Exercises in Functional Analysis

Kluwer Texts in the Mathematical Sciences

VOLUME 26

A Graduate-Level Book Series

The titles published in this series are listed at the end of this volume.

Exercises in Functional Analysis

by

Constantin Costara
Department of Mathematics,
Ovidius University of Constanta, Romania

and

Dumitru Popa
Department of Mathematics,
Ovidius University of Constanta, Romania

KLUWER ACADEMIC PUBLISHERS
DORDRECHT / BOSTON / LONDON

A C.I.P. Catalogue record for this book is available from the Library of Congress.

ISBN 1-4020-1560-7

Published by Kluwer Academic Publishers,
P.O. Box 17, 3300 AA Dordrecht, The Netherlands.

Sold and distributed in North, Central and South America
by Kluwer Academic Publishers,
101 Philip Drive, Norwell, MA 02061, U.S.A.

In all other countries, sold and distributed
by Kluwer Academic Publishers,
P.O. Box 322, 3300 AH Dordrecht, The Netherlands.

Printed on acid-free paper

All Rights Reserved
© 2003 Kluwer Academic Publishers
No part of this work may be reproduced, stored in a retrieval system, or transmitted
in any form or by any means, electronic, mechanical, photocopying, microfilming, recording
or otherwise, without written permission from the Publisher, with the exception
of any material supplied specifically for the purpose of being entered
and executed on a computer system, for exclusive use by the purchaser of the work.

Printed in the Netherlands.

Contents

Preface vii

Some Standard Notations and Conventions ix

Part I: Normed spaces 1

Chapter 1. Open, closed, and bounded sets in normed spaces 3
1.1 Exercises 4
1.2 Solutions 11

Chapter 2. Linear and continuous operators on normed spaces 36
2.1 Exercises 37
2.2 Solutions 42

Chapter 3. Linear and continuous functionals. Reflexive spaces 68
3.1 Exercises 68
3.2 Solutions 72

Chapter 4. The distance between sets in Banach spaces 86
4.1 Exercises 86
4.2 Solutions 92

Chapter 5. Compactness in Banach spaces. Compact operators 107
5.1 Exercises 108
5.2 Solutions 115

Chapter 6. The Uniform Boundedness Principle 147
6.1 Exercises 147
6.2 Solutions 155

Chapter 7. The Hahn–Banach theorem 175
7.1 Exercises 175
7.2 Solutions 180

Chapter 8. Applications for the Hahn–Banach theorem 195
8.1 Exercises 196
8.2 Solutions 199

Chapter 9. Baire's category. The open mapping and closed graph theorems . . . **213**
    9.1   Exercises 214
    9.2   Solutions 220

Part II: Hilbert spaces **241**

Chapter 10. Hilbert spaces, general theory **243**
    10.1 Exercises 245
    10.2 Solutions 250

Chapter 11. The projection in Hilbert spaces **271**
    11.1 Exercises 271
    11.2 Solutions 281

Chapter 12. Linear and continuous operators on Hilbert spaces **305**
    12.1 Exercises 306
    12.2 Solutions 318

Part III: General topological spaces **366**

Chapter 13. Linear topological and locally convex spaces **368**
    13.1 Exercises 369
    13.2 Solutions 377

Chapter 14. The weak topologies **403**
    14.1 Exercises 405
    14.2 Solutions 412

Bibliography **444**

List of Symbols **447**

Index **449**

Preface

The understanding of results and notions for a student in mathematics requires solving exercises. The exercises are also meant to test the reader's understanding of the text material, and to enhance the skill in doing calculations. This book is written with these three things in mind. It is a collection of more than 450 exercises in Functional Analysis, meant to help a student understand much better the basic facts which are usually presented in an introductory course in Functional Analysis.

Another goal of this book is to help the reader to understand the richness of ideas and techniques which Functional Analysis offers, by providing various exercises, from different topics, from simple ones to, perhaps, more difficult ones. We also hope that some of the exercises herein can be of some help to the teacher of Functional Analysis as seminar tools, and to anyone who is interested in seeing some applications of Functional Analysis. To what extent we have managed to achieve these goals is for the reader to decide.

We have also tried to show the reader, on the one hand, the various and sometimes unexpected connections between the usual parts of a course in Functional Analysis and other topics in mathematics, and, on the other hand, the fine interplay between the topics in the field of Functional Analysis. As a consequence we have decided to include in our chapters some exercises which are connected with the respective chapter, even if the solution we give requires the use of notions and results which are usually studied later in a course in Functional Analysis. We hope that this option will be to the benefit of the reader. When we refer to an exercise from the same chapter we only indicate the number for that exercise, and when we refer to an exercise from another chapter we also indicate the chapter's number.

Naturally, many of the exercises presented here are borrowed from different books on Functional Analysis, where they usually appear as unsolved exercises, and sometimes with short indications. Some of them are, in fact, lemmas, propositions, and theorems: we have included some theoretical results as *exercises*, and in this case their proper citations are given. Others can be found in various papers, some of them are just folklore results, and some of them are, perhaps, invented by us, meeting the need of giving concrete examples for the abstract notions and results presented in a course in Functional Analysis. When possible we indicate at the beginning of a solution one of the possible sources for the exercise, not necessarily the original one, so that the reader can see other ideas which do not appear in the solution given here. If the source of some exercises are not given, we sincerely regret it. Any such omission is not made intentionally. It must also be emphasized that the lack of a citation does not imply a claim of originality on our part for the statement of an exercise.

The book is divided into three parts. The first one deals with the theory of normed spaces. It contains exercises on the general properties of sets in normed spaces, linear bounded operators on normed spaces, the dual of a normed space, compactness in normed spaces, and on the basic principles of Functional Analysis: the Hahn–Banach theorem (regarded as an extension theorem and as a separation theorem), the uniform boundedness principle, the open mapping and the closed graph theorems on Banach spaces. The second part deals with Hilbert spaces and contains exercises on the general theory of Hilbert spaces (inner products, orthogonality, orthonormal basis in Hilbert spaces), the projection theorem (on closed convex sets and on closed linear subspaces), and linear and continuous operators on Hilbert spaces. The third part deals with linear topological spaces. A large number of exercises on the weak topologies are presented in the final chapter. For the convenience of the reader a brief glossary of notations is given at the beginning of the book, whilst a list of symbols and an index are supplied in the final pages.

At the beginning of each chapter we include a summary of the main notions, notations, and theoretical results needed in order to solve many of the exercises in that chapter. All the notations and notions used and not defined are standard. The reader is assumed to be familiar with some basic results in topology, measure theory and complex functions theory. For some background results on these topics, we refer the reader to J. Kelley [34], P. Halmos [27] and W. Rudin [45].

The sets of definitions and theorems recalled at the beginning of each chapter are not complete and they are meant to increase the readability of the book. For a serious background in Functional Analysis we refer to some books which we have used, in which the reader can find not only the proofs for the results merely stated in this book, but also all the basic results in Functional Analysis: S. Banach [5], N. Bourbaki [9], N. Dunford and J. Schwartz [20], E. Hille and R.S. Phillips [30], L.V. Kantorovich and G.P. Akilov [33], J. Kelley and I. Namioka [35], W. Rudin [44], H. Schaefer [46], K. Yoshida [51].

Dumitru Popa thanks his wife Maria for her support during the writing of this book, and for suggesting the submission of the manuscript to the Kluwer Academic Publishers for possible publication.

Finally, we thank the three anonymous referees whose comments and remarks improved the initial version of the manuscript.

> C. Costara and D. Popa
> University Ovidius,
> Faculty of Mathematics and Informatics
> Constanta, Romania, 2003

Some Standard Notations and Conventions

Convention. All the linear spaces used in the book are considered over the scalar field $\mathbb{K} = \mathbb{R}$ (or $\mathbb{C}$). It will be specified or it will be clear from the context when the scalar field is $\mathbb{R}$.

Spaces of sequences

For $1 \leq p < \infty$ we denote

$$l_p = \left\{ (x_n)_{n \in \mathbb{N}} \subseteq \mathbb{K} \,\bigg|\, \sum_{n=1}^{\infty} |x_n|^p < \infty \right\}.$$

Then l_p is a linear space with respect to the natural operations for addition and scalar multiplication, and a Banach space under the norm

$$\|x\|_p = \left(\sum_{n=1}^{\infty} |x_n|^p \right)^{1/p} \quad \forall x = (x_n)_{n \in \mathbb{N}} \in l_p.$$

We denote

$$l_\infty = \left\{ (x_n)_{n \in \mathbb{N}} \subseteq \mathbb{K} \mid (x_n)_{n \in \mathbb{N}} \text{ is bounded} \right\}.$$

Then l_∞ is a linear space with respect to the natural operations for addition and scalar multiplication and a Banach space under the norm

$$\|x\|_\infty = \sup_{n \in \mathbb{N}} |x_n| \quad \forall x = (x_n)_{n \in \mathbb{N}} \in l_\infty.$$

We denote

$$c_0 = \left\{ (x_n)_{n \in \mathbb{N}} \subseteq \mathbb{K} \,\Big|\, \lim_{n \to \infty} x_n = 0 \right\}$$

and

$$c = \left\{ (x_n)_{n \in \mathbb{N}} \subseteq \mathbb{K} \mid (x_n)_{n \in \mathbb{N}} \text{ converges} \right\}.$$

Then c_0 and c are linear spaces with respect to the natural operations for addition and scalar multiplication and Banach spaces with respect to the norm $\|x\|_\infty = \sup_{n \in \mathbb{N}} |x_n| \; \forall x = (x_n)_{n \in \mathbb{N}} \in c_0$, or c.

For $1 \leq p \leq \infty$ and $n \in \mathbb{N}$ we denote $e_n = (0, ..., 0, 1, 0, ...) \in l_p$ (1 in the nth position). The same notation is used for the space c_0. Also, for $X = l_p$ with $1 \leq p \leq \infty$, or c_0, we define the canonical projections $(p_n)_{n \in \mathbb{N}}$, $p_n : X \to \mathbb{K}$, $p_n(x_1, ..., x_n, ...) = x_n \; \forall (x_1, ..., x_n, ...) \in X$.

For $n \in \mathbb{N}$ and $1 \leq p \leq \infty$ we denote $l_p^n = \left(\mathbb{K}^n, \|\cdot\|_p \right)$, where $\|(x_1, ..., x_n)\|_p = \left(\sum_{k=1}^{n} |x_k|^p \right)^{1/p}$ for $1 \leq p < \infty$ and $\|(x_1, ..., x_n)\|_\infty = \max_{1 \leq k \leq n} |x_k|$.

Spaces of functions

If T is a compact Hausdorff space we denote

$$C(T) = \{f : T \to \mathbb{K} \mid f \text{ continuous}\}.$$

Then $C(T)$ is a linear space with respect to the usual pointwise operations for addition and scalar multiplication, and a Banach space under the norm $\|\cdot\|_\infty : C(T) \to \mathbb{R}$,

$$\|f\|_\infty = \sup_{x \in T} |f(x)| \quad \forall f \in C(T).$$

Let (S, Σ, μ) be a measure space. For $1 \leq p < \infty$ we denote

$$\mathcal{L}_p(\mu) = \left\{ f : S \to \mathbb{K} \,\bigg|\, f \text{ is } \mu\text{-measurable and } \int_S |f|^p \, d\mu < \infty \right\}.$$

Then $\mathcal{L}_p(\mu)$ is a linear space with respect to the usual pointwise operations for addition and scalar multiplication. The map $\|\cdot\|_p : \mathcal{L}_p(\mu) \to \mathbb{R}$ defined by

$$\|f\|_p = \left(\int_S |f|^p \, d\mu \right)^{1/p} \quad \forall f \in \mathcal{L}_p(\mu)$$

is a seminorm. Factorizing this linear space by the equivalence relation given by the equality μ-almost everywhere, we obtain the linear space $L_p(\mu)$, which is a Banach space with respect to the norm $\|\tilde{f}\|_p = \left(\int_S |f|^p \, d\mu \right)^{1/p} \forall \tilde{f} \in L_p(\mu), f \in \tilde{f}$. We identify a function $f \in \mathcal{L}_p(\mu)$ with its equivalence class $\tilde{f}$ in $L_p(\mu)$, and we write $f \in L_p(\mu)$, as usual.

We denote

$$\mathcal{L}_\infty(\mu) = \{f : S \to \mathbb{K} \mid f \text{ is } \mu\text{-measurable}, \exists M > 0 \text{ such that } |f| \leq M, \mu\text{-a.e.}\}.$$

Then $\mathcal{L}_\infty(\mu)$ is a linear space with respect to the usual pointwise operations for addition and scalar multiplication. The map $\|\cdot\|_\infty : \mathcal{L}_\infty(\mu) \to \mathbb{R}$ defined by

$$\|f\|_\infty = \inf\{M > 0 \mid |f| \leq M, \mu\text{-a.e.}\} \quad \forall f \in \mathcal{L}_\infty(\mu)$$

is a seminorm. Factorizing this linear space by the equivalence relation given by the equality μ-a.e., we obtain the linear space $L_\infty(\mu)$, which is a Banach space with respect to the norm $\|\tilde{f}\|_\infty = \|f\|_\infty \, \forall \tilde{f} \in L_\infty(\mu), f \in \tilde{f}$. We identify a function $f \in \mathcal{L}_\infty(\mu)$ with its equivalence class $\tilde{f}$ in $L_\infty(\mu)$, and we write $f \in L_\infty(\mu)$, as usual.

If $n \in \mathbb{N}$ and the measure space is (S, Σ, μ), where $S \subseteq \mathbb{R}^n$ is a Lebesgue measurable set, Σ is the σ-algebra of all the Lebesgue measurable subsets of S, and $\mu : \Sigma \to [0, \infty]$ is the Lebesgue measure, we write simply $L_p(S)$.

Convention. On all the Banach spaces defined above, if not otherwise specified, we always consider these canonical norms, and when there is no risk of confusion we will write simply $\|\cdot\|$ for the norm.

Part I

Normed spaces

Chapter 1

Open, closed, and bounded sets in normed spaces

Definition. Let X be a linear space. A map $p : X \to \mathbb{R}$ is called:

i) *subadditive* if $p(x+y) \leq p(x) + p(y) \ \forall x, y \in X$.
ii) *positive homogeneous* if $p(\lambda x) = \lambda p(x) \ \forall x \in X, \forall \lambda \geq 0$.
iii) *absolutely homogeneous* if $p(\lambda x) = |\lambda| p(x) \ \forall x \in X, \forall \lambda \in \mathbb{K}$.
iii) *sublinear* if it is subadditive and positive homogeneous.
iv) *supra-linear* if $-p$ is sublinear.
v) *seminorm* if it is subadditive and absolutely homogeneous.

Definition. Let X be a linear space. A map $\|\cdot\| : X \to \mathbb{R}_+$ is called a *norm* if:

i) $\|x\| = 0 \Leftrightarrow x = 0$.
ii) $\|\lambda x\| = |\lambda| \|x\| \ \forall \lambda \in \mathbb{K}, \forall x \in X$.
iii) $\|x + y\| \leq \|x\| + \|y\| \ \forall x, y \in X$.

The couple $(X, \|\cdot\|)$, or simply X when the norm is understood, is called a *normed space*.

For a normed space $(X, \|\cdot\|)$, $B_X = \{x \in X \mid \|x\| \leq 1\}$ is called the *closed unit ball* of X, $S_X = \{x \in X \mid \|x\| = 1\}$ is called the *unit sphere* of X and for $\varepsilon > 0$ and $x \in X$, the sets $B(x, \varepsilon) = \{y \in X \mid \|y - x\| < \varepsilon\}$, respectively $\overline{B}(x, \varepsilon) = \{y \in X \mid \|y - x\| \leq \varepsilon\}$ are called the *open*, respectively the *closed ball* with center at x and radius ε.

Definition. Let X be a normed space. A sequence $(x_n)_{n \in \mathbb{N}} \subseteq X$ is called:

i) *a Cauchy sequence* if $\forall \varepsilon > 0 \ \exists n_\varepsilon \in \mathbb{N}$ such that $\forall n \geq n_\varepsilon$ and $\forall m \geq n_\varepsilon$ it follows that $\|x_n - x_m\| < \varepsilon$, or, equivalently, $\forall \varepsilon > 0 \ \exists n_\varepsilon \in \mathbb{N}$ such that $\forall n \geq n_\varepsilon$ and $\forall p \in \mathbb{N}$ it follows that $\|x_{n+p} - x_n\| < \varepsilon$.

ii) *a convergent sequence* if there is $x \in X$ such that: $\forall \varepsilon > 0 \ \exists n_\varepsilon \in \mathbb{N}$ such that $\forall n \geq n_\varepsilon$ it follows that $\|x_n - x\| < \varepsilon$. In this case we write $x_n \to x$.

If the norm on X is complete (that is, any Cauchy sequence in X is convergent), then X is called a *Banach space*.

Definition. For a normed space $(X, \|\cdot\|)$ the map $d : X \times X \to \mathbb{R}$ defined by $d(x,y) = \|x-y\|$ is a metric (or a distance). This metric is called the *metric associated with the norm*. Hence normed spaces are metric spaces.

The topology generated by a norm is by definition the topology generated by the metric associated with the norm, that is,
$$\tau = \{D \subseteq X \mid \forall x \in D \; \exists \varepsilon > 0 \text{ such that } B(x,\varepsilon) \subseteq D\}.$$

Convention. All the topological notions in a normed space (if not otherwise specified) are understood to be defined with respect to the metric associated to the norm.

Definition. Two norms $\|\cdot\|_1$, $\|\cdot\|_2$ on a linear space X are called *equivalent* if and only if they generate the same topology, or, equivalently, there are $c_1 > 0$, $c_2 > 0$ such that $\|x\|_1 \leq c_2 \|x\|_2 \; \forall x \in X$ and $\|x\|_2 \leq c_1 \|x\|_1 \; \forall x \in X$.

Theorem. On a finite-dimensional linear space any two norms are equivalent.

Notation. If X is a topological space, and $A \subseteq X$ is a subset, by $\overset{\circ}{A}$, respectively $\overline{A}$, we denote the *interior*, respectively the *closure* of A.

Definition. Let X be a topological space.

i) A subset $A \subseteq X$ is called *dense* if and only if $\overline{A} = X$.

ii) X is called *separable* if and only if there is a dense countable subset of X.

Definition. Let X be a normed space. A subset $A \subseteq X$ is called *bounded* if and only if there is $M > 0$ such that $\|x\| \leq M \; \forall x \in A$.

Definition. Let X be a normed space. A series $\sum_{n=1}^{\infty} x_n$ in X is called:

i) *convergent* if and only if the sequence $\left(\sum_{k=1}^{n} x_k\right)_{n \in \mathbb{N}}$ is convergent.

ii) *absolutely convergent* if and only if the series $\sum_{n=1}^{\infty} \|x_n\|$ is convergent.

Theorem. Let X be a normed space. Then X is a Banach space if and only if any absolutely convergent series in X is convergent.

Theorem. Let X be a normed space. Then any finite-dimensional linear subspace of X is closed.

Definition. Let X be a normed space. A sequence $(x_n)_{n \in \mathbb{N}} \subseteq X$ is called a *Schauder basis* if and only if for every $x \in X$ there is a unique sequence of scalars $(\alpha_n)_{n \in \mathbb{N}}$ such that $x = \sum_{n=1}^{\infty} \alpha_n x_n$ (that is, $\lim_{n \to \infty} \left\| x - \sum_{i=1}^{n} \alpha_i x_i \right\| = 0$).

Definition. Let X be a normed space and $Y \subseteq X$ a closed linear subspace. Let X/Y be the quotient linear space. Then X/Y is a normed space with respect to the quotient norm $\|\overline{x}\| = \inf\{\|y\| \mid x - y \in Y\}$, where $\overline{x} = x + Y$.

1.1 Exercises

Any linear space can be normed

1. Let $X \neq \{0\}$ be a real or complex linear space. Prove that there is at least one norm on X.

1.1 Exercises

The triangle axiom and the convexity of the closed unit ball are equivalent

2. Prove that the triangle axiom from the definition for a norm is equivalent with the convexity of the closed unit ball. More precisely, if X is a linear space on which is given a function $p : X \to [0, \infty)$ with the properties:
  i) $p(x) = 0 \Leftrightarrow x = 0$;
  ii) $p(\lambda x) = |\lambda| p(x) \; \forall x \in X, \forall \lambda \in \mathbb{K}$,
then p is a norm if and only if $B_X = \{x \in X \mid p(x) \leq 1\}$ is convex.

The closed unit sphere in a normed space of dimension ≥ 2 is path connected

3. Let X be a normed space of dimension ≥ 2. Prove that S_X is path connected, hence connected.

Finite additivity and countable additivity for norms

4. Let $\varphi : [0, \infty) \to \mathbb{R}$ be a C^1 convex function with $\varphi(0) = 0$ and $\varphi'(0) = 0$, and let X be a normed space of dimension ≥ 2.
  i) Prove that for any $x \in X \setminus \{0\}$ and for any $\alpha > 0$, there is $y \in X$ such that $\|y\| = \alpha$ and $\varphi(\|x + y\|) = \varphi(\|x\|) + \varphi(\|y\|)$.
  ii) Deduce that if, in addition, X is a Banach space, and $\varphi(t) = 0$ if and only if $t = 0$, then for any sequence of strictly positive real numbers $(\alpha_n)_{n \in \mathbb{N}}$ for which the series $\sum_{n=1}^{\infty} \alpha_n$ converges and for any $x \in X \setminus \{0\}$ there is a sequence $(y_n)_{n \in \mathbb{N}} \subseteq X$ such that $\|y_n\| = \alpha_n$ $\forall n \in \mathbb{N}$ and $\varphi \left(\left\| x + \sum_{n=1}^{\infty} y_n \right\| \right) = \varphi(\|x\|) + \sum_{n=1}^{\infty} \varphi(\|y_n\|)$.

The minimum of two norms

5. Let $a > 0$. On $C[0, 1]$ we consider the following norms: $\|f\|_\infty = \sup_{t \in [0,1]} |f(t)|$, $\|f\|_1 = a \int_0^1 |f(t)| \, dt$ $\forall f \in C[0, 1]$. Prove that $\|f\| = \min\{\|f\|_\infty, \|f\|_1\}$ is a norm on $C[0, 1]$ if and only if $a \leq 1$.

Open sets in l_2

6. Prove that the parallelepiped $P = \{(x_n)_{n \in \mathbb{N}} \in l_2 \mid |x_n| < 1 \; \forall n \in \mathbb{N}\}$ is an open set in l_2.

7. Let $(a_n)_{n \in \mathbb{N}} \subseteq (0, \infty)$. Find a necessary and sufficient condition on the sequence $(a_n)_{n \in \mathbb{N}}$ such that the parallelepiped $P = \{(x_n)_{n \in \mathbb{N}} \in l_2 \mid |x_n| < a_n \; \forall n \in \mathbb{N}\}$ be an open set in l_2.

The set of all polynomials in $C[-1, 1]$

8. Is the set of all polynomials open in $C[-1, 1]$?

Closed linear subspaces in l_p and L_p

9. Let $1 \leq p < \infty$ and $G_p = \left\{ (x_n)_{n \in \mathbb{N}} \in l_p \;\Big|\; \sum_{n=1}^{\infty} x_n = 0 \right\}$.
i) Prove that $G_p \subseteq l_p$ is a linear subspace.
ii) Is $G_p \subseteq l_p$ a closed set?

10. For $1 \leq p < \infty$, consider $E_p = \{ f \in L_p[0, \infty) \mid \int_0^{\infty} f(x)\, dx = 0 \}$.
i) Prove that $E_p \subseteq L_p[0, \infty)$ is a linear subspace.
ii) Prove that $E_p \subseteq L_p[0, \infty)$ is a closed set if and only if $p = 1$.

Algebraic sets which are closed

11. Let X be a normed space, and $A \in L(X)$. Prove that the following linear subspaces are closed in $L(X)$:
i) $\{B \in L(X) \mid AB = 0\}$, i.e., the set of all the right divisors for A.
ii) $\{B \in L(X) \mid AB = BA\}$, i.e., the centralizer of A.

12. Let $G \subseteq X$ be a subset in a normed space X, and Y a normed space. Prove that the set $M = \{U \in L(X, Y) \mid G \subseteq \ker U\}$ is a closed linear subspace in $L(X, Y)$.

Bounded sets in l_p

13. Let $1 \leq p < \infty$ and consider a sequence $(a_n)_{n \in \mathbb{N}} \subseteq (0, \infty)$. Find a necessary and sufficient condition on the sequence $(a_n)_{n \in \mathbb{N}}$ such that:
i) The parallelepiped $\{(x_n)_{n \in \mathbb{N}} \in l_p \mid |x_n| < a_n \; \forall n \in \mathbb{N}\}$ be a bounded set in l_p.
ii) The ellipsoid
$$\left\{ (x_n)_{n \in \mathbb{N}} \in l_p \;\Big|\; \sum_{n=1}^{\infty} \frac{|x_n|^p}{a_n^p} < 1 \right\}$$
be a bounded set in l_p.

Topological properties for linear subspaces

14. Let X be a normed space. Find all the linear subspaces $L \subseteq X$ which are contained in a ball.

15. Let X be a normed space. Find all the linear subspaces $L \subseteq X$ which contain a ball.

16. Let X and Y be two normed spaces, $T : X \to Y$ is a linear operator and $T_n : X \to Y$, $n \in \mathbb{N}$, is a sequence of linear operators. Prove that the sets
$$A = \{x \in X \mid T_n x \text{ does not converge towards } Tx\}$$
and
$$B = \{x \in X \mid (T_n x)_{n \in \mathbb{N}} \text{ is not a Cauchy sequence}\}$$
are either empty, or dense in X.

17. Let X be a normed space and $A \subseteq X$ a set with the property that $X \setminus A$ is a linear subspace. Prove that A is either dense, or empty.

18. Let X be a normed space and $G \subseteq X$ a linear subspace. Prove that either $\overline{G} = X$, or $\overset{\circ}{G} = \emptyset$.

Closed convex sets

19. Let $A = \{f \in L_2[0,1] \mid \exists I_f \text{ interval} \subseteq [0,1], 1/2 \in I_f, f = 0 \text{ a.e. on } I_f\}$. Is A closed in $L_2[0,1]$?

20. Let $M = \{f \in L_2[0,1] \mid f([0,1]) \subseteq [0,1] \text{ a.e.}\}$. Prove that M is a closed convex set in $L_2[0,1]$.

A dense linear subspace in l_∞

21. For a subset $M \subseteq \mathbb{N}$ we denote $\chi_M \in l_\infty$ as being the characteristic function for M: $\chi_M(k) = \begin{cases} 1, & k \in M, \\ 0, & k \notin M, \end{cases}$ $\forall k \in \mathbb{N}$. Prove that $\text{Sp}\{\chi_M \mid M \in \mathcal{P}(\mathbb{N})\} \subseteq l_\infty$ is dense.

The closure of l_p and c_0 in l_∞

22. Prove that $l_p \subseteq c_0$ and $l_p \neq c_0$, where $1 \leq p < \infty$ is fixed.

23. Find the closure of l_p and c_0 in l_∞, where $1 \leq p < \infty$ is fixed.

A characterization for the closed convex sets in $\mathbb{R}^n$

24. Let $n \in \mathbb{N}$ and $A \subseteq \mathbb{R}^n$ a convex set such that the intersection between A and any line from $\mathbb{R}^n$ is a closed set in $\mathbb{R}^n$. Prove that A is closed.

Closed and open sets in the space $\mathcal{M}_n$

25. For $n \in \mathbb{N}$ let $\mathcal{M}_n$ be the space of all the $n \times n$ real matrices. On $\mathcal{M}_n$ we define the following metric: $d(A,B) = \sum_{i,j=1}^{n} |a_{ij} - b_{ij}|$, where $A = (a_{ij})_{i,j}$, $B = (b_{ij})_{i,j}$. Prove that the set of all nilpotent matrices is closed in $\mathcal{M}_n$ with respect to this metric.

26. Let $m, n \in \mathbb{N}$ with $m < n$. Let $\mathcal{M}(m,n)$ be the space of all the linear transformations from $\mathbb{R}^m$ into $\mathbb{R}^n$ and let L be the set of all the linear transformations from $\mathcal{M}(m,n)$ of rank m.

i) Prove that L is an open set in $\mathcal{M}(m,n)$.

ii) Prove that there is a continuous function $T: L \to \mathcal{M}(n,m)$ such that $T(A)A = I_m$, for any $A \in L$, where I_m is the identity on $\mathbb{R}^m$.

The Banach space of all the scalar convergent series

27. Denote
$$S_c = \left\{(x_n)_{n \in \mathbb{N}} \subseteq \mathbb{K} \;\middle|\; \left(\sum_{k=1}^{n} x_k\right)_{n \in \mathbb{N}} \text{ converges}\right\}$$

and for any $x = (x_n)_{n \in \mathbb{N}} \in S_c$, define $\|x\| = \sup_{n \in \mathbb{N}} \left|\sum_{k=1}^{n} x_k\right|$.

i) Prove that $(S_c, \|\cdot\|)$ is a Banach space.

ii) Prove that $l_1 \subseteq S_c$ is dense.

iii) Prove that $U : S_c \to c_0$, $U((x_n)_{n\in\mathbb{N}}) = \left(\sum_{k=n}^{\infty} x_k\right)_{n\in\mathbb{N}}$ is an isomorphism of Banach spaces.

iv) Prove that $T : S_c \to c_0$, $T((x_n)_{n\in\mathbb{N}}) = \left(\sum_{k=n+1}^{\infty} x_k\right)_{n\in\mathbb{N}}$ is a surjective linear and continuous operator, that $V : c_0 \to S_c$, $V((x_n)_{n\in\mathbb{N}}) = (-x_1, x_1 - x_2, x_2 - x_3, \ldots)$ is a linear isometry, and that $TV = I_{c_0}$ but $VT \neq I_{S_c}$.

A Cantor type of characterization for a Banach space

28. Let X be a normed space. Prove that if $\overline{B}(x, \alpha) \subseteq \overline{B}(y, \beta)$, then $\beta \geq \alpha$ and $\|x - y\| \leq \beta - \alpha$. In particular, if $\overline{B}(x, \alpha) = \overline{B}(y, \beta)$ then $\alpha = \beta$ and $x = y$.

29. Let X be a normed space. Prove that X is a Banach space if and only if any decreasing sequence of closed balls from X with the sequence of radii converging to 0 has non-empty intersection.

30. Let X be a normed space.

i) If $\overline{B}(x_1, r_1) \supseteq \overline{B}(x_2, r_2) \supseteq \cdots \supseteq \overline{B}(x_n, r_n) \supseteq \overline{B}(x_{n+1}, r_{n+1}) \supseteq \cdots$ is a decreasing sequence of closed balls in X and $r_n \geq r > 0$ $\forall n \in \mathbb{N}$, prove that for any $r > \varepsilon > 0$, there is an $x \in X$ such that $\overline{B}(x, r - \varepsilon) \subseteq \bigcap_{n\in\mathbb{N}} \overline{B}(x_n, r_n)$.

ii) Under the same hypothesis as in (i), if X is Banach prove that there is an $x \in X$ such that $\overline{B}(x, r) \subseteq \bigcap_{n\in\mathbb{N}} \overline{B}(x_n, r_n)$.

iii) Give an example of a normed space X and a sequence $\left(\overline{B}(x_n, r_n)\right)_{n\in\mathbb{N}}$ as in (i) such that we cannot find an $x \in X$ with $\overline{B}(x, r) \subseteq \bigcap_{n\in\mathbb{N}} \overline{B}(x_n, r_n)$.

iv) Let X be a normed space for which there is an $r \geq 0$ such that from $\overline{B}(x_1, r_1) \supseteq \overline{B}(x_2, r_2) \supseteq \cdots \supseteq \overline{B}(x_n, r_n) \supseteq \overline{B}(x_{n+1}, r_{n+1}) \supseteq \cdots$, with $(x_n)_{n\in\mathbb{N}} \subseteq X$, and $r_n \geq r$ for all $n \in \mathbb{N}$, it follows that there is an $x \in X$ such that $\overline{B}(x, r) \subseteq \bigcap_{n\in\mathbb{N}} \overline{B}(x_n, r_n)$. Prove that X is a Banach space.

Differences between norms and metrics

31. i) If X is a linear space on which we have two norms which generate the same topology on X, prove that either both of them are complete, or none of them is complete.

ii) We define: $d_1 : \mathbb{R} \times \mathbb{R} \longrightarrow \mathbb{R}_+$, $d_1(x, y) = |x - y|$, $d_2 : \mathbb{R} \times \mathbb{R} \longrightarrow \mathbb{R}_+$, $d_2(x, y) = |\phi(x) - \phi(y)|$, where $\phi : \mathbb{R} \to \mathbb{R}$, $\phi(x) = x/(1 + |x|)$ $\forall x \in \mathbb{R}$. Prove that d_1 and d_2 are distances on $\mathbb{R}$, which generate the same topology, but $(\mathbb{R}, d_1)$ is complete and $(\mathbb{R}, d_2)$ is not complete.

32. Let (X_1, d_1) and (X_2, d_2) be two metric spaces and $f : X_1 \to X_2$ a continuous surjective function such that $d_1(p, q) \leq d_2(f(p), f(q))$ $\forall p, q \in X_1$.

i) If X_1 is complete then X_2 is complete! Give a proof or a counterexample.

ii) If X_2 is complete then X_1 is complete! Give a proof or a counterexample.

iii) Analyse (i) and (ii) if, in addition, X_1 and X_2 are two normed spaces, d_1 and d_2 are the distances associated to the norms, and f is linear.

1.1 Exercises

Equivalent norms

33. Let c_{00} be the space of all the scalar sequences with finite support: for $x = (x_n)_{n \in \mathbb{N}} \subseteq \mathbb{K}$, $x \in c_{00}$ if and only if there is an $m \in \mathbb{N}$, depending on x, such that $x_n = 0\ \forall n \geq m$.

i) Let $a = (a_n)_{n \in \mathbb{N}}$ be a scalar sequence. On c_{00} we consider the following map: $\|x\|_a = \sum_{n=1}^{\infty} |a_n|\,|x_n|\ \forall x \in c_{00}$. Prove that $\|\cdot\|_a$ is a norm if and only if $a_n \neq 0\ \forall n \in \mathbb{N}$.

ii) Let $a = (a_n)_{n \in \mathbb{N}}$ and $b = (b_n)_{n \in \mathbb{N}}$ be two scalar sequences such that $a_n \neq 0$ and $b_n \neq 0\ \forall n \in \mathbb{N}$. Prove that $\|\cdot\|_a$ and $\|\cdot\|_b$ are equivalent norms if and only if $0 < \inf_{n \in \mathbb{N}}(|a_n|/|b_n|) \leq \sup_{n \in \mathbb{N}}(|a_n|/|b_n|) < \infty$.

A tree type of property for the completeness

34. Let X be a normed space, $Y \subseteq X$ be a closed linear subspace such that Y (with the norm from X) is complete and X/Y (with the quotient norm) is also complete. Prove that X is complete.

A stability property for finite linearly independent systems

35. Let $n \in \mathbb{N}$, X is a normed space and $x_1, x_2, ..., x_n$ are n linearly independent elements in X. Prove that there is $\varepsilon > 0$ such that: if $y_1, y_2, ..., y_n \in X$ are such that $\|y_i\| < \varepsilon$, $i = \overline{1,n}$, then $x_1 + y_1, x_2 + y_2, ..., x_n + y_n$ are also n linearly independent elements in X.

The projection onto finite-dimensional linear subspaces in a normed space

36. Let X be a normed space and $x, y \in X$. Prove that the function $\varphi : \mathbb{R} \to \mathbb{R}$, $\varphi(t) = \|x - ty\|$ attains its infimum on $\mathbb{R}$.

37. i) If $(X, \|\cdot\|)$ is a normed space, and $Y \subseteq X$ is a finite-dimensional linear subspace, prove that any element of X has a projection on Y, i.e., $\forall x \in X\ \exists y_0 \in Y$ such that $\|x - y_0\| = d(x, Y) = \inf\{\|x - y\| \mid y \in Y\}$

ii) Is this projection unique? Give a proof or a counterexample.

A property for the set of all polynomials of degree at most k

38. Prove that for any $k \in \mathbb{N}$ there is a real constant $c_k > 0$ such that $|P(0)| \leq c_k \int_0^1 |P(t)|\,dt$ for any real polynomial of degree at most k.

Finite-dimensional linear subspaces in function spaces

39. i) Let $n \in \mathbb{N}$, X be a set, and V be a linear subspace of dimension n in the real linear space of all functions from X to $\mathbb{R}$. Prove that there are n elements $x_1, x_2, ..., x_n \in X$ such that the operator $f \to (f(x_1), f(x_2), ..., f(x_n))^t$, defined on V with values in $\mathbb{R}^n$, is a linear isomorphism.

ii) Let X be a topological Hausdorff space and V a linear subspace of $C(X; \mathbb{R})$, with $\dim_{\mathbb{R}} V = n \in \mathbb{N}$.

a) Prove that there are n open pairwise disjoint sets $U_i \subseteq X$ such that: if $f \in V$ is such that for any $1 \leq i \leq n$ there is an $x_i \in U_i$ with $f(x_i) = 0$, then $f = 0$.

b) Suppose that X is a compact space. Consider the sets U_i with the property from (a). Let $x_i \in U_i, i = \overline{1,n}$. Deduce that there is a constant $c > 0$ such that: $\forall f \in V$, $\sup_{x \in X} |f(x)| \leq c \sum_{i=1}^n |f(x_i)|$.

c) Let X be a compact Hausdorff space and $V \subseteq C(X; \mathbb{R})$ a finite-dimensional linear subspace. Using (b) prove that if $(f_n)_{n \in \mathbb{N}} \subseteq V$, $f_n \to 0$ pointwise, then $f_n \to 0$ uniformly.

Continuous and discontinuous mappings on a normed space

40. Let $f : C[0,1] \to \mathbb{R}$, $f(x) = x(1) \; \forall x \in C[0,1]$.

i) Prove that f is continuous, if on $C[0,1]$ we consider the usual norm.

ii) Prove that is not continuous, if on $C[0,1]$ we consider the p-norm, that is, $\|f\| = \left(\int_0^1 |f(x)|^p dx\right)^{1/p}$, where $1 \leq p < \infty$.

41. Let $U : C[0,1] \to C[0,1]$, $U(f) = f^2$. Prove that:

i) U is continuous but not uniformly continuous, if on $C[0,1]$ we consider the usual norm.

ii) U is not continuous if on both spaces we consider the p-norm.

iii) U is continuous but is not uniformly continuous if on the domain of definition we consider the usual norm of $C[0,1]$ and on the range we consider the p-norm.

The spaces L_1 and L_p, $1 < p < \infty$, are homeomorphic

42. i) Let $1 < p < \infty$, $(X, \|\cdot\|)$ is a normed space, and $a, b \in X$ with $\|a\| = \|b\| = 1$. Prove that $\|a - tb\|^p \leq 2^p \|a - t^p b\|$ and $\|a - t^p b\| \leq p \|a - tb\| \; \forall t \in [0,1]$, and then deduce that for any $x, y \in X$, we have $\|x - y\|^p \leq 2^p \| \|x\|^{p-1} x - \|y\|^{p-1} y \|$ and $\| \|x\|^{p-1} x - \|y\|^{p-1} y \| \leq p \|x - y\| (\|x\| + \|y\|)^{p-1}$.

ii) Prove that, given a measurable space (S, Σ, μ), and $1 < p < \infty$, then the function $T : L_1(\mu) \to L_p(\mu)$, $T(f) = |f|^{1/p-1} f$ is bijective and uniformly continuous.

iii) Prove that, given a measurable space (S, Σ, μ), and $1 < p < \infty$, then the function $U : L_p(\mu) \to L_1(\mu)$, $U(f) = |f|^{p-1} f$ is bijective and uniformly continuous on any bounded set.

iv) Deduce that, given a measurable space (S, Σ, μ), and $1 < p < \infty$, then the topological spaces $L_1(\mu)$ and $L_p(\mu)$ are homeomorphic.

Properties for convex sets in finite or infinite-dimensional normed spaces

43. i) Let $n \in \mathbb{N}$. Prove that if $A \subseteq \mathbb{R}^n$ is a convex set which contains n linearly independent elements, then A cannot have an empty interior. Deduce from here that if $A \subseteq \mathbb{R}^n$ is convex and dense, then $A = \mathbb{R}^n$.

ii) Let $P = \{(x_n)_{n \in \mathbb{N}} \in l_1 \mid x_n \geq 0 \; \forall n \in \mathbb{N}\}$. Prove that $P \subseteq l_1$ is convex, $\mathrm{Sp}(P) = l_1$, but $\overset{\circ}{P} = \emptyset$.

iii) Let X be a Hausdorff real linear topological space on which there is a linear discontinuous functional $f : X \to \mathbb{R}$. Let $A = [f \geq 0]$, $B = [f < 0]$. Prove that:

a) A, B are convex, non-empty, and $A \cup B = X$.
b) $\mathrm{Sp}(A) = X$ and $\mathrm{Sp}(B) = X$.
c) $\overline{A} = X$, $\overline{B} = X$.

A norm on an infinite-dimensional linear spaces is never unique

44. Let $(X, \|\cdot\|)$ be an infinite-dimensional normed space and τ its topology. Prove that there are two norms $\|\cdot\|_1$ and $\|\cdot\|_2$ on X such that if τ_1 and τ_2 are the associated topologies for $\|\cdot\|_1$ and $\|\cdot\|_2$, then $\tau_2 \subseteq \tau \subseteq \tau_1$, $\tau_2 \neq \tau$, $\tau_1 \neq \tau$, i.e., τ_1 is (strictly) stronger than τ, and τ_2 is (strictly) weaker than τ.

1.2 Solutions

1. As is well known from linear algebra, every linear space has a basis. Let $B = \{e_i \mid i \in I\}$ be an algebraic basis of X over $\mathbb{K}$ ($= \mathbb{R}$ or $\mathbb{C}$). Then any $x \in X \setminus \{0\}$ can be written, uniquely, in the form $x = x_{i_1} e_{i_1} + \cdots + x_{i_n} e_{i_n}$ with $n \in \mathbb{N}$, $x_{i_j} \in \mathbb{K} \setminus \{0\}$ and $i_j \in I$, $j = 1, \ldots, n$, pairwise distinct. Define $\|x\| = \sum_{j=1}^{n} |x_{i_j}|$. Also let $\|x\| = 0$ if $x = 0$. We will prove that $\|\cdot\|$ is indeed a norm on X.

i) Let $x \in X$, $x \neq 0$. Then $x = x_{i_1} e_{i_1} + \cdots + x_{i_n} e_{i_n}$, with $n \in \mathbb{N}$, $x_{i_j} \in \mathbb{K}$, $i_j \in I$ pairwise distinct, $j = 1, \ldots, n$. Since $x \neq 0$ at least one of the x_{i_j} is not zero, which implies that $\|x\| = \sum_{j=1}^{n} |x_{i_j}| > 0$.

ii) Let $x \in X$ and $\lambda \in \mathbb{K}$. If $x = 0$ or $\lambda = 0$ then $\lambda x = 0$ and so $\|\lambda x\| = |\lambda| \|x\|$. Suppose now that $\lambda \in \mathbb{K} \setminus \{0\}$, $x \in X \setminus \{0\}$. If x has a decomposition as above, then λx has the decomposition $\lambda x = \lambda x_{i_1} e_{i_1} + \cdots + \lambda x_{i_n} e_{i_n}$. Then $\|\lambda x\| = \sum_{j=1}^{n} |\lambda x_{i_j}| = |\lambda| \|x\|$.

iii) Let now $x, y \in X$. If x or y is zero then $\|x + y\| = \|x\| + \|y\|$. Let us suppose now that x and y are not zero. Consider for x a decomposition as above and let $y = \sum_{s=1}^{m} y_{t_s} e_{t_s}$, with $m \in \mathbb{N}$, $y_{t_s} \in \mathbb{K} \setminus \{0\}$, $t_s \in I$, pairwise distinct, $s = 1, \ldots, m$. Let $A_x = \{i_1, \ldots, i_n\}$, $A_y = \{t_1, \ldots, t_m\}$, $A_x, A_y \subseteq I$. If $A_x \cap A_y = \emptyset$ then $x + y = \sum_{j=1}^{n} x_{i_j} e_{i_j} + \sum_{s=1}^{m} y_{t_s} e_{t_s}$, and this is the only decomposition of $x + y$ with respect to the basis B, with coefficients in $\mathbb{K} \setminus \{0\}$. Then $\|x + y\| = \sum_{j=1}^{n} |x_{i_j}| + \sum_{s=1}^{m} |y_{t_s}| = \|x\| + \|y\|$.

Let us suppose now that $A_{xy} := A_x \cap A_y$ is non-empty. Suppose, for example, that $i_n = t_m$, $i_{n-1} = t_{m-1}, \ldots, i_{n-k} = t_{m-k}$, so $A_{xy} = \{i_n, \ldots, i_{n-k}\} = \{t_m, \ldots, t_{m-k}\}$. Then $x + y$ has the decomposition: $x + y = \sum_{j=1}^{n-k-1} x_{i_j} e_{i_j} + \sum_{s=1}^{m-k-1} y_{t_s} e_{t_s} + \left[\sum_{l=1}^{k} \left(x_{i_{n-l}} + y_{t_{m-l}} \right) e_{i_{n-l}} \right]^*$,

where by $\sum^*$ we understand that in this summation appear only those elements for which $x_{i_{n-l}} + y_{t_{m-l}} \neq 0$. If $x + y = 0$ then clearly $\|x + y\| \leq \|x\| + \|y\|$. If $x + y \neq 0$ then

$$\|x + y\| = \sum_{j=1}^{n-k-1} |x_{i_j}| + \sum_{s=1}^{m-k-1} |y_{t_s}| + \left[\sum_{l=1}^{k} |x_{i_{n-l}} + y_{t_{m-l}}| \right]^* \leq \sum_{j=1}^{n-k-1} |x_{i_j}| + \sum_{s=1}^{m-k-1} |y_{t_s}| + \sum_{l=1}^{k} \left(|x_{i_{n-l}}| + |y_{t_{m-l}}| \right) = \|x\| + \|y\|.$$

Hence $(X, \|\cdot\|)$ is a normed space.

Remark. The reader can prove that if X is infinite-dimensional, then the above norm is not a complete one. Also, using the same kind of reasoning, one can prove that on any linear space we can construct an inner product.

2. See [40, exercise 1.56].

If $(X, \|\cdot\|)$ is a normed space, for $x, y \in B_X$ and $t \in [0, 1]$ we have $\|tx + (1-t)y\| \leq t\|x\| + (1-t)\|y\| \leq 1$, i.e., B_X is convex.

Conversely, let us assume that B_X is a convex set. We must show that p is a norm, i.e., we must prove the triangle inequality: $p(x+y) \leq p(x) + p(y)$ $\forall x, y \in X$. Let $x, y \in X$. If, for example, $p(x) = 0$, it follows from (i) that $x = 0$, and then $p(x+y) = p(y) = p(x) + p(y)$. If $p(x) \neq 0$ and $p(y) \neq 0$, the elements $a = x/p(x)$ and $b = y/p(y)$ have the properties that $p(a) = 1$, $p(b) = 1$. Then, since B_X is a convex set,

$$p\left(\frac{p(x)}{p(x)+p(y)}a + \frac{p(y)}{p(x)+p(y)}b\right) \leq 1,$$

i.e.,

$$p\left(\frac{x+y}{p(x)+p(y)}\right) \leq 1.$$

Applying now (ii) we obtain that

$$\frac{p(x+y)}{p(x)+p(y)} \leq 1,$$

i.e., $p(x+y) \leq p(x) + p(y)$.

3. Let $x, y \in S_X$. We have the following cases:

i) $y \neq -x$. In this case $(1-t)x + ty \neq 0$ $\forall t \in [0, 1]$. Indeed, if there is a $t \in [0, 1]$ such that $(1-t)x + ty = 0$, then $t \neq 0$ and $t \neq 1$ ($x, y \in S_X$). Hence $t \in (0, 1)$ and $(1-t)x + ty = 0$, $(1-t)x = -ty$, and then $|1-t|\|x\| = |t|\|y\|$, $1-t = t$, $t = 1/2$, from whence $(1 - 1/2)x = -y/2$, $y = -x$, contradiction. Therefore we can define

$$\gamma : [0, 1] \to X, \gamma(t) = \frac{(1-t)x + ty}{\|(1-t)x + ty\|}.$$

Obviously γ is continuous, $\gamma([0,1]) \subseteq S_X$, $\gamma(0) = x$ and $\gamma(1) = y$, i.e., γ is a path in S_X connecting x with y.

ii) $y = -x$. We must prove that x and $-x$ can be connected by a continuous path contained in S_X. Since $\dim_{\mathbb{K}} X \geq 2$, there is a $z \in X$ such that the system $\{x, z\}$ is linearly independent. Then $x \neq z/\|z\|$ and $-x \neq z/\|z\|$. From (i), x and $z/\|z\|$ can be connected by a continuous path $\gamma_1 : [0, 1] \to X : \gamma_1([0, 1]) \subseteq S_X, \gamma_1(0) = x, \gamma_1(1) = z/\|z\|$. Also, $-x$ and $z/\|z\|$ can be connected by a continuous path $\gamma_2 : [0, 1] \to X : \gamma_2([0, 1]) \subseteq S_X$, $\gamma_2(0) = -x$, $\gamma_2(1) = z/\|z\|$. As is well known, the union $\gamma_1 \vee \gamma_2$ of γ_1 and γ_2 connects x with $-x$ and its range is contained in S_X: we consider $\gamma_1 \vee \gamma_2 : [0, 2] \to X$,

$$(\gamma_1 \vee \gamma_2)(t) = \begin{cases} \gamma_1(t) & \text{if } 0 \leq t \leq 1, \\ \gamma_2(2-t) & \text{if } 1 \leq t \leq 2. \end{cases}$$

4. We first prove that for every convex function $\varphi : [0, \infty) \to \mathbb{R}$ of class C^1, with $\varphi'(0) = 0$, we have:

1.2 Solutions

(1) $\varphi(t+\alpha) - \varphi(t) - \varphi(\alpha) \geq -\varphi(0),$
(2) $\varphi(|t-\alpha|) - \varphi(t) - \varphi(\alpha) \leq -\varphi(0), \forall t \geq 0, \forall \alpha > 0.$

Indeed, for $\alpha > 0$ fixed, let $h : [0, \infty) \to \mathbb{R}$, $h(t) = \varphi(t+\alpha) - \varphi(t) - \varphi(\alpha) + \varphi(0)$. We have $h'(t) = \varphi'(t+\alpha) - \varphi'(t)$. Since φ is convex and of class C^1, φ' is increasing, i.e., $\varphi'(t+\alpha) \geq \varphi'(t)$, $h'(t) \geq 0$, i.e., h is increasing, and therefore $h(t) \geq h(0) = 0 \ \forall t \geq 0$, i.e., the first inequality. For the second one let $g : [0, \infty) \to \mathbb{R}$, $g(t) = \varphi(|t-\alpha|) - \varphi(t) - \varphi(\alpha) + \varphi(0)$. Then using the condition $\varphi'(0) = 0$ it is easy to see that g is differentiable at α and $g'(\alpha) = -\varphi'(\alpha)$. It follows that g is differentiable on $[0, \infty)$ and that

$$g'(t) = \begin{cases} \varphi'(t-\alpha) - \varphi'(t) & \text{if } t \geq \alpha, \\ -(\varphi'(\alpha - t) + \varphi'(t)) & \text{if } 0 \leq t < \alpha. \end{cases}$$

Since φ' is increasing we obtain that $g'(t) \leq 0 \ \forall t \geq 0$, i.e., g is decreasing, from whence $g(t) \leq g(0) = 0 \ \forall t \geq 0$, i.e., the second inequality.

i) Let $x \in X \setminus \{0\}$ and $\alpha > 0$. Consider the function $f : X \to \mathbb{R}$, $f(y) = \varphi(\|x+y\|) - \varphi(\|x\|) - \varphi(\alpha)$. Obviously, f is continuous. Moreover,

$$f\left(\frac{\alpha x}{\|x\|}\right) = \varphi\left(\left\|x + \frac{\alpha x}{\|x\|}\right\|\right) - \varphi(\|x\|) - \varphi(\alpha)$$
$$= \varphi(\|x\| + \alpha) - \varphi(\|x\|) - \varphi(\alpha).$$

From the inequality (1), for $t = \|x\|$ it follows that $f((\alpha x)/\|x\|) \geq -\varphi(0) = 0$. Also,

$$f\left(-\frac{\alpha x}{\|x\|}\right) = \varphi\left(\left\|x - \frac{\alpha x}{\|x\|}\right\|\right) - \varphi(\|x\|) - \varphi(\alpha)$$
$$= \varphi(|\|x\| - \alpha|) - \varphi(\|x\|) - \varphi(\alpha),$$

and from the inequality (2) for $t = \|x\|$ it follows that $f(-(\alpha x)/\|x\|) \leq -\varphi(0) = 0$. But $\alpha x/\|x\|, -\alpha x/\|x\| \in S_\alpha$, where $S_\alpha = \{x \in X \mid \|x\| = \alpha\}$. Since $\dim_\mathbb{K} X \geq 2$, from exercise 3 it follows that $S_\alpha \subseteq X$ is connected, from whence, since f is continuous, $f(S_\alpha) \subseteq \mathbb{R}$ is connected, hence $f(S_\alpha) \subseteq \mathbb{R}$ is an interval. Now $f(\alpha x/\|x\|), f(-\alpha x/\|x\|) \in f(S_\alpha)$, and since $f(S_\alpha) \subseteq \mathbb{R}$ is an interval, we obtain that $[f(-\alpha x/\|x\|), f(\alpha x/\|x\|)] \subseteq f(S_\alpha)$. But as we have already showed, $0 \in [f(-\alpha x/\|x\|), f(\alpha x/\|x\|)]$, i.e., $0 \in f(S_\alpha)$, from whence it follows that there is $y \in S_\alpha$ such that $f(y) = 0$, i.e., the statement.

ii) From the additional hypothesis on φ, that is $\varphi(t) = 0$ if and only if $t = 0$, it follows that $\varphi(t) > 0 \ \forall t > 0$.

From (i) it follows that for an $x \in X \setminus \{0\}$ and an $\alpha_1 > 0$, there is $y_1 \in X$ such that $\|y_1\| = \alpha_1$ and $\varphi(\|x + y_1\|) = \varphi(\|x\|) + \varphi(\|y_1\|)$. Since $\|x\| > 0$ it follows that $\varphi(\|x\|) > 0$, from whence $\varphi(\|x\|) + \varphi(\|y_1\|) > 0$, that is, $\varphi(\|x+y_1\|) > 0$, $\|x+y_1\| > 0$, i.e., $x+y_1 \in X$ is not zero. Applying again (i) for the couple $(x+y_1, \alpha_2)$, we deduce that there is $y_2 \in X$ such that $\|y_2\| = \alpha_2$ and $\varphi(\|x + y_1 + y_2\|) = \varphi(\|x + y_1\|) + \varphi(\|y_2\|)$. In this way we construct, by induction, a sequence $(y_n)_{n \in \mathbb{N}} \subseteq X$ such that $\|y_n\| = \alpha_n \ \forall n \in \mathbb{N}$ and $\varphi(\|x + y_1\|) = \varphi(\|x\|) + \varphi(\|y_1\|)$, $\varphi\left(\left\|x + \sum_{i=1}^n y_i\right\|\right) = \varphi\left(\left\|x + \sum_{i=1}^{n-1} y_i\right\|\right) + \varphi(\|y_n\|)$, for all $n \geq 2$. Then $\varphi\left(\left\|x + \sum_{i=1}^n y_i\right\|\right) = \varphi(\|x\|) + \sum_{i=1}^n \varphi(\|y_i\|) \ \forall n \geq 1$. Since

$\|y_n\| = \alpha_n \ \forall n \in \mathbb{N}$, and the series $\sum_{n=1}^{\infty} \alpha_n$ is convergent, it follows that the series $\sum_{n=1}^{\infty} y_n$ is absolutely convergent in X, and therefore convergent, since X is a Banach space. Then $\left\|x + \sum_{i=1}^{n} y_i\right\| \to \left\|x + \sum_{n=1}^{\infty} y_n\right\|$, from whence, using the continuity of the function φ, we obtain that $\varphi\left(\left\|x + \sum_{n=1}^{\infty} y_n\right\|\right) = \varphi(\|x\|) + \lim_{n \to \infty} \sum_{i=1}^{n} \varphi(\|y_i\|)$, i.e., the series $\sum_{n=1}^{\infty} \varphi(\|y_n\|)$ converges and $\varphi\left(\left\|x + \sum_{n=1}^{\infty} y_n\right\|\right) = \varphi(\|x\|) + \sum_{n=1}^{\infty} \varphi(\|y_n\|)$, i.e., the statement.

Remarks. 1. If we have an arbitrary convex C^1 function $\psi : [0, \infty) \to \mathbb{R}$, if we consider $\varphi : [0, \infty) \to \mathbb{R}$, $\varphi(t) = \psi(t) - \psi(0) - t\psi'(0)$, and if is true that $\varphi(t) = 0$ if and only if $t = 0$, then φ satisfies the conditions from the statement.

2. For $\varphi(t) = t^2$ the part (i) of the above exercise is the part (a) from [9, exercise 7, chapter V, section 1].

5. Let $f_n(t) = t^n \ \forall t \in [0, 1], \ \forall n \geq 0$. Then $\|f_0\|_\infty = 1$, $\|f_0\|_1 = a$, hence $\|f_0\| = \min(1, a)$. We also have: $\|f_n\|_\infty = 1$, $\|f_n\|_1 = a/(n+1)$, i.e., $\|f_n\| = \min(1, a/(n+1))$ $\forall n \in \mathbb{N}$. For any $n \in \mathbb{N}$, $\|f_0 + f_n\|_\infty = 2$, $\|f_0 + f_n\|_1 = a(1 + 1/(n+1))$ so $\|f_0 + f_n\| = \min(2, a(1 + 1/(n+1))) \ \forall n \in \mathbb{N}$. If $\|\cdot\|$ is a norm on $C[0, 1]$, then $\|f_0 + f_n\| \leq \|f_0\| + \|f_n\| \ \forall n \in \mathbb{N}$, i.e., $\min(2, a(1 + 1/(n+1))) \leq \min(1, a) + \min(1, a/(n+1)) \ \forall n \in \mathbb{N}$, from whence passing to the limit for $n \to \infty$, $\min(2, a) \leq \min(1, a) + \min(1, 0)$, $\min(2, a) \leq \min(1, a)$, i.e., $a \leq 1$.

Conversely, if $a \leq 1$ then $\|f\|_1 = a \int_0^1 |f(x)| \, dx \leq \|f\|_\infty$, and therefore $\|f\| = \min(\|f\|_\infty, \|f\|_1) = \|f\|_1$, which is clearly a norm.

Remark. The above exercise shows that the minimum of two norms is not a norm, in general. The reader can prove that the maximum of two norms is indeed a norm.

6. See [40, exercise 1.47].

Let $x = (x_n)_{n \in \mathbb{N}} \in P$. Then $\sum_{n=1}^{\infty} |x_n|^2 < \infty$, from whence it follows that there is a $k \in \mathbb{N}$ such that $\sum_{n=k+1}^{\infty} |x_n|^2 < 1/4$. Let $\delta = \min(1/2, 1 - |x_1|, ..., 1 - |x_k|)$, $\delta > 0$ since $x \in P$. We will prove that $B(x, \delta) \subseteq P$, which by the definition of the topology associated with a norm assures that P is open.

Let $y = (y_n)_{n \in \mathbb{N}} \in B(x, \delta)$, that is $\|y - x\| < \delta$. If $1 \leq n \leq k$, then $|y_n| \leq |y_n - x_n| + |x_n| \leq \|y - x\| + |x_n| < \delta + |x_n| \leq 1 - |x_n| + |x_n| = 1$. Therefore $|y_n| < 1 \ \forall 1 \leq n \leq k$. We also have $\|(y_n)_{n \geq k+1}\| \leq \|(y_n - x_n)_{n \geq k+1}\| + \|(x_n)_{n \geq k+1}\| \leq \|y - x\| + \left(\sum_{n=k+1}^{\infty} |x_n|^2\right)^{1/2} < \delta + 1/2 \leq 1$. Hence $|y_n| < 1 \ \forall n \geq k+1$, so $|y_n| < 1$ $\forall n \in \mathbb{N}$, i.e., $y \in P$.

Remark. One can prove that if $1 \leq p < \infty$, then the parallelepiped $P = \{(x_n)_{n \in \mathbb{N}} \in l_p \mid |x_n| < 1 \ \forall n \in \mathbb{N}\}$ is an open set in l_p.

7. See [40, exercise 1.48].

We will prove that P is an open set in l_2 if and only if $\inf_{n \in \mathbb{N}} a_n > 0$.

1.2 Solutions

Let us suppose that $P \subseteq l_2$ is an open set. Since $a_1 > 0$, there is $\varepsilon > 0$ such that $a_1 > \varepsilon > 0$. The element $x = (a_1 - \varepsilon, 0, 0, ...)$ belongs to P, and since P is open there is a $\delta > 0$ such that $B(x, \delta) \subseteq P$. Let $\eta = \min(\varepsilon, \delta/2) > 0$. Then we have $\overline{B}(x, \eta) \subseteq B(x, \delta) \subseteq P$. Let $n \in \mathbb{N}$, $n \geq 2$. The element $y = (a_1 - \varepsilon, 0, ..., 0, \eta, 0, ...)$ (η on the nth position) is in $\overline{B}(x, \eta)$, and therefore $y \in P$, i.e., from the definition of P, $\eta < a_n$. Obviously, $a_1 > \varepsilon \geq \eta$. Hence $\exists \eta > 0$ such that $a_n > \eta \ \forall n \in \mathbb{N}$, from whence $\inf_{n \in \mathbb{N}} a_n \geq \eta > 0$.

Conversely, if $\inf_{n \in \mathbb{N}} a_n > 0$, there is an $\varepsilon > 0$ such that $a_n > \varepsilon \ \forall n \in \mathbb{N}$. Let $x = (x_n)_{n \in \mathbb{N}} \in P$. Then $\sum_{n=1}^{\infty} |x_n|^2 < \infty$, and then it follows that there is a $k \in \mathbb{N}$ such that $\sum_{n=k+1}^{\infty} |x_n|^2 < \varepsilon^2/4$. Let $\delta = \min(\varepsilon/2, a_1 - |x_1|, ..., a_k - |x_k|)$, $\delta > 0$ since $x \in P$. We shall prove that $B(x, \delta) \subseteq P$. Let $y \in B(x, \delta)$, $y = (y_n)_{n \in \mathbb{N}}$. Then $\|y - x\| < \delta$. Also, $|y_n| \leq |y_n - x_n| + |x_n| \leq \|y - x\| + |x_n| < \delta + |x_n| \leq a_n - |x_n| + |x_n| = a_n$, $|y_n| < a_n$, $1 \leq n \leq k$. We also have $\|(y_n)_{n \geq k+1}\| \leq \|(y_n - x_n)_{n \geq k+1}\| + \|(x_n)_{n \geq k+1}\| \leq \|y - x\| + \left(\sum_{n=k+1}^{\infty} |x_n|^2\right)^{1/2} < \delta + \varepsilon/2 \leq \varepsilon$. Then $|y_n| \leq \|(y_n)_{n \geq k+1}\| < \varepsilon < a_n$, i.e., $|y_n| < a_n \ \forall n \geq k+1$. Hence $|y_n| < a_n \ \forall n \in \mathbb{N}$, i.e., $y = (y_n)_{n \in \mathbb{N}} \in P$.

8. See [40, exercise 1.39].

The answer is no. The function $f : [-1, 1] \to \mathbb{R}$, $f(x) = x$ is a polynomial. If the set P of all polynomials were open in $C[-1, 1]$, then there is an $\varepsilon > 0$ such that $B(f, \varepsilon) \subseteq P$. Let $g : [-1, 1] \to \mathbb{R}$, $g(x) = x + \varepsilon x/(2(1+|x|))$. We have $|g(x) - f(x)| = \varepsilon |x|/(2(1+|x|)) \leq \varepsilon/2 \ \forall x \in [-1, 1]$, i.e., $\|g - f\| \leq \varepsilon/2 < \varepsilon$, i.e., $g \in B(f, \varepsilon)$, and therefore $g \in P$, i.e., g is a polynomial. But the polynomials are C^{∞} functions, and an elementary calculus shows that g is not twice differentiable at zero. Hence the set of all polynomials is not an open set in $C[-1, 1]$.

9. See [40, exercise 1.68].

i) Let $x = (x_n)_{n \in \mathbb{N}}$, $y = (y_n)_{n \in \mathbb{N}}$, $x, y \in G_p$, and $\alpha, \beta \in \mathbb{R}$. Then $\alpha x + \beta y = (\alpha x_n + \beta y_n)_{n \in \mathbb{N}} \in l_p$ and $\sum_{n=1}^{\infty}(\alpha x_n + \beta y_n) = \alpha \sum_{n=1}^{\infty} x_n + \beta \sum_{n=1}^{\infty} y_n = 0$, i.e., $\alpha x + \beta y \in G_p$.

ii) Let us suppose that $1 < p < \infty$. We will prove that in this case G_p is not closed. Indeed, for $n \geq 1$, the element $(-1, 1/n, 1/n, ..., 1/n, 0, ...)$ (n times $1/n$) belongs to G_p and $x_n \to x = (-1, 0, 0, ...)$, since $\|x_n - x\| = (1/n^p + 1/n^p + \cdots + 1/n^p)^{1/p} = (n/n^p)^{1/p} = n^{1/p - 1} \to 0$ ($1 < p < \infty$!). Since x is not in G_p we obtain that G_p is not closed.

For $p = 1$, $G_1 = \left\{(x_n)_{n \in \mathbb{N}} \in l_1 \,\bigg|\, \sum_{n=1}^{\infty} x_n = 0\right\}$. Let $f : l_1 \to \mathbb{R}$, $f(x) = \sum_{n=1}^{\infty} x_n \ \forall x = (x_n)_{n \in \mathbb{N}} \in l_1$. We have $|f(x)| = \left|\sum_{n=1}^{\infty} x_n\right| \leq \sum_{n=1}^{\infty} |x_n| = \|x\| \ \forall x \in l_1$. Since f is also linear, it follows that f is continuous. Since $G_1 = \ker f = f^{-1}(\{0\})$, $\{0\} \subseteq \mathbb{R}$ is a closed set and f is continuous, we obtain that $G_1 \subseteq l_1$ is a closed set.

10. i) It is obvious.

ii) Let $1 < p < \infty$. For $n \in \mathbb{N}$, let us consider the function $f_n = -\chi_{[0,1]} + (1/n)\chi_{[1,n+1]}$. Then $\int_0^{\infty} f_n(x)dx = -1 + (1/n)\mu([1, n+1]) = 0$, i.e., $f_n \in E_p$. Since $p > 1$, it is easy to see that $f_n \to -\chi_{[0,1]}$ in L_p. Since $-\chi_{[0,1]} \notin E_p$ we obtain that E_p is not closed.

If $p = 1$, let $U : L_1[0,\infty) \to \mathbb{R}$, $U(f) = \int_0^\infty f(x)\,dx$. Obviously U is linear, and $|U(f)| = |\int_0^\infty f(x)\,dx| \le \int_0^\infty |f(x)|\,dx = \|f\|_1 \; \forall f \in L_1$, i.e., U is continuous. Since $E_1 = \ker U = U^{-1}(\{0\})$, $\{0\} \subseteq \mathbb{R}$ is closed, and U continuous, then $U^{-1}(\{0\}) = E_1$ is closed in L_1. Hence $E_p \subseteq L_p$ is a closed set if and only if $p = 1$.

11. i) Let $h : L(X) \to L(X)$, $h(B) = AB$. Obviously h is linear and $\|h(B)\| = \|AB\| \le \|A\|\|B\| \; \forall B \in L(X)$, i.e., h is continuous. The set from the statement is $\ker h = h^{-1}(\{0\})$ and since $\{0\} \subseteq L(X)$ is closed, it follows that $\ker h$ is a closed set.

ii) Let $g : L(X) \to L(X)$, $g(B) = AB - BA \; \forall B \in L(X)$. Then g is linear and $\|g(B)\| \le 2\|A\|\|B\| \; \forall B \in L(X)$, i.e., g is continuous. Again, the set from the statement is $\ker g$, and therefore is closed.

12. Let $U, V \in M$ and $\alpha, \beta \in \mathbb{K}$. Then $G \subseteq \ker U$, $G \subseteq \ker V$, from whence $G \subseteq \ker(\alpha U + \beta V)$, i.e., $\alpha U + \beta V \in M$ and M is a linear subspace. Let now $(U_n)_{n \in \mathbb{N}} \subseteq M$ and $U \in L(X,Y)$ such that $U_n \to U$. Since $U_n \in M$ we have $G \subseteq \ker U_n \; \forall n \in \mathbb{N}$. Since $U_n \to U$ it follows that $\bigcap_{n \in \mathbb{N}} \ker U_n \subseteq \ker U$ (indeed, if $x \in \bigcap_{n \in \mathbb{N}} \ker U_n$, then $U_n(x) = 0 \; \forall n \in \mathbb{N}$, and from $U_n(x) \to U(x)$ it follows that $U(x) = 0$, i.e., $x \in \ker U$), from whence $G \subseteq \bigcap_{n \in \mathbb{N}} \ker U_n \subseteq \ker U$, i.e., $G \subseteq \ker U$, $U \in M$. Therefore M is a closed set.

13. See [40, exercise 1.46].

i) Let P be the parallelepiped from the statement. Let us suppose that $P \subseteq l_p$ is a bounded set. Then there is an $M > 0$ such that $\|x\| \le M \; \forall x \in P$. Let $n \in \mathbb{N}$ be fixed. Then $\forall 0 < \varepsilon < \max(a_1, ..., a_n)$, the element $(a_1 - \varepsilon, ..., a_n - \varepsilon, 0, ...)$ belongs to P and then $\left(\sum_{k=1}^n (a_k - \varepsilon)^p\right)^{1/p} \le M$. Passing to the limit for $\varepsilon \to 0$, $\varepsilon > 0$ we obtain that $\sum_{k=1}^n a_k^p \le M^p$, and then $\sum_{n=1}^\infty a_n^p < \infty$, $a = (a_n)_{n \in \mathbb{N}} \in l_p$.

Conversely, if $a = (a_n)_{n \in \mathbb{N}} \in l_p$, since for every $x = (x_n)_{n \in \mathbb{N}} \in P$ we have $|x_n| < a_n \; \forall n \in \mathbb{N}$, then $\sum_{n=1}^\infty |x_n|^p < \sum_{n=1}^\infty a_n^p$, so $\|x\| \le \|a\| \; \forall x \in P$.

Hence the parallelepiped P is a bounded set in l_p if and only if $(a_n)_{n \in \mathbb{N}} \in l_p$.

ii) Let E be the ellipsoid from the statement. Let us suppose that $E \subseteq l_p$ is a bounded set, i.e. there is an $M > 0$ such that $\|x\| \le M \; \forall x \in E$. Let n be fixed. Since $0 < a_n$, let $0 < \varepsilon < a_n$. The element $x = (0, ..., 0, a_n - \varepsilon, 0, ...)$ ($n - 1$ times 0) belongs to E and then $\|x\| \le M$, i.e., $a_n - \varepsilon \le M \; \forall 0 < \varepsilon < a_n$. For $\varepsilon \to 0$ we obtain that $a_n \le M \; \forall n \in \mathbb{N}$, i.e., the sequence $(a_n)_{n \in \mathbb{N}}$ is in l_∞.

Conversely, let us suppose that the sequence $(a_n)_{n \in \mathbb{N}}$ is bounded: $\exists M > 0$ such that $a_n \le M \; \forall n \in \mathbb{N}$. If $x \in E$, then

$$\|x\|^p = \sum_{n=1}^\infty |x_n|^p = \sum_{n=1}^\infty \frac{|x_n|^p}{a_n^p} a_n^p \le M^p \sum_{n=1}^\infty \frac{|x_n|^p}{a_n^p} < M^p,$$

i.e., $\|x\| \le M \; \forall x \in E$.

Hence the ellipsoid $E \subseteq l_p$ is bounded if and only if $(a_n)_{n \in \mathbb{N}} \in l_\infty$.

1.2 Solutions

14. See [40, exercise 1.63].
Let L be a linear subspace and $B(a,\varepsilon) \subseteq X$ an open ball such that $L \subseteq B(a,\varepsilon)$. Let $x \in L$. Then, since L is a linear subspace, $nx \in L \ \forall n \in \mathbb{N}$, and since $L \subseteq B(a,\varepsilon)$, $nx \in B(a,\varepsilon) \ \forall n \in \mathbb{N}$, i.e., $||nx - a|| < \varepsilon \ \forall n \in \mathbb{N}$, from whence $||nx|| \leq ||nx - a|| + ||a|| < \varepsilon + ||a||$. Therefore $||x|| \leq (\varepsilon + ||a||)/n \ \forall n \in \mathbb{N}$ and passing to the limit for $n \to \infty$ we deduce that $||x|| \leq 0$, i.e., $x = 0$. Hence $L = \{0\}$.

15. Let $B(a,\varepsilon) \subseteq L$. Then $a \in B(a,\varepsilon) \subseteq L$, $a \in L$. Let $x \in B(0,\varepsilon)$, i.e., $||x|| < \varepsilon$. Then $a + x \in B(a,\varepsilon) \subseteq L$, $a + x \in L$, $a \in L$ and L is a linear subspace. Then $x \in L$, i.e., $B(0,\varepsilon) \subseteq L$. But $\forall x \in X$ with $x \neq 0$ we have $\varepsilon x/(2||x||) \in B(0,\varepsilon) \subseteq L$, i.e., $\varepsilon x/(2||x||) \in L$. Since L is a linear subspace we obtain that $x \in L$. Hence $X \subseteq L$, and therefore $L = X$.

16. See [24, exercise 3, section 1].
We have $\mathbf{C}_A = X \setminus A = \{x \in X \mid T_n x \to Tx\}$, from whence using the linearity we obtain that $\mathbf{C}_A$ is a linear subspace. Suppose that $x_0 \in A$. If $x \in \mathbf{C}_A$, then $x + \lambda x_0 \in A \ \forall \lambda \in \mathbb{K}$, $\lambda \neq 0$. Indeed, if $x + \lambda x_0 \in \mathbf{C}_A$, as $\mathbf{C}_A$ is a linear subspace and $x \in \mathbf{C}_A$, it follows that $\lambda x_0 \in \mathbf{C}_A$, from whence $x_0 \in \mathbf{C}_A$, which is false! Then: $x + (1/n) x_0 \in A \ \forall x \in \mathbf{C}_A$, $\forall n \geq 1$, and $x + (1/n) x_0 \to x$, from whence it follows that $x \in \overline{A}$, i.e., $\mathbf{C}_A \subseteq \overline{A}$. Hence $X = A \cup \mathbf{C}_A \subseteq \overline{A}$, i.e., $\overline{A} = X$, that is A is dense in X. Analogously for B.

17. The solution is the same as the one for exercise 16 and we will omit it.

18. If $\overset{\circ}{G} \neq \varnothing$, then using exercise 15 we obtain that $X = G$.
Remark. If $A \subseteq X$ has the property that $\mathbf{C}_A \subseteq X$ is a linear subspace, then it follows that $\mathbf{C}_A = X$ or $\overset{\circ}{\mathbf{C}_A} = \varnothing$, i.e., $A = \varnothing$ or $\mathbf{C}_{\overline{A}} = \varnothing$, that is $A = \varnothing$ or $\overline{A} = X$. Therefore A is either empty, or dense. We obtain a new solution for exercise 17.

19. Let $E_n = [1/2 - 1/n, 1/2 + 1/n]$ and $f_n = 1 - \chi_{E_n}$. Then $f_n = 0$ on E_n, $1/2 \in E_n$. We have $f_n \to f = 1$ in $L_2[0,1]$, since $||f_n - f||_2 = ||\chi_{E_n}||_2 = \sqrt{\mu(E_n)} = \sqrt{2/n} \to 0$. We obtain that the set A is not closed in $L_2[0,1]$.

20. Let $f_n \in M$, $f_n \to f \in L_2[0,1]$. We will prove that $f \in M$, which will assure that M is closed. Since $f_n \in M$, $f_n([0,1]) \subseteq [0,1]$ a.e., i.e., there is a set $A_n \subseteq [0,1]$ with $\mu(\mathbf{C}_{A_n}) = 0$ and $f_n(A_n) \subseteq [0,1]$. But from $f_n \to f$ in $L_2[0,1]$, as is well known from measure theory, there is a subsequence $(f_{k_n})_{n \in \mathbb{N}}$ such that $f_{k_n} \to f$ a.e.. Let $A \subseteq [0,1]$ with $\mu(\mathbf{C}_A) = 0$ such that $f_{k_n}(t) \to f(t) \ \forall t \in A$. Let $B = A \cap \bigcap_{n \in \mathbb{N}} A_{k_n}$. Then $f_{k_n}(B) \subseteq f_{k_n}(A_{k_n}) \subseteq [0,1] \ \forall n \in \mathbb{N}$ and since $f_{k_n}(t) \to f(t) \ \forall t \in B (\subseteq A)$, it follows that $f(t) \in [0,1] \ \forall t \in B$. But $\mathbf{C}_B = \mathbf{C}_A \cup \bigcup_{n \in \mathbb{N}} \mathbf{C}_{A_{k_n}}$ and since μ is countably additive we obtain that $\mu(\mathbf{C}_B) \leq \mu(\mathbf{C}_A) + \sum_{n=1}^{\infty} \mu(\mathbf{C}_{A_{k_n}}) = 0$, $\mu(\mathbf{C}_B) = 0$, i.e., $f([0,1]) \subseteq [0,1]$ a.e., i.e., $f \in M$. The fact that the set M is convex is obvious.

Remark. The same calculations as above assure that if $A \subseteq \mathbb{R}$ is a closed set, then the set $M = \{f \in L_2[0,1] \mid f([0,1]) \subseteq A \text{ a.e.}\}$ is closed in $L_2[0,1]$.

21. Let $\lambda = (\lambda_k)_{k \in \mathbb{N}} \in l_\infty$ and $\varepsilon > 0$. Let us consider the interval $[-\|\lambda\|, \|\lambda\|]$. We can find N intervals $I_1, ..., I_N$, closed to the left and opened to the right, each of length less than ε, pairwise disjoint, such that their union contains $[-\|\lambda\|, \|\lambda\|]$. For each $1 \leq i \leq N$, we denote by $M_i = \{n \in \mathbb{N} \mid \lambda_n \in I_i\}$. Then M_i are pairwise disjoint sets and $\bigcup_{i=1}^{N} M_i = \mathbb{N}$. Let $\alpha_i \in I_i$, $i = \overline{1, N}$, and define $x \in l_\infty$, $x = \sum_{i=1}^{N} \alpha_i \chi_{M_i}$. Then $x = (x_k)_{k \in \mathbb{N}} \in \mathrm{Sp}\{\chi_M \mid M \in \mathcal{P}(\mathbb{N})\}$.

For each $k \in \mathbb{N}$, there is an $i \in \{1, ..., N\}$ such that $\lambda_k \in I_i$. Since $\alpha_i \in I_i$, and I_i has its length less than ε, it follows that $|\lambda_k - \alpha_i| < \varepsilon$, i.e., $|\lambda_k - x_k| \leq \varepsilon$. Since $k \in \mathbb{N}$ is arbitrary, from the definition for the norm in l_∞ it follows that $\|\lambda - x\|_\infty \leq \varepsilon$.

Remark. Above, we have considered l_∞ as a linear space over $\mathbb{R}$. Using what is proved above, the reader can easily solve the same exercise, working with $\mathbb{K}$ ($= \mathbb{C}$ or $\mathbb{R}$) instead of $\mathbb{R}$.

22. See [40, exercise 1.32].

Let $1 \leq p < \infty$ and $x = (x_n)_{n \in \mathbb{N}} \in l_p$, i.e., $\sum_{n=1}^{\infty} |x_n|^p$ converges. Then $|x_n|^p \to 0$, $|x_n| \to 0$, and therefore $x = (x_n)_{n \in \mathbb{N}} \in c_0$. Let now $x = (1, 1/\ln 2, ..., 1/\ln n, ...)$. Obviously, $x \in c_0$, $(1/\ln n \to 0)$. Suppose that there is $1 \leq p < \infty$ such that $x \in l_p$. We obtain that $1 + \sum_{n=2}^{\infty} 1/(\ln n)^p < \infty$. From the well known Cauchy test we obtain that the series $\sum_{n=2}^{\infty} 2^n/(\ln 2^n)^p$ converges, i.e., the series $\sum_{n=2}^{\infty} 2^n/(n \ln 2)^p$ converges, which is false, since $\lim_{n \to \infty} 2^n/n^p = \infty$ if $1 \leq p < \infty$.

23. We have $l_p \subseteq c_0 \subseteq l_\infty$. We prove first that c_0 is a closed set in l_∞. Let $\xi \in \overline{c_0}$. Then we can find a sequence $(\zeta_n)_{n \in \mathbb{N}} \subseteq c_0$ such that $\|\zeta_n - \xi\| \to 0$. Let $\xi = (\xi_k)_{k \in \mathbb{N}} \in l_\infty$ and $\zeta_n = (\zeta_n^k)_{k \in \mathbb{N}} \in c_0$. Then $\forall \varepsilon > 0\ \exists n_\varepsilon \in \mathbb{N}$ such that $\sup_{k \in \mathbb{N}} |\zeta_{n_\varepsilon}^k - \xi_k| < \varepsilon/2$. Since $\zeta_{n_\varepsilon} = (\zeta_{n_\varepsilon}^k)_{k \in \mathbb{N}} \in c_0$, i.e., $\lim_{k \to \infty} |\zeta_{n_\varepsilon}^k| = 0$, there is a $k_\varepsilon \in \mathbb{N}$ such that $|\zeta_{n_\varepsilon}^k| < \varepsilon/2\ \forall k \geq k_\varepsilon$. Then $\forall k \geq k_\varepsilon$, we have $|\xi_k| \leq |\zeta_{n_\varepsilon}^k| + |\xi_k - \zeta_{n_\varepsilon}^k| < \varepsilon/2 + \sup_{n \in \mathbb{N}} |\xi_n - \zeta_{n_\varepsilon}^n| < \varepsilon$, i.e., $\xi_k \to 0$, i.e., $\xi = (\xi_k)_{k \in \mathbb{N}} \in c_0$. Hence $\overline{c_0} \subseteq c_0$, and since always $c_0 \subseteq \overline{c_0}$, we obtain that $\overline{c_0} = c_0$. Since $l_p \subseteq c_0$ we obtain that $\overline{l_p} \subseteq \overline{c_0} = c_0$. Let now $x = (x_n)_{n \in \mathbb{N}} \in c_0$. We denote by $\xi_n = (x_1, x_2, ..., x_n, 0, ...) \in l_p$. Then $\|\xi_n - x\|_\infty = \sup_{k \geq n+1} |x_k|$. Since $x = (x_n)_{n \in \mathbb{N}} \in c_0$, then $\forall \varepsilon > 0\ \exists n_\varepsilon \in \mathbb{N}$ such that $|x_k| < \varepsilon\ \forall k \geq n_\varepsilon$, and therefore $\|\xi_n - x\|_\infty \leq \varepsilon\ \forall n \geq n_\varepsilon$. Then $c_0 \subseteq \overline{l_p}$ and hence $\overline{l_p} = c_0$.

We give now a second solution for the fact that l_p is dense in c_0. Obviously $l_p \subseteq c_0$ is a linear subspace. Let $f \in c_0^*$ such that $f(x) = 0$ for all $x \in l_p$. Using the fact that $c_0^* = l_1$, there is a sequence $\xi = (\xi_n)_{n \in \mathbb{N}} \in l_1$ such that $f(x) = \sum_{n=1}^{\infty} x_n \xi_n\ \forall x = (x_n)_{n \in \mathbb{N}} \in c_0$. Then $f(e_k) = 0\ \forall k \in \mathbb{N}$, since $e_k \in l_p$ for each $k \in \mathbb{N}$, and therefore $\xi_k = 0\ \forall k \in \mathbb{N}$. Hence $f(x) = 0\ \forall x \in c_0$ and using now a separation result (see chapter 8) it follows that l_p is dense in $(c_0, \|\cdot\|_\infty)$.

24. See [9, exercise 15, chapter II, section 1].

We will solve the exercise by induction on $n \in \mathbb{N}$.

1.2 Solutions

If $n = 1$, since $A \subseteq \mathbb{R}$ is convex we obtain that $A \subseteq \mathbb{R}$ is an interval. Since by our hypothesis $A \cap \mathbb{R}$ (we consider the line $\mathbb{R}$) is closed it follows that $A \subseteq \mathbb{R}$ is a closed interval.

Let us suppose now that for a fixed $n \geq 2$, for each $1 \leq m \leq n-1$, if $B \subseteq \mathbb{R}^m$ is a convex set such that the intersection of B with each line from $\mathbb{R}^m$ is a closed set in $\mathbb{R}^m$, then $B \subseteq \mathbb{R}^m$ is closed. Consider $A \subseteq \mathbb{R}^n$ convex such that $A \cap d \subseteq \mathbb{R}^n$ is closed for each line $d \subseteq \mathbb{R}^n$. If A does not contain n linearly independent elements, then there are $m \leq n-1$ linearly independent elements $x_1, ..., x_m \in \mathbb{R}^n$ such that $A \subseteq \mathrm{Sp}\{x_1, ..., x_m\}$, (see the solution for exercise 43(i)). Let $Y = \mathrm{Sp}\{x_1, ..., x_m\}$, $\dim Y = m \leq n-1$. Let $T : Y \to \mathbb{R}^m$, $T\left(\sum_{i=1}^m \alpha_i x_i\right) = (\alpha_1, ..., \alpha_m)^t$ for $(\alpha_1, ..., \alpha_m)^t \in \mathbb{R}^m$. Then T is linear and bijective. Then $T : Y \to \mathbb{R}^m$, $T^{-1} : \mathbb{R}^m \to Y$ are continuous. Let $B = T(A) \subseteq \mathbb{R}^m$. Since T is linear, $B \subseteq \mathbb{R}^m$ is convex. Let d be a line in $\mathbb{R}^m$. Since T^{-1} is linear, $T^{-1}(d) \subseteq Y$ is a line, so $T^{-1}(d) \subseteq \mathbb{R}^n$ is a line. Then $T^{-1}(d) \cap A \subseteq \mathbb{R}^n$ is closed and since T is homeomorphism, $T(T^{-1}(d) \cap A) = d \cap T(A) = d \cap B \subseteq \mathbb{R}^m$ is closed. By our hypothesis of induction, $B \subseteq \mathbb{R}^m$ is closed, and since T is continuous, $A = T^{-1}(B)$ is closed in $\mathbb{R}^n$.

Let us suppose now that A contains n linearly independent elements. Then by exercise 43(i), $\overset{\circ}{A} \neq \varnothing$. Without loss of generality we can and do assume that $0 \in \overset{\circ}{A}$ (otherwise consider a translation), and then there is an $R > 0$ such that $\overline{B}(0, R) \subseteq A$, where $\overline{B}(0, R) = \{x \in \mathbb{R}^n | \|x\| \leq R\}$ ($\|\cdot\|$ is a norm on $\mathbb{R}^n$). Suppose that there is an $x \in \mathbb{R}^n$ such that $x \in \overline{A}$, but $x \notin A$. Then $x \neq 0$. We consider now the line which contains the points 0 and x, i.e., $d = \{tx | t \in \mathbb{R}\}$. Since $d \cap A \subseteq \mathbb{R}^n$ is closed and convex, there is a closed interval $[a, b] \subseteq \mathbb{R}$ such that $d \cap A = \{tx | t \in [a, b]\}$. Since $0 \in d \cap A$ it follows that $0 \in [a, b]$. Since $x \notin d \cap A$ it follows that $b < 1$. Put $t_0 = b$, and then: $tx \in A$ $\forall t \in [0, t_0]$, $tx \notin A$ $\forall t > t_0$, and $t_0 \in (0, 1)$. Let $r = (1 - t_0)R/t_0$, $r > 0$. We will show that $B(x, r) \cap A = \varnothing$ and then $x \notin \overline{A}$, which is contradictory.

Suppose that there is $y \in B(x, r)$ such that $y \in A$. Consider now the line which contains the elements y and $t_0 x$. We observe that $y \neq t_0 x$. If $y = t_0 x$, then $\|t_0 x - x\| < r$, i.e., $\|x(1 - t_0)\| < ((1-t_0)/t_0)R$, $\|x\| < R/t_0$, $\|t_0 x\| < R$, so there is a $t > t_0$, $\|tx\| < R$, and then $tx \in A$, false! So there is a (unique) line which contains the elements y and $t_0 x$, $d = \{t_0 x + t(y - t_0 x) | t \in \mathbb{R}\}$. For $t = -t_0/(1 - t_0)$, we obtain the element $(t_0 x - t_0 y)/(1 - t_0) \in d$. But

$$\left\| \frac{t_0 x - t_0 y}{1 - t_0} \right\| = \frac{t_0}{1 - t_0} \|x - y\| < \frac{t_0}{1 - t_0} r = R,$$

so there is $s > 0$ such that

$$B\left(\frac{t_0 x - t_0 y}{1 - t_0}, s\right) \subseteq B(0, R),$$

and so $B((t_0 x - t_0 y)/(1 - t_0), s) \subseteq A$. Then

$$t_0 y + (1 - t_0) B\left(\frac{t_0 x - t_0 y}{1 - t_0}, s\right)$$

is included in A, since A is convex, and is open. But

$$t_0 x = t_0 y + (1 - t_0) \frac{t_0 x - t_0 y}{1 - t_0}$$

and so
$$t_0 x \in t_0 y + (1-t_0) B\left(\frac{t_0 x - t_0 y}{1-t_0}, s\right).$$
Hence there is a $q > 0$ such that $B(t_0 x, q) \subseteq A$, and so there is a $t > t_0$ such that $tx \in A$, which is false! Hence $\overline{A} = A$.

25. See [1, exercise 16, Fall 1991].
Let $X, Y, X', Y' \in \mathcal{M}_n$, with $X = (x_{ij})$, $Y = (y_{ij})$, $X' = (x'_{ij})$, $Y' = (y'_{ij})$. Then:

$$\begin{aligned}
d\left(X'Y', XY\right) &= \sum_{i,j=1}^{n} \left| \sum_{k=1}^{n} x'_{ik} y'_{kj} - \sum_{k=1}^{n} x_{ik} y_{kj} \right| \\
&\leq \sum_{i,j=1}^{n} \sum_{k=1}^{n} \left| x'_{ik} y'_{kj} - x_{ik} y_{kj} \right| \\
&= \sum_{i,j=1}^{n} \sum_{k=1}^{n} \left| x'_{ik} \left(y'_{kj} - y_{kj} \right) + \left(x'_{ik} - x_{ik} \right) y_{kj} \right| \\
&\leq \sum_{i,j=1}^{n} \sum_{k=1}^{n} \left(\left| x'_{ik} \left(y'_{kj} - y_{kj} \right) \right| + \left| \left(x'_{ik} - x_{ik} \right) y_{kj} \right| \right) \\
&\leq \sum_{i,j=1}^{n} \sum_{k=1}^{n} \left(d\left(X', 0_n\right) d\left(Y, Y'\right) + d\left(Y, 0_n\right) d\left(X, X'\right) \right) \\
&= n^3 \left(d\left(X', 0_n\right) d\left(Y, Y'\right) + d\left(Y, 0_n\right) d\left(X, X'\right) \right).
\end{aligned}$$

For $X = Y$ and $X' = Y'$ we obtain that:

$$\begin{aligned}
d\left(X^2, \left(X'\right)^2\right) &\leq n^3 \left(d\left(X', 0_n\right) + d\left(X, 0_n\right) \right) d\left(X, X'\right) \\
&\leq n^3 \left(d\left(X, X'\right) + d\left(X, 0_n\right) + d\left(X, 0_n\right) \right) d\left(X, X'\right) \\
&= n^3 \left(d\left(X, X'\right) + 2 d\left(X, 0_n\right) \right) d\left(X, X'\right).
\end{aligned}$$

Now if $X_k \xrightarrow{d} X_0$ it follows that $X_k^2 \xrightarrow{d} X_0^2$, from whence $X_k^4 \xrightarrow{d} X_0^4$, etc.. We obtain that $X_k^{2^s} \xrightarrow{d} X_0^{2^s}$, for any $s \in \mathbb{N}$. Let now $A_k \xrightarrow{d} A$, such that for any k the matrix A_k is nilpotent. Then $A_k^n = 0_n$, for any k. Choosing now $s \in \mathbb{N}$ such that $N := 2^s \geq n$, we deduce that $A_k^N \xrightarrow{d} A^N$, and $A_k^N = 0_n$, for any k. It follows then that $A^N = 0_n$ i.e., A is nilpotent (here 0_n is the $n \times n$ null matrix).

26. See [1, exercise 20, Fall 1983].
For $p, q \in \mathbb{N}$ on the set $\mathcal{M}(p, q) = \mathcal{M}_{q \times p}(\mathbb{R})$ we consider the distance $d(A, B) = \sum_{i=1}^{q} \sum_{j=1}^{p} |a_{ij} - b_{ij}|$, where $A = (a_{ij})_{i=\overline{1,q},\, j=\overline{1,p}}$, $B = (b_{ij})_{i=\overline{1,q},\, j=\overline{1,p}}$. It is clear the fact that a sequence of matrices $(A_n)_{n \in \mathbb{N}} \subseteq \mathcal{M}_{q \times p}$ converges with respect to this metric to a matrix $A \in \mathcal{M}_{q \times p}$ if and only if $[A_n]_{i,j} \to A_{i,j}\ \forall i, j$ (with obvious notations). From this it follows

1.2 Solutions

that if $A_n \xrightarrow{d} A$ and $B_n \xrightarrow{d} B$, with $A, A_n \in \mathcal{M}_{q\times p}$ and $B, B_n \in \mathcal{M}_{p\times r}$, then $A_n B_n \xrightarrow{d} AB$ in $\mathcal{M}_{q\times r}$. (1)

In the case $p = q$ we consider the determinant function $\det : \mathcal{M}_p \to \mathbb{R}$, given by $\det A = \sum_{\sigma \in S_p} \varepsilon(\sigma) a_{1\sigma(1)} \cdots a_{p\sigma(p)}$, where $A = (a_{ij})_{i,j=\overline{1,p}}$. From the above remarks it follows that the determinant function is continuous if we consider it as a function between two metric spaces. Consider a sequence of invertible matrices $(A_n)_{n\in\mathbb{N}} \subseteq \mathcal{M}_p$, with $A_n = \left(a_{ij}^{(n)}\right)_{i,j=\overline{1,p}}$ such that $A_n \xrightarrow{d} A \in \mathcal{M}_p$, with A invertible. Consider $A_n^* = \begin{pmatrix} A_{11}^{(n)} & A_{21}^{(n)} & \ldots & A_{p1}^{(n)} \\ \ldots & \ldots & \ldots & \ldots \\ A_{1p}^{(n)} & A_{2p}^{(n)} & \ldots & A_{pp}^{(n)} \end{pmatrix}$, the adjoint for the matrix A_n, ($A_{ij}^{(n)}$ is the algebraic complement of the element $a_{ij}^{(n)}$ in the matrix A_n), and consider also $A^* = \begin{pmatrix} A_{11} & \ldots & A_{p1} \\ \ldots & \ldots & \ldots \\ A_{1p} & \ldots & A_{pp} \end{pmatrix}$.

Since the determinant function is continuous it follows that $A_{ij}^{(n)} \to A_{ij}$, for $i, j = \overline{1,p}$ and that $\det A_n \to \det A$. Then $A_n^{-1} = (1/\det A_n) A_n^* \xrightarrow{d} (1/\det A) A^* = A^{-1}$, for $n \to \infty$. (2)

i) Let $(A_k)_{k\in\mathbb{N}}$ be a sequence of matrices from $\mathcal{M}(m,n)$ with $A_k \xrightarrow{d} A \in \mathcal{M}(m,n)$ and such that $\text{rank}(A_k) < m$ $\forall k \in \mathbb{N}$. It follows that $(A_k^t) A_k \in \mathcal{M}_m$ has its rank at most the rank of A_k, i.e., strictly less than m, for any $k \in \mathbb{N}$. So $\det(A_k^t A_k) = 0$ for $k \in \mathbb{N}$. But $A_k \xrightarrow{d} A$ and $A_k^t \xrightarrow{d} A^t$. From (1) it follows that $(A_k^t) A_k \xrightarrow{d} A^t A$, hence by the continuity of the determinant it follows that $\det(A_k^t) A_k \to \det(A^t A)$, i.e., $\det(A^t A) = 0$, which means that $\text{rank}(A^t A) < m$. From here it follows that $\text{rank}(A) < m$. Hence $\mathcal{M}(m,n) \setminus L$ is closed in $(\mathcal{M}(m,n), d)$, which means that the set L is open in $(\mathcal{M}(m,n), d)$.

ii) We define $T : L \to \mathcal{M}(n,m)$ given by $T(A) = (A^t A)^{-1} A^t$. Then T is well defined and $T(A) A = I_m$, for any $A \in \mathcal{M}(m,n)$. We must prove that $T : (L, d) \to (\mathcal{M}(m,n), d)$ is continuous. Let $(A_k)_{k\in\mathbb{N}} \subseteq L$, with $A_k \xrightarrow{d} A \in L$. Then $(A_k^t A_k)_{k\in\mathbb{N}} \subseteq \mathcal{M}_m$, with $A_k^t A_k \xrightarrow{d} A^t A$. But $A_k^t A_k$ is invertible for any $k \in \mathbb{N}$, and $A^t A$ is also invertible. From (2) it follows that $(A_k^t A_k)^{-1} \xrightarrow{d} (A^t A)^{-1}$ and since $A_k^t \xrightarrow{d} A^t$ from (1) we obtain that $(A_k^t A_k)^{-1} A_k^t \xrightarrow{d} (A^t A)^{-1} A^t$, i.e., T is continuous.

27. For (i) and (ii), see [9, exercise 10, chapter III, section 4].

i) If $x = (x_n)_{n\in\mathbb{N}} \in S_c$, $y = (y_n)_{n\in\mathbb{N}} \in S_c$, then $x + y = (x_n + y_n)_{n\in\mathbb{N}}$ and the sequence $\left(\sum_{k=1}^n (x_k + y_k)\right)_{n\in\mathbb{N}} = \left(\sum_{k=1}^n x_k\right)_{n\in\mathbb{N}} + \left(\sum_{k=1}^n y_n\right)_{n\in\mathbb{N}}$ converges, so $x + y \in S_c$. Obviously if $x \in S_c$ and $\alpha \in \mathbb{K}$, then $\alpha x \in S_c$, so S_c is a linear space. Let us observe now that if $x = (x_n)_{n\in\mathbb{N}} \in S_c$, then the sequence $\left(\sum_{k=1}^n x_k\right)_{n\in\mathbb{N}}$ is bounded, since is convergent, so $\|x\| < \infty$. If $\|x\| = 0$, then $\sum_{k=1}^n x_k = 0$ $\forall n \in \mathbb{N}$, from whence $x_1 = 0$ and $x_n =$

$\sum_{k=1}^{n} x_k - \sum_{k=1}^{n-1} x_k = 0$ for $n \geq 2$, i.e., $x = 0$. Also

$$\|\alpha x\| = \sup_{n \in \mathbb{N}} \left|\sum_{k=1}^{n} \alpha x_k\right| = \sup_{n \in \mathbb{N}} |\alpha| \left|\sum_{k=1}^{n} x_k\right| = |\alpha| \|x\|.$$

If $y = (y_n)_{n \in \mathbb{N}}$, $z = (z_n)_{n \in \mathbb{N}}$, then

$$\|y + z\| = \sup_{n \in \mathbb{N}} \left|\sum_{k=1}^{n}(y_k + z_k)\right| \leq \sup_{n \in \mathbb{N}} \left|\sum_{k=1}^{n} y_k\right| + \sup_{n \in \mathbb{N}} \left|\sum_{k=1}^{n} z_k\right| = \|y\| + \|z\|.$$

We obtain that $\|\cdot\|$ is a norm on S_c.

We prove now that this norm is a complete one. For this, let $(w_n)_{n \in \mathbb{N}} \subseteq S_c$ be a Cauchy sequence with respect to this norm, i.e., $\forall \varepsilon > 0 \ \exists n_\varepsilon \in \mathbb{N}$ such that $\|w_n - w_m\| < \varepsilon$ for $n > m \geq n_\varepsilon$. If we denote by w_n^k the kth component for w_n, $w_n = (w_n^k)_{k \in \mathbb{N}}$, then: $\forall \varepsilon > 0$ $\exists n_\varepsilon \in \mathbb{N}$ such that

$$\left|w_n^k - w_m^k\right| \leq \left|\sum_{i=1}^{k}(w_n^i - w_m^i)\right| + \left|\sum_{i=1}^{k-1}(w_n^i - w_n^i)\right| \leq 2\|w_n - w_m\| < 2\varepsilon,$$

$\forall n > m \geq n_\varepsilon$, $\forall k \in \mathbb{N}$. This shows that for any $k \in \mathbb{N}$, the sequence $(w_n^k)_{n \in \mathbb{N}} \subseteq \mathbb{K}$ is Cauchy. Since $\mathbb{K}$ is complete there is $w = (w^k)_{k \in \mathbb{N}}$ such that $w_n^k \to w^k \ \forall k \in \mathbb{N}$. For $n > m \geq n_\varepsilon$, we have $\sup_{i \in \mathbb{N}} \left|\sum_{k=1}^{i}(w_n^k - w_m^k)\right| < \varepsilon$, and then for any $i \in \mathbb{N}$, $\left|\sum_{k=1}^{i}(w_n^k - w_m^k)\right| < \varepsilon$ for $n > m \geq n_\varepsilon$. Now passing to the limit for $n \to \infty$ we obtain that $\forall \varepsilon > 0 \ \exists n_\varepsilon \in \mathbb{N}$ such that $\forall m \geq n_\varepsilon$, $\left|\sum_{k=1}^{i}(w^k - w_m^k)\right| \leq \varepsilon$ for all $i \in \mathbb{N}$. Therefore: $\forall \varepsilon > 0 \ \exists n_\varepsilon \in \mathbb{N}$, such that $\forall m \geq n_\varepsilon$, $\|w - w_m\| \leq \varepsilon$, i.e., $w \in S_c$ and $w_m \to w$.

ii) Let $x = (x_n)_{n \in \mathbb{N}} \in S_c$. Then $\forall \varepsilon > 0 \ \exists n'_\varepsilon \in \mathbb{N}$ such that $\left|\sum_{i=n+1}^{n+p} x_i\right| < \varepsilon/2 \ \forall n \geq n'_\varepsilon$, $\forall p \geq 1$ (1). Let $n''_\varepsilon \in \mathbb{N}$ such that $1/2^n < \varepsilon/2 \ \forall n \geq n''_\varepsilon$ (2). Let $k = \max(n'_\varepsilon, n''_\varepsilon)$. The element $\xi = (x_1, ..., x_k, 1/2^{k+1}, 1/2^{k+2}, ...)$ belongs to l_1, and $\xi - x = (0, ..., 0, 1/2^{k+1} - x_{k+1}, ...)$, $\|\xi - x\| = \sup_{p \geq k+1} \left|\sum_{i=k+1}^{p}(1/2^i - x_i)\right|$. But for $p \geq k + 1$, using (1) and (2) we have

$$\left|\sum_{i=k+1}^{p}\left(\frac{1}{2^i} - x_i\right)\right| \leq \sum_{i=k+1}^{p} \frac{1}{2^i} + \left|\sum_{i=k+1}^{p} x_i\right| \leq \frac{1}{2^k} + \frac{\varepsilon}{2} < \varepsilon,$$

i.e., $\|\xi - x\| \leq \varepsilon$, and therefore l_1 is dense in S_c.

iii) Since for $(x_n)_{n \in \mathbb{N}} \in S_c$ the series $\sum_{n=1}^{\infty} x_n$ converges, it follows that $\lim_{n \to \infty} \sum_{k=n}^{\infty} x_k = 0$, and therefore U is well defined. Obviously, U is linear. For any $x = (x_n)_{n \in \mathbb{N}} \in S_c$, we have

$$\left|\sum_{k=1}^{\infty} x_k\right| = \left|\lim_{n \to \infty} \sum_{k=1}^{n} x_k\right| \leq \lim_{n \to \infty} \left|\sum_{k=1}^{n} x_k\right| \leq \sup_{n \in \mathbb{N}} \left|\sum_{k=1}^{n} x_k\right| = \|x\|.$$

1.2 Solutions

Then for any $n \in \mathbb{N}$ with $n \geq 2$,

$$\left|\sum_{k=n}^{\infty} x_k\right| \leq \left|\sum_{k=1}^{\infty} x_k\right| + \left|\sum_{k=1}^{n-1} x_k\right| \leq 2\|x\|,$$

so $\sup_{n \in \mathbb{N}} \left|\sum_{k=n}^{\infty} x_k\right| \leq 2\|x\|$, i.e., $\|Ux\| \leq 2\|x\| \ \forall x \in S_c$, hence $U \in L(S_c, c_0)$ and $\|U\| \leq 2$. Also, we have $U(1, -2, 0, 0, \ldots) = (-1, -2, 0, \ldots)$ and $2 = \|(-1, -2, 0, \ldots)\| \leq \|U\| \|(1, -2, 0, 0, \ldots)\| = \|U\|$, i.e., $\|U\| = 2$. Let $x = (x_n)_{n \in \mathbb{N}} \in S_c$ with $Ux = 0$, i.e., $\sum_{k=n}^{\infty} x_k = 0 \ \forall n \in \mathbb{N}$. Then for any $n \in \mathbb{N}$ we have $x_n = \sum_{k=n}^{\infty} x_k - \sum_{k=n+1}^{\infty} x_k = 0 - 0 = 0$, so $x = 0$, from whence it follows that U is injective. Let now $(y_n)_{n \in \mathbb{N}} \in c_0$. We define $x = (y_1 - y_2, y_2 - y_3, y_3 - y_4, \ldots)$. Then $(y_1 - y_2) + (y_2 - y_3) + \cdots + (y_n - y_{n+1}) = y_1 - y_{n+1} \to y_1$, that is, $x \in S_c$. We also have $Ux = y$, so $U \in L(S_c, c_0)$ is linear, continuous and bijective. Since $U^{-1} : c_0 \to S_c$ is defined by $U^{-1}(y) = x$ if and only if $Ux = y$ it follows that $U^{-1}(y) = (y_1 - y_2, y_2 - y_3, y_3 - y_4, \ldots)$ and $\|U^{-1}(y)\| = \sup_{n \in \mathbb{N}} |y_1 - y_{n+1}| \leq 2\|y\| \ \forall y \in c_0$ i.e., U^{-1} is continuous with $\|U^{-1}\| \leq 2$. Also, $U^{-1}(1, -1, 0, 0, \ldots) = (2, -1, 0, \ldots)$, thus $2 = \|(2, -1, 0, \ldots)\| \leq \|U^{-1}\| \|(1, -1, 0, 0, \ldots)\| = \|U^{-1}\|$, i.e., $\|U^{-1}\| = 2$.

iv) As in (iii), T is well defined, linear, and continuous, with $\|T\| \leq 2$. We also have $T(1, -2, 0, 0, \ldots) = (-2, 0, 0, \ldots)$ and $2 = \|(-2, 0, \ldots)\| \leq \|T\| \|(1, -2, 0, 0, \ldots)\| = \|T\|$, i.e., $\|T\| = 2$. Now let $(y_n)_{n \in \mathbb{N}} \in c_0$. We define $x = (-y_1, y_1 - y_2, y_2 - y_3, y_3 - y_4, \ldots)$. Then $-y_1 + (y_1 - y_2) + (y_2 - y_3) + \cdots + (y_n - y_{n+1}) = -y_{n+1} \to 0$, so $x \in S_c$. We also have $Tx = y$, so T is surjective. Clearly V is well defined, linear, and $\|V(x)\| = \sup_{n \in \mathbb{N}} |x_n| = \|x\| \ \forall x \in c_0$, i.e., V is an isometry, and hence $\|V\| = 1$. We also have $T(V(x)) = T(-x_1, x_1 - x_2, x_2 - x_3, \ldots) = (x_1, x_2, x_3, \ldots) = x$ for all $x \in c_0$, and $V(T(x)) = V\left(\sum_{n=2}^{\infty} x_n, \sum_{n=3}^{\infty} x_n, \ldots\right) = \left(-\sum_{n=2}^{\infty} x_n, x_2, x_3, \ldots\right)$ for all $x \in S_c$.

Remark. For (iii), using the open mapping theorem, from $U \in L(S_c, c_0)$ being bijective it follows directly that U is an isomorphism, since the norms on S_c and c_0 are complete.

28. The case $x = y$ is obvious. If $x \neq y$, let us consider the element $z = x + \alpha(x - y)/\|x - y\|$. Then $\|x - z\| = \alpha$, so $z \in \overline{B}(x, \alpha)$, hence, by our hypothesis, $z \in \overline{B}(y, \beta)$, i.e., $\|z - y\| \leq \beta$. Then

$$\left\|(x - y)\left(1 + \frac{\alpha}{\|x - y\|}\right)\right\| \leq \beta,$$

so

$$\|x - y\|\left(1 + \frac{\alpha}{\|x - y\|}\right) \leq \beta,$$

that is, $\|x - y\| \leq \beta - \alpha$ and $\beta \geq \alpha$.

Remark. If we eliminate the second axiom from the definition for a norm, the statement from this exercise is no longer true. For example, if we consider $q : \mathbb{R} \to \mathbb{R}_+$, $q(x) = |x|/(1 + |x|)$, then $q(x) = 0 \Leftrightarrow x = 0$, $q(-x) = q(x)$, $q(x + y) \leq q(x) + q(y) \ \forall x, y \in \mathbb{R}$, and if we denote $\overline{B}(0, \alpha) = \{x \in \mathbb{R} \mid q(x) \leq \alpha\}$, then $\overline{B}(0, 2) = \overline{B}(0, 1)$.

29. Suppose that X is a Banach space. For any $n \in \mathbb{N}$, let $B_n = \overline{B}(x_n, r_n)$, $x_n \in X$, $r_n \geq 0$, such that $B_{n+1} \subseteq B_n$ $\forall n \in \mathbb{N}$, and $\lim_{n\to\infty} r_n = 0$. Then for any $m \geq n \geq 1$, $x_m \in B_m \subseteq B_n = \overline{B}(x_n, r_n)$, so $\|x_n - x_m\| \leq r_n$. Since $\lim_{n\to\infty} r_n = 0$ it follows that the sequence $(x_n)_{n\in\mathbb{N}}$ is Cauchy. Since X is complete, there is an $x \in X$ such that $x_n \to x$. For any $k \in \mathbb{N}$, the sequence $(x_n)_{n\geq k}$ is contained in B_k and converges to x. Since B_k is closed we obtain that $x \in B_k$, and therefore $x \in \bigcap_{k\in\mathbb{N}} B_k$.

Let us suppose now that for any decreasing sequence of closed balls in X with the sequence of radii convergent to zero, their intersection is not the empty set. Let $(x_n)_{n\in\mathbb{N}} \subseteq X$ be a Cauchy sequence. To prove that the sequence $(x_n)_{n\in\mathbb{N}}$ converges, it is sufficient to find a convergent subsequence. Since $(x_n)_{n\in\mathbb{N}} \subseteq X$ is a Cauchy sequence, then there is a subsequence $(n_k)_{k\in\mathbb{N}}$ of $\mathbb{N}$ such that $\|x_{n_{k+1}} - x_{n_k}\| \leq 1/2^k$ for all $k \in \mathbb{N}$. Let now $B_k = \overline{B}(x_{n_{k+1}}, 1/2^k)$ $\forall k \in \mathbb{N}$. For $k \in \mathbb{N}$, let $x \in B_{k+1}$. Then $\|x - x_{n_{k+1}}\| \leq \|x - x_{n_{k+2}}\| + \|x_{n_{k+2}} - x_{n_{k+1}}\| \leq 1/2^k$, so $x \in B_k$. Therefore $B_{k+1} \subseteq B_k$ for all $k \in \mathbb{N}$ and then, by hypothesis, there is an $x \in \bigcap_{k\in\mathbb{N}} B_k$. Then $\|x - x_{n_{k+1}}\| \leq 1/2^k$ for all $k \in \mathbb{N}$, and therefore $\lim_{k\to\infty} x_{n_k} = x$, which is exactly what we wanted to prove.

30. i) See [6, exercise 2, chapter 2].

Let us denote $B_n = \overline{B}(x_n, r_n)$. From exercise 28 it follows that the sequence $(r_n)_{n\in\mathbb{N}}$ is decreasing, and also $\|x_n - x_m\| \leq |r_n - r_m|$ $\forall m, n \in \mathbb{N}$. Since $r_n \geq r > 0$ $\forall n \in \mathbb{N}$, it follows that $\inf_{n\in\mathbb{N}} r_n = \lim_{n\to\infty} r_n = R \geq r$. Let $r > \varepsilon > 0$. The sequence $(r_n)_{n\in\mathbb{N}}$ converges, so it is a Cauchy sequence, hence there is an $N \in \mathbb{N}$, depending on ε, such that $|r_n - r_m| \leq \varepsilon$ for all $m, n \geq N$. We will prove that $\overline{B}(x_N, R - \varepsilon) \subseteq \bigcap_{k\in\mathbb{N}} B_k$. If $k \leq N$, since $R - \varepsilon < R \leq r_N$, it follows that $\overline{B}(x_N, R - \varepsilon) \subseteq \overline{B}(x_N, r_N) = B_N \subseteq B_k$. Let now $k > N$ and $y \in X$ with $\|y - x_N\| \leq R - \varepsilon$. Since $k > N$ we have $|r_k - r_N| \leq \varepsilon$ and therefore $\|x_k - x_N\| \leq \varepsilon$. Then $\|y - x_k\| \leq \|y - x_N\| + \|x_N - x_k\| \leq R - \varepsilon + \varepsilon = R \leq r_k$, i.e., $y \in B_k$. We obtain that $\overline{B}(x_N, R - \varepsilon) \subseteq \bigcap_{n\in\mathbb{N}} B_n$, thus $\overline{B}(x_N, r - \varepsilon) \subseteq \bigcap_{n\in\mathbb{N}} B_n$.

ii) Let us denote again $B_n = \overline{B}(x_n, r_n)$. Suppose now that the norm on X is complete. The sequence $(x_n)_{n\in\mathbb{N}}$ is a Cauchy sequence, hence there is an $x \in X$ such that $x_n \to x$. We will prove first that $B(x, r) \subseteq \bigcap_{n\in\mathbb{N}} B_n$. Indeed, let $y \in B(x, r)$, that is, $\|y - x\| < r$. Using (i) for any $r > \varepsilon > 0$ there is an $x_{N_\varepsilon} \in X$ which can be chosen such that $\|x - x_{N_\varepsilon}\| < \varepsilon/2$ and $\overline{B}(x_{N_\varepsilon}, r - \varepsilon/2) \subseteq \bigcap_{n\in\mathbb{N}} B_n$. Let $\varepsilon = r - \|y - x\|$, $r > \varepsilon > 0$. Then $\|y - x_{N_\varepsilon}\| \leq \|y - x\| + \|x - x_{N_\varepsilon}\| < r - \varepsilon + \varepsilon/2 = r - \varepsilon/2$, so $y \in B(x_{N_\varepsilon}, r - \varepsilon/2)$, i.e., $y \in \bigcap_{n\in\mathbb{N}} B_n$. We have proved that $B(x, r) \subseteq \bigcap_{n\in\mathbb{N}} B_n$. But B_n is closed $\forall n \in \mathbb{N}$, hence $\bigcap_{n\in\mathbb{N}} B_n$ is closed. It follows that $\overline{B}(x, r) = \overline{B(x, r)} \subseteq \bigcap_{n\in\mathbb{N}} B_n$.

iii) Let $X = c_{00}$, the space of all sequences with finite support, with the supremum norm. For any $n \in \mathbb{N}$, let $x_n = (1, 1/2, 1/2^2, ..., 1/2^n, 0, 0, ...) \in c_{00}$, $r_n = 1 + 1/2^n$ and $B_n = \overline{B}(x_n, r_n)$. As it is easy to see, we have $B_1 \supseteq B_2 \supseteq \cdots \supseteq B_n \supseteq \cdots$, and let us suppose that there is an $x \in c_{00}$ such that $\overline{B}(x, 1) \subseteq \bigcap_{n\in\mathbb{N}} B_n$. Then $B(x, 1) \subseteq B_n$ for all $n \in \mathbb{N}$, hence by exercise 28 it follows that $\|x - x_n\| \leq 1/2^n$ $\forall n \in \mathbb{N}$, from whence $x_n \to x$. From here it follows that $x = (1, 1/2, 1/2^2, ..., 1/2^n, 1/2^{n+1}, ...)$, and therefore

$x \notin c_{00}$!

iv) Let $(x_n)_{n \in \mathbb{N}}$ be a Cauchy sequence. It follows that there is a subsequence $(x_{n_k})_{k \in \mathbb{N}}$ such that $\|x_{n_{k+1}} - x_{n_k}\| \leq 1/2^{k+1}$ for all $k \in \mathbb{N}$. Let $B_k = \overline{B}(x_{n_k}, r + 1/2^k)$ for $k \in \mathbb{N}$. For $y \in B_{k+1}$, we have $\|y - x_{n_k}\| \leq \|y - x_{n_{k+1}}\| + \|x_{n_{k+1}} - x_{n_k}\| \leq r + 1/2^{k+1} + 1/2^{k+1} = r + 1/2^k$, so $y \in B_k$, i.e., $B_{k+1} \subseteq B_k$ for all $k \in \mathbb{N}$. By our hypothesis, there is an $x \in X$ such that $\overline{B}(x, r) \subseteq \bigcap_{k \in \mathbb{N}} \overline{B}(x_{n_k}, r + 1/2^k)$. Using exercise 28, it follows that $\|x - x_{n_k}\| \leq 1/2^k$ for all $k \in \mathbb{N}$, hence $x_{n_k} \to x$. Since $(x_n)_{n \in \mathbb{N}}$ is a Cauchy sequence, and it has a subsequence convergent towards x, it follows that $\lim_{n \to \infty} x_n = x \in X$. Therefore X is complete.

31. i) If $\|\cdot\|_1$ and $\|\cdot\|_2$ are two norms which generate the same topology, they are equivalent, that is, we can find $m, M > 0$ such that $m\|x\|_1 \leq \|x\|_2 \leq M\|x\|_1$ for all $x \in X$.

If $\|\cdot\|_1$ is complete, then from $m\|x\|_1 \leq \|x\|_2 \; \forall x \in X$ it follows that if $(x_n)_{n \in \mathbb{N}} \subseteq X$ is a Cauchy sequence with respect to the norm $\|\cdot\|_2$, then the sequence is also Cauchy with respect to the norm $\|\cdot\|_1$ and then there is an $x \in X$ such that $\|x_n - x\|_1 \to 0$. Using now that $\|x_n - x\|_2 \leq M\|x_n - x\|_1$ it follows that $\|x_n - x\|_2 \to 0$. Analogously, if $\|\cdot\|_2$ is complete then the norm $\|\cdot\|_1$ is also complete.

ii) See [44, exercise 12, chapter 1].

Obviously $(\mathbb{R}, d_1)$ is a metric space, and, since ϕ is injective, $(\mathbb{R}, d_2)$ is also a metric space. The function $\phi : \mathbb{R} \to (-1, 1) \subseteq \mathbb{R}$ is bijective and continuous on $(\mathbb{R}, |\cdot|)$, and $\phi^{-1} : (-1, 1) \subseteq \mathbb{R} \to \mathbb{R}$, $\phi^{-1}(y) = y/(1 - |y|)$ is also continuous on $(\mathbb{R}, |\cdot|)$. It is easy to see now that the two metric generate the same topology on $\mathbb{R}$.

To prove that $(\mathbb{R}, d_2)$ is not a complete metric space, we consider the sequence $x_n = n$ $\forall n \in \mathbb{N}$. Then $x_n/(1 + |x_n|) = n/(1 + n) \to 1$, from whence $d_2(x_n, x_m) \to 0$ for $m, n \to \infty$, i.e., $(x_n)_{n \in \mathbb{N}} \subseteq (\mathbb{R}, d_2)$ is a Cauchy sequence. Let us suppose that there is an $x_0 \in \mathbb{R}$ with $x_n \xrightarrow{d_2} x_0$. Then $|\phi(x_n) - \phi(x_0)| \to 0$, and since $\phi(x_n) \to 1$ we will obtain that $\phi(x_0) = 1$, i.e., $x_0/(1 + |x_0|) = 1$, $x_0 > 0$, $x_0 = 1 + x_0$, a contradiction.

32. For (i) and (ii), see [1, exercise 6, Fall 1992].

i) Suppose that X_1 is complete. We will prove that X_2 is complete. Let $(y_n)_{n \in \mathbb{N}} \subseteq X_2$ be a Cauchy sequence. Since f is surjective, there is a sequence $(x_n)_{n \in \mathbb{N}} \subseteq X_1$ such that $f(x_i) = y_i \; \forall i \in \mathbb{N}$. Then $d_1(x_i, x_j) \leq d_2(f(x_i), f(x_j)) = d_2(y_i, y_j)$ for all $i, j \in \mathbb{N}$ and we deduce that $(x_n)_{n \in \mathbb{N}} \subseteq X_1$ is a Cauchy sequence in X_1, and so there is an $x \in X_1$ such that $x_n \to x$. Let $y = f(x)$. Since f is continuous, $f(x_n) \to f(x)$, i.e., $y_n \to y$ and therefore X_2 is complete.

ii) We will construct a counterexample. Let $X_1 = \{1, 1/2, 1/3, ...\}$, $d_1 : X_1 \times X_1 \to \mathbb{R}_+$, $d_1(x, y) = |x - y|$ and $X_2 = \{1, 2, 3, ...\}$, $d_2 : X_2 \times X_2 \to \mathbb{R}_+$, $d_2(x, y) = |x - y|/(1 + |x - y|)$. Obviously (X_1, d_1) and (X_2, d_2) are metric spaces. The sequence $(x_n)_{n \in \mathbb{N}} \subseteq X_1$, $x_n = 1/n \; \forall n \in \mathbb{N}$ is Cauchy in X_1 and does not converge in X_1, since $0 \notin X_1$. Let now $(y_n)_{n \in \mathbb{N}} \subseteq X_2$ be a Cauchy sequence in (X_2, d_2). Then $d_2(y_n, y_m) \to 0$ for $n, m \to \infty$, i.e., $|y_n - y_m|/(1 + |y_n - y_m|) \to 0$ for $n, m \to \infty$. Since for any $k \in \mathbb{N}$ we have $k/(1 + k) \geq 1/2$ it follows then that the sequence $(y_n)_{n \in \mathbb{N}}$ is stationary, therefore convergent. Hence (X_1, d_1) is a metric space which is not complete and (X_2, d_2) is a complete metric space. Define now $f : X_1 \to X_2$, $f(1/n) = n$ for $n \geq 1$. Obviously f is surjective, and

$$d_1(1/n, 1/m) \leq d_2(f(1/n), f(1/m))$$
$$\Leftrightarrow \left|\frac{1}{n} - \frac{1}{m}\right| \leq \frac{|n-m|}{1+|n-m|}$$
$$\Leftrightarrow 1 + |n-m| \leq mn,$$

which is clearly true for all $m, n \in \mathbb{N}$. Hence $d_1(p,q) \leq d_2(f(p), f(q))$ $\forall p, q \in X_1$. We must also prove that f is continuous. Let $(x_n)_{n \in \mathbb{N}} \subseteq X_1$, $x_n \to x \in X_1$. If the sequence $(x_n)_{n \in \mathbb{N}}$ is not stationary, then we must have $x = 0$, and $0 \notin X_1$. It follows then that there is a $k \in \mathbb{N}$ such that $x_n = x_k = x$ for $n \geq k$, hence $f(x_n) = f(x_k)$ for $n \geq k$ so $f(x_n) \to f(x_k) = f(x)$, i.e., f is continuous.

iii) We will prove that in this case (ii) is also true. Since f is in addition linear, and by the initial hypothesis f is continuous, it follows that $\|f(p)\|_2 \leq \|f\| \|p\|_1$ for $p \in X_1$. Also, by hypothesis, $\|f(p) - f(q)\|_2 \geq \|p - q\|_1$ for $p, q \in X_1$. In particular, $\|f(p)\|_2 \geq \|p\|_1$ for all $p \in X_1$.

Let $(x_n)_{n \in \mathbb{N}} \subseteq X_1$ be a Cauchy sequence and denote $y_n = f(x_n)$. Since f is linear and continuous, from $\|f(x_n) - f(x_m)\|_2 \leq \|f\| \|x_n - x_m\|_1$ it follows that $(y_n)_{n \in \mathbb{N}} \subseteq X_2$ is a Cauchy sequence. Since X_2 is complete there is $y \in X_2$ such that $\|y_n - y\|_2 \to 0$. Because f is surjective there is $x \in X_1$ such that $f(x) = y$. Then $\|x_n - x\|_1 \leq \|f(x_n) - f(x)\|_2 = \|y_n - y\|_2$, hence $\|x_n - x\|_1 \to 0$.

33. i) Suppose that $\|\cdot\|_a$ is a norm. Let $A = \{n \in \mathbb{N} \mid a_n = 0\}$. If $A \neq \emptyset$, then there is $k \in \mathbb{N}$ such that $a_k = 0$. We have $\|e_k\|_a = |a_k| = 0$, hence since $\|\cdot\|_a$ is a norm it follows that $e_k = 0$, which is false. Hence $A = \emptyset$, i.e., $a_n \neq 0$ $\forall n \in \mathbb{N}$. The converse is clear.

ii) By our hypothesis and (i), $\|\cdot\|_a$ and $\|\cdot\|_b$ are norms. If $\|\cdot\|_a$ and $\|\cdot\|_b$ are equivalent norms, then there are $c_1, c_2 > 0$ such that $\|x\|_a \leq c_1 \|x\|_b$ and $\|x\|_b \leq c_2 \|x\|_a$ for all $x \in c_{00}$. In particular for all $n \in \mathbb{N}$ we have $\|e_n\|_a \leq c_1 \|e_n\|_b$ and $\|e_n\|_b \leq c_2 \|e_n\|_a$, i.e., $|a_n| \leq c_1 |b_n|$ and $|b_n| \leq c_2 |a_n|$. The converse is clear.

34. See [20, exercise 6, chapter II, section 4].

Let $(x_n)_{n \in \mathbb{N}} \subseteq X$ be a Cauchy sequence. For any $x \in X$ we have that $\|\overline{x}\| = \inf \{\|z\| \mid z - x \in Y\}$, thus $\|\overline{x}\| \leq \|x\|$ for all $x \in X$. Since $(x_n)_{n \in \mathbb{N}} \subseteq X$ is a Cauchy sequence from this inequality it follows that the sequence $(\overline{x_n})_{n \in \mathbb{N}} \subseteq X/Y$ is a Cauchy sequence, hence there is a $z \in X$ such that $\lim_{n \to \infty} \|\overline{x_n} - \overline{z}\| = 0$, i.e., $\lim_{n \to \infty} \|\overline{x_n - z}\| = 0$. But if $\|a_n\| \to 0$, then there is a subsequence $(n_k)_{k \in \mathbb{N}}$ of $\mathbb{N}$ such that $\|a_{k_n}\| < 1/2^n$ $\forall n \in \mathbb{N}$. Using this fact we obtain that there is a subsequence $(k_n)_{n \in \mathbb{N}}$ of $\mathbb{N}$ such that $\|\overline{x_{k_n} - z}\| < 1/2^n$ $\forall n \in \mathbb{N}$. From the definition of the quotient norm we deduce that there is $y_n \in Y$ such that $\|(x_{k_n} - z) + y_n\| < 1/2^n$ $\forall n \in \mathbb{N}$. For any $m, n \in \mathbb{N}$ we have $\|y_n - y_m\| \leq \|y_n + (x_{k_n} - z)\| + \|x_{k_n} - x_{k_m}\| + \|(x_{k_m} - z) + y_m\|$, and therefore $\lim_{m,n \to \infty} \|y_n - y_m\| = 0$. Since Y is complete it follows that there is $y \in Y$ such that $\lim_{n \to \infty} y_n = y$. Since $\lim_{n \to \infty} \|(x_{k_n} - z) + y_n\| = 0$ we deduce that $\lim_{n \to \infty} x_{k_n} = -y + z$. Hence the Cauchy sequence $(x_n)_{n \in \mathbb{N}}$ contains a convergent subsequence, and therefore $(x_n)_{n \in \mathbb{N}}$ converges to $-y + z$.

1.2 Solutions

35. See [25, exercise 288, chapter I, section 11].

First solution. (See [3, Lemma 2.2.1].) We will prove the assertion from the exercise by induction on $n \in \mathbb{N}$. For $n = 1$, if $\{x\} \subseteq X$ is a linearly independent system, then $x \neq 0$, so $\|x\| > 0$. For $y \in X$ with $\|y\| < \|x\|$, we have $x + y \neq 0$ and so $\{x + y\} \subseteq X$ is a linearly independent system. Suppose now that the result is true for $n - 1$ vectors. Let $x_1, x_2, ..., x_n$ be n linearly independent vectors in X. Suppose that we cannot find $\varepsilon > 0$ as in the statement, and then there are sequences $(z_1^k)_{k \in \mathbb{N}}, ..., (z_n^k)_{k \in \mathbb{N}} \subseteq X$, $z_j^k \to 0$, $j = \overline{1, n}$, and $(\lambda_1^k)_{k \in \mathbb{N}}, ..., (\lambda_{n-1}^k)_{k \in \mathbb{N}} \subseteq \mathbb{K}$ such that

$$(*) \quad x_n + z_n^k = \lambda_1^k (x_1 + z_1^k) + \cdots + \lambda_{n-1}^k (x_{n-1} + z_{n-1}^k) \; \forall k \in \mathbb{N}.$$

If all the sequences $(\lambda_j^k)_{k \in \mathbb{N}}, j = \overline{1, n-1}$, are bounded, then passing if necessary to subsequences, we can suppose that $\lambda_j^k \to \lambda_j, j = \overline{1, n-1}$, so passing to the limit for $k \to \infty$ in $(*)$ we obtain that $x_n = \lambda_1 x_1 + \cdots + \lambda_{n-1} x_{n-1}$, which is false, since the elements $x_1, x_2, ..., x_n$ are linearly independent in X. Then without loss of generality we suppose that $\lim_{k \to \infty} |\lambda_1^k| = \infty$. Then there is a rank from where all the components of the sequence $(\lambda_1^k)_{k \in \mathbb{N}}$ are not zero, hence we can suppose that $\lambda_1^k \neq 0 \; \forall k \in \mathbb{N}$. It follows then that $x_1 + w_1^k + \alpha_2^k (x_2 + z_2^k) + \cdots + \alpha_{n-1}^k (x_{n-1} + z_{n-1}^k) = 0$, where $\alpha_j^k = \lambda_j^k / \lambda_1^k, j = \overline{2, n-1}$, $k \in \mathbb{N}$ and $w_1^k = z_1^k - (x_n + z_n^k)/\lambda_1^k \; \forall k \in \mathbb{N}$. But $w_1^k \to 0$ for $k \to \infty$, and this means that we can find the sequences $(w_1^k)_{k \in \mathbb{N}}, (z_2^k)_{k \in \mathbb{N}}, ..., (z_{n-1}^k)_{k \in \mathbb{N}} \subseteq X$, all convergent to $0 \in X$, such that the system $\{x_1 + w_1^k, x_2 + z_2^k, ..., x_{n-1} + z_{n-1}^k\} \subseteq X$ is linearly dependent, for every $k \in \mathbb{N}$. This contradicts our hypothesis of induction.

Second solution. The assertion from exercise is that for any n linearly independent elements $x_1, ..., x_n \in X$ there is an $\varepsilon > 0$ such that: $\forall y_1 \in B(x_1, \varepsilon), ..., \forall y_n \in B(x_n, \varepsilon)$, the set $\{y_1, ..., y_n\} \subseteq X$ is linearly independent. Let us assume, for a contradiction, that this fact is not true. This means that: $\forall \varepsilon > 0 \; \exists y_1 \in B(x_1, \varepsilon), ..., \exists y_n \in B(x_n, \varepsilon)$, such that $y_1, ..., y_n \in X$ are not linearly independent, i.e., $\forall \varepsilon > 0 \; \exists y_1 \in B(x_1, \varepsilon), ..., \exists y_n \in B(x_n, \varepsilon)$ and $\alpha_1, ..., \alpha_n \in \mathbb{K}$ not all zero (that is, $|\alpha_1| + \cdots + |\alpha_n| > 0$) such that $\alpha_1 y_1 + \cdots + \alpha_n y_n = 0$. Then $\alpha_1(y_1 - x_1) + \cdots + \alpha_n(y_n - x_n) = -(\alpha_1 x_1 + \cdots + \alpha_n x_n)$, from whence $\|\alpha_1 x_1 + \cdots + \alpha_n x_n\| = \|\alpha_1(y_1 - x_1) + \cdots + \alpha_n(y_n - x_n)\| \leq |\alpha_1| \|y_1 - x_1\| + \cdots + |\alpha_n| \|y_n - x_n\| < \varepsilon |\alpha_1| + \cdots + \varepsilon |\alpha_n|$, i.e., denoting by $c_i = \alpha_i/(|\alpha_1| + \cdots + |\alpha_n|)$, we have: $|c_1| + \cdots + |c_n| = 1$ and $\|c_1 x_1 + \cdots + c_n x_n\| < \varepsilon$.

Hence: $\forall \varepsilon > 0 \; \exists (c_1, ..., c_n) \in \mathbb{K}^n$, such that $|c_1| + \cdots + |c_n| = 1$ and $\|c_1 x_1 + \cdots + c_n x_n\| < \varepsilon$ (1). Let $E = \{c = (c_1, ..., c_n) \in \mathbb{K}^n \mid |c_1| + \cdots + |c_n| = 1\} \subseteq \mathbb{K}^n$ and $f : E \to \mathbb{R}$, $f(c_1, ..., c_n) = \|c_1 x_1 + \cdots + c_n x_n\|$. Obviously f is continuous and the set E is compact, since it is closed and bounded. The relation (1) says that: $\forall \varepsilon > 0$ $\exists c_\varepsilon \in E$, such that $f(c_\varepsilon) < \varepsilon$. For $\varepsilon = 1/k$, $k \in \mathbb{N}$, we obtain that there is $c_k \in E$, such that $f(c_k) < 1/k$. Therefore $\inf f(E) = 0$, and, using the Weierstrass theorem, we obtain that there is $c \in E$ such that $f(c) = 0$, i.e., $\|c_1 x_1 + \cdots + c_n x_n\| = 0$, $c_1 x_1 + \cdots + c_n x_n = 0$. Since $x_1, ..., x_n \in X$ are linearly independent vectors, it follows that $c_1 = 0, ..., c_n = 0$, which contradicts the fact that $c \in E$, i.e., $|c_1| + \cdots + |c_n| = 1$.

36. If $y = 0$ then $\varphi(t) = \|x\| \; \forall t \in \mathbb{R}$ and $\inf_{t \in \mathbb{R}} \varphi(t) = \varphi(0)$. If $y \neq 0$, let $m = \inf_{t \in \mathbb{R}} \varphi(t)$. Let us observe that $\varphi(t) \geq 0 \; \forall t \in \mathbb{R}$ and hence $m \in [0, \infty)$. Let $(t_n)_{n \in \mathbb{N}} \subseteq \mathbb{R}$ be such that $\varphi(t_n) \to m$. Then $\|t_n y\| = \|(x - t_n y) - x\| \leq \|x - t_n y\| + \|x\| = \varphi(t_n) + \|x\| \leq$

$M + \|x\|$ ($(\varphi(t_n))_{n\in\mathbb{N}}$ being convergent, it is bounded) and since $\|y\| > 0$, $|t_n| \leq (M + \|x\|)/\|y\|$ $\forall n \in \mathbb{N}$, i.e., the sequence $(t_n)_{n\in\mathbb{N}} \subseteq \mathbb{R}$ is bounded. By the Cesaro lemma this sequence contains a convergent subsequence, i.e., there is a $t_0 \in \mathbb{R}$ and a subsequence $(k_n)_{n\in\mathbb{N}}$ of $\mathbb{N}$ such that $t_{k_n} \to t_0$. Since $|\varphi(t_1) - \varphi(t_2)| = |\|x - t_1 y\| - \|x - t_2 y\|| \leq \|(x - t_1 y) - (x - t_2 y)\| = |t_1 - t_2| \|y\|$ we obtain that φ is Lipschitz, hence continuous. Therefore $\varphi(t_{k_n}) \to \varphi(t_0)$. Since $\varphi(t_n) \to m$ it follows that $\varphi(t_{k_n}) \to m$, and therefore $m = \varphi(t_0)$.

37. i) Let $\delta = d(x, Y)$. Then there is a sequence $(y_n)_{n\in\mathbb{N}} \subseteq Y$ such that $\|x - y_n\| \to \delta$. It follows then that the sequence $(x - y_n)_{n\in\mathbb{N}} \subseteq X$ is bounded, and therefore, using the triangle inequality, we obtain that the sequence $(y_n)_{n\in\mathbb{N}}$ is bounded in X, and so it is bounded in Y. Since Y is finite-dimensional, there is a subsequence $(y_{n_k})_{k\in\mathbb{N}}$ such that $y_{n_k} \to y_0 \in Y$. Then $\|x - y_{n_k}\| \to \|x - y_0\|$. Since $\|x - y_n\| \to \delta$ we deduce that $\delta = \|x - y_0\|$, $y_0 \in Y$.

ii) For the general case the answer is no! Indeed, for any $n \in \mathbb{N}$, let $E_n = \mathrm{Sp}\{e_1, ..., e_n\} \subseteq l_\infty$ and $x = e_{n+1} \in l_\infty$. It is easy to verify that $d(x, E_n) = 1$ and that $\|x - y\|_\infty = 1$ for all $y = (\alpha_1, ..., \alpha_n, 0, 0, ...) \in E_n$ with $|\alpha_i| \leq 1$, $i = \overline{1,n}$.

Remark. It is proved in exercise 30 of chapter 4 a more general result in this direction.

38. Let P_k be the space of all the real polynomials of degree at most k. If we consider on P_k the norms $\|P\|_\infty := \sup_{t\in[0,1]} |P(t)|$ and $\|P\|_1 := \int_0^1 |P(t)|\, dt$ $\forall P \in P_k$, then since P_k is finite-dimensional the norms $\|\cdot\|_1$ and $\|\cdot\|_\infty$ are equivalent, so there is a $c_k > 0$ such that $\|P\|_\infty \leq c_k \|P\|_1$ $\forall P \in P_k$, and then $|P(0)| \leq \sup_{t\in[0,1]} |P(t)| = \|P\|_\infty \leq c_k \|P\|_1 = c_k \int_0^1 |P(t)|\, dt$ for all $P \in P_k$.

39. i) See [1, exercise 16, Spring 1988].

Let $\{f_i : X \to \mathbb{R},\ i = \overline{1,n}\}$ be a basis for V. Then $\forall f \in V$, we can find a unique vector $(\alpha_1, \alpha_2, ..., \alpha_n)^t \in \mathbb{R}^n$ such that $f = \sum_{i=1}^n \alpha_i f_i$. Then the operator $h : V \to \mathbb{R}^n$, defined by $h(f) = (\alpha_1, \alpha_2, ..., \alpha_n)^t \in \mathbb{R}^n$ is a linear isomorphism. We prove that there are $x_1, x_2, ..., x_n \in X$ such that $D(x_1, x_2, ..., x_n) = \begin{vmatrix} f_1(x_1) & f_2(x_1) & \cdots & f_n(x_1) \\ f_1(x_2) & f_2(x_2) & \cdots & f_n(x_2) \\ \cdots & \cdots & \cdots & \cdots \\ f_1(x_n) & f_2(x_n) & \cdots & f_n(x_n) \end{vmatrix} \neq 0$.

Let us assume, for a contradiction, that $\forall (x_1, x_2, ..., x_n) \in X^n$, $D(x_1, x_2, ..., x_n) = 0$. Let R be the maximum over those r for which there are $x_1, ..., x_n \in X$, with

$$\mathrm{rank} \begin{pmatrix} f_1(x_1) & f_2(x_1) & \cdots & f_n(x_1) \\ f_1(x_2) & f_2(x_2) & \cdots & f_n(x_2) \\ \cdots & \cdots & \cdots & \cdots \\ f_1(x_n) & f_2(x_n) & \cdots & f_n(x_n) \end{pmatrix} = r.$$

Then, by our supposition, $R < n$. Let $x_1, ..., x_n \in X$, with rank $\begin{pmatrix} f_1(x_1) & f_2(x_1) & \cdots & f_n(x_1) \\ f_1(x_2) & f_2(x_2) & \cdots & f_n(x_2) \\ \cdots & \cdots & \cdots & \cdots \\ f_1(x_n) & f_2(x_n) & \cdots & f_n(x_n) \end{pmatrix} = R$ and suppose, for example, that $\begin{vmatrix} f_1(x_1) & \dots & f_R(x_1) \\ \dots & \dots & \dots \\ f_1(x_R) & \dots & f_R(x_R) \end{vmatrix} \neq 0$. Then it follows that $\begin{vmatrix} f_1(x_1) & \dots & f_{R+1}(x_1) \\ \dots & \dots & \dots \\ f_1(x_R) & \dots & f_{R+1}(x_R) \\ f_1(x) & \dots & f_{R+1}(x) \end{vmatrix} =$

$0 \ \forall x \in X$. Developing this determinant over the line $R+1$ we obtain that $\sum_{i=1}^{R+1} \alpha_i f_i(x) = 0$, where α_i does not depend on x, and $\alpha_{R+1} \neq 0$. Then f_{R+1} is a linear combination of $f_1, ..., f_R$, which is false!

So there are $x_1, x_2, \ldots, x_n \in X$ such that $D(x_1, x_2, \ldots, x_n) \neq 0$. We have the diagram $V \xrightarrow{h} \mathbb{R}^n \xrightarrow{A} \mathbb{R}^n$, $f \longrightarrow \alpha = (\alpha_1, \alpha_2, \ldots, \alpha_n)^t \xrightarrow{A} A\alpha$, where the matrix $A : \mathbb{R}^n \longrightarrow \mathbb{R}^n$ is defined by: $A = \begin{pmatrix} f_1(x_1) & f_2(x_1) & \cdots & f_n(x_1) \\ f_1(x_2) & f_2(x_2) & \cdots & f_n(x_2) \\ \cdots & \cdots & \cdots & \cdots \\ f_1(x_n) & f_2(x_n) & \cdots & f_n(x_n) \end{pmatrix}$. Since $\det(A) \neq 0$, $A : \mathbb{R}^n \longrightarrow \mathbb{R}^n$ is an isomorphism. Let us consider the composition $\sigma : V \longrightarrow \mathbb{R}^n$, $\sigma = A \circ h$. Since A and h are isomorphisms it follows that σ is an isomorphism. We have

$\sigma(f) = \sigma\left(\sum_{i=1}^n \alpha_i f_i\right) = A \begin{pmatrix} \alpha_1 \\ \alpha_2 \\ \vdots \\ \alpha_n \end{pmatrix} = \begin{pmatrix} f_1(x_1) & f_2(x_1) & \cdots & f_n(x_1) \\ f_1(x_2) & f_2(x_2) & \cdots & f_n(x_2) \\ \cdots & \cdots & \cdots & \cdots \\ f_1(x_n) & f_2(x_n) & \cdots & f_n(x_n) \end{pmatrix} \begin{pmatrix} \alpha_1 \\ \alpha_2 \\ \vdots \\ \alpha_n \end{pmatrix} =$

$\left(\sum_{i=1}^n \alpha_i f_i(x_1), \ldots, \sum_{i=1}^n \alpha_i f_i(x_n)\right)^t = (f(x_1), f(x_2), \ldots, f(x_n))^t$ for all $f \in V$.

ii) For (a) and (b), see [13, exercise 9, chapter I, section 2].

a) Using (i) we obtain that there are n pairwise distinct elements $x_1, \ldots, x_n \in X$ and a basis $\{f_1, \ldots, f_n\}$ of V such that $\begin{vmatrix} f_1(x_1) & \dots & \dots & f_n(x_1) \\ \dots & \dots & \dots & \dots \\ f_1(x_n) & \dots & \dots & f_n(x_n) \end{vmatrix} \neq 0$. Now from the continuity of the determinant and the continuity for the functions $f_1, \ldots, f_n$ it follows that there are $U_1, \ldots, U_n \subseteq X$, open neighborhoods respectively for $x_1, \ldots, x_n$, such that $\begin{vmatrix} f_1(y_1) & \dots & \dots & f_n(y_1) \\ \dots & \dots & \dots & \dots \\ f_1(y_n) & \dots & \dots & f_n(y_n) \end{vmatrix} \neq 0 \ \forall y_1 \in U_1, \ldots, \forall y_n \in U_n$. Since X is Hausdorff and the points $x_1, \ldots, x_n$ are pairwise distinct, then we can find $V_1, \ldots, V_n \subseteq X$, open neighborhoods respectively for $x_1, \ldots, x_n$, such that $V_i \cap V_j = \emptyset$, for $i \neq j$. Hence if we replace $U_1, \ldots, U_n$ with $U_1 \cap V_1, \ldots, U_n \cap V_n$ we can assume that the sets U_i are also pairwise disjoint. Let $f \in V$ with the property that for any $1 \leq i \leq n$ there is an $y_i \in U_i$ with $f(y_i) = 0$. Let $f = \sum_{i=1}^n \alpha_i f_i$ be the decomposition of f with respect to the basis $\{f_1, \ldots, f_n\}$ of V. Then $0 = f(y_j) = \sum_{i=1}^n \alpha_i f_i(y_j)$, $j = \overline{1,n}$, i.e., $\begin{pmatrix} f_1(y_1) & \dots & \dots & f_n(y_1) \\ \dots & \dots & \dots & \dots \\ f_1(y_n) & \dots & \dots & f_n(y_n) \end{pmatrix} \begin{pmatrix} \alpha_1 \\ \vdots \\ \alpha_n \end{pmatrix} = 0$ and since $\begin{vmatrix} f_1(y_1) & \dots & \dots & f_n(y_1) \\ \dots & \dots & \dots & \dots \\ f_1(y_n) & \dots & \dots & f_n(y_n) \end{vmatrix} \neq$ 0 it follows that $\alpha_1 = \cdots = \alpha_n = 0$, so $f = 0$.

b) Let $x_i \in U_i$, $i = \overline{1,n}$. We consider $f_1, \ldots, f_n$ as in (a). We denote by $A =$

$$\begin{pmatrix} f_1(x_1) & \ldots & \ldots & f_n(x_1) \\ \ldots & \ldots & \ldots & \ldots \\ f_1(x_n) & \ldots & \ldots & f_n(x_n) \end{pmatrix}.$$ If $\det A = 0$, there are $\alpha_1, \ldots, \alpha_n \in \mathbb{R}$ not all zero such that $A(\alpha_1, \ldots, \alpha_n)^t = 0$. Let $f = \sum_{i=1}^{n} \alpha_i f_i$. Then it follows that $f \in V$ and $f(x_1) = \ldots = f(x_n) = 0$. From (a) it follows that $f = 0$, i.e., $\sum_{i=1}^{n} \alpha_i f_i = 0$, and since $f_1, \ldots, f_n$ is a basis, $\alpha_1 = 0, \ldots, \alpha_n = 0$, which is false! Hence $\det A \neq 0$ and then $\forall 1 \leq i \leq n$ there are $\alpha_1^i, \ldots, \alpha_n^i \in \mathbb{R}$ such that $A(\alpha_1^i, \ldots, \alpha_n^i)^t = e_i$. Denoting by $g_i = \sum_{k=1}^{n} \alpha_k^i f_k$ we obtain that $g_i \in V, i = \overline{1,n}$, with $g_i(x_j) = \delta_{ij}, i,j = \overline{1,n}$. Let now $f \in V$ and let $h = f - \sum_{i=1}^{n} g_i f(x_i)$. Then $h(x_k) = f(x_k) - \sum_{i=1}^{n} g_i(x_k) f(x_i) = f(x_k) - \sum_{i=1}^{n} \delta_{ik} f(x_i) = f(x_k) - f(x_k) = 0$, $k = \overline{1,n}$, and then $f = \sum_{i=1}^{n} g_i f(x_i)$.

Consider $c = \max_{i=\overline{1,n}} \sup_{x \in X} |g_i(x)|$. Then $c < \infty$ since X is compact and the functions g_i are continuous. Then for every element $f \in V$ we obtain the fact that $\sup_{x \in X} |f(x)| \leq c \sum_{i=1}^{n} |f(x_i)|$.

c) It is clearly a consequence of (b).

Remark. For example, if $k \in \mathbb{N}$ and if $(P_n)_{n \in \mathbb{N}}$ is a sequence of polynomials all of degree $\leq k$, which converges pointwise to 0 on $[0,1]$, then $(P_n)_{n \in \mathbb{N}}$ is uniformly convergent to 0 on $[0,1]$.

40. See [40, exercise 1.92].
Obviously f is linear.
i) We have $|f(x)| = |x(1)| \leq \sup_{t \in [0,1]} |x(t)| = \|x\|$ for all $x \in C[0,1]$ and from here it follows that f is continuous.

ii) If f were continuous, since is linear there is an $M > 0$ such that $|f(x)| \leq M\|x\|$ $\forall x \in C[0,1]$, that is, $|x(1)| \leq M \left(\int_0^1 |x(t)|^p dt \right)^{1/p}$ $\forall x \in C[0,1]$. Taking $x(t) = t^n$, we must have $1 \leq M/(np+1)^{1/p}$ $\forall n \in \mathbb{N}$ and for $n \to \infty$ we obtain a contradiction!

41. See [40, exercise 1.93].
i) Let $V : C[0,1] \times C[0,1] \to C[0,1]$, $V(f,g) = fg$. Obviously V is a bilinear operator. In addition: $\|V(f,g)\| = \|fg\| \leq \|f\|\|g\|$ $\forall (f,g) \in C[0,1] \times C[0,1]$. Therefore V is continuous. Since $U(f) = V(f,f)$ $\forall f \in C[0,1]$, i.e., U is the quadratic form associated with V, it follows that U is continuous.

Suppose that U is uniformly continuous. Then for $\varepsilon = 1 > 0$ $\exists \delta > 0$ such that: $\|f - g\| \leq \delta \Rightarrow \|U(f) - U(g)\| < 1$, i.e., $\|f - g\| \leq \delta$, $f, g \in C[0,1] \Rightarrow \|f^2 - g^2\| < 1$. Taking $f = n + \delta$, $g = n$ (constant functions), $n \in \mathbb{N}$, we obtain that $\|(\delta + n)^2 - n^2\| < 1$, i.e., $2n\delta + \delta^2 < 1$ $\forall n \in \mathbb{N}$, $n < (1 - \delta^2)/(2\delta)$ $\forall n \in \mathbb{N}$, a contradiction.

1.2 Solutions

ii) Let $0 < a < 1$ and $b > 0$. We consider the function $f : [0, 1] \to \mathbb{R}$,

$$f(x) = \begin{cases} b(1 - x/a) & 0 \le x \le a, \\ 0 & a < x \le 1. \end{cases}$$

Then $\|f\|_p^p = \int_0^a b^p (1 - x/a)^p \, dx = (-ab^p/(p+1))(1 - x/a)^{p+1} \big|_0^a = ab^p/(p+1)$, i.e., $\|f\|_p = ba^{1/p}/(p+1)^{1/p}$. Also, $\|f^2\|_p^p = \int_0^a b^{2p} (1 - x/a)^{2p} \, dx = ab^{2p}/(2p+1)$, that is, $\|f^2\|_p = b^2 a^{1/p}/(2p+1)^{1/p}$.

We choose now a and b depending on $n \in \mathbb{N}$ such that $a^{1/p}b = 1/n$ and $b^2 a^{1/p} = 1$ ($a = 1/n^{2p}$ and $b = n$). For these values of a and b we obtain $f_n \in C[0, 1]$ with $\|f_n\|_p = 1/(n(p+1)^{1/p}) \to 0$ and $\|U(f_n)\|_p = 1/(2p+1)^{1/p}$, i.e., $f_n \to 0$ and the sequence $(U(f_n))_{n \in \mathbb{N}}$ does not converge towards zero, i.e., U is not continuous at the origin.

iii) Let $T : C[0, 1] \times C[0, 1] \to C[0, 1]$, $T(f, g) = fg$. Clearly T is a bilinear operator and $\|T(f, g)\| = \|fg\|_p = \left(\int_0^1 |f(x)g(x)|^p \, dx \right)^{1/p} \le \|f\|_\infty \|g\|_\infty$, i.e., T is continuous. Since $U(f) = T(f, f)$ it follows that U is continuous. Analogously with (i) one can prove that U is not uniformly continuous.

42. See [12, exercise 10, chapter IV, section 6].

i) Using the axioms for a norm, we have $\|a - tb\| \le \|a - t^p b\| + \|t^p b - tb\| = \|a - t^p b\| + |t^p - t| = \|a - t^p b\| + (t - t^p) \le \|a - t^p b\| + 1 - t^p \; \forall t \in [0, 1]$, from whence $\|a - t^p b\| \ge \|a - tb\| + t^p - 1$ (1). Also, $1 = \|a\| \le \|a - t^p b\| + \|t^p b\| = \|a - t^p b\| + t^p \; \forall t \in [0, 1]$, from whence $\|a - t^p b\| \ge 1 - t^p$ (2). Adding (1) and (2) we obtain that $\|a - t^p b\| \ge \|a - tb\|/2$ (3). Now $0 \le \|a - tb\|/2 \le (\|a\| + t\|b\|)/2 \le 1$, from whence $\|a - tb\|/2 \ge (\|a - tb\|/2)^p$, hence (3) gives $\|a - t^p b\| \ge \|a - tb\|/2 \ge (\|a - tb\|/2)^p$, i.e., $\|a - tb\|^p \le 2^p \|a - t^p b\|$, that is, the first inequality from the statement.

For the second one, we have $\|a - t^p b\| \le \|a - tb\| + \|tb - t^p b\| = \|a - tb\| + t - t^p \; \forall t \in [0, 1]$ (4). If we will prove that $t - t^p \le (p - 1)(1 - t) \le (p - 1)\|a - tb\|$ (5), then using (5) from inequality (4) we deduce that $\|a - t^p b\| \le p\|a - tb\|$, i.e., the second inequality. To show (5) we will prove that $t - t^p \le (p - 1)(1 - t) \; \forall 0 \le t \le 1$, and $1 - t \le \|a - tb\| \; \forall 0 \le t \le 1$. Let $\varphi : [0, 1] \to \mathbb{R}$, $\varphi(t) = t - t^p - (p - 1)(1 - t)$. Then the derivative of φ is $\varphi'(t) = p(1 - t^{p-1}) \ge 0$, hence φ is increasing, from whence $\varphi(t) \le \varphi(1) = 0 \; \forall 0 \le t \le 1$. Also, we have $1 = \|a\| \le \|a - tb\| + \|tb\| = \|a - tb\| + t$, i.e., $1 - t \le \|a - tb\| \; \forall 0 \le t \le 1$.

For the remaining inequalities, we can suppose that x and y are not zero, since if one of them is zero then the inequalities are trivial. Without loss of generality we can assume that $\|x\| \le \|y\|$. Then if we take $a = y/\|y\|$, $b = x/\|x\|$ and $t = \|x\|/\|y\|$ in the first inequality which we have already proved, we obtain that

$$\left\| \frac{y}{\|y\|} - \frac{\|x\|}{\|y\|} \frac{x}{\|x\|} \right\|^p \le 2^p \left\| \frac{y}{\|y\|} - \left(\frac{\|x\|}{\|y\|} \right)^p \frac{x}{\|x\|} \right\|,$$

i.e., $\|x - y\|^p \le 2^p \|\|x\|^{p-1} x - \|y\|^{p-1} y\|$. For the same values for a, b and t, using the second inequality we obtain that

$$\left\| \frac{y}{\|y\|} - \left(\frac{\|x\|}{\|y\|} \right)^p \frac{x}{\|x\|} \right\| \le p \left\| \frac{y}{\|y\|} - \frac{\|x\|}{\|y\|} \frac{x}{\|x\|} \right\|,$$

i.e.,
$$\left\|\frac{y}{\|y\|} - \frac{\|x\|^{p-1}}{\|y\|^p} x\right\| \le p \left\|\frac{y}{\|y\|} - \frac{x}{\|y\|}\right\|,$$

and then
$$\left\|\|x\|^{p-1} x - \|y\|^{p-1} y\right\| \le p \|x-y\| \|y\|^{p-1} \le p \|x-y\| (\|x\| + \|y\|)^{p-1}.$$

ii) We have
$$\|Tf - Tg\|^p = \int_S \left| |f(s)|^{1/p-1} f(s) - |g(s)|^{1/p-1} g(s) \right|^p d\mu.$$

But for any $s \in S$ using (i) we have
$$\left| |f(s)|^{1/p-1} f(s) - |g(s)|^{1/p-1} g(s) \right|^p \le 2^p |f(s) - g(s)|,$$

so we obtain that
$$\|Tf - Tg\|^p \le 2^p \int_S |f(s) - g(s)| d\mu = 2^p \|f - g\|,$$

i.e., $\|Tf - Tg\| \le 2 \|f-g\|^{1/p}$, from whence it follows that T is uniformly continuous, since $\forall \varepsilon > 0 \ \exists \delta_\varepsilon = (\varepsilon/2)^p > 0$ such that for any $f, g \in L_1(\mu)$ with $\|f-g\| < \delta_\varepsilon$ we have $\|Tf - Tg\| < \varepsilon$.

iii) For any $f, g \in L_p(\mu)$ we have
$$\|Uf - Ug\| = \int_S \left| |f(s)|^{p-1} f(s) - |g(s)|^{p-1} g(s) \right|^p d\mu.$$

But for any $s \in S$ using (i) we have that
$$\left| |f(s)|^{p-1} f(s) - |g(s)|^{p-1} g(s) \right|^p \le p |f(s) - g(s)| (|f(s)| + |g(s)|)^{p-1},$$

so we obtain that
$$\begin{aligned}
\int_S \left| |f(s)|^{p-1} f(s) - |g(s)|^{p-1} g(s) \right|^p d\mu &\le p \int_S |f(s) - g(s)| (|f(s)| + |g(s)|)^{p-1} d\mu \\
&\le p \left(\int_S |f(s) - g(s)|^p d\mu \right)^{1/p} \\
&\quad \left(\int_S (|f(s)| + |g(s)|)^{(p-1)q} d\mu \right)^{1/q} \\
&= p \|f - g\| \, \|f\| + \|g\|\|^{p/q} \, (1/p + 1/q = 1).
\end{aligned}$$

Thus $\|Uf - Ug\| \le p \|f-g\| \, \|f\| + |g|\|^{p/q}$, and this implies that $\|Uf - Ug\| \le p \|f-g\| (\|f\| + \|g\|)^{p/q}$. Let $A \subseteq L_p(\mu)$ be a bounded set and let $M = \sup_{f \in A} \|f\| < \infty$. Consider $f, g \in A$. Then $\|Uf - Ug\| \le p \|f-g\| (2M)^{p/q}$, from whence it follows that U is uniformly continuous on A, since $\forall \varepsilon > 0 \ \exists \delta_\varepsilon = \varepsilon/(p(2M)^{p/q}) > 0$ such that for any $f, g \in A$ with $\|f-g\| < \delta_\varepsilon$ we have that $\|Uf - Ug\| < \varepsilon$.

1.2 Solutions

We have $TU = I$ and $UT = I$ where I is the identity operator, and therefore T and U are bijective, and $T^{-1} = U$.

iv) By (ii) it remains to be proved that $T^{-1} = U$ is continuous. Let $f_n \to f$ in the norm of $L_p(\mu)$. Then the set $A = \{f_n \mid n \in \mathbb{N}\} \cup \{f\}$ is bounded and let $M = \sup_{g \in A} \|g\| < \infty$. By (iii) we have $\|T^{-1}f_n - T^{-1}f\| \leq p\|f_n - f\|(2M)^{p/q} \ \forall n \in \mathbb{N}$, and since $f_n \to f$ in the norm of $L_p(\mu)$ we deduce that $T^{-1}f_n \to T^{-1}f$ in the norm of $L_1(\mu)$, i.e. T^{-1} is continuous.

43. See [13, exercise 11, chapter II, section 2].

i) Using, if necessary, a translation, we can assume that $0 \in A$. Let $x_1, \ldots, x_n \in A$, linearly independent elements in $\mathbb{R}^n$. Then $\{x_1, \ldots, x_n\} \subseteq \mathbb{R}^n$ is a basis and let us define the linear operator $T : \mathbb{R}^n \to \mathbb{R}^n$ by $T(e_i) = x_i$, $i = \overline{1, n}$. Then $T, T^{-1} : \mathbb{R}^n \to \mathbb{R}^n$ are continuous (on $\mathbb{R}^n$ we consider the Euclidean norm), so $T : \mathbb{R}^n \to \mathbb{R}^n$ is a homeomorphism. Let $B = \left\{(\alpha_1, \ldots, \alpha_n)^t \in \mathbb{R}^n \ \middle| \ \alpha_i \geq 0, i = \overline{1,n}, \sum_{i=1}^n \alpha_i \leq 1\right\}$. For $(\alpha_1, \ldots, \alpha_n)^t \in B$ we have

$$T(\alpha_1, \ldots, \alpha_n)^t = \sum_{i=1}^n \alpha_i x_i = \sum_{i=1}^n \alpha_i x_i + \left(1 - \sum_{i=1}^n \alpha_i\right) 0 \in A,$$

so $T(B) \subseteq A$. Let $x = (1/(2n), \ldots, 1/(2n)) \in B$. Then $B(x, 1/(2n)) \subseteq B$. Indeed, if $y \in B(x, 1/(2n))$, $y = (y_1, \ldots, y_n)^t$, then $\left(\sum_{i=1}^n (y_i - 1/(2n))^2\right)^{1/2} \leq 1/(2n)$, from whence $|y_i - 1/(2n)| \leq 1/(2n)$, $i = \overline{1, n}$, so $y_i \in [0, 1/n]$, $i = \overline{1, n}$. Then $y_i \geq 0$, $i = \overline{1, n}$, $\sum_{i=1}^n y_i \leq \sum_{i=1}^n 1/n = 1$, i.e., $y \in B$, so $B(x, 1/(2n)) \subseteq B$, from whence $T(B(x, 1/(2n))) \subseteq T(B)$. But $T(B) \subseteq A$, hence we obtain that $T(B(x, 1/(2n))) \subseteq A$. Since T is a homeomorphism, $T(B(x, 1/(2n)))$ is open and non-empty, and then it follows that $\overset{\circ}{A} \neq \varnothing$.

Let now $A \subseteq \mathbb{R}^n$ be a convex set which is dense in $\mathbb{R}^n$. Let $R = \max\{1 \leq r \leq n \mid \exists \{x_1, \ldots, x_r\} \subseteq A \text{ linearly independent vectors}\}$. Let us suppose that $R < n$, and let $\{x_1, \ldots, x_R\} \subseteq A$ be a linearly independent system. For any $x \in A$, by the definition of R, the set $\{x, x_1, \ldots, x_R\}$ is linearly dependent, so there are $\alpha, \alpha_1, \alpha_2, \ldots, \alpha_R \in \mathbb{R}$, not all zero, such that $\alpha x + \sum_{i=1}^R \alpha_i x_i = 0$. Since $x_1, \ldots, x_R$ are linearly independent, it follows that $\alpha \neq 0$, so $x \in \text{Sp}\{x_1, \ldots, x_R\}$. We proved that $A \subseteq \text{Sp}\{x_1, \ldots, x_R\}$, from whence $\overline{A} \subseteq \overline{\text{Sp}\{x_1, \ldots, x_R\}} = \text{Sp}\{x_1, \ldots, x_R\} \neq \mathbb{R}^n$ since $R < n$, which is false. So there are n linearly independent elements in A and by what we have proved above it follows that $\overset{\circ}{A} \neq \varnothing$. Now since A is convex, using exercise 12 of chapter 13 we obtain that $\overset{\circ}{\overline{A}} = \overset{\circ}{A}$, i.e., $\overset{\circ}{A} = \mathbb{R}^n$, that is, $A = \mathbb{R}^n$.

ii) Obviously $P \subseteq l_1$ is convex. For $x = (x_n)_{n \in \mathbb{N}}$, we define $y = (x_n^+)_{n \in \mathbb{N}}$, $z = (x_n^-)_{n \in \mathbb{N}}$, where $x_n^+ = \max(x_n, 0)$, $x_n^- = \max(-x_n, 0)$. We have $0 \leq x_n^-, x_n^+ \leq |x_n|$ $\forall n \in \mathbb{N}$, from whence $\sum_{n=1}^\infty x_n^+ \leq \sum_{n=1}^\infty |x_n| < \infty$, $\sum_{n=1}^\infty x_n^- < \infty$, i.e., $y, z \in P$, and $y - z = x$, so $\text{Sp}(P) = l_1$.

Let us suppose now that we can find $x = (x_n)_{n \in \mathbb{N}} \in P$ and $\varepsilon > 0$ such that $B(x, \varepsilon) \subseteq P$. Since $\sum_{n=1}^{\infty} |x_n| < \infty$ there is an $n \geq 2$ such that $|x_n| = x_n < \varepsilon/3$. Let $y = (x_1, \ldots, x_{n-1}, -(x_n + \varepsilon/3), x_{n+1}, \ldots)$. Then $y \in l_1$. We also have $\|y - x\|_{l_1} = |2x_n + \varepsilon/3| = 2x_n + \varepsilon/3 < \varepsilon$, so $y \in B(x, \varepsilon)$, and therefore $y \in P$. It follows that $-(x_n + \varepsilon/3) \geq 0$, so $x_n + \varepsilon/3 \leq 0$, which is false, since $x_n \geq 0, \varepsilon > 0$.

iii) a) Obviously, since f is linear, A and B are convex. We clearly have $A \cap B = \varnothing$, $A \cup B = X$. If we suppose that B is empty, then $A = X$, i.e., $f(x) \geq 0$, $f(x) = f(-(-x)) = -f(-x) \leq 0 \; \forall x \in X$, so $f(x) = 0$, $f \equiv 0$, f continuous, which is false! Analogously $A \neq \varnothing$.

b) Let $x \in X$. If $f(x) \geq 0$ then $x \in A$. If $f(x) < 0$, then $f(-x) = -f(x) > 0$, i.e., $-x \in A$, so $x \in \text{Sp}(A)$. Therefore $\text{Sp}(A) = X$ and, analogously, $\text{Sp}(B) = X$.

c) We will prove that $\overset{\circ}{B} = \varnothing$, and from here it follows that $\overline{A} = X$. Let us suppose, for a contradiction, that $\overset{\circ}{B} \neq \varnothing$, i.e., there are $a \in X$ and a balanced neighborhood V of 0 such that $a + V \subseteq B$. Since V is balanced, $\forall x \in V$ we have $a \pm x \in B$, i.e., by the definition of B, $f(a + x) < 0$ and $f(a - x) < 0$, from whence $|f(x)| < -f(a)$, i.e., f is bounded on a neighborhood of 0. From exercise 16 of chapter 13 it follows that f is continuous, which contradicts our hypothesis on f. Analogously we obtain the property $\overline{B} = X$.

44. See [9, exercise 12 (a), chapter II, section 5], or [35, exercise I, chapter 3, section 11].

Since X is infinite-dimensional there is a linearly independent system $(x_n)_{n \in \mathbb{N}} \subseteq X$ such that $\|x_n\| = 1 \; \forall n \in \mathbb{N}$. The system $(x_n)_{n \in \mathbb{N}}$ can be extended to an algebraic basis of X. Let $(y_i)_{i \in I} \subseteq X$ be a linearly independent system with $\|y_i\| = 1$ for all $i \in I$ such that $\{x_n \mid n \in \mathbb{N}\} \cup \{y_i \mid i \in I\}$ is an algebraic basis for X. We take $A = \text{eco}(\{(1/n)x_n \mid n \in \mathbb{N}\} \cup \{y_i \mid i \in I\})$, the balanced convex hull. Recall that if $E \subseteq X$ then $\text{eco}(E) = \{\lambda_1 a_1 + \cdots + \lambda_n a_n \mid a_1, \ldots, a_n \in E, |\lambda_1| + \cdots + |\lambda_n| \leq 1, n \in \mathbb{N}\}$. We will prove that A is absorbing. Let $x \in X$, $x \neq 0$. Since $\{x_n \mid n \in \mathbb{N}\} \cup \{y_i \mid i \in I\}$ is an algebraic basis for X, x can be written in the form $x = \alpha_1 x_1 + \cdots + \alpha_p x_p + \beta_1 y_{i_1} + \cdots + \beta_m y_{i_m}$, that is,

$$x = \alpha_1 \frac{x_1}{1} + \cdots + p \alpha_p \frac{x_p}{p} + \beta_1 y_{i_1} + \cdots + \beta_m y_{i_m}.$$

Let $\varepsilon = |\alpha_1| + 2|\alpha_2| + \cdots + p|\alpha_p| + |\beta_1| + \cdots + |\beta_m|$. Then since $x \neq 0$, $\varepsilon > 0$. Then

$$\frac{x}{\varepsilon} = \frac{\alpha_1}{\varepsilon} \frac{x_1}{1} + \cdots + \frac{p\alpha_p}{\varepsilon} \frac{x_p}{p} + \frac{\beta_1}{\varepsilon} y_{i_1} + \cdots + \frac{\beta_m}{\varepsilon} y_{i_m}$$
$$\in \text{eco}(\{x_n/n \mid n \in \mathbb{N}\} \cup \{y_i \mid i \in I\}) = A,$$

i.e., $x \in \varepsilon A$. Hence A is absorbing, and also A is balanced and convex. Let $p_A : X \to \mathbb{R}_+$ be the Minkowski seminorm associated with A. Since $\|x_n\| = 1 \; \forall n \in \mathbb{N}$ and $\|y_i\| = 1$ $\forall i \in I$, it follows that $\{(1/n)x_n \mid n \in \mathbb{N}\} \cup \{y_i \mid i \in I\} \subseteq B_X$, and since B_X is balanced and convex, it follows that $A = \text{eco}(\{(1/n)x_n \mid n \in \mathbb{N}\} \cup \{y_i \mid i \in I\}) \subseteq B_X$. But $[p_A < 1] \subseteq A$ from whence $[p_A < 1] \subseteq B_X$ and from here it follows in a standard way that $\|x\| \leq p_A(x)$ for all $x \in X$. If $p_A(x) = 0$, then $\|x\| = 0$, $x = 0$. Therefore p_A is a norm.

We will prove now that $p_A(x_n) = n$ for all $n \in \mathbb{N}$. Since $x_n/n \in A$ it follows that $p_A(x_n) \leq n$. Let now $\mu > 0$ such that $x_n \in \mu A$, i.e., $x_n/\mu \in A$. Using the form for

1.2 Solutions

the balanced convex hull, it follows that there are $p, m \in \mathbb{N}$, $i_1, ..., i_m \in I$, $\lambda_1, ..., \lambda_p \in \mathbb{K}$, $\beta_1, ..., \beta_m \in \mathbb{K}$ such that

$$\frac{x_n}{\mu} = \lambda_1 \frac{x_1}{1} + \lambda_2 \frac{x_2}{2} + \cdots + \lambda_p \frac{x_p}{p} + \beta_1 y_{i_1} + \cdots + \beta_m y_{i_m},$$

and $|\lambda_1| + \cdots + |\lambda_p| + |\beta_1| + \cdots + |\beta_m| \leq 1$. If $p < n$, i.e., $p \leq n - 1$, since $(\lambda_1/1)x_1 + \cdots + (\lambda_p/p)x_p - (1/\mu)x_n + \beta_1 y_{i_1} + \cdots + \beta_m y_{i_m} = 0$ and the system $\{x_n \mid n \in \mathbb{N}\} \cup \{y_i \mid i \in I\}$ is linearly independent, it follows that $\lambda_1 = \cdots = \lambda_n = 0$, $-1/\mu = 0$, $\beta_1 = \cdots = \beta_m = 0$, which is false! Hence $p \geq n$, i.e., we have $x_n/\mu = (\lambda_1/1)x_1 + \cdots + (\lambda_n/n)x_n + \cdots + (\lambda_p/p)x_p + \beta_1 y_{i_1} + \cdots + \beta_m y_{i_m}$ and $p \geq n$, from whence, using the fact that the system $\{x_n \mid n \in \mathbb{N}\} \cup \{y_i \mid i \in I\}$ is linearly independent we deduce that $\lambda_k = 0$ for $k \neq n$, $1/\mu = \lambda_n/n$, and $\beta_1 = \cdots = \beta_m = 0$. But from $|\lambda_1| + \cdots + |\lambda_p| + |\beta_1| + \cdots + |\beta_m| \leq 1$ it follows that $|\lambda_n| \leq 1$, and then $n/\mu \leq 1$, that is, $\mu \geq n$. Hence $\{\mu > 0 \mid x_n \in \mu A\} \subseteq [n, \infty)$ and then $p_A(x_n) \geq n$. So $p_A(x_n) = n$ for all $n \in \mathbb{N}$.

Let $\|\cdot\|_1 = p_A$. Then from $\|x\| \leq \|x\|_1 = p_A(x)$ it follows that the identity operator $I : (X, \|\cdot\|_1) \to (X, \|\cdot\|)$ is continuous, i.e., $\tau \subseteq \tau_1$. If $\tau = \tau_1$ then $I : (X, \|\cdot\|) \to (X, \|\cdot\|_1)$ is also continuous, i.e., there is an $M > 0$ such that $\|x\|_1 \leq M \|x\|$ $\forall x \in X$. In particular, $p_A(x_n) \leq M \|x_n\|$ $\forall n \in \mathbb{N}$, i.e., $n \leq M$ $\forall n \in \mathbb{N}$, which is false.

The idea to construct $\|\cdot\|_2$ is to use the same technique as above on the dual space X^*. Since X is infinite-dimensional, X^* is infinite-dimensional, and therefore from the first part there is a norm $p : X^* \to \mathbb{R}$ and a sequence $(x_n^*)_{n \in \mathbb{N}} \subseteq X^*$ such that $\|x_n^*\| = 1$ $\forall n \in \mathbb{N}$, $\|x^*\| \leq p(x^*)$ $\forall x^* \in X^*$ and $p(x_n^*) = n$ $\forall n \in \mathbb{N}$. Define now $\|\cdot\|_2 : X \to \mathbb{R}_+$, $\|x\|_2 = \sup\{|f(x)| \mid f \in X^*, p(f) \leq 1\}$. If $\|x\|_2 = 0$ then $|f(x)| = 0$ for all $f \in X^*$ with $p(f) \leq 1$. If $f \in X^*$ then $p(f/(p(f) + 1)) \leq 1$, from whence $|f(x)|/(p(f)+1) = 0$ $\forall f \in X^*$, and therefore $x = 0$. The other axioms for the norm $\|\cdot\|_2$ are easy to verify. The properties $\tau \subseteq \tau_2$ and $\tau_2 \neq \tau$ can be proved as in the first case.

The norm for some linear and continuous operators

7. Let $(\xi_n)_{n\geq 0} \in l_\infty$. We define $U : l_1 \to C[0,1]$, $(Ua)(x) = \sum_{n=0}^{\infty} a_n \xi_n x^n$ $\forall x \in [0,1]$, $\forall a = (a_n)_{n\geq 0} \in l_1$. Prove that U is well defined, linear, and continuous, with $\|U\| = \sup_{n\geq 0} |\xi_n|$.

8. Let $1 \leq p < \infty$ and $U : l_\infty \to L_p[0,1]$, $U(x_1, x_2, ...) = \sum_{n=1}^{\infty} x_n \chi_{[\frac{1}{2^n}, \frac{1}{2^{n-1}})}$. Prove that U is a linear and continuous operator, and calculate $\|U\|$.

Norm convergence and pointwise convergence for sequences of operators

9. We consider the linear operators $A_n : l_2 \to l_2$, $B_n : l_2 \to l_2$, $A_n(x) = (x_1/n, x_2/(2n), x_3/(3n), ...)$, $B_n(x) = (0, 0, ..., 0, x_{n+1}, x_{n+2}, ...)$ $\forall x = (x_1, x_2, ...) \in l_2$, $\forall n \in \mathbb{N}$. Prove that $\|A_n\| \to 0$ and $B_n(x) \to 0$ $\forall x \in l_2$, but that $(B_n)_{n\in\mathbb{N}}$ does not converge towards 0 in the operator norm.

If $L(X,Y)$ is a Banach space, then Y is a Banach space

10. Let X and Y be two normed spaces such that $X \neq \{0\}$. If $L(X,Y)$ is a Banach space, prove that Y is a Banach space.

Isometries

11. Let $1 \leq p < \infty$. Find an isometry $j : l_\infty \to L(L_p[0,1])$.

12. Let $1 \leq p < \infty$ and consider $U : l_p \to L_p[0, \infty)$, $U(x) = \sum_{n=1}^{\infty} x_n \chi_{[n-1,n)}$ $\forall x = (x_1, x_2, ...) \in l_p$. Prove that U is an isometry.

The Hardy operators on the space l_p

13. i) Let $1 < p < \infty$, $n \in \mathbb{N}$, $a_1 \geq 0, ..., a_n \geq 0$. Prove the inequalities

$$\sum_{k=1}^{n} \left(\frac{a_1 + \cdots + a_k}{k} \right)^p \leq \frac{p}{p-1} \left(\sum_{k=1}^{n} \left(\frac{a_1 + \cdots + a_k}{k} \right)^{p-1} a_k \right)$$

and

$$\left(\sum_{k=1}^{n} \left(\frac{a_1 + \cdots + a_k}{k} \right)^p \right)^{1/p} \leq \frac{p}{p-1} \left(\sum_{k=1}^{n} a_k^p \right)^{1/p}.$$

ii) Let $1 < p < \infty$ and let $H : l_p \to l_p$ be the Hardy operator defined by

$$H((x_n)_{n\in\mathbb{N}}) = \left(\frac{x_1 + x_2 + \cdots + x_n}{n} \right)_{n\in\mathbb{N}}.$$

Prove that H is linear and continuous, and that $\|H\| = p/(p-1)$.

2.1 Exercises

iii) Let $f : [0, \infty) \to [0, \infty)$ be a continuous differentiable convex function, with $f(0) = 0$, and let $(a_n)_{n \in \mathbb{N}} \subseteq [0, \infty)$. Prove that

$$f\left(\sum_{k=n}^{n+p} a_k\right) \leq \sum_{k=n}^{n+p} a_k f'\left(\sum_{j=k}^{n+p} a_j\right) \quad \forall n, p \in \mathbb{N}.$$

iv) Consider $1 < p < \infty$ and let $H : l_p \to l_p$ be the Hardy operator defined by

$$H((x_n)_{n \in \mathbb{N}}) = \left(\sum_{k=n}^{\infty} \frac{x_k}{k}\right)_{n \in \mathbb{N}}.$$

Prove that H is linear and continuous, and that $\|H\| = p$.

The Hilbert, Hardy, and Schur operators on the space $L_p(0, \infty)$

14. i) Let $1 < p < \infty$ and $K : (0, \infty) \to \mathbb{R}$ be a Hardy kernel, i.e., a Lebesgue measurable function with the property that $\int_0^\infty x^{-1/p} |K(x)| \, dx < \infty$. We consider the operators $\mathcal{I}_K : L_p(0, \infty) \to L_p(0, \infty)$ defined by $\mathcal{I}_K f(x) = (1/x) \int_0^\infty K(y/x) f(y) \, dy$ and $\mathcal{M}_K : L_p(0, \infty) \to L_p(0, \infty)$ defined by $\mathcal{M}_K f(x) = x^{1-2/p} \int_0^\infty K(xy) f(y) \, dy$. Prove that $\mathcal{I}_K$ and $\mathcal{M}_K$ are linear and continuous, and that

$$\left|\int_0^\infty y^{-1/p} K(y) \, dy\right| \leq \|\mathcal{I}_K\| \leq \int_0^\infty y^{-1/p} |K(y)| \, dy,$$

$$\left|\int_0^\infty y^{-1/p} K(y) \, dy\right| \leq \|\mathcal{M}_K\| \leq \int_0^\infty y^{-1/p} |K(y)| \, dy.$$

In particular, if $K \geq 0$, then $\|\mathcal{M}_K\| = \|\mathcal{I}_K\| = \int_0^\infty y^{-1/p} K(y) \, dy$.

ii) Prove that for the Hardy operator $H : L_p(0, \infty) \to L_p(0, \infty)$ defined by $Hf(x) = (1/x) \int_0^x f(y) \, dy$, we have $\|H\| = p/(p-1)$.

iii) Prove that for the Hilbert operator $H : L_p(0, \infty) \to L_p(0, \infty)$ defined by $Hf(x) = \int_0^\infty (f(y)/(x+y)) \, dy$ we have $\|H\| = \pi / \sin \frac{\pi}{p}$.

iv) Prove that for the Schur operator $S : L_p(0, \infty) \to L_p(0, \infty)$ defined by $Sf(x) = \int_0^\infty (f(y)/\max(x,y)) \, dy$ we have $\|S\| = p^2/(p-1)$.

v) Prove that for any $\alpha > 0$, the modified Riemann–Liouville integral operator of order α, $\mathcal{R}_\alpha : L_p(0, \infty) \to L_p(0, \infty)$ defined by

$$\mathcal{R}_\alpha f(x) = \frac{1}{\Gamma(\alpha)} \int_0^x \frac{(x-y)^{\alpha-1}}{x^\alpha} f(y) dy$$

is linear and continuous, with $\|\mathcal{R}_\alpha\| = \Gamma(1 - 1/p) / \Gamma(\alpha + 1 - 1/p)$.

vi) Prove that for any $\alpha > 0$, the operator $\mathcal{I}_\alpha : L_p(0, \infty) \to L_p(0, \infty)$ defined by

$$\mathcal{I}_\alpha f(x) = \frac{1}{x^\alpha} \int_0^x y^{\alpha-1} f(y) dy$$

is linear and continuous, with $\|\mathcal{I}_\alpha\| = B(1 - 1/p, \alpha)$.

Here, Γ and B are the Euler functions.

The Schauder basis

15. Let $\alpha, \beta \in \mathbb{C}\setminus\{0\}$ such that $0 < |\beta/\alpha| < 1$, and let $x_1 = (\alpha, \beta, 0, ...)$, $x_2 = (0, \alpha, \beta, 0, ...)$, $x_3 = (0, 0, \alpha, \beta, 0, ...)$, ... Prove that $(x_n)_{n \in \mathbb{N}}$ is a Schauder basis in l_2.

16. Let X be a Banach space and $(x_n)_{n \in \mathbb{N}} \subseteq X$ be a Schauder basis. We denote by $Y = \left\{ y = (\alpha_n)_{n \in \mathbb{N}} \subseteq \mathbb{K} \ \Big| \ \sum_{n=1}^{\infty} \alpha_n x_n \text{ converges} \right\}$ and for any $y \in Y$, we define $\|y\|_1 = \sup_{n \in \mathbb{N}} \left\| \sum_{i=1}^{n} \alpha_i x_i \right\|$. Prove that $(Y, \|\cdot\|_1)$ is a Banach space.

17. Let X be a Banach space which has a Schauder basis $(x_n)_{n \in \mathbb{N}}$. For every $x \in X$, $x = \sum_{k=1}^{\infty} \alpha_k x_k$, and for every $n \in \mathbb{N}$ we define $x_n^*(x) = \alpha_n$. Prove that $x_n^* \in X^*$ for all $n \in \mathbb{N}$.

18. Let X be a Banach space which has a Schauder basis $(x_n)_{n \in \mathbb{N}}$. Prove that $x_n \notin \overline{\operatorname{Sp}\{x_1, ..., x_{n-1}, x_{n+1}, x_{n+2}, ...\}}^{\|\cdot\|} \ \forall n \in \mathbb{N}$, i.e., the system $(x_n)_{n \in \mathbb{N}}$ is topologically linearly independent.

Separable Banach spaces are quotient of l_1

19. Prove that any separable Banach space is a quotient of l_1, i.e., if X is a separable Banach space, then there is a linear continuous and surjective operator from l_1 to X. More precisely, if $(x_n)_{n \in \mathbb{N}} \subseteq B_X$ is a dense set one can consider $Q : l_1 \to X$, $Q\left((\alpha_n)_{n \in \mathbb{N}}\right) = \sum_{n=1}^{\infty} \alpha_n x_n$.

A characterization for separable Banach spaces

20. Prove that a Banach space X is separable if and only if it is isometrically isomorphic with a closed linear subspace of a $C(K)$ space, where K is a compact metric space.

Non-separable Banach spaces

21. Let (X, d) be a metric space with the property that there are $A \subseteq X$, A uncountable, and $\varepsilon > 0$, such that: $\forall a, b \in A$, $a \neq b$, we have $d(a, b) \geq \varepsilon$. Prove that X is not a separable space.

22. Prove that l_∞ is not a separable space.

23. Prove that for any $1 \leq p < \infty$, the space $L(L_p[0,1])$ is not separable.

24. Find:
i) an isometry $j : l_\infty \to L(l_1, C[0,1])$.
ii) for $1 \leq p < \infty$, an isometry $j : l_\infty \to L(l_p)$.
Then deduce that $L(l_1, C[0,1])$ and $L(l_p)$, $1 \leq p < \infty$, are not separable spaces.

An extension result

25. Let X and Y be two normed spaces, and $T : X \to Y$ a linear continuous and surjective operator. Prove that the operator $\widetilde{T} : X/\ker T \to Y$, $\widetilde{T}(x + \ker T) = Tx$ $\forall x \in X$, is a well defined bijective linear and continuous operator, and that $\|T\| = \|\widetilde{T}\|$.

2.1 Exercises

Operators with closed range

26. Let X be a complex normed space and $T : X \to X$ a linear and continuous operator.

i) Prove that if there is $S : X \to X$ linear and continuous such that $T = TST$, then the range of T in X is a closed set.

ii) Deduce that if there is $P \in \mathbb{C}[z]$, $P(z) = \sum_{n=0}^{N} a_n z^n$, such that $P(T) = 0$, then:
a) If $a_0 \neq 0$, prove that the range of T is a closed set in X.
b) If $\min \{k = \overline{0, N} \mid a_k \neq 0\} = m > 0$, prove that T^m has closed range.

The dual for the inclusion operator

27. Let X be a normed space, $Y \subseteq X$ a linear subspace, and $i : Y \to X$ the inclusion operator. Prove that $i^* : X^* \to Y^*$ is the restriction operator, that is, $i^*(x^*) = x^*|_Y$, $\forall x^* \in X^*$.

Isometric isomorphisms

28. Let X and Y be two Banach spaces, and $T \in L(X, Y)$. Prove that T is an isometric isomorphism if and only if $T^* \in L(Y^*, X^*)$ is an isometric isomorphism.

Biorthogonal systems

29. Let X be a Banach space and $(x_i)_{i \in \mathbb{N}} \subseteq X$, $(x_i^*)_{i \in \mathbb{N}} \subseteq X^*$ a biorthogonal system for X, i.e., $x_i^*(x_j) = \delta_{ij}$ for all $i, j \in \mathbb{N}$.

i) If $\sup_{n \in \mathbb{N}} \left\| \sum_{i=1}^{n} x^*(x_i) x_i^* \right\| < \infty$ for any $x^* \in X^*$, prove that the representation $x = \sum_{i=1}^{\infty} x_i^*(x) x_i$ is true for every $x \in \overline{\mathrm{Sp}\{x_n \mid n \in \mathbb{N}\}}^{\|\cdot\|}$.

ii) If $\sup_{n \in \mathbb{N}} \left\| \sum_{i=1}^{n} x_i^*(x) x_i \right\| < \infty$ for any $x \in X$, prove that the representation $x^* = \sum_{i=1}^{\infty} x^*(x_i) x_i^*$ is true for every $x^* \in \overline{\mathrm{Sp}\{x_n^* \mid n \in \mathbb{N}\}}^{\|\cdot\|}$.

The lifting property for l_1

30. Let X be a Banach space, and $T : l_1 \to X$ a linear and continuous operator. Prove that for any Banach space Y, for any metric surjection $Q : Y \to X$, and for any $\varepsilon > 0$, there is a linear and continuous operator $\overline{T} : l_1 \to Y$ such that $\|\overline{T}\| \leq (1+\varepsilon)\|T\|$, $T = Q \circ \overline{T}$.

An operator $Q \in L(X, Y)$ is called a metric surjection if it is surjective and $\|Qx\| = \inf\{\|y\| \mid Qy = Qx\}$ $\forall x \in X$.

31. Let X be a Banach space, and Y a closed linear subspace of X. Prove that for any $\varepsilon > 0$ and for any $T \in L(l_1, X/Y)$, there is $\overline{T} \in L(l_1, X)$ such that $Q \circ \overline{T} = T$ and $\|\overline{T}\| \leq (1+\varepsilon)\|T\|$, where $Q : X \to X/Y$ is the canonical projection.

The extension property for l_∞

32. i) Let X be a normed space. If $T : X \to l_\infty$ is a linear and continuous operator, prove that there is a bounded sequence $(x_n^*)_{n\in\mathbb{N}} \subseteq X^*$ such that $T(x) = (x_n^*(x))_{n\in\mathbb{N}}$. Much more, $\|T\| = \sup_{n\in\mathbb{N}} \|x_n^*\|$.

ii) If X is a normed space, $Y \subseteq X$ is a linear subspace, and $T : Y \to l_\infty$ is a linear and continuous operator, prove that there is $\widetilde{T} : X \to l_\infty$ linear and continuous such that $\widetilde{T}\big|_Y = T$, $\|\widetilde{T}\| = \|T\|$.

An operator with the above properties is called a Phillips extension for T.

2.2 Solutions

1. See [42, exercise 37, chapter IV].

Let $f \in C(T)$ with $\|f\| \leq 1$, i.e., $|f(t)| \leq 1 \ \forall t \in T$, that is, $-1 \leq f \leq 1$. Therefore $f + 1 \geq 0$, $U(f+1) \geq 0$, $-U(1) \leq U(f)$. Analogously, $U(f) \leq U(1)$. Thus $-U(1) \leq U(f) \leq U(1)$, and therefore $\|U(f)\| \leq \|U(1)\| \ \forall \|f\| \leq 1$. Thus U is continuous and $\|U\| \leq \|U(1)\|$. Also $\|U(1)\| \leq \|U\| \|1\|$. But $\|1\| = \sup_{t\in T} |1(t)| = 1$, and therefore $\|U(1)\| \leq \|U\|$. We obtain that $\|U(1)\| = \|U\|$.

2. (i) If $f \in C[0,1]$, i.e., $f : [0,1] \to \mathbb{R}$ is continuous, then since $\varphi : [0,1] \times [0,1] \to [0,\infty)$ is continuous we obtain that $h(x,t) = \varphi(x,t) f(t)$, $h : [0,1] \times [0,1] \to \mathbb{R}$ is also continuous. Then the function $x \mapsto \int_0^1 \varphi(x,t) f(t)\, dt$ is continuous, i.e., $U(f) \in C[0,1]$. Evidently, U is a linear operator. We prove that U is a positive operator. Let $f \in C[0,1]$, $f \geq 0$, i.e., $f(t) \geq 0 \ \forall t \in [0,1]$. Then since $\varphi(x,t) \geq 0 \ \forall (x,t) \in [0,1] \times [0,1]$ it follows that $\int_0^1 \varphi(x,t) f(t)\, dt \geq 0 \ \forall x \in [0,1]$, i.e., $(Uf)(x) \geq 0 \ \forall x \in [0,1]$, $Uf \geq 0$. Using exercise 1 we obtain that $\|U\| = \|U(1)\|$. Let $h = U(1)$. Then $h(x) = \int_0^1 \varphi(x,t)\, dt$ $\forall x \in [0,1]$. Since $\partial\varphi/\partial x : [0,1] \times [0,1] \to [0,\infty)$ is continuous, from the theorem on differentiation in a Riemann integral with parameter it follows that h is differentiable on $[0,1]$ and $h'(x) = \int_0^1 \frac{\partial \varphi}{\partial x}(x,t)\, dt \ \forall x \in [0,1]$. Also, since $\partial\varphi/\partial x$ is positive it follows that $h'(x) = \int_0^1 \frac{\partial \varphi}{\partial x}(x,t)\, dt \geq 0 \ \forall x \in [0,1]$, i.e., h is increasing. Then $\|h\| = \sup_{x\in[0,1]} h(x) = h(1) = \int_0^1 \varphi(1,t)\, dt$.

(ii) We take $\varphi : [0,1] \times [0,1] \to [0,\infty)$, $\varphi(x,t) = e^{xt}$, with $\frac{\partial \varphi}{\partial x}(x,t) = te^{xt}$. Then by (i) U is a linear and continuous operator and $\|U\| = \int_0^1 \varphi(1,t)\, dt = \int_0^1 e^t dt = e - 1$.

3. See [40, exercise 8.22].

i) Let $f \in C[0,1]$. Using the Weierstrass theorem, for every $n \in \mathbb{N}$ there is a $c_n \in [0,1]$ such that $\|U_n f - f\| = \left| f\left(c_n^{1+1/n}\right) - f(c_n) \right|$. Suppose now that $\|U_n f - f\|$ does not converge towards 0, and therefore there are an $\varepsilon > 0$ and a subsequence $(n_k)_{k\in\mathbb{N}}$ of $\mathbb{N}$ such that $\|U_{n_k} f - f\| \geq \varepsilon \ \forall k \in \mathbb{N}$, i.e., $\left| f\left(c_{n_k}^{1+1/n_k}\right) - f(c_{n_k}) \right| \geq \varepsilon \ \forall k \in \mathbb{N}$ (1). Since $(c_{n_k})_{k\in\mathbb{N}} \subseteq [0,1]$ we use Cesaro's lemma and without loss of generality we assume that $c_{n_k} \to c_0 \in [0,1]$. Then $c_{n_k}^{1+1/n_k} \to c_0$ and $f\left(c_{n_k}^{1+1/n_k}\right) \to f(c_0)$, $f(c_{n_k}) \to f(c_0)$. We

2.2 Solutions

obtain that $f\left(c_{n_k}^{1+1/n_k}\right) - f(c_{n_k}) \to 0$, which contradicts (1).

ii) We have: $\|U_n - I\| = \sup_{\|f\|\leq 1} \|U_n(f) - f\| = \sup_{\|f\|\leq 1} \sup_{x\in[0,1]} \left|f\left(x^{1+1/n}\right) - f(x)\right| \leq \sup_{\|f\|\leq 1} \sup_{x\in[0,1]} \left(\left|f\left(x^{1+1/n}\right)\right| + |f(x)|\right) \leq 2$. For every $n \in \mathbb{N}$ we consider $f_n \in C[0,1]$ such that the graph of f connects, linearly, the points $(0,0)$, $\left((1/2)^{1+1/n}, 1\right)$, $(1/2, -1)$ and $(1, 1)$. Evidently, $\|f_n\|_\infty = 1$ and

$$\|U_n(f_n) - f_n\|_\infty = \sup_{x\in[0,1]} \left|f_n\left(x^{1+1/n}\right) - f_n(x)\right| \geq \left|f_n\left((1/2)^{1+1/n}\right) - f_n(1/2)\right| = 2.$$

We obtain that $\|U_n - I\| \geq 2$, and therefore $\|U_n - I\| = 2 \ \forall n \in \mathbb{N}$.

Remark. The same type of reasoning shows that if $g : [0,1] \to [0,1]$ is a continuous function which is not the identity map, and we define $T : C[0,1] \to C[0,1]$, $(Tf)(x) = f(g(x))$, we obtain a linear and continuous operator such that $\|T - I\| = 2$.

4. See [40, exercise 8.21].

We know (see exercise 2) that $Uf, U_n f, V_n f \in C[0,1] \ \forall f \in C[0,1]$, $\forall n \in \mathbb{N}$. Using the Taylor expansion $e^x = \sum_{n=0}^\infty x^n/n!$ we have

$$(Uf - U_n f)(x) = \int_0^1 \left(\sum_{k=n+1}^\infty \frac{(tx)^k}{k!}\right) f(t)\, dt.$$

By exercise 1 we have

$$\|U_n - U\| = \sup_{x\in[0,1]} \int_0^1 \left(\sum_{k=n+1}^\infty \frac{(tx)^k}{k!}\right) dt = \int_0^1 \left(\sum_{k=n+1}^\infty \frac{t^k}{k!}\right) dt \to 0.$$

We also have

$$(Uf - V_n f)(x) = \int_0^1 e^{xt} f(t)\, dt - \int_{\varepsilon_n}^{1-\varepsilon_n} e^{xt} f(t)\, dt = \int_0^{\varepsilon_n} e^{xt} f(t)\, dt + \int_{1-\varepsilon_n}^1 e^{xt} f(t)\, dt.$$

This equality shows that $U - V_n$ is a positive operator, and using exercise 1 we obtain that

$$\|U - V_n\| = \|(U - V_n)(1)\| = \sup_{x\in[0,1]} \left(\int_0^{\varepsilon_n} e^{xt}\, dt + \int_{1-\varepsilon_n}^1 e^{xt}\, dt\right)$$
$$= \int_0^{\varepsilon_n} e^t\, dt + \int_{1-\varepsilon_n}^1 e^t\, dt \to 0,$$

since $\varepsilon_n \to 0$.

5. See [40, exercise 8.20].

From standard analysis we know that if $f : [0,1] \to \mathbb{R}$ is a continuous function, then $x \longmapsto \int_0^x f(t)\, dt$ is also a continuous function, i.e., U and U_n take their values in $C[0,1]$.

Using the Taylor expansion $e^t = \sum_{n=0}^{\infty} \frac{t^n}{n!}$ we obtain that

$$(Uf - U_n f)(x) = \int_0^x \left(\sum_{k=n+1}^{\infty} \frac{t^k}{k!} \right) f(t)\, dt \quad \forall x \in [0,1].$$

Then

$$|(Uf - U_n f)(x)| \leq \int_0^x \left(\sum_{k=n+1}^{\infty} \frac{t^k}{k!} \right) |f(t)|\, dt \leq \|f\| \int_0^1 \left(\sum_{k=n+1}^{\infty} \frac{t^k}{k!} \right) dt,$$

for all $x \in [0,1]$, thus

$$\|U_n - U\| \leq \|f\| \int_0^1 \left(\sum_{k=n+1}^{\infty} \frac{t^k}{k!} \right) dt \quad \forall f \in C[0,1],$$

i.e.,

$$\|U_n - U\| \leq \int_0^1 \left(\sum_{k=n+1}^{\infty} \frac{t^k}{k!} \right) dt.$$

But if $u_n(t) = \sum_{k=n+1}^{\infty} t^k/k!$, $u_n : [0,1] \to \mathbb{R}$, then $|u_n(t)| \leq \sum_{k=n+1}^{\infty} 1/k! \to 0$, i.e., $u_n \to 0$ uniformly on $[0,1]$. Then $\int_0^1 u_n(t)\, dt \to 0$. Thus $\|U_n - U\| \to 0$.

To calculate $\|U_n - U\|$, we can also use exercise 1, if we observe that $U - U_n : C[0,1] \to C[0,1]$ is positive and therefore

$$\|U_n - U\| = \|(U_n - U)(1)\| = \sup_{x \in [0,1]} \int_0^x \left(\sum_{k=n+1}^{\infty} \frac{t^k}{k!} \right) dt = \int_0^1 \left(\sum_{k=n+1}^{\infty} \frac{t^k}{k!} \right) dt.$$

6. See [36, exercise 1, chapter 3, section 9].

i) Let us observe first that U is well defined, since the composition of two continuous functions is a continuous function, and that U is linear. Since U is positive and $U(1) = 1$, using exercise 1 it follows that $\|U\| = \|U(1)\| = 1$.

ii) Suppose that U is surjective. Let $s_1, s_2 \in S$ with $s_1 \neq s_2$. By the well known Urysohn lemma it follows that there is an $f \in C(S)$ such that $0 \leq f(s) \leq 1 \, \forall s \in S$, with $f(s_1) = 1$ and $f(s_2) = 0$. Since U is surjective there is a $\varphi \in C(T)$ such that $U(\varphi) = f$, from whence $1 = f(s_1) = \varphi(h(s_1))$ and $0 = f(s_2) = \varphi(h(s_2))$, thus $h(s_1) \neq h(s_2)$, i.e., h is injective.

Conversely, suppose that h is injective. Since h is continuous and S is compact, it follows that the co-restriction of h, i.e., $h : S \to h(S)$ is a homeomorphism (see the remark following the solution for exercise 29 of chapter 14). Let now $g \in C(S)$. Then the function $g \circ h^{-1} : h(S) \to \mathbb{R}$ is continuous. Now by the Tietze extension theorem it follows that there is an $f \in C(T)$ such that $f(t) = (g \circ h^{-1})(t) \, \forall t \in h(S)$, and then $f(h(s)) = g(s) \, \forall s \in S$, i.e., $U(f) = g$.

2.2 Solutions

iii) Suppose that U is an isometry. Let $t_0 \in T$. If $h(S) \cap \{t_0\} = \varnothing$, since $h(S)$ is compact, by the Urysohn lemma we obtain that there is an $f \in C(T)$ with $0 \leq f(t) \leq 1$ $\forall t \in T$, $f = 0$ on $h(S)$ and $f(t_0) = 1$. Since U is an isometry, we have $\|U(f)\| = \|f\|$. But $f = 0$ on $h(S)$, i.e., $f(h(s)) = 0$ $\forall s \in S$, i.e., $U(f) = 0$, so $\|f\| = 0$, i.e. $f = 0$, which is false, because $f(t_0) = 1$. Hence for each $t_0 \in T$, $h(S) \cap \{t_0\} \neq \varnothing$, i.e., there is $s \in S$ such that $h(s) = t_0$. Therefore h is surjective.

Conversely, if h is surjective then $h(S) = T$, hence $\sup_{s \in S}|f(h(s))| = \sup_{t \in T}|f(t)|$, i.e., $\|U(f)\| = \|f\|$ $\forall f \in C(T)$.

7. We have $|a_n \xi_n x^n| \leq |a_n| |\xi_n|$ $\forall x \in [0,1]$, $\forall n \geq 0$ and $\sum_{n=0}^{\infty} |a_n| |\xi_n| \leq \|\xi\| \|a\| < \infty$. Using the Weierstrass criterium we obtain that the series $\sum_{n=0}^{\infty} a_n \xi_n x^n$ is uniformly convergent on $[0,1]$ and that the function $x \longmapsto \sum_{n=0}^{\infty} a_n \xi_n x^n$ is a continuous one, i.e., $U(a) \in C[0,1]$ $\forall a \in l_1$. Clearly U is linear. Also $\|U(a)\| = \sup_{x \in [0,1]} \left|\sum_{n=0}^{\infty} a_n \xi_n x^n\right| \leq \|a\| \|\xi\|$, i.e., U is continuous, $\|U\| \leq \|\xi\| = \sup_{n \geq 0} |\xi_n|$. But $(Ue_n)(x) = \xi_n x^n$ $\forall x \in [0,1]$ and therefore $|\xi_n| \leq \|U\|$ $\forall n \geq 0$. We obtain that $\sup_{n \geq 0} |\xi_n| = \|U\|$.

8. If $x \in l_\infty$, then

$$\left|\sum_{n=1}^{\infty} x_n \chi_{[\frac{1}{2^n}, \frac{1}{2^{n-1}})}\right|^p = \sum_{n=1}^{\infty} |x_n|^p \chi_{[\frac{1}{2^n}, \frac{1}{2^{n-1}})},$$

and therefore

$$\|U(x)\|_p = \left(\int_0^1 \sum_{n=1}^{\infty} |x_n|^p \chi_{[\frac{1}{2^n}, \frac{1}{2^{n-1}})} d\mu\right)^{1/p} = \left(\sum_{n=1}^{\infty} |x_n|^p \frac{1}{2^n}\right)^{1/p} \leq \|x\| < \infty,$$

i.e., $U(x) \in L_p[0,1]$. Since U is linear, from the above inequality we obtain that U is continuous and $\|U\| \leq 1$. But for any $n \in \mathbb{N}$, $x = (\underbrace{1,1,\dots,1}_{n},0,\dots) \in l_\infty$ (n times 1), $\|x\| = 1$, and $U(x) = \sum_{k=1}^{n} \chi_{[\frac{1}{2^k},\frac{1}{2^{k-1}})} = \chi_{[\frac{1}{2^n},1)}$. Using the fact that $\|U(x)\| \leq \|U\| \|x\|$, we obtain that $(1 - 1/2^n)^{1/p} \leq \|U\|$ $\forall n \in \mathbb{N}$, and, for $n \to \infty$, $1 \leq \|U\|$, and therefore $\|U\| = 1$.

9. See [40, exercise 8.19].
We have

$$\|A_n(x)\| = \left(\sum_{k=1}^{\infty} \frac{|x_k|^2}{k^2 n^2}\right)^{1/2} = \frac{1}{n}\left(\sum_{k=1}^{\infty} \frac{|x_k|^2}{k^2}\right)^{1/2} \leq \frac{1}{n}\left(\sum_{k=1}^{\infty} |x_k|^2\right)^{1/2} = \frac{1}{n}\|x\|,$$

i.e., $\|A_n\| \leq 1/n \to 0$. Also, $\|B_n(x)\| = \left(\sum_{k=n+1}^{\infty} |x_k|^2\right)^{1/2}$ $\forall x \in l_2$. But $x \in l_2$, i.e., $s = \sum_{n=1}^{\infty} |x_n|^2 < \infty$, and therefore $s - \sum_{k=1}^{n} |x_k|^2 \to 0$, i.e., $\sum_{k=n+1}^{\infty} |x_k|^2 \to 0$. Thus

$\|B_n(x)\| \to 0 \ \forall x \in l_2$. Also $\|B_n(x)\| = \left(\sum\limits_{k=n+1}^{\infty} |x_k|^2\right)^{1/2} \leq \left(\sum\limits_{k=1}^{\infty} |x_k|^2\right)^{1/2} = \|x\|$, i.e., $\|B_n\| \leq 1$. For $e_{n+1} = (0,0,...,0,1,0,...) \in l_2$, $B_n(e_{n+1}) = e_{n+1}$, thus $1 = \|e_{n+1}\| = \|B_n(e_{n+1})\| \leq \|B_n\| \|e_{n+1}\| = \|B_n\|$, and therefore $\|B_n\| = 1 \ \forall n \in \mathbb{N}$.

10. Consider $x_0 \in X$, with $\|x_0\| = 1$. Using the Hahn–Banach theorem we obtain that there is an $x^* \in X^*$, $\|x^*\| = 1$, such that $x^*(x_0) = \|x_0\| = 1$. Consider now a Cauchy sequence $(y_n)_{n \in \mathbb{N}} \subseteq Y$. We define the operators $A_n : X \to Y$, $A_n(x) = x^*(x)y_n$. Evidently $A_n \in L(X,Y) \ \forall n \in \mathbb{N}$. Also: $\|A_n - A_m\| = \sup\limits_{\|x\| \leq 1} \|x^*(x)(y_n - y_m)\| = \|y_n - y_m\| \sup\limits_{\|x\| \leq 1} |x^*(x)| = \|y_n - y_m\| \ \forall n,m \in \mathbb{N}$, that is, $(A_n)_{n \in \mathbb{N}} \subseteq L(X,Y)$ is a Cauchy sequence. Since $L(X,Y)$ is a Banach space, there is an operator $A \in L(X,Y)$ such that $A_n \to A$ in norm. Let $Ax_0 = y_0$. Then $\|y_n - y_0\| = \|A_n(x_0) - A(x_0)\| = \|(A_n - A)(x_0)\| \leq \|A_n - A\| \|x_0\| \to 0$, therefore $y_n \to y_0$, and Y is a Banach space.

Remark. The converse is also true. If Y is a Banach space, and X is a normed space, then $L(X,Y)$ is a Banach space.

11. Let $A_n = [\frac{1}{2^n}, \frac{1}{2^{n-1}})$, $n \in \mathbb{N}$. For $x = (x_n)_{n \in \mathbb{N}} \in l_\infty$ we consider $U_x : L_p[0,1] \to L_p[0,1]$, $U_x(f) = \sum\limits_{n=1}^{\infty} x_n \chi_{A_n} f$. Evidently, U_x is a linear operator. Since $(A_n)_{n \in \mathbb{N}}$ are pairwise disjoint sets, we have $|U_x(f)|^p = \sum\limits_{n=1}^{\infty} |x_n|^p \chi_{A_n} |f|^p$, and therefore

$$\int_0^1 |U_x(f)|^p \, d\mu = \sum_{n=1}^{\infty} |x_n|^p \int_{A_n} |f|^p \, d\mu$$
$$\leq \|x\|^p \sum_{n=1}^{\infty} \int_{A_n} |f|^p = \|x\|^p \int_0^1 |f|^p \, d\mu < \infty$$

(we have used the property $\bigcup\limits_{n \in \mathbb{N}} A_n = (0,1)$ and that if g is a positive measurable function then $A \mapsto \int_A g \, d\mu$ is a countably additive measure). Thus $U_x(f) \in L_p[0,1]$ and $\|U_x(f)\| \leq \|x\| \|f\| \ \forall f \in L_p[0,1]$, i.e., U_x is linear and continuous, $\|U_x\| \leq \|x\| \ \forall x \in l_\infty$. For any $k \in \mathbb{N}$ we have: $\|U_x(\chi_{A_k})\| \leq \|U_x\| \|\chi_{A_k}\|$. Since $(A_n)_{n \in \mathbb{N}}$ are pairwise disjoint sets, we have $U_x(\chi_{A_k}) = x_k \chi_{A_k}$, and therefore $\|U_x(\chi_{A_k})\| = \|x_k \chi_{A_k}\| = |x_k| (\mu(A_k))^{1/p}$. Thus $|x_k| (\mu(A_k))^{1/p} \leq \|U_x\| (\mu(A_k))^{1/p}$, $\mu(A_k) > 0$, and therefore $|x_k| \leq \|U_x\| \ \forall k \in \mathbb{N}$, hence $\|x\| = \sup\limits_{k \in \mathbb{N}} |x_k| \leq \|U_x\|$. Thus $\|U_x\| = \|x\| \ \forall x \in l_\infty$. Let $j : l_\infty \to L(L_p[0,1])$, $j(x) = U_x$. Evidently j is a linear operator, and $\|j(x)\| = \|U_x\| = \|x\| \ \forall x \in l_\infty$, i.e., j is an isometry.

12. We have $|U(x)|^p = \sum\limits_{n=1}^{\infty} |x_n|^p \chi_{[n-1,n)}$, and therefore

$$\int_0^\infty |U(x)|^p \, d\mu = \sum_{n=1}^{\infty} |x_n|^p \mu([n-1,n)) = \sum_{n=1}^{\infty} |x_n|^p,$$

i.e., $\|U(x)\| = \|x\| \ \forall x \in l_p$.

2.2 Solutions

13. See [28, p. 328].

Let q be the conjugate of p. Let $M_0 = 0$ and $M_k = (a_1 + \cdots + a_k)/k$, $1 \leq k \leq n$. For $1 \leq k \leq n$ we have

$$M_k^p - \frac{p}{p-1} M_k^{p-1} a_k = M_k^p - \frac{p}{p-1} M_k^{p-1} (kM_k - (k-1)M_{k-1})$$

$$= \left(1 - \frac{pk}{p-1}\right) M_k^p + \frac{k-1}{p-1} p M_k^{p-1} M_{k-1}.$$

Since the function $x \longmapsto \log x$ is concave we have the inequality

$$\log \frac{\alpha x^p + \beta y^p}{\alpha + \beta} \geq \frac{\alpha \log x^p + \beta \log y^p}{\alpha + \beta}$$

for any $\alpha, \beta, x, y > 0$. In particular, for $\alpha = p - 1$, $\beta = 1$, we obtain that $px^{p-1}y \leq (p-1)x^p + y^p$, for any $x \geq 0$, $y \geq 0$. From this we deduce that

$$M_k^p - \frac{p}{p-1} M_k^{p-1} a_k = \left(1 - \frac{pk}{p-1}\right) M_k^p + \frac{k-1}{p-1} p M_k^{p-1} M_{k-1}$$

$$\leq \left(1 - \frac{pk}{p-1}\right) M_k^p + \frac{k-1}{p-1} \left((p-1) M_k^p + M_{k-1}^p\right)$$

$$= \frac{1}{p-1} \left((k-1) M_{k-1}^p - k M_k^p\right).$$

Hence we have the inequality

$$M_k^p - \frac{p}{p-1} M_k^{p-1} a_k \leq \frac{1}{p-1} \left((k-1) M_{k-1}^p - k M_k^p\right).$$

Adding these inequalities for $k = 1, \ldots, n$ we deduce that

$$\sum_{k=1}^n M_k^p - \frac{p}{p-1} \sum_{k=1}^n M_k^{p-1} a_k \leq -\frac{n M_n^p}{p-1} \leq 0,$$

i.e., the first inequality from the statement. Applying the Hölder inequality we deduce that

$$\sum_{k=1}^n M_k^p \leq \frac{p}{p-1} \sum_{k=1}^n M_k^{p-1} a_k \leq \frac{p}{p-1} \left(\sum_{k=1}^n M_k^{(p-1)q}\right)^{1/q} \left(\sum_{k=1}^n a_k^p\right)^{1/p}$$

$$= \frac{p}{p-1} \left(\sum_{k=1}^n M_k^p\right)^{1/q} \left(\sum_{k=1}^n a_k^p\right)^{1/p}$$

and from here

$$\left(\sum_{k=1}^n M_k^p\right)^{1/p} \leq \frac{p}{p-1} \left(\sum_{k=1}^n a_k^p\right)^{1/p}.$$

The inequalities from the exercise are called the Elliott's inequalities.

ii) Let $x = (x_n)_{n\in\mathbb{N}} \in l_p$. From (i) and the properties for the modulus we deduce that for any $n \in \mathbb{N}$ we have

$$\sum_{k=1}^{n} \left(\frac{|x_1 + \cdots + x_k|}{k}\right)^p \leq \left(\frac{p}{p-1}\right)^p \left(\sum_{k=1}^{n} |x_k|^p\right)$$

and passing to the limit for $n \to \infty$ we deduce that

$$\sum_{n=1}^{\infty} \left(\frac{|x_1 + \cdots + x_n|}{n}\right)^p \leq \left(\frac{p}{p-1}\right)^p \left(\sum_{n=1}^{\infty} |x_n|^p\right),$$

i.e., $\|H(x)\| \leq (p/(p-1)) \|x\|$ $\forall x \in l_p$, so since H is linear it follows that H is continuous with $\|H\| \leq p/(p-1)$. To prove the converse inequality, consider the element $x = \left(1, 1/2^{1/p}, \ldots, 1/n^{1/p}, 0, \ldots\right) \in l_p$. We have $\|H(x)\|^p \leq \|H\|^p \|x\|^p$, $\|x\|^p = \sum_{k=1}^{n} 1/k$ and

$$\|H(x)\|^p \geq \sum_{k=1}^{n} \left(\frac{1}{k}\left(1 + \frac{1}{2^{1/p}} + \cdots + \frac{1}{k^{1/p}}\right)\right)^p.$$

It follows that

$$\|H\|^p \geq \frac{\sum_{k=1}^{n} \left(\frac{1}{k}\left(1 + \frac{1}{2^{1/p}} + \cdots + \frac{1}{k^{1/p}}\right)\right)^p}{\sum_{k=1}^{n} \frac{1}{k}} \quad \forall n \in \mathbb{N}.$$

Using the Stolz–Cesaro lemma we obtain that

$$\lim_{n\to\infty} \frac{\sum_{k=1}^{n} \left(\frac{1}{k}\left(1 + \frac{1}{2^{1/p}} + \cdots + \frac{1}{k^{1/p}}\right)\right)^p}{\sum_{k=1}^{n} \frac{1}{k}} = \lim_{n\to\infty} \frac{\left(\frac{1}{n}\left(1 + \frac{1}{2^{1/p}} + \cdots + \frac{1}{n^{1/p}}\right)\right)^p}{\frac{1}{n}}$$

$$= \lim_{n\to\infty} \left(\frac{1 + \frac{1}{2^{1/p}} + \cdots + \frac{1}{n^{1/p}}}{n^{1/q}}\right)^p.$$

Again by the Stolz–Cesaro lemma,

$$\lim_{n\to\infty} \frac{1 + \frac{1}{2^{1/p}} + \cdots + \frac{1}{n^{1/p}}}{n^{1/q}} = \lim_{n\to\infty} \frac{\frac{1}{(n+1)^{1/p}}}{(n+1)^{1/q} - n^{1/q}} = \lim_{n\to\infty} \frac{\frac{1}{n}\left(\frac{n}{n+1}\right)^{1/p}}{\left(1 + \frac{1}{n}\right)^{1/q} - 1} = \frac{p}{p-1}.$$

Thus $\|H\| \geq p/(p-1)$.

iii) Using Lagrange's theorem, $f(b) - f(a) = (b-a)f'(c)$, $a < c < b$, and since f' is increasing (f is convex), we obtain that $f(b) - f(a) \leq (b-a)f'(b)$. Therefore $f(x+a) - f(a) \leq xf'(x+a)$ $\forall 0 \leq x, a < \infty$, that is, $f(x+a) - xf'(x+a) \leq f(a)$. If $x = a_n$ and $a = \sum_{k=n+1}^{n+p} a_k$, we obtain that

$$f\left(\sum_{k=n}^{n+p} a_k\right) - a_n f'\left(\sum_{k=n}^{n+p} a_k\right) \leq f\left(\sum_{k=n+1}^{n+p} a_k\right)$$

2.2 Solutions

Consider now $\varphi : [0, \infty)^{p+1} \to \mathbb{R}$ given by

$$\varphi(a_n, a_{n+1}, \ldots, a_{n+p}) = f\left(\sum_{k=n}^{n+p} a_k\right) - \sum_{k=n}^{n+p} a_k f'\left(\sum_{j=k}^{n+p} a_j\right).$$

We have that

$$\begin{aligned}
\varphi(a_n, a_{n+1}, \ldots, a_{n+p}) &= f\left(\sum_{k=n}^{n+p} a_k\right) - a_n f'\left(\sum_{j=n}^{n+p} a_j\right) - \sum_{k=n+1}^{n+p} a_k f'\left(\sum_{j=k}^{n+p} a_j\right) \\
&\leq f\left(\sum_{k=n+1}^{n+p} a_k\right) - \sum_{k=n+1}^{n+p} a_k f'\left(\sum_{j=k}^{n+p} a_j\right) \\
&= \varphi(0, a_{n+1}, a_{n+2}, \ldots, a_{n+p}).
\end{aligned}$$

Then $\varphi(0, a_{n+1}, a_{n+2}, \ldots, a_{n+p}) \leq \varphi(0, 0, a_{n+2}, \ldots, a_{n+p}) \leq \cdots \leq \varphi(0, 0, 0, \ldots, a_{n+p})$. But $\varphi(0, 0, 0, \ldots, x) = f(x) - xf'(x) \leq 0$ $\forall 0 \leq x < \infty$, since f is convex on $[0, \infty)$ and $f(0) = 0$. Thus $\varphi(a_n, a_{n+1}, \ldots, a_{n+p}) \leq 0$, i.e., the inequality we wanted.

Taking $f(x) = x^p$, $p > 1$, we obtain the Davies–Petersen inequalities, i.e.,

$$\left(\sum_{k=n}^{n+m} a_k\right)^p \leq p \sum_{k=n}^{n+m} a_k \left(\sum_{j=k}^{n+m} a_j\right)^{p-1} \quad \forall n, m \in \mathbb{N},$$

and, for $m \to \infty$,

$$\left(\sum_{k=n}^{\infty} a_k\right)^p \leq p \sum_{k=n}^{\infty} a_k \left(\sum_{j=k}^{\infty} a_j\right)^{p-1} \quad \forall n \in \mathbb{N}.$$

iv) For any positive sequence $(a_n)_{n \in \mathbb{N}} \in l_p$, let $b_n = \sum_{k=n}^{\infty} a_k/k$. Then for any $n \in \mathbb{N}$ we have $b_n - b_{n+1} = a_n/n$.

First solution. Fix $n \in \mathbb{N}$. Using the Davies–Petersen inequality, we have

$$\left(\sum_{k=n}^{\infty} \frac{a_k}{k}\right)^p \leq p \sum_{k=n}^{\infty} \frac{a_k}{k} \left(\sum_{j=k}^{\infty} \frac{a_j}{j}\right)^{p-1} = p \sum_{k=n}^{\infty} \frac{a_k}{k} b_k^{p-1},$$

i.e., $b_n^p \leq p \sum_{k=n}^{\infty} (a_k/k) b_k^{p-1}$, hence

$$\begin{aligned}
\sum_{i=1}^{n} b_i^p &\leq p \sum_{i=1}^{n} \sum_{k=i}^{\infty} \frac{a_k}{k} b_k^{p-1} = p \left(\sum_{k=1}^{\infty} \frac{a_k}{k} b_k^{p-1} + \sum_{k=2}^{\infty} \frac{a_k}{k} b_k^{p-1} + \cdots + \sum_{k=n}^{\infty} \frac{a_k}{k} b_k^{p-1}\right) \\
&= p \left(\sum_{k=1}^{n} \frac{a_k}{k} b_k^{p-1} + \sum_{k=2}^{n} \frac{a_k}{k} b_k^{p-1} + \cdots + \sum_{k=n-1}^{n} \frac{a_k}{k} b_k^{p-1} + n \sum_{k=n}^{\infty} \frac{a_k}{k} b_k^{p-1}\right) \\
&= p \left(a_1 b_1^{p-1} + a_2 b_2^{p-1} + \cdots + a_n b_n^{p-1} + n \sum_{k=n}^{\infty} \frac{a_k}{k} b_k^{p-1}\right).
\end{aligned}$$

By the Hölder inequality, we have

$$a_1 b_1^{p-1} + a_2 b_2^{p-1} + \cdots + a_n b_n^{p-1} \leq \left(\sum_{i=1}^n a_i^p\right)^{1/p} \left(\sum_{i=1}^n b_i^{q(p-1)}\right)^{1/q}$$

$$= \left(\sum_{i=1}^n a_i^p\right)^{1/p} \left(\sum_{i=1}^n b_i^p\right)^{1/q},$$

thus

$$\sum_{i=1}^n b_i^p \leq p\left(\left(\sum_{i=1}^n a_i^p\right)^{1/p} \left(\sum_{i=1}^n b_i^p\right)^{1/q} + n\sum_{k=n}^\infty \frac{a_k}{k} b_k^{p-1}\right).$$

Since $(b_n)_{n\in\mathbb{N}}$ is decreasing we have $\sum_{k=n}^\infty (a_k/k) b_k^{p-1} \leq b_n^{p-1} \sum_{k=n}^\infty (a_k/k) = b_n^p$, hence $n\sum_{k=n}^\infty (a_k/k) b_k^{p-1} \leq nb_n^p$. Now again by the Hölder inequality we have $nb_n^p \leq n\left(\sum_{k=n}^\infty a_k^p\right) \left(\sum_{k=n}^\infty 1/k^q\right)^{p/q}$. But

$$\sum_{k=n}^\infty \frac{1}{k^q} \leq \int_{n-1}^\infty \frac{1}{x^q} dx = \frac{1}{(q-1)(n-1)^{q-1}},$$

so

$$0 \leq nb_n^p \leq \left(\sum_{k=n}^\infty a_k^p\right) \frac{n}{(q-1)^{p/q}(n-1)} \to 0.$$

Using that

$$\sum_{k=1}^n b_k^p \leq p\left(\sum_{k=1}^n a_k^p\right)^{1/p} \left(\sum_{k=1}^n b_k^p\right)^{1/q} + pnb_n^p,$$

that is,

$$\left(\sum_{k=1}^n b_k^p\right)^{1/q} \left(\left(\sum_{k=1}^n b_k^p\right)^{1/p} - p\left(\sum_{k=1}^n a_k^p\right)^{1/p}\right) \leq pnb_n^p,$$

from here it follows easily that the increasing sequence $\left(\sum_{k=1}^n b_k^p\right)_{n\in\mathbb{N}}$ is upper bounded, i.e., the series $\sum_{n=1}^\infty b_n^p$ is convergent. Passing to the limit for $n \to \infty$ we obtain that $\sum_{n=1}^\infty b_n^p \leq p\left(\sum_{n=1}^\infty a_n^p\right)^{1/p} \left(\sum_{n=1}^\infty b_n^p\right)^{1/q}$, or $\left(\sum_{n=1}^\infty b_n^p\right)^{1/p} \leq p\left(\sum_{n=1}^\infty a_n^p\right)^{1/p}$. Now if $x = (x_n)_{n\in\mathbb{N}} \in l_p$ and we write $r_n = \sum_{k=n}^\infty (x_k/k)$, then $|r_n| \leq \sum_{k=n}^\infty (|x_k|/k)$ and by the above proved inequality, applied for $\{|x_k|\}$, we have

$$\left(\sum_{n=1}^\infty |r_n|^p\right)^{1/p} \leq \left(\sum_{n=1}^\infty \left(\sum_{k=n}^\infty \frac{|x_k|}{k}\right)^p\right)^{1/p} \leq p\left(\sum_{n=1}^\infty |x_n|^p\right)^{1/p},$$

2.2 Solutions

i.e., $\|H(x)\| \leq p\|x\| \ \forall x \in l_p$. To prove the reverse inequality let $\alpha > 1/p$ and for $n \in \mathbb{N}$ consider the element $x = (0, ..., 0, 1/n^\alpha, 1/(n+1)^\alpha, ...) \in l_p$. We have

$$H(x) = \left(\sum_{k=n}^{\infty} \frac{1}{k^{\alpha+1}}, ..., \sum_{k=n}^{\infty} \frac{1}{k^{\alpha+1}}, \sum_{k=n+1}^{\infty} \frac{1}{k^{\alpha+1}}, ...\right),$$

$\|H(x)\|^p \leq \|H\|^p \|x\|^p$, $\|x\|^p = \sum_{k=n}^{\infty} \frac{1}{k^{\alpha p}}$, and

$$\|H(x)\|^p \geq \left(\sum_{k=n}^{\infty} \frac{1}{k^{\alpha+1}}\right)^p + \left(\sum_{k=n+1}^{\infty} \frac{1}{k^{\alpha+1}}\right)^p + \cdots.$$

It follows that

$$\|H\|^p \geq \frac{\left(\sum_{k=n}^{\infty} \frac{1}{k^{\alpha+1}}\right)^p + \left(\sum_{k=n+1}^{\infty} \frac{1}{k^{\alpha+1}}\right)^p + \cdots}{\sum_{k=n}^{\infty} \frac{1}{k^{\alpha p}}} \quad \forall n \in \mathbb{N}.$$

Using the Stolz–Cesaro lemma we obtain that

$$\lim_{n \to \infty} \frac{\left(\sum_{k=n}^{\infty} \frac{1}{k^{\alpha+1}}\right)^p + \left(\sum_{k=n+1}^{\infty} \frac{1}{k^{\alpha+1}}\right)^p + \cdots}{\sum_{k=n}^{\infty} \frac{1}{k^{\alpha p}}} = \lim_{n \to \infty} \frac{-\left(\sum_{k=n}^{\infty} \frac{1}{k^{\alpha+1}}\right)^p}{-\frac{1}{n^{\alpha p}}}$$

$$= \lim_{n \to \infty} \frac{\left(\sum_{k=n}^{\infty} \frac{1}{k^{\alpha+1}}\right)^p}{\frac{1}{n^{\alpha p}}}$$

$$= \lim_{n \to \infty} \left(n^\alpha \sum_{k=n}^{\infty} \frac{1}{k^{\alpha+1}}\right)^p,$$

hence $\|H\|^p \geq \left(\lim_{n \to \infty} n^\alpha \sum_{k=n}^{\infty} (1/k^{\alpha+1})\right)^p$, i.e., $\|H\| \geq \lim_{n \to \infty} \left(n^\alpha \sum_{k=n}^{\infty} (1/k^{\alpha+1})\right)$. Now again by the Stolz–Cesaro lemma we have

$$\lim_{n \to \infty} \frac{\sum_{k=n}^{\infty} \frac{1}{k^{\alpha+1}}}{\frac{1}{n^\alpha}} = \lim_{n \to \infty} \frac{-\frac{1}{n^{\alpha+1}}}{\frac{1}{(n+1)^\alpha} - \frac{1}{n^\alpha}} = \lim_{n \to \infty} \frac{n^{\alpha-1}}{(n+1)^\alpha - n^\alpha} = \frac{1}{\alpha}.$$

Thus $\|H\| \geq 1/\alpha$, $\alpha > 1/p$, and passing to the limit for $\alpha \to 1/p$, $\alpha > 1/p$ we obtain that $\|H\| \geq p$.

Second solution. Consider a fixed $n \in \mathbb{N}$. For $1 \leq k \leq n$ we have $b_k^p - pb_k^{p-1} a_k = b_k^p - pb_k^{p-1}(kb_k - kb_{k+1}) = (1-pk)b_k^p + kpb_k^{p-1}b_{k+1}$. Since $px^{p-1}y \leq (p-1)x^p + y^p$, for any $x \geq 0$, $y \geq 0$, from this we deduce that $b_k^p - pb_k^{p-1} a_k \leq (1-pk)b_k^p + k\left((p-1)b_k^p + b_{k+1}^p\right) = kb_{k+1}^p - (k-1)b_k^p$. Hence we have that $b_k^p - pb_k^{p-1} a_k \leq$

$kb_{k+1}^p - (k-1)b_k^p$. Adding these inequalities for $k = 1, ..., n$, we deduce that $\sum_{k=1}^n b_k^p - p\sum_{k=1}^n b_k^{p-1}a_k \le nb_{n+1}^p$. By the Hölder inequality, we have $\sum_{k=1}^n b_k^p \le p\sum_{k=1}^n b_k^{p-1}a_k + nb_{n+1}^p \le p\left(\sum_{k=1}^n a_k^p\right)^{1/p}\left(\sum_{k=1}^n b_k^{(p-1)q}\right)^{1/q} + nb_{n+1}^p$, i.e., $\sum_{k=1}^n b_k^p \le p\left(\sum_{k=1}^n a_k^p\right)^{1/p}\left(\sum_{k=1}^n b_k^p\right)^{1/q} + nb_{n+1}^p$.
From here we continue as in the first solution.

Third solution. If we identify l_q^* with l_p then the operator from (iv) is the adjoint of the operator from (ii). The details are the following.

Let $H^* : l_q^* \to l_q^*$ be the dual of the operator H from the part (ii). We know that $l_q^* = l_p$. In fact, the map $J : l_q^* \to l_p$ defined by $J(x^*) = (a_n)_{n \in \mathbb{N}}$ if and only if $x^*(x) = \sum_{n=1}^\infty a_n x_n$ $\forall x = (x_n)_{n \in \mathbb{N}} \in l_q$, is an isomorphic isometry. We consider the diagram

$$\begin{array}{ccc} l_q^* & \xrightarrow{H^*} & l_q^* \\ J^{-1} \uparrow & & \downarrow J \\ l_p & \xrightarrow{JH^*J^{-1}} & l_p \end{array}$$

and we will prove that $(JH^*J^{-1})(a_n)_{n \in \mathbb{N}} = \left(\sum_{k=n}^\infty (a_k/k)\right)_{n \in \mathbb{N}}$ $\forall (a_n)_{n \in \mathbb{N}} \in l_p$, i.e., JH^*J^{-1} is the operator from the statement. Since $\|H^*\| = \|H\|$ and J is an isomorphic isometry, we obtain that $\|JH^*J^{-1}\| = \|H\|$, and by (ii), $\|H\| = p$.

Let $a = (a_n)_{n \in \mathbb{N}} \in l_p$. Let $J^{-1}(a) = x^*$, i.e., $J(x^*) = (a_n)_{n \in \mathbb{N}}$, which means that $x^*((x_n)_{n \in \mathbb{N}}) = \sum_{n=1}^\infty a_n x_n$ $\forall (x_n)_{n \in \mathbb{N}} \in l_q$. In particular, $x^*(0, ...0, 1/n, 1/(n+1), ...) = \sum_{k=n}^\infty (a_k/k)$. Let also $J^{-1}\left(\left(\sum_{k=n}^\infty (a_k/k)\right)_{n \in \mathbb{N}}\right) = y^*$. Then $J(y^*) = \left(\sum_{k=n}^\infty (a_k/k)\right)_{n \in \mathbb{N}}$, which means that

$$y^*((x_n)_{n \in \mathbb{N}}) = \sum_{n=1}^\infty \left(\sum_{k=n}^\infty (a_k/k)\right) x_n \quad \forall (x_n)_{n \in \mathbb{N}} \in l_q.$$

In particular, $y^*(e_n) = \sum_{k=n}^\infty (a_k/k)$ $\forall n \in \mathbb{N}$. We must prove that

$$(H^*J^{-1})(a) = J^{-1}\left(\left(\sum_{k=n}^\infty (a_k/k)\right)_{n \in \mathbb{N}}\right).$$

We have $(H^*J^{-1})(a) = H^*(x^*) = x^* \circ H$, the last equality by the definition of the dual for a linear and continuous operator. The equality which we must prove is now $x^* \circ H = y^*$. Since both members are linear and continuous, it is enough to prove that $(x^* \circ H)(e_n) = y^*(e_n)$ $\forall n \in \mathbb{N}$. We have $H(e_n) = H(0, ..., 0, 1, 0, ...) = (0, ...0, 1/n, 1/(n+1), ...)$, and then

$$(x^* \circ H)(e_n) = x^*(0, ...0, 1/n, 1/(n+1), ...) = \sum_{k=n}^\infty (a_k/k) = y^*(e_n).$$

2.2 Solutions

14. See [28, p. 326], or [20, exercises 20, 21, chapter VI, section 11].

i) Let $g \in L_q(0, \infty)$ where $1/p + 1/q = 1$. We use the Fubini theorem to write

$$\int_0^\infty \left(\frac{1}{x} \int_0^\infty K(y/x) f(y) \, dy \right) g(x) \, dx = \iint_{(0,\infty)^2} \frac{1}{x} K(y/x) f(y) g(x) \, dx dy$$

$$= \iint_{(0,\infty)^2} \left((y/x)^\alpha \left(\frac{1}{x} K(y/x) \right)^{1/p} f(y) \right)$$

$$\left((x/y)^\alpha \left(\frac{1}{x} K(y/x) \right)^{1/q} g(x) \right) dx dy,$$

and therefore, by the Hölder inequality we have

$$\left| \int_0^\infty \left(\frac{1}{x} \int_0^\infty K\left(\frac{y}{x}\right) f(y) \, dy \right) g(x) \, dx \right| \leq \left(\iint_{(0,\infty)^2} \left(\frac{y}{x}\right)^{\alpha p} \left| K\left(\frac{y}{x}\right) \right| |f(y)|^p \frac{dx}{x} dy \right)^{1/p}$$

$$\left(\iint_{(0,\infty)^2} \left(\frac{x}{y}\right)^{\alpha q} \left| K\left(\frac{y}{x}\right) \right| |g(x)|^q \frac{dx}{x} dy \right)^{1/q}$$

for some $\alpha \in \mathbb{R}$ which is to be calculated. For the first term we can again use the Fubini theorem to write

$$\iint_{(0,\infty)^2} (y/x)^{\alpha p} \frac{1}{x} |K(y/x)| |f(y)|^p \, dx dy = \int_0^\infty \left(\int_0^\infty (y/x)^{\alpha p} \frac{1}{x} |K(y/x)| \, dx \right) |f(y)|^p \, dy.$$

Now we evaluate the integral $\int_0^\infty (y/x)^{\alpha p} (1/x) |K(y/x)| \, dx$. We consider the change of variables $y/x = t$, and then

$$\int_0^\infty (y/x)^{\alpha p} \frac{1}{x} |K(y/x)| \, dx = \int_0^\infty t^{\alpha p - 1} |K(t)| \, dt.$$

We obtain that

$$\iint_{(0,\infty)^2} (y/x)^{\alpha p} \frac{1}{x} |K(y/x)| |f(y)|^p \, dx dy = \left(\int_0^\infty t^{\alpha p - 1} |K(t)| \, dt \right) \left(\int_0^\infty |f(y)|^p \, dy \right).$$

For the second term we proceed in the same way. We have

$$\iint_{(0,\infty)^2} (x/y)^{\alpha q} \frac{1}{x} |K(y/x)| |g(x)|^q \, dx dy = \int_0^\infty \left(\int_0^\infty (x/y)^{\alpha q} \frac{1}{x} |K(y/x)| \, dy \right) |g(x)|^q \, dx.$$

Now we evaluate the integral $\int_0^\infty (x/y)^{\alpha q} (1/x) |K(y/x)| \, dy$. We make the change of variables $y/x = t$, and then

$$\int_0^\infty (x/y)^{\alpha q} \frac{1}{x} |K(y/x)| \, dy = \int_0^\infty t^{-\alpha q} |K(t)| \, dt.$$

We obtain that
$$\iint_{(0,\infty)^2} (x/y)^{\alpha q} \frac{1}{x} |K(y/x)| |g(x)|^q \, dxdy = \left(\int_0^\infty t^{-\alpha q} |K(t)| \, dt \right) \left(\int_0^\infty |g(x)|^q \, dx \right).$$

Using this relations we obtain that
$$\left| \int_0^\infty \left(\frac{1}{x} \int_0^\infty K(y/x) f(y) \, dy \right) g(x) \, dx \right| \leq \left(\int_0^\infty t^{\alpha p - 1} |K(t)| \, dt \right)^{1/p}$$
$$\left(\int_0^\infty |f(y)|^p \, dy \right)^{1/p}$$
$$\left(\int_0^\infty t^{-\alpha q} |K(t)| \, dt \right)^{1/q}$$
$$\left(\int_0^\infty |g(x)|^q \, dx \right)^{1/q}$$
$$= \|f\|_p \|g\|_q \left(\int_0^\infty t^{\alpha p - 1} |K(t)| \, dt \right)^{1/p}$$
$$\left(\int_0^\infty t^{-\alpha q} |K(t)| \, dt \right)^{1/q}.$$

Now we choose $\alpha \in \mathbb{R}$ such that $\alpha p - 1 = -\alpha q$, i.e., $\alpha = 1/(p+q)$ and $\alpha p - 1 = -q/(p+q) = -1/p$, and we obtain that
$$\left| \int_0^\infty \left(\frac{1}{x} \int_0^\infty K(y/x) f(y) \, dy \right) g(x) \, dx \right| \leq \|f\|_p \|g\|_q \left(\int_0^\infty t^{-1/p} |K(t)| \, dt \right).$$

But, as is well known, if we have a σ-finite measure space (S, Σ, μ), f is a measurable function and $M > 0$ is a constant such that $\left| \int_S fg \, d\mu \right| \leq M \|g\|_q \; \forall g \in L_q(\mu)$, then $f \in L_p(\mu)$ and $\|f\|_p \leq M$. In our situation it follows that the function $\mathcal{I}_K f(x) = (1/x) \int_0^\infty K(y/x) f(y) \, dy$ belongs to $L_p(0, \infty)$ and
$$\|\mathcal{I}_K f\| \leq \|f\|_p \left(\int_0^\infty t^{-1/p} |K(t)| \, dt \right) \quad \forall f \in L_p(0, \infty).$$

This implies $\|\mathcal{I}_K\| \leq \int_0^\infty t^{-1/p} |K(t)| \, dt$.

For the second operator we will proceed in the same way. Let $g \in L_q(0, \infty)$. Then we use the Fubini theorem to write
$$\int_0^\infty \left(x^{1-2/p} \int_0^\infty K(xy) f(y) \, dy \right) g(x) \, dx = \iint_{(0,\infty)^2} x^{1/q} x^{-1/p} K(xy) f(y) g(x) \, dxdy$$
$$= \iint_{(0,\infty)^2} \left((xy)^\alpha \left(\frac{1}{x} K(xy) \right)^{1/p} f(y) \right)$$
$$\left(\left(\frac{1}{xy} \right)^\alpha (xK(xy))^{1/q} g(x) \right) dxdy.$$

2.2 Solutions

By Hölder's inequality,

$$\left| \iint_{(0,\infty)^2} \left((xy)^\alpha \left(\frac{1}{x} K(xy) \right)^{1/p} f(y) \right) \left(\left(\frac{1}{xy} \right)^\alpha (xK(xy))^{1/q} g(x) \right) dx dy \right|$$

is lesser than

$$\left(\iint_{(0,\infty)^2} (xy)^{\alpha p} \frac{1}{x} |K(xy)| |f(y)|^p \, dxdy \right)^{1/p} \left(\iint_{(0,\infty)^2} \left(\frac{1}{xy} \right)^{\alpha q} x |K(xy)| |g(x)|^q \, dxdy \right)^{1/q}$$

for some $\alpha \in \mathbb{R}$ which is to be calculated. For the first term we can use again the Fubini theorem to write

$$\iint_{(0,\infty)^2} (xy)^{\alpha p} \frac{1}{x} |K(xy)| |f(y)|^p \, dxdy = \int_0^\infty \left(\int_0^\infty (xy)^{\alpha p} \frac{1}{x} |K(xy)| \, dx \right) |f(y)|^p \, dy.$$

Now we evaluate the integral $\int_0^\infty (xy)^{\alpha p} (1/x) |K(xy)| \, dx$. We consider the change of variables $x = t/y$, and then

$$\int_0^\infty (xy)^{\alpha p} \frac{1}{x} |K(xy)| \, dx = \int_0^\infty t^{\alpha p - 1} |K(t)| \, dt.$$

We obtain that

$$\iint_{(0,\infty)^2} (xy)^{\alpha p} \frac{1}{x} |K(xy)| |f(y)|^p \, dxdy = \left(\int_0^\infty t^{\alpha p - 1} |K(t)| \, dt \right) \left(\int_0^\infty |f(y)|^p \, dy \right).$$

For the second term we proceed in the same way. We have

$$\iint_{(0,\infty)^2} \left(\frac{1}{xy} \right)^{\alpha q} x |K(xy)| |g(x)|^q \, dxdy = \int_0^\infty \left(\int_0^\infty \left(\frac{1}{xy} \right)^{\alpha q} x |K(xy)| \, dy \right) |g(x)|^q \, dx.$$

Now we evaluate the integral $\int_0^\infty (1/xy)^{\alpha q} x |K(xy)| \, dy$. We make the change of variables $xy = t$, and then

$$\int_0^\infty \left(\frac{1}{xy} \right)^{\alpha q} x |K(xy)| \, dy = \int_0^\infty t^{-\alpha q} |K(t)| \, dt.$$

We obtain that

$$\iint_{(0,\infty)^2} \left(\frac{1}{xy} \right)^{\alpha q} x |K(xy)| |g(x)|^q \, dxdy = \left(\int_0^\infty t^{-\alpha q} |K(t)| \, dt \right) \left(\int_0^\infty |g(x)|^q \, dx \right).$$

Using this relations we obtain that

$$\left| \int_0^\infty \left(x^{1-2/p} \int_0^\infty K(xy) f(y) \, dy \right) g(x) \, dx \right| \leq \left(\int_0^\infty t^{\alpha p - 1} |K(t)| \, dt \right)^{1/p}$$

$$\left(\int_0^\infty |f(y)|^p \, dy \right)^{1/p}$$

$$\left(\int_0^\infty t^{-\alpha q} |K(t)| \, dt \right)^{1/q}$$

$$\left(\int_0^\infty |g(x)|^q \, dx \right)^{1/q}$$

$$= \|f\|_p \|g\|_q \left(\int_0^\infty t^{\alpha p - 1} K(t) \, dt \right)^{1/p}$$

$$\left(\int_0^\infty t^{-\alpha q} |K(t)| \, dt \right)^{1/q}.$$

Now we choose $\alpha \in \mathbb{R}$ such that $\alpha p - 1 = -\alpha q$, i.e., $\alpha = 1/(p+q)$ and $\alpha p - 1 = -q/(p+q) = -1/p$, and we obtain that

$$\left| \int_0^\infty \left(x^{1-2/p} \int_0^\infty K(xy) f(y) \, dy \right) g(x) \, dx \right| \leq \|f\|_p \|g\|_q \left(\int_0^\infty t^{-1/p} |K(t)| \, dt \right).$$

Then, as above, it follows that the function $\mathcal{M}_K f(x) = x^{1-2/p} \int_0^\infty K(xy) f(y) \, dy$ belongs to $L_p(0, \infty)$ and that $\|\mathcal{M}_K f\| \leq \|f\|_p \left(\int_0^\infty t^{-1/p} |K(t)| \, dt \right) \forall f \in L_p(0, \infty)$, and then $\|\mathcal{M}_K\| \leq \int_0^\infty t^{-1/p} |K(t)| \, dt$.

Remark. We tried to choose $\alpha \in \mathbb{R}$ such that $\int_0^\infty t^{\alpha p - 1} K(t) \, dt < \infty$ and $\int_0^\infty t^{-\alpha q} K(t) \, dt < \infty$, i.e., $t^{\alpha - 1/p} K^{1/p}(t) \in L_p(0, \infty)$ and $t^{-\alpha} K^{1/q}(t) \in L_q(0, \infty)$. Then by the Hölder inequality it follows that $t^{-\alpha} K^{1/q}(t) t^{\alpha - 1/p} K^{1/p}(t) \in L_1(0, \infty)$, i.e., $t^{-1/p} K(t) \in L_1(0, \infty)$, that is, $\int_0^\infty t^{-1/p} |K(t)| \, dt < \infty$. Hence the condition from the hypothesis is a very natural one.

In the remainder, we will use the well known results on the change of variables formula, integration by parts for the Lebesgue integral and the Lebesgue differentiation theorem. For a proof for these results the reader can consult [29, chapter V].

For any $n \in \mathbb{N}$, consider the functions $f_n, g_n : (0, \infty) \to \mathbb{R}$ defined by $f_n(x) = x^{-1/p + 1/(pn)} \chi_{(0,1)}(x)$ and $g_n(x) = x^{-1/q + 1/(qn)} \chi_{(0,1)}(x) \, \forall x \in (0, \infty)$. Then $\mathcal{I}_K f_n(x) = (1/x) \int_0^1 K(y/x) y^{-1/p + 1/(pn)} dy$ and

$$\int_0^\infty \mathcal{I}_K f_n(x) g_n(x) \, dx = \int_0^1 \mathcal{I}_K f_n(x) x^{-1/q + 1/(qn)} dx$$

$$= \int_0^1 \left(\frac{1}{x} \int_0^1 K(y/x) y^{-1/p + 1/(pn)} dy \right) x^{-1/q + 1/(qn)} dx$$

$$= \int_0^1 x^{-1 - 1/q + 1/(qn)} \left(\int_0^1 K(y/x) y^{-1/p + 1/pn} dy \right) dx.$$

2.2 Solutions

Making the change of variables $y/x = t$, we obtain that

$$\int_0^1 K(y/x) y^{-1/p+1/(pn)} dy = \int_0^{1/x} K(t) t^{-1/p+1/(pn)} x^{1-1/p+1/(pn)} dt,$$

and then, since $1 - 1/p = 1/q$, we obtain that

$$\int_0^\infty \mathcal{I}_K f_n(x) g_n(x) dx = \int_0^1 x^{-1-1/q+1/(qn)} \left(\int_0^{1/x} K(t) t^{-1/p+1/(pn)} x^{1/q+1/(pn)} dt \right) dx$$

$$= \int_0^1 x^{-1+1/n} \left(\int_0^{1/x} K(t) t^{-1/p+1/(pn)} dt \right) dx.$$

Now if we denote $G(y) = \int_0^y K(t) t^{-1/p+1/(pn)} dt$ we have

$$\int_0^\infty \mathcal{I}_K f_n(x) g_n(x) dx = \int_0^1 x^{-1+1/n} G(1/x) dx.$$

For a fixed $y \in (0, \infty)$, we have $\left| K(t) t^{-1/p+1/(pn)} \right| \leq y^{1/(pn)} \left| K(t) t^{-1/p} \right|$ $\forall t \in (0, y]$, and from here $|G(y)| \leq y^{1/(pn)} \int_0^\infty t^{-1/p} |K(t)| dt$, hence $y^{-1/n} |G(y)| \leq y^{-1/(qn)} \int_0^\infty t^{-1/p} |K(t)| dt \to 0$ for $y \to \infty$. Moreover by the Lebesgue differentiation theorem $G'(y) = K(y) y^{-1/p+1/(pn)}$ for almost all $y \in (0, \infty)$. Making the change of variables $u = 1/x$ and integrating by parts, we have

$$\int_0^1 x^{-1+1/n} G(1/x) dx = \int_1^\infty u^{-1-1/n} G(u) du = -n \int_1^\infty \left(u^{-1/n} \right)' G(u) du$$

$$= -n u^{-1/n} G(u) \Big|_1^\infty + n \int_1^\infty u^{-1/n} G'(u) du$$

$$= nG(1) + n \int_1^\infty u^{-1/n} K(u) u^{-1/p+1/(pn)} du$$

$$= n \left(\int_0^1 y^{-1/p+1/(pn)} K(y) dy + \int_1^\infty y^{-1/p-1/(qn)} K(y) dy \right).$$

Also $\|f_n\| = \left(\int_0^1 x^{-1+1/n} dx \right)^{1/p} = n^{1/p}$ and $\|g_n\| = \left(\int_0^1 x^{-1+1/n} dx \right)^{1/q} = n^{1/q}$. Since $\mathcal{I}_K$ is continuous, we have $\|\mathcal{I}_K f_n\| \leq \|\mathcal{I}_K\| \|f_n\|$ and also by the Hölder inequality

$$\left| \int_0^\infty \mathcal{I}_K f_n(x) g_n(x) dx \right| \leq \|\mathcal{I}_K f_n\| \|g_n\| \leq \|\mathcal{I}_K\| \|f_n\| \|g_n\|.$$

Resuming, we have

$$n \left| \int_0^1 y^{-1/p+1/(pn)} K(y) dy + \int_1^\infty y^{-1/p-1/(qn)} K(y) dy \right| \leq n \|\mathcal{I}_K\|,$$

that is,

$$\left| \int_0^1 y^{-1/p+1/(pn)} K(y) dy + \int_1^\infty y^{-1/p-1/(qn)} K(y) dy \right| \leq \|\mathcal{I}_K\|,$$

for any $n \in \mathbb{N}$.

For any $0 < y \leq 1$ we have $y^{-1/p+1/(pn)} |K(y)| \leq y^{-1/p} |K(y)|$. Also $\int_0^1 y^{-1/p} |K(y)| \, dy \leq \int_0^\infty y^{-1/p} |K(y)| \, dy < \infty$ and $y^{-1/p+1/(pn)} K(y) \to y^{-1/p} |K(y)|$ for any $y \in (0, 1)$, hence by the Lebesgue dominated convergence theorem

$$\int_0^1 y^{-1/p+1/(pn)} K(y) \, dy \to \int_0^1 y^{-1/p} K(y) \, dy.$$

In the same way for any $y \in [1, \infty)$ we have $y^{-1/p-1/(qn)} |K(y)| \leq y^{-1/p} |K(y)|$, $\int_1^\infty y^{-1/p} |K(y)| \, dy \leq \int_0^\infty y^{-1/p} |K(y)| \, dy < \infty$ and $y^{-1/p-1/(qn)} K(y) \to y^{-1/p} K(y)$ for any $y \in [1, \infty)$, hence by the Lebesgue dominated convergence theorem,

$$\int_1^\infty y^{-1/p-1/(qn)} K(y) \, dy \to \int_1^\infty y^{-1/p} K(y) \, dy.$$

We obtain that $\left| \int_0^\infty y^{-1/p} K(y) \, dy \right| \leq \|\mathcal{I}_K\|$.

For $\mathcal{M}_K$ we proceed in the same way. For any $n \in \mathbb{N}$, let $f_n, g_n : (0, \infty) \to \mathbb{R}$ defined by $f_n(x) = x^{-1/p+1/(pn)} \chi_{(0,1)}(x)$ and $g_n(x) = x^{-1/q-1/(qn)} \chi_{(1,\infty)}(x)$. Then $\mathcal{M}_K f_n(x) = x^{1-2/p} \int_0^1 K(xy) \, y^{-1/p+1/(pn)} \, dy$ and

$$\begin{aligned}
\int_0^\infty \mathcal{M}_K f_n(x) g_n(x) \, dx &= \int_1^\infty \mathcal{M}_K f_n(x) x^{-1/q-1/(qn)} \, dx \\
&= \int_1^\infty \left(x^{1-2/p} \int_0^1 K(xy) \, y^{-1/p+1/(pn)} \, dy \right) x^{-1/q-1/(qn)} \, dx \\
&= \int_1^\infty x^{1-2/p-1/q-1/(qn)} \left(\int_0^1 K(xy) \, y^{-1/p+1/(pn)} \, dy \right) dx.
\end{aligned}$$

Then we use the change of variables $xy = t$, and we obtain that

$$\begin{aligned}
\int_0^1 K(xy) \, y^{-1/p+1/(pn)} \, dy &= \int_0^x (1/x) K(t) (t/x)^{-1/p+1/(pn)} \, dt \\
&= \int_0^x K(t) t^{-1/p+1/(pn)} x^{-1+1/p-1/(pn)} \, dt,
\end{aligned}$$

and then, since $1 - 1/p = 1/q$, we obtain that

$$\begin{aligned}
\int_0^\infty \mathcal{M}_K f_n(x) g_n(x) \, dx &= \int_1^\infty x^{1-2/p-1/q-1/(qn)} \\
&\qquad \left(\int_0^x K(t) t^{-1/p+1/(pn)} x^{-1/q-1/(pn)} \, dt \right) dx \\
&= \int_1^\infty x^{-1-1/n} \left(\int_0^x K(t) t^{-1/p+1/(pn)} \, dt \right) dx.
\end{aligned}$$

If we denote now $G(y) = \int_0^y K(t) t^{-1/p+1/(pn)} \, dt$, we have

$$\int_0^\infty \mathcal{M}_K f_n(x) g_n(x) \, dx = \int_1^\infty x^{-1-1/n} G(x) \, dx.$$

2.2 Solutions

It is proved above that

$$\int_1^\infty x^{-1-1/n} G(x)\, dx = n \left(\int_0^1 y^{-1/p+1/(pn)} K(y)\, dy + \int_1^\infty y^{-1/p-1/(qn)} K(y)\, dy \right).$$

We have $\|f_n\| = \left(\int_0^1 x^{-1+1/n} dx \right)^{1/p} = n^{1/p}$ and $\|g_n\| = \left(\int_1^\infty x^{-1-1/n} dx \right)^{1/q} = n^{1/q}$. Since $\mathcal{M}_K$ is continuous we have $\|\mathcal{M}_K f_n\| \leq \|\mathcal{M}_K\| \|f_n\|$, and by the Hölder inequality we obtain that

$$\left| \int_0^\infty \mathcal{I}_K f_n(x) g_n(x)\, dx \right| \leq \|\mathcal{M}_K f_n\| \|g_n\| \leq \|\mathcal{M}_K\| \|f_n\| \|g_n\|.$$

Resuming, we have

$$n \left| \int_0^1 y^{-1/p+1/(pn)} K(y)\, dy + \int_1^\infty y^{-1/p-1/(qn)} K(y)\, dy \right| \leq n \|\mathcal{M}_K\| \quad \forall n \in \mathbb{N},$$

that is,

$$\left| \int_0^1 y^{-1/p+1/(pn)} K(y)\, dy + \int_1^\infty y^{-1/p-1/(qn)} K(y)\, dy \right| \leq \|\mathcal{M}_K\| \quad \forall n \in \mathbb{N}.$$

From here we continue as above to deduce that $\left| \int_0^\infty y^{-1/p} K(y)\, dy \right| \leq \|\mathcal{M}_K\|$.

ii) We take $K(t) = \begin{cases} 1, & \text{if } t \leq 1, \\ 0, & \text{if } t > 1. \end{cases}$ Then $H = \mathcal{I}_K$ and

$$\|\mathcal{I}_K\| = \int_0^\infty y^{-1/p} K(y)\, dy = \int_0^1 y^{-1/p} dy = \frac{p}{p-1}.$$

iii) We take $K(t) = 1/(1+t)$. Then $H = \mathcal{I}_K$ and

$$\|\mathcal{I}_K\| = \int_0^\infty y^{-1/p} K(y)\, dy = \int_0^\infty \frac{y^{-1/p}}{y+1} dy = \frac{\pi}{\sin \frac{\pi}{p}}.$$

iv) We take $K(t) = 1/\max(t, 1)$. Then $S = \mathcal{I}_K$ and

$$\|\mathcal{I}_K\| = \int_0^\infty y^{-1/p} K(y)\, dy = \int_0^1 y^{-1/p} dy + \int_1^\infty y^{-1/p-1} dy = \frac{p}{p-1} + p = \frac{p^2}{p-1}.$$

v) We take

$$K(t) = \frac{1}{\Gamma(\alpha)} (1-t)^{\alpha-1} \chi_{(0,1)}(t).$$

Then

$$\begin{aligned}
\mathcal{I}_K f(x) &= \frac{1}{x} \int_0^\infty K(y/x) f(y)\, dy = \frac{1}{\Gamma(\alpha)} \frac{1}{x} \int_0^\infty (1-y/x)^{\alpha-1} \chi_{(0,1)}(y/x) f(y)\, dy \\
&= \frac{1}{\Gamma(\alpha)} \int_0^x \frac{(x-y)^{\alpha-1}}{x^\alpha} f(y)\, dy = \mathcal{R}_\alpha f(x).
\end{aligned}$$

Moreover,

$$\|\mathcal{R}_\alpha\| = \int_0^\infty y^{-1/p} K(y) dy = \frac{1}{\Gamma(\alpha)} \int_0^1 y^{-1/p} (1-y)^{\alpha-1} dy = \frac{1}{\Gamma(\alpha)} B(1-1/p, \alpha)$$
$$= \frac{\Gamma(1-1/p)\Gamma(\alpha)}{\Gamma(\alpha+1-1/p)\Gamma(\alpha)} = \frac{\Gamma(1-1/p)}{\Gamma(\alpha+1-1/p)}.$$

vi) We take $K(t) = t^{\alpha-1} \chi_{(0,1)}(t)$. Then

$$\mathcal{I}_K f(x) = \frac{1}{x} \int_0^\infty K(y/x) f(y) dy = \frac{1}{x} \int_0^\infty (y/x)^{\alpha-1} \chi_{(0,1)}(y/x) f(y) dy$$
$$= \frac{1}{x^\alpha} \int_0^x y^{\alpha-1} f(y) dy = \mathcal{I}_\alpha f(x).$$

Moreover,

$$\|\mathcal{I}_\alpha\| = \int_0^\infty y^{-1/p} K(y) dy = \int_0^1 y^{-1/p} y^{\alpha-1} dy = B(1-1/p, \alpha).$$

15. See [26, exercise 30, chapter IX].

If $\{x_n \mid n \in \mathbb{N}\}$ is a Schauder basis in l_2, we have: $\forall x = (\xi_n)_{n \in \mathbb{N}} \in l_2$ there is a unique sequence $(\lambda_n)_{n \in \mathbb{N}} \subseteq \mathbb{K}$ such that $x = \sum_{n=1}^\infty \lambda_n x_n$, the series being convergent in l_2. We obtain that $\forall k \in \mathbb{N}$, $\langle x, e_k \rangle = \sum_{n=1}^\infty \lambda_n \langle x_n, e_k \rangle$, i.e., the sequence $(\lambda_n)_{n \in \mathbb{N}}$ satisfies the equations: $\lambda_1 \alpha = \xi_1$, $\lambda_1 \beta + \lambda_2 \alpha = \xi_2$, $\lambda_2 \beta + \lambda_3 \alpha = \xi_3$, $\lambda_3 \beta + \lambda_4 \alpha = \xi_4$, ..., and therefore $\lambda_1 = \xi_1/\alpha$, $\lambda_2 = (1/\alpha)(\xi_2 - (\beta/\alpha)\xi_1)$,..., $\lambda_n = (1/\alpha)\left(\xi_n + \sum_{k=1}^{n-1}(-1)^{n-k}(\beta/\alpha)^{n-k}\xi_k\right)$ $\forall n \geq 2$. Thus the sequence $(\lambda_n)_{n \in \mathbb{N}}$, if it exists, has the above value. To have a Schauder basis, we must show that $x = \sum_{n=1}^\infty \lambda_n x_n$ in l_2, i.e., $\|\lambda_1 x_1 + \cdots + \lambda_n x_n - x\| \to 0$. But $\lambda_1 x_1 + \cdots + \lambda_n x_n = (\lambda_1 \alpha, \lambda_1 \beta + \lambda_2 \alpha, ..., \lambda_{n-1}\beta + \lambda_n \alpha, \lambda_n \beta, 0, ...)$, and therefore we obtain that $\|\lambda_1 x_1 + \cdots + \lambda_n x_n - x\|^2 = |\lambda_1 \alpha - \xi_1|^2 + |\lambda_1 \beta + \lambda_2 \alpha - \xi_2|^2 + \cdots + |\lambda_{n-1}\beta + \lambda_n \alpha - \xi_n|^2 + |\lambda_n \beta - \xi_{n+1}|^2 + |\xi_{n+2}|^2 + \cdots$. Taking into account the expression for $(\lambda_n)_{n \in \mathbb{N}}$, we obtain that $\|\lambda_1 x_1 + \cdots + \lambda_n x_n - x\|^2 = |\lambda_n \beta - \xi_{n+1}|^2 + \sum_{k=n+2}^\infty |\xi_k|^2$. But $\sum_{k=n+2}^\infty |\xi_k|^2 \to 0$, since the series $\sum_{k=1}^\infty |\xi_k|^2$ converges. Also $\xi_{n+1} \to 0$. We know that if $\xi_k \to 0$ and the series $\sum_{k=1}^\infty a_k$ is absolutely convergent, then $a_{n-1}\xi_1 + a_{n-2}\xi_2 + \cdots + a_2 \xi_{n-2} + a_1 \xi_{n-1} \to 0$ (see the solution for exercise 14 of chapter 6). Applying this fact for $\lambda_n = (1/\alpha)\left(\xi_n + \sum_{k=1}^{n-1}(-1)^{n-k}(\beta/\alpha)^{n-k}\xi_k\right)$ we obtain that $\lambda_n \to 0$ (here $\sum_{k=1}^\infty (-1)^k (\beta/\alpha)^k$ is absolutely convergent, $\sum_{k=1}^\infty \left|(-1)^k (\beta/\alpha)^k\right| = \sum_{k=1}^\infty |\beta/\alpha|^k < \infty$, since $|\beta/\alpha| < 1$). Therefore $\|\lambda_1 x_1 + \cdots + \lambda_n x_n - x\|^2 \to 0$ for $n \to \infty$.

2.2 Solutions

16. See [20, exercise 8, chapter II, section 4].

Evidently Y is a linear space. If we have $\|y\|_1 = 0$, then $\sum_{i=1}^{n} \alpha_i x_i = 0 \ \forall n \in \mathbb{N}$, therefore
$$\alpha_n x_n = \left(\sum_{i=1}^{n} \alpha_i x_i\right) - \left(\sum_{i=1}^{n-1} \alpha_i x_i\right) = 0 \ \forall n \in \mathbb{N}, \text{ and since } x_n \neq 0 \ \forall n \in \mathbb{N} \text{ (from the}$$
definition for a Schauder basis, more precisely from the uniqueness of the representation with respect to the basis), we obtain that $\alpha_n = 0 \ \forall n \in \mathbb{N}$, thus $y = 0$. The other axioms for a norm are easy to verify. We observe that if $y = (\alpha_n)_{n \in \mathbb{N}} \in Y$, then $\|\alpha_n x_n\| = \left\|\left(\sum_{i=1}^{n} \alpha_i x_i\right) - \left(\sum_{i=1}^{n-1} \alpha_i x_i\right)\right\| \leq \left\|\sum_{i=1}^{n} \alpha_i x_i\right\| + \left\|\sum_{i=1}^{n-1} \alpha_i x_i\right\| \leq 2 \sup_{k \in \mathbb{N}} \left\|\sum_{i=1}^{k} \alpha_i x_i\right\| = 2 \|y\|_1$
$\forall n \in \mathbb{N}$ (1).

We prove now that the norm considered on Y is complete. Let $(y^{(k)})_{k \in \mathbb{N}} \subseteq Y$, $y^{(k)} = (\alpha_n^k)_{n \in \mathbb{N}} \ \forall k \in \mathbb{N}$, such that: $\forall \varepsilon > 0 \ \exists N_\varepsilon \in \mathbb{N}$ such that $\sup_{n \in \mathbb{N}} \left\|\sum_{i=1}^{n} (\alpha_i^m - \alpha_i^p) x_i\right\| < \varepsilon$
$\forall m, p \geq N_\varepsilon$ (2). For any fixed $k \in \mathbb{N}$, using (1) we have $\|(\alpha_k^m - \alpha_k^p) x_k\| \leq 2 \sup_{n \in \mathbb{N}} \left\|\sum_{i=1}^{n} (\alpha_i^m - \alpha_i^p) x_i\right\|$, and therefore $|\alpha_k^m - \alpha_k^p| \leq 2\varepsilon / \|x_k\| \ \forall m, p \geq N_\varepsilon$. We obtain that for any $k \in \mathbb{N}$, the sequence $(\alpha_k^n)_{n \in \mathbb{N}}$ is Cauchy, thus convergent, and let $y = (\alpha_k)_{k \in \mathbb{N}}$ a sequence of scalars such that $\lim_{n \to \infty} \alpha_k^n = \alpha_k \ \forall k \in \mathbb{N}$. We prove now that $y \in Y$ and that $\lim_{k \to \infty} \|y^{(k)} - y\|_1 = 0$. Let $n \in \mathbb{N}$. Using (2) we have $\left\|\sum_{i=1}^{n} (\alpha_i^m - \alpha_i^p) x_i\right\| < \varepsilon$,
$\forall m, p \geq N_\varepsilon$. For $p \to \infty$ we obtain that $\left\|\sum_{i=1}^{n} (\alpha_i^m - \alpha_i) x_i\right\| \leq \varepsilon \ \forall m \geq N_\varepsilon$. Then $\sup_{n \in \mathbb{N}} \left\|\sum_{i=1}^{n} (\alpha_i^m - \alpha_i) x_i\right\| \leq \varepsilon \ \forall m \geq N_\varepsilon$, thus $\|y^{(m)} - y\|_1 \leq \varepsilon \ \forall m \geq N_\varepsilon$, and we obtain that $y \in Y$ and $\lim_{k \to \infty} y^{(k)} = y$.

17. See [20, exercise 9, chapter II, section 4].

For each $x \in X$ we have a unique sequence of scalars $(\alpha_i)_{i \in \mathbb{N}}$ such that $x = \sum_{i=1}^{\infty} \alpha_i x_i$. We consider the Banach space Y given by exercise 16 and we have the operator $T : X \to Y$,
$$T\left(\sum_{i=1}^{\infty} \alpha_i x_i\right) = (\alpha_i)_{i \in \mathbb{N}}.$$
Then T is linear, bijective, and
$$\|T^{-1}((\alpha_i)_{i \in \mathbb{N}})\| = \left\|\sum_{i=1}^{\infty} \alpha_i x_i\right\| = \lim_{n \to \infty} \left\|\sum_{i=1}^{n} \alpha_i x_i\right\| \leq \sup_{n \in \mathbb{N}} \left\|\sum_{i=1}^{n} \alpha_i x_i\right\| = \|(\alpha_i)_{i \in \mathbb{N}}\|_1.$$

This means that T^{-1} is continuous. Since X and Y are Banach spaces, by the inverse mapping theorem (see chapter 9) we obtain that T is also continuous. For any $i \in \mathbb{N}$ we define $e_i^* : Y \to \mathbb{K}$, $e_i^*((\alpha_n)_{n \in \mathbb{N}}) = \alpha_i$. Evidently e_i^* is linear for every $i \in \mathbb{N}$. Also $|e_i^*(\alpha)| = |\alpha_i| = \|\alpha_i x_i\| / \|x_i\| \leq (2 \|\alpha\|) / \|x_i\|$, using the relation (1) from the solution for exercise 16. Therefore $e_i^* \in Y^* \ \forall i \in \mathbb{N}$. We observe that $x_i^* = T^* e_i^* \ \forall i \in \mathbb{N}$, and therefore $x_i^* \in X^* \ \forall i \in \mathbb{N}$.

18. See [20, exercise 10, chapter II, section 4].

Let us suppose that we can find a positive integer n such that $x_n \in \overline{\mathrm{Sp}\{x_1, x_2, ..., x_{n-1}, x_{n+1}, x_{n+2}, ...\}}^{\|\cdot\|}$. Consider then $x_n^* \in X^*$ given by exercise 17. We have $x_n^*(x_m) = \delta_{nm} \ \forall m \in \mathbb{N}$, thus $x_n^*(x_m) = 0 \ \forall m \neq n$. Since x_n^* is linear, we obtain that $x_n^*(x) = 0 \ \forall x \in \mathrm{Sp}\{x_1, x_2, ..., x_{n-1}, x_{n+1}, x_{n+2}, ...\}$ and, since x_n^* is continuous, we obtain that $x_n^*(x) = 0 \ \forall x \in \overline{\mathrm{Sp}\{x_1, x_2, ..., x_{n-1}, x_{n+1}, x_{n+2}, ...\}}^{\|\cdot\|}$. But $x_n \in \overline{\mathrm{Sp}\{x_1, x_2, ..., x_{n-1}, x_{n+1}, x_{n+2}, ...\}}^{\|\cdot\|}$, and we obtain that $x_n^*(x_n) = 0$, a contradiction, since $x_n^*(x_n) = 1$.

19. See [18, chapter 7, the proof of theorem 5].

We first prove that Q is well defined, that is, the series $\sum_{n=1}^{\infty} \alpha_n x_n$ converges. We have $\sum_{n=1}^{\infty} \|\alpha_n x_n\| = \sum_{n=1}^{\infty} |\alpha_n| \|x_n\| \leq \sum_{n=1}^{\infty} |\alpha_n| < \infty$, and therefore the series $\sum_{n=1}^{\infty} \alpha_n x_n$ converges absolutely in X. Since X is a Banach space the series converges, and Q is well defined. Evidently Q is linear and, for every $(\alpha_n)_{n \in \mathbb{N}} \in l_1$, $\|Q((\alpha_n)_{n \in \mathbb{N}})\| = \left\|\sum_{n=1}^{\infty} \alpha_n x_n\right\| \leq \sum_{n=1}^{\infty} \|\alpha_n x_n\| \leq \|(\alpha_n)_{n \in \mathbb{N}}\|_{l_1}$, and then Q is continuous. It remains to be proved that Q is surjective. Let $x \in B_X$. Then there is an $n_1 \in \mathbb{N}$ such that $\|x - x_{n_1}\| < 1/2$. If there is $k = 1, ..., n_1$ such that $2(x - x_{n_1}) - x_k = 0$, then $x = x_k/2 + x_{n_1}$, and then $x = Q(e_k/2 + e_{n_1}) \in Q(l_1)$. If not, let $\delta > 0$ such that $\delta \leq 1/2^2$ and $\delta < \|2(x - x_{n_1}) - x_k\|$ for $k = 1, ..., n_1$. Then we can find $n_2 \in \mathbb{N}$ such that $\|2(x - x_{n_1}) - x_{n_2}\| \leq \delta$, and then $\|2(x - x_{n_1}) - x_{n_2}\| \leq 1/2^2$ and $n_2 > n_1$. In this way we obtain that either $x \in Q(l_1)$, or we can construct a subsequence $(n_k)_{k \in \mathbb{N}}$ of $\mathbb{N}$ such that $\|2^{k-1}(x - x_{n_1}) - 2^{k-2}x_{n_2} - \cdots - 2x_{n_{k-1}} - x_{n_k}\| \leq 1/2^k$ for all $k \in \mathbb{N}$. Then $\|x - x_{n_1} - x_{n_2}/2 - \cdots - x_{n_k}/2^{k-1}\| \leq 1/2^{2k-1}$, and therefore $x = x_{n_1} + x_{n_2}/2 + \cdots + x_{n_k}/2^{k-1} + \cdots = Q(\alpha)$, where $\alpha = (\alpha_n)_{n \in \mathbb{N}}$, $\alpha_n = 1/2^{k-1}$ if $n = n_k$ for a $k \in \mathbb{N}$, $\alpha_n = 0$ if not, $\alpha \in l_1$, since $\sum_{k=0}^{\infty} 1/2^k < \infty$. Then $B_X \subseteq Q(l_1)$, and since Q is linear we obtain that $X \subseteq Q(l_1)$.

20. See [6, theorem 4, chapter 8].

Suppose that X is a separable Banach space. Using exercise 29 of chapter 14 we obtain that $(B_{X^*}, \mathrm{weak}^*)$ is a compact metric space. Define then $T : X \to C(B_{X^*})$, $T(x) = \widehat{x}(\cdot)$. Evidently T is linear, and for every $x \in X$ we have $\|Tx\| = \|\widehat{x}(\cdot)\|_{B_{X^*}} = \sup_{x^* \in B_{X^*}} |\widehat{x}(x^*)| = \|x\|$, thus T is an isometry. Since T is linear we obtain that T is injective. Denote by $Y = T(X) \subseteq C(B_{X^*})$, and we obtain the isometric isomorphism $T : X \to Y \subseteq C(B_{X^*})$. Since X is complete and T is an isometry it follows that $T(X) = Y \subseteq C(B_{X^*})$ is a closed linear subspace. Indeed, let $(x_n)_{n \in \mathbb{N}} \subseteq X$, $Tx_n \to y \in C(B_{X^*})$. Then the sequence $(T(x_n))_{n \in \mathbb{N}} \subseteq Y$ is Cauchy, hence since T is a linear isometry the sequence $(x_n)_{n \in \mathbb{N}}$ is Cauchy. Since X is a Banach space there is $x \in X$ such that $x_n \to x$ and by the continuity of T, $Tx_n \to Tx$, and therefore $y = Tx$.

Suppose now that X is a Banach space for which there is a compact metric space K, a closed linear subspace $Y \subseteq C(K)$, and an isometric isomorphism $T : X \to Y$. We know that $C(K)$ is a separable space, and therefore Y is separable. Let $(y_k)_{k \in \mathbb{N}} \subseteq Y$ be

2.2 Solutions

dense. Let $(x_k)_{k\in\mathbb{N}} \subseteq X$ be such that $Tx_k = y_k$ $\forall k \in \mathbb{N}$. Consider $x \in X$ arbitrary. For $\varepsilon > 0$ there is $k \in \mathbb{N}$ such that $\|y_k - Tx\| < \varepsilon$ ($(y_k)_{k\in\mathbb{N}} \subseteq Y$ dense), thus $\|Tx_k - Tx\| < \varepsilon$, $\|T(x_k - x)\| < \varepsilon$, therefore since T is an isometry $\|x_k - x\| < \varepsilon$. We obtain that $(x_k)_{k\in\mathbb{N}} \subseteq X$ is dense, thus X is separable.

21. Recall that a topological space X is separable if there is a countable dense set $B \subseteq X$. Suppose, for a contradiction, that X is separable and consider $\{x_n \mid n \in \mathbb{N}\} \subseteq X$ a countable dense set. For every $a \in A$, $B(a, \varepsilon/2)$ is a neighborhood of a, and from the density we obtain that $B(a, \varepsilon/2) \cap \{x_n \mid n \in \mathbb{N}\} \neq \emptyset$, i.e., there is an $n \in \mathbb{N}$ such that $x_n \in B(a, \varepsilon/2)$, i.e., $d(x_n, a) < \varepsilon/2$. Let $n_a = \min\{n \in \mathbb{N} \mid d(x_n, a) < \varepsilon/2\}$, and then $d(x_{n_a}, a) < \varepsilon/2$. Consider the function $f : A \to \mathbb{N}$, $f(a) = n_a$. We shall prove that f is injective, and we shall obtain that A has the same cardinality as the set $f(A) \subseteq \mathbb{N}$, and since $\mathbb{N}$ is countable we obtain that A is countable, a contradiction. Let $a, b \in A$ such that $a \neq b$ and $f(a) = f(b)$. Then $n_a = n_b$. We have that $d(x_{n_a}, a) < \varepsilon/2$ and $d(x_{n_b}, b) < \varepsilon/2$, i.e., $d(x_{n_a}, b) < \varepsilon/2$, and therefore $d(a, b) \leq d(a, x_{n_a}) + d(x_{n_a}, b) < \varepsilon$, i.e., $d(a, b) < \varepsilon$, which contradicts our hypothesis: $\forall a, b \in A$, $a \neq b \Rightarrow d(a, b) \geq \varepsilon$.

22. For $A \subseteq \mathbb{N}$ we consider the sequence $\chi_A : \mathbb{N} \to \mathbb{K}$, $\chi_A(n) = \begin{cases} 1, & n \in A, \\ 0, & n \notin A. \end{cases}$ Then $|\chi_A(n)| \leq 1$ $\forall n \in \mathbb{N}$, i.e., $\chi_A \in l_\infty$. If $A, B \subseteq \mathbb{N}$, $A \neq B$, then $\|\chi_A - \chi_B\| \geq 1$. Indeed, since $A \neq B$ there is a $k \in \mathbb{N}$ such that $k \in A \setminus B$, or $k \in B \setminus A$. Suppose, for example, that $k \in A \setminus B$. Then: $\|\chi_A - \chi_B\| = \sup_{n\in\mathbb{N}} |\chi_A(n) - \chi_B(n)| \geq |\chi_A(k) - \chi_B(k)| = |1 - 0| = 1$. (In fact, $\|\chi_A - \chi_B\| = 1$.) Since the set $(\chi_A)_{A\in\mathcal{P}(\mathbb{N})} \subseteq l_\infty$ is not countable, using exercise 21 we obtain that l_∞ is not separable.

23. From exercise 11 we obtain that there is $j : l_\infty \to L(L_p[0,1])$, $\|j(x)\| = \|x\|$ $\forall x \in l_\infty$ and j linear. The set $\{j(\chi_A) \mid A \in \mathcal{P}(\mathbb{N})\} \subseteq L(L_p[0,1])$ is uncountable and: $\forall A, B \in \mathcal{P}(\mathbb{N})$, $A \neq B$, $\|j(\chi_A) - j(\chi_B)\| = \|j(\chi_A - \chi_B)\| = \|\chi_A - \chi_B\| \geq 1$. Using exercise 21 we obtain that $L(L_p[0,1])$ is not separable.

24. i) Let $\xi = (\xi_n)_{n\geq 0} \in l_\infty$. Define $U_\xi : l_1 \to C[0,1]$, $(U_\xi)(a)(x) = \sum_{n=0}^\infty a_n \xi_n x^n$ $\forall a = (a_n)_{n\geq 0} \in l_1$, $\forall x \in [0,1]$. Using exercise 7 we have $\|U_\xi\| = \|\xi\|$, i.e., $\xi \to U_\xi$ is an isometry from l_∞ to $L(l_1, C[0,1])$.

ii) Let $\xi = (\xi_n)_{n\in\mathbb{N}} \in l_\infty$. We define $U_\xi : l_p \to l_p$, $U_\xi(x) = (\xi_n x_n)_{n\in\mathbb{N}}$ $\forall x = (x_n)_{n\in\mathbb{N}} \in l_p$. Evidently

$$\|U_\xi(x)\|^p = \sum_{n=1}^\infty |\xi_n|^p |x_n|^p \leq \|\xi\|^p \sum_{n=1}^\infty |x_n|^p < \infty \quad \forall x = (x_n)_{n\in\mathbb{N}} \in l_p,$$

i.e., U_ξ is well defined, and $\|U_\xi(x)\| \leq \|\xi\| \|x\|$ $\forall x \in l_p$. Since U_ξ is linear we obtain that U_ξ is a linear and continuous operator, and that $\|U_\xi\| \leq \|\xi\|$. But for every $k \in \mathbb{N}$ $\|U_\xi(e_k)\| \leq \|U_\xi\| \|e_k\|$. Since $U_\xi(e_k) = (0, ..., 0, \xi_k, 0, ...)$ we obtain that $|\xi_k| \leq \|U_\xi\|$ $\forall k \in \mathbb{N}$, and therefore $\|\xi\| = \sup_{k\in\mathbb{N}} |\xi_k| \leq \|U_\xi\|$. Thus $\|U_\xi\| = \|\xi\|$ $\forall \xi \in l_\infty$. Let $j : l_\infty \to L(l_p)$, $j(\xi) = U_\xi$. Evidently j is linear, and the above relation shows that $\|j(\xi)\| = \|\xi\|$ $\forall \xi \in l_\infty$, i.e., j is an isometry.

Analogously to the solution of exercise 23 we obtain that $L(l_1, C[0,1])$ and $L(l_p)$, $1 \leq p < \infty$, are not separable spaces.

25. Since T is a linear and continuous operator, $\ker T \subseteq X$ is a closed linear subspace, and therefore $X/\ker T$ is a normed space. We prove now that $\widetilde{T}$ is well defined. If $x + \ker T = y + \ker T$, then $x - y \in \ker T$, thus $T(x-y) = 0$, that is, $Tx = Ty$. Evidently, $\widetilde{T}$ is linear. Since T is surjective, we obtain that $\widetilde{T}$ is also a surjective operator. If $\widetilde{T}(x + \ker T) = 0$ then $Tx = 0$, that is, $x \in \ker T$, thus $x + \ker T = 0 + \ker T$, and $\widetilde{T}$ is injective. We calculate now the norm for $\widetilde{T}$. Let $\overline{x} = x + \ker T \in X/\ker T$. For any $z \in \overline{x}$, $\|\widetilde{T}\overline{x}\| = \|Tz\| \leq \|T\| \|z\|$, thus $\|\widetilde{T}\overline{x}\| \leq \|T\| \inf\{\|z\| \mid z \in \overline{x}\}$, and therefore $\|\widetilde{T}\overline{x}\| \leq \|T\| \|\overline{x}\|$ $\forall \overline{x} \in X/\ker T$. We obtain that $\widetilde{T} \in L(X/\ker T, Y)$, $\|\widetilde{T}\| \leq \|T\|$. Consider now $x \in X$, $\|x\| \leq 1$. Then $\|\overline{x}\| \leq 1$, thus $\|Tx\| = \|\widetilde{T}\overline{x}\| \leq \sup_{\|\overline{y}\| \leq 1}\|\widetilde{T}\overline{y}\| = \|\widetilde{T}\|$. Then $\sup_{\|x\| \leq 1}\|Tx\| \leq \|\widetilde{T}\|$, that is, $\|T\| \leq \|\widetilde{T}\|$, and therefore $\|T\| = \|\widetilde{T}\|$.

26. i) See [44, exercise 14, chapter 5].
Let $x \in \overline{T(X)}$. Then there is $(x_n)_{n \in \mathbb{N}} \subseteq X$ with $Tx_n \to x \in X$. Then $TSTx_n \to TSx$, thus, by hypothesis, $Tx_n \to TSx$, hence $x = TSx$, i.e., $x \in T(X)$.

ii) Since $P(T) = 0$ we have $\sum_{n=0}^{N} a_n T^n = 0$, where $T^n = T \circ \cdots \circ T$, n-fold composition, $T^0 = I$.

a) If $a_0 \neq 0$ then $I = \sum_{n=1}^{N} b_n T^n$, thus $T = \sum_{n=1}^{N} b_n T^{n+1}$. We obtain that $T = T\left(\sum_{n=1}^{N} b_n T^{n-1}\right)T$, and therefore by (i) the range of T is closed.

b) Suppose now that $\sum_{n=m}^{N} a_n T^n = 0$, with $m > 0$, and $a_m \neq 0$. We obtain that $T^m = \sum_{n=m+1}^{N} b_n T^n$. If all the b_n are 0 then $T^m = 0$, and the range of T^m is closed. If not, let k be the smallest n between $m+1$ and N such that $b_n \neq 0$. Then $T^m = \sum_{n=k}^{N} b_n T^n$ with $b_k \neq 0$. We obtain that $T^k = \sum_{n=k}^{N} b_n T^{n+k-m}$, thus T^k is a linear combination of T^n, $n > k$. Then T^m is a linear combination of T^n, $n > k$. After a finite number of steps we obtain that T^m is a linear combination of T^n, $n \geq 2m$. Thus $T^m = \sum_{n=2m}^{M} c_n T^n$. We obtain that $T^m = T^m\left(\sum_{n=2m}^{M} c_n T^{n-2m}\right) T^m$, and therefore by (i) the range of T^m is a closed set.

27. We have $(i^*(x^*))(y) = x^*(i(y)) = x^*(y)$ $\forall y \in Y$, and therefore $i^*(x^*) = x^*|_Y$ for all $x^* \in X^*$.

28. Suppose that $T \in L(X,Y)$ is a bijective operator, and that $\|Tx\| = \|x\|$ $\forall x \in X$. For any $y^* \in Y^*$, since T is a surjective isometry we have $\|T^*y^*\| = \sup_{\|x\| \leq 1}|(T^*y^*)(x)| =$

2.2 Solutions

$\sup\limits_{\|x\|\leq 1} |y^*(Tx)| = \sup\limits_{\|y\|\leq 1} |y^*(y)| = \|y^*\|$, and therefore $T^* \in L(Y^*, X^*)$ is an isometry. Then T^* is injective and we must prove that T^* is surjective. Let $x^* \in X^*$. Since $\|Tx\| = \|x\|$ $\forall x \in X$, and T is bijective, we obtain that $T^{-1} : Y \to X$ is well defined, linear, $\|T^{-1}y\| = \|y\|$ $\forall y \in Y$, and therefore $T^{-1} \in L(Y, X)$ and T^{-1} is also an isometry. We define then $y^* \in Y^*$, $Y \xrightarrow{T^{-1}} X \xrightarrow{x^*} \mathbb{K}$, $y^* = x^* \circ T^{-1}$. We have $(T^*y^*)(x) = y^*(Tx) = x^*(T^{-1}(Tx)) = x^*(x)$ $\forall x \in X$, and therefore $T^*y^* = x^*$.

Suppose now that $T^* \in L(Y^*, X^*)$ is an isometric isomorphism. For all $x \in X$, since T^* is a surjective isometry, we have $\|Tx\| = \sup\limits_{\|y^*\|\leq 1} |y^*(Tx)| = \sup\limits_{\|y^*\|\leq 1} |(T^*(y^*))(x)| = \sup\limits_{\|x^*\|\leq 1} |x^*(x)| = \|x\|$, thus $T \in L(X, Y)$ is an isometry. Since T is linear, we obtain that T is injective. It remains to be proved that T is surjective. Since X is complete and T is an isometry it follows that $T(X) \subseteq Y$ is closed. This is proved in the solution for exercise 20. Suppose that $T(X) \neq Y$, and let $y_0 \in Y \setminus T(X)$. Since $T(X) \subseteq Y$ is closed, using the Hahn–Banach theorem (see chapter 8) there is $y^* \in Y^*$ such that $y^*|_{T(X)} = 0$, $y^*(y_0) \neq 0$. Then $y^*(Tx) = 0$ $\forall x \in X$, thus $(T^*y^*)(x) = 0$ $\forall x \in X$, and then $T^*y^* = 0 \in X^*$. But T^* is injective, and therefore $y^* = 0$, $y^*(y_0) = 0$, a contradiction.

29. See [20, exercise 12 (a), (b), chapter II, section 4].

i) Define $T : X^* \to l_\infty(X^*)$, $T(x^*) = \left(\sum\limits_{i=1}^{n} x^*(x_i) x_i^*\right)_{n \in \mathbb{N}}$. Here, $l_\infty(X^*)$ is the space of all norm bounded sequences $(x_n^*)_{n \in \mathbb{N}} \subseteq X^*$, with the norm given by: $\|(x_n^*)_{n \in \mathbb{N}}\| = \sup\limits_{n \in \mathbb{N}} \|x_n^*\|$ $\forall (x_n^*)_{n \in \mathbb{N}} \in l_\infty(X^*)$. In the usual way one can prove that, since X^* is a Banach space, then $l_\infty(X^*)$ is a Banach space. By our hypothesis T is a well defined operator. Since X^* and $l_\infty(X^*)$ are Banach spaces and T is linear, in order to prove that T is continuous we will use the closed graph theorem (see chapter 9). Let $(y_k^*)_{k \geq 0} \subseteq X^*$, $y_k^* \to y_0^*$, $T(y_k^*) \to \lambda = (\lambda_n^*)_{n \in \mathbb{N}} \in l_\infty(X^*)$. Then $\lim\limits_{k \to \infty} \sum\limits_{i=1}^{n} y_k^*(x_i) x_i^* = \lambda_n^*$ $\forall n \in \mathbb{N}$, and therefore $\sum\limits_{i=1}^{n} y_0^*(x_i) x_i^* = \lambda_n^*$ $\forall n \in \mathbb{N}$, that is, $T(y_0^*) = \lambda$. Thus $T \in L(X^*, l_\infty(X^*))$, and then $\sup\limits_{\|x^*\|\leq 1} \|T(x^*)\|_{l_\infty(X^*)} = \|T\| < \infty$, that is, $\forall \|x^*\| \leq 1$, $\forall n \in \mathbb{N}$, $\left\|\sum\limits_{i=1}^{n} x^*(x_i) x_i^*\right\| \leq \|T\|$.

Therefore $\forall \|x^*\| \leq 1$, $\forall n \in \mathbb{N}$, $\forall \|x\| \leq 1$, $\left|x^*\left(\sum\limits_{i=1}^{n} x_i^*(x) x_i\right)\right| \leq \|T\|$. Taking the supremum for $\|x^*\| \leq 1$ it follows that $\forall n \in \mathbb{N}$, $\forall \|x\| \leq 1$, $\left\|\sum\limits_{i=1}^{n} x_i^*(x) x_i\right\| \leq \|T\|$, and from here $\forall n \in \mathbb{N}$, $\forall x \in X$, $\left\|\sum\limits_{i=1}^{n} x_i^*(x) x_i\right\| \leq \|T\| \|x\|$, thus $\sup\limits_{n \in \mathbb{N}} \left\|\sum\limits_{i=1}^{n} x_i^*(x) x_i\right\| \leq \|T\| \|x\|$ $\forall x \in X$.

Let $E = \left\{x \in X \mid \sum\limits_{i=1}^{\infty} x_i^*(x) x_i \text{ converges}, x = \sum\limits_{i=1}^{\infty} x_i^*(x) x_i\right\}$. We prove now that $E \subseteq X$ is a closed linear subspace. The fact that $E \subseteq X$ is a linear subspace is obvious. Let now $(y_k)_{k \in \mathbb{N}} \subseteq E$, $y_k \to y \in X$ in the norm of X. Let $\varepsilon > 0$. There is $k \in \mathbb{N}$ such that $\|y_k - y\| < \varepsilon$. Since $y_k \in E$ there is $n_\varepsilon \in \mathbb{N}$ such that $\left\|\sum\limits_{i=1}^{n} x_i^*(y_k) x_i - y_k\right\| < \varepsilon$

$\forall n \geq n_\varepsilon$. Then, for $n \geq n_\varepsilon$ we have

$$\left\|\sum_{i=1}^n x_i^*(y) x_i - y\right\| \leq \left\|\sum_{i=1}^n x_i^*(y) x_i - \sum_{i=1}^n x_i^*(y_k) x_i\right\|$$
$$+ \left\|\sum_{i=1}^n x_i^*(y_k) x_i - y_k\right\| + \|y_k - y\|$$
$$\leq \|T\|\varepsilon + \varepsilon + \varepsilon = (2 + \|T\|)\varepsilon,$$

thus the series $\sum_{i=1}^\infty x_i^*(y) x_i$ converges, and y is its limit. Thus $E \subseteq X$ is a closed linear subspace. We observe that by biorthogonality $x_i \in E\ \forall i \in \mathbb{N}$, and then $\overline{\text{Sp}\{x_i \mid i \in \mathbb{N}\}}^{\|\cdot\|} \subseteq E$.

ii) Exercise! Evidently one can use the technique from (i).

30. See [38, C.3.6, chapter 0].

For each $n \in \mathbb{N}$ let $x_n = Te_n$, $x_n \in X$. Since $Q : Y \to X$ is a metric surjection, for each $n \in \mathbb{N}$ there is an $y_n \in Y$ such that $Qy_n = x_n$, $\|y_n\| \leq (1+\varepsilon)\|x_n\|$. Define $\overline{T} : l_1 \to Y$, given by $\overline{T}(\alpha) = \sum_{n=1}^\infty \alpha_n y_n\ \forall \alpha = (\alpha_n)_{n\in\mathbb{N}} \in l_1$. Then $\sum_{n=1}^\infty \|\alpha_n y_n\| = \sum_{n=1}^\infty |\alpha_n|\|y_n\| \leq (1+\varepsilon) \sum_{n=1}^\infty |\alpha_n|\|Te_n\| \leq (1+\varepsilon)\|T\|\|\alpha\|_{l_1}$. Thus $\overline{T} : l_1 \to Y$ is well defined (the series $\sum_{n=1}^\infty \alpha_n y_n$ converges absolutely and, since Y is a Banach space, the series converges). Evidently $\overline{T}$ is linear. Since $\|\overline{T}(\alpha)\| \leq \sum_{n=1}^\infty \|\alpha_n y_n\| \leq (1+\varepsilon)\|T\|\|\alpha\|_{l_1}$ we obtain that $\overline{T}$ is also continuous, $\|\overline{T}\| \leq (1+\varepsilon)\|T\|$. Also $\forall \alpha = (\alpha_n)_{n\in\mathbb{N}} \in l_1$, since Q and T are linear and continuous operators,

$$Q(\overline{T}(\alpha)) = Q\left(\sum_{n=1}^\infty \alpha_n y_n\right) = \sum_{n=1}^\infty \alpha_n Q(y_n) = \sum_{n=1}^\infty \alpha_n x_n$$
$$= \sum_{n=1}^\infty \alpha_n T(e_n) = T\left(\sum_{n=1}^\infty \alpha_n e_n\right) = T(\alpha),$$

that is, $Q \circ \overline{T} = T$.

31. See [16, exercise 3.21 (a), chapter I].

We know that if X is a Banach space, and $Y \subseteq X$ is a closed linear subspace, then X/Y becomes a Banach space, with the quotient norm: $\|\overline{x}\| = \inf\{\|y\| \mid y \in \overline{x}\}\ \forall \overline{x} \in X/Y$. We shall show now that $Q : X \to X/Y$ is a metric surjection and then we will use exercise 30. Evidently $Q : X \to X/Y$ is a surjection. Then $Qx = \overline{x}\ \forall x \in X$ and by the definition of the quotient norm we have

$$\|Qx\| = \inf\{\|y\| \mid y \in \overline{x}\} = \inf\{\|y\| \mid Qy = \overline{x}\} = \inf\{\|y\| \mid Qy = Qx\}.$$

2.2 Solutions 67

32. See [38, C.3.2, chapter 0].

i) We consider the adjoint of T, $T^* : l_\infty^* \to X^*$, which is a linear and continuous operator. For every $n \in \mathbb{N}$ let $p_n : l_\infty \to \mathbb{K}$, $p_n(x) = x_n$ $\forall x = (x_1, x_2, ...) \in l_\infty$. Then $p_n \in l_\infty^*$ $\forall n \in \mathbb{N}$ and we define $x_n^* = T^*(p_n)$. Since T^* is continuous and $||p_n|| = 1$ $\forall n \in \mathbb{N}$, we obtain that the sequence $(x_n^*)_{n \in \mathbb{N}} \subseteq X^*$ is bounded. For $x \in X$, if $Tx = (y_1, y_2, ...) \in l_\infty$, then $y_n = p_n(Tx) = (T^*p_n)(x) = x_n^*(x)$ for all $n \in \mathbb{N}$, and therefore $Tx = (x_n^*(x))_{n \in \mathbb{N}}$ $\forall x \in X$. Then $||Tx||_\infty = \sup_{n \in \mathbb{N}} |x_n^*(x)| \leq \left(\sup_{n \in \mathbb{N}} ||x_n^*||\right) ||x||$ and therefore $||T|| \leq \sup_{n \in \mathbb{N}} ||x_n^*||$. For $\varepsilon > 0$ there is an $n_\varepsilon \in \mathbb{N}$ such that $||x_{n_\varepsilon}^*|| + \varepsilon \geq \sup_{n \in \mathbb{N}} ||x_n^*||$. For $x_{n_\varepsilon}^* \in X^*$, since $||x_{n_\varepsilon}^*|| = \sup_{||x|| \leq 1} |x_{n_\varepsilon}^*(x)|$ there is an $x \in X$, $||x|| \leq 1$, such that $|x_{n_\varepsilon}^*(x)| + \varepsilon \geq ||x_{n_\varepsilon}^*||$. Then $|x_{n_\varepsilon}^*(x)| \geq \sup_{n \in \mathbb{N}} ||x_n^*|| - 2\varepsilon$, and therefore $||T|| \geq \sup_{n \in \mathbb{N}} |x_n^*(x)| \geq \sup_{n \in \mathbb{N}} ||x_n^*|| - 2\varepsilon$. Since $\varepsilon > 0$ was arbitrary we obtain that $||T|| \geq \sup_{n \in \mathbb{N}} ||x_n^*||$, and therefore $||T|| = \sup_{n \in \mathbb{N}} ||x_n^*||$.

ii) Since $T : Y \to l_\infty$ is a bounded linear operator we use (i) and we obtain $(y_n^*)_{n \in \mathbb{N}} \subseteq Y^*$ such that $Ty = (y_n^*(y))_{n \in \mathbb{N}}$ $\forall y \in Y$ and $||T|| = \sup_{n \in \mathbb{N}} ||y_n^*||$. Using the Hahn–Banach theorem (see chapter 7) there is a sequence $(x_n^*)_{n \in \mathbb{N}} \subseteq X^*$ such that $x_n^*|_Y = y_n^*$, $||x_n^*|| = ||y_n^*||$ $\forall n \in \mathbb{N}$. Define then $\widetilde{T} : X \to l_\infty$, $\widetilde{T}x = (x_n^*(x))_{n \in \mathbb{N}}$ $\forall x \in X$. Then by (i) $\widetilde{T} : X \to l_\infty$ is a well defined bounded linear operator, and $||\widetilde{T}|| = \sup_{n \in \mathbb{N}} ||x_n^*|| = \sup_{n \in \mathbb{N}} ||y_n^*|| = ||T||$. For any $y \in Y$ we have $\widetilde{T}y = (x_n^*(y))_{n \in \mathbb{N}} = (y_n^*(y))_{n \in \mathbb{N}} = Ty$.

Chapter 3

Linear and continuous functionals. Reflexive spaces

Definition. For a normed space X its *dual* $L(X, \mathbb{K})$ is denoted by X^*, and for an element $x^* \in X^*$ its norm is given by one of the formulas:

$$\|x^*\| = \inf \{M \geq 0 \mid |x^*(x)| \leq M \|x\| \ \forall x \in X\} = \sup_{\|x\| \leq 1} |x^*(x)| = \sup_{\|x\| = 1} |x^*(x)|.$$

For a normed space X its *bidual* $L(X^*, \mathbb{K})$ is denoted by X^{**}.

Definition. Let X be a normed space. For $x \in X$ consider $\widehat{x} : X^* \to \mathbb{K}$ defined by $\widehat{x}(x^*) = x^*(x) \ \forall x^* \in X^*$. Then $\widehat{x} \in X^{**}$ and $\|\widehat{x}\|_{X^{**}} = \|x\|$, and let $K_X : X \to X^{**}$ be defined by $K_X(x) = \widehat{x} \ \forall x \in X$. K_X is called the *canonical embedding* into the bidual.

Proposition. If X is a Banach space then $K_X(X) \subseteq X^{**}$ is a closed linear subspace.

Definition. A Banach space is called *reflexive* if and only if the canonical embedding into its bidual is surjective.

3.1 Exercises

A discontinuous additive functional

1. Prove that there is a discontinuous additive function $f : \mathbb{R} \to \mathbb{R}$. ($f$ additive means $f(x + y) = f(x) + f(y) \ \forall x, y \in \mathbb{R}$.)

Characterizations for finite-dimensional normed spaces

2. Let X be a normed space. Prove that the following assertions are equivalent:

i) X is finite-dimensional;

3.1 Exercises

ii) Any linear functional on X is continuous;
iii) Any linear subspace of X is closed.

Dense sets defined by discontinuous linear functionals

3. i) Let X be a linear topological space and $f : X \to \mathbb{K}$ a linear functional. Let $G \subseteq X$ be a dense linear subspace, f being discontinuous on G. Prove that $G \cap \ker f$ is dense in X. Thus if $f : X \to \mathbb{K}$ is linear and discontinuous on X then $\ker f$ is dense in X.

ii) Let X be a real linear topological space and $A \subseteq X$ a dense convex set in X. Prove that for any closed hyperplane $H \subseteq X$ the set $H \cap A$ is dense in H.

4. Let $E = \{P \mid P \text{ polynomial}, P(1) = 0\} \subseteq L_2([0,1]; \mathbb{R})$. Prove that E is dense in $L_2([0,1]; \mathbb{R})$.

5. For $\lambda \in \mathbb{R}$ we denote $E_\lambda = \{x \in C([0,1]; \mathbb{R}) \mid x(0) = \lambda\}$. Prove that E_λ is a dense convex set in $L_2([0,1]; \mathbb{R})$.

Linearly independent functionals and norms

6. Let $n \in \mathbb{N}$. Let V be a linear space and $f_1, ..., f_n$ be n linear functionals on V. Prove that there is a norm $\|\cdot\|$ on V such that all the functionals $f_1, ..., f_n$ are continuous from $(V, \|\cdot\|)$ to $\mathbb{K}$. What happens in the case when we have an infinite number of linear functionals? If we have an infinite number of linearly independent functionals?

The Auerbach Lemma

7. Let $(X, \|\cdot\|)$ be a finite-dimensional normed space with $\dim_\mathbb{K} X = n \in \mathbb{N}$. Prove that there are $x_1, ..., x_n \in X$, $x_1^*, ..., x_n^* \in X^*$, such that $\|x_i\| = \|x_i^*\| = 1$ $\forall i = \overline{1,n}$ and
$$x_i^*(x_j) = \delta_{ij} = \begin{cases} 1, & i = j, \\ 0, & i \neq j, \end{cases} \forall i, j = \overline{1,n}.$$

Norm attaining and non-norm attaining linear and continuous functionals

8. Consider $L = \left\{ f \in C[a,b] \mid \int_a^{(a+b)/2} f(x)dx = \int_{(a+b)/2}^b f(x)dx \right\}$.
i) Prove that L is a closed linear subspace in $C[a,b]$ and find $x^* \in (C[a,b])^*$ such that $L = \ker x^*$.
ii) Prove that $x^* \in (C[a,b])^*$ does not attain its norm on the closed unit ball.

9. Let $x^* : c_0 \to \mathbb{R}$, $x^*(x) = \sum_{n=1}^\infty x_n/2^{n-1}$ $\forall x = (x_1, x_2, ...) \in c_0$.
i) Prove that x^* is a linear continuous functional and that $\|x^*\| = 2$.
ii) Prove that $x^* \in (c_0)^*$ does not attain its norm on the closed unit ball.

10. Find the form for the linear and continuous functionals on c_0 which achieve their norm on the closed unit ball of c_0.

11. Let $x^* : l_1 \to \mathbb{R}$, $x^*(x) = \sum_{n=1}^\infty (1 - 1/n)x_n$ $\forall x = (x_n)_{n \in \mathbb{N}}$.
i) Prove that x^* is a linear continuous functional on l_1 and that $\|x^*\| = 1$.
ii) Prove that x^* does not achieve its norm on the closed unit ball of l_1.

12. Let $(\xi_n)_{n \in \mathbb{N}} \in l_\infty$, $M = \sup_{n \in \mathbb{N}} |\xi_n|$, such that $M \notin \{|\xi_n| \mid n \in \mathbb{N}\}$. Then let $x^* : l_1 \to \mathbb{R}$ given by $x^*(x) = \sum_{n=1}^\infty \xi_n x_n$ $\forall x = (x_n)_{n \in \mathbb{N}} \in l_1$.

i) Prove that x^* is a linear and continuous functional and that $\|x^*\| = M$.

ii) Prove that x^* does not achieve its norm on the closed unit ball of l_1.

13. Prove that the functional $x^* : l_1 \to \mathbb{R}$, $x^*(x) = \sum_{n=1}^{\infty} (1/n) x_n$ achieve its norm on the closed unit ball of l_1.

The James Theorem

14. Let X be a Banach space. Prove that the following assertions are equivalent:

i) X is reflexive;

ii) Any linear and continuous functional on X achieve its norm on the closed unit ball of X.

Non reflexive spaces

15. Prove that c_0, l_1 and $C[a,b]$ are not reflexive spaces.

The weak convergence in reflexive spaces

16. Let $(x_n)_{n \in \mathbb{N}}$ be a bounded sequence in a normed space X. For every $n \in \mathbb{N}$ we denote by $K_n = \overline{co} \{x_n, x_{n+1}, ...\}$.

i) Prove that if $x_n \to x_0 \in X$ weak then $\bigcap_{n=1}^{\infty} K_n = \{x_0\}$.

ii) If X is in addition reflexive and $\bigcap_{n=1}^{\infty} K_n = \{x_0\}$, where $x_0 \in X$, then $x_n \to x_0$ weak.

iii) Give a counterexample to prove that the assertion from (ii) is no longer true if X is not supposed to be reflexive.

A hereditary property for reflexive spaces

17. Let X be a reflexive Banach space and $Y \subseteq X$ a closed linear subspace. Prove that Y is a reflexive Banach space.

A Banach space is reflexive if and only if its dual is reflexive

18. Let X be a Banach space. Prove that X is reflexive if and only if X^* is reflexive.

The annihilator for a set in a normed space

19. For any normed space X and for any subset $Z \subseteq X$ we denote by
$$Z^\perp = \{x^* \in X^* \mid x^*|_Z = 0\},$$
the annihilator of Z.

i) If X is a normed space, and $Y \subseteq X$ a linear subspace, prove that the operator $T : X^*/Y^\perp \to Y^*$, $x^* + Y^\perp \xrightarrow{T} y^*$, where $y^* = x^*|_Y$, is an isometric isomorphism.

ii) If X is a normed space, and $Y \subseteq X$ a closed linear subspace, prove that the operator $S : Y^\perp \to (X/Y)^*$, $x^* \xrightarrow{S} \overline{x}^*$, where $\overline{x}^*(x+Y) = x^*(x) \, \forall x \in X$, is an isometric isomorphism.

iii) Let X be a reflexive space, and consider a closed linear subspace $Y \subseteq X$. Prove that $Y^{\perp\perp} = K_X(Y)$.

iv) Is the assertion from (iii) still true if we eliminate the hypothesis that X is reflexive?

3.1 Exercises

Tree type of properties for reflexive spaces

20. Using exercise 19 prove that if X is a reflexive space and $Y \subseteq X$ is a closed linear subspace then X/Y is also a reflexive space (with the quotient norm).

21. Let X be a Banach space and $Y \subseteq X$ a closed linear subspace. If Y and X/Y are reflexive spaces, Y with the norm from X and X/Y with the quotient norm, prove that X is also a reflexive space.

Separable linear subspaces into the dual

22. Let X be a normed space, and $Z \subseteq X^*$ a separable linear subspace. Prove that there is a separable linear subspace $Y \subseteq X$ such that Z is isometrically isomorphic to a linear subspace of Y^*.

A Cantor type of theorem in reflexive spaces

23. Let X be a Banach space. Prove that the following assertions are equivalent:
i) X is a reflexive space;
ii) For any sequence $(K_n)_{n\in\mathbb{N}}$ of non-empty bounded closed and convex sets of X, with $K_{n+1} \subseteq K_n \; \forall n \in \mathbb{N}$, we have that $\bigcap_{n\in\mathbb{N}} K_n$ is non-empty.

Failure for the Cantor type of theorem in non-reflexive spaces

24. i) For each $n \in \mathbb{N}$ let $K_n \subseteq c_0$,
$$K_n = \left\{ x \in c_0 \mid \|x\| \leq 1, x = (x_k)_{k\in\mathbb{N}}, x_k = 1, k = \overline{1,n} \right\}.$$
Prove that K_n is a non-empty bounded closed and convex set for each $n \in \mathbb{N}$, that $K_{n+1} \subseteq K_n \; \forall n \in \mathbb{N}$, but $\bigcap_{n\in\mathbb{N}} K_n = \emptyset$.

ii) For each $n \in \mathbb{N}$ let $K_n \subseteq l_1$,
$$K_n = \left\{ x \in l_1 \;\middle|\; \|x\| \leq 1, x = (x_k)_{k\in\mathbb{N}}, \sum_{k=1}^{\infty} (1 - 1/k) x_k \geq 1 - 1/n \right\}.$$
Prove that K_n is a non-empty bounded closed and convex set for each $n \in \mathbb{N}$, that $K_{n+1} \subseteq K_n \; \forall n \in \mathbb{N}$, but $\bigcap_{n\in\mathbb{N}} K_n = \emptyset$.

iii) Let $a \in [0,1)$, $(a_n)_{n\in\mathbb{N}} \subseteq (a,1]$, with $a_n \searrow a$. For every $n \in \mathbb{N}$ let
$$K_n = \{f \in C[0,1] \mid f(a) = 0, \|f\| \leq 1, f(x) = 1 \; \forall x \in [a_n, 1]\}.$$
Prove that K_n is a non-empty bounded closed and convex set for any $n \in \mathbb{N}$, that $K_{n+1} \subseteq K_n \; \forall n \in \mathbb{N}$, but $\bigcap_{n\in\mathbb{N}} K_n = \emptyset$.

The dual of l_∞

25. Prove that for any $x^* \in l_\infty^*$ there is a unique finitely additive measure $\mu : \mathcal{P}(\mathbb{N}) \to \mathbb{K}$, with bounded variation, such that $\mu(A) = x^*(\chi_A) \ \forall A \subseteq \mathbb{N}$, where $\chi_A = (x_n)_{n\in\mathbb{N}}$, $x_n = \begin{cases} 1 & \text{if } n \in A, \\ 0 & \text{if } n \notin A. \end{cases}$ We also have $|\mu|(\mathbb{N}) \leq \|x^*\|$. ($|\mu|$ is the total variation measure of μ.)

Conversely, for every finitely additive measure $\mu : \mathcal{P}(\mathbb{N}) \to \mathbb{K}$, with bounded variation, there is a unique element $x^* \in l_\infty^*$ such that $\mu(A) = x^*(\chi_A) \ \forall A \subseteq \mathbb{N}$. We also have $\|x^*\| \leq |\mu|(\mathbb{N})$.

3.2 Solutions

1. We consider $\mathbb{R}$ as a linear space over $\mathbb{Q}$. Evidently, since $\mathbb{R}$ is not countable and $\mathbb{Q}$ is countable, we obtain that the dimension of $\mathbb{R}$ over $\mathbb{Q}$ is infinite. Analogously to the solution for exercise 2 (the part '(ii) $\Rightarrow$ (i)') we obtain that there is $f : \mathbb{R} \to \mathbb{R}$, $\mathbb{Q}$-linear and discontinuous. The fact that $f : \mathbb{R} \to \mathbb{R}$ is $\mathbb{Q}$-linear is equivalent with the fact that $f : \mathbb{R} \to \mathbb{R}$ is additive. We have $f(x+y) = f(x) + f(y)$ for every x and y from $\mathbb{R}$, and f is discontinuous.

2. '(i) $\Rightarrow$ (ii)' If X is finite-dimensional let $\dim_\mathbb{K} X = n \in \mathbb{N}$. Consider a basis $\{e_1, ..., e_n\}$ for X. Consider a linear functional $f : X \to \mathbb{K}$. Thus $f(x) = \sum_{i=1}^{n} \alpha_i f(e_i) = \sum_{i=1}^{n} \alpha_i a_i$, where $x = \sum_{i=1}^{n} \alpha_i e_i$, $a_i := f(e_i)$, $i = \overline{1,n}$. Then

$$|f(x)| \leq \max_{i=\overline{1,n}} |a_i| \sum_{i=1}^{n} |\alpha_i|.$$

Since $x = \sum_{i=1}^{n} \alpha_i e_i \longmapsto \|x\| = \sum_{i=1}^{n} |\alpha_i|$ is a norm on X, and any two norms on X (X is finite-dimensional) are equivalent, we obtain that f is continuous.

'(ii) $\Rightarrow$ (i)' Suppose, for a contradiction, that X is not finite-dimensional. We can find then a linearly independent system $\{x_n \mid n \in \mathbb{N}\} \subseteq X$. Then $x_n \neq 0 \ \forall n \in \mathbb{N}$, and the family $\{x_n/\|x_n\| \mid n \in \mathbb{N}\}$ is again a linearly independent one. Thus, there is a linearly independent family $\{x_n \mid n \in \mathbb{N}\} \subseteq X$ such that $\|x_n\| = 1 \ \forall n \in \mathbb{N}$. Then we can obtain a basis B for X, $B = \{x_n \mid n \in \mathbb{N}\} \cup \{x_i \mid i \in I\}$. We define $f : X \to \mathbb{K}$ linear, given on B by $\begin{cases} f(x_n) = n & \forall n \in \mathbb{N}, \\ f(x_i) = 0 & \forall i \in I. \end{cases}$ We can extend f, given on B, to a linear functional $f : X \to \mathbb{K}$. Then $|f(x_n)| = n$, $\|x_n\| = 1 \ \forall n \in \mathbb{N}$, and we obtain a linear functional on X which is not continuous (if f were continuous, then $|f(x_n)| \leq \|f\| \|x_n\| \ \forall n \in \mathbb{N}$, that is, $n \leq \|f\| \ \forall n \in \mathbb{N}$).

'(ii) $\Rightarrow$ (iii)' Consider a linear subspace $Y \subseteq X$. We consider a basis $\{e_j \mid j \in J\}$ for Y, and we can complete this linearly independent system (in X) to a basis $\{e_i \mid i \in I\}$ for X ($J \subseteq I$). For every $i \in I$ we define $l_i^* : X \to \mathbb{K}$, $\begin{cases} l_i^*(e_k) = 0 & \text{if } k \neq i, \\ l_i^*(e_k) = 1 & \text{if } k = i, \end{cases}$ l_i linear. By our hypothesis (ii), $l_i^* : X \to \mathbb{K}$ is a linear and continuous functional, for any $i \in I$.

3.2 Solutions 73

We prove now that $Y = \bigcap_{i \in I \setminus J} \ker l_i^*$ and this solves the exercise, because $\ker l_i^*$ is a closed set $\forall i \in I$ (for the case $I = J$, that is, $Y = X$, it is obvious that Y is closed). Since $l_i^* |_Y = 0$ for $i \notin J$, we have $Y \subseteq \ker l_i^* \; \forall i \in I \setminus J$, that is, $Y \subseteq \bigcap_{i \in I \setminus J} \ker l_i^*$. Consider $x \in \bigcap_{i \in I \setminus J} \ker l_i^*$. We have $x = \sum_{i \in I} \alpha_i e_i$ (only a finite number of $\{\alpha_i \mid i \in I\}$ are non-null). For $i \in I \setminus J$, since $l_i^*(x) = 0$ ($x \in \ker l_i^*$, by our supposition) we obtain that $\alpha_i = 0$. Then

$$x = \sum_{i \in J} \alpha_i e_i + \sum_{i \in I \setminus J} \alpha_i e_i = \sum_{i \in J} \alpha_i e_i \in Y,$$

and therefore $\bigcap_{i \in I \setminus J} \ker l_i^* \subseteq Y$.

'(iii) $\Rightarrow$ (ii)' Let $f : X \to \mathbb{K}$ be a linear functional. Its kernel $\ker f$ is a linear subspace of X, and therefore by our hypothesis is a closed linear subspace. We prove that if $\ker f \subseteq X$ is a closed set then f is continuous. Since f is linear it is sufficient to prove that f is continuous at 0. Let $\varepsilon > 0$, and denote $V = \{x \subset X \mid |f(x)| < \varepsilon\}$. If $f = 0$, then $V = X$, and therefore V is an open neighborhood of $0 \in X$. If $f \neq 0$, there is $a \in X$ such that $f(a) \neq 0$, and let $x_0 = a/f(a) \in X$, $f(x_0) = 1$. Then $x_0 \notin \ker f$, $\varepsilon x_0 \notin \ker f$, and since $\ker f \subseteq X$ is a closed set, $\varepsilon x_0 \in \mathbf{C}_{\ker f}$, an open set in a normed space. Therefore there is an $r > 0$ such that $B(\varepsilon x_0, r) \subseteq \mathbf{C}_{\ker f}$, i.e., $B(\varepsilon x_0, r) \cap \ker f = \varnothing$. Consider $x \in B(0, r)$. If $x \notin V$ then $|f(x)| \geq \varepsilon$, and, if we denote $y = -(\varepsilon x)/f(x)$, then $\|y\| \leq \|x\| < r$, that is, $y + \varepsilon x_0 \in B(\varepsilon x_0, r)$. But $f(y + \varepsilon x_0) = 0$, and therefore $y + \varepsilon x_0 \in B(\varepsilon x_0, r) \cap \ker f = \varnothing$, a contradiction. Thus $\forall x \in B(0, r)$ we obtain that $x \in V$, i.e., $B(0, r) \subseteq V$, that is, V is an open neighborhood of $0 \in X$. Thus f is continuous at 0. Since f is linear, f is continuous on X.

3. i) We show that for any open set $D \neq \varnothing$, we have $f(D \cap G) = \mathbb{K}$. Suppose that there exists D such that $f(D \cap G) \neq \mathbb{K}$. Since $D \neq \varnothing$ and G is dense in X, it results that $D \cap G \neq \varnothing$. Let $a \in D \cap G$. Since $a \in D$ we obtain that $D - a = V$ is a neighborhood of 0. We prove now that $f(V \cap G) \neq \mathbb{K}$. If, for a contradiction, $f(V \cap G) = \mathbb{K}$, we consider $\lambda \in \mathbb{K}$, and, for $\lambda - f(a) \in \mathbb{K}$, $\exists x \in V \cap G$ such that $f(x) = -f(a) + \lambda$. Then $x + a \in V + a = D$ and $x \in G$, $a \in G$, $x + a \in G$, thus $x + a \in D \cap G$ and $f(x + a) = \lambda$, i.e., $f(D \cap G) = \mathbb{K}$, a contradiction. Since $f(V \cap G) \neq \mathbb{K}$, there is $\alpha \in \mathbb{K}$ such that $\alpha \notin f(V \cap G)$. Let $U \subseteq V$, U a balanced neighborhood of 0. Since $f(U \cap G) \subseteq f(V \cap G)$ we obtain that $\alpha \notin f(U \cap G)$. We shall prove that $|f(x)| \leq \alpha$ $\forall x \in U \cap G$, and we shall obtain that f is continuous at 0, on G ($\forall \varepsilon > 0 \; \exists (\varepsilon/\alpha) U$ neighborhood of 0 such that $\forall x \in ((\varepsilon/\alpha) U) \cap G$, $|f(x)| \leq \varepsilon$), and using the fact that f is linear we obtain that f is continuous on G, a contradiction. If $|\alpha| < |f(x)|$, $x \in U \cap G$, then $|\alpha/f(x)| < 1$. Let $\lambda = \alpha/f(x)$. Since G is a linear subspace and U balanced, $\lambda f(U \cap G) = f(\lambda U \cap \lambda G) \subseteq f(U \cap G)$, and therefore $\lambda f(x) \in f(U \cap G)$, i.e., $\alpha \in f(U \cap G)$, a contradiction!

Thus for any non-empty open set D we have $f(D \cap G) = \mathbb{K}$, and since $0 \in \mathbb{K}$, $\exists x \in D \cap G$ such that $f(x) = 0$, i.e., $x \in D \cap G \cap \ker f$. We obtain that $D \cap G \cap \ker f \neq \varnothing$ $\forall D$ open, and therefore $\overline{G \cap \ker f} = X$.

ii) See [9, exercise 18, chapter II, section 1].

Recall that a closed hyperplane $H \subseteq X$ is a set of the form $H = \{x \in X \mid x^*(x) = \alpha\}$, where $x^* : X \to \mathbb{R}$ is a linear continuous functional and $\alpha \in \mathbb{R}$.

Consider a closed hyperplane $H \subseteq X$. If $x^* = 0$, then $H = \begin{cases} X & \text{if } \alpha = 0, \\ \varnothing & \text{if } \alpha \neq 0. \end{cases}$ Then $H \cap A = \begin{cases} A & \text{if } \alpha = 0, \\ \varnothing & \text{if } \alpha \neq 0, \end{cases}$ a set which is dense in H. Suppose now that $x^* \neq 0$. Let $x_0 \in H$ and consider V, a balanced neighborhood of 0. We prove that $(x_0 + V) \cap [x^* < \alpha] \neq \varnothing$. If, for a contradiction, $(x_0 + V) \cap [x^* < \alpha] = \varnothing$, i.e., $x_0 + V \subseteq [x^* \geq \alpha]$, then, $\forall x \in V$, $x + x_0 \in x_0 + V \subseteq [x^* \geq \alpha]$, i.e., $x^*(x + x_0) \geq \alpha$. Since $x_0 \in H$, $x^*(x_0) = \alpha$, we obtain that $x^*(x) \geq 0$. Thus $\forall x \in V$ we have that $x^*(x) \geq 0$. But V being balanced, $\forall x \in V$ we have $-x \in V$, and therefore $x^*(-x) \geq 0$, $x^*(x) \leq 0$. Thus $\forall x \in V$, $x^*(x) = 0$. But V is absorbing, i.e., $\forall x \in X$ there is $\lambda \neq 0$ such that $\lambda x \in V$. Therefore $x^*(\lambda x) = 0$, and $x^*(x) = 0$, i.e., $x^* = 0$, a contradiction. The set $[x^* < \alpha] = (x^*)^{-1}(-\infty, \alpha)$ is open (x^* is continuous) and therefore $(x_0 + V) \cap [x^* < \alpha]$ is a non-empty open set. Since A is dense we obtain that $A \cap (x_0 + V) \cap [x^* < \alpha] \neq \varnothing$. Therefore there is $a_1 \in V$ such that $x_0 + a_1 \in A$ and $x^*(x_0 + a_1) < \alpha$. Analogously, $A \cap (x_0 + V) \cap [x^* > \alpha] \neq \varnothing$, i.e., there is $a_2 \in V$ such that $x_0 + a_2 \in A$ and $x^*(x_0 + a_2) > \alpha$. Let now $f : [0, 1] \to \mathbb{R}$, $f(t) = x^*(t(x_0 + a_1) + (1 - t)(x_0 + a_2))$. Then f is continuous, $f(0) > \alpha$, $f(1) < \alpha$. Therefore there is $t \in (0, 1)$ such that $f(t) = \alpha$, i.e., $x = t(x_0 + a_1) + (1-t)(x_0 + a_2) \in H$. Since A is convex, $x \in A$, as a convex combination of $x_0 + a_1, x_0 + a_2 \in A$. Also, $x = x_0 + ta_1 + (1-t)a_2 \in x_0 + V + V$ (V is balanced, $a_1, a_2 \in V$, $0 < t, 1-t < 1$, and then $ta_1 \in V$, $(1-t)a_2 \in V$). Thus, $x \in H \cap A \cap (x_0 + V + V)$, i.e., $H \cap A \cap (x_0 + V + V) \neq \varnothing$. Therefore we have proved that $\forall x_0 \in H$, $\forall V$ a balanced neighborhood of 0, we have $H \cap A \cap (x_0 + V + V) \neq \varnothing$. Now let $x_0 \in H$ and let W be a neighborhood of 0. There is a balanced neighborhood V of 0 such that $V + V \subseteq W$. From what we have proved above, we have $H \cap A \cap (x_0 + V + V) \neq \varnothing$, and therefore $H \cap A \cap (x_0 + W) \neq \varnothing$. This implies that $H \cap A$ is dense in H.

4. See [40, exercise 4.16].

Evidently E is a linear subspace of $L_2[0, 1]$. We consider a linear continuous functional $f : L_2[0, 1] \to \mathbb{R}$ such that $f|_E = 0$. There is $g \in L_2[0, 1]$ such that $f(x) = \int_0^1 g(t) x(t) dt$ for all $x \in L_2[0, 1]$. If we define $h(t) = (1 - t)g(t)$ then $h \in L_2[0, 1]$, and for every polynomial P we have

$$f(P(t)(1-t)) = 0, \quad \int_0^1 g(t)(1-t) P(t) dt = 0, \quad \int_0^1 h(t) P(t) dt = 0.$$

Thus h is orthogonal on the set of all polynomials. But for any $u \in C[0, 1]$ there is a sequence of polynomials $(P_n)_{n \in \mathbb{N}}$ such that $\|P_n - u\|_\infty \to 0$ (the Bernstein theorem), and since

$$\left| \int_0^1 h(t) u(t) dt \right| = \left| \int_0^1 h(t)(u(t) - P_n(t)) dt \right| \leq \int_0^1 |h(t)| |u(t) - P_n(t)| dt$$
$$\leq \|P_n - u\|_\infty \left(\int_0^1 |h(t)|^2 dt \right)^{1/2} = \|P_n - u\|_\infty \|h\|_2,$$

for $n \to \infty$ we obtain that $\int_0^1 h(t) u(t) dt = 0$ $\forall u \in C[0, 1]$, thus h is orthogonal on $C[0, 1] \subseteq L_2[0, 1]$. But $C[0, 1]$ is dense in $L_2[0, 1]$ and therefore $h = 0$, μ-a.e., that is,

3.2 Solutions

$g = 0$, μ-a.e., and then $f = 0$. We have proved that if $f \in (L_2[0,1])^*$, $f|_E = 0$, then $f = 0$. We obtain that E is dense in $L_2[0,1]$ (see the theorems from chapter 8).

We can obtain another solution by using exercise 3(i) (see the second solution for exercise 5).

5. See [40, exercise 4.15].

First solution. Let $E = \{x \in C[0,1] \mid x(0) = 0\}$. Then $E \subseteq L_2[0,1]$ is dense. Indeed, let $x \in C[0,1]$. We consider $x_n(t) = x(t)(1-(1-t)^n)$, and therefore $x_n(0) = 0$, $x_n \in E$ $\forall n \in \mathbb{N}$, and

$$\|x_n - x\|_2^2 = \int_0^1 |x(t)|^2 (1-t)^{2n} dt \leq M^2 \int_0^1 (1-t)^{2n} dt = \frac{M^2}{2n+1} \to 0,$$

$M = \sup_{t \in [0,1]} |x(t)| < \infty$. Thus $\overline{E}^{\|\cdot\|_2} \supseteq C[0,1]$. But $\overline{C[0,1]}^{\|\cdot\|_2} = L_2[0,1]$, and therefore $\overline{E}^{\|\cdot\|_2} = L_2[0,1]$, i.e., $\overline{E}^{\|\cdot\|_2} = L_2[0,1]$. Since $E_\lambda = \lambda + E$ we obtain that $\overline{E_\lambda} = \overline{\lambda + E} = \lambda + L_2[0,1] = L_2[0,1]$, i.e., E_λ is dense in $L_2[0,1]$. Obviously E_λ is also convex.

Second solution. We consider $\mathcal{L}_2[0,1]$ and not $L_2[0,1]$, thus we do not identify a function $f \in \mathcal{L}_2[0,1]$ with its equivalence class $\tilde{f} \in L_2[0,1]$ (see Notations). $\mathcal{L}_2[0,1]$ is a seminormed space with respect to the seminorm $f \mapsto \left(\int_0^1 |f(x)|^2 dx\right)^{1/2}$, hence a linear topological space. Then $C[0,1] \subseteq \mathcal{L}_2[0,1]$ is dense (with respect to the topology given by this seminorm).

We consider $f : \mathcal{L}_2[0,1] \to \mathbb{R}$, $f(x) = x(0)$. Then $E = C[0,1] \cap \ker f$. We prove that f is discontinuous on $(C[0,1], \|\cdot\|_2)$. Let $x_n(t) = (1-t) - (1-t)^{n+1}$, $x(t) = 1-t$. Then

$$\|x_n - x\|_2^2 = \int_0^1 (1-t)^{2n+2} dt = \frac{1}{2n+3} \to 0,$$

i.e., $x_n \to x$. But $f(x_n) = x_n(0) = 0$ $\forall n \in \mathbb{N}$, and $f(x) = 1$, i.e., f is discontinuous on $C[0,1]$. Then, by exercise 3(i), $\overline{E}^{\|\cdot\|_2} = \mathcal{L}_2[0,1]$, and therefore, if we consider E as a subset in $L_2[0,1]$, we have $\overline{E}^{\|\cdot\|_2} = L_2[0,1]$. Since $E_\lambda = \lambda + E$, we obtain that $\overline{E_\lambda}^{\|\cdot\|_2} = \overline{\lambda + E}^{\|\cdot\|_2} = \lambda + L_2[0,1] = L_2[0,1]$, i.e., E_λ is dense in $L_2[0,1]$.

6. See [6, exercise 23, chapter 2].

Using exercise 1 of chapter 1 we obtain that there is a norm $\|\cdot\|_1$ on V. Consider now the set $\{f_1, ..., f_n\} \subseteq V'$. Let $\|\cdot\| : V \to \mathbb{R}_+$, $\|x\| = \|x\|_1 + \max_{1 \leq i \leq n} |f_i(x)|$. If $\|x\| = 0$, then $\|x\|_1 = 0$, and so $x = 0$. Obviously $\|\lambda x\| = |\lambda| \|x\|$ $\forall \lambda \in \mathbb{K}$, $\forall x \in V$, and $\|x+y\| \leq \|x\| + \|y\|$ $\forall x, y \in V$. Hence $(V, \|\cdot\|)$ is a normed space. For any $1 \leq i \leq n$ we have $|f_i(x)| \leq \|x\|$ $\forall x \in V$, so $f_i : (V, \|\cdot\|) \to \mathbb{K}$ is continuous.

In the case when we have an infinite linearly independent set of functionals on V, we cannot say that there is always a norm on V with respect to which all the functionals from our set are continuous. For example, one can take, if V is infinite-dimensional, our set of functionals as being a (linear) basis $\mathcal{B}$ for V', the algebraic dual of V. Then using exercise 2 we obtain that we cannot find a norm $\|\cdot\|$ on V with respect to which all the functionals from $\mathcal{B}$ are continuous, since then it follows that all the functionals from V' are continuous.

7. See [38, B.4.8, chapter 0] or [19, chapter 6].
Let $e_1, ..., e_n$ be an algebraic basis for X and let us define $d : (B_{X^*})^n \to [0, \infty)$,

$$d(a_1, ..., a_n) = \left|\det(a_i(e_j))_{i,j=\overline{1,n}}\right| \quad \forall a_1, ..., a_n \in B_{X^*}.$$

Since X^* is finite-dimensional B_{X^*} is compact. Hence $(B_{X^*})^n$ is compact, and since d is continuous by the Weierstrass theorem we can find $x_1^*, ..., x_n^* \in X^*$ such that d attains its maximum on $(B_{X^*})^n$ at $(x_1^*, ..., x_n^*)$. Let $\delta_0 > 0$ be this maximum. Since $\delta_0 \neq 0$, $\det(x_i^*(e_j))_{i,j=\overline{1,n}} \neq 0$, hence there are $x_1, ..., x_n \in X$, linearly independent, such that $\sum_{j=1}^n x_j^*(e_i) x_j = e_i, i = \overline{1,n}$. Applying now x_k^* we obtain that $\sum_{j=1}^n x_j^*(e_i) x_k^*(x_j) = x_k^*(e_i)$, i.e., $BA = A$, where $B = (x_k^*(x_j))_{k,j}$, $A = (x_j^*(e_i))_{j,i}$. The matrix A is invertible, and therefore B is the unit matrix, that is, $x_k^*(x_j) = \delta_{kj}, k,j = \overline{1,n}$. From $\sum_{j=1}^n x_j^*(e_i) a_k(x_j) = a_k(e_i), \forall k, i$, it follows that $\det((x_j^*(e_i))) \det(a_k(x_j)) = \det((a_k(e_i)))$, hence, by our choice for $x_1^*, ..., x_n^*$, $|\det(a_k(x_j))| \leq 1 \; \forall a_1, ..., a_n \in B_{X^*}$. For $a_i := a, a_k := x_k^* \; \forall k \neq i$, we obtain that $|a(x_i)| \leq 1 \; \forall a \in B_{X^*}$, and therefore $\|x_i\| \leq 1$. Then $1 = x_i^*(x_i) \leq \|x_i^*\| \|x_i\| \leq 1$, and hence $\|x_i^*\| = \|x_i\| = 1, i = \overline{1,n}$.

8. See [40, exercise 11.25].
i) Let $x^* : C[a,b] \to \mathbb{R}$, $x^*(f) = \int_a^{(a+b)/2} f(x)dx - \int_{(a+b)/2}^b f(x)dx$. Then x^* is linear, and

$$|x^*(f)| \leq \int_a^{(a+b)/2} |f(x)| \, dx + \int_{(a+b)/2}^b |f(x)| \, dx$$
$$= \int_a^b |f(x)| \, dx \leq \|f\| \int_a^b 1 dx = (b-a) \|f\|,$$

i.e., x^* is continuous, $\|x^*\| \leq b - a$. Since $L = \ker x^*$, we obtain that L is a closed linear subspace.
Let us now calculate the norm for x^*. There is n_0 such that $1/n < (b-a)/2 \; \forall n \geq n_0$. Let $f_n : [a,b] \to \mathbb{R}$,

$$f_n(x) = \begin{cases} 1, & x \in [a, (a+b)/2 - 1/n], \\ n((a+b)/2 - x), & x \in ((a+b)/2 - 1/n, (a+b)/2 + 1/n), \quad \forall n \geq n_0. \\ -1, & x \in [(a+b)/2 + 1/n, b], \end{cases}$$

Then $|f_n(x)| \leq 1 \; \forall x \in [a,b]$, i.e., $\|f_n\| \leq 1$, and $x^*(f_n) = b - a + 2/n \leq \|x^*\| \|f_n\| \leq \|x^*\| \; \forall n \geq n_0$. For $n \to \infty$ we obtain that $b - a \leq \|x^*\|$, and therefore $\|x^*\| = b - a$.

ii) Suppose that x^* achieve its norm on the unit ball of $C[a,b]$, i.e., there is $f \in C[a,b]$ with $\|f\| \leq 1$ such that $x^*(f) = \|x^*\| = b - a$, i.e., f is continuous, $|f(x)| \leq 1 \; \forall x \in [a,b]$, that is, $-1 \leq f(x) \leq 1 \; \forall x \in [a,b]$ and

$$\int_a^{(a+b)/2} f(x)dx - \int_{(a+b)/2}^b f(x)dx = \int_a^b 1 dx.$$

3.2 Solutions

Then
$$\int_a^{(a+b)/2} (f(x)-1)dx = \int_{(a+b)/2}^b (f(x)+1)dx.$$

But $f(x) - 1 \leq 0 \ \forall x \in [a,b]$ and $f(x) + 1 \geq 0 \ \forall x \in [a,b]$. Thus
$$0 \geq \int_a^{(a+b)/2} (f(x)-1)dx = \int_{(a+b)/2}^b (f(x)+1)dx \geq 0,$$

i.e.,
$$\int_a^{(a+b)/2} (f(x)-1)dx = \int_{(a+b)/2}^b (f(x)+1)dx = 0,$$

and therefore $f(x) - 1 = 0 \ \forall x \in [a, (a+b)/2]$, $f(x) + 1 = 0 \ \forall x \in [(a+b)/2, b]$, and then $f((a+b)/2) = 1$, $f((a+b)/2) = -1$, a contradiction.

9. See [40, exercise 11.26].
i) Obviously x^* is linear. Also
$$|x^*(x)| \leq \sum_{n=1}^\infty \frac{|x_n|}{2^{n-1}} \leq \|x\| \sum_{n=1}^\infty \frac{1}{2^{n-1}} = 2\|x\| \quad \forall x \in c_0,$$

i.e., x^* is continuous and $\|x^*\| \leq 2$. For $n \in \mathbb{N}$ let $x = (1,1,\ldots,1,0,0,\ldots) \in c_0$ (n times 1). Then $x^*(x) = \sum_{k=1}^n \frac{1}{2^{k-1}}$, and then
$$\sum_{k=1}^n \frac{1}{2^{k-1}} = |x^*(x)| \leq \|x^*\| \|x\| = \|x^*\|.$$

For $n \to \infty$, $\sum_{n=1}^\infty \frac{1}{2^{n-1}} \leq \|x^*\|$, i.e., $2 \leq \|x^*\|$.

ii) Suppose that x^* achieve its norm on the unit ball of c_0, i.e., there is $x = (x_n)_{n \in \mathbb{N}} \in c_0$ such that $\|x\| \leq 1$, $x^*(x) = \|x^*\| = 2$. Then $\sum_{n=1}^\infty \frac{x_n}{2^{n-1}} = 2$, and therefore
$$2 = \left|\sum_{n=1}^\infty \frac{x_n}{2^{n-1}}\right| \leq \sum_{n=1}^\infty \frac{|x_n|}{2^{n-1}} \leq \sum_{n=1}^\infty \frac{1}{2^{n-1}} = 2.$$

Thus $\sum_{n=1}^\infty \frac{|x_n|}{2^{n-1}} = \sum_{n=1}^\infty \frac{1}{2^{n-1}}$, that is, $\sum_{n=1}^\infty \frac{1-|x_n|}{2^{n-1}} = 0$. But $1 - |x_n| \geq 0 \ \forall n \in \mathbb{N}$, and therefore
$$0 \leq \frac{1-|x_n|}{2^{n-1}} \leq \sum_{k=1}^\infty \frac{1-|x_k|}{2^{k-1}} = 0,$$

i.e., $1 - |x_n| = 0$, $|x_n| = 1 \ \forall n \in \mathbb{N}$, a contradiction, because $(x_n)_{n \in \mathbb{N}} \in c_0$, i.e., $|x_n| \to 0$.

10. See [23, exercise 5.7, section 5].

Let $f \in c_0^*$, $f \neq 0$. Since $c_0^* = l_1$ there is $\xi = (\xi_n)_{n \in \mathbb{N}} \in l_1$ such that $f(x) = \sum_{n=1}^{\infty} x_n \xi_n$ $\forall x \in c_0$ and $\|f\| = \sum_{n=1}^{\infty} |\xi_n|$. If f achieve its norm on the closed unit ball of c_0, i.e., $\exists x \in c_0$ with $\|x\| \leq 1$ such that $f(x) = \|f\|$, then

$$\|f\| = |f(x)| \leq \sum_{n=1}^{\infty} |x_n| |\xi_n| \leq \sum_{n=1}^{\infty} |\xi_n| = \|f\|.$$

Thus $\|f\| = \sum_{n=1}^{\infty} |x_n| |\xi_n| = \sum_{n=1}^{\infty} |\xi_n|$, and therefore $\sum_{n=1}^{\infty} |\xi_n| (1 - |x_n|) = 0$. Since

$$0 \leq (1 - |x_n|) |\xi_n| \leq \sum_{k=1}^{\infty} |\xi_k| (1 - |x_k|) = 0$$

we obtain that $(1 - |x_n|) |\xi_n| = 0 \; \forall n \in \mathbb{N}$. Let $A = \{n \in \mathbb{N} \mid \xi_n \neq 0\}$. Suppose that A is infinite. Then there is a subsequence $(k_n)_{n \in \mathbb{N}}$ of $\mathbb{N}$ such that $k_n \in A \; \forall n \in \mathbb{N}$. Then $|x_{k_n}| = 1 \; \forall n \in \mathbb{N}$. Since $x_n \to 0$, we obtain that $x_{k_n} \to 0$, a contradiction. Therefore A is finite ($\xi \in c_{00}$) and suppose, for example, that $A = \{1, 2, 3, ..., n\}$. Then $\xi_k = 0 \; \forall k \notin A$ and therefore $f(x) = x_1 \xi_1 + \cdots + x_n \xi_n \; \forall x \in c_0$. Then $\|f\| = |\xi_1| + \cdots + |\xi_n|$ and f achieve its norm at $(\text{sgn}(\xi_1), ..., \text{sgn}(\xi_n), 0, ...)$, element from the unit ball of c_0.

11. i) Let $\xi_n = 1 - 1/n \; \forall n \in \mathbb{N}$. Then $|\xi_n| = \xi_n \nearrow 1$, and $1 = \lim_{n \to \infty} \xi_n = \sup_{n \in \mathbb{N}} \xi_n = \sup_{n \in \mathbb{N}} |\xi_n|$, i.e., $(\xi_n)_{n \in \mathbb{N}} \in l_\infty$. We obtain that $x^* : l_1 \to \mathbb{R}$ is a linear and continuous functional and that $\|x^*\| = \sup_{n \in \mathbb{N}} |\xi_n| = 1$.

ii) See [23, exercise 5.3, section 5].
Suppose that there is $x \in l_1$ with $\|x\| \leq 1$ such that $\|x^*\| = x^*(x) = 1$. Then

$$1 = \left| \sum_{n=1}^{\infty} \left(1 - \frac{1}{n}\right) x_n \right| \leq \sum_{n=1}^{\infty} \left(1 - \frac{1}{n}\right) |x_n| \leq \sum_{n=1}^{\infty} |x_n| = \|x\| = 1.$$

Thus $\sum_{n=1}^{\infty} (1 - 1/n) |x_n| = \sum_{n=1}^{\infty} |x_n| = 1$, therefore $\sum_{n=1}^{\infty} (1/n) |x_n| = 0$, and we obtain that $|x_n| = 0 \; \forall n \in \mathbb{N}$, i.e., $\|x\| = \sum_{n=1}^{\infty} |x_n| = 0$, $x = 0$. Thus $x^*(x) = x^*(0) = 0$, a contradiction.

12. i) It is obvious from the general theory.
ii) Suppose that there is $x \in l_1$ with $\|x\| \leq 1$ such that $x^*(x) = \|x^*\| = M$. Then

$$M = \sum_{n=1}^{\infty} \xi_n x_n = \left| \sum_{n=1}^{\infty} \xi_n x_n \right| \leq \sum_{n=1}^{\infty} |\xi_n| |x_n| \leq M \sum_{n=1}^{\infty} |x_n| = M \|x\| \leq M.$$

Thus $M \sum_{n=1}^{\infty} |x_n| = \sum_{n=1}^{\infty} |\xi_n| |x_n|$, that is, $\sum_{n=1}^{\infty} (M - |\xi_n|) |x_n| = 0$. Since $M - |\xi_n| \geq 0$ $\forall n \in \mathbb{N}$, we obtain that $(M - |\xi_n|) |x_n| = 0 \; \forall n \in \mathbb{N}$. But $M \neq |\xi_n| \; \forall n \in \mathbb{N}$, and

3.2 Solutions

therefore $|x_n| = 0 \ \forall n \in \mathbb{N}$, i.e., $\|x\| = \sum_{n=1}^{\infty} |x_n| = 0$, $x = 0$, $x^*(x) = \|x^*\| = M = 0$, a contradiction.

13. $\|x^*\| = \sup_{n \in \mathbb{N}} 1/n = 1$, and $x^*(e_1) = 1 = \|x^*\|$ for $e_1 = (1, 0, 0, ...) \in l_1$.

14. '(i) $\Rightarrow$ (ii)' See [40, exercise 13.16].
Let $x^* \in X^*$, $x^* \neq 0$. Using a consequence of the Hahn–Banach theorem we obtain that there is $x^{**} \in X^{**}$ such that $\|x^{**}\| = 1$ and $x^{**}(x^*) = \|x^*\|$. But X is reflexive and therefore the canonical embedding $K_X : X \to X^{**}$, $K_X(x) = \widehat{x}$ is surjective, and then there is $x \in X$ such that $\widehat{x} = x^{**}$. Then $1 = \|x^{**}\| = \|\widehat{x}\| = \|x\|$, and $x^*(x) = \widehat{x}(x^*) = x^{**}(x^*) = \|x^*\|$, i.e., x^* achieve its norm at x.
'(ii) $\Rightarrow$ (i)' For a proof, see [17, chapter I, section 3, theorem 3.2].

15. See [40, exercise 13.17].
We use exercises 8, 9 and 12, and the easy part of the James theorem (see exercise 14).

16. See [20, exercise 43, chapter V].
i) We know that if $A \subseteq X$ is a convex set then $\overline{A}^{\text{weak}} = \overline{A}^{\|\cdot\|}$ (the Mazur theorem). Suppose that $x_n \to x_0$ weak. For any $m \in \mathbb{N}$, $(x_n)_{n \geq m}$ converges weak to x_0. Since $x_n \in K_m \ \forall n \geq m$, we obtain that $x_0 \in \overline{K_m}^{\text{weak}} = \overline{K_m}^{\|\cdot\|} = K_m$. Thus $x_0 \in K_m \ \forall m \in \mathbb{N}$, i.e., $\{x_0\} \subseteq \bigcap_{n=1}^{\infty} K_n$.

Let now $y \in \bigcap_{n=1}^{\infty} K_n$. Consider $\varepsilon > 0$ and an element $x^* \in X^*$. Since $x_n \to x_0$ weak there is an $m \in \mathbb{N}$ such that $|x^*(x_n - x_0)| \leq \varepsilon$, for any $n \geq m$. Let then $A = \{x \in X \mid |x^*(x - x_0)| \leq \varepsilon\}$. We obtain that A is a closed convex set, $x_n \in A \ \forall n \geq m$, and therefore $K_m \subseteq A$. Since $y \in K_m$ we obtain that $y \in A$, and therefore $|x^*(y - x_0)| \leq \varepsilon$. Since $\varepsilon > 0$ was arbitrary, passing to the limit for $\varepsilon \to 0$ we obtain that $x^*(y - x_0) = 0$ $\forall x^* \in X^*$, and therefore $y = x_0$.

ii) We first make the following general remark. If X is a linear topological space and $(x_n)_{n \in \mathbb{N}} \subseteq X$, $x_0 \in X$ are such that for any subsequence $(k_n)_{n \in \mathbb{N}}$ of $\mathbb{N}$ we can find a subsequence $(p_n)_{n \in \mathbb{N}}$ such that $x_{k_{p_n}} \to x_0$, then $x_n \to x_0$. Indeed, if we suppose that x_n does not converge towards x_0 then there is a neighborhood V of 0 such that $\forall n \in \mathbb{N} \ \exists k \geq n$ such that $x_k - x_0 \notin V$. Then in a standard way we can construct a subsequence $(k_n)_{n \in \mathbb{N}}$ of $\mathbb{N}$ such that for any n, $x_{k_n} - x_0 \notin V$. By hypothesis there is a subsequence $(p_n)_{n \in \mathbb{N}}$ such that $x_{k_{p_n}} \to x_0$, hence there is an $m \in \mathbb{N}$ such that $x_{k_{p_m}} - x_0 \in V$, and this contradicts our choice for the subsequence $(k_n)_{n \in \mathbb{N}}$.

In order to prove that $x_n \to x_0$ weak we will use the above remark. Consider a subsequence $(k_n)_{n \in \mathbb{N}}$ of $\mathbb{N}$. Since the sequence $(x_{k_n})_{n \in \mathbb{N}} \subseteq X$ is bounded and X is a reflexive space, using the Eberlein–Smulian theorem we obtain that we can find $y \in X$ and a subsequence $a_n = x_{k_{p_n}}$ such that $a_n \to y$ weak. Using (i), $\bigcap_{n \in \mathbb{N}} A_n = \{y\}$, where $A_n = \overline{co}\{a_n, a_{n+1}, ...\}$. We have $k_{p_n} \geq p_n \geq n \ \forall n \in \mathbb{N}$, and therefore $\{a_n, a_{n+1}, ...\} \subseteq \{x_n, x_{n+1}, ...\}$. We obtain that $A_n = \overline{co}\{a_n, a_{n+1}, ...\} \subseteq \overline{co}\{x_n, x_{n+1}, ...\} = K_n$, and therefore $\{y\} = \bigcap_{n \in \mathbb{N}} A_n \subseteq \bigcap_{n \in \mathbb{N}} K_n$. By our hypothesis, $\bigcap_{n \in \mathbb{N}} K_n = \{x_0\}$, and therefore $y = x_0$, i.e., $x_{k_{p_n}} \to x_0$ weak. Now we apply the above remark.

iii) Let $(e_n)_{n\in\mathbb{N}} \subseteq l_1$ be the standard basis. Then the set $\text{co}\{e_n, e_{n+1}, ...\}$ is included in the set of elements from l_1 which have 0 at the first $n-1$ positions, and therefore K_n is included in the set of elements from l_1 which have 0 at the first $n-1$ positions. Then $\bigcap_{n=1}^{\infty} K_n = \{0\}$. But if $x^* : l_1 \to \mathbb{K}$, $x^*(x_1, x_2, ...) = \sum_{n=1}^{\infty} x_n$, i.e., $x^* = (1, 1, ...) \in l_1^* = l_\infty$, then $x^*(e_n) = 1\ \forall n \in \mathbb{N}$ and therefore $(e_n)_{n\in\mathbb{N}}$ does not converge *weak* towards 0.

17. See [20, theorem 23, chapter II, section 4].

Let $y^{**} : Y^* \to \mathbb{K}$ be a linear and continuous functional. For any $x^* : X \to \mathbb{K}$ linear and continuous, we consider its restriction on Y, $x^*|_Y : Y \to \mathbb{K}$. Then $x^*|_Y$ is linear and

$$\|x^*|_Y\|_{Y^*} = \sup_{y \in Y, \|y\| \le 1} |x^*(y)| \le \sup_{x \in X, \|x\| \le 1} |x^*(x)| = \|x^*\|,$$

i.e., $x^*|_Y \in Y^*$, $\|x^*|_Y\|_{Y^*} \le \|x^*\|$. We define $\widetilde{y^{**}} : X^* \to \mathbb{K}$, $\widetilde{y^{**}}(x^*) = y^{**}(x^*|_Y)$ $\forall x^* \in X^*$. Then $\widetilde{y^{**}}$ is well defined and linear. We have

$$\left|\widetilde{y^{**}}(x^*)\right| = |y^{**}(x^*|_Y)| \le \|y^{**}\|\,\|x^*|_Y\|_{Y^*} \le \|y^{**}\|\,\|x^*\|,$$

i.e., $\widetilde{y^{**}} \in X^{**}$. Since X is a reflexive space, there is $x \in X$ such that $\widehat{x} = \widetilde{y^{**}}$, i.e., $\widetilde{y^{**}}(x^*) = x^*(x)\ \forall x^* \in X^*$, and therefore $y^{**}(x^*|_Y) = x^*(x)\ \forall x^* \in X^*$. Suppose that $x \notin Y$. Using a corollary for the Hahn–Banach theorem (see chapter 7), we obtain that there is $x^* \in X^*$ such that $x^*|_Y = 0$, $x^*(x) = 1$, $\|x^*\| = 1/\delta$, $\delta = d(x, Y) > 0$ (Y is closed). But $y^{**}(x^*|_Y) = x^*(x)$, and we obtain a contradiction. Thus $x \in Y$, and therefore there is an $y = x \in Y$ such that $y^{**}(x^*|_Y) = x^*(y)\ \forall x^* \in X^*$. Let now $y^* \in Y^*$, $y^* : Y \to \mathbb{K}$ linear and continuous. Using the Hahn–Banach theorem we obtain an extension $x^* : X \to \mathbb{K}$ for y^*, and therefore $y^{**}(y^*) = y^{**}(x^*|_Y) = x^*(y) = y^*(y)$. We obtain that $y^{**}(y^*) = y^*(y)\ \forall y^* \in Y^*$, i.e., Y is reflexive.

18. See [18, exercise 3 (iv), chapter III].

Suppose that X is a reflexive space, and consider a linear and continuous functional $x^{***} : X^{**} \to \mathbb{K}$. We then construct $x^* : X \to \mathbb{K}$, given by $x^*(x) = x^{***}(\widehat{x})$. Then x^* is linear, and $|x^*(x)| \le \|x^{***}\|\,\|\widehat{x}\| = \|x^{***}\|\,\|x\|$, that is, $x^* \in X^*$, $\|x^*\| \le \|x^{***}\|$. If we prove that $x^{***}(x^{**}) = x^{**}(x^*)\ \forall x^{**} \in X^{**}$, we will obtain that X^* is reflexive. Let $x^{**} \in X^{**}$. Since X is reflexive there is $x \in X$ such that $\widehat{x} = x^{**}$. Then $x^{***}(x^{**}) = x^{***}(\widehat{x}) = x^*(x) = x^{**}(x^*)$.

Suppose now that X^* is reflexive. Since X is a Banach space then $K_X(X) \subseteq X^{**}$ is a closed linear subspace, and suppose that $K_X(X) \ne X^{**}$. Then there is $\widetilde{x}^{**} \in X^{**}$, $\widetilde{x}^{**} \notin K_X(X)$. Since $d(\widetilde{x}^{**}, K_X(X)) > 0$, using a corollary of the Hahn–Banach theorem we obtain that there is $\widetilde{x}^{***} \in X^{***}$ such that $\widetilde{x}^{***}(\widehat{x}) = 0\ \forall x \in X$ and $\widetilde{x}^{***}(\widetilde{x}^{**}) = 1$. But X^* is a reflexive space, and therefore there is $\widetilde{x}^* \in X^*$ such that $\widetilde{x}^{***}(x^{**}) = x^{**}(\widetilde{x}^*)$ $\forall x^{**} \in X^{**}$. Then $\widetilde{x}^*(x) = \widehat{x}(\widetilde{x}^*) = \widetilde{x}^{***}(\widehat{x}) = 0\ \forall x \in X$, and therefore $\widetilde{x}^* = 0$. We obtain that $\widetilde{x}^{***}(\widetilde{x}^{**}) = \widetilde{x}^{**}(0) = 0$, and we contradict $\widetilde{x}^{***}(\widetilde{x}^{**}) = 1$.

19. See [20, exercise 17, chapter II, section 4].

i) We observe that for any $Z \subseteq X$, $Z^\perp \subseteq X^*$ is a closed linear subspace, since $Z^\perp = \bigcap_{x \in Z} \ker \widehat{x}$, where $\widehat{x} : X^* \to \mathbb{K}$, $\widehat{x}(x^*) = x^*(x)$. Then $X^*/Y^\perp$ is well defined as a normed

3.2 Solutions

space. We prove now that $T : X^*/Y^\perp \to Y^*$ is well defined. If $x^* + Y^\perp = z^* + Y^\perp$, then $x^* - z^* \in Y^\perp$, thus $(x^* - z^*)(y) = 0 \; \forall y \in Y$, that is, $x^*(y) = z^*(y) \; \forall y \in Y$, i.e., $x^* |_Y = z^* |_Y$, that is, $T(x^* + Y^\perp) = T(z^* + Y^\perp)$. Evidently, T is linear. We now prove that T is an isometry. We must prove that $\inf\{\|z^*\| \mid (z^* - x^*)|_Y = 0\} = \|x^* |_Y\|_{Y^*} \; \forall x^* \in X^*$. If $z^* \in X^*$, $(z^* - x^*)|_Y = 0$, then $z^* |_Y = x^* |_Y$, and therefore $\|x^* |_Y\| = \|z^* |_Y\|$. But $\|z^* |_Y\| \le \|z^*\|$, and therefore $\|x^* |_Y\| \le \|z^*\|$. Also, for $x^* |_Y \in Y^*$, using the Hahn–Banach theorem (see chapter 7), there is $z^* \in X^*$, $z^* |_Y = x^* |_Y$, $\|z^*\|_{X^*} = \|x^* |_Y\|_{Y^*}$. Thus T is an isometry, and since T is linear we obtain that T is injective. We now prove that T is surjective. Let $y^* \in Y^*$. Again, using the Hahn–Banach theorem we obtain that there is $x^* \in X^*$ such that $x^* |_Y = y^*$, thus $T(x^* + Y^\perp) = y^*$.

ii) As above, S is well defined, that is, $Sx^* \in (X/Y)^* \; \forall x^* \in Y^\perp$. Evidently, S is a linear operator. We prove that S is an isometry, that is, $\sup_{\|x\|<1} |x^*(x)| = \sup_{\|x+Y\|<1} |x^*(x)|$ $\forall x^* \in Y^\perp$. For any $x \in X$, from the definition of the quotient norm, $\|x + Y\| = \inf_{y \in Y} \|x + y\| \le \|x\|$, and therefore we obtain the inequality '$\le$'. Let now $x \in X$ such that $\inf_{y \in Y} \|x + y\| = \|x + Y\| < 1$. There is then $z \in X$ such that $\|z\| < 1$, $z - x \in Y$. Then $x^*(z - x) = 0$, that is, $x^*(z) = x^*(x)$. We obtain that $|x^*(x)| = |x^*(z)| \le \sup_{\|w\|<1}|x^*(w)|$, and since $x + Y \in X/Y$ was arbitrary of norm < 1 we obtain the inequality '$\ge$'. Thus S is an isometry, and since S is linear we obtain that S is injective. Let $\overline{x}^* : X/Y \to \mathbb{K}$ linear and continuous. Define $x^* : X \to \mathbb{K}$, $x^*(x) := \overline{x}^*(x + Y)$, $\forall x \in X$. We have $X \xrightarrow{\pi} X/Y \xrightarrow{\overline{x}^*} \mathbb{K}$, $x^* = \overline{x}^* \circ \pi$, where $\pi : X \to X/Y$ is the canonical surjection, $\pi(x) = x + Y \; \forall x \in X$, and therefore $x^* \in X^*$ since π and $\overline{x}^*$ are linear and continuous operators. Evidently $Sx^* = \overline{x}^*$.

iii) Let $\widehat{y} \in K_X(Y)$. For $x^* \in Y^\perp$, $\widehat{y}(x^*) = x^*(y)$, and $x^*(y) = 0$, since $y \in Y$ and $x^* \in Y^\perp$, and therefore $K_X(Y) \subseteq Y^{\perp\perp}$. Now let $x^{**} \in Y^{\perp\perp}$. Since X is reflexive there is $x \in X$ such that $\widehat{x} = x^{**}$. Thus $\widehat{x} \in Y^{\perp\perp}$, i.e., from the definition of $Y^{\perp\perp}$, $\widehat{x} = 0$ on $Y^\perp$, and, from the definition of $Y^\perp$, $\widehat{x}(x^*) = 0 \; \forall x^* \in X^*$ with $x^* |_Y = 0$. Thus $x^* \in X^*$ and $x^* |_Y = 0$ implies that $x^*(x) = 0$. Using a corollary of the Hahn–Banach theorem we obtain that $x \in \overline{Y}$. Since Y is closed we obtain that $x \in Y$, and therefore $x^{**} \in K_X(Y)$.

iv) Evidently the assertion from (iii) is no longer true if X is not a reflexive space. Consider $X = c_0$, $Y = X$. An element $x^* \in l_1 = c_0^*$ belongs to $Y^\perp$ if and only if $x^* |_{c_0} = 0$, that is, $x^* = 0$. Thus $Y^\perp = \{0\}$, and therefore $Y^{\perp\perp} = \{x^* \in l_1^* \mid x^*(0) = 0\} = l_1^* = l_\infty = c_0^{**}$. From exercise 15 we know that c_0 is not a reflexive space, i.e., $K_{c_0}(c_0) \ne l_\infty$, $K_X(Y) \ne Y^{\perp\perp}$.

20. See [20, exercise 19, chapter II, section 4].
We use the notations from exercise 19. We have the following bijections

$$X/Y \xrightarrow{U} X^{**}/Y^{\perp\perp} \xrightarrow{T} (Y^\perp)^* \xrightarrow{(S^{-1})^*} (X/Y)^{**},$$

where $U(x + Y) = \widehat{x} + Y^{\perp\perp} \; \forall x \in X$. Let $\overline{x}^{**} \in (X/Y)^{**}$, and therefore there is $\overline{x} = x + Y \in X/Y$ such that $\overline{x}^{**} = (S^{-1})^*(T(\widehat{x} + Y^{\perp\perp}))$. For any $\overline{x}^* \in (X/Y)^*$ we have

$$(\overline{x}^{**})(\overline{x}^*) = (S^{-1})^*(T(\widehat{x} + Y^{\perp\perp}))(\overline{x}^*) = (T(\widehat{x} + Y^{\perp\perp}))(S^{-1}\overline{x}^*).$$

We denote $y^* = S^{-1}\overline{x}^*, y^* \in Y^\perp$. Then

$$\overline{x}^{**}(\overline{x}^*) = \left(T\left(\widehat{x} + Y^{\perp\perp}\right)\right)(y^*) = \widehat{x}(y^*) = y^*(x),$$

by the definition of T. But $Sy^* = \overline{x}^*$ and therefore, by our definition for S, $\overline{x}^*(x+Y) = y^*(x)$, and therefore $\overline{x}^{**}(\overline{x}^*) = \overline{x}^*(\overline{x}) \ \forall \overline{x}^* \in (X/Y)^*$, where $\overline{x} = x + Y \in X/Y$. We obtain that X/Y with the quotient norm is a reflexive space.

21. See [20, exercise 20, chapter II, section 4].

First we prove that if $x_0^{**} \in X^{**}$ then there is an $x_0 \in X$ such that $x_0^{**}(y^*) = y^*(x_0)$ $\forall y^* \in Y^\perp$. Let $\pi : X \to X/Y$ be the canonical projection, a linear and continuous operator. Then $\pi^{**} : X^{**} \to (X/Y)^{**}$, thus $\pi^{**}(x_0^{**}) \in (X/Y)^{**}$. But X/Y is reflexive, and therefore there is $\overline{x}_0 \in X/Y$, $\overline{x}_0 = x_0 + Y$, $x_0 \in X$, such that $\pi^{**}(x_0^{**}) = \widehat{\overline{x}_0}$, i.e., $(\pi^{**}(x_0^{**}))(\overline{x}^*) = (\overline{x}^*)(\overline{x}_0) \ \forall \overline{x}^* \in (X/Y)^*$. Using exercise 19 we have the isometrical isomorphism $S : Y^\perp \to (X/Y)^*$. Then $(\pi^{**}(x_0^{**}))(Sy^*) = (Sy^*)(\overline{x}_0) \ \forall y^* \in Y^\perp$, thus $x_0^{**}(\pi^*(Sy^*)) = y^*(x_0) \ \forall y^* \in Y^\perp$. If we denote $\pi^*(Sy^*) = x^* \in X^*$ then for any $x \in X$ we have

$$x^*(x) = [\pi^*(Sy^*)](x) = (Sy^*)(\pi(x)) = y^*(x)$$

by the definition of S, and therefore $x^* = y^*$, that is, $\pi^*(Sy^*) = y^*$. Then $x_0^{**}(y^*) = y^*(x_0) \ \forall y^* \in Y^\perp$. We have proved that for any $x^{**} \in X^{**}$ there is an $x_0 \in X$ such that $x^{**}(y^*) = y^*(x_0) \ \forall y^* \in Y^\perp$. We obtain that $(x^{**} - \widehat{x}_0) \in Y^{\perp\perp}$. Using exercise 19, we have

$$Y^{\perp\perp} \xrightarrow{S} \left(X^*/Y^\perp\right)^* \xrightarrow{(T^{-1})^*} Y^{**},$$

where the operators S and $(T^{-1})^*$ are isometrical isomorphisms. Since Y is a reflexive space, there is $y_0 \in Y$ such that $\left[(T^{-1})^*(S(x^{**} - \widehat{x}_0))\right](y^*) = y^*(y_0) \ \forall y^* \in Y^*$, that is, $[S(x^{**} - \widehat{x}_0)](T^{-1}y^*) = y^*(y_0) \ \forall y^* \in Y^*$. Let now $x^* \in X^*$ arbitrary. Then $T(x^* + Y^\perp) \in Y^*$, thus $[S(x^{**} - \widehat{x}_0)](x^* + Y^\perp) = \left[T(x^* + Y^\perp)\right](y_0) \ \forall x^* \in X^*$. Using the definitions for S and T, we have

$$[S(x^{**} - \widehat{x}_0)](x^* + Y^\perp) = (x^{**} - \widehat{x}_0)(x^*), \quad \left[T(x^* + Y^\perp)\right](y_0) = x^*(y_0),$$

and therefore $x^{**}(x^*) = x^*(x_0 + y_0)$. If we denote $x = x_0 + y_0$, we obtain $x \in X$, depending only on $x^{**} \in X^{**}$, such that $x^{**}(x^*) = x^*(x) \ \forall x^* \in X^*$, and therefore X is a reflexive space.

22. See [20, lemma 8, chapter VI, section 8].

Let $(x_n^*)_{n \in \mathbb{N}} \subseteq Z \setminus \{0\}$ be a countable dense set in Z. For every $n \in \mathbb{N}$ and for every $m \in \mathbb{N}$, $(m/(m+1)) \|x_n^*\| < \|x_n^*\| = \sup_{\|x\|=1} |x_n^*(x)|$, and therefore there is an $x_{n,m} \in X$ such that $\|x_{n,m}\| = 1$ and $|x_n^*(x_{n,m})| \geq (m/(m+1)) \|x_n^*\|$. Let $Y = \text{Sp}\{x_{n,m} \mid n, m \in \mathbb{N}\}$, $Y \subseteq X$. Then Y is a separable space, and let $T : Z \to Y^*$, $T(x^*) = x^*|_Y \ \forall x^* \in Z \subseteq X^*$. For any $x^* \in Z$,

$$\|Tx^*\|_{Y^*} = \sup_{\|y\| \leq 1, y \in Y} |x^*(y)| \leq \sup_{\|x\| \leq 1, x \in X} |x^*(x)| = \|x^*\|_{X^*},$$

3.2 Solutions

thus T is well defined, linear, and continuous. For any $n \in \mathbb{N}$,

$$\|Tx_n^*\|_{Y^*} = \sup_{\|y\|\leq 1,\, y \in Y} |x_n^*(y)| \geq \sup_{m \in \mathbb{N}} |x_n^*(x_{n,m})| \geq \sup_{m \in \mathbb{N}} \frac{m}{m+1} \|x_n^*\|_{X^*} = \|x_n^*\|_{X^*}.$$

Thus $\|Tx_n^*\|_{Y^*} = \|x_n^*\|_{X^*}$ $\forall n \in \mathbb{N}$, i.e., $\|Tx^*\|_{Y^*} = \|x^*\|_{X^*}$ $\forall x^* \in \{x_n^* \mid n \in \mathbb{N}\}$. Let $x^* \in Z$ arbitrary. Since $(x_n^*)_{n \in \mathbb{N}} \subseteq Z$ is a dense set there is a sequence $(a_k)_{k \in \mathbb{N}}$ of elements from the set $(x_n^*)_{n \in \mathbb{N}}$ such that $a_k \to x^*$ in the norm of Z, thus in the norm of X^*. Since T is continuous, $Ta_k \to Tx^*$ in the norm of Y^*. Thus $\|a_k\|_{X^*} \to \|x^*\|_{X^*}$, $\|Ta_k\|_{Y^*} \to \|Tx^*\|_{Y^*}$, $\|Ta_k\|_{Y^*} = \|a_k\|_{X^*}$ $\forall k \in \mathbb{N}$, and we obtain that $\|Tx^*\|_{Y^*} = \|x^*\|_{X^*}$ $\forall x^* \in Z$. Thus Z is isometrically isomorphic with $T(Z) \subseteq Y^*$.

23. See [23, theorem 5.5, section 5].

'(i) $\Rightarrow$ (ii)' For any $n \in \mathbb{N}$ we have $K_n \neq \emptyset$ and we can choose $x_n \in K_n$. Then $(x_n)_{n \in \mathbb{N}} \subseteq K_1$, and since K_1 is bounded we obtain that the sequence $(x_n)_{n \in \mathbb{N}}$ is bounded. Since X is a reflexive space, using the Eberlein–Smulian theorem we obtain that there is $x_0 \in X$ and a subsequence $(k_n)_{n \in \mathbb{N}} \subseteq \mathbb{N}$ such that $x_{k_n} \to x_0$ weak in X. For any $n \in \mathbb{N}$ the sequence $(x_{k_s})_{s \geq n}$ is included in K_{k_n} and since $x_{k_n} \to x_0$ weak we obtain that $x_0 \in \overline{K_{k_n}}^{weak}$ $\forall n \in \mathbb{N}$. But K_{k_n} is a convex and norm closed set, and therefore K_{k_n} is closed in the weak topology (the Mazur theorem). Thus $x_0 \in K_{k_n}$ $\forall n \in \mathbb{N}$. But $k_n \geq n$ $\forall n \in \mathbb{N}$, and therefore $K_{k_n} \subseteq K_n$ $\forall n \in \mathbb{N}$, i.e., $\bigcap_{n=1}^{\infty} K_{k_n} \subseteq \bigcap_{n=1}^{\infty} K_n$. Since $x_0 \in \bigcap_{n=1}^{\infty} K_{k_n}$ we obtain that $x_0 \in \bigcap_{n=1}^{\infty} K_n$, and therefore $\bigcap_{n=1}^{\infty} K_n \neq \emptyset$.

'(ii) $\Rightarrow$ (i)' Let $x^* \in X^* \setminus \{0\}$. For $n \in \mathbb{N}$, we define

$$K_n = \{x \in X \mid \|x\| \leq 1, x^*(x) \geq \|x^*\| - 1/n\}.$$

Since $\|x^*\| - 1/n < \|x^*\|$, there is $a \in X$ with $\|a\| \leq 1$ such that $\|x^*\| - 1/n < |x^*(a)| = x^*(x)$, where $x = a\,\mathrm{sgn}(x^*(a))$, i.e., $x \in K_n \neq \emptyset$. Evidently K_n is a closed convex and bounded set, and $K_{n+1} \subseteq K_n$ $\forall n \in \mathbb{N}$. From (ii) $\bigcap_{n=1}^{\infty} K_n \neq \emptyset$, i.e., there is $x \in \bigcap_{n=1}^{\infty} K_n$, i.e., $\|x\| \leq 1$ and $x^*(x) \geq \|x^*\| - 1/n$ $\forall n \in \mathbb{N}$. For $n \to \infty$, it results that $x^*(x) \geq \|x^*\|$. Then $\|x^*\| \leq x^*(x) = |x^*(x)| \leq \|x^*\|\|x\| \leq \|x^*\|$, i.e., there is $\|x\| = 1$ with $\|x^*\| = x^*(x)$. Using the difficult part of the James theorem (see exercise 14) we obtain that X is a reflexive space.

24. i) $K_n = B_{c_0} \cap H_n$, where B_{c_0} is the closed unit ball in c_0, and

$$H_n = \left\{x \in c_0 \mid x = (x_k)_{k \in \mathbb{N}}, x_k = 1, k = \overline{1,n}\right\} = \bigcap_{k=1}^{n} [p_k = 1],$$

with usual notations, i.e., $p_k(x_1, x_2, \ldots) = x_k$ $\forall x = (x_n)_{n \in \mathbb{N}} \in c_0$. Since $p_k \in c_0^*$ $\forall k \in \mathbb{N}$, H_n is a closed convex set $\forall n \in \mathbb{N}$. Thus K_n is a closed convex bounded set $\forall n \in \mathbb{N}$. The fact that K_n is non-empty $\forall n \in \mathbb{N}$, is obvious (for example $(1, \ldots, 1, 0, 0, \ldots) \in K_n$, 1 on the first n positions). Evidently $K_{n+1} \subseteq K_n$ $\forall n \in \mathbb{N}$, and suppose that $x_0 \in \bigcap_{n=1}^{\infty} K_n$. Then

$x_0 \in K_n$ $\forall n \in \mathbb{N}$, thus $p_n(x_0) = 1$ $\forall n \in \mathbb{N}$, and therefore $x_0 = (1, 1, ...)$. This contradicts $x_0 \in c_0$.

ii) We observe that $e_n \in K_n$, i.e., $K_n \neq \emptyset$. The other properties for K_n are obvious. If, for a contradiction, there is $x = (x_1, x_2, ...) \in \bigcap_{n=1}^{\infty} K_n$, we obtain that $\sum_{k=1}^{\infty} |x_k| \leq 1$ and $\sum_{k=1}^{\infty} (1 - 1/k) x_k \geq 1 - 1/n$ $\forall n \in \mathbb{N}$, and, for $n \to \infty$, $\sum_{k=1}^{\infty} (1 - 1/k) x_k \geq 1$. But

$$1 \leq \sum_{k=1}^{\infty} \left(1 - \frac{1}{k}\right) x_k = \left|\sum_{k=1}^{\infty} \left(1 - \frac{1}{k}\right) x_k\right| \leq \sum_{k=1}^{\infty} \left(1 - \frac{1}{k}\right) |x_k| \leq \sum_{k=1}^{\infty} |x_k| \leq 1,$$

and therefore $x_k = 0$ $\forall k \in \mathbb{N}$, a contradiction.

iii) See [40, exercise 3.45].

For $n \in \mathbb{N}$, for example, the function $f : [0, 1] \to \mathbb{R}$,

$$f(x) = \begin{cases} 0 & \text{if } 0 \leq x \leq a, \\ (x - a)/(a_n - a) & \text{if } a < x \leq a_n, \\ 1 & \text{if } a_n < x \leq 1 \end{cases}$$

belongs to K_n. It is obvious that K_n is a closed convex set, and, since $K_n \subseteq B_{C[0,1]}$, K_n is also bounded. If, for a contradiction, there is $f \in \bigcap_{n=1}^{\infty} K_n$, then $f(a) = 0$ and $f(x) = 1$ $\forall x \in [a_n, 1]$, $\forall n \in \mathbb{N}$. We obtain that $f(a_n) = 1$ $\forall n \in \mathbb{N}$. Since $a_n \to a$ for $n \to \infty$ and f is continuous we obtain that $f(a) = 1$, a contradiction.

25. See [18, theorem 7, chapter 7].

Consider $x^* \in l_\infty^*$ and define $\mu : \mathcal{P}(\mathbb{N}) \to \mathbb{K}$, $\mu(A) = x^*(\chi_A)$ $\forall A \subseteq \mathbb{N}$. If $A_1, ..., A_n \in \mathcal{P}(\mathbb{N})$ are pairwise disjoint sets, then $\chi_{A_1} + \cdots + \chi_{A_n} = \chi_{A_1 \cup \cdots \cup A_n}$, and therefore

$$\mu\left(\bigcup_{i=1}^{n} A_i\right) = x^*(\chi_{A_1 \cup \cdots \cup A_n}) = \sum_{i=1}^{n} x^*(\chi_{A_i}) = \sum_{i=1}^{n} \mu(A_i).$$

Thus $\mu : \mathcal{P}(\mathbb{N}) \to \mathbb{K}$ is finitely additive. The variation $|\mu| : \mathcal{P}(\mathbb{N}) \to [0, \infty]$ is defined by

$$|\mu|(E) = \sup\left\{\sum_{i=1}^{n} |\mu(A_i)| \,\Big|\, \bigcup_{i=1}^{n} A_i = E,\ A_i \cap A_j = \emptyset\ \forall i \neq j\right\} \forall E \subseteq \mathcal{P}(\mathbb{N}).$$

If $|\mu|(\mathbb{N}) < \infty$, we say that μ is a measure with bounded variation. Consider $A_1, ..., A_n \subseteq \mathbb{N}$, pairwise disjoint sets. Then we have

$$\sum_{i=1}^{n} |\mu(A_i)| = \sum_{i=1}^{n} |x^*(\chi_{A_i})| = \sum_{i=1}^{n} x^*(\chi_{A_i}) \operatorname{sgn}(x^*(\chi_{A_i}))$$

$$= x^*\left(\sum_{i=1}^{n} \operatorname{sgn}(x^*(\chi_{A_i})) \chi_{A_i}\right) \leq \|x^*\|,$$

since $\left\|\sum_{i=1}^{n} \operatorname{sgn}(x^*(\chi_{A_i})) \chi_{A_i}\right\|_\infty \leq 1$. We take the supremum over $A_1, ..., A_n \subseteq \mathbb{N}$ and we obtain that $|\mu|(\mathbb{N}) \leq \|x^*\|$.

3.2 Solutions

Consider now a finitely additive measure $\mu : \mathcal{P}(\mathbb{N}) \to \mathbb{K}$, with bounded variation $|\mu|(\mathbb{N}) < \infty$. For any $x \in l_\infty$ of the form $\sum_{i=1}^{n} \alpha_i \chi_{E_i}$, with $\alpha_i \in \mathbb{K}$, $E_i \subseteq \mathbb{N}$ pairwise disjoint sets, we define $x^*(x) = \sum_{i=1}^{n} \alpha_i \mu(E_i)$. If we denote by $X \subseteq l_\infty$ the linear subspace of elements of the form $\sum_{i=1}^{n} \alpha_i \chi_{E_i}$, with $\alpha_i \in \mathbb{K}$, $E_i \subseteq \mathbb{N}$ pairwise disjoint sets, using exercise 21 of chapter 1 we obtain that $X \subseteq (l_\infty, \|\cdot\|_\infty)$ is a dense linear subspace. Evidently x^* is well defined and linear on X, and

$$|x^*(x)| \leq \sum_{i=1}^{n} |\alpha_i| |\mu(E_i)| \leq \|x\|_\infty \sum_{i=1}^{n} |\mu(E_i)| \leq \|x\|_\infty |\mu|(\mathbb{N}) \quad \forall x \in X.$$

Since $\mathbb{K}$ is a complete space, we can extend x^* to a linear continuous functional on l_∞, denoted again by $x^* : l_\infty \to \mathbb{K}$, such that $\|x^*\| \leq |\mu|(\mathbb{N})$. By our construction $x^*(\chi_E) = \mu(E) \; \forall E \in \mathcal{P}(\mathbb{N})$.

Chapter 4

The distance between sets in Banach spaces

Definition. Let X be a normed space, $A \subseteq X$ a non-empty subset, and let $x \in X$. Then
$$d(x, A) = \text{dist}(x, A) = \inf_{a \in A} d(x, a) = \inf_{a \in A} \|x - a\|$$
is called the *distance* from the element x to the subset A.

Definition. Let X be a normed space, and $A, B \subseteq X$ two non-empty subsets. Then
$$d(A, B) = \text{dist}(A, B) = \inf_{a \in A,\, b \in B} d(a, b) = \inf_{a \in A,\, b \in B} \|a - b\|$$
is called the *distance* from the set A to the set B.

The Riesz Lemma. Let X be a normed space, and $G \subseteq X$ a closed linear subspace with $G \neq X$. Then for any $0 < \varepsilon < 1$ there is $x_\varepsilon \notin G$ with $\|x_\varepsilon\| = 1$ and $d(x_\varepsilon, G) \geq 1 - \varepsilon$.

4.1 Exercises

Properties for the distance from a point to a set

1. Let X be a normed space, $A \subseteq X$ a non-empty set, and $f : X \to \mathbb{R}_+$, $f(x) = d(x, A) = \inf_{a \in A} \|x - a\|$. Prove that:

i) $d(x, A) = d(x, \overline{A})\ \forall x \in X$. In particular $d(x, A) = 0 \Leftrightarrow x \in \overline{A}$.

ii) $|d(x, A) - d(y, A)| \leq d(x, y)\ \forall x, y \in X$.

iii) $d(\lambda x, \lambda A) = |\lambda|\, d(x, A)\ \forall \lambda \in \mathbb{K}, \forall x \in X$, and $d(x+y, A+B) \leq d(x, A) + d(y, B)$ $\forall x, y \in X, \forall A, B \subseteq X$ non-empty. In particular, if A is a linear subspace then f is a seminorm, and $f(x) \leq \|x\|\ \forall x \in X$.

iv) If A is convex then f is convex.

4.1 Exercises

The distance from a function to a set of polynomials in $C[0,1]$

2. Calculate the following distances in $C([0,1]; \mathbb{R})$.
i) From $x_0(t) = t$ to the linear subspace of all constant functions.
ii) From $x_1(t) = t^2$ to the linear subspace of all the polynomials of degree at most 1.

The distance from a point to the kernel of a norm-one projector

3. i) Let X be a normed space and $P \in L(X)$ a non-null projector (that is, $P^2 = P$, $P \neq 0$). Let $L = \ker P$. Prove that

$$\frac{\|P(x)\|}{\|P\|} \leq d(x, L) \leq \|P(x)\| \quad \forall x \in X.$$

In particular, if $P \in L(X)$ is a norm-one projector, then $d(x, L) = \|P(x)\| \, \forall x \in X$.

ii) Consider $L = \{f \in C[-1,1] \mid f(x) = f(-x) \, \forall x \in [-1,1]\}$, and $S = \{f \in C[-1,1] \mid f(x) = -f(-x) \, \forall x \in [-1,1]\}$. Using (1) calculate $d(e^x, L)$ and $d(e^x, S)$.

iii) Let $a = (a_n)_{n \in \mathbb{N}} \in l_\infty$, and $M_a : l_2 \to l_2$ be the multiplication operator $M_a(x_1, x_2, ...) = (a_1 x_1, a_2 x_2, ...)$. Let $L = \{U \in L(l_2) \mid U = U^*\}$ and $S = \{U \in L(l_2) \mid U = -U^*\}$. Using (i) calculate $d(M_a, L)$ and $d(M_a, S)$.

iv) Let $L = \{U \in L(L_2[0,1]) \mid U = U^*\}$, $S = \{U \in L(L_2[0,1]) \mid U = -U^*\}$, $\varphi \in L_\infty[0,1]$ and $M_\varphi : L_2[0,1] \to L_2[0,1]$, the multiplication operator $M_\varphi(f) = f\varphi$ $\forall f \in L_2[0,1]$. Using (i) calculate $d(M_\varphi, L)$ and $d(M_\varphi, S)$.

The distance from a point to the kernel of a linear and continuous functional

4. Let X be a normed space, $x^* \in X^*$, $x^* \neq 0$ and $L = \ker x^*$. Prove that

$$d(x, L) = \frac{|x^*(x)|}{\|x^*\|} \quad \forall x \in X.$$

A geometric characterization for reflexive spaces

5. i) Let $x^* \in X^*$, $x^* \neq 0$, $c \in \mathbb{K}$, and consider the closed hyperplane $A = \{x \in X \mid x^*(x) = c\}$. Prove that $A \neq \emptyset$ and that

$$d(x, A) = \frac{|x^*(x) - c|}{\|x^*\|} \quad \forall x \in X.$$

ii) Prove that if X is a Banach space then the following assertions are equivalent:
a) X is reflexive;
b) Each closed hyperplane contains a minimal norm element.

The distance from a point to an open half-plane

6. Let X be a real normed space, $x^* \in X^* \setminus \{0\}$, and let $x \in X$ such that $x^*(x) \geq 0$. Prove that

$$d(x, [x^* < 0]) = \frac{x^*(x)}{\|x^*\|}.$$

The distance from an element to a linear subspace

7. Let L be a one-dimensional linear subspace in a Hilbert space H, and $a \in L$, $a \neq 0$. Prove that
$$d(x, L^\perp) = \frac{|\langle x, a \rangle|}{\|a\|} \quad \forall x \in H.$$

8. i) Let $n \in \mathbb{N}$. In the space l_2, calculate $d(x, L_n)$, the distance from the element $x = (1, 0, 0, ...) \in l_2$ to the linear subspace
$$L_n = \left\{ (x_k)_{k \in \mathbb{N}} \in l_2 \;\bigg|\; \sum_{k=1}^n x_k = 0 \right\}.$$
What is $\lim_{n \to \infty} d(x, L_n)$?

ii) Let $(a_k)_{k \in \mathbb{N}} \subseteq \mathbb{K} \setminus \{0\}$ be a sequence for which there exists $\lim_{n \to \infty} |a_{n+1}/a_n|$, and this limit is not equal to 1. Let $n \in \mathbb{N}$ and let
$$L_n = \left\{ (x_k)_{k \in \mathbb{N}} \in l_2 \;\bigg|\; \sum_{k=1}^n a_k x_k = 0 \right\}.$$
Calculate $d(e_n, L_n)$ and $\lim_{n \to \infty} d(e_n, L_n)$.

9. Let $x \in L_1[0, \infty) \cap L_2[0, \infty)$. For $n \in \mathbb{N}$ we denote by
$$L_n = \left\{ f \in L_2[0, \infty) \;\bigg|\; \int_0^n f(t)\,dt = 0 \right\}.$$
Calculate $d(x, L_n)$ and $\lim_{n \to \infty} d(x, L_n)$.

10. Let (S, Σ, μ) be a measure space, $1 < p < \infty$, $1/p + 1/q = 1$, and let $g \in L_q(\mu)$ be a fixed function. Calculate the distance from a function $f \in L_p(\mu)$ to the linear subspace
$$L = \left\{ x \in L_p(\mu) \;\bigg|\; \int_S g(s)\,x(s)\,d\mu(s) = 0 \right\}.$$

11. In the space $L_2[0, 1]$, calculate the distance from the function $f(t) = t^2$ to the linear subspace
$$L = \left\{ x \in L_2[0, 1] \;\bigg|\; \int_0^1 x(t)\,dt = 0 \right\}.$$

The distance between two parallel linear manifolds

12. i) Let X be a normed space, and $V_1 = L + x_1$, $V_2 = L + x_2$ two parallel linear manifolds, i.e., $L \subseteq X$ is a linear subspace and $x_1, x_2 \in X$. Prove that $d(V_1, V_2) = d(x_1 - x_2, L)$.

ii) Using (i) prove that if X is a normed space and $H_1 = \{x \in X \mid x^*(x) = \alpha_1\}$, $H_2 = \{x \in X \mid x^*(x) = \alpha_2\}$ are two closed hyperplanes ($x^* \in X^* \setminus \{0\}$, $\alpha_1, \alpha_2 \in \mathbb{K}$), then
$$d(H_1, H_2) = \frac{|\alpha_1 - \alpha_2|}{\|x^*\|}.$$

4.1 Exercises

iii) Consider two parallel planes in $\mathbb{R}^3$, $\pi_1 : a_1 x + b_1 y + c_1 z + d_1 = 0$, $\pi_2 : a_1 x + b_1 y + c_1 z + d_2 = 0$. Using (ii) deduce the following well known formula from the analytic geometry:
$$d(\pi_1, \pi_2) = \frac{|d_1 - d_2|}{\sqrt{a_1^2 + b_1^2 + c_1^2}}.$$

iv) Using (ii) calculate the distance between the closed hyperplanes
$$H_1 = \left\{ f \in C[0,1] \,\bigg|\, \int_0^1 f(x)\,dx = 1 \right\}$$

and

$$H_2 = \left\{ f \in C[0,1] \,\bigg|\, \int_0^1 f(x)\,dx = -1 \right\}.$$

v) Let H be a Hilbert space, and $a \in H$, $a \neq 0$. Let $A_1 = \{x \in H \mid \langle x, a \rangle = 1\}$ and $A_2 = \{x \in H \mid \langle x, a \rangle = 2\}$. Using (ii) prove that $d(A_1, A_2) = 1/\|a\|$.

The structure for the set of all minimal norm elements

13. Let X be a normed space and $A \subseteq X$ a convex set which has at least two elements of minimal norm. Prove that A contains an infinity of elements of minimal norms. More precisely: $\{x \in A \mid x$ is an element of minimal norm in $A\}$ is convex.

Effective calculation for the set of all minimal norm elements

14. Let
$$H = \left\{ (x_n)_{n \in \mathbb{N}} \in l_1 \,\bigg|\, \sum_{n=1}^{\infty} \frac{1}{n} x_n = 1 \right\}.$$
Prove that in H there is a unique element of minimal norm, and find this element.

15. Let
$$H = \left\{ (x_n)_{n \in \mathbb{N}} \in l_1 \,\bigg|\, x_1 + x_2 + \frac{x_3}{2} + \frac{x_4}{3} + \cdots = 1 \right\}.$$
Prove that H has an infinity of elements of minimal norm, and find these elements.

16. Let
$$H = \left\{ f \in C([0,1];\mathbb{R}) \,\bigg|\, \int_0^1 \frac{f(x)}{x+1}\,dx = 1 \right\}.$$
Prove that H has only one element of minimal norm, and find this element.

17. Let
$$H = \left\{ f \in C([0,1];\mathbb{R}) \,\bigg|\, \int_0^{1/2} x f(x)\,dx = 1 \right\}.$$
Prove that H has an infinity of elements of minimal norm, and find these elements.

18. Let
$$M = \left\{ f \in C([0,1];\mathbb{R}) \,\bigg|\, \int_0^{1/2} f(t)\,dt - \int_{1/2}^1 f(t)\,dt = 1 \right\}.$$

Prove that M is a closed, convex non-empty subset of $C[0,1]$ which has no minimal norm element.

19. Let
$$M = \left\{ f \in L_1[0,1] \,\bigg|\, \int_0^1 f(t)\,dt = 1 \right\}.$$
Prove that $M \subseteq L_1[0,1]$ is a closed, convex non-empty set which has an infinity of elements of minimal norm.

Limitations for the Riesz lemma

20. i) Let X be the closed linear subspace of $C[0,1]$ given by all the functions taking the value 0 at 0. Let $Y \subseteq X$ be the closed linear subspace of all functions $x \in X$ for which $\int_0^1 x(t)\,dt = 0$. Prove that we cannot find an $x \in S_X$ such that $d(x,Y) \geq 1$.
ii) If X is a Hilbert space, and $Y \subseteq X$ is a proper closed linear subspace, prove that there is an $x \in S_X$ such that $d(x, S_Y) = \sqrt{2}$.

21. Let X be a Banach space. Prove that the following assertions are equivalent:
i) X is reflexive;
ii) For each closed linear subspace $Y \subseteq X$, $Y \neq X$, there is an $x \in S_X$ such that $d(x,Y) \geq 1$.

22. Let
$$X = \{(x_n)_{n\in\mathbb{N}} \in c_0 \mid x_{2n} = 0 \,\forall n \in \mathbb{N}\}$$
and
$$Y = \left\{ (x_n)_{n\in\mathbb{N}} \in X \,\bigg|\, \sum_{n=1}^{\infty} \frac{x_n}{2^n} = 0 \right\}.$$
Prove that we cannot find an $x \in S_X$ such that $d(x,Y) \geq 1$.

A proof of the Riesz theorem

23. Let X be a normed space such that B_X is norm compact.
i) Prove that there are $n \in \mathbb{N}$ and $x_1, ..., x_n \in B_X$ such that $B_X \subseteq \bigcup_{i=1}^{n}(x_i + (1/2)B_X)$.
ii) Let $Y = \mathrm{Sp}\{x_1, ..., x_n\}$. Prove that $B_X \subseteq Y + (1/2^n)B_X \,\forall n \in \mathbb{N}$.
iii) Deduce that $X = Y$ and hence that X is finite-dimensional (**The Riesz Theorem**).

Sparse sequences in the unit sphere of an infinite-dimensional normed space

24. Let X be an infinite-dimensional normed space. Using exercise 21 prove that there is a sequence $(x_n)_{n\in\mathbb{N}} \subseteq S_X$ such that $\|x_n - x_m\| \geq 1 \,\forall n \neq m$.

25. i) Let $n \in \mathbb{N}$ and $x_1, ..., x_n$ be n linearly independent elements in a normed space X of dimension $n+1$. Prove that there is $x_{n+1} \in S_X$ such that $\|x_{n+1} - x_i\| > 1 \,\forall 1 \leq i \leq n$.
ii) Using (i) prove that if X is an infinite-dimensional normed space, then there is a sequence $(x_n)_{n\in\mathbb{N}} \subseteq X$ such that $\|x_n\| = 1 \,\forall n \in \mathbb{N}$ and $\|x_n - x_m\| > 1 \,\forall m, n \in \mathbb{N}$, $m \neq n$.

4.1 Exercises

The distance to the unit sphere of a finite-dimensional subspace

26. Let X be a real infinite-dimensional normed space, and $Y \subseteq X$ a finite-dimensional linear subspace.

i) For each $\varepsilon \in (0,1)$, let $y_1^*, ..., y_n^* \in S_{Y^*}$ be an ε-net for S_{Y^*}. Prove that

$$\max_{i=\overline{1,n}} y_i^*(y) \geq (1-\varepsilon)\|y\| \quad \forall y \in Y.$$

ii) Prove that for each $\varepsilon \in (0,1)$ there is an $x \in X$ with $\|x\| = 1$ such that $(1-\varepsilon)\|y\| \leq \|x+y\| \ \forall y \in Y$. Deduce that $d(\text{Sp}\{x\}, S_Y) \geq 1 - \varepsilon$.

An evaluation for the norm of a bijective operator on l_∞^n

27. i) Let X be a real normed space and $x_1, ..., x_n \in X$ with $\|x_i\| = 1$ for each $1 < i \leq n$. Let $\varepsilon \in (0,1)$ and let us suppose that

$$\left\|\sum_{i=1}^n \lambda_i x_i\right\| \leq (1+\varepsilon) \max_{1 \leq i \leq n} |\lambda_i| \quad \forall \lambda_1, ..., \lambda_n \in \mathbb{R}.$$

Prove that

$$\left\|\sum_{i=1}^n \lambda_i x_i\right\| \geq (1-\varepsilon) \max_{1 \leq i \leq n} |\lambda_i| \quad \forall \lambda_1, ..., \lambda_n \in \mathbb{R}.$$

ii) Let $\varepsilon \in (0,1)$ and let X be a real normed space. If $T \in L(l_\infty^n, X)$ is bijective, $\|T\| \leq 1+\varepsilon$, and $\|Te_i\| = 1$, $i = \overline{1,n}$, prove that $\|T^{-1}\| \geq 1/(1-\varepsilon)$.

A finite-dimensional normed space is close to a subspace of an l_∞^n-space

28. Let $(X, \|\cdot\|)$ be a finite-dimensional normed space. Then B_{X^*} is compact and for $0 < \varepsilon < 1$ let $\{x_1^*, ..., x_n^*\} \subseteq B_{X^*}$ be an ε-net for B_{X^*}. Define then $T : X \to l_\infty^n$, $T(x) = (x_1^*(x), ..., x_n^*(x))$. Prove that:

i) T is injective, linear, continuous, and $\|T\| \leq 1$.

ii) Denoting by $G = T(X) \subseteq l_\infty^n$, prove that $T^{-1} : G \to X$ is linear, continuous, and $\|T^{-1}\| \leq 1/(1-\varepsilon)$.

A surjectivity property for l_∞

29. i) Let X be a normed space, and consider a closed linear subspace $G \subseteq X$, $G \neq X$. Prove that $\forall \varepsilon > 0 \ \exists x_\varepsilon \notin G$ such that $d(x_\varepsilon, G) \geq \varepsilon$.

ii) Using (i) prove that if X is an infinite-dimensional normed space then there is a sequence $(x_n)_{n \in \mathbb{N}} \subseteq X$ with the property that: $\forall (\lambda_n)_{n \in \mathbb{N}} \in l_\infty \ \exists x^* \in X^*$ such that $x^*(x_n) = \lambda_n \ \forall n \in \mathbb{N}$.

Geometric characterizations for reflexive spaces

30. i) Let X be a Banach space. Prove that for each closed convex non-empty set $A \subseteq X$, the function $x \longmapsto d(x, A)$ is lower semicontinuous on X with respect to the *weak* topology, i.e., $\forall \lambda \in \mathbb{R}$, $\{x \in X \mid d(x, A) \leq \lambda\}$ is *weak* closed.

ii) Let X be a Banach space. Prove that the following assertions are equivalent:
a) X is reflexive;
b) Each closed convex non-empty subset $A \subseteq X$ has a minimal norm element, i.e., there is an $x_0 \in A$ such that $\|x_0\| = d(0, A) = \inf_{x \in A} \|x\|$.

If, in addition, each point of the unit sphere S_X is an extremal point of B_X then the element x_0 is unique.

iii) Using (i) and (ii) prove that the following assertions are equivalent:
a) X is reflexive;
b) For each closed convex non-empty set $A \subseteq X$ and each closed convex non-empty and bounded set $B \subseteq X$ we can find $x \in A$ and $y \in B$ such that $\|x - y\| = d(A, B)$.

4.2 Solutions

1. i) We will prove that (i) and (ii) are true in any metric space. If (X, d) is a metric space and $A \subseteq X$ a non-empty set, then the distance from the point $x \in X$ to the set A is $d(x, A) = \inf_{a \in A} d(x, a)$.

Since $A \subseteq \overline{A}$ it follows that $d(x, \overline{A}) \leq d(x, A)$. Let $a \in \overline{A}$. Then there is a sequence $(a_n)_{n \in \mathbb{N}} \subseteq A$ such that $a_n \to a$, and therefore $d(x, a_n) \to d(x, a)$. But $d(x, a_n) \geq d(x, A)$ $\forall n \in \mathbb{N}$. Passing to the limit for $n \to \infty$ we obtain that $d(x, a) \geq d(x, A)$ $\forall a \in \overline{A}$, hence $d(x, \overline{A}) \geq d(x, A)$.

If $d(x, A) = 0$, then $d(x, A) < 1/n$ $\forall n \in \mathbb{N}$, i.e., there is a sequence $(a_n)_{n \in \mathbb{N}} \subseteq A$ such that $d(a_n, x) < 1/n$ $\forall n \in \mathbb{N}$, i.e., $d(a_n, x) \to 0$, $x \in \overline{A}$. Conversely, if $x \in \overline{A}$, then it is clear that $d(x, \overline{A}) = 0$, and therefore $d(x, A) = 0$.

ii) We have
$$\begin{aligned} d(x, A) &= \inf_{a \in A} d(x, a) \leq \inf_{a \in A} (d(x, y) + d(y, a)) \\ &= \inf_{a \in A} d(y, a) + d(x, y) = d(y, A) + d(x, y), \end{aligned}$$

hence $d(x, A) - d(y, A) \leq d(x, y)$. Changing x with y, we obtain that $d(y, A) - d(x, A) \leq d(y, x) = d(x, y)$. Then $|d(x, A) - d(y, A)| \leq d(x, y)$.

iii) Let $x \in X$, $\lambda \in \mathbb{K}$. We have
$$d(\lambda x, \lambda A) = \inf_{a \in A} \|\lambda x - \lambda a\| = \inf_{a \in A} |\lambda| \|x - a\| = |\lambda| \inf_{a \in A} \|x - a\| = |\lambda| d(x, A).$$

Let $a \in A$, $b \in B$. Then $a + b \in A + B$, from whence
$$d(x + y, A + B) \leq \|(x + y) - (a + b)\| \leq \|x - a\| + \|y - b\|.$$

Taking the infimum over $a \in A$ and $b \in B$ we obtain that $d(x + y, A + B) \leq d(x, A) + d(y, B)$.

4.2 Solutions

If A is a linear subspace then $\lambda A = A \ \forall \lambda \neq 0$, from whence we obtain that $d(\lambda x, A) = |\lambda| d(x, A) \ \forall \lambda \neq 0$. For $\lambda = 0$, we have $d(\lambda x, A) = d(0, A) = 0 = |\lambda| d(x, A)$, since $0 \in A$. Since A is a linear subspace, $A + A = A$, from whence $d(x+y, A) \leq d(x, A) + d(y, A)$. In addition, we have $d(x, A) \leq d(x, 0) = \|x\|$ because $0 \in A$.

iv) Let $x_1, x_2 \in X$, and consider $0 < \lambda < 1$. Let $a_1, a_2 \in A$. Then since A is convex, $\lambda a_1 + (1 - \lambda) a_2 \in A$, from whence

$$\begin{aligned} d(\lambda x_1 + (1 - \lambda) x_2, A) &\leq d(\lambda x_1 + (1 - \lambda) x_2, \lambda a_1 + (1 - \lambda) a_2) \\ &= \|\lambda x_1 + (1 - \lambda) x_2 - \lambda a_1 - (1 - \lambda) a_2\| \\ &\leq \lambda \|x_1 - a_1\| + (1 - \lambda) \|x_2 - a_2\|, \end{aligned}$$

i.e.,

$$d(\lambda x_1 + (1 - \lambda) x_2, A) \leq \lambda \|x_1 - a_1\| + (1 - \lambda) \|x_2 - a_2\| \ \forall a_1, a_2 \in A.$$

Taking the infimum over $a_1 \in A$ and $a_2 \in A$, we obtain that

$$d(\lambda x_1 + (1 - \lambda) x_2, A) \leq \lambda d(x_1, A) + (1 - \lambda) d(x_2, A).$$

2. We denote by P_0 the linear subspace of all the polynomials of degree zero and with P_1 the linear subspace of all polynomials of degree ≤ 1.

i) We have: $d(x_0, P_0) = \inf \{\|x_0 - k\|_\infty \mid k \in \mathbb{R}\} = \inf_{k \in \mathbb{R}} \sup_{t \in [0,1]} |t - k|$. Since $\sup_{t \in [0,1]} |t - 1/2| = 1/2$ and $\sup_{t \in [0,1]} |t - k| > 1/2 \ \forall k \neq 1/2$, we obtain that $d(x_0, P_0) = 1/2$, and this distance is attained only at one element of P_0, the constant function $1/2$.

ii) We have:

$$d(x_1, P_1) = \inf \{\|x_1 - (at + b)\|_\infty \mid a, b \in \mathbb{R}\} = \inf_{a, b \in \mathbb{R}} \sup_{t \in [0,1]} |t^2 - at - b|.$$

Let us denote $T_{a,b} = \sup_{t \in [0,1]} |t^2 - at - b|$, $a, b \in \mathbb{R}$. Then:

a) If $a \in (-\infty, 0]$ then the vertex of the parabole $t \longmapsto t^2 - at - b$ is not in the interval $(0, 1)$, and then $T_{a,b} = \max \{|1 - a - b|, |b|\}$. Since $a \in (-\infty, 0]$ it follows that $T_{a,b} \geq 1/2$ in this case.

b) If $a \in [2, \infty)$ then again the vertex of the parabole $t \longmapsto t^2 - at - b$ is not in the interval $(0, 1)$, and then $T_{a,b} = \max \{|1 - a - b|, |b|\}$. Since $a \in [2, \infty)$, again $T_{a,b} \geq 1/2$.

c) Let us suppose now that $a \in (0, 2)$. Then the vertex of the parabole is in the interval $(0, 1)$, and hence

$$T_{a,b} = \max \{|1 - a - b|, |b|, |a^2/4 + b|\}.$$

If $a = 1$, then $T_{1,b} = \max \{|b|, |1/4 + b|\}$. For $b = -1/8$, $T_{1,-1/8} = 1/8$. If $b > -1/8$, then $1/4 + b > 1/8$ and $T_{a,b} > 1/8$, and if $b < -1/8$ then $|b| > 1/8$ and $T_{1,b} > 1/8$. Let us suppose now that there is an $a \in (0, 2)$, $a \neq 1$, and a $b \in \mathbb{R}$ such that $T_{a,b} \leq 1/8$. Then $|1 - a - b| \leq 1/8$, $|b| \leq 1/8$ and $|a^2/4 + b| \leq 1/8$. Since $\max \{|b|, |a^2/4 + b|\} \geq a^2/8$, it follows that $1/8 \geq a^2/8$, and since $a \in (0, 2)$, $a \neq 1$, we obtain that $a \in (0, 1)$. If $b > 0$, then $a^2/4 \leq 1/8$, i.e., $a \leq 1/\sqrt{2}$, and since $b \leq 1/8$, it follows that $a + b \leq 1/\sqrt{2} + 1/8$.

Then $1/8 \geq |1-(a+b)| = 1-(a+b) \geq 1-\left(1/\sqrt{2}+1/8\right)$, which is false. If $b \leq 0$, then if $a^2/4+b < 0$, we have $a^2/4 < -b = |b| \leq 1/8$, i.e., $a \leq 1/\sqrt{2}$, which is false, by the above calculations. Hence $a^2/4+b \geq 0$, and then $a^2/4+b \leq \frac{1}{8}$ (1). We have $a \in (0,1)$, $b \leq 0$ and then $|1-a-b| = 1-a-b \leq 1/8$ (2). Adding (1) and (2) we obtain that $1-a+a^2/4 \leq 1/4$ and then $a^2-4a+3 \leq 0$, which is false, since $a \in (0,1)$.

We obtain that $d(x_1, P_1) = 1/8$.

3. i) Let $y \in L = \ker P$. Then $P(x-y) = P(x) - P(y) = P(x)$, from whence $\|P(x)\| = \|P(x-y)\| \leq \|P\| \|x-y\|$, i.e., $\|P(x)\|/\|P\| \leq \|x-y\| \; \forall y \in L$. Taking the infimum over $y \in L$ we obtain that $\|P(x)\|/\|P\| \leq d(x,L)$. Since P is a projector, $P^2 = P$, $x - P(x) \in \ker P$, from whence $\|x - (x - P(x))\| \geq d(x,L)$, i.e., $d(x,L) \leq \|P(x)\|$.

ii) Let $P : C[-1,1] \to C[-1,1]$, $(Pf)(x) = (f(x) - f(-x))/2$. Then $\ker P = L$ and P is a projector with $\|P\| = 1$ (this is easy to prove, see exercise 16 of chapter 5). Using (i) we have $d(e^x, L) = \|P(e^x)\|$. Since $P(e^x) = (e^x - e^{-x})/2$ we have $\|P(e^x)\| = (e^2-1)/(2e)$.

Let now $P : C[-1,1] \to C[-1,1]$, $(Pf)(x) = (f(x) + f(-x))/2$. Then again P is a norm-one projector. Using (i) we have that $d(e^x, S) = \|P(e^x)\|$. But $P(e^x) = (e^x + e^{-x})/2$, and hence $\|P(e^x)\| = (e^2+1)/(2e)$.

For (iii) and (iv) we use the same notations as for exercises 14 and 26 from chapter 12.

iii) Let $P : L(l_2) \to L(l_2)$, $P(U) = (U - U^*)/2$. We have $P^2(U) = P(V) = (V - V^*)/2$, where $V = P(U)$, $V^* = (U^* - U^{**})/2 = (U^* - U)/2 = -P(U)$, and therefore P is a projector. We have $\|P(U)\| \leq (\|U\| + \|U^*\|)/2 = \|U\|$, i.e., $\|P\| \leq 1$, and then since P is a non-null projector we have $\|P\| = 1$. Also $\ker P = L$, and using (i) we obtain that $d(M_a, L) = \|P(M_a)\|$. We have

$$P(M_a)(x_1, x_2, ...) = \left(\frac{M_a - M_a^*}{2}\right)(x_1, x_2, ...) = (ib_1 x_1, ..., ib_n x_n, ...),$$

where $b_k = \Im a_k$, i.e., $P(M_a)$ is a multiplication operator. Therefore, $\|P(M_a)\| = \sup_{k \in \mathbb{N}} |b_k| = \sup_{k \in \mathbb{N}} |\Im a_k|$, i.e., $d(M_a, L) = \sup_{k \in \mathbb{N}} |\Im a_k|$.

We take now $P(U) = (U + U^*)/2$ and let $S = \ker P$. Then from (i) since P is a norm-one projector we have $d(M_a, S) = \|P(M_a)\|$. Since $P(M_a)(x_1, x_2, ...) = (x_1 \Re a_1, ..., x_n \Re a_n, ..)$ is a multiplication operator, then $\|P(M_a)\| = \sup_{k \in \mathbb{N}} |\Re a_k|$, and therefore $d(M_a, S) = \sup_{k \in \mathbb{N}} |\Re a_k|$.

iv) By (i), $d(M_\varphi, L) = \|P(M_\varphi)\|$, where $P(V) = (V - V^*)/2$ for all $V \in L(L_2[0,1])$. Therefore $(P(M_\varphi))(f) = i(\Im\varphi) f$ for all $f \in L_2[0,1]$. Then

$$\|P(M_\varphi)\| = \|\Im\varphi\|_{L_\infty[0,1]}.$$

Analogously we obtain that $d(M_\varphi, L) = \|\Re\varphi\|_{L_\infty[0,1]}$.

4. Let $x \in X$ and consider $\ell \in L = \ker x^*$. Then $x^*(\ell) = 0$ and $|x^*(x - \ell)| \leq \|x^*\| \|x - \ell\|$, and therefore $|x^*(x)|/\|x^*\| \leq \|x - \ell\| \; \forall \ell \in L$, from whence $d(x, L) \geq |x^*(x)|/\|x^*\|$. Since $x^* \neq 0$ then $\|x^*\| > 0$, and let $0 < \varepsilon < \|x^*\|$. Then

$$0 < \|x^*\| - \varepsilon < \|x^*\| = \sup_{\|x\|=1} |x^*(x)|,$$

4.2 Solutions

from whence it follows that there is $x_\varepsilon \in X$, $\|x_\varepsilon\| = 1$, such that $\|x^*\| - \varepsilon < |x^*(x_\varepsilon)| \leq \|x^*\|$. Let $y_\varepsilon = x_\varepsilon/x^*(x_\varepsilon)$ and $a_\varepsilon = x - y_\varepsilon x^*(x)$. Then $a_\varepsilon \in L = \ker x^*$, from whence

$$d(x, L) \leq \|x - a_\varepsilon\| = \|y_\varepsilon\| \, |x^*(x)| = \frac{\|x_\varepsilon\|}{|x^*(x_\varepsilon)|} \, |x^*(x)| = \frac{|x^*(x)|}{|x^*(x_\varepsilon)|},$$

i.e., $d(x, L) \leq |x^*(x)| / |x^*(x_\varepsilon)| \ \forall \varepsilon > 0$. Since $\lim_{\varepsilon>0, \varepsilon\to 0} |x^*(x_\varepsilon)| = \|x^*\|$, we obtain that $d(x, L) \leq |x^*(x)| / \|x^*\|$.

5. i) Since $x^* \neq 0$ there is an $x \in X$ such that $x^*(x) \neq 0$, from whence $a = cx/x^*(x) \in A$, and therefore $A \neq \emptyset$. Then $A = a + \ker x^*$. Using exercise 4 we obtain that

$$\begin{aligned} d(x, A) &= \inf\{\|x - a - y\| \mid y \in \ker x^*\} = d(x - a, \ker x^*) \\ &= \frac{|x^*(x - a)|}{\|x^*\|} = \frac{|x^*(x) - c|}{\|x^*\|}. \end{aligned}$$

ii) '(b) $\Rightarrow$ (a)' Let $x^* \in X^* \setminus \{0\}$ and let $\alpha \in \mathbb{K} \setminus \{0\}$. Consider $H = [x^* = \alpha]$, $H \neq \emptyset$. Then

$$\inf_{x \in H} \|x\| = d(0, H) = \frac{|x^*(0) - \alpha|}{\|x^*\|} = \frac{|\alpha|}{\|x^*\|}$$

(the last equality by (i)). From our hypothesis (b), there is an $a \in H \setminus \{0\}$ such that $\|a\| = \inf_{x \in H} \|x\| = |\alpha| / \|x^*\|$. But $a \in H$, from whence $x^*(a) = \alpha$, and then $\|a\| = |x^*(a)| / \|x^*\|$. Then $\|x^*\| = x^*(u)$, where $u = (\mathrm{asgn}(x^*(a))) / \|a\|$, i.e., x^* attains its norm on B_X. Using the James theorem (the difficult part of the James theorem, see exercise 14 of chapter 3) we obtain that X is reflexive.

'(a) $\Rightarrow$ (b)' Let $x^* \in X^*$, $x^* \neq 0$, and let $H = [x^* = \alpha] \neq \emptyset$, where $\alpha \neq 0$. Since by hypothesis X is reflexive, using the James theorem (the easy part of the James theorem) there is $u \in S_X$ such that $x^*(u) = \|x^*\|$. Then $x^*(u/\|x^*\|) = 1$, and then $a = \alpha u / \|x^*\| \in H$. In addition, by (i)

$$\|a\| = \frac{|\alpha| \|u\|}{\|x^*\|} = \frac{|\alpha|}{\|x^*\|} = d(0, H),$$

i.e., a is a minimal norm element in H. If $\alpha = 0$, then $0 \in H$ is a minimal norm element for H.

Remark. If $h : ax + by - c = 0$ is a line in the Euclidean plane and $M(x_0, y_0) \in \mathbb{R}^2$, then

$$d(M, h) = \frac{|ax_0 + by_0 - c|}{\sqrt{a^2 + b^2}}$$

and if $\pi : ax + by + cz - d = 0$ is a plane in the Euclidean space and $M_0(x_0, y_0, z_0) \in \mathbb{R}^3$, then

$$d(M_0, \pi) = \frac{|ax_0 + by_0 + cz_0 - d|}{\sqrt{a^2 + b^2 + c^2}},$$

i.e., we obtain the well known formulas from the analytic geometry.

Indeed, let $x^* : \mathbb{R}^2 \to \mathbb{R}$, $x^*(x, y) = ax + by$. Then x^* is linear and continuous, with $\|x^*\| = \sqrt{a^2 + b^2}$. In addition $h = \{(x, y) \in \mathbb{R}^2 \mid x^*(x, y) = c\}$ and then

$$d(M, h) = \frac{|x^*(x_0, y_0) - c|}{\|x^*\|} = \frac{|ax_0 + by_0 - c|}{\sqrt{a^2 + b^2}}.$$

The same for the second case.

6. Obviously, $d(x, [x^* \leq 0]) \leq d(x, [x^* = 0])$. Let now $y \in [x^* \leq 0]$, that is, $x^*(y) \leq 0$. Then

$$\|x - y\| = \sup_{\|y^*\|=1} |y^*(x-y)| = \sup_{\|y^*\|=1} y^*(x-y) = \sup_{y^* \neq 0} \frac{y^*(x-y)}{\|y^*\|}$$
$$\geq \frac{x^*(x-y)}{\|x^*\|} = \frac{x^*(x)}{\|x^*\|} - \frac{x^*(y)}{\|x^*\|} \geq \frac{x^*(x)}{\|x^*\|}.$$

From the definition of the distance from a point to a set it follows that $d(x, [x^* \leq 0]) \geq x^*(x)/\|x^*\|$. Since $x^*(x) \geq 0$, using exercise 4 we obtain that $d(x, [x^* = 0]) = x^*(x)/\|x^*\|$ from whence we obtain that

$$d(x, [x^* \leq 0]) = d(x, [x^* = 0]) = \frac{x^*(x)}{\|x^*\|}.$$

Also, since x^* is continuous and $x^* \neq 0$ it follows that $\overline{[x^* < 0]} = [x^* \leq 0]$. Indeed, it is clear that $\overline{[x^* < 0]} \subseteq [x^* \leq 0]$ and $[x^* < 0] \subseteq \overline{[x^* < 0]}$. Let $x \in X$ such that $x^*(x) = 0$. Let $x_0 \in X$ such that $x^*(x_0) \neq 0$, and for every $n \in \mathbb{N}$ let $x_n = x - x_0/(nx^*(x_0))$. Then $x_n \in [x^* < 0]$ and $x_n \to x$, and therefore $\overline{[x^* < 0]} = [x^* \leq 0]$. Using exercise 1(i), we have $d(x, [x^* < 0]) = d(x, [x^* \leq 0])$, from whence we obtain that $d(x, [x^* < 0]) = x^*(x)/\|x^*\|$.

7. Let $f : H \to \mathbb{K}$, $f(x) = \langle x, a \rangle$. Then $L^\perp = \ker f$ and using exercise 4 we obtain that

$$d(x, L^\perp) = d(x, \ker f) = \frac{|f(x)|}{\|f\|} = \frac{|\langle x, a \rangle|}{\|a\|}$$

since $\|f\| = \|a\|$.

8. i) Let $a_n = (1, 1, ..., 1, 0, 0, ...) \in l_2$ (n times 1). Then $L_n = \{x \in l_2 | \langle x, a_n \rangle = 0\} = \{a_n\}^\perp$. Then by exercise 7, we have

$$d(x, L_n) = \frac{|\langle x, a_n \rangle|}{\|a_n\|} = \frac{1}{\sqrt{n}} \to 0.$$

ii) We have $L_n = \{x_n\}^\perp$, where $x_n = (\overline{a_1}, ..., \overline{a_n}, 0, ...)$. Then

$$d(e_n, L_n) = \frac{|\langle e_n, x_n \rangle|}{\|x_n\|} = \frac{|a_n|}{\sqrt{|a_1|^2 + \cdots + |a_n|^2}} = \sqrt{\frac{|a_n|^2}{|a_1|^2 + \cdots + |a_n|^2}}.$$

Let $l = \lim_{n \to \infty} |a_{n+1}/a_n|$. The following cases can occur:

1) $l > 1$. From the well known ratio test for series it follows that the series $\sum_{n=1}^{\infty} |a_n|^2$ diverges. Using now the Stolz–Cesaro lemma, we obtain that

$$\lim_{n \to \infty} \frac{|a_n|^2}{|a_1|^2 + \cdots + |a_n|^2} = \lim_{n \to \infty} \frac{|a_{n+1}|^2 - |a_n|^2}{|a_{n+1}|^2} = 1 - \frac{1}{l^2},$$

4.2 Solutions

and then $\lim_{n\to\infty} d(e_n, L_n) = \sqrt{1 - 1/l^2}$.

2) $l < 1$. Again by the ratio test for series it follows that the series $\sum_{n=1}^{\infty} |a_n|^2$ converges. In particular $a_n \to 0$. Then

$$\lim_{n\to\infty} \frac{|a_n|^2}{|a_1|^2 + \cdots + |a_n|^2} = \frac{0}{s} = 0,$$

where $s = \sum_{n=1}^{\infty} |a_n|^2$.

9. Let $a_n = \chi_{[0,n]} \in L_2[0, \infty)$. Then $L_n = \{a_n\}^\perp$ and using exercise 7 we obtain that

$$d(x, L_n) = \frac{|\langle x, a_n \rangle|}{\|a_n\|} = \frac{|\int_{\mathbb{R}} x(t)\chi_{[0,n]}(t)dt|}{\|\chi_{[0,n]}\|_2} = \frac{|\int_0^n x(t)dt|}{\sqrt{n}}.$$

We have

$$\left|\int_0^n x(t)dt\right| \leq \int_0^n |x(t)|\, dt \leq \int_0^\infty |x(t)|\, dt = \|x\|_1,$$

hence $d(x, L) \leq \|x\|_1/\sqrt{n} \to 0$.

10. Let $x^* : L_p(\mu) \to \mathbb{K}$, $x^*(x) = \int_S gx\, d\mu$. Obviously x^* is linear, and from the Hölder inequality,

$$|x^*(x)| = \left|\int_S gx\, d\mu\right| \leq \int_S |g|\, |x|\, d\mu \leq \left(\int_S |g|^q\, d\mu\right)^{1/q} \left(\int_S |x|^p\, d\mu\right)^{1/p}$$
$$= \|g\|_q \|x\|_p \quad \forall x \in L_p(\mu)$$

it follows that x^* is continuous, i.e., $x^* \in (L_p(\mu))^*$. Using exercise 4 it follows that $d(f, L) = |x^*(f)|/\|x^*\|$ (clearly $\ker x^* = L$). We have $\|x^*\| \leq \|g\|_q$. Taking $x = |g|^{q-1} \operatorname{sgn}(g)$, where $\operatorname{sgn}(g) = \begin{cases} |g|/g & g \neq 0, \\ 0 & g = 0, \end{cases}$ we have

$$x^*(x) = \int_S g\operatorname{sgn}(g) |g|^{q-1}\, d\mu = \int_S |g|^q\, d\mu.$$

Then

$$\int_S |g|^q\, d\mu = |x^*(x)| \leq \|x^*\|\, \|x\|_p = \|x^*\| \left(\int_S |x|^p\, d\mu\right)^{1/p}$$
$$= \|x^*\| \left(\int_S |g|^{p(q-1)}\, d\mu\right)^{1/p} = \|x^*\| \left(\int_S |g|^q\, d\mu\right)^{1/p}.$$

Using the fact that $1 - 1/p = 1/q$ we obtain that $\|g\|_q \leq \|x^*\|$ and therefore $\|x^*\| = \|g\|_q$. Hence

$$d(f, L) = \frac{|\int_S gf\, d\mu|}{\left(\int_S |g|^q\, d\mu\right)^{1/q}}.$$

11. Take $g(t) = 1 \in L_2[0,1]$. Using exercise 10, we obtain that

$$d(f, L) = \frac{\left|\int_0^1 1 \cdot f(t)dt\right|}{\|g\|_2} = \frac{\left|\int_0^1 t^2 dt\right|}{\left(\int_0^1 1 dt\right)^{1/2}} = \frac{1}{3}.$$

12. i) We have

$$d(V_1, V_2) = \inf_{y \in V_2, x \in V_1} \|x - y\| = \inf_{x, y \in L} \|x + x_1 - (y + x_2)\|$$
$$= \inf_{z \in L} \|z - (x_1 - x_2)\| = d(x_1 - x_2, L).$$

ii) Let $H_1 = \ker x^* + x_1$ where $x^*(x_1) = \alpha_1$, and $H_2 = \ker x^* + x_2$ where $x^*(x_2) = \alpha_2$. Then by (i) and exercise 4 we have

$$d(H_1, H_2) = d(x_1 - x_2, \ker x^*) = \frac{|x^*(x_1 - x_2)|}{\|x^*\|} = \frac{|\alpha_1 - \alpha_2|}{\|x^*\|}.$$

iii) We have $x^* : \mathbb{R}^3 \to \mathbb{R}$, $x^*(x, y, z) = a_1 x + b_1 y + c_1 z$, $\|x^*\| = \sqrt{a_1^2 + b_1^2 + c_1^2}$ and we use (ii).

iv) Since for $x^* \in (C[0,1])^*$, $x^*(f) = \int_0^1 f(x)\,dx$, we have $\|x^*\| = 1$, then using (ii) we obtain that $d(H_1, H_2) = 2$.

v) Using (ii), $d(A_1, A_2) = 1/\|a\|$.

13. Let $a_1, a_2 \in A$, $a_1 \neq a_2$ be such that $\|a_1\| = \|a_2\| = \inf_{x \in A} \|x\| = d(0, A) =: d$. Let $0 \leq \lambda \leq 1$ and $a_\lambda = \lambda a_1 + (1-\lambda) a_2 \in A$, since A is convex. Then

$$\|a_\lambda\| \leq \lambda \|a_1\| + (1-\lambda)\|a_2\| = \lambda d + (1-\lambda)d = d = \inf_{x \in A} \|x\| \leq \|a_\lambda\|,$$

i.e., $\|a_\lambda\| = \inf_{a \in A} \|a\|$. Let us now observe that if $\lambda_1 \neq \lambda_2$ then $a_{\lambda_1} \neq a_{\lambda_2}$, since $a_1 \neq a_2$.

14. Using exercise 5(i), $\min_{x \in H} \|x\| = 1/\|x^*\|$, where $x^*(x_1, x_2 \ldots) = \sum_{n=1}^{\infty} x_n/n$, and $\|x^*\| = \sup\{1/n \mid n \in \mathbb{N}\} = 1$. We have $e_1 \in H$ and $\|e_1\| = \min_{x \in H} \|x\| = 1$.

Let now $a = (a_n)_{n \in \mathbb{N}} \in H$ be such that $\|a\| = \min_{x \in H} \|x\| = 1$, i.e., $\sum_{n=1}^{\infty} a_n/n = 1$ and $\sum_{n=1}^{\infty} |a_n| = 1$. Then

$$1 = \sum_{n=1}^{\infty} \frac{1}{n} a_n = \left|\sum_{n=1}^{\infty} \frac{1}{n} a_n\right| \leq \sum_{n=1}^{\infty} \frac{1}{n} |a_n| \leq \sum_{n=1}^{\infty} |a_n| = 1,$$

$$\sum_{n=2}^{\infty} |a_n| \frac{n-1}{n} = 0,$$

from whence $|a_n| = 0 \ \forall n \geq 2$, i.e., $a_n = 0 \ \forall n \geq 2$. Then $a_1 = 1$, hence $a = e_1$.

4.2 Solutions   99

15. We have (see exercise 5(i)) $\inf_{x \in H} \|x\| = 1 = \|e_1\| = \|e_2\|$. Using exercise 13 we obtain that $\lambda e_1 + (1 - \lambda) e_2 = (\lambda, 1 - \lambda, 0, 0, ...)$, $\lambda \in [0, 1]$, are minimal norm elements. Let now $x = (x_n)_{n \in \mathbb{N}} \in H$ be a minimal norm element. Then

$$x_1 + x_2 + \frac{x_3}{2} + \frac{x_4}{3} + \cdots = 1$$

and $|x_1| + |x_2| + |x_3| + \cdots = 1$. We then have

$$\begin{aligned} 1 &= x_1 + x_2 + \frac{x_3}{2} + \cdots = \left| x_1 + x_2 + \frac{x_3}{2} + \cdots \right| \\ &\leq |x_1| + |x_2| + \frac{|x_3|}{2} + \cdots \leq |x_1| + |x_2| + \cdots = 1, \end{aligned}$$

so it follows that $|x_3| = 0, |x_4| = 0, ...$, i.e., $x = (x_1, x_2, 0, ...)$. In addition $x_1 + x_2 = 1$ and $|x_1| + |x_2| = 1$. This easily implies that $x_2 = 1 - \lambda, x_1 = \lambda$, with $\lambda \in [0, 1]$.

16. We will use in the solutions for exercises 16–18 the following well known property: if $f : [a, b] \to [0, \infty)$ is a continuous functions and $\int_a^b f(x)\, dx = 0$, then $f(x) = 0$ $\forall x \in [a, b]$.

Let $x^* : C[0, 1] \to \mathbb{R}$, $x^*(f) = \int_0^1 (f(x)/(x+1))\, dx$. We have $|x^*(f)| \leq \|f\| \int_0^1 (1/(x+1))\, dx = \ln 2 \|f\|$ $\forall f \in C[0, 1]$ and from $|x^*(1)| \leq \|x^*\| \|1\| = \|x^*\|$, $\|x^*\| \geq \int_0^1 (1/(x+1))\, dx = \ln 2$, we obtain that $\|x^*\| = \ln 2$. Let $f_0(x) = 1/\ln 2$. Obviously $f_0 \in H$ and using exercise 5(i) we have

$$\inf_{g \in H} \|g\| = d(0, H) = \frac{|x^*(0) - 1|}{\|x^*\|} = \frac{1}{\ln 2} = \|f_0\|,$$

i.e., f_0 is a minimal norm element.

Let now $f \in H$ be a minimal norm element, i.e., $\|f\| = \min_{g \in H} \|g\| = 1/\ln 2$. Then $|f(x)| \leq 1/\ln 2$ $\forall 0 \leq x \leq 1$, i.e., $-1/\ln 2 \leq f(x) \leq 1/\ln 2$. But $f \in H$, that is, $\int_0^1 (f(x)/(x+1))\, dx = 1$, i.e., $\int_0^1 ((f(x) - 1/\ln 2)/(x+1))\, dx = 0$. Since $1/\ln 2 - f(x) \geq 0$ $\forall x \in [0, 1]$, we obtain that $(1/\ln 2 - f(x))/(x+1) = 0$ $\forall 0 \leq x \leq 1$, that is, $f(x) = 1/\ln 2$ $\forall 0 \leq x \leq 1$.

17. Let $x^* : C[0, 1] \to \mathbb{R}$, $x^*(f) = \int_0^{1/2} x f(x)\, dx$. We have $|x^*(f)| \leq \|f\| \int_0^{1/2} x\, dx = \|f\|/8$, i.e., $\|x^*\| \leq 1/8$. Taking $f_0 : [0, 1] \to \mathbb{R}$, $f_0(x) = 1$ $\forall x \in [0, 1]$, we have $|x^*(f_0)| \leq \|x^*\| \|f_0\| = \|x^*\|$, that is, $\int_0^{1/2} x\, dx \leq \|x^*\|$, and therefore $\|x^*\| = 1/8$. Now using exercise 5(i) we obtain that

$$\inf_{f \in H} \|f\| = d(0, H) = \frac{|x^*(0) - 1|}{\|x^*\|} = 8.$$

If we consider $g = 8 f_0$ then $g \in H$ and $\|g\| = \inf_{f \in H} \|f\|$.

Let now $f \in H$ be a minimal norm element, i.e., $\|f\| = \inf_{g \in H} \|g\| = 8$. Then $|f(x)| \leq 8$ $\forall x \in [0, 1]$, that is, $-8 \leq f(x) \leq 8$ $\forall x \in [0, 1]$. Since $f \in H$ we have $\int_0^{1/2} x f(x)\, dx =$

$1 = \int_0^{1/2} 8x\,dx$, that is, $\int_0^{1/2} x\,(8 - f(x))\,dx = 0$. Then $x\,(8 - f(x)) = 0 \ \forall 0 \leq x \leq 1/2$, i.e., $f(x) = 8 \ \forall 0 < x \leq 1/2$, from whence, since f is continuous, $f(x) = 8 \ \forall 0 \leq x \leq 1/2$. Hence if $f \in H$ is a minimal norm element, then $f(x) = 8 \ \forall 0 \leq x \leq 1/2$, and $|f(x)| \leq 8 \ \forall x \in [0,1]$.

Conversely, let $f \in C[0,1]$ with $f(x) = 8 \ \forall 0 \leq x \leq 1/2$ and $|f(x)| \leq 8 \ \forall 0 \leq x \leq 1$. Then $\|f\| = 8$ and $x^*(f) = \int_0^{1/2} xf(x)\,dx = \int_0^{1/2} 8x\,dx = 1$, i.e., $f \in H$ is a minimal norm element.

18. Let us consider $x^* : C[0,1] \to \mathbb{R}$, $x^*(f) = \int_0^{1/2} f(t)\,dt - \int_{1/2}^1 f(t)\,dt$. Then x^* is linear and from the property

$$|x^*(f)| \leq \int_0^{1/2} |f(t)|\,dt + \int_{1/2}^1 |f(t)|\,dt \leq \|f\|_\infty$$

we obtain that x^* is continuous. Hence M is a hyperplane in $C[0,1]$ and therefore $M \subseteq C[0,1]$ is a closed, convex, and non-empty set. But it is proved in exercise 8 of chapter 3 that $\|x^*\| = 1$, and then, by exercise 5(i), $\inf\{\|f\| \mid f \in M\} = 1$. Suppose that we can find an $f \in M$ such that $\|f\| = 1$. Then

$$\int_0^{1/2} (1 - f(t))\,dt + \int_{1/2}^1 (1 + f(t))\,dt = 0,$$

and the functions $t \longmapsto 1 - f(t)$ and $t \longmapsto 1 + f(t)$ are continuous and positive. It follows that $f(t) = 1$ on $[0,1/2]$, $f(t) = -1$ on $[1/2,1]$, so $f(1/2) = 1$ and $f(1/2) = -1$, contradiction!

19. Let $x^* : L_1[0,1] \to \mathbb{K}$, $x^*(f) = \int_0^1 f(t)\,dt$. Then x^* is linear and since $|x^*(f)| \leq \int_0^1 |f(t)|\,dt = \|f\|_1 \ \forall f \in L_1[0,1]$, we obtain that x^* is continuous with $\|x^*\| \leq 1$. Also, taking $f = 1$ we deduce that $\|x^*\| = 1$. Hence M is a hyperplane in $L_1[0,1]$, therefore a closed, convex, and non-empty set. By exercise 5(i), $\inf\{\|f\|_1 \mid f \in M\} = 1$. For each $a, b \in [0, \infty)$, with $a/2 + b = 1$, define $f_{a,b} : [0,1] \to \mathbb{K}$, $f_{a,b}(t) = at + b$. Then $x^*(f_{a,b}) = 1$, i.e., $f_{a,b} \in M$. Also $\|f_{a,b}\| = \int_0^1 |f_{a,b}(t)|\,dt = 1$.

20. i) See [18, exercise 3 (i), chapter 1].

Suppose that there is $x \in C[0,1]$, $x(0) = 0$ and $|x(t)| \leq 1 \ \forall t \in [0,1]$, such that $d(x,Y) \geq 1$. For each $\alpha > 1$, define $x_\alpha \in C[0,1]$ by

$$x_\alpha(t) = x(t) - \alpha t^{\alpha-1} \int_0^1 x(s)\,ds.$$

Then $x_\alpha(0) = 0 \ \forall \alpha > 1$, and

$$\int_0^1 x_\alpha(t)\,dt = \int_0^1 x(t)\,dt \left(1 - \alpha \int_0^1 t^{\alpha-1}\,dt\right) = 0 \ \forall \alpha > 1.$$

Hence $x_\alpha \in Y \ \forall \alpha > 1$, from whence $\|x_\alpha - x\|_\infty \geq d(x,Y) \geq 1 \ \forall \alpha > 1$, i.e., $\forall \alpha > 1$ there is a $t_\alpha \in [0,1]$ such that $\left|\alpha t_\alpha^{\alpha-1} \int_0^1 x(t)\,dt\right| \geq 1$ (we use the Weierstrass theorem: any real continuous function on a compact set attains its bounds), i.e.,

$\left|\int_0^1 x(t)\,dt\right| \geq 1/(\alpha t_\alpha^{\alpha-1})$, and hence $\left|\int_0^1 x(t)\,dt\right| \geq 1/\alpha \ \forall \alpha > 1$ (obviously, $t_\alpha \neq 0$ $\forall \alpha > 1$). Passing to the limit for $\alpha \to 1$, we obtain that $\left|\int_0^1 x(t)\,dt\right| \geq 1$. Then

$$\int_0^1 (1 - |x(t)|)\,dt = 1 - \int_0^1 |x(t)|\,dt \leq 0$$

and since $1 - |x(t)| \geq 0 \ \forall t \in [0,1]$ and x is continuous it follows that $x(t) = 1 \ \forall t \in [0,1]$, which is false, because $x(0) = 0$.

ii) See [18, exercise 3 (ii), chapter 1].

Since Y is a closed proper linear subspace in X then $Y^\perp \neq \{0\}$, and then there is an element $x \in Y^\perp$ such that $\|x\| = 1$. Then for each $y \in Y$, $\|y\| = 1$ we have $\|x-y\|^2 = \langle x-y, x-y\rangle = \langle x,x\rangle - \langle x,y\rangle - \langle y,x\rangle + \langle y,y\rangle = 2$, i.e., $\|x-y\| = \sqrt{2}$, and therefore $d(x, S_Y) = \sqrt{2}$.

21. See [18, exercise 3 (iii), chapter 1].

'(i) $\Rightarrow$ (ii)' Since $Y \subseteq X$ is a proper linear subspace, there is $x_0 \in X \setminus \{0\}$ such that $x_0 \notin Y$. Using the Hahn–Banach theorem (see chapter 7) and that Y is closed we obtain that there is $x_0^* \in X$ such that $x_0^*|_Y = 0$ and $x_0^*(x_0) \neq 0$. Since X is reflexive, by the easy part of the James theorem (see exercise 14 of chapter 3) we obtain that x_0^* attains its norm on the unit sphere of X, i.e., there is an $x \in X$, $\|x\| = 1$ such that $\|x_0^*\| = x_0^*(x)$. For each $y \in Y$, we have $x_0^*(x-y) = x_0^*(x) = \|x_0^*\|$. Then

$$\|x - y\| = \sup_{\|x^*\| \leq 1} |x^*(x-y)| \geq \frac{x_0^*(x-y)}{\|x_0^*\|} = 1.$$

Hence $\|x - y\| \geq 1 \ \forall y \in Y$ and then it follows that $d(x, Y) \geq 1$.

'(ii) $\Rightarrow$ (i)' Let $x^* \in X^*$, $x^* \neq 0$. Then $Y = \ker x^* \neq X$. Using the hypothesis (ii) there is an $x \in S_X$ such that $d(x, Y) \geq 1$, i.e., $d(x, \ker x^*) \geq 1$. Using now exercise 4 we obtain that $d(x, \ker x^*) = |x^*(x)|/\|x^*\|$. Hence $|x^*(x)| \geq \|x^*\|$, or $\|x^*\| \leq |x^*(x)| \leq \|x^*\| \|x\| = \|x^*\|$, i.e., $u = x \operatorname{sgn}(x^*(x)) \in B_X$ and $x^*(u) = \|x^*\|$. From the difficult part of the James theorem it follows that X is reflexive.

22. Let $x^* : X \to \mathbb{K}$, $x^*((x_n)_{n \in \mathbb{N}}) = \sum_{n=1}^\infty \frac{x_n}{2^n}$. We have $Y = \ker x^*$. We will prove that $\|x^*\| = 2/3$. Indeed, let $x = (x_1, x_2, ...) \in X$. Then $x = (x_1, 0, x_3, 0, ...) \in c_0$ and $x^*(x) = \sum_{n=1}^\infty \frac{x_{2n-1}}{2^{2n-1}}$, from whence

$$|x^*(x)| \leq \sum_{n=1}^\infty \frac{|x_{2n-1}|}{2^{2n-1}} \leq \|x\| \sum_{n=1}^\infty \frac{1}{2^{2n-1}} = \frac{2}{3} \|x\|,$$

that is, $\|x^*\| \leq 2/3$. For any $n \in \mathbb{N}$ let $x = (1, 0, 1, 0, ..., 1, 0, 0, 0, ...) \in X$ (n occasions of 1). Then $x^*(x) = \sum_{k=1}^n 1/2^{2k-1}$ and $\sum_{k=1}^n 1/2^{2k-1} \leq \|x^*\| \|x\| = \|x^*\|$, hence passing to the limit for $n \to \infty$ we obtain that $\sum_{n=1}^\infty 1/2^{2n-1} \leq \|x^*\|$, i.e., $\|x^*\| = 2/3$. Let us suppose that there is an $x \in S_X$ such that $d(x, Y) \geq 1$. Then by exercise 4 we have $|x^*(x)| / \|x^*\| \geq 1$ and therefore there is $u = \text{sgn}(x^*(x)) x \in S_X$ such that $x^*(u) = \|x^*\|$. If $u = (x_1, 0, x_3, 0, ...) \in c_0$, then $\sup_{n \in \mathbb{N}} |x_{2n-1}| = 1$ and $\sum_{n=1}^{\infty} \frac{x_{2n-1}}{2^{2n-1}} = x^*(u) = 2/3$. We have

$$\frac{2}{3} = \left|\sum_{n=1}^{\infty} \frac{x_{2n-1}}{2^{2n-1}}\right| \leq \sum_{n=1}^{\infty} \frac{|x_{2n-1}|}{2^{2n-1}} \leq \sum_{n=1}^{\infty} \frac{1}{2^{2n-1}} = \frac{2}{3},$$

from whence

$$\sum_{n=1}^{\infty} \frac{|x_{2n-1}|}{2^{2n-1}} = \sum_{n=1}^{\infty} \frac{1}{2^{2n-1}},$$

i.e., $|x_{2n-1}| = 1 \ \forall n \in \mathbb{N}$, which is false, since $u = (x_1, 0, x_3, 0, ...) \in c_0$, i.e., $|x_{2n-1}| \to 0$.

Remark. If we take $f : c_0 \to \mathbb{K}$, $f((x_n)_{n \in \mathbb{N}}) = \sum_{n=1}^{\infty} x_n/2^n$, then we have $x^* = f|_X$ and we observe that $\|f|_X\| = 2/3 < \|f\| = 1$.

23. See [18, Notes and Remarks, chapter 1].

i) Denoting by B the open unit ball of X, since $B_X \subseteq \bigcup_{x \in B_X} (x + (1/2) B)$ and using the compactness of B_X it follows that there are $n \in \mathbb{N}$ and $x_1, ..., x_n \in B_X$ such that $B_X \subseteq \bigcup_{i=1}^{n} (x_i + (1/2) B) \subseteq \bigcup_{i=1}^{n} (x_i + (1/2) B_X)$.

ii) By (i) we have $B_X \subseteq Y + (1/2) B_X$. Then using (i) again and that Y is a linear subspace, we deduce that

$$B_X \subseteq Y + \frac{1}{2}\left(Y + \frac{1}{2}B_X\right) = Y + \frac{1}{2^2}B_X.$$

By induction we obtain that $B_X \subseteq Y + (1/2^n) B_X \ \forall n \in \mathbb{N}$.

iii) Let $x \in B_X$. For each $n \in \mathbb{N}$, by (ii) there are $x_n \in B_X$ and $y_n \in Y$ such that $x = y_n + (1/2^n) x_n$. Since $\|x_n\| \leq 1 \ \forall n \in \mathbb{N}$, it follows that $(1/2^n) x_n \to 0$ and hence $y_n \to x$. Thus $x \in \overline{Y}$. But $Y \subseteq X$ is a finite-dimensional linear subspace, thus Y is closed. It follows that $B_X \subseteq Y$ and since $Y \subseteq X$ is a linear subspace, it follows that $X = Y$.

24. See [18, Notes and Remarks, chapter 1].

Let $x_1 \in X$, $\|x_1\| = 1$. Since X is infinite-dimensional, it follows that $\text{Sp}\{x_1\} \neq X$, and let $\widetilde{x}_2 \in X \setminus \text{Sp}\{x_1\}$. We denote by $\widetilde{X}_2 = \text{Sp}\{x_1, \widetilde{x}_2\}$, $X_1 = \text{Sp}\{x_1\}$. We obtain that X_1 is a closed linear subspace (since is finite-dimensional) of a closed linear subspace $\widetilde{X}_2$ of X ($\widetilde{X}_2$ is also finite-dimensional). So we have the following situation: X_1 is a closed proper linear subspace of $\widetilde{X}_2$, where $\widetilde{X}_2$ is finite-dimensional, hence reflexive. Using exercise 21 there is $x_2 \in S_{\widetilde{X}_2}$ such that $\text{dist}(x_2, S_{X_1}) \geq 1$. Hence we can find $x_2 \in X$, $\|x_2\| = 1$ such that $\|x_2 - x_1\| \geq 1$. Let us consider now $X_2 = \text{Sp}\{x_1, x_2\}$. X is infinite-dimensional, so there is $\widetilde{x}_3 \in X \setminus X_2$. Denote by $\widetilde{X}_3 = \text{Sp}\{x_1, x_2, \widetilde{x}_3\}$. Now the same reasoning as above can be applied, and we can find $x_3 \in S_{\widetilde{X}_3}$ such that $\text{dist}(x_3, S_{X_2}) \geq 1$, therefore $\|x_3 - x_1\| \geq 1$, $\|x_3 - x_2\| \geq 1$. It is clear now the inductive process to construct a sequence $(x_n)_{n \in \mathbb{N}} \subseteq S_X$ such that $\|x_n - x_m\| \geq 1 \ \forall n \neq m$.

4.2 Solutions

25. i) See [6, lemma 11, chapter 4].
Since $\dim_\mathbb{K} X = n+1$ there is a $z \in X$ such that $\{x_1, ..., x_n, z\} \subseteq X$ is a basis for X. Because $z \notin \mathrm{Sp}\{x_1, ..., x_n\}$, by the Hahn–Banach theorem there is an $x^* \in X^*$ such that $x^*(x_i) = 0 \; \forall 1 \leq i \leq n$ and $x^*(z) = 1$. Normalizing, i.e., replacing x^* with $x^*/\|x^*\|$ it follows that there is an $x^* \in S_{X^*}$ such that $x^*(x_i) = 0 \; \forall i = \overline{1, n}$. In the same way there is $y^* \in X^*$ such that $y^*(x_i) = 1 \; \forall i = \overline{1, n}$ (for example, for i between 1 and n, since $x_i \notin \mathrm{Sp}\{x_1, ..., x_{i-1}, x_{i+1}, ..., x_n\}$ we can find $x_i^* \in X^*$ such that $x_i^*(x_i) = 1$ and $x_i^* = 0$ on $\mathrm{Sp}\{x_1, ..., x_{i-1}, x_{i+1}, ..., x_n\}$, and then consider $y^* = x_1^* + \cdots + x_n^*$). Since $\|x^*\| = 1$ and S_X is compact it follows that $K := (x^*)^{-1}(\{1\}) \cap S_X$ is a compact (as a closed subset of a compact set) and non-empty set (the Weierstrass theorem!). Now, again from the Weierstrass theorem, there is an $x_{n+1} \in K$ such that $y^*(x_{n+1}) = \min\limits_{x \in K} y^*(x)$. We have $\|x_{n+1} - x_i\| \geq x^*(x_{n+1} - x_i) = 1$, and since $y^*(x_{n+1} - x_i) = y^*(x_{n+1}) - 1 < y^*(x_{n+1}) = \min\limits_{x \in K} y^*(x) \; \forall 1 \leq i \leq n$, it follows that $x_{n+1} - x_i \notin K \; \forall 1 \leq i \leq n$. Then $\|x_{n+1} - x_i\| \neq 1, i = \overline{1, n}$, and hence $\|x_{n+1} - x_i\| > 1, i = \overline{1, n}$.
ii) See [18, Notes and Remarks, chapter 1].
It follows immediately from (i), by induction.

26. i) Since Y is finite-dimensional Y^* is also finite-dimensional and hence $S_{Y^*} = \{y \in Y^* \mid \|y^*\| = 1\}$ is compact. It follows that there is an ε-net $\{y_1^*, ..., y_n^*\}$ for S_{Y^*}, i.e., $S_{Y^*} \subseteq \bigcup\limits_{i=1}^n B(y_i^*, \varepsilon)$. Then for each $y^* \in S_{Y^*}$ there is $1 \leq i \leq n$ such that $\|y^* - y_i^*\| \leq \varepsilon$. If $y \in Y$ we have $|(y^* - y_i^*)(y)| \leq \varepsilon \|y\|$ and hence

$$y^*(y) \leq y_i^*(y) + \varepsilon \|y\| \leq \max_{j=\overline{1,n}} y_j^*(y) + \varepsilon \|y\|.$$

Thus we have proved that $\forall y \in Y$ and $\forall y^* \in S_{Y^*}$ we have

$$y^*(y) \leq \max_{i=\overline{1,n}} y_i^*(y) + \varepsilon \|y\|.$$

Now taking the supremum with respect to $y^* \in S_{Y^*}$ we obtain that $\|y\| \leq \max\limits_{i=\overline{1,n}} y_i^*(y) + \varepsilon \|y\| \; \forall y \in Y$, and hence $\max\limits_{i=\overline{1,n}} y_i^*(y) \geq (1-\varepsilon) \|y\| \; \forall y \in Y$.

ii) See [6, exercise 12, chapter 4].
With the same notations as in (i), for each $1 \leq i \leq n$ let $x_i^* \in X^*$ be a Hahn–Banach extension for y_i^*, i.e., $\|x_i^*\| = \|y_i^*\| = 1$, $x_i^*|_Y = y_i^*$. Since X is infinite-dimensional, there is an $x \in X$ with $\|x\| = 1$ such that $x \in \bigcap\limits_{i=1}^n \ker x_i^*$ (see exercise 16(ii) of chapter 14). Then $\forall y \in Y$ we have

$$\|x + y\| \geq \max_{i=\overline{1,n}} x_i^*(x+y) = \max_{i=\overline{1,n}} x_i^*(y) = \max_{i=\overline{1,n}} y_i^*(y) \geq (1-\varepsilon) \|y\|.$$

From this inequality we deduce that $\forall \lambda \in \mathbb{R} \setminus \{0\}$ and $\forall y \in S_Y$ we have

$$\|\lambda x + y\| = |\lambda| \left\| x + \frac{y}{\lambda} \right\| \geq |\lambda|(1-\varepsilon) \left\| \frac{y}{\lambda} \right\| = 1 - \varepsilon.$$

Hence, $d\left(\mathrm{Sp}(\{x\}), S_Y\right) \geq 1 - \varepsilon$.

27. See [6, exercise 8, chapter 4].

i) Let $T : l_\infty^n \to X$, $T(\lambda_1, ..., \lambda_n) = \sum_{i=1}^n \lambda_i x_i \ \forall \lambda_1, ..., \lambda_n \in \mathbb{R}$. Then, by hypothesis, $T \in L(l_\infty^n, X)$, $\|T\| \leq 1 + \varepsilon$ and hence $\|T^* : X^* \to l_1^n\| \leq 1 + \varepsilon$. Since $T^*(x^*) = (x^*(x_1), ..., x^*(x_n))$ it follows that $\forall x^* \in X^*$ with $\|x^*\| = 1$ we have $\sum_{i=1}^n |x^*(x_i)| \leq 1 + \varepsilon$. Let now $\lambda_1, ..., \lambda_n \in \mathbb{R}$ and suppose, for example, that $|\lambda_1| = \max_{i=\overline{1,n}} |\lambda_i|$. By the Hahn–Banach theorem there is $x^* \in X^*$, $\|x^*\| = 1$ such that $|x^*(x_1)| = \|x_1\| = 1$. Then

$$\left\| \sum_{i=1}^n \lambda_i x_i \right\| \geq \left| x^* \left(\sum_{i=1}^n \lambda_i x_i \right) \right| \geq |\lambda_1| \, |x^*(x_1)| - \sum_{i=2}^n |\lambda_i| \, |x^*(x_i)|$$

$$\geq \max_{i=\overline{1,n}} |\lambda_i| \left(1 - \sum_{i=2}^n |x^*(x_i)| \right).$$

Since $\sum_{i=1}^n |y^*(x_i)| \leq 1 + \varepsilon \ \forall y^* \in S_{X^*}$, and $|x^*(x_1)| = 1$, we obtain that $\sum_{i=2}^n |x^*(x_i)| \leq \varepsilon$. Then $\left\| \sum_{i=1}^n \lambda_i x_i \right\| \geq (1-\varepsilon) \max_{i=\overline{1,n}} |\lambda_i|$.

ii) It follows immediately from (i), taking $x_i = Te_i$, $i = \overline{1,n}$.

28. i) Obviously, T is linear, and

$$\|Tx\|_\infty = \max_{i=\overline{1,n}} |x_i^*(x)| \leq \|x\| \max_{i=\overline{1,n}} \|x_i^*\| \leq \|x\|,$$

hence T is continuous, with $\|T\| \leq 1$. Let $x \in X$ such that $Tx = 0$. For each $x^* \in B_{X^*}$, since $\{x_1^*, ..., x_n^*\} \subseteq B_{X^*}$ is an ε-net, there is an i such that $\|x^* - x_i^*\| \leq \varepsilon$. Then $|(x^* - x_i^*)(x)| \leq \varepsilon \|x\|$. Since $Tx = 0$, that is, $(x_1^*(x), ..., x_n^*(x)) = 0$, it follows that $x_i^*(x) = 0$, and therefore $|x^*(x)| \leq \varepsilon \|x\|$. Taking the supremum over $x^* \in B_{X^*}$, it follows that $\|x\| \leq \varepsilon \|x\|$, and therefore $x = 0$ since $\varepsilon < 1$. Thus T is injective.

ii) Let $T^{-1} : G \to X$, which is well defined and linear. Let $y \in G$ and $x \in X$ such that $T^{-1}(y) = x$, i.e., $Tx = y$. For $x^* \in B_{X^*}$ there is an i such that $\|x_i^* - x^*\| < \varepsilon$. Then

$$|x^*(x)| \leq |(x^* - x_i^*)(x)| + |x_i^*(x)| \leq \|x_i^* - x^*\| \, \|x\| + \max_{i=\overline{1,n}} |x_i^*(x)|$$

$$\leq \varepsilon \|x\| + \|Tx\| = \varepsilon \|x\| + \|y\|.$$

Taking the supremum over $x^* \in B_{X^*}$ we obtain that $\|x\| \leq \varepsilon \|x\| + \|y\|$, and therefore $\|T^{-1}y\| \leq \|y\|/(1-\varepsilon) \ \forall y \in G$. We obtain that T^{-1} is continuous, with $\|T^{-1}\| \leq 1/(1-\varepsilon)$.

Remark. For two finite-dimensional Banach spaces X and Y of the same dimension, the real number $d(X,Y) = \inf \{\|T\| \, \|T^{-1}\| \mid T : X \to Y \text{ is an isomorphism}\}$ is called *the Banach–Mazur distance* between X and Y. Hence the exercise shows that for a finite-dimensional normed space X, the Banach–Mazur distance between X and some closed linear subspace of some l_∞^n-space is as close to 1 as we wish.

For more information concerning the Banach–Mazur distance the reader can consult [49].

4.2 Solutions

29. i) Let us suppose, for a contradiction, that there is an $\varepsilon_0 > 0$ such that $\forall x \notin G$ we have $d(x, G) < \varepsilon_0$, i.e., $\mathbf{C}_G \subseteq \{x \in X \mid d(x, G) < \varepsilon_0\}$. Since $\forall x \in G, d(x, G) = 0 < \varepsilon_0$, then $G \subseteq \{x \in X \mid d(x, G) < \varepsilon_0\}$ and we deduce that $X = G \cup \mathbf{C}_G \subseteq \{x \in X \mid d(x, G) < \varepsilon_0\} \subseteq X$, i.e., $X = \{x \in X \mid d(x, G) < \varepsilon_0\}$. Let $x \in X$ and $n \in \mathbb{N}$. Then $nx \in X$, from whence $d(nx, G) < \varepsilon_0$, i.e., $\exists g_n \in G$ such that $\|nx - g_n\| < \varepsilon_0$, that is, $\|x - g_n/n\| < \varepsilon_0/n$ $\forall n \in \mathbb{N}$. For $n \to \infty$, $\|x - g_n/n\| \to 0$, i.e., $g_n/n \to x$. Using that $(g_n/n)_{n \in \mathbb{N}} \subseteq G$ it follows that $x \in \overline{G} = G$, since G is closed. Hence $X = G$, which is false!

ii) See [9, exercise 19 (a), chapter IV, section 5].

Let $x_1 \in X$, $\|x_1\| = 2$. By the Hahn–Banach theorem there is an $x_1^* \in X^*$ such that $x_1^*(x_1) = 1$ and $\|x_1^*\| \leq 1/2$. Let us suppose now that we have $x_1, ..., x_{n-1} \in X$ and $x_1^*, ..., x_{n-1}^* \in X^*$ such that $\|x_i^*\| \leq 1/2^i$, $i = \overline{1, n-1}$ and $x_i^*(x_j) = \delta_{ij}$ for $i, j = 1, ..., n-1$. Let $G_n = \mathrm{Sp}\{x_1, ..., x_{n-1}\}$. Then G_n is a finite-dimensional linear subspace and $G_n \neq X$, since X is infinite-dimensional. From (i) there is an $x_n \notin G_n$ such that $d(x_n, G_n) \geq 2^n$. Now, from the Hahn–Banach theorem there is an $x_n^* \in X^*$ such that $x_n^*|_{G_n} = 0$, $x_n^*(x_n) = 1$ and $\|x_n^*\| = 1/d(x_n, G_n)$. Then $x_n^*(x_i) = 0$, $i = \overline{1, n-1}$, and $\|x_n^*\| \leq 1/2^n$. In this way we obtain by induction two sequences $(x_n)_{n \in \mathbb{N}} \subseteq X$, $(x_n^*)_{n \in \mathbb{N}} \subseteq X^*$ such that $x_i^*(x_j) = \delta_{ij}$ $\forall i, j \in \mathbb{N}$ and $\|x_n^*\| \leq 1/2^n$ $\forall n \in \mathbb{N}$. Let now $(\lambda_n)_{n \in \mathbb{N}} \in l_\infty$. Since X^* is a Banach space and

$$\sum_{n=1}^\infty \|\lambda_n x_n^*\| \leq M \sum_{n=1}^\infty \frac{1}{2^n} < \infty, \quad M = \sup_{n \in \mathbb{N}} |\lambda_n| < \infty,$$

the series $\sum_{n=1}^\infty \lambda_n x_n^*$ converges. Let $x^* = \sum_{n=1}^\infty \lambda_n x_n^* \in X^*$. Then

$$x^*(x_i) = \sum_{n=1}^\infty \lambda_n x_n^*(x_i) = \sum_{n=1}^\infty \lambda_n \delta_{ni} = \lambda_i \quad \forall i \in \mathbb{N}.$$

30. i) See [9, exercise 1 (a), chapter IV, section 5].

Let $\lambda \in \mathbb{R}$. Since A is convex, by exercise 1 (ii) and (iv), $x \longmapsto d(x, A)$ is norm continuous and convex, and we obtain that the set $E_\lambda = \{x \in X \mid d(x, A) \leq \lambda\}$ is norm closed and convex. But the norm closed convex sets are $weak$ closed (the Mazur theorem), and then E_λ is $weak$ closed.

ii) '(a) $\Rightarrow$ (b)' See [9, exercise 1 (b), chapter IV, section 5].

Let us suppose that X is reflexive. Let $d = d(0, A) = \inf_{a \in A} \|a\|$. For each $n \in \mathbb{N}$ there is an $a_n \in A$ such that $d \leq \|a_n\| < d + 1/n$. Let $A_n = \{x \in A \mid \|x\| \leq d + 1/n\}$. Then $A_n \neq \emptyset$ $\forall n \in \mathbb{N}$. Obviously $A_n = A \cap B_{d+1/n}$ is convex and closed, and $A_{n+1} \subseteq A_n$ $\forall n \in \mathbb{N}$ (B_r is the closed ball of center 0 and radius r in X). Then A_n is $weak$ closed $\forall n \in \mathbb{N}$. X is reflexive, hence by the Alaoglu–Bourbaki theorem B_X is $weak$ compact and therefore $B_{d+1/n}$ is $weak$ compact. Hence $A_n \subseteq B_{d+1/n}$ is $(weak)$ closed $\subseteq (weak)$ compact, and it follows that A_n is $weak$ compact $\forall n \in \mathbb{N}$. Since $A_{n+1} \subseteq A_n$ $\forall n \in \mathbb{N}$, and A_n is $weak$ compact and non-empty $\forall n \in \mathbb{N}$, we obtain that $\bigcap_{n \in \mathbb{N}} A_n \neq \emptyset$, that is, $\exists x_0 \in \bigcap_{n \in \mathbb{N}} A_n$, i.e., $x_0 \in A_n$ $\forall n \in \mathbb{N}$, and therefore $x_0 \in A$ and $d \leq \|x_0\| \leq d + 1/n$ $\forall n \in \mathbb{N}$. Then, for $n \to \infty$, $d \leq \|x_0\| \leq d$, i.e., $d = \|x_0\|$.

'(b) $\Rightarrow$ (a)' Let $x^* \in X^*$, $x^* \neq 0$ and let $A = \{x \in X \mid x^*(x) = 1\}$. Then A is closed convex and non-empty. By (b) there is $a \in A$ a minimal norm element, i.e.,

$$\|a\| = \inf_{x \in A} \|x\| = d(0, A) = \frac{|x^*(0) - 1|}{\|x^*\|}$$

(the last equality, by exercise 5(i)), that is, $\|a\| = 1/\|x^*\|$, i.e., $x^*(a) = 1$ and $\|a\| = 1/\|x^*\|$. Then $\|a\| = x^*(a)/\|x^*\|$, that is, $\|x^*\| = x^*(x)$, where $x = a/\|a\|$. From the James theorem (the difficult part) it follows that X is reflexive.

Let us suppose now that each point of the unit sphere S_X is an extremal point in B_X. Let $a_1, a_2 \in A$ such that $\|a_1\| = \|a_2\| = d = \text{dist}(0, A)$. If $d = 0$, then $\|a_1\| = \|a_2\| = 0$, $a_1 = a_2 = 0$. If $d \neq 0$, let $x_1 = a_1/d$, $x_2 = a_2/d$ and $x = (x_1 + x_2)/2 = (a_1 + a_2)/(2d)$. Since A is convex, $(a_1 + a_2)/2 \in A$, and therefore $d \leq \|(a_1 + a_2)/2\|$, and then $d \leq \|(a_1 + a_2)/2\| \leq (\|a_1\| + \|a_2\|)/2 = d$, and therefore $\|x\| = 1$. Hence $\|x\| = 1$, $\|x_1\| = \|x_2\| = 1$ and $x = (x_1 + x_2)/2$. Now using the hypothesis that each point of the unit sphere S_X is an extremal point in B_X it follows that $x = x_1 = x_2$, and then $a_1 = a_2$.

iii) '(a) $\Rightarrow$ (b)' See [9, exercise 1 (c), chapter IV, section 5].

First solution. We have

$$d(A, B) = \inf_{a \in A, b \in B} d(a, b) = \inf_{b \in B} \left(\inf_{a \in A} d(b, a) \right) = \inf_{b \in B} d(b, A).$$

Let $f : X \to [0, \infty)$, $f(x) = d(x, A)$. We know from (i) that f is lower semicontinuous with respect to the *weak* topology. Since B is closed and convex, B is *weak* closed. Also, by hypothesis X is reflexive, hence B_X is *weak* compact, and since B is bounded, there is an $M \geq 0$ such that $B \subseteq MB_X$. Then B is (*weak*) closed $\subseteq$ (*weak*) compact and hence B is *weak* compact. Now, as is well known, each lower semicontinuous function on a non-empty compact set attains its infimum, i.e., $\exists y \in B$ such that $d(A, B) = \inf_{b \in B} f(b) = f(y)$, that is, $d(A, B) = d(y, A) = d(0, A - y)$. Using (ii) there is $a \in A - y$ such that $d(0, A - y) = \|a\|$. Since $a \in A - y$, there is an $x \in A$ such that $a = x - y$, and therefore $\|x - y\| = d(A, B)$.

Second solution. We have $d(A, B) = \inf_{a \in A, b \in B} \|a - b\| = d(0, A - B)$. Since by hypothesis X is reflexive and B is closed, convex, and bounded, it follows that B is *weak* compact. Also, by the Mazur theorem, since A is norm closed and convex it follows that A is *weak* closed. Then $A - B$ is *weak* closed (see exercise 10(iv) of chapter 13, for a proof). Also, since A, B are convex it follows that $A - B$ is convex. Hence $A - B$ is *weak* closed, convex, and non-empty. Again, since X is reflexive, by (ii) there exists $x \in A - B$ such that $d(0, A - B) = \|x\|$, i.e., (b).

'(b) $\Rightarrow$ (a)' Let $x^* \in X^*$, $x^* \neq 0$, and consider $A = [x^* = 1]$, $B = \{0\}$. There is then, by hypothesis, an $a \in A$ such that $d(A, B) = \|a - 0\| = \|a\|$, i.e., $x^*(a) = 1$ and $\|a\| = 1/\|x^*\|$. Then $\|a\| = x^*(a)/\|x^*\|$, and therefore $\|x^*\| = x^*(x)$, where $x = a/\|a\|$. Then from the James theorem (the difficult part) it follows that X is reflexive.

Remark. From (ii) we deduce that if X is reflexive and $A \subseteq X$ is closed convex and non-empty, then $\forall x \in X \; \exists a \in A$ such that $d(x, a) = d(x, A)$, i.e., for example, if X is a Hilbert space we have a new proof, much more complicated, for the existence part in the orthogonal projection theorem.

Chapter 5

Compactness in Banach spaces. Compact operators

Theorem. Let (X, d) be a metric space and consider a set $A \subseteq X$. Then the following assertions are equivalent:

i) A is *totally bounded*, i.e., $\forall \varepsilon > 0\ \exists$ a finite system $x_1, ..., x_n \in X$ called an ε-net for A such that $A \subseteq \bigcup_{i=1}^{n} B(x_i, \varepsilon)$;

ii) Any sequence $(x_n)_{n \in \mathbb{N}} \subseteq A$ contains a Cauchy subsequence.

If in addition (X, d) is a complete metric space then A is totally bounded if and only if A is *relatively compact*, i.e., $\overline{A}$ is compact.

The Arzela–Ascoli Theorem. Let (T, d) be a compact metric space and $A \subseteq C(T)$. Then the following assertions are equivalent:

i) A is relatively norm compact;

ii) A is uniformly bounded, i.e., $\exists M > 0$ such that $|f(t)| \leq M\ \forall f \in A, \forall t \in T$, and A is equicontinuous, i.e., $\forall \varepsilon > 0\ \exists \delta_\varepsilon > 0$ such that $\forall x, y \in T$ with $d(x, y) < \delta_\varepsilon$ it follows that $|f(x) - f(y)| < \varepsilon\ \forall f \in A$;

iii) Any sequence $(f_n)_{n \in \mathbb{N}} \subseteq A$ contains a uniformly convergent subsequence.

Relative compactness in l_p, $1 \leq p < \infty$. A subset $A \subseteq l_p$ is relatively norm compact if and only if A is bounded and $\forall \varepsilon > 0\ \exists n_\varepsilon \in \mathbb{N}$ such that $\sum_{k=n_\varepsilon}^{\infty} |p_k(x)|^p < \varepsilon\ \forall x \in A$.

Relative compactness in c_0. A set $A \subseteq c_0$ is relatively norm compact if and only if there is a sequence $\lambda = (\lambda_n)_{n \in \mathbb{N}} \in c_0$ such that $|p_n(x)| \leq |\lambda_n|\ \forall x \in A, \forall n \in \mathbb{N}$.

Definition. Let X, Y be normed spaces. An operator $U \in L(X, Y)$ is called *compact* if and only if $U(B_X) \subseteq Y$ is totally bounded.

If, in addition, Y is a Banach space then $U \in L(X, Y)$ is a *compact* operator if and only if $U(B_X) \subseteq Y$ is relatively compact.

We denote $K(X,Y) = \{U \in L(X,Y) \mid U \text{ is compact}\}$ and $K(X) = K(X,X)$.

Theorem. i) Let X, Y be normed spaces. Then $K(X,Y)$ is a closed linear subspace of $L(X,Y)$.

ii) Let X, Y, Z, T be normed spaces. If $A \in L(X,Y)$, $B \in K(Y,Z)$, $C \in L(Z,T)$, then the composition CBA is a compact operator (the ideal property for the compact operators).

Definition. Let X, Y be normed spaces. A linear and continuous operator $U : X \to Y$ is called a *finite rank operator* if and only if the range $U(X) \subseteq Y$ is finite-dimensional, or, equivalently, if there are $n \in \mathbb{N}$, $x_1^*, ..., x_n^* \in X^*$ and $y_1, ..., y_n \in Y$ such that $U(x) = x_1^*(x)y_1 + \cdots + x_n^*(x)y_n \ \forall x \in X$.

The Schauder Theorem. Let X, Y be Banach spaces and $U \in L(X,Y)$. Then U is compact if and only if U^* is compact.

The Mazur Theorem. Let X be a Banach space and $A \subseteq X$ a subset. Then the following assertions are equivalent:

i) A is relatively compact.
ii) co (A) is relatively compact.
iii) ec (A) is relatively compact.
iv) eco (A) is relatively compact.

The Grothendieck Theorem. Let X be a Banach space and $A \subseteq X$ a subset. Then A is relatively compact if and only if there is $(x_n)_{n \in \mathbb{N}} \subseteq X$ such that $x_n \to 0$ and $A \subseteq \overline{\text{co}}\{x_n \mid n \in \mathbb{N}\}$.

5.1 Exercises

Examples of compact sets in l_2

1. Under what conditions on the sequence $(\lambda_n)_{n \in \mathbb{N}} \subseteq (0, \infty)$ are the following sets compact in l_2?

i) The parallelepiped $\{(x_1, x_2, ...) \in l_2 \mid |x_n| \leq \lambda_n \ \forall n \in \mathbb{N}\}$.

ii) The ellipsoid $\left\{(x_1, x_2, ...) \in l_2 \mid \sum_{n=1}^{\infty} \left(|x_n|^2 / \lambda_n^2\right) \leq 1\right\}$.

Relatively compact sets in $C(T)$

2. Prove that a set $M \subseteq C[a,b]$ for which there are $m, L > 0$ and $x_0 \in [a,b]$ such that $|f(x_0)| \leq m \ \forall f \in M$ and $|f(x) - f(y)| \leq L|x-y| \ \forall f \in M, \ \forall x, y \in [a,b]$, is relatively compact in $C[a,b]$.

3. Prove that a set M of C^1 functions f on the interval $[a,b]$ which satisfy the conditions $|f(a)| \leq k_1$, $\int_a^b |f'(x)|^2 dx \leq k_2$, $(k_1, k_2 > 0$ are constants) is relatively compact in $C[a,b]$.

4. Prove that a set M of C^1 functions f on the interval $[a,b]$ which satisfy the condition $\int_a^b \left(|f(x)|^2 + |f'(x)|^2\right) dx \leq k$, $(k > 0$ is a constant), is relatively compact in the space $C[a,b]$.

5.1 Exercises

5. Let M be a bounded set in the space $C[a,b]$. Prove that the set A containing all functions of the form $y(t) = \int_0^t x(s)ds$, where $x \in M$, is relatively compact in $C[a,b]$.

6. Let $(g_n)_{n \in \mathbb{N}}$ be a sequence of twice continuous differentiable functions on a fixed open neighborhood of $[0, 1]$ such that $g_n(0) = g_n'(0) = 0\ \forall n \in \mathbb{N}$. Suppose also that $|g_n''(x)| \leq 1\ \forall n \in \mathbb{N},\ \forall x \in [0, 1]$. Prove that there is a subsequence of $(g_n)_{n \in \mathbb{N}}$ which is uniformly convergent on $[0, 1]$.

7. Let M be a set of C^1 functions on the interval $[a, b]$ which satisfy the following conditions:

i) There is $L > 0$ such that $|f'(x)| \leq L\ \forall f \in M,\ \forall x \in [a, b]$;
ii) For each function $f \in M$, the equation $f(x) = 0$ has at least one solution.

Prove that M is a relatively compact set in $C[a, b]$.

8. Establish if the following sets of functions are relatively compact in $C[0, 1]$.

i) $f_n(x) = x^n,\ n \in \mathbb{N}$.
ii) $f_n(x) = \sin nx,\ n \in \mathbb{N}$.
iii) $f_\alpha(x) = \sin \alpha x,\ \alpha \in \mathbb{R}$.
iv) $f_\alpha(x) = \sin(x + \alpha),\ \alpha \in \mathbb{R}$.
v) $f_\alpha(x) = \arctan(\alpha x),\ \alpha \in \mathbb{R}$.
vi) $f_\alpha(x) = e^{x-\alpha},\ \alpha \in \mathbb{R}, \alpha \geq 0$.

9. i) Let X be a Banach space, and consider a Cauchy sequence $(x_n)_{n \in \mathbb{N}} \subseteq X$. Prove that the set $A = \{x_n \mid n \in \mathbb{N}\} \subseteq X$ is relatively compact.

ii) Let T be a compact metric space, $(f_n)_{n \in \mathbb{N}} \subseteq C(T)$ a uniformly bounded sequence, and let $g : T \to l_\infty$, $g(x) = (f_n(x))_{n \in \mathbb{N}}$. Prove that the set $(f_n)_{n \in \mathbb{N}} \subseteq C(T)$ is relatively compact if and only if the function g is continuous.

10. i) Find the continuous functions $\varphi : [0, 1] \to [0, \infty)$ such that the set $A = \{f \in C[0, 1] \mid |f(x)| \leq \varphi(x)\ \forall x \in [0, 1]\}$ be relatively compact in the space $C[0, 1]$.

ii) Let T be a compact metric space and $\varphi : T \to [0, \infty)$ continuous. Find φ for which the set $A = \{f \in C(T) \mid |f(x)| \leq \varphi(x)\ \forall x \in T\}$ is relatively compact in the space $C(T)$.

Limitations for the Arzela–Ascoli theorem

11. The Arzela–Ascoli theorem implies that a sequence $(f_n)_{n \in \mathbb{N}}$ of real valued continuous functions defined on a metric space Ω is relatively compact (i.e., has an uniformly convergent subsequence) if:

i) Ω is compact;
ii) $\sup_{n \in \mathbb{N}} \|f_n\| < \infty$;
iii) The sequence is equicontinuous.

Give examples of sequences which are not relatively compact as a set, such that: (i) and (ii) are true, but (iii) is false; (i) and (iii) are true, but (ii) is false; (ii) and (iii) are true, but (i) is false. Take Ω as a subset of $\mathbb{R}$.

The Arzela–Ascoli theorem, a more general case

12. Let (Ω, τ) be a compact space and (X, d) a metric space. On the space $C(\Omega, X) = \{f : \Omega \to X \mid f \text{ continuous}\}$ we will consider two topologies: the pointwise convergence topology, τ_p, i.e., the weaker topology for which the operators $f \to f(t)$ are continuous

for every $t \in \Omega$, and the uniform convergence topology, τ_u, given by the distance: $\rho : C(\Omega, X) \times C(\Omega, X) \to \mathbb{R}_+$, $\rho(f,g) = \sup_{t \in \Omega} d(f(t), g(t))$ $\forall f, g \in C(\Omega, X)$. A subset $A \subseteq C(\Omega, X)$ is called equicontinuous if for every $t_0 \in \Omega$ and $\varepsilon > 0$ there is an open neighborhood U of t_0 in Ω such that $d(f(t), f(t_0)) < \varepsilon$ $\forall t \in U, \forall f \in A$.

i) If $A \subseteq C(\Omega, X)$ is equicontinuous, prove that τ_p and τ_u coincide on A.

ii) If $A \subseteq C(\Omega, X)$ is equicontinuous and for any $t \in \Omega$, the set $\{f(t) \mid f \in A\} \subseteq X$ is relatively compact, prove that A is relatively compact in the τ_u topology.

Compact composition operators

13. Let X be a Banach space. For each $A, B \in L(X)$ we define $T_{A,B} : L(X) \to L(X)$, $T_{A,B}(S) = ASB$ $\forall S \in L(X)$.

i) If A, B are compact operators prove that $T_{A,B}$ is also compact.

ii) If A, B are not zero and if $T_{A,B}$ is compact prove that A and B are compact operators.

Compact multiplication operators between $C(T)$ spaces

14. i) For a function $\varphi \in C[0,1]$, we consider the multiplication operator $M_\varphi : C[0,1] \to C[0,1]$, $(M_\varphi f)(x) = \varphi(x)f(x)$ $\forall x \in [0,1], \forall f \in C[0,1]$.

a) Find the norm $\|M_\varphi\|$.

b) Under what conditions on the function φ is the operator M_φ compact?

ii) Let T be a compact metric space and $\varphi \in C(T)$. Under what conditions on the function φ is the multiplication operator $M_\varphi : C(T) \to C(T)$, $(M_\varphi f)(x) = \varphi(x)f(x)$ $\forall x \in T, \forall f \in C(T)$ compact?

Effective examples of compact and non-compact operators

15. Given $U : C[0,1] \to C[0,1]$ as below, find its norm and establish if U is compact.

i) $(Uf)(x) = xf(x)$.
ii) $(Uf)(x) = f(0) + xf(1)$.
iii) $(Uf)(x) = \int_0^1 e^{tx} f(t)\,dt$.
iv) $(Uf)(x) = f(x^2)$.

16. Prove that the operators $U, V : C[-1,1] \to C[-1,1]$ given by $(Uf)(x) = (f(x) + f(-x))/2$ $\forall x \in [-1,1]$ and $(Vf)(x) = (f(x) - f(-x))/2$ $\forall x \in [-1,1]$ are linear and continuous and find their norms. Are U and V compact operators?

17. For $1 \le p < \infty$, find the norm for the operators $U : l_p \to l_p$ below, and if they are compact.

i) $U(x_1, x_2, \ldots) = (0, x_1, x_2, \ldots)$ $\forall (x_1, x_2, \ldots) \in l_p$.
ii) $U(x_1, x_2, \ldots) = (x_1, x_2/2, x_3/3, \ldots)$ $\forall (x_1, x_2, \ldots) \in l_p$.
iii) $U(x_1, x_2, \ldots) = (0, x_1, x_2/2, x_3/3, \ldots)$ $\forall (x_1, x_2, \ldots) \in l_p$.

18. Is the canonical inclusion $J : l_1 \to l_2$, $J(x) = x$ $\forall x \in l_1$ a compact operator?

5.1 Exercises

The Volterra and Hardy operators

19. i) Let $V : C[0,1] \to C[0,1]$, $(Vf)(x) = \int_0^x f(t)\,dt$ $\forall f \in C[0,1]$, $\forall x \in [0,1]$ (V is the Volterra operator). Prove that V is linear, continuous, and compact. Find its norm.

ii) Let $1 < p < \infty$. Let $H : L_p(0,\infty) \to L_p(0,\infty)$, $(Hf)(x) = (1/x)\int_0^x f(t)\,dt$ $\forall f \in L_p(0,\infty)$, $\forall x \in (0,\infty)$ (H is the Hardy operator). Prove that H is well defined, linear and continuous but not compact.

An example of a non-compact operator for which its square is compact

20. Let X be one of the spaces c_0 or l_p with $1 \leq p \leq \infty$. We consider the operator $U : X \to X$, $U(x_1, x_2, \ldots) = (0, x_1, 0, x_3, 0, x_5, \ldots)$. Prove that U is not a compact operator but U^2 is compact.

Compact multiplication operators on l_p

21. Let $1 \leq p < \infty$, and $\lambda = (\lambda_n)_{n\in\mathbb{N}} \subseteq \mathbb{K}$ with $\sup_{n\in\mathbb{N}} |\lambda_n| < \infty$. We define the multiplication operator $M_\lambda : l_p \to l_p$, $M_\lambda(x) = (\lambda_1 x_1, \lambda_2 x_2, \ldots)$ $\forall x = (x_1, x_2, \ldots) \in l_p$. Prove that:

i) M_λ is linear and continuous and $\|M_\lambda\| = \sup_{n\in\mathbb{N}} |\lambda_n|$.

ii) M_λ is a compact operator if and only if $\lambda \in c_0$.

The summation operator is not compact

22. i) Prove that the summation operator $\Sigma : l_1 \to l_\infty$ given by $\Sigma\left((t_n)_{n\in\mathbb{N}}\right) = \left(\sum_{i=1}^n t_i\right)_{n\in\mathbb{N}}$ is linear, continuous, but not compact.

ii) Prove that the inclusion $i : l_1 \to l_\infty$ is a linear and continuous operator which has Σ defined at (i) as a factor, and that i is not compact.

iii) Let $U : L_1(0,\infty) \to L_\infty(0,\infty)$, $Uf(x) = \int_0^x f(t)\,dt$. Prove that U is linear and continuous but not compact.

Compact operators on infinite-dimensional Banach spaces are not surjective

23. i) Let X be an infinite-dimensional Banach space, and $A \in K(X)$. Prove that there is $y \in X$ such that the equation $A(x) = y$ has no solution, i.e., A is not surjective.

ii) Let $1 \leq p < \infty$, and $U : l_p \to l_p$, $U(x_1, x_2, \ldots) = (x_1, x_2/2, x_3/3, \ldots)$. Find an element $y \in l_p$ for which the equation $U(x) = y$ has no solution.

iii) Let $A : c_{00} \to c_{00}$, $A\left((x_n)_{n\in\mathbb{N}}\right) = (x_n/n)_{n\in\mathbb{N}}$ $\forall (x_n)_{n\in\mathbb{N}} \in c_{00}$. Prove that A is compact and bijective. (On c_{00} we have the norm from l_∞.)

Compact non-injective operators on infinite-dimensional normed spaces

24. Let X, Y be normed spaces and $A \in L(X,Y)$ which has the property: $\exists c > 0$ such that $\forall x \in X$, $\|Ax\| \geq c\|x\|$. Can A be a compact operator?

25. i) Let X, Y be normed spaces, X infinite-dimensional, and $A \in K(X,Y)$. Prove that there is a sequence $(x_n)_{n\in\mathbb{N}} \subseteq X$ with $\|x_n\| = 1$ $\forall n \in \mathbb{N}$ and $Ax_n \to 0$. Therefore $0 \in \overline{A(S_X)}$.

ii) Using (i) prove that if X, Y are normed spaces, X infinite-dimensional, then $\{U \in K(X,Y) \mid U$ is not injective$\}$ is dense in $K(X,Y)$.

Compact projection operators are finite rank operators

26. Let X be a normed space and $T \in L(X)$ which is compact and $T^2 = T$. Prove that T is a finite rank operator.

The range of a bounded, closed, and convex set by a compact operator

27. Let X be a reflexive space, Y a Banach space, and $A : X \to Y$ linear and continuous. If $M \subseteq X$ is closed, convex, and bounded, prove that $A(M) \subseteq Y$ is closed.
If, in addition, A is a compact operator, prove that $A(M) \subseteq Y$ is compact.

28. Let H be a Hilbert space and $T : H \to H$ a compact operator. Prove that T attains its norm, i.e., there is an $x \in H$ with $\|x\| \leq 1$ such that $\|Tx\| = \|T\|$.

29. i) Let X be a reflexive space, Y a Banach space, and $A : X \to Y$ a compact operator. If $M \subseteq X$ is a non-empty bounded closed and convex set and $y \in Y$ prove that there is an $x_0 \in M$ such that $\inf_{x \in M} \|Ax - y\| = \|Ax_0 - y\|$. If, in addition, Y is a Hilbert space prove that $x_0 \in M$ obtained as above is a solution for the equation $A(x) = \Pr_{A(M)}(y)$, where Pr denotes the orthogonal projection onto a closed, convex, and non-empty set in a Hilbert space.

ii) Let $1 < p < 2$, $A : l_p \to l_2$ is the multiplication operator $A(x_1, x_2, \ldots) = (x_1/2, x_2/3, \ldots, x_n/(n+1), \ldots)$, and

$$M = \left\{ (x_1, x_2, \ldots) \in l_p \;\middle|\; \sum_{n=1}^{\infty} \frac{n^p |x_n|^p}{(n+1)^p} \leq 1 \right\}.$$

Find an explicit element $x_0 \in M$ such that $\inf_{x \in M} \|Ax - 2e_1\| = \|Ax_0 - 2e_1\|$.

Relatively compact sets and weak* null convergent sequences in separable Banach spaces

30. Let X be a separable Banach space and $A \subseteq X$ a subset. Prove that A is relatively compact if and only if for every sequence $(x_n^*)_{n \in \mathbb{N}} \subseteq X^*$, weak* convergent to zero, it follows that $x_n^* \to 0$ uniformly on A.

A compact set in a Banach space

31. Let X be a Banach space, and $(x_n)_{n \in \mathbb{N}} \subseteq X$ such that $\lim_{n \to \infty} \|x_n\| = 0$. Prove that the set $K = \left\{ x \in X \;\middle|\; x = \sum_{n=1}^{\infty} a_n x_n, \sum_{n=1}^{\infty} |a_n| \leq 1 \right\} \subseteq X$ is norm compact.

Limitations for the Mazur Theorem

32. Give an example of a normed space X, a sequence $(x_n)_{n \in \mathbb{N}}$ in X, norm convergent to 0, for which there is a sequence $(\lambda_n)_{n \in \mathbb{N}} \in l_1$ such that the series $\sum_{n=1}^{\infty} \lambda_n x_n$ is not convergent in X. Then deduce that there is a normed space (which is not a Banach space!) in which the closed convex hull of a compact set is not always compact and hence that the Mazur theorem is no longer true without the hypothesis on the completeness of the norm.

A general property for a compact operator

33. Let X and Y be Banach spaces, and $T : X \to Y$ a linear, continuous, and compact operator. Prove that there is a sequence $(y_m^*)_{m \in \mathbb{N}} \subseteq B_{Y^*}$ such that $(T^*(y_m^*))_{m \in \mathbb{N}} \subseteq T^*(B_{Y^*})$ is norm dense, and for any fixed $1 \leq p < \infty$, we have: $\forall \varepsilon > 0 \; \exists \delta_\varepsilon > 0$ such that

$$\|Tx\|^p \leq \varepsilon \|x\|^p + \delta_\varepsilon \sum_{m=1}^{\infty} \frac{|y_m^*(Tx)|^p}{2^m} \quad \forall x \in X.$$

Compact operators factor through closed linear subspaces of c_0

34. Let X and Y be two Banach spaces. Prove that:
i) A linear and continuous operator $T : X \to Y$ is compact if and only if there is a sequence $(x_n^*)_{n \in \mathbb{N}} \subseteq X^*$ norm convergent to zero such that

$$\|Tx\| \leq \sup_{n \in \mathbb{N}} |x_n^*(x)| \quad \forall x \in X.$$

Hence T is compact if and only if there are $\lambda = (\lambda_n)_{n \in \mathbb{N}} \in c_0$ and $(y_n^*)_{n \in \mathbb{N}} \subseteq X^*$ bounded such that $\|Tx\| \leq \sup_{n \in \mathbb{N}} (|\lambda_n|^2 |y_n^*(x)|) \; \forall x \in X$.

ii) Each linear, continuous, and compact operator between two Banach spaces factors compactly through a closed linear subspace of c_0, i.e., if $T \in L(X, Y)$ is compact, then we can find a closed linear subspace $Z \subseteq c_0$ and two compact operators $A \in L(X, Z)$, $B \in L(Z, Y)$ such that $T = BA$.

Operators on c_0

35. Prove that the linear and continuous operators from c_0 into a Banach space X correspond to $weak$ Cauchy series in X, with the correspondence given by $T \in L(c_0, X) \longmapsto (Te_n)_{n \in \mathbb{N}}$.
The series $\sum_{n=1}^{\infty} x_n$ in X is said to be $weak$ Cauchy ($(x_n)_{n \in \mathbb{N}} \in w_1(X)$) if and only if $\sum_{n=1}^{\infty} |x^*(x_n)| < \infty \; \forall x^* \in X^*$.

36. Let X be a Banach space. Prove that the operator $T \in L(c_0, X)$ is compact if and only if every subseries of the series $\sum_{n=1}^{\infty} Te_n$ is norm convergent.

37. Let X be a Banach space. Prove that $T : c_0 \to X$ is $weak$ compact if and only if $T : c_0 \to X$ is compact.

38. Let $(\xi_n)_{n \in \mathbb{N}}$ be a scalar sequence. Prove that the function $\sum_{n=1}^{\infty} a_n \xi_n \chi_{[2^{n-1}, 2^n)}$ is in $L_1[1, \infty)$ for any $(a_n)_{n \in \mathbb{N}} \in c_0$ if and only if the series $\sum_{n=1}^{\infty} 2^{n-1} \xi_n$ is absolutely convergent, and in this case prove that the operator $U : c_0 \to L_1[1, \infty)$ defined by $U((a_n)_{n \in \mathbb{N}}) = \sum_{n=1}^{\infty} a_n \xi_n \chi_{[2^{n-1}, 2^n)}$ is compact.

Operators on l_1

39. Prove that the linear and continuous operators from l_1 to a Banach space X correspond to the bounded sequences from X, with the correspondence given by $T \in L(l_1, X) \longmapsto (Te_n)_{n \in \mathbb{N}} \subseteq X$.

40. Let X be a Banach space. Prove that the operator $T \in L(l_1, X)$ is *weak* compact if and only if the set $\{Te_n \mid n \in \mathbb{N}\} \subseteq X$ is relatively *weak* compact.

41. i) Let X be a Banach space. Prove that the operator $T \in L(l_1, X)$ is compact if and only if the set $\{Te_n \mid n \in \mathbb{N}\} \subseteq X$ is relatively norm compact.

ii) Let $(a_n)_{n \in \mathbb{N}} \in l_\infty$. Using (i) prove that the operator $U : l_1 \to l_\infty$, $U(x_1, x_2, ...) = \left(\sum_{k=1}^{n} a_k x_k\right)_{n \in \mathbb{N}}$ is compact if and only if $(a_n)_{n \in \mathbb{N}} \in c_0$.

iii) Let $(\xi_n)_{n \geq 0} \in l_\infty$. We define $U : l_1 \to C[0,1]$, $(Ua)(x) = \sum_{n=0}^{\infty} a_n \xi_n x^n$ $\forall a = (a_n)_{n \geq 0} \in l_1$, $\forall x \in [0,1]$. Using (i) prove that U is compact if and only if $(\xi_n)_{n \geq 0} \in c_0$.

Operators into c_0

42. Prove that the linear and continuous operators from a Banach space X to the space c_0 correspond to *weak** null sequences in X^*, with the correspondence given by $T \in L(X, c_0) \longmapsto (T^* p_n)_{n \in \mathbb{N}} \subseteq X^*$. (Here, $p_n : c_0 \to \mathbb{K}$ are the canonical projections.)

More precisely, if $T : X \to c_0$ is a linear and continuous operator, we can find $(x_n^*)_{n \in \mathbb{N}} \subseteq X^*$ such that $x_n^* \to 0$ *weak** and $T(x) = (x_n^*(x))_{n \in \mathbb{N}}$ $\forall x \in X$, and conversely, if $x_n^* \to 0$ *weak**, and $T : X \to c_0$, $T(x) = (x_n^*(x))_{n \in \mathbb{N}}$ $\forall x \in X$, then T is linear and continuous.

43. Let X be a Banach space. Prove that $T \in L(X, c_0)$ is *weak* compact if and only if $T^* p_n \to 0$ *weak* in X^*. More precisely, if $T \in L(X, c_0)$, $T(x) = (x_n^*(x))_{n \in \mathbb{N}}$ with $x_n^* \to 0$ *weak**, prove that T is *weak* compact if and only if $x_n^* \to 0$ *weak*. Therefore the *weak* compact operators with values in c_0 correspond to *weak* null sequences.

44. i) Let X be a Banach space. Prove that $T \in L(X, c_0)$ is compact if and only if $T^* p_n \to 0$ in the norm of X^*.

More precisely, let $T \in L(X, c_0)$, $T(x) = (x_n^*(x))_{n \in \mathbb{N}}$ with $x_n^* \to 0$ *weak**. Prove that T is compact if and only if $x_n^* \to 0$ in norm, i.e., the compact operators with values in c_0 correspond to norm null sequences.

ii) Let $(a_n)_{n \in \mathbb{N}} \in l_\infty$, and let $U : c_0 \to c_0$, $U(x_1, x_2, ...) = (a_1 x_1, a_2 x_2, ...)$. Prove that U is *weak* compact $\Leftrightarrow$ U is compact $\Leftrightarrow$ $(a_n)_{n \in \mathbb{N}} \in c_0$.

iii) Let H be a Hilbert space, and $(x_n)_{n \in \mathbb{N}} \subseteq H \setminus \{0\}$ an orthogonal system. Prove that the sequence $(\langle x, x_n \rangle)_{n \in \mathbb{N}}$ converges towards 0 for any $x \in H$ if and only if $\sup_{n \in \mathbb{N}} \|x_n\| < \infty$, and in this case the operator $U : H \to c_0$, $U(x) = (\langle x, x_n \rangle)_{n \in \mathbb{N}}$ is linear and continuous. If U is linear and continuous, prove that U is compact if and only if $\|x_n\| \to 0$.

Operators into l_1

45. Prove that the linear and continuous operators from a Banach space X to the space l_1 correspond to *weak* Cauchy series $\sum_{n=1}^{\infty} x_n^*$, with the correspondence given by $T \in$

$L(X, l_1) \longmapsto (T^* p_n)_{n \in \mathbb{N}} \in w_1(X^*)$ (see exercise 35 for the definition for *weak* Cauchy series). (Here $p_n : l_1 \to \mathbb{K}$ are the canonical projections.)

46. Let X be a Banach space and $T \in L(X, l_1)$. Prove that T is compact if and only if T is *weak* compact.

47. Let X be a Banach space. Prove that an operator $T \in L(X, l_1)$ is compact if and only if all the subseries for the series $\sum_{n=1}^{\infty} T^* p_n$ are norm convergent.

48. i) Let $(a_n)_{n \in \mathbb{N}}$ be a scalar sequence. Prove that the series $\sum_{n=1}^{\infty} a_n \int_0^{1/n} f(x)\, dx$ is absolutely convergent for any $f \in C[0, 1]$ if and only if the series $\sum_{n=1}^{\infty} a_n/n$ is absolutely convergent, and in this case the operator $U : C[0, 1] \to l_1$ defined by $U(f) = \left(a_n \int_0^{1/n} f(x)\, dx \right)_{n \in \mathbb{N}}$ is compact.

ii) Let H be a Hilbert space, $(x_n)_{n \in \mathbb{N}} \subseteq H \setminus \{0\}$ an orthogonal system, and $(a_n)_{n \in \mathbb{N}}$ a scalar sequence. Prove that the series $\sum_{n=1}^{\infty} a_n \langle x, x_n \rangle$ is absolutely convergent for any $x \in H$ if and only if $(a_n \|x_n\|)_{n \in \mathbb{N}} \in l_2$, and in this case the operator $U : H \to l_1$ defined by $U(x) = (a_n \langle x, x_n \rangle)_{n \in \mathbb{N}}$ is compact.

Operators on l_p, $1 < p < \infty$

49. Let $1 < p < \infty$, and let X be a Banach space. We denote by $w_p(X)$ the set of all sequences $(x_n)_{n \in \mathbb{N}} \subseteq X$ with the property that $(x^*(x_n))_{n \in \mathbb{N}} \in l_p$ $\forall x^* \in X^*$. For each $x = (x_n)_{n \in \mathbb{N}} \in w_p(X)$, we define $\|x\|_p^{\text{weak}} = \sup_{\|x^*\| \leq 1} \left(\sum_{n=1}^{\infty} |x^*(x_n)|^p \right)^{1/p}$. Prove that $\left(w_p(X), \|\cdot\|_p^{\text{weak}} \right)$ is a Banach space.

50. Let $1 < p < \infty$ and let X be a Banach space. Prove that there exists an isomorphic isomorphism from $L(l_p, X)$ onto $w_q(X)$ given by the correspondence $T \in L(l_p, X) \longmapsto (Te_n)_{n \in \mathbb{N}} \in w_q(X)$, where q is the conjugate of p, i.e., $1/p + 1/q = 1$.

Operators into l_p, $1 < p < \infty$

51. Consider $1 < p < \infty$ and let X be a Banach space. Prove that there exists an isometric isomorphism from $L(X, l_p)$ to $w_p(X^*)$, given by the correspondence $T \in L(X, l_p) \longmapsto (T^* p_n)_{n \in \mathbb{N}} \in w_p(X^*)$. (Here $p_n : l_p \to \mathbb{K}$ are the canonical projections.)

5.2 Solutions

1. See [40, exercise 15.51].

i) Let P be the parallelepiped from the statement. We have $P = \bigcap_{n \in \mathbb{N}} A_n$, $A_n = \{x \mid |x_n| \leq \lambda_n\}$. Since $p_n : l_2 \to \mathbb{R}$ is continuous, it follows that A_n is closed, $A_n = p_n^{-1}([-\lambda_n, \lambda_n])$ $\forall n \in \mathbb{N}$, and therefore P is a closed set. Suppose that P is compact. Then $\forall \varepsilon > 0$ $\exists n_\varepsilon \in \mathbb{N}$ such that $\sum_{k=n_\varepsilon}^{\infty} |p_k(x)|^2 \leq \varepsilon$ $\forall x \in A$. Let $n \geq n_\varepsilon$ and $p \geq 0$.

The element $x = (0,...,0,\lambda_n,...,\lambda_{n+p},0,...)$ belongs to A and therefore $\sum_{k=n_\varepsilon}^{\infty} |p_k(x)|^2 \leq \varepsilon$, $\sum_{k=n}^{n+p} |\lambda_k|^2 \leq \varepsilon$, i.e., by the Cauchy test for series, $\sum_{n=1}^{\infty} |\lambda_n|^2$ converges.

Conversely, let us suppose that the series $\sum_{n=1}^{\infty} |\lambda_n|^2$ converges. Then $\forall \varepsilon > 0 \; \exists n_\varepsilon \in \mathbb{N}$ such that $\sum_{k=n_\varepsilon}^{\infty} |\lambda_k|^2 \leq \varepsilon$. Since $|p_n(x)| \leq \lambda_n \; \forall x \in P, \forall n \in \mathbb{N}$, it follows that $\sum_{k=n_\varepsilon}^{\infty} |p_k(x)|^2 \leq \varepsilon \; \forall x \in P$, i.e., P is relatively compact in l_2.

Hence $P \subseteq l_2$ is compact $\Leftrightarrow (\lambda_n)_{n\in\mathbb{N}} \in l_2$.

Remark. Since the series $\sum_{n=1}^{\infty} 1/n^2$ converges, it follows that the parallelepiped

$$P = \{(x_1, x_2, ...) \in l_2 \mid |x_n| \leq 1/n \; \forall n \in \mathbb{N}\}$$

is a compact set in l_2.

ii) Let E be the ellipsoid from the statement. We have $E = \bigcap_{n\in\mathbb{N}} E_n$, $E_n = \left\{ x \mid \sum_{k=1}^{n} (|x_k|^2/\lambda_k^2) \leq 1 \right\}$. Since $T : l_2 \to \mathbb{R}$, $T(x) = \sum_{k=1}^{n} (|x_k|^2/\lambda_k^2)$ is obviously continuous (T is not linear!), we obtain that $E_n = T^{-1}((-\infty, 1])$ is a closed set for every $n \in \mathbb{N}$, hence E is closed. Suppose that E is compact. Then $\forall \varepsilon > 0 \; \exists n_\varepsilon \in \mathbb{N}$ such that $\sum_{k=n_\varepsilon}^{\infty} |p_k(x)|^2 \leq \varepsilon^2 \; \forall x \in E$. Let $n \geq n_\varepsilon$. Then $\lambda_n e_n = (0,...,0,\lambda_n,0,...) \in E$, and then $\sum_{k=n_\varepsilon}^{\infty} |p_k(\lambda_n e_n)|^2 \leq \varepsilon^2$, $|\lambda_n|^2 \leq \varepsilon^2$, $|\lambda_n| \leq \varepsilon$. Therefore $\lambda_n \to 0$.

Conversely, if $\lambda_n \to 0$ then $\forall \varepsilon > 0 \; \exists n_\varepsilon \in \mathbb{N}$ such that $|\lambda_n| \leq \varepsilon \; \forall n \geq n_\varepsilon$. If $x = (x_n)_{n\in\mathbb{N}} \in E$ then

$$\sum_{k=n_\varepsilon}^{\infty} |p_k(x)|^2 = \sum_{k=n_\varepsilon}^{\infty} |x_k|^2 = \sum_{k=n_\varepsilon}^{\infty} \frac{|x_k|^2}{\lambda_k^2} \lambda_k^2$$
$$\leq \varepsilon^2 \sum_{k=n_\varepsilon}^{\infty} \frac{|x_k|^2}{\lambda_k^2} \leq \varepsilon^2 \sum_{k=1}^{\infty} \frac{|x_k|^2}{\lambda_k^2} \leq \varepsilon^2,$$

and then E is relatively compact, since E is also bounded (see exercise 13 of chapter 1).

Hence the ellipsoid is compact $\Leftrightarrow (\lambda_n)_{n\in\mathbb{N}} \in c_0$.

Remark. Since $(1/n)_{n\in\mathbb{N}}$ converges towards 0, it follows that the ellipsoid

$$\left\{ (x_1, x_2, ...) \in l_2 \;\middle|\; \sum_{n=1}^{\infty} n^2 x_n^2 \leq 1 \right\}$$

is a compact set in l_2. The same reasoning as above shows that if $1 \leq p < \infty$, the set

$$\left\{ (x_1, x_2, ...) \in l_p \;\middle|\; \sum_{n=1}^{\infty} \frac{|x_n|^p}{\lambda_n^p} \leq 1 \right\}$$

5.2 Solutions

is compact in l_p if and only if $(\lambda_n)_{n\in\mathbb{N}} \in c_0$.

2. See [40, exercise 15.33].

We have $|f(x) - f(y)| \leq L|x - y|$ $\forall f \in M$, $\forall x, y \in [a, b]$. This assures that M is equicontinuous. We also have $|f(x_0)| \leq m$ $\forall f \in M$. Using the uniform Lipschitz condition we obtain that

$$\begin{aligned} |f(x)| &\leq |f(x) - f(x_0)| + |f(x_0)| \leq L|x - x_0| + m \\ &\leq L(|x| + |x_0|) + m \\ &\leq L(\max(|b|, |a|) + |x_0|) + m \quad \forall f \in M, \forall x \in [a, b], \end{aligned}$$

i.e., M is uniformly bounded. The Arzela–Ascoli theorem assures that M is relatively compact.

3. See [40, exercise 15.34].

Let $x, y \in [a, b]$, $x < y$. We have $f(y) - f(x) = \int_x^y f'(t)\,dt$ and using the Hölder inequality we obtain

$$\begin{aligned} |f(y) - f(x)| &\leq \int_x^y |f'(t)|\,dt \leq \left(\int_x^y |f'(t)|^2\,dt\right)^{1/2} \left(\int_x^y 1\,dt\right)^{1/2} \\ &\leq \sqrt{y-x}\left(\int_a^b |f'(t)|^2\,dt\right)^{1/2} \leq \sqrt{k_2}\sqrt{y-x}. \end{aligned}$$

Hence $|f(y) - f(x)| \leq \sqrt{k_2}\sqrt{|y-x|}$ $\forall f \in M$. From here it follows that M is equicontinuous: $\forall \varepsilon > 0 \, \exists \delta_\varepsilon = \varepsilon^2/\sqrt{k_2} > 0$ such that $\forall x, y \in [a, b]$ with $|x - y| < \delta_\varepsilon \Rightarrow |f(x) - f(y)| < \varepsilon$ $\forall f \in M$. Let now $x \in [a, b]$. Then $|f(x) - f(a)| \leq \sqrt{k_2}\sqrt{b-a}$ $\forall f \in M$ from whence using the hypothesis,

$$|f(x)| \leq |f(x) - f(a)| + |f(a)| \leq k_1 + \sqrt{k_2(b-a)} \quad \forall f \in M, \forall x \in [a, b],$$

i.e., M is uniformly bounded. The Arzela–Ascoli theorem assures that M is relatively compact.

4. See [40, exercise 15.35].

Obviously, $\int_a^b |f'(x)|^2\,dx \leq \int_a^b \left(|f(x)|^2 + |f'(x)|^2\right)dx \leq k$ $\forall f \in M$. Then as in the solution for exercise 3, we obtain that $|f(x) - f(a)| \leq \sqrt{k(b-a)} = k_1$ $\forall f \in M$, $\forall x \in [a, b]$, from whence $|f(a)| \leq |f(a) - f(x)| + |f(x)| \leq k_1 + |f(x)|$ $\forall f \in M$. Then

$$\begin{aligned} (b-a)|f(a)| &= \int_a^b |f(a)|\,dt \leq \int_a^b k_1\,dt + \int_a^b |f(x)|\,dx \\ &\leq k_1(b-a) + \left(\int_a^b |f(x)|^2\,dx\right)^{1/2}\left(\int_a^b 1\,dx\right)^{1/2} \\ &\leq k_1(b-a) + (b-a)^{1/2}\left(\int_a^b \left(|f(x)|^2 + |f'(x)|^2\right)dx\right)^{1/2} \\ &\leq k_1(b-a) + \sqrt{b-a}\sqrt{k}, \end{aligned}$$

i.e., $|f(a)| \le c_1 \; \forall f \in M$. Using exercise 3 it follows that the set M is relatively compact in $C[a,b]$.

5. See [40, exercise 15.31].
Let $t_1 < t_2$, $t_1, t_2 \in [a,b]$. We have

$$|y(t_2) - y(t_1)| = \left|\int_{t_1}^{t_2} x(s)ds\right| \le \int_{t_1}^{t_2} |x(s)|\,ds \le \left(\sup_{t \in [a,b]} |x(t)|\right)(t_2 - t_1) \le L(t_2 - t_1),$$

i.e., $|y(t_2) - y(t_1)| \le L|t_2 - t_1| \; \forall t_1, t_2 \in [a,b]$, and therefore the set A is uniformly Lipschitz ($L = \sup_{x \in M} \|x\| < \infty$, by hypothesis). Since $y(0) = 0 \; \forall y \in A$, using exercise 2 it follows that the set A is relatively compact.

6. See [1, exercise 5, Spring 1982].
By the Taylor formula, $\forall x, x_0 \in [0,1]$ we have $g_n(x) = g_n(x_0) + (x - x_0)g_n'(x_0) + ((x - x_0)^2/2) g_n''(c)$, $c \in (0,1)$. Taking $x_0 = 0$ we have by hypothesis $g_n(x) = (x^2/2) g_n''(c)$, $|g_n(x)| = (x^2/2) |g_n''(c)| \le 1/2 \; \forall n \in \mathbb{N}, \forall x \in [0,1]$, i.e., $(g_n)_{n \in \mathbb{N}}$ is uniformly bounded. In the same way

$$|g_n(x) - g_n(x_0)| \le |x - x_0| |g_n'(x_0)| + \frac{|x - x_0|^2}{2}.$$

But $|g_n'(x) - g_n'(0)| = |x| |g_n''(d)| \le |x|$, i.e., $|g_n'(x)| \le |x| \; \forall x \in [0,1]$, from whence

$$|g_n(x) - g_n(x_0)| \le |x - x_0| + \frac{|x - x_0|^2}{2} \; \forall n \in \mathbb{N}, \; \forall x, x_0 \in [0,1],$$

and then $(g_n)_{n \in \mathbb{N}}$ is equicontinuous. We apply now the Arzela–Ascoli theorem.

7. See [40, exercise 15.36].
Using the Lagrange formula it follows that $|f(t_1) - f(t_2)| = |t_1 - t_2| |f'(c)| \le L|t_1 - t_2| \; \forall f \in M, \forall t_1, t_2 \in [a,b]$ and therefore M is equicontinuous, i.e., $\forall \varepsilon > 0 \; \exists \delta_\varepsilon = \varepsilon/L > 0$ such that $\forall t_1, t_2 \in [a,b]$ with $|t_1 - t_2| < \delta_\varepsilon$, we have $|f(t_1) - f(t_2)| < \varepsilon \; \forall f \in M$. Let $f \in M$ and $t \in [a,b]$. Then, by our hypothesis, there is $u \in [a,b]$ such that $f(u) = 0$ (u depends on the function $f \in M$). Then

$$|f(t)| = |f(t) - f(u)| = |t - u| |f'(c)| \le L|t - u| \le L(|t| + |u|) \le 2L \max(b, -a)$$

($t \in [a,b] \Rightarrow |t| = \max(t, -t) \le \max(b, -a)$), i.e., M is uniformly bounded. From the Arzela–Ascoli theorem it follows that M is relatively compact in $C[a,b]$.

8. See [40, exercise 15.44].
i) The answer is no, since $\{f_n \mid n \in \mathbb{N}\}$ is not equicontinuous. If the set were equicontinuous then $\forall \varepsilon > 0 \; \exists \delta_\varepsilon > 0$ such that $\forall x, y \in [0,1]$ with $|x - y| < \delta_\varepsilon \Rightarrow |f_n(x) - f_n(y)| < \varepsilon \; \forall n \in \mathbb{N}$. There is $n_\varepsilon \in \mathbb{N}$ such that $1/n < \delta_\varepsilon \; \forall n \ge n_\varepsilon$. Then for $x = 1 - 1/n$, $y = 1$, $|x - y| = 1/n < \delta_\varepsilon$, we have $|f_n(1 - 1/n) - f_n(1)| < \varepsilon \; \forall n \ge n_\varepsilon$, i.e., $|(1 - 1/n)^n - 1| < \varepsilon \; \forall n \ge n_\varepsilon$ and for $n \to \infty$, $|e^{-1} - 1| < \varepsilon \; \forall \varepsilon > 0$, which is false!

5.2 Solutions

Let us observe that $|f_n(x)| \leq 1 \ \forall n \in \mathbb{N}, \forall x \in [0,1]$.

ii) The answer is no, since the family $\{f_n \mid n \in \mathbb{N}\}$ is not equicontinuous. The proof is similar to (i), taking for example $x_n = \pi/(2n)$, $y_n = (2\pi)/n$.

iii) The answer is no, since $\{f_n \mid n \in \mathbb{N}\} \subseteq \{f_\alpha \mid \alpha \in \mathbb{R}\}$.

iv) The answer is yes! We have $|f'_\alpha(x)| = |\cos(x+\alpha)| \leq 1 \ \forall x \in [0,1], \forall \alpha \in \mathbb{R}$, and from the Lagrange formula it follows that $|f_\alpha(x) - f_\alpha(y)| = |x-y||f'_\alpha(c)| \leq |x-y|$ $\forall \alpha \in \mathbb{R}, \forall x, y \in [0,1]$, i.e., $(f_\alpha)_{\alpha \in \mathbb{R}}$ is uniformly Lipschitz. Since $|f_\alpha(x)| \leq 1 \ \forall \alpha \in \mathbb{R}$, $\forall x \in [0,1]$, from exercise 2 it follows that $(f_\alpha)_{\alpha \in \mathbb{R}}$ is relatively compact.

v) The answer is no, since the considered set is not equicontinuous. If the set were equicontinuous, then $\forall \varepsilon > 0 \ \exists \delta_\varepsilon > 0$ such that: $|x-y| < \delta_\varepsilon \Rightarrow |f_\alpha(x) - f_\alpha(y)| < \varepsilon$ $\forall \alpha \in \mathbb{R}$. Let $n \in \mathbb{N}$ such that $1/n < \delta_\varepsilon$. For $x = 1/n$, $y = 0$, $|x-y| = 1/n < \delta_\varepsilon$, from whence $|f_n(1/n) - f_n(0)| < \varepsilon$. But $f_n(1/n) = \arctan(1) = \pi/4$, $f_n(0) = 0$, and we obtain that $\pi/4 < \varepsilon \ \forall \varepsilon > 0$, which is false!

vi) The answer is yes! We have $|f_\alpha(0)| = e^{-\alpha} \leq 1 \ \forall \alpha \geq 0$, and
$$\int_0^1 |f'_\alpha(x)|^2 \, dx = \int_0^1 e^{2x-2\alpha} dx = \frac{e^2-1}{2} e^{-2\alpha} \leq \frac{e^2-1}{2} \ \forall \alpha \geq 0.$$

From exercise 3 it follows that the set $\{f_\alpha \mid \alpha \geq 0\}$ is relatively compact.

9. i) Suppose that $(x_n)_{n \in \mathbb{N}} \subseteq X$ is a Cauchy sequence. Since X is a Banach space, there is an $x \in X$ such that $x_n \to x$, i.e., $\forall \varepsilon > 0 \ \exists n_\varepsilon \in \mathbb{N}$ such that $\|x_n - x\| < \varepsilon \ \forall n \geq n_\varepsilon$. We will show that $A \subseteq \bigcup_{i=1}^{n_\varepsilon} B(x_i, \varepsilon) \bigcup B(x, \varepsilon)$, which will assure that A is relatively compact.

Let $n \in \mathbb{N}$. If $n \leq n_\varepsilon$ then $x_n \in B(x_n, \varepsilon)$. If $n \geq n_\varepsilon$ then $\|x_n - x\| < \varepsilon$, i.e., $x_n \in B(x, \varepsilon)$.

In fact, from $x_n \to x$ it follows that the set $B = \{x_n \mid n \in \mathbb{N}\} \bigcup \{x\}$ is compact, hence relatively compact. Since $A \subseteq B$ it follows that A is relatively compact.

ii) Let us observe that by hypothesis g takes its values in l_∞. Since T is a compact metric space, g is continuous if and only if it is uniformly continuous, i.e., $\forall \varepsilon > 0 \ \exists \delta_\varepsilon > 0$ such that $\forall x, y \in T$ with $d(x, y) < \delta_\varepsilon$ it follows that $\|g(x) - g(y)\| < \varepsilon$. Using the definition for the norm in the space l_∞, this is equivalent to $\forall \varepsilon > 0 \ \exists \delta_\varepsilon > 0$ such that $\forall x, y \in T$ with $d(x, y) < \delta_\varepsilon$ it follows that $|f_n(x) - f_n(y)| < \varepsilon \ \forall n \in \mathbb{N}$. Now, by the Arzela–Ascoli theorem this is equivalent to the family $(f_n)_{n \in \mathbb{N}} \subseteq C(T)$ being relatively compact.

10. i) See [40, exercise 15.45].

We show that A is relatively compact if and only if $\varphi(x) = 0 \ \forall x \in [0,1]$. Let us suppose that A is relatively compact. Let $a \in (0,1)$. For $n \in \mathbb{N}$ with $1/n < a$, we define $f_n : [0,1] \to \mathbb{R}$, $f_n(x) = \begin{cases} 0 & 0 \leq x < a - 1/n, \\ n(x-a)+1 & a - 1/n \leq x \leq a, \\ 1 & a < x \leq 1. \end{cases}$ Then $f_n \varphi \in A$ $\forall n > 1/a$. Since A is relatively compact, it is equicontinuous, and therefore $\forall \varepsilon > 0$, $\exists \delta_\varepsilon > 0$ such that: $\forall x, y \in [0,1]$ with $|x-y| < \delta_\varepsilon \Rightarrow |f_n(x)\varphi(x) - f_n(y)\varphi(y)| < \varepsilon$ $\forall n \in \mathbb{N}$ with $1/n < a$. There is an $n_\varepsilon \in \mathbb{N}$ such that $1/n < \min(a, \delta_\varepsilon) \ \forall n \geq n_\varepsilon$. For $x = a - 1/n$, $y = a$, $|x-y| = 1/n < \delta_\varepsilon$, we have $|f_n(a - 1/n)\varphi(a - 1/n) - f_n(a)\varphi(a)| < \varepsilon$ $\forall n \geq n_\varepsilon$, that is, $|\varphi(a)| < \varepsilon \ \forall \varepsilon > 0$, and passing to the limit for $\varepsilon \to 0$, $|\varphi(a)| \leq 0$, i.e., $\varphi(a) = 0 \ \forall a \in (0,1)$. Since φ is continuous we have $\varphi(0) = \lim_{a \searrow 0} \varphi(a) = 0$ and $\varphi(1) = \lim_{a \nearrow 1} \varphi(a) = 0$. Hence $\varphi(a) = 0 \ \forall a \in [0,1]$.

The converse is obvious, since $A = \{0\}$.

ii) Let T' be the set of all the accumulation points in T. We will prove that if $T' = \varnothing$ then A is relatively compact for any $\varphi : T \to [0, \infty)$, and that if $T' \neq \varnothing$ then A is relatively compact if and only if $\varphi(a) = 0 \ \forall a \in T'$. Indeed if $T' = \varnothing$ then since T is a compact metric space it follows that T is a finite set, say with n elements, and in this case we can identify in a natural way $C(T)$ with l_∞^n and the set A with a set of the form $\{(x_1, ..., x_n) \in l_\infty^n \mid |x_i| \leq a_i \ \forall 1 \leq i \leq n\}$ for some fixed $(a_1, ..., a_n) \in \mathbb{R}_+^n$, which is clearly relatively norm compact.

Let $a \in T$. For $n \in \mathbb{N}$ define $f_n : T \to \mathbb{R}$, $f_n(x) = \min(n\mathrm{d}(x, a), 1)$. We will show that
$$f_n(x) \to \begin{cases} 1, & \text{if } x \neq a, \\ 0, & \text{if } x = a, \end{cases} \quad \forall x \in T.$$
Indeed, if $x = a$ then $f_n(a) = \min(n\mathrm{d}(a, a), 1) = 0$ $\forall n \in \mathbb{N}$. If $x \neq a$ then $\mathrm{d}(x, a) > 0$, hence there exists $n_x \in \mathbb{N}$ such that $1/n < \mathrm{d}(x, a)$ $\forall n \geq n_x$, i.e., $f_n(x) = \min(n\mathrm{d}(x, a), 1) = 1$ $\forall n \geq n_x$, and then $f_n(x) \to 1$. Suppose that $T' \neq \varnothing$ and let $a \in T'$. If A is relatively compact then $\forall \varepsilon > 0$ $\exists \delta_\varepsilon > 0$ such that $\forall x, y \in T$ with $\mathrm{d}(x, y) < \delta_\varepsilon$ it follows that $|f(x) - f(y)| < \varepsilon$ $\forall f \in A$. Since $f_n \varphi \in A$ $\forall n \in \mathbb{N}$ then $\forall x \in T$, $x \neq a$, with $\mathrm{d}(x, a) < \delta_\varepsilon$, it follows that $|f_n(x)\varphi(x) - f_n(a)\varphi(a)| < \varepsilon$ $\forall n \in \mathbb{N}$, i.e., $|f_n(x)\varphi(x)| < \varepsilon$ $\forall n \in \mathbb{N}$. For $n \to \infty$ we obtain that $|\varphi(x)| \leq \varepsilon$. Hence $\forall \varepsilon > 0$ $\exists \delta_\varepsilon > 0$ such that $\forall x \in T$, $x \neq a$ with $\mathrm{d}(x, a) < \delta_\varepsilon$, it follows that $|\varphi(x)| \leq \varepsilon$, i.e., $\lim_{x \to a} \varphi(x) = 0$. As φ is continuous, $\lim_{x \to a} \varphi(x) = \varphi(a)$, and then $\varphi(a) = 0$.

The converse for the case when $T' = T$ is obvious, since $A = \{0\}$. If $T' \neq T$ then $T \setminus T'$ is finite (in a compact metric space an infinite subset has an accumulation point). If $T \setminus T'$ has at least two elements let $\delta = \min(\delta_1, \delta_2) > 0$ where $\delta_1 = d(T \setminus T', T')$ and $\delta_2 = \min\{d(a, b) \mid a, b \in T \setminus T', a \neq b\}$. Then as is easy to see, $\forall x, y \in T$ with $d(x, y) < \delta$ it follows that either x and y belong to T', or $x = y \in T \setminus T'$, and then for any $f \in A$ we have $f(x) = f(y)$ (since $\varphi = 0$ on T'). From this and the Arzela–Ascoli theorem it follows that A is relatively compact. If $T \setminus T'$ has one element, $T \setminus T' = \{a\}$, we take $\delta = d(a, T')$. Then as it is easy to see, $\forall x, y \in T$ with $d(x, y) < \delta$ it follows that either x and y belong to T', or $x = y = a$. In both cases we have that $f(x) = f(y)$ for any $f \in A$. From this and the Arzela–Ascoli theorem it follows that A is relatively compact.

11. See [1, exercise 1, Fall 1986].

i) Let $\Omega = [0, 1]$ and let $f_n : [0, 1] \to \mathbb{R}$, $f_n(x) = x^n$. Then Ω is compact and $\|f_n\| = \sup_{x \in [0,1]} |x^n| = 1$ $\forall n \in \mathbb{N}$, i.e., the conditions (i) and (ii) are satisfied. Suppose that there is a subsequence $(f_{n_k})_{k \in \mathbb{N}}$ such that $f_{n_k} \to f$ uniformly $[0, 1]$. Then f is continuous. Since $f_{n_k}(x) \to f(x)$ $\forall x \in [0, 1]$, we obtain that $f(x) = \begin{cases} 0 & x \in [0, 1), \\ 1 & x = 1, \end{cases}$ which is not a continuous function. Hence $(f_n)_{n \in \mathbb{N}}$ is not relatively compact and therefore it is not equicontinuous (by the Arzela–Ascoli theorem!).

ii) Let $\Omega = [0, 1]$ and $f_n : [0, 1] \to \mathbb{R}$, $f_n(x) = n$ $\forall x \in [0, 1]$, $\forall n \in \mathbb{N}$. Then Ω is compact, the sequence $(f_n)_{n \in \mathbb{N}}$ is equicontinuous and since $\|f_n\| = n$ $\forall n \in \mathbb{N}$ it follows that $\sup_{n \in \mathbb{N}} \|f_n\| = \infty$.

iii) Let $\Omega = \mathbb{R}$ and $f_n : \mathbb{R} \to \mathbb{R}$, $f_n(x) = \begin{cases} 0 & x \leq n, \\ \arctan(x - n) & x > n. \end{cases}$ It is clear that f_n is continuous $\forall n \in \mathbb{N}$. We also have $\|f_n\| = \sup_{x \in \mathbb{R}} |f_n(x)| = \pi/2$ $\forall n \in \mathbb{N}$. Hence Ω

5.2 Solutions

is not compact, and $\sup_{n\in\mathbb{N}} \|f_n\| < \infty$. We will show that the set $(f_n)_{n\in\mathbb{N}}$ is equicontinuous. For this, let $x_0 \in \mathbb{R}$. We will show that for each $x \in \mathbb{R}$ with $|x - x_0| < \varepsilon$ we have $|f_n(x) - f_n(x_0)| < \varepsilon$ $\forall n \in \mathbb{N}$. Let $n \in \mathbb{N}$. Without loss of generality, we can suppose that $x \leq x_0$. For x and x_0 we can have one of the following situations:

1) $x \leq n$ and $x_0 \leq n$. Then $f_n(x) = f_n(x_0) = 0$, thus $|f_n(x) - f_n(x_0)| < \varepsilon$.
2) $x \leq n < x_0$. Then $f_n(x) = 0$ and $f_n(x_0) = \arctan(x_0 - n)$. Since

$$|(\arctan(t - n))'| = \frac{1}{1 + (t - n)^2} \leq 1 \quad \forall t \in \mathbb{R},$$

using the Lagrange formula it follows that

$$|f_n(x_0) - f_n(x)| = |\arctan(x_0 - n) - \arctan(n - n)| \leq |x_0 - n| \leq x_0 - x < \varepsilon.$$

3) $n < x \leq x_0$. Then we have $f_n(x) = \arctan(x - n)$ and $f_n(x_0) = \arctan(x_0 - n)$. Analogously with (2), we obtain the fact that

$$|f_n(x) - f_n(x_0)| = |\arctan(x - n) - \arctan(x_0 - n)| \leq x_0 - x < \varepsilon.$$

Hence the sequence $(f_n)_{n\in\mathbb{N}}$ is equicontinuous.

We show now that the set $(f_n)_{n\in\mathbb{N}}$ is not relatively compact. For $x \in \mathbb{R}$, there is an $N \in \mathbb{N}$ such that $x \leq n$ $\forall n \geq N$. Then $f_n(x) = 0$ $\forall n \geq N$, i.e., $(f_n)_{n\in\mathbb{N}}$ converges pointwise to the zero function on $\mathbb{R}$. If $(f_n)_{n\in\mathbb{N}}$ has a convergent subsequence, then this subsequence must converge to the zero function. Suppose, therefore, that there is a subsequence $(n_k)_{k\in\mathbb{N}}$ of $\mathbb{N}$ such that $f_{n_k} \to 0$ uniformly on $\mathbb{R}$. Then, there is $k_1 \in \mathbb{N}$ such that $|f_{n_k}(x)| < 1$ $\forall k \geq k_1$, $\forall x \in \mathbb{R}$. Hence, $|f_{n_{k_1}}(x)| < 1$ $\forall x \in \mathbb{R}$. For $x \to \infty$, $f_{n_{k_1}}(x) \to \pi/2$, and then $\pi/2 \leq 1$, false!

12. See [8, lemma 1 and theorem 2, chapter IV].

i) For each $f \in C(\Omega, X)$, a basis of neighborhoods for f in the τ_p topology is given by $W(f; t_1, ..., t_n; \varepsilon) = \{g \in C(\Omega, X) \mid d(f(t_i), g(t_i)) < \varepsilon, i = \overline{1, n}\}$, with $\varepsilon > 0$, $t_1, ..., t_n \in \Omega$. Let $\varepsilon > 0$. Since by hypothesis A is equicontinuous, for each $t \in \Omega$, there is an open neighborhood $V(t)$ of t such that $d(f(s), f(t)) < \varepsilon/3$ $\forall s \in V(t)$, $\forall f \in A$. Then $\Omega = \bigcup_{t\in\Omega} V(t)$, and since Ω is compact there are $t_1, ..., t_n \in \Omega$ such that $\Omega \subseteq \bigcup_{i=1}^{n} V(t_i)$. For an element $f_0 \in A$, let us consider the set $W(f_0; t_1, ..., t_n; \varepsilon/3)$ defined above. Let $f \in W(f_0; t_1, ..., t_n; \varepsilon/3) \cap A$. Let $t \in \Omega$. Since $\Omega \subseteq \bigcup_{i=1}^{n} V(t_i)$, there is an i such that $t \in V(t_i)$, and then $d(f(t), f(t_i)) < \varepsilon/3$, $d(f_0(t), f_0(t_i)) < \varepsilon/3$. Then $d(f(t), f_0(t)) \leq d(f(t), f(t_i)) + d(f(t_i), f_0(t_i)) + d(f_0(t_i), f_0(t)) < \varepsilon$. Since $t \in \Omega$ is arbitrary, it follows that $\rho(f, f_0) \leq \varepsilon$. Hence: for every $f_0 \in A$ and $\varepsilon > 0$ there is $U \in \tau_p$, $f_0 \in U$, such that $U \cap A \subseteq \overline{B_\rho(f_0, \varepsilon)} \cap A$, with an obvious notation. It follows that $\tau_u|_A \subseteq \tau_p|_A$. But obviously, we have $\tau_p|_A \subseteq \tau_u|_A$, and then $\tau_u|_A = \tau_p|_A$.

ii) We consider the product space X^Ω, endowed with the product topology. The elements of X^Ω are of the form $\varphi = (\varphi_t)_{t\in\Omega}$, with $\varphi_t \in X$ for each $t \in \Omega$. We define $\psi : C(\Omega, X) \to X^\Omega$, $\psi(f) = (f(t))_{t\in\Omega}$. Let us consider on $C(\Omega, X)$ the topology τ_p and on X^Ω the product topology. We show that ψ is continuous. By the definition of the product topology it is sufficient to prove that for each $t \in \Omega$, the function $f \to f(t)$ from $(C(\Omega, X), \tau_p)$ to X is continuous, which is obviously true. Clearly ψ is injective, and let us prove now that $\psi^{-1} : \psi(C(\Omega, X)) \to C(\Omega, X)$ is continuous. Again this is equivalent to the continuity for the functions $\delta_t \circ \psi^{-1} : \psi(C(\Omega, X)) \to X$ where $\delta_t(f) = f(t)$ $\forall f \in C(\Omega, X)$. But $\delta_t \circ \psi^{-1} = \mathrm{pr}_t$ on $\psi(C(\Omega, X))$, where pr_t is the canonical projection $\varphi = (\varphi_t)_{t \in \Omega} \longmapsto \varphi_t$ from X^Ω to X, which, by the definition of the product topology is continuous. Hence $\psi : (C(\Omega, X), \tau_p) \to \psi(C(\Omega, X)) \subseteq X^\Omega$ is a homeomorphism (1).

Let $F = \overline{A}^{\tau_p}$, $A \subseteq C(\Omega, X)$ equicontinuous. We show that F is also equicontinuous. Let $t_0 \in \Omega$, $\varepsilon > 0$. Since A is equicontinuous, there is U an open neighborhood of t_0 such that $d(f(t), f(t_0)) < \varepsilon/3$ $\forall t \in U$, $\forall f \in A$. Let $t \in U$, $g \in F$. Since $g \in F = \overline{A}^{\tau_p}$ and $W(g; t, t_0; \varepsilon/3)$ is a neighborhood of g with respect to the τ_p topology, from the definition of the closure it follows that $W(g; t, t_0; \varepsilon/3) \cap A \neq \emptyset$, hence there is $f \in A$, with $d(f(t), g(t)) < \varepsilon/3$, $d(f(t_0), g(t_0)) < \varepsilon/3$. Then $d(g(t), g(t_0)) \leq d(g(t), f(t)) + d(f(t), f(t_0)) + d(f(t_0), g(t_0)) < \varepsilon$, and hence $\forall g \in F$, $\forall t \in U$ we have $d(g(t), g(t_0)) < \varepsilon$, hence F is equicontinuous (2).

By (1), $\psi\left(\overline{A}^{\tau_p}\right) \subseteq X^\Omega$ is closed in the product topology. Since X is Hausdorff (every metric space is Hausdorff) it follows that X^Ω is also Hausdorff. If we will show that $\psi\left(\overline{A}^{\tau_p}\right) \subseteq X^\Omega$ is contained in a compact set, then $\psi\left(\overline{A}^{\tau_p}\right)$ will also be compact, and then by (1) it follows that $\overline{A}^{\tau_p}$ is compact. By (2), $\overline{A}^{\tau_p}$ is equicontinuous. Then by (i), τ_u and τ_p coincide on $\overline{A}^{\tau_p}$, and then $\overline{A}^{\tau_p} = \overline{A}^{\tau_u}$. Since $\left(\overline{A}^{\tau_p}, \tau_p\right)$ is compact it follows that $\left(\overline{A}^{\tau_p}, \tau_u\right)$ is compact, hence $\left(\overline{A}^{\tau_u}, \tau_u\right)$ is compact, which is just what we want to prove.

Hence it remains to be proved that $\psi\left(\overline{A}^{\tau_p}\right)$ is contained in a compact set of X^Ω. By our hypothesis, for each $t \in \Omega$ there is a compact set $X_t \subseteq X$ such that $f(t) \in X_t$ $\forall f \in A$. For $\varphi = \psi(f)$, $f \in \overline{A}^{\tau_p}$, $\varphi_t = f(t)$ $\forall t \in \Omega$. Since $f \in \overline{A}^{\tau_p}$ there is a net $(f_\delta)_{\delta \in \Delta} \subseteq A$ such that $f_\delta \xrightarrow{\tau_p} f$. Hence $f_\delta(t) \to f(t)$. Then $f(t) \in \overline{X_t}^d = X_t$, and hence $\varphi \in \prod_{t \in \Omega} X_t$, i.e., $\{\psi(f) | f \in \overline{A}^{\tau_p}\} \subseteq \prod_{t \in \Omega} X_t$. Since X_t is compact for each $t \in \Omega$, by the well known Tychonoff theorem it follows that $\prod_{t \in \Omega} X_t$ is compact.

13. i) See [8, theorem 3, chapter IV].

Obviously $T_{A,B} \in L(L(X), L(X))$, $\|T_{A,B}\| \leq \|A\| \|B\|$. Let us denote as usual with B_X the closed unit ball in X. Since B is compact, $\overline{B(B_X)} \subseteq X$ is compact. Let $\Omega = \overline{B(B_X)}$. Let $\mathcal{A} = \{AS \mid S \in L(X), \|S\| \leq 1\} \subseteq C(\Omega, X)$. For $x_0 \in \Omega$, $\varepsilon > 0$, let $x \in \Omega$ with $\|x - x_0\| < \varepsilon / (\|A\| + 1)$. Then $\|ASx - ASx_0\| \leq \|A\| \|S\| \|x - x_0\| \leq (\varepsilon \|A\|) / (\|A\| + 1) < \varepsilon$ $\forall S \in L(X), \|S\| \leq 1$ and hence $\mathcal{A} \subseteq C(\Omega, X)$ is equicontinuous. For each $x \in \Omega$ the set $\{ASx \mid S \in L(X), \|S\| \leq 1\}$ is contained in $A(\{z \in X \mid \|z\| \leq \|x\|\})$, which, since A is compact, is relatively compact.

By the Arzela–Ascoli theorem, the form from exercise 12, it follows that $\mathcal{A} \subseteq C(\Omega, X)$ is relatively compact with respect to the uniform convergence topology. If we consider a sequence $(S_n)_{n \in \mathbb{N}} \subseteq B_{L(X)}$, there is a subsequence $(n_k)_{k \in \mathbb{N}}$ of $\mathbb{N}$ such that the sequence $(AS_{n_k})_{k \in \mathbb{N}}$ is uniformly convergent in $C(\Omega, X)$. Then $(AS_{n_k}B)_{k \in \mathbb{N}}$ is uniformly convergent on B_X, hence in the operator norm.

ii) Since $B \neq 0$ we also have $B^* \neq 0$ (we know that $\|B\| = \|B^*\|$). Let then $x_0^* \in X^*$, with $B^* x_0^* \neq 0$. For $x^* \in X^*$ and $x \in X$ we have $T_{A,B}(x^* \otimes x) = B^* x^* \otimes Ax$, since $[T_{A,B}(x^* \otimes x)](y) = [A \circ (x^* \otimes x)](By) = x^*(By)Ax = [(B^* x^*) \otimes Ax](y)$ $\forall y \in X$.

5.2 Solutions

Let $(x_n)_{n\in\mathbb{N}} \subseteq B_X$. Then $\|x_0^* \otimes x_n\| = \|x_0^*\|\|x_n\| \leq \|x_0^*\|$, and since $T_{A,B}$ is compact, there is a subsequence $(n_k)_{k\in\mathbb{N}}$ of $\mathbb{N}$ such that $(T_{A,B}(x_0^* \otimes x_{n_k}))_{k\in\mathbb{N}}$ converges. Then $\lim_{k,s\to\infty} \|T_{A,B}(x_0^* \otimes x_{n_k}) - T_{A,B}(x_0^* \otimes x_{n_s})\| = 0$, hence $\lim_{k,s\to\infty} \|B^*x_0^* \otimes (Ax_{n_k} - Ax_{n_s})\| = 0$, i.e., $\lim_{k,s\to\infty} \|B^*x_0^*\|\|Ax_{n_k} - Ax_{n_s}\| = 0$. Since $B^*x_0^* \neq 0$ it follows that the sequence $(Ax_{n_k})_{k\in\mathbb{N}} \subseteq X$ is Cauchy, hence convergent (X is a Banach space). Therefore A is compact and in the same way one can prove that B is also compact.

Remark. For $x^* \in X^*$, $x \in X$ we denote as usual with $x^* \otimes x \in L(X)$ the operator $(x^* \otimes x)(y) = x^*(y)x \; \forall y \in X$. Obviously, $\|x^* \otimes x\| = \|x^*\|\|x\|$.

14. i) a) $M_\varphi f \in C[0,1]$, since the product of two continuous functions is a continuous function. Obviously M_φ is linear. Also

$$\|M_\varphi f\| = \sup_{x\in[0,1]} |(M_\varphi f)(x)| = \sup_{x\in[0,1]} |\varphi(x)f(x)| \leq \|\varphi\|\|f\|,$$

from whence $\|M_\varphi\| \leq \|\varphi\|$. Since $M_\varphi(1) = \varphi$ and $\|\varphi\| = \|M_\varphi(1)\| \leq \|M_\varphi\|\|1\| = \|M_\varphi\|$, it follows that $\|M_\varphi\| = \|\varphi\| = \sup_{x\in[0,1]} |\varphi(x)|$.

b) If M_φ is compact then $\{M_\varphi f \mid \|f\| \leq 1\}$ is relatively compact, i.e., $\{f\varphi \mid |f(x)| \leq 1 \; \forall x \in [0,1]\}$ is relatively compact. Reasoning as in the solution for exercise 10(i) we obtain that $\varphi(x) = 0 \; \forall x \in [0,1]$, i.e., $M_\varphi = 0$.

ii) Let T' be the set of all the accumulation points in T. Reasoning as in the solution for exercise 10(ii) we obtain that if $T' = \varnothing$ then M_φ is compact for any $\varphi \in C(T)$, while if $T' \neq \varnothing$ then M_φ is compact if and only if $\varphi(a) = 0 \; \forall a \in T'$.

15. See [40, exercise 16.1].

i) Taking $\varphi(x) = x$, we have $Uf = \varphi f$ and from exercise 14, $\|U\| = \|\varphi\| = \sup_{x\in[0,1]} |x| = 1$. Since $\varphi \neq 0$ we obtain that U is not compact.

ii) Since U is positive using exercise 1 of chapter 2 we obtain that $\|U\| = \|U(1)\| = \sup_{x\in[0,1]} (1+x) = 2$. Let $A = \{Uf \mid \|f\| \leq 1\}$. If $g \in A$ then $g(x) = f(0) + xf(1)$ $\forall x \in [0,1]$, from whence $|g(0)| = |f(0)| \leq \|f\| \leq 1$, i.e., $|g(0)| \leq 1 \; \forall g \in A$. Also, $g'(x) = f(1)$, and $\int_0^1 |g'(x)|^2 dx = |f(1)|^2 \leq \|f\|^2 \leq 1 \; \forall g \in A$. From exercise 3 it follows that A is relatively compact, i.e., U is a compact operator. (In fact, U is a finite rank operator.)

iii) For $\|f\| \leq 1$ let $g(x) = \int_0^1 e^{tx} f(t)\,dt \; \forall x \in [0,1]$. Using the differentiation theorem for the Riemann integral with a parameter, we have

$$g'(x) = \int_0^1 \frac{\partial}{\partial x}(e^{tx}f(t))dt = \int_0^1 te^{tx}f(t)\,dt,$$

$$|g'(x)| \leq \int_0^1 te^{tx}|f(t)|\,dt \leq \int_0^1 te^{tx}dt \leq e \; \forall x \in [0,1].$$

Using the Lagrange formula it follows that the set of all functions g is uniformly Lipschitz. We also have

$$|g(0)| = \left|\int_0^1 f(t)\,dt\right| \leq \int_0^1 |f(t)|\,dt \leq \|f\| \leq 1$$

for all g. From exercise 2 it follows that $\{Uf \mid \|f\| \leq 1\}$ is relatively compact. The norm of $\|U\|$ is calculated at exercise 2 of chapter 2.

iv) Since $\varphi : [0,1] \to [0,1]$, $\varphi(x) = x^2$ is surjective, we have

$$\|Uf\| = \sup_{x\in[0,1]} |Uf(x)| = \sup_{x\in[0,1]} |f(x^2)| = \sup_{x\in[0,1]} |f(\varphi(x))|$$
$$= \sup_{x\in[0,1]} |f(x)| = \|f\| \quad \forall f \in C[0,1],$$

i.e., $\|U\| = 1$. See also exercise 6 of chapter 2. For the compactness we have that $\{Uf \mid \|f\| \leq 1\} = \{f \mid \|f\| \leq 1\} = B_{C[0,1]}$, which is not relatively compact, since $C[0,1]$ is infinite-dimensional.

16. See [40, exercise 16.2].

Let $g = Uf$. Then $(Ug)(x) = (g(x) + g(-x))/2 = g(x) = (Uf)(x)$, since $g(x) = g(-x) \ \forall x \in [-1,1]$. Hence $U^2 = U$, i.e., U is a projector. Then $\|U\| = \|U^2\| \leq \|U\|^2$ and since $U \neq 0$ we obtain that $1 \leq \|U\|$. But $\|Uf\| = \sup_{x\in[0,1]} |Uf(x)|$, and

$$|Uf(x)| \leq \frac{|f(x)| + |f(-x)|}{2} \leq \frac{\|f\| + \|f\|}{2} = \|f\|,$$

i.e., $\|Uf\| \leq \|f\| \ \forall f \in C[-1,1]$, $\|U\| \leq 1$. Therefore $\|U\| = 1$. Also, U is not compact: we consider the set $\{Uf_n \mid n \in \mathbb{N}\} \subseteq \{Uf \mid \|f\| \leq 1\}$, where $f_n(x) = x^{2n} \ \forall x \in [-1,1]$, and we use exercise 8(i).

The same type of reasoning can also be applied for V.

17. See [40, exercise 16.8].

i) We have $\|Ux\| = \left(\sum_{n=1}^{\infty} |x_n|^p\right)^{1/p} = \|x\| \ \forall x \in l_p$, from whence $\|U\| = 1$. Let us observe that $U(x_1, x_2, ...) = (0, x_1, x_2, ...)$ and hence if we denote $V : l_p \to l_p$, $V(x_1, x_2, ...) = (x_2, x_3, ...)$ then $(VU)(x_1, x_2, ...) = V(U(x_1, x_2, ...)) = V(0, x_1, x_2, ...) = (x_1, x_2, ...)$, i.e., $VU = I$. Obviously, V is linear and continuous, since

$$\|Vx\| = \left(\sum_{n=2}^{\infty} |x_n|^p\right)^{1/p} \leq \left(\sum_{n=1}^{\infty} |x_n|^p\right)^{1/p} = \|x\|.$$

If U were compact, by the ideal property for the compact operators, we would obtain that VU is compact, i.e., I is compact, i.e., the set

$$\{I(x) \mid \|x\| \leq 1\} = \left\{ x = (x_1, x_2, ...) \ \Big| \ \sum_{n=1}^{\infty} |x_n|^p \leq 1 \right\}$$

must be relatively compact in l_p. But from exercise 1(ii) (see the remark following the solution), this set corresponds to the sequence $(1, 1, ...) \notin c_0$, and therefore it is not relatively compact.

The above technique can also be used in other situations. We can obtain a much shorter solution, observing that $A = \{U(x) \mid \|x\| \leq 1\}$ is not relatively compact, since $e_n \in A$

5.2 Solutions

$\forall n \geq 2$ and $\|e_{n+p} - e_n\| = 2^{1/p}$ $\forall n \geq 2$, $\forall p \in \mathbb{N}$, i.e., $\{e_n \mid n \geq 2\} \subseteq A$ does not contain Cauchy subsequences. We can also solve the exercise by using the Riesz theorem, since l_p is not finite-dimensional.

ii) We have

$$\|Ux\| = \left(\sum_{n=1}^{\infty} \frac{|x_n|^p}{n^p}\right)^{1/p} \leq \left(\sum_{n=1}^{\infty} |x_n|^p\right)^{1/p} = \|x\| \quad \forall x \in l_p,$$

i.e., $\|U\| \leq 1$. In addition, $\|e_1\| = \|U(e_1)\| \leq \|U\| \|e_1\|$, i.e., $\|U\| \geq 1$, from whence $\|U\| = 1$. We have

$$\{Ux \mid \|x\| \leq 1\} = \left\{y = (y_1, y_2, \ldots) \in l_p \;\middle|\; \sum_{n=1}^{\infty} n^p |y_n|^p \leq 1\right\}.$$

Since the sequence $(1/n)_{n \in \mathbb{N}}$ is in c_0, from exercise 1(ii) the set $\{Ux \mid \|x\| \leq 1\}$ is relatively compact in l_p, and therefore U is a compact operator.

We give now another solution for the fact that U is compact. Let $U_n : l_p \to l_p$, $U_n(x_1, x_2, \ldots) = (x_1, x_2/2, \ldots, x_n/n, 0, 0, \ldots)$. Then for $\|x\| \leq 1$ we have

$$\|U(x) - U_n(x)\| = \left\|\left(0, \ldots, 0, \frac{x_{n+1}}{n+1}, \ldots\right)\right\| = \left(\sum_{k=n+1}^{\infty} \frac{|x_k|^p}{k^p}\right)^{1/p}$$

$$\leq \frac{1}{n}\left(\sum_{k=n+1}^{\infty} |x_k|^p\right)^{1/p} \leq \frac{1}{n},$$

from whence $\|U_n - U\| \to 0$. But U_n are finite rank operators, since $U_n(x) = \sum_{k=1}^{n} p_k(x) e_k/k$ $\forall x \in l_p$, where $p_k : l_p \to \mathbb{K}$ are the canonical projections. Since every finite rank operator is compact it follows that U_n is compact $\forall n \in \mathbb{N}$, and since $U_n \to U$ in norm it follows that U is compact.

iii) Let $A : l_p \to l_p$, $A(x_1, x_2, \ldots) = (0, x_1, x_2, \ldots)$ and $B : l_p \to l_p$, $B(x_1, x_2, \ldots) = (x_1, x_2/2, \ldots)$, i.e., the operators from (i) and (ii). Then $U = AB$. Since B is compact, by the ideal property of the compact operators it follows that U is compact. Analogously with (ii) one can prove that $\|U\| = 1$. (We use that $U(e_1) = e_2$.)

18. See [40, exercise 16.10].
If $x \in l_1$, $x = (x_1, x_2, \ldots)$, then

$$\|x\|_2 = \left(\sum_{n=1}^{\infty} |x_n|^2\right)^{1/2} \leq \sum_{n=1}^{\infty} |x_n| = \|x\|_1,$$

i.e., $J(x) \in l_2$ and $\|J(x)\| \leq \|x\|$ $\forall x \in l_2$. Therefore $\|J\| \leq 1$. In addition, $\|e_1\| = \|J(e_1)\| \leq \|J\| \|e_1\|$. Then $\|J\| \geq 1$, and therefore $\|J\| = 1$. Since

$$\{e_n \mid n \in \mathbb{N}\} = \{J(e_n) \mid n \in \mathbb{N}\} \subseteq J(B_{l_1}),$$

it follows that J is not compact.

19. i) $V : C[0,1] \to C[0,1]$, $(Vf)(x) = \int_0^x f(t)\,dt \; \forall f \in C[0,1], \; \forall x \in [0,1]$, is clearly a linear operator. Also, $\|Vf\| = \sup_{x\in[0,1]} \left|\int_0^x f(t)\,dt\right| \leq \|f\| \; \forall f \in C[0,1]$, and therefore V is continuous with $\|V\| \leq 1$. If we consider the constant function $\mathbf{1}(x) = 1 \; \forall x \in [0,1]$, with $\|\mathbf{1}\| = 1$, we obtain that $\|V(\mathbf{1})\| = \sup_{x\in[0,1]} \left|\int_0^x dt\right| = \sup_{x\in[0,1]} |x| = 1$, and therefore $\|V\| = 1$. For the compactness see exercise 5.

ii) It is proved in the solution for exercise 14(ii) of chapter 2 that H is linear and continuous.

It is proved in the solution for exercise 38(i) of chapter 14 that the following general result is true: if X, Y are Banach spaces and $U \in L(X,Y)$ is a compact operator, then from $x_n \to 0$ weak in X it follows that $Ux_n \to 0$ in the norm of Y. For the sequence $f_n = n\chi_{(0,1/n^p)}$, we have $f_n \to 0$ weak (the reasoning used in the solution for exercise 12 of chapter 14 is also true in $L_p(0,\infty)$) and if, for a contradiction, H is compact, then by the above result we must have $Hf_n \to 0$ in norm. Now, if $a,b > 0$ and $f = a\chi_{(0,b)}$ then

$$Hf(x) = \frac{1}{x}\int_0^x a\chi_{(0,b)}(t)\,dt = \frac{a}{x}\mu((0,x)\cap(0,b)) = \begin{cases} a & \text{if } 0 < x < b, \\ ab/x & \text{if } x \geq b, \end{cases}$$

from whence

$$\|Hf\|^p = a^p b + (ab)^p \int_b^\infty x^{-p}\,dx = \frac{p}{p-1}a^p b.$$

Thus $\|Hf_n\|^p = p/(p-1) \; \forall n \in \mathbb{N}$, and therefore H is not compact.

20. See [40, exercise 16.13].

We have $U^2(x) = U(U(x)) = U(0,x_1,0,x_3,0,x_5,\ldots) = (0,0,\ldots)$, i.e., $U^2 = 0$, and therefore U^2 is compact. We have $U(e_{2n-1}) = e_{2n} \; \forall n \in \mathbb{N}$. Since for $n,p \in \mathbb{N}$, $\|e_{n+k} - e_n\| = 1$ if $X = c_0$ or l_∞, and $\|e_{n+k} - e_n\| = 2^{1/p}$ if $X = l_p$, $1 \leq p < \infty$, it follows that the set $\{U(e_{2n-1}) \mid n \in \mathbb{N}\} = \{e_{2n} \mid n \in \mathbb{N}\} \subseteq U(B_X)$ is not relatively compact, from whence it follows that $U(B_X)$ is not relatively compact, i.e., U is not a compact operator.

21. See [40, exercise 16.21].

i) Let $x = (x_n)_{n\in\mathbb{N}} \in l_p$. We have $|\lambda_n x_n|^p \leq \|\lambda\|_\infty^p |x_n|^p \; \forall n \in \mathbb{N}$ and the series $\sum_{n=1}^\infty |x_n|^p$ converges, hence by the comparison test for series it follows that the series $\sum_{n=1}^\infty |\lambda_n x_n|^p$ converges. Moreover,

$$\|M_\lambda(x)\| = \left(\sum_{n=1}^\infty |\lambda_n x_n|^p\right)^{1/p} \leq \|\lambda\|_\infty \left(\sum_{n=1}^\infty |x_n|^p\right)^{1/p} = \|\lambda\|_\infty \|x\| \quad \forall x \in l_p.$$

Since M_λ is linear it follows that M_λ is continuous and $\|M_\lambda\| \leq \|\lambda\|_\infty$. Also, for any $n \in \mathbb{N}$,

$$|\lambda_n| = \|M_\lambda(e_n)\| \leq \|M_\lambda\|\|e_n\| = \|M_\lambda\|,$$

5.2 Solutions

so $\|\lambda\|_\infty = \sup_{n\in\mathbb{N}} |\lambda_n| \leq \|M_\lambda\|$.

ii) Suppose that M_λ is compact, i.e., $M_\lambda(B_{l_p})$ is relatively compact. Then $\forall \varepsilon > 0$ $\exists n_\varepsilon \in \mathbb{N}$ such that $\sum_{k=n_\varepsilon}^\infty |p_k(M_\lambda(x))|^p \leq \varepsilon^p \ \forall x \in B_{l_p}$. Let $n \geq n_\varepsilon$. Then for $e_n \in B_{l_p}$ we have $\sum_{k=n_\varepsilon}^\infty |p_k(M_\lambda(e_n))|^p \leq \varepsilon^p$, that is, $|\lambda_n| \leq \varepsilon$. Therefore $\lambda_n \to 0$.

Conversely, if $\lambda_n \to 0$, then $\forall \varepsilon > 0 \ \exists n_\varepsilon \in \mathbb{N}$ such that $|\lambda_n| \leq \varepsilon \ \forall n \geq n_\varepsilon$. If $x = (x_k)_{k\in\mathbb{N}} \in B_{l_p}$, then

$$\sum_{k=n_\varepsilon}^\infty |p_k(M_\lambda(x))|^p = \sum_{k=n_\varepsilon}^\infty |\lambda_k x_k|^p \leq \varepsilon^p \sum_{k=n_\varepsilon}^\infty |x_k|^p \leq \varepsilon^p \sum_{k=1}^\infty |x_k|^p \leq \varepsilon^p,$$

and therefore $M_\lambda(B_{l_p})$ is relatively compact, i.e., M_λ is compact.

22. See [18, exercise 7 (i), chapter VII].
Obviously Σ is well defined and linear. We have

$$\|\Sigma((t_n)_{n\in\mathbb{N}})\|_\infty = \sup_{n\in\mathbb{N}} \left|\sum_{i=1}^n t_i\right| \leq \sup_{n\in\mathbb{N}} \sum_{i=1}^n |t_i| = \|(t_n)_{n\in\mathbb{N}}\|_1,$$

hence Σ is a continuous operator. To show that Σ is not compact, let $\{e_n \mid n \in \mathbb{N}\}$ be the standard basis from l_1. We have $\|e_n\|_1 = 1 \ \forall n \in \mathbb{N}$ and $\|\Sigma(e_n) - \Sigma(e_m)\|_\infty = 1 \ \forall n \neq m$. Hence the sequence $(\Sigma(e_n))_{n\in\mathbb{N}}$ cannot contain Cauchy subsequences, and therefore Σ is not compact.

ii) Clearly i is well defined and linear. We also have

$$\|i((t_n)_{n\in\mathbb{N}})\|_\infty = \sup_{n\in\mathbb{N}} |t_n| \leq \sum_{n=1}^\infty |t_n| = \|(t_n)_{n\in\mathbb{N}}\|_1,$$

hence i is continuous. We define $j : l_\infty \to l_\infty$,

$$j((t_n)_{n\in\mathbb{N}}) = (t_1, t_2 - t_1, t_3 - t_2, ..., t_{n+1} - t_n, ...).$$

Obviously j is well defined and linear. We have $\|j((t_n)_{n\in\mathbb{N}})\|_\infty \leq 2\|(t_n)_{n\in\mathbb{N}}\|_\infty$ $\forall (t_n)_{n\in\mathbb{N}} \in l_\infty$, and therefore j is continuous. Also, for each $(t_n)_{n\in\mathbb{N}} \in l_1$ we have

$$\begin{aligned} j(\Sigma((t_n)_{n\in\mathbb{N}})) &= j(t_1, t_1 + t_2, t_1 + t_2 + t_3, ...) \\ &= (t_1, (t_1 + t_2) - t_1, (t_1 + t_2 + t_3) - (t_1 + t_2), ...) = (t_1, t_2, t_3, ...), \end{aligned}$$

i.e., $j \circ \Sigma = i$.

Let us consider again the elements $\{e_n \mid n \in \mathbb{N}\}$ in l_1. We have $\|i(e_n) - i(e_m)\|_\infty = 1$ $\forall m \neq n$, and i cannot be a compact operator. The property $j \circ \Sigma = i$ now implies (this is already proved at (i)) that Σ is not compact.

In connection with this exercise, see exercise 43 of chapter 14.

iii) We have

$$\left|\int_0^x f(t)dt\right| \leq \int_0^x |f(t)|\,dt \leq \int_0^\infty |f(t)|\,dt = \|f\|_1 \quad \forall x \in (0, \infty),$$

i.e., U takes its values in $L_\infty(0,\infty)$, and $\|Uf\|_\infty \leq \|f\|_1 \ \forall f \in L_1(0,\infty)$. Since obviously U is linear, it follows that U is continuous, and $\|U\| \leq 1$. From

$$U(\chi_{(0,1)})(x) = \int_0^x \chi_{(0,1)}(t)dt = \begin{cases} x & \text{if } 0 < x < 1, \\ 1 & \text{if } x \geq 1, \end{cases}$$

$\|U(\chi_{(0,1)})\|_\infty = 1$ and $\|U(\chi_{(0,1)})\|_\infty \leq \|U\| \|\chi_{(0,1)}\|_1$, it follows that $\|U\| \geq 1$, and therefore $\|U\| = 1$.

Let now $f_n = \chi_{[n,n+1)}$, $\|f_n\|_1 = 1 \ \forall n \in \mathbb{N}$. If we denote by $g_n = Uf_n$, we have

$$g_n(x) = \begin{cases} 0 & \text{if } 0 < x < n, \\ x - n & \text{if } n \leq x < n+1, \\ 1 & \text{if } x \geq n+1, \end{cases}$$

and therefore $\|g_{n+k} - g_n\|_\infty = 1 \ \forall n \in \mathbb{N}, \ \forall k \in \mathbb{N}$. The sequence $(Uf_n)_{n \in \mathbb{N}}$ cannot contain Cauchy subsequences and therefore U is not compact.

23. i) See [40, exercise 16.38].

Let us suppose, for a contradiction, that A is surjective. From the open mapping theorem it follows that A is an open operator, in particular $A(B(0,1)) \subseteq X$ is an open set, i.e., there is $\varepsilon > 0$ such that $\overline{B}(0,\varepsilon) = \varepsilon\overline{B}(0,1) \subseteq A(B(0,1))$, where $B(0,1) = \{x \in X \mid \|x\| < 1\}$. Since A is compact it follows that $A(B(0,1))$ is relatively compact. Hence there is $\varepsilon > 0$ such that $\overline{B}(0,\varepsilon) = \varepsilon\overline{B}(0,1)$ is relatively compact, from whence $\overline{B}(0,1)$ is relatively compact, and therefore compact. But if $\overline{B}(0,1)$ is compact then by the Riesz theorem X is finite-dimensional, which is false! Hence, A is not surjective.

ii) Choose $y = (1/n^\alpha)_{n \in \mathbb{N}} \in l_p$, i.e., the generalized harmonic series $\sum_{n=1}^\infty 1/n^{\alpha p}$ converges, that is, $\alpha p > 1$, $\alpha > 1/p$. If $x = (x_n)_{n \in \mathbb{N}} \in l_p$ has the property that $U(x) = y$, then $(1/n)x_n = 1/n^\alpha \ \forall n \in \mathbb{N}$, i.e., $x_n = 1/n^{\alpha-1} \ \forall n \in \mathbb{N}$. Since $x \in l_p$, the generalized harmonic series $\sum_{n=1}^\infty 1/n^{(\alpha-1)p}$ converges, and therefore $(\alpha-1)p > 1$, $\alpha > 1 + 1/p$. From here it follows that for $y = (1/n^{1/p+1})_{n \in \mathbb{N}} \in l_p$, the equation $U(x) = y$ has no solution.

iii) For $n \in \mathbb{N}$, let $A_n : c_{00} \to c_{00}$, $A_n(x_1, x_2, ...) = (x_1, x_2/2, ..., x_n/n, 0, ...)$ $\forall (x_k)_{k \in \mathbb{N}} \in c_{00}$. Then $\|A - A_n\| = 1/(n+1) \ \forall n \in \mathbb{N}$, and therefore $A_n \to A$. Since A_n are finite rank operators, we obtain that A is compact. The fact that A is bijective is obvious.

24. See [40, exercise 16.41].

Let $Z = A(X) \subseteq Y$. The condition from our hypothesis assures that A is injective (if $Ax = 0$ then $x = 0$). Hence $U : X \to Z$, $U(x) = A(x)$ is bijective. Let us consider $U^{-1} : Z \to X$. Clearly U^{-1} is linear. Also $\|U^{-1}(y)\| \leq \|y\|/c \ \forall y \in Z$. Indeed, let $y \in Z = A(X)$. There is then an $x \in X$ such that $A(x) = y$, i.e., $U(x) = y$, $U^{-1}(y) = x$. By our hypothesis, $\|Ax\| \geq c\|x\|$, i.e., $\|y\| \geq c\|U^{-1}(y)\|$, $\|U^{-1}(y)\| \leq \|y\|/c \ \forall y \in Z$, i.e., $U^{-1} : Z \to X$ is linear and continuous. If U is a compact operator, then by the ideal property for the compact operators it follows that $U^{-1}U : X \to X$ is compact, i.e., $I : X \to X$ is compact, which means that X is finite-dimensional (the Riesz theorem).

5.2 Solutions

If X is finite-dimensional then every $A \in L(X,Y)$ is compact (in a finite-dimensional normed space a set is compact if and only if it is closed and bounded). Hence A is compact if and only if X is finite-dimensional.

25. i) See [40, exercise 16.25].

Since X is infinite-dimensional, using exercise 24 it follows that $\forall c > 0 \; \exists x_c \in X$ such that $\|Ax_c\| < c \|x_c\|$. In particular, $\forall n \in \mathbb{N} \; \exists x_n \in X$ such that $\|Ax_n\| < \|x_n\|/n$. Let us observe that $x_n \neq 0 \; \forall n \in \mathbb{N}$. Let

$$a_n = \frac{x_n}{\|x_n\|} \in X, \; \|a_n\| = 1 \; \forall n \in \mathbb{N}.$$

Then $\|A(a_n)\| < 1/n \; \forall n \in \mathbb{N}$, and therefore $A(a_n) \to 0$.

ii) Let $U \in K(X,Y)$ and $\varepsilon > 0$. From (i) it follows that there is $x_\varepsilon \in X$ with $\|x_\varepsilon\| = 1$ such that $\|U(x_\varepsilon)\| < \varepsilon$. Using now the Hahn–Banach theorem it follows that there is $x_\varepsilon^* \in X^*$ with $x_\varepsilon^*(x_\varepsilon) = \|x_\varepsilon\| = 1$ and $\|x_\varepsilon^*\| = 1$. We define $U_\varepsilon : X \to Y$ by $U_\varepsilon(x) = U(x) - x_\varepsilon^*(x)U(x_\varepsilon)$. Since U is compact and the space of all the compact operators is a linear subspace it follows that $U_\varepsilon \in K(X,Y)$. Clearly, $U_\varepsilon(x_\varepsilon) = 0$, i.e., U_ε is not injective, since $x_\varepsilon \neq 0$. In addition, $\|U - U_\varepsilon\| \leq \|x_\varepsilon^*\| \|U(x_\varepsilon)\| < \varepsilon$, and (ii) is proved.

26. Let $Y = T(X)$ be the range of T in X. Then Y with the norm given by X is a normed space and consider $T|_Y : Y \to Y$. If $y \in Y$ there is an $x \in X$ such that $T(x) = y$ and therefore $T(Tx) = T(y)$, $T(x) = T(y)$, $y = T(y)$. Hence $T|_Y$ is the identity of Y. Since T is compact $T|_Y$ is also compact, hence the identity of Y is a compact operator. By the Riesz theorem it follows that $\dim_\mathbb{K} Y < \infty$, i.e., T is a finite rank operator.

27. See [40, exercise 16.33 (a)].

Let $(y_n)_{n \in \mathbb{N}} \subseteq A(M)$, $y_n \to y \in Y$. Let $y_n = Ax_n$, $x_n \in M \; \forall n \in \mathbb{N}$. Since M is bounded it follows that the sequence $(x_n)_{n \in \mathbb{N}}$ is bounded, and using the fact that X is reflexive, from the Eberlein–Smulian theorem it follows that there is a subsequence $(x_{k_n})_{n \in \mathbb{N}}$ such that $x_{k_n} \to x \in X$ weak. Then $x \in \overline{M}^{\text{weak}}$. Since M is convex, $\overline{M}^{\text{weak}} = \overline{M}^{\|\cdot\|} = M$, M being norm closed. Hence $x \in M$. Since $A : X \to Y$ is continuous, $A : X \to Y$ is weak-to-weak continuous (see exercise 18 of chapter 14). Since $x_{k_n} \to x$ weak it follows that $Ax_{k_n} \to Ax$ weak. But $Ax_n \to y$ in norm, and hence $Ax_n \to y$ weak. Since the weak topology is Hausdorff it follows that $Ax = y$ and thus $y \in A(M)$, i.e., $A(M) \subseteq Y$ is closed.

If $A : X \to Y$ is in addition compact, since $M \subseteq X$ is bounded it follows that $A(M) \subseteq Y$ is relatively compact. Since $A(M)$ is closed it follows that $A(M) \subseteq Y$ is a compact set.

28. See [40, exercise 16.36].

From exercise 27 we have that $T(B_H) \subseteq H$ is a compact set. Since $\|\cdot\|$ is continuous, from the Weierstrass theorem it follows that there is an $x \in B_H$ such that

$$\|Tx\| = \sup_{\|y\| \leq 1} \|Ty\| = \|T\|.$$

29. i) See [40, exercise 16.33 (b)].

Denote by $d = \inf_{x \in M} \|Ax - y\|$. There is a sequence $(x_n)_{n \in \mathbb{N}} \subseteq M$ such that $\|Ax_n - y\| \to d$. Using exercise 27, the set $A(M) \subseteq Y$ is a compact one, and therefore there is a subsequence $(x_{k_n})_{n \in \mathbb{N}}$ and an element $x_0 \in M$ such that $Ax_{k_n} \to Ax_0$. Then $Ax_{k_n} - y \to Ax_0 - y$, $\|Ax_{k_n} - y\| \to \|Ax_0 - y\|$. We obtain that $d = \|Ax_0 - y\|$, i.e., $\|Ax_0 - y\| = \inf_{x \in M} \|Ax - y\|$.

Suppose now that Y is a Hilbert space. We have $\inf_{x \in M} \|Ax - y\| = d(y, A(M))$, and since Y is a Hilbert space, by the projection theorem (see chapter 11), $d(y, A(M)) = \|y - \text{Pr}_{A(M)}(y)\|$ and $\text{Pr}_{A(M)}(y)$ is the only element for which the infimum is attained. Then $A(x_0) = \text{Pr}_{A(M)}(y)$.

ii) Let $\lambda = (\lambda_n)_{n \in \mathbb{N}} \in c_0$, and consider the multiplication operator $A : l_p \to l_2$, $A((x_n)_{n \in \mathbb{N}}) = (\lambda_n x_n)_{n \in \mathbb{N}}$. It is easy to see that under our hypothesis ($1 < p < 2$) the operator is well defined, linear and continuous (see also exercise 24 of chapter 9 for a more general result). Also, if we consider the sequence of finite rank operators defined by $A_n(x_1, x_2, ...) = (\lambda_1 x_1, \lambda_2 x_2, ..., \lambda_n x_n, 0, ...)$, we have $\|A_n - A\| = \sup_{k \geq n+1} |\lambda_k| \to 0$, hence A is compact. A simple calculation shows that if, for a sequence $a = (a_n)_{n \in \mathbb{N}} \in l_\infty \setminus c_0$ with $a_n \neq 0 \ \forall n \in \mathbb{N}$, we consider the ellipsoid

$$E_a = \left\{ (x_n)_{n \in \mathbb{N}} \in l_p \ \bigg| \ \sum_{n=1}^{\infty} \frac{|x_n|^p}{|a_n|^p} \leq 1 \right\}$$

(which by exercise 13(ii) of chapter 1 is a bounded set and by exercise 1(ii) is not a compact set), then

$$A(E_a) = \left\{ (x_n)_{n \in \mathbb{N}} \in l_2 \ \bigg| \ \sum_{n=1}^{\infty} \frac{|x_n|^p}{|\lambda_n|^p |a_n|^p} \leq 1 \right\}.$$

By (i) $x_0 \in E_a$ with the properties from the statement is a solution for the equation $A(x) = \text{Pr}_{A(E_a)}(2e_1)$. For our exercise we have $\lambda_n = 1/(n+1)$, $a_n = (n+1)/n$, $E_a = M$, and

$$A(M) = A(E_a) = \left\{ (x_n)_{n \in \mathbb{N}} \in l_2 \ \bigg| \ \sum_{n=1}^{\infty} n^p |x_n|^p \leq 1 \right\}.$$

As in the solution for exercise 6 of chapter 11 we obtain that $\text{Pr}_{A(E_a)}(2e_1) = e_1$. The equation $A(x_0) = e_1$ has $x_0 = 2e_1$ as a solution.

30. This result is known as the Gelfand–Phillips theorem, see [33].

First we prove the easy implication. Let $A \subseteq X$ be a relatively compact set and $x_n^* \to 0$ *weak** in X^*. Let $B = \overline{A}$, $B \subseteq X$ is compact, and if we show that $x_n^* \to 0$ uniformly on B, it follows that $x_n^* \to 0$ uniformly on A. Let us suppose that x_n^* does not converge uniformly to 0 on B. Then there is $\varepsilon_0 > 0$ such that for each $N \in \mathbb{N}$ there is an $n \geq N$ and an $x_n \in B$ such that $|x_n^*(x_n)| \geq \varepsilon_0$. We can then construct in a standard way a subsequence $(n_k)_{k \in \mathbb{N}}$ of $\mathbb{N}$ such that $|x_{n_k}^*(x_{n_k})| \geq \varepsilon_0$, $x_{n_k} \in B \ \forall k \in \mathbb{N}$. Since the sequence $(x_{n_k}^*)_{k \in \mathbb{N}}$ is *weak** convergent, it is bounded, i.e., there is an $M > 0$ such that $\|x_{n_k}^*\| \leq M \ \forall k \in \mathbb{N}$. Since $(x_{n_k})_{k \in \mathbb{N}} \subseteq B$, and B is norm compact, there is a subsequence $(n_{k_p})_{p \in \mathbb{N}}$ of the sequence

5.2 Solutions

$(n_k)_{k\in\mathbb{N}}$ and an $x_0 \in B$ such that $x_{n_{k_p}} \to x_0$ in norm. Hence there is $p_{\varepsilon_0} \in \mathbb{N}$ such that $\|x_{n_{k_p}} - x_0\| \leq \varepsilon_0/(2M)\ \forall p \geq p_{\varepsilon_0}$. Then, for $p \geq p_{\varepsilon_0}$ we have

$$\left|x^*_{n_{k_p}}(x_0)\right| \geq \left|x^*_{n_{k_p}}(x_{n_{k_p}})\right| - \left|x^*_{n_{k_p}}(x_{n_{k_p}} - x_0)\right| \geq \varepsilon_0 - \left\|x^*_{n_{k_p}}\right\|\left\|x_{n_{k_p}} - x_0\right\| \geq \varepsilon_0 - \frac{\varepsilon_0}{2} = \frac{\varepsilon_0}{2},$$

so $|x^*_{n_{k_p}}(x_0)| \geq \varepsilon_0/2\ \forall p \geq p_{\varepsilon_0}$, which is false, since $x^*_{n_{k_p}} \to 0\ weak^*$.

Conversely, let us suppose now that $x^*_n \to 0\ weak^*$ implies $x^*_n \to 0$ uniformly on A. Let $T : X \to T(X) \subseteq C(K)$ be the isometric isomorphism given by exercise 20 of chapter 2. Recall that $K = (B_{X^*}, weak^*)$ and $T(x) = \widehat{x}(\cdot)$, where $\widehat{x} : K \to \mathbb{K}$ is given by $\widehat{x}(x^*) = x^*(x)\ \forall x^* \in B_{X^*}$. We will show that $T(A) \subseteq C(K)$ is relatively compact, and the fact that $T : X \to T(X) \subseteq C(K)$ is an isometric isomorphism will assure that A is relatively compact. In order to prove that $T(A) \subseteq C(K)$ is relatively compact, we will use the Arzela–Ascoli theorem, i.e., we show that $T(A) \subseteq C(K)$ is equicontinuous and uniformly bounded. If A is not bounded, then for any $n \in \mathbb{N}$ we can find $x_n \in A$ such that $\|x_n\| > n$. Then we can find $y^*_n \in X^*$ such that $\|y^*_n\| = 1$ and $|y^*_n(x_n)| > n$. Let $x^*_n = y^*_n/\sqrt{n}\ \forall n \subset \mathbb{N}$. Then $\|x^*_n\| \to 0$ and therefore $x^*_n \to 0\ weak^*$. By our hypothesis, $x^*_n \to 0$ uniformly on A. But $|x^*_n(x_n)| > \sqrt{n}\ \forall n \in \mathbb{N}$, and we obtain a contradiction. Therefore A is norm bounded, i.e., there is an $M > 0$ such that $\|x\| \leq M$ $\forall x \in A$. Then for each $x \in A$, $\|Tx\| \leq \|T\|\|x\| \leq M$, hence $T(A)$ is uniformly bounded. We must also prove that the family $(\widehat{x}(\cdot))_{x\in A}$ is equicontinuous. Let $x^*_n \in K = B_{X^*}$ $\forall n \geq 0$, and suppose that $x^*_n \to x^*_0$ with respect to the metric on $(B_{X^*}, weak^*)$. We must prove that $\forall \varepsilon > 0\ \exists n_\varepsilon \in \mathbb{N}$ such that $\sup_{x\in A}|\widehat{x}(x^*_n) - \widehat{x}(x^*_0)| < \varepsilon\ \forall n \geq n_\varepsilon$. Let us suppose, for a contradiction, that this is not true, i.e., there is an $\varepsilon_0 > 0$, $(n_k)_{k\in\mathbb{N}}$ a subsequence of $\mathbb{N}$, and $x_{n_k} \in A\ \forall k \in \mathbb{N}$ such that $\left|\widehat{x_{n_k}}(x^*_{n_k}) - \widehat{x_{n_k}}(x^*_0)\right| \geq \varepsilon_0\ \forall k \in \mathbb{N}$, i.e., $|x^*_{n_k}(x_{n_k}) - x^*_0(x_{n_k})| \geq \varepsilon_0\ \forall k \in \mathbb{N}$. Now for any $k \in \mathbb{N}$ we have $\varphi_k := x^*_{n_k} - x^*_0 \in X^*$. Also, since $x^*_n \to x^*_0$ with respect to the metric on $(B_{X^*}, weak^*)$ it follows that $x^*_{n_k} \to x^*_0$ with respect to the metric on $(B_{X^*}, weak^*)$, and then $x^*_{n_k} \to x^*_0\ weak^*$, that is, $\varphi_k \to 0$ $weak^*$. Then by hypothesis $\varphi_k \to 0$ uniformly on A, i.e., $\sup_{x\in A}|\varphi_k(x)| \to 0$, and this is not true, since we have $\sup_{x\in A}|\varphi_k(x)| \geq |x^*_{n_k}(x_{n_k}) - x^*_0(x_{n_k})| \geq \varepsilon_0\ \forall k \in \mathbb{N}$.

31. See [37, Proposition 1.e.2].

Define the operator $T : l_1 \to X$, $T((a_n)_{n\in\mathbb{N}}) = \sum_{n=1}^{\infty} a_n x_n$. Let us show first that T is well defined, i.e., we must show that the series converges. Since $\|x_n\| \to 0$, $\sup_{n\in\mathbb{N}}\|x_n\| = M < \infty$. Then for each $(a_n)_{n\in\mathbb{N}} \in l_1$, i.e., $\sum_{n=1}^{\infty}|a_n| < \infty$, we have $\sum_{n=1}^{\infty}\|a_n x_n\| \leq M\sum_{n=1}^{\infty}|a_n| < \infty$, i.e., the series $\sum_{n=1}^{\infty} a_n x_n$ is absolutely convergent. Since X is complete it follows that the series $\sum_{n=1}^{\infty} a_n x_n$ converges. Hence T is well defined and it is easy to see that T is linear, continuous, with $\|T\| \leq M$. For $n \in \mathbb{N}$, define $T_n : l_1 \to X$, $T_n((a_k)_{k\in\mathbb{N}}) = \sum_{k=1}^{n} a_k x_k$. We have $T - T_n = \sum_{k=n+1}^{\infty} a_k x_k$, and from what we have just proved, $\|T - T_n\| \leq \sup_{k\geq n+1}\|x_k\| \to 0$, i.e., $\|T_n - T\| \to 0$. Since T_n is a finite rank operator $\forall n \in \mathbb{N}$, it follows that T is compact, hence $T(B_{l_1}) \subseteq X$ is relatively compact.

If we will show that $T(B_{l_1}) \subseteq X$ is norm closed, the exercise will be solved. Let therefore $a^k \in B_{l_1}$ $\forall k \in \mathbb{N}$, $a^k = (a_n^k)_{n \in \mathbb{N}}$ $\forall k \in \mathbb{N}$, such that $\lim\limits_{k \to \infty} \sum\limits_{n=1}^{\infty} a_n^k x_n = y \in X$ in norm. Since $|a_1^k| \leq 1$ for each $k \in \mathbb{N}$, there is a subsequence $(a_1^{k_n})_{n \in \mathbb{N}}$ of the sequence $(a_1^k)_{k \in \mathbb{N}}$ and a scalar a_1 such that $\lim\limits_{n \to \infty} a_1^{k_n} = a_1$. Again, $|a_2^{k_n}| \leq 1$ for each $n \in \mathbb{N}$ and so there is a subsequence $(a_2^{k_{n_p}})_{p \in \mathbb{N}}$ of the sequence $(a_2^{k_n})_{n \in \mathbb{N}}$ and a scalar a_2 such that $\lim\limits_{p \to \infty} a_2^{k_{n_p}} = a_2$. Then also $\lim\limits_{p \to \infty} a_1^{k_{n_p}} = a_1$. In this way we can construct, by induction, a sequence of scalars $(a_n)_{n \in \mathbb{N}}$ such that: for each $N \in \mathbb{N}$ there is a subsequence $(m_p)_{n \in \mathbb{N}} \subseteq \mathbb{N}$ such that $\lim\limits_{p \to \infty} a_j^{m_p} = a_j$, for each $1 \leq j \leq N$. Let us show that $\sum\limits_{n=1}^{\infty} |a_n| \leq 1$. For this it is sufficient to show that for each $N \in \mathbb{N}$, $\sum\limits_{n=1}^{N} |a_n| \leq 1$. Let $N \in \mathbb{N}$, and let us consider the subsequence $(m_p)_{p \in \mathbb{N}}$ as above. Then $(a_1^{m_j}, ..., a_N^{m_j}) \to (a_1, ..., a_N)$ for $j \to \infty$ and since $\sum\limits_{i=1}^{N} |a_i^{m_j}| \leq \sum\limits_{i=1}^{\infty} |a_i^{m_j}| \leq 1$ for each $j \in \mathbb{N}$, it follows that $\sum\limits_{i=1}^{N} |a_i| \leq 1$. Hence $\sum\limits_{n=1}^{\infty} |a_n| \leq 1$, so we can consider the element $\sum\limits_{n=1}^{\infty} a_n x_n \in X$. We will show that $y = \sum\limits_{n=1}^{\infty} a_n x_n$, and the exercise will be solved. Indeed, let $\varepsilon > 0$. Since $\|x_n\| \to 0$ there is an $N'_\varepsilon \in \mathbb{N}$ such that $\|x_n\| \leq \varepsilon$ $\forall n \geq N'_\varepsilon$. Since $\lim\limits_{k \to \infty} \sum\limits_{n=1}^{\infty} a_n^k x_n = y$ there is an $N''_\varepsilon \in \mathbb{N}$ such that $\left\| \sum\limits_{n=1}^{\infty} a_n^k x_n - y \right\| \leq \varepsilon$ $\forall k \geq N''_\varepsilon$. Let $N_\varepsilon = \max(N'_\varepsilon, N''_\varepsilon)$. Then for $N, k \geq N_\varepsilon$ we have

$$\left\| \sum_{n=1}^{N} a_n^k x_n - y \right\| \leq \left\| \sum_{n=1}^{N} a_n^k x_n - \sum_{n=1}^{\infty} a_n^k x_n \right\| + \left\| \sum_{n=1}^{\infty} a_n^k x_n - y \right\|$$

$$\leq \left\| \sum_{n=N+1}^{\infty} a_n^k x_n \right\| + \varepsilon \leq \sum_{n=N+1}^{\infty} |a_n^k| \|x_n\| + \varepsilon \leq \varepsilon \left(\sum_{n=N+1}^{\infty} |a_n^k| \right) + \varepsilon$$

$$\leq \varepsilon \left(\sum_{n=1}^{\infty} |a_n^k| \right) + \varepsilon \leq 2\varepsilon.$$

Let $N \geq N_\varepsilon$. By our construction, there is a subsequence $(m_p)_{p \in \mathbb{N}}$ of $\mathbb{N}$ such that $\lim\limits_{j \to \infty} a_i^{m_j} = a_i$, $i = \overline{1, N}$, hence there is $j \in \mathbb{N}$ such that $m_j \geq N_\varepsilon$ and $|a_i^{m_j} - a_i| \leq \varepsilon/2^i$, $i = \overline{1, N}$. Then

$$\left\| \sum_{n=1}^{N} a_n x_n - y \right\| \leq \left\| \sum_{n=1}^{N} (a_n - a_n^{m_j}) x_n \right\| + \left\| \sum_{n=1}^{N} a_n^{m_j} x_n - y \right\|$$

$$\leq \sum_{n=1}^{N} |a_n - a_n^{m_j}| \|x_n\| + 2\varepsilon$$

$$\leq M \sum_{n=1}^{N} \frac{\varepsilon}{2^i} + 2\varepsilon = (2 + M)\varepsilon,$$

5.2 Solutions

where $M = \sup\limits_{n \in \mathbb{N}} \|x_n\|$. So, $\forall \varepsilon > 0 \; \exists N_\varepsilon \in \mathbb{N}$ such that $\left\|\sum\limits_{n=1}^{N} a_n x_n - y\right\| \leq (2+M)\varepsilon$ $\forall N \geq N_\varepsilon$, i.e., $y = \sum\limits_{n=1}^{\infty} a_n x_n$.

32. See [23, exercise 6.19].

Let c_{00} be the space of all scalar sequences with finite support, with the norm given by l_1. Let $x_n = e_n/n \; \forall n \in \mathbb{N}$, where $\{e_n \mid n \in \mathbb{N}\}$ is the standard basis for c_{00}. Let us observe that $\|x_n\| = 1/n \to 0$, i.e., $x_n \to 0$.

Let now $\lambda_n = 1/n^2$. Then $\sum\limits_{n=1}^{\infty} |\lambda_n| = \sum\limits_{n=1}^{\infty} 1/n^2 < \infty$. Let us denote by $s_n = \sum\limits_{i=1}^{n} \lambda_i x_i$ $\forall n \in \mathbb{N}$, and suppose that there is $y = (y_n)_{n \in \mathbb{N}} \in c_{00}$ such that $s_n \to y$. Then $s_n^k \to y_k$ $\forall k \in \mathbb{N}$ (the convergence on the components), and therefore $y_k = 1/k^3 \; \forall k \in \mathbb{N}$, so $x \notin c_{00}$, contradiction!

If we consider $K = \{x_n \mid n \in \mathbb{N}\} \cup \{0\}$ then $K \subseteq c_{00}$ is compact. If we denote by $\lambda_n = 1/n^{3/2}$, $s_n = \sum\limits_{i=1}^{n} \lambda_i x_i$ and $\lambda = \sum\limits_{n=1}^{\infty} 1/n^{3/2}$, then

$$\frac{s_n}{\lambda} = \sum_{i=1}^{n} \frac{\lambda_i}{\lambda} x_n + \left(1 - \sum_{i=1}^{n} \frac{\lambda_i}{\lambda}\right) \cdot 0,$$

so $s_n/\lambda \in \operatorname{co}(K) \; \forall n \in \mathbb{N}$. Therefore $s_n/\lambda \in \overline{\operatorname{co}}(K) \; \forall n \in \mathbb{N}$. As above, one can prove that the sequence $(s_n/\lambda)_{n \in \mathbb{N}}$ has no convergent subsequence in c_{00}, and therefore $\overline{\operatorname{co}}(K)$ is not a compact set.

33. See [15, Lemma 2].

Since T is compact, from the Schauder theorem it follows that T^* is compact, i.e., $\overline{T^*(B_{Y^*})}$ is compact. We shall prove that if A is a set in a metric space such that $\overline{A}$ is compact then there is a countable subset $E \subseteq A$ such that $\overline{A} \subseteq \overline{E}$. Applying this fact for $T^*(B_{Y^*})$ it follows that there is a sequence $(y_m^*)_{m \in \mathbb{N}} \subseteq B_{Y^*}$ such that $(T^*(y_m^*))_{m \in \mathbb{N}} \subseteq T^*(B_{Y^*})$ is norm dense.

Indeed, if $\overline{A}$ is a compact set in a metric space (X, d) then since $\forall \varepsilon > 0$ we have $\overline{A} \subseteq \bigcup\limits_{x \in \overline{A}} B(x, \varepsilon/2)$ it follows that we can find $x_1 \in \overline{A}, ..., x_n \in \overline{A}$ such that $\overline{A} \subseteq \bigcup\limits_{i=1}^{n} B(x_i, \varepsilon/2)$. Then for any $1 \leq i \leq n$ we have $x_i \in \overline{A}$ and $B(x_i, \varepsilon/2)$ is a neighborhood of x_i, thus by the definition of a closure point we have $B(x_i, \varepsilon/2) \cap A \neq \emptyset$, i.e., there is $a_i \in A$ such that $d(x_i, a_i) < \varepsilon/2$. From here we obtain that $\overline{A} \subseteq \bigcup\limits_{i=1}^{n} B(a_i, \varepsilon)$. Then it follows that for any $m \in \mathbb{N}$ we can find $k_m \in \mathbb{N}$ and $\{x_i^m \mid 1 \leq i \leq k_m\} \subseteq A$ such that $\overline{A} \subseteq \bigcup\limits_{i=1}^{k_m} B(x_i^m, 1/m)$ (1). Let $E = \{x_i^m \mid m \in \mathbb{N}, 1 \leq i \leq k_m\} \subseteq A$, which is obviously a countable set. We shall prove that $\overline{A} \subseteq \overline{E}$. Indeed, let $x \in \overline{A}$ and $\varepsilon > 0$. Then there is $m \in \mathbb{N}$ such that $1/m < \varepsilon$. Then by (1), $x \in \bigcup\limits_{i=1}^{k_m} B(x_i^m, 1/m)$ hence there is $1 \leq i \leq k_m$ such that $x \in B(x_i^m, 1/m)$, i.e., $\|x - x_i^m\| < 1/m < \varepsilon$, that is, $x_i^m \in E \cap B(x, \varepsilon)$, and then $x \in \overline{E}$.

The second assertion from the exercise is the following: $\forall \varepsilon > 0 \, \exists \delta_\varepsilon > 0$ such that

$$\|Tx\|^p \leq \varepsilon + \delta_\varepsilon \sum_{m=1}^\infty \frac{|y_m^*(Tx)|^p}{2^m} \quad \forall \|x\| = 1.$$

Let us suppose, for a contradiction, that this is not true, i.e., $\exists \varepsilon_0 > 0$ such that $\forall n \in \mathbb{N}$ $\exists \|x_n\| = 1$ with the property that

$$\|Tx_n\|^p > \varepsilon_0 + n \sum_{m=1}^\infty \frac{|y_m^*(Tx_n)|^p}{2^m}.$$

Then

$$\frac{|y_m^*(Tx_n)|}{2^{m/p}} \leq \frac{\|Tx_n\|}{n^{1/p}} \leq \frac{\|T\|\|x_n\|}{n^{1/p}} = \frac{\|T\|}{n^{1/p}} \to 0,$$

that is, $\lim_{n\to\infty} y_m^*(Tx_n) = 0 \, \forall m \in \mathbb{N}$, i.e., $x^*(x_n) \to 0 \, \forall x^* \in \{T^*(y_m^*) \mid m \in \mathbb{N}\}$, which, by our hypothesis, is norm dense in $T^*(B_{Y^*})$. Therefore $x^*(x_n) \to 0 \, \forall x^* \in T^*(B_{Y^*})$. Then $y^*(Tx_n) \to 0 \, \forall y^* \in B_{Y^*}$. But since T is compact we can find $y \in Y$ and a subsequence $(x_{k_n})_{n\in\mathbb{N}}$ such that $Tx_{k_n} \to y$ in norm. Let $y^* \in B_{Y^*}$. Then $y^*(Tx_{k_n}) \to y^*(y)$. But $y^*(Tx_{k_n}) \to 0$, from whence $y^*(y) = 0$. Therefore $y = 0$, i.e., $Tx_{k_n} \to 0$ in norm, and then $\|Tx_{k_n}\| \to 0$. But $\|Tx_{k_n}\|^p > \varepsilon_0 \, \forall n \in \mathbb{N}$, and we obtain a contradiction.

34. See [18, exercise 6 (ii) and (iii), chapter I].

i). Let $T \in L(X, Y)$ compact. Then $T^* \in L(Y^*, X^*)$ is compact, by the Schauder theorem, hence $T^*(B_{Y^*}) \subseteq X^*$ is relatively norm compact. Then by the Grothendieck theorem there is a sequence $(x_n^*)_{n\in\mathbb{N}} \subseteq X^*$, $\lim_{n\to\infty} \|x_n^*\| = 0$, such that $T^*(B_{Y^*}) \subseteq \overline{\operatorname{co}}(\{x_n^*\}_{n\in\mathbb{N}})$. For each $x \in X$, we have $\|Tx\| = \sup_{\|y^*\|\leq 1} |y^*(Tx)| = \sup_{\|y^*\|\leq 1} |(T^*(y^*))(x)|$. Consider A, the set of all elements from X^* which are limit of convex combinations of elements from the sequence $(x_n^*)_{n\in\mathbb{N}}$, and we obtain a closed, convex set with $\{x_n^*\}_{n\in\mathbb{N}} \subseteq A$. It follows that $\overline{\operatorname{co}}(\{x_n^*\}_{n\in\mathbb{N}}) \subseteq A$, and so $T^*(B_{Y^*}) \subseteq A$. Hence for every $y^* \in B_{Y^*}$, there is a sequence $(z_n^*)_{n\in\mathbb{N}} \subseteq X^*$ such that $z_n^* \to T^*y^*$ in the norm of X^*, each z_n^* being a convex combination of elements from the set $\{x_j^* \mid j \in \mathbb{N}\}$. Let $x \in X$. For $n \in \mathbb{N}$,

$$z_n^* = \sum_{j=1}^{k_n} \alpha_{jn} x_j^*, \alpha_{jn} \geq 0, j = \overline{1, k_n}, \sum_{j=1}^{k_n} \alpha_{jn} = 1, \text{ and then } |z_n^*(x)| = \left|\sum_{j=1}^{k_n} (\alpha_{jn} x_j^*)(x)\right| \leq$$

$\sup_{j\in\mathbb{N}} |x_j^*(x)| \, \forall n \in \mathbb{N}$. Since $z_n^* \to T^*y^*$ in the norm of X^*, $z_n^*(x) \to (T^*y^*)(x)$, and then $|(T^*y^*)(x)| \leq \sup_{n\in\mathbb{N}} |x_n^*(x)| \, \forall \|y^*\| \leq 1$, from whence $\|Tx\| \leq \sup_{n\in\mathbb{N}} |x_n^*(x)| \, \forall x \in X$.

Let us suppose now that $x_n^* \to 0$ in X^* and that $\|Tx\| \leq \sup_{n\in\mathbb{N}} |x_n^*(x)| \, \forall x \in X$. Using exercise 44, or using the characterization of relatively norm compact sets in c_0, it follows that the operator $S : X \to c_0$, $S(x) = (x_n^*(x))_{n\in\mathbb{N}}$ is compact. We have $\|Tx\| \leq \|Sx\|$ $\forall x \in X$ and since S is compact, it follows that T is also compact (if $(x_n)_{n\in\mathbb{N}} \subseteq B_X$, since S is compact there is a subsequence $(k_n)_{n\in\mathbb{N}}$ of $\mathbb{N}$ such that $(Sx_{k_n})_{n\in\mathbb{N}}$ is a Cauchy sequence, so from $\|Tx_{k_n} - Tx_{k_m}\| \leq \|Sx_{k_n} - Sx_{k_m}\| \, \forall n, m \in \mathbb{N}$ it follows that $(Tx_{k_n})_{n\in\mathbb{N}}$ is a sequence Cauchy, and therefore T is a compact operator).

The last assertion from (i) is obvious.

ii) Let $T \in L(X,Y)$ be compact. By (i), we can find $\lambda = (\lambda_n)_{n\in\mathbb{N}} \in c_0$ and $(y_n^*)_{n\in\mathbb{N}} \subseteq X^*$ bounded such that $\|Tx\| \leq \sup_{n\in\mathbb{N}}\left(|\lambda_n|^2 |y_n^*(x)|\right) \forall x \in X$. Let $W \subseteq c_0$, $W = \{(\lambda_n y_n^*(x))_{n\in\mathbb{N}} \mid x \in X\}$. Since $\lambda_n \to 0$ and $(y_n^*)_{n\in\mathbb{N}} \subseteq X^*$ is bounded, it follows that $W \subseteq c_0$, and from the fact that y_n^* is linear for each $n \in \mathbb{N}$ it follows that $W \subseteq c_0$ is a linear subspace. We define $A : X \to W$, $A(x) = (\lambda_n y_n^*(x))_{n\in\mathbb{N}} \forall x \in X$. Then A is linear and from $\|Ax\| = \sup_{n\in\mathbb{N}}|\lambda_n y_n^*(x)| \leq \left(\|\lambda\|_\infty \sup_{n\in\mathbb{N}}\|y_n^*\|\right)\|x\|$ it follows that A is continuous. Denoting $\tau_n = |\lambda_n|\|y_n^*\| \forall n \in \mathbb{N}$, we have $\tau_n \to 0$ and $|(Ax)_n| \leq \tau_n \forall n \in \mathbb{N}$, $\forall \|x\| \leq 1$, hence by the characterization of relatively norm compact sets in c_0 we obtain that $A(B_X) \subseteq c_0$ is relatively compact and therefore A is compact.

Define now $B : W \to Y$, $B\left((\lambda_n y_n^*(x))_{n\in\mathbb{N}}\right) = Tx \ \forall x \in X$. Let us observe first that B is well defined, since if $(\lambda_n y_n^*(x))_{n\in\mathbb{N}} = (\lambda_n y_n^*(y))_{n\in\mathbb{N}}$, then $\sup_{n\in\mathbb{N}}|\lambda_n|^2 |y_n^*(x-y)| = 0$, hence $\|T(x-y)\| = 0$, i.e., $Tx = Ty$. Obviously B is linear, from the linearity of T. We also have

$$\left\|B\left((\lambda_n y_n^*(x))_{n\in\mathbb{N}}\right)\right\| = \|Tx\| \leq \sup_{n\in\mathbb{N}}|\lambda_n|^2 |y_n^*(x)| \leq \|\lambda\|_\infty \left\|(\lambda_n y_n^*(x))_{n\in\mathbb{N}}\right\|_{c_0},$$

hence B is continuous. Define $S : W \to c_0$, $S((\alpha_n)_{n\in\mathbb{N}}) = (\lambda_n \alpha_n)_{n\in\mathbb{N}}$. Since $(\lambda_n)_{n\in\mathbb{N}} \in c_0$, S is linear, continuous, and compact (we use again the characterization for the relatively norm compact sets in c_0). We have, in addition, $\|B((\alpha_n)_{n\in\mathbb{N}})\| \leq \|S((\alpha_n)_{n\in\mathbb{N}})\|$ $\forall (\alpha_n)_{n\in\mathbb{N}} \in W$, and then (see the final part of the solution for (i)) B is also compact. We have $BA = T$, the only inconvenience is that W is only a linear subspace of c_0, and not necessarily a closed one.

Let $Z = \overline{W} \subseteq c_0$. Then A can be considered with values in Z, and we obtain $A \in L(X,Z)$ compact. We have $B \in L(W,Y)$ and since Y is complete we can extend B in the usual way to an operator $C \in L(Z,Y)$. Then $T = CA$, and we must prove that C is compact. Let $(z_n)_{n\in\mathbb{N}} \subseteq Z$, $\|z_n\| \leq 1 \ \forall n \in \mathbb{N}$. Since $Z = \overline{W}$ there are $(w_n)_{n\in\mathbb{N}} \subseteq W$, $\|w_n - z_n\| \leq 1/n \ \forall n \in \mathbb{N}$. It follows that the sequence $(w_n)_{n\in\mathbb{N}} \subseteq W$ is bounded and from the compactness of $B \in L(W,Y)$ we can suppose that $Bw_n \to y \in Y$. We have $\|Cw_n - Cz_n\| \leq \|C\|/n$, hence $\|Bw_n - Cz_n\| \leq \|C\|/n \to 0$, from whence $Cz_n \to y \in Y$, i.e., C is also compact.

35. See [18, exercise 2, chapter VII].

Let $(x_n)_{n\in\mathbb{N}} \subseteq X$, with $\sum_{n=1}^\infty |x^*(x_n)| < \infty \ \forall x^* \in X^*$. Then the operator $S : X^* \to l_1$, $S(x^*) = (x^*(x_n))_{n\in\mathbb{N}}$ is well defined. Obviously S is linear and since X^* and l_1 are Banach spaces we will use the closed graph theorem in order to prove that S is continuous. Let $x_n^* \to x_0^*$ in X^*, $Sx_n^* \to y_0$ in l_1. Then $x_n^*(x_k) \to x_0^*(x_k) \ \forall k \in \mathbb{N}$, and $x_n^*(x_k) \to y_0^k$ $\forall k \in \mathbb{N}$. It follows that $y_0^k = x_0^*(x_k) \ \forall k \in \mathbb{N}$, i.e., $Sx_0^* = y_0$. The graph of S is closed, and therefore S is continuous. Then

$$\|S\| = \sup_{\|x^*\|\leq 1} \sum_{n=1}^\infty |x^*(x_n)| < \infty.$$

We show now that there is $T : c_0 \to X$ linear and continuous such that $T^* = S$. Using exercise 19 of chapter 14, it is sufficient to show that $S : (X^*, weak^*) \to (c_0^*, weak^*)$

is continuous. For this we will use the characterization of the continuity with nets. Let $(x_\delta^*)_{\delta \in \Delta} \subseteq X^*$ be a net such that $x_\delta^* \to x_0^*$ weak*. Then by the Uniform Boundedness Principle it follows that there is an $M > 0$ such that $\|x_\delta^*\| \leq M \ \forall \delta \in \Delta$, and since $x_\delta^* \to x_0^*$ weak*, we also have $\|x_0^*\| \leq M$. We must show that $Sx_\delta^* \to Sx_0^*$ weak*, or, equivalently, $(Sx_\delta^*)(\alpha) \to (Sx_0^*)(\alpha) \ \forall \alpha \in c_0$. Let $\alpha = (\alpha_n)_{n \in \mathbb{N}} \in c_0$. For any $\varepsilon > 0$ there is an $n_\varepsilon \in \mathbb{N}$ such that $|\alpha_n| < \varepsilon/(4M\|S\|) \ \forall n \geq n_\varepsilon$. Since $x_\delta^* \to x_0^*$ weak*, it follows that $x_\delta^*(x_k) \to x_0^*(x_k)$, $k = \overline{1, n_\varepsilon}$, and then there is a $\delta_\varepsilon \in \Delta$ such that

$$|x_\delta^*(x_k) - x_0^*(x_k)| < \frac{\varepsilon}{2\left(1 + \sum_{k=1}^{n_\varepsilon} |\alpha_k|\right)} \ \forall \delta \geq \delta_\varepsilon, \ \forall k = \overline{1, n_\varepsilon}.$$

Then for $\delta \geq \delta_\varepsilon$ we have

$$|(Sx_\delta^*)(\alpha) - (Sx_0^*)(\alpha)| = \left|\sum_{i=1}^\infty \alpha_i(x_\delta^* - x_0^*)(x_i)\right| \leq \sum_{i=1}^{n_\varepsilon} |\alpha_i|\,|x_\delta^*(x_i) - x_0^*(x_i)|$$
$$+ \sum_{i=n_\varepsilon+1}^\infty |\alpha_i|\,|x_\delta^*(x_i) - x_0^*(x_i)|.$$

But

$$\sum_{i=n_\varepsilon+1}^\infty |x_\delta^*(x_i) - x_0^*(x_i)| \leq \sum_{i=1}^\infty |x_\delta^*(x_i)| + \sum_{i=1}^\infty |x_0^*(x_i)| = \|Sx_\delta^*\| + \|Sx_0^*\| \leq 2M\|S\|$$

and then

$$|(Sx_\delta^*)(\alpha) - (Sx_0^*)(\alpha)| \leq \frac{\varepsilon}{2\left(1 + \sum_{k=1}^{n_\varepsilon} |\alpha_k|\right)} \sum_{i=1}^{n_\varepsilon} |\alpha_i| + \frac{\varepsilon}{4M\|S\|} \cdot 2M\|S\| \leq \varepsilon,$$

hence $Sx_\delta^* \to Sx_0^*$ weak*. Then by exercise 19 of chapter 14 there is a linear and continuous operator $T : c_0 \to X$ such that $T^* = S$. We have $(x^*)(Te_n) = (T^*x^*)(e_n) \ \forall n \in \mathbb{N}$, $\forall x^* \in X^*$, hence $x^*(Te_n) = (Sx^*)(e_n)$, i.e., $x^*(Te_n) = x^*(x_n) \ \forall n \in \mathbb{N}, \ \forall x^* \in X^*$, that is, $Te_n = x_n \ \forall n \in \mathbb{N}$.

Let now $T : c_0 \to X$ be linear and continuous, and let $Te_n = x_n \ \forall n \in \mathbb{N}$. Then for each $x^* \in X^*$,

$$\sum_{n=1}^\infty |x^*(x_n)| = \sum_{n=1}^\infty |x^*(Te_n)| = \sum_{n=1}^\infty |(T^*x^*)(e_n)| = \|T^*x^*\| < \infty,$$

hence $(x_n)_{n \in \mathbb{N}} \in w_1(X)$.

36. See [18, exercise 2 (ii), chapter VII].

Let $T : c_0 \to X$ be a compact operator. Then, by the Schauder theorem, $T^* : X^* \to l_1$ is compact, i.e., $T^*(B_{X^*}) \subseteq l_1$ is relatively compact. Then, using the characterization for relatively compact sets in l_1, $\forall \varepsilon > 0 \ \exists N_\varepsilon \in \mathbb{N}$ such that $\sum_{k=N_\varepsilon+1}^\infty |(T^*x^*)_k| \leq \varepsilon \ \forall x^* \in X^*$, $\|x^*\| \leq 1$. We denote $x_n = Te_n \ \forall n \in \mathbb{N}$. We have $(T^*x^*)_k = (T^*x^*)(e_k) = x^*(Te_k) =$

$x^{*}(x_{k}) \forall k \in \mathbb{N}$, and then $\sum_{k=N_{\varepsilon}+1}^{\infty} |x^{*}(x_{k})| \leq \varepsilon \ \forall x^{*} \in X^{*}, \|x^{*}\| \leq 1$. Let now $(n_{k})_{k \in \mathbb{N}}$ be a subsequence of $\mathbb{N}$. Then for $m > p \geq N_{\varepsilon}$ we have

$$\left\| \sum_{k=p+1}^{m} Te_{n_k} \right\| = \sup_{\|x^*\| \leq 1} \left| \sum_{k=p+1}^{m} x^*(Te_{n_k}) \right| \leq \sup_{\|x^*\| \leq 1} \left(\sum_{k=p+1}^{m} |x^*(x_{n_k})| \right) \leq \varepsilon,$$

and since X is a Banach space the series $\sum_{k=1}^{\infty} Te_{n_k}$ converges.

Let us suppose now that all the subseries for $\sum_{n=1}^{\infty} x_n$ are norm convergent, where $x_n = Te_n \ \forall n \in \mathbb{N}$. Since T takes its values in $\overline{\text{Sp}\{x_n \mid n \in \mathbb{N}\}}^{\|\cdot\|} \subseteq X$, we can and do suppose, replacing if necessary X with $Y = \overline{\text{Sp}\{x_n \mid n \in \mathbb{N}\}}^{\|\cdot\|}$, that X is separable. We will show that $T \in L(c_0, X)$ is compact, and for this, by the Schauder theorem, it is enough to show that $T^* : X^* \to l_1$ is compact. For each $x^* \in X^*$, $(T^*x^*)_k = (T^*x^*)(e_k) = x^*(Te_k) = x^*(x_k)$, hence $T^*x^* = (x^*(x_k))_{k \in \mathbb{N}} \ \forall x^* \in X^*$. Let $(x_n^*)_{n \in \mathbb{N}} \subseteq B_{X^*}$. Since X is separable, by exercise 29 of chapter 14 the $weak^*$ topology on B_{X^*} is metrizable, and since by the Alaoglu–Bourbaki theorem B_{X^*} is $weak^*$ compact, there are $x_0^* \in B_{X^*}$ and a subsequence $(x_{n_k}^*)_{k \in \mathbb{N}}$ of the sequence $(x_n^*)_{n \in \mathbb{N}}$ such that $x_{n_k}^* \to x_0^*$ $weak^*$ in X^*. We will show that $T^*x_{n_k}^* \to T^*x_0^*$ in l_1. Using the Schur theorem, exercise 3(iv) of chapter 14, it is enough to show that $T^*x_{n_k}^* \to T^*x_0^*$ $weak$ in l_1. Since the sequence $(x_{n_k}^*)_{k \in \mathbb{N}}$ is bounded, we obtain that the sequence $(T^*x_{n_k}^*)_{k \in \mathbb{N}} \subseteq l_1$ is bounded. Since $l_1^* = l_\infty$ and by exercise 21 of chapter 1, $\{\chi_M \mid M \subseteq \mathbb{N}\}$ is a fundamental set, to show that $T^*x_{n_k}^* \to T^*x_0^*$ $weak$ in l_1 it is enough to show that $\chi_M(T^*x_{n_k}^*) \to \chi_M(T^*x_0^*) \ \forall M \subseteq \mathbb{N}$ (see exercise 10 of chapter 14). Let M be a subset of $\mathbb{N}$. If M is finite then

$$\chi_M(T^*x_0^*) = \sum_{i \in M} x_0^*(x_i) = x_0^* \left(\sum_{i \in M} x_i \right) = \lim_{k \to \infty} x_{n_k}^* \left(\sum_{i \in M} x_i \right)$$
$$= \lim_{k \to \infty} \sum_{i \in M} x_{n_k}^*(x_i) = \lim_{k \to \infty} \chi_M(T^*x_{n_k}^*).$$

If M is not finite, $M = \{m_1, m_2, ...\}$, then

$$\chi_M(T^*x_0^*) = \sum_{j=1}^{\infty} x_0^*(x_{m_j}) = x_0^* \left(\sum_{j=1}^{\infty} x_{m_j} \right) = \lim_{k \to \infty} x_{n_k}^* \left(\sum_{j=1}^{\infty} x_{m_j} \right)$$
$$= \lim_{k \to \infty} \sum_{j=1}^{\infty} x_{n_k}^*(x_{m_j}) = \lim_{k \to \infty} \chi_M(T^*x_{n_k}^*)$$

(all the subseries $\sum_{j=1}^{\infty} x_{m_j}$ are convergent).

37. We prove first that if Y is a Banach space then $U : Y \to l_1$ is $weak$ compact if and only if U is compact. Indeed, suppose that U is $weak$ compact, i.e., $U(B_Y)$ is relatively $weak$ compact in l_1, and let $(y_n)_{n \in \mathbb{N}} \subseteq B_Y$. Then $(U(y_n))_{n \in \mathbb{N}} \subseteq U(B_Y)$ and then by the

Eberlein–Smulian theorem we can find a subsequence $(k_n)_{n\in\mathbb{N}}$ of $\mathbb{N}$ and an element $\xi \in l_1$ such that $U(y_{k_n}) \to \xi$ weak. Now by the Schur theorem, exercise 3(iv) of chapter 14, it follows that $U(y_{k_n}) \to \xi$ in norm. Thus $U(B_Y)$ is relatively norm compact, i.e., U is a compact operator. Conversely, if X and Y are Banach spaces and $U \in L(Y, X)$ is compact, then U is weak compact. Indeed, since U is compact, then $\overline{U(B_Y)}^{\|\cdot\|}$ is norm compact. But $U(B_Y)$ is convex, and using the Mazur theorem we obtain that $\overline{U(B_X)}^{\|\cdot\|} = \overline{U(B_X)}^{weak}$. Since the identity $I : (X, \|\cdot\|) \to (X, weak)$ is continuous we obtain that $\overline{U(B_X)}^{weak}$ is weak compact and then U is weak compact.

Using the Gantmacher theorem, $T : c_0 \to X$ is weak compact if and only if $T^* : X^* \to l_1$ is weak compact. Using what we have proved above, this is equivalent to $T^* : X^* \to l_1$ being compact, which by the Schauder theorem is equivalent to $T : c_0 \to X$ being compact.

38. We will prove the following more general result. Let (S, Σ, μ) be a measure space, and $(A_n)_{n\in\mathbb{N}} \subseteq \Sigma$ a sequence of pairwise disjoint sets. For a scalar sequence $(\xi_n)_{n\in\mathbb{N}}$, we have $\sum_{n=1}^{\infty} a_n \xi_n \chi_{A_n} \in L_1(\mu)$ $\forall (a_n)_{n\in\mathbb{N}} \in c_0$ if and only if the series $\sum_{n=1}^{\infty} \xi_n \mu(A_n)$ is absolutely convergent, and in this case the operator $U : c_0 \to L_1(\mu)$ defined by $U((a_n)_{n\in\mathbb{N}}) = \sum_{n=1}^{\infty} a_n \xi_n \chi_{A_n}$ is compact.

Suppose that $\sum_{n=1}^{\infty} a_n \xi_n \chi_{A_n} \in L_1(\mu)$ $\forall (a_n)_{n\in\mathbb{N}} \in c_0$. Using the countable additivity for the integral of positive measurable functions, the fact that the function $\sum_{n=1}^{\infty} a_n \xi_n \chi_{A_n}$ is in $L_1(\mu)$ means that

$$\sum_{n=1}^{\infty} |a_n \xi_n \mu(A_n)| = \sum_{n=1}^{\infty} \int_S |a_n \xi_n| \chi_{A_n} d\mu = \int_S \left|\sum_{n=1}^{\infty} a_n \xi_n \chi_{A_n}\right| d\mu < \infty,$$

hence the multiplication operator $M : c_0 \to l_1$, $M((a_n)_{n\in\mathbb{N}}) = (a_n \xi_n \mu(A_n))_{n\in\mathbb{N}}$ is well defined, linear, and, as it is easy to prove, with its closed graph. Since c_0, l_1 are Banach spaces, by the closed graph theorem M is a linear and continuous operator. For any fixed $n \in \mathbb{N}$ we have $\sum_{k=1}^{n} |\xi_k \mu(A_k)| = \|U(1, 1, ..., 1, 0, ...)\| \leq \|U\|$, and therefore $\sum_{k=1}^{\infty} |\xi_k \mu(A_k)| \leq \|U\|$. If $\sum_{k=1}^{\infty} |\xi_k \mu(A_k)| < \infty$, it is easy to see that $\sum_{n=1}^{\infty} a_n \xi_n \chi_{A_n} \in L_1(\mu)$ $\forall (a_n)_{n\in\mathbb{N}} \in c_0$. To prove the compactness of the operator U we consider for any $n \in \mathbb{N}$ the operator $U_n : c_0 \to L_1(\mu)$ defined by $U_n((a_k)_{k\in\mathbb{N}}) = \sum_{k=1}^{n} a_k \xi_k \chi_{A_k}$. Then U_n is a finite rank operator, thus compact, and we observe that by the countable additivity of the integral for positive measurable functions we obtain that

$$\|(U_n - U)(a)\| = \sum_{k=n+1}^{\infty} |a_k \xi_k \mu(A_k)| \leq \|a\| \sum_{k=n+1}^{\infty} |\xi_k \mu(A_k)| \quad \forall a \in c_0,$$

so $\|U_n - U\| \leq \sum_{k=n+1}^{\infty} |\xi_k \mu(A_k)| \to 0$, that is, $U_n \to U$. Since the space of all the compact operators is closed, it follows that U is compact.

5.2 Solutions

39. See [18, exercise 5, chapter VII].

Let $(x_n)_{n \in \mathbb{N}} \subseteq X$ for which there is an $M > 0$ such that $\|x_n\| \leq M \ \forall n \in \mathbb{N}$. Define $T : l_1 \to X$, $T\left((\alpha_n)_{n \in \mathbb{N}}\right) = \sum_{n=1}^{\infty} \alpha_n x_n$. For each $\alpha = (\alpha_n)_{n \in \mathbb{N}} \in l_1$, we have

$$\sum_{n=1}^{\infty} \|\alpha_n x_n\| \leq \sum_{n=1}^{\infty} |\alpha_n| \|x_n\| \leq M \sum_{n=1}^{\infty} |\alpha_n| = M \|\alpha\|.$$

Hence the series $\sum_{n=1}^{\infty} \alpha_n x_n$ is absolutely convergent, and since X is a Banach space, it converges, so T is well defined. Obviously T is linear and

$$\|T(\alpha)\| = \left\| \sum_{n=1}^{\infty} \alpha_n x_n \right\| \leq \sum_{n=1}^{\infty} \|\alpha_n x_n\| \leq M \|\alpha\|_1 \ \forall \alpha = (\alpha_n)_{n \in \mathbb{N}} \in l_1,$$

which implies that T is continuous. We also have $Te_n = x_n \ \forall n \in \mathbb{N}$.

Conversely, if $T : l_1 \to X$ is linear and continuous, then denoting $x_n = Te_n \ \forall n \in \mathbb{N}$, since $\|e_n\|_1 = 1 \ \forall n \in \mathbb{N}$ it follows that $\|Te_n\| \leq \|T\| \ \forall n \in \mathbb{N}$, i.e., $\|x_n\| \leq \|T\| \ \forall n \in \mathbb{N}$, hence the sequence $(x_n)_{n \in \mathbb{N}} \subseteq X$ is bounded.

40. See [18, exercise 5 (i), chapter VII].

If $T : l_1 \to X$ is *weak* compact, then $T(B_{l_1}) \subseteq X$ is relatively compact in the *weak* topology. As $e_n \in B_{l_1} \ \forall n \in \mathbb{N}$ we have $\{Te_n \mid n \in \mathbb{N}\} \subseteq T(B_{l_1})$, and hence $\{Te_n \mid n \in \mathbb{N}\} \subseteq X$ is relatively compact in the *weak* topology.

Let us suppose now that $\{Te_n \mid n \in \mathbb{N}\} \subseteq X$ is relatively compact in the *weak* topology. Then there is a *weak* compact set $A \subseteq X$ such that $\{Te_n \mid n \in \mathbb{N}\} \subseteq A$. Since A is *weak* compact, by the Krein theorem, the set $\overline{\text{eco}(A)}^{\text{weak}} \subseteq X$ is also *weak* compact. If we show that $T(B_{l_1}) \subseteq \overline{\text{eco}(A)}^{\text{weak}}$, the exercise will be solved. Let $x_n = Te_n \ \forall n \in \mathbb{N}$. Let $\alpha = (\alpha_n)_{n \in \mathbb{N}} \in B_{l_1}$, that is, $\sum_{n=1}^{\infty} |\alpha_n| \leq 1$. Let us denote $y_k = \sum_{n=1}^{k} \alpha_n x_n \ \forall k \in \mathbb{N}$. Since $\sum_{n=1}^{k} |\alpha_n| \leq 1$, we have $y_k \in \text{eco}(A) \ \forall k \in \mathbb{N}$. For each $k \in \mathbb{N}$ we have $\|y_k - T(\alpha)\| \leq \sum_{n=k+1}^{\infty} |\alpha_n| \|x_n\| \leq \|T\| \sum_{n=k+1}^{\infty} |\alpha_n| \to 0$, hence $T(\alpha) \in \overline{\text{eco}(A)}^{\|\cdot\|} = \overline{\text{eco}(A)}^{\text{weak}}$.

41. i) See [18, exercise 2 (ii), chapter VII].

The solution is the same as the one given at exercise 40, working with the norm topology instead of the *weak* topology and using the Mazur theorem (the closure for the balanced convex hull of a norm compact set is a norm compact set) instead of the Krein theorem.

ii) Clearly, U is linear and continuous, with $\|U\| \leq \|(a_n)_{n \in \mathbb{N}}\|_{\infty}$. Suppose that the set $A = \{U(e_n) \mid n \in \mathbb{N}\} \subseteq l_\infty$ is relatively norm compact. If, for a contradiction, the sequence $(a_n)_{n \in \mathbb{N}}$ does not converge towards zero, then there are $\varepsilon > 0$ and a subsequence $(k_n)_{n \in \mathbb{N}}$ of $\mathbb{N}$ such that $|a_{k_n}| \geq \varepsilon \ \forall n \in \mathbb{N}$ (1). We have $\{U(e_{k_n}) \mid n \in \mathbb{N}\} \subseteq A$, therefore we can find a convergent subsequence $\left(U(e_{k_{n_p}})\right)_{p \in \mathbb{N}}$ of $(U(e_{k_n}))_{n \in \mathbb{N}}$. Then

$\|U(e_{k_{n_p}}) - U(e_{k_{n_s}})\| \to 0$ for $p, s \to \infty$. But for $p > s$ we have

$$\|U(e_{k_{n_p}}) - U(e_{k_{n_s}})\| = \max\{|a_{k_{n_s}}|, |a_{k_{n_p}} - a_{k_{n_s}}|\},$$

and then we can find an s such that $|a_{k_{n_s}}| < \varepsilon$, and this contradicts (1).

Conversely, if $(a_n)_{n \in \mathbb{N}} \in c_0$ then using that

$$\|U(e_n) - U(e_m)\| = \max(|a_n|, |a_m - a_n|) \to 0$$

for $m > n$, $m, n \to \infty$, we obtain that the set $\{U(e_n) \mid n \in \mathbb{N}\} \subseteq l_\infty$ is relatively norm compact.

iii) Clearly U is linear and continuous. By (i) U is compact if and only if $\{U(e_n) \mid n \geq 0\} \subseteq C[0,1]$ is relatively compact, i.e., if and only if the set $\{\xi_n x^n \mid n \geq 0\} \subseteq C[0,1]$ is relatively compact.

Suppose that U is compact. From the Arzela–Ascoli theorem it follows that $\forall \varepsilon > 0$ $\exists \delta_\varepsilon > 0$ such that $\forall x, y \in [0,1]$ with $|x - y| < \delta_\varepsilon$ we have that $|\xi_n(x^n - y^n)| < \varepsilon$ $\forall n \in \mathbb{N}$. There is $n_\varepsilon \in \mathbb{N}$, such that $1/(2n) < \delta_\varepsilon$ $\forall n \geq n_\varepsilon$. Let $n \geq n_\varepsilon$. For $x = 1 - 1/n$, $y = 1 - 1/(2n)$ we have $|x - y| < \delta_\varepsilon$, and then $|\xi_n||(1 - 1/n)^n - (1 - 1/(2n))^n| < \varepsilon$ $\forall n \geq n_\varepsilon$. For $n \to \infty$ we obtain that $\limsup_{n \to \infty} |\xi_n| |1/e - 1/\sqrt{e}| \leq \varepsilon$, i.e., $\limsup_{n \to \infty} |\xi_n| \leq (\varepsilon e)/(\sqrt{e} - 1)$ $\forall \varepsilon > 0$, i.e., $\limsup_{n \to \infty} |\xi_n| = 0$, $\xi_n \to 0$.

Conversely, we have $\|\xi_n x^n\|_\infty = |\xi_n| \to 0$, and therefore the set $\{\xi_n x^n \mid n \geq 0\} \subseteq C[0,1]$ is relatively compact.

42. See [18, exercise 4, chapter VII].

First solution. Let $T : X \to c_0$ be linear and continuous. Then $T^* : l_1 \to X^*$ is linear and continuous, and hence $weak^*$-to-$weak^*$ continuous (see exercise 19 of chapter 14). Consider the set $(p_n)_{n \in \mathbb{N}} \subseteq c_0^* = l_1$. For each $\alpha = (\alpha_n)_{n \in \mathbb{N}} \in c_0$, $p_n(\alpha) = \alpha_n \to 0$, hence $p_n \to 0$ $weak^*$ in c_0^*. Then it follows that $T^* p_n \to 0$ $weak^*$ in X^*. Now let $(x_n^*)_{n \in \mathbb{N}} \subseteq X^*$ with $x_n^* \to 0$ $weak^*$. Define $T : X \to c_0$, $T(x) = (x_n^*(x))_{n \in \mathbb{N}}$. Since $x_n^* \to 0$ $weak^*$ it follows that $x_n^*(x) \to 0$ $\forall x \in X$, hence T takes its values in c_0. T is linear and to show that T is continuous since X and c_0 are Banach spaces we can use the closed graph theorem. Let $x_n \to x$, $Tx_n \to y = (y_k)_{k \in \mathbb{N}} \in c_0$. Then $x_k^*(x_n) \to x_k^*(x)$ $\forall k \in \mathbb{N}$. Since $Tx_n \to y$ it follows that $x_k^*(x_n) \to y_k$ $\forall k \in \mathbb{N}$, and then $x_k^*(x) = y_k$ $\forall k \in \mathbb{N}$, i.e., $Tx = y$. Let us observe that $(T^* p_n)(x) = p_n(Tx) = x_n^*(x)$ $\forall x \in X$, hence $T^* p_n = x_n^*$ $\forall n \in \mathbb{N}$.

Second solution. Let $(x_n^*)_{n \in \mathbb{N}} \subseteq X^*$ with $x_n^* \to 0$ $weak^*$, i.e., $x_n^*(x) \to 0$ $\forall x \in X$. Then $(x_n^*(x))_{n \in \mathbb{N}} \in c_0$, i.e., $T : X \to c_0$, $T(x) = (x_n^*(x))_{n \in \mathbb{N}}$ is well defined. Obviously, T is linear. We also have

$$\|T(x)\| = \sup_{n \in \mathbb{N}} |x_n^*(x)| \leq \left(\sup_{n \in \mathbb{N}} \|x_n^*\|\right) \|x\|.$$

Since $x_n^*(x) \to 0$ $\forall x \in X$, from the Uniform Boundedness Principle it follows that $\sup_{n \in \mathbb{N}} \|x_n^*\| < \infty$. Therefore T is linear and continuous. Let us also observe that

$$\|T\| = \sup_{\|x\| \leq 1} \|T(x)\| = \sup_{\|x\| \leq 1} \sup_{n \in \mathbb{N}} |x_n^*(x)| = \sup_{n \in \mathbb{N}} \sup_{\|x\| \leq 1} |x_n^*(x)| = \sup_{n \in \mathbb{N}} \|x_n^*\|.$$

Conversely, suppose that $T : X \to c_0$ is a linear and continuous operator. Then $p_n \circ T : X \to \mathbb{K}$ is linear and continuous as a composition of two linear and continuous operators. Let $x_n^* = p_n \circ T \in X^*$. If $x \in X$, $T(x) = (p_n(Tx))_{n \in \mathbb{N}} = (x_n^*(x))_{n \in \mathbb{N}}$. For $x \in X$ we have $Tx \in c_0$, that is, $p_n(Tx) \to 0$, $x_n^*(x) \to 0$, and therefore $x_n^* \to 0$ $weak^*$.

43. See [18, exercise 4 (i), chapter VII].

By the Gantmacher theorem, $T : X \to c_0$ is $weak$ compact if and only if $T^* : l_1 \to X^*$ is $weak$ compact, which by exercise 40 is equivalent with the fact that $\{T^*p_n \mid n \in \mathbb{N}\} \subseteq X^*$ is relatively compact in the $weak$ topology.

In the sequel we will use the following general result. Let (X, τ) be a topological Hausdorff space and $x_n \to x$ in X. Then the set $A = \{x_n \mid n \in \mathbb{N}\} \cup \{x\}$ is compact. Since X is Hausdorff, A is closed, and from $B = \{x_n \mid n \in \mathbb{N}\} \subseteq A$ it follows that $\overline{B} \subseteq \overline{A} = A$. Since A is compact and $\overline{B}$ closed it follows that $\overline{B}$ is compact, i.e., $B = \{x_n \mid n \in \mathbb{N}\}$ is relatively compact.

If we suppose that $T^*p_n \to 0$ $weak$ in X^* then by the general result proved above it follows that $\{T^*p_n \mid n \in \mathbb{N}\}$ is relatively compact in the $weak$ topology. Conversely, let us suppose that $T : X \to c_0$ is $weak$ compact. Then $\{T^*p_n \mid n \in \mathbb{N}\} \subseteq X^*$ is relatively $weak$ compact. Suppose that T^*p_n does not converge towards 0 in the $weak$ topology. Then there is an $x^{**} \in X^{**}$ such that $x^{**}(T^*p_n)$ does not converge towards 0. Then there are $\varepsilon > 0$ and a subsequence $(n_k)_{k \in \mathbb{N}}$ of $\mathbb{N}$ such that $|x^{**}(T^*p_{n_k})| \geq \varepsilon$ $\forall k \in \mathbb{N}$. Using the relative compactness in the $weak$ topology for the set $\{T^*p_{n_k} \mid k \in \mathbb{N}\}$, by the Eberlein–Smulian theorem it follows that there are $x^* \in X^*$ and a subsequence $(n_{k_j})_{j \in \mathbb{N}}$ of $(n_k)_{k \in \mathbb{N}}$ such that $T^*p_{n_{k_j}} \to x^*$ $weak$. Since $T \in L(X, c_0)$, T^* is $weak^*$-to-$weak^*$ continuous (see exercise 19 of chapter 14), and therefore $T^*p_{n_{k_j}} \to 0$ $weak^*$. But $T^*p_{n_{k_j}} \to x^*$ $weak$, and therefore $T^*p_{n_{k_j}} \to x^*$ $weak^*$. Then $x^* = 0$, $x^{**}\left(T^*p_{n_{k_j}}\right) \to 0$, and we obtain a contradiction! Hence $T^*p_n \to 0$ $weak$.

44. i) See [18, exercise 4 (ii), chapter VII].

First solution. From the Schauder theorem a linear and continuous operator $T : X \to c_0$ is compact if and only if $T^* : l_1 \to X^*$ is compact. Using exercise 41, $T^* : l_1 \to X^*$ is compact if and only if $\{T^*p_n \mid n \in \mathbb{N}\} \subseteq X^*$ is relatively compact in norm. If we denote by $x_n^* = T^*p_n$, since $T \in L(X, c_0)$ and $p_n \to 0$ $weak^*$, by exercise 19 of chapter 14 it follows that $x_n^* \to 0$ $weak^*$. Now we show that if X is a Banach space and $(x_n^*)_{n \in \mathbb{N}} \subseteq X^*$ is $weak^*$ convergent to 0, then $\{x_n^* \mid n \in \mathbb{N}\} \subseteq X^*$ is relatively norm compact if and only if $x_n^* \to 0$ in the norm on X^*.

Let $\{x_n^* \mid n \in \mathbb{N}\} \subseteq X^*$ relatively compact, with $x_n^* \to 0$ $weak^*$. Let us suppose that $(x_n^*)_{n \in \mathbb{N}}$ does not converge towards 0 in the norm of X^*. Then there are $\varepsilon > 0$ and a subsequence $(n_k)_{k \in \mathbb{N}}$ such that $\|x_{n_k}^*\| \geq \varepsilon$ $\forall k \in \mathbb{N}$. But from the relative compactness, there are a subsequence $(n_{k_j})_{j \in \mathbb{N}}$ of the sequence $(n_k)_{k \in \mathbb{N}}$ and $x^* \in X^*$ such that $x_{n_{k_j}}^* \to x^*$ in the norm of X^*. It follows that $x_{n_{k_j}}^* \to x^*$ $weak^*$, and since the $weak^*$ topology is Hausdorff it follows that $x_{n_{k_j}}^* \to 0$, which contradicts the above inequality.

The other implication is obvious (see the solution for exercise 43).

Second solution. Suppose that T is compact, i.e., $A = T(B_X) \subseteq c_0$ is relatively compact. Then there is a sequence $\lambda = (\lambda_n)_{n \in \mathbb{N}} \in c_0$ such that $|p_n(\xi)| \leq |\lambda_n|$ $\forall n \in \mathbb{N}$, $\forall \xi \in A$, and therefore $|p_n(Tx)| \leq |\lambda_n|$ $\forall n \in \mathbb{N}$, $\forall \|x\| \leq 1$. Then $\|x_n^*\| \leq |\lambda_n|$ $\forall n \in \mathbb{N}$

and since $\lambda_n \to 0$ it follows that $\|x_n^*\| \to 0$, i.e., $x_n^* \to 0$ in norm.

Conversely, let us suppose that $x_n^* \to 0$ in norm. Let $T_n : X \to c_0$ given by $T_n(x) = (x_1^*(x), ..., x_n^*(x), 0, 0, ...)$. We obtain that

$$\|Tx - T_n x\|_{c_0} = \|(0, ..., 0, x_{n+1}^*(x), ...)\| = \sup_{k \geq n+1} |x_k^*(x)| \leq \left(\sup_{k \geq n+1} \|x_k^*\| \right) \|x\|,$$

i.e., $\|T - T_n\| \leq \sup\limits_{k \geq n+1} \|x_k^*\| \to 0$, since $\|x_n^*\| \to 0$. Hence $T_n \to T$ in norm. Since for every $n \in \mathbb{N}$, T_n is a finite rank operator, it follows that T is compact ($K(X, c_0) \subseteq L(X, c_0)$ is a closed linear subspace).

ii) Let $x_n^* : c_0 \to \mathbb{K}$, $x_n^*(x_1, x_2, ...) = a_n x_n$. Then $U(x) = (x_n^*(x))_{n \in \mathbb{N}}$ and from exercise 43, U is weak compact if and only if $x_n^* \to 0$ weak. Since $x_n^* \in c_0^* = l_1$, $x_n^* \to 0$ weak in $l_1 \Leftrightarrow x_n^* \to 0$ in norm, by the Schur theorem (see exercise 3(iv) of chapter 14). But $\|x_n^*\| = |a_n| \; \forall n \in \mathbb{N}$. Hence by the above and (i) U is weak compact $\Leftrightarrow U$ compact $\Leftrightarrow x_n^* \to 0$ in norm $\Leftrightarrow (a_n)_{n \in \mathbb{N}} \in c_0$.

iii) For any $n \in \mathbb{N}$, let $x_n^* : H \to \mathbb{K}$ be defined by $x_n^*(x) = \langle x, x_n \rangle$. Then $x_n^* \to 0$ weak* if and only if U is linear and continuous, with $\|U\| = \sup\limits_{n \in \mathbb{N}} \|x_n^*\|$. Since $\|x_n^*\| = \|x_n\|$, if U is linear and continuous it follows that $\sup\limits_{n \in \mathbb{N}} \|x_n\| < \infty$. Moreover by (i) U is compact if and only if $\|x_n^*\| \to 0$, i.e., the statement.

45. See [18, exercise 3, chapter VII].

Let $T : X \to l_1$ be linear and continuous. Then $T^* : l_\infty \to X^*$ is linear and continuous, and for each $x \in X$ we have

$$\sum_{n=1}^\infty |(T^* p_n)(x)| = \sum_{n=1}^\infty |p_n(Tx)| = \|Tx\| < \infty.$$

Let $x_n^* = T^* p_n \in X^* \; \forall n \in \mathbb{N}$. Then we have $\sum\limits_{n=1}^\infty |x_n^*(x)| = \|Tx\| \leq \|T\| \; \forall x \in X$ with $\|x\| \leq 1$. Consider now $x^{**} \in X^{**}$ with $\|x^{**}\| \leq 1$. Using the Goldstine theorem, there is a net $(x_\delta)_{\delta \in \Delta} \subseteq B_X$ such that $x^*(x_\delta) \to x^{**}(x^*) \; \forall x^* \in X^*$. For any $n \in \mathbb{N}$ we have $\sum\limits_{k=1}^n |x_k^*(x_\delta)| \leq \|T\| \; \forall \delta \in \Delta$ and passing to the limit for $\delta \in \Delta$ it follows that $\sum\limits_{k=1}^n |x^{**}(x_k^*)| \leq \|T\|$, hence $\sum\limits_{n=1}^\infty |x^{**}(x_n^*)| \leq \|T\|$, i.e., $(x_n^*)_{n \in \mathbb{N}} \in w_1(X^*)$.

Let now $(x_n^*)_{n \in \mathbb{N}} \subseteq X^*$ such that $\sum\limits_{n=1}^\infty |x^{**}(x_n^*)| < \infty \; \forall x^{**} \in X^{**}$. Then $\sum\limits_{n=1}^\infty |x_n^*(x)| < \infty \; \forall x \in X$. This inequality shows that $T : X \to l_1$, $T(x) = (x_n^*(x))_{n \in \mathbb{N}}$ is well defined. The operator T is obviously linear, and since X and l_1 are Banach spaces, in order to show that T is continuous we can use the closed graph theorem. Let therefore $(x_n)_{n \in \mathbb{N}} \subseteq X$, $x_n \to x$, $y = (y_k)_{k \in \mathbb{N}} \in l_1$, $Tx_n \to y$. Then $x_k^*(x_n) \to x_k^*(x) \; \forall k \in \mathbb{N}$, and $x_k^*(x_n) \to y_k$ $\forall k \in \mathbb{N}$. It follows that $y_k = x_k^*(x) \; \forall k \in \mathbb{N}$, hence $T(x) = y$. Then $T \in L(X, l_1)$ and for each $x \in X$ we have

$$(T^* p_n)(x) = p_n(Tx) = x_n^*(x),$$

i.e., $T^* p_n = x_n^* \; \forall n \in \mathbb{N}$.

5.2 Solutions

46. See the solution for exercise 37.

47. See [18, exercise 3 (ii), chapter VII].
Let us suppose that $T : X \to l_1$ is linear, continuous, and compact. By the Schauder theorem, $T^* : l_\infty \to X^*$ is also compact. Consider the operator $S : c_0 \to X^*$, $S(x) = T^*(x)$ $\forall x \in c_0$, i.e., $S = T^* \circ K_{c_0}$, where $K_{c_0} : c_0 \hookrightarrow l_\infty$ is the canonical embedding into the bidual. Since K_{c_0} is linear and continuous, by the ideal property for compact operators it follows that S is compact and using exercise 36 we obtain that all the subseries for the series $\sum_{n=1}^{\infty} S(e_n)$ are norm convergent. We also have $S(e_n) = T^*(K_{c_0}(e_n)) = T^* p_n$ $\forall n \in \mathbb{N}$.

Let us suppose now that all the subseries for the series $\sum_{n=1}^{\infty} T^* p_n$ are convergent. We can then define a compact operator $S : c_0 \to X^*$ by considering $S(e_n) = T^* p_n$ $\forall n \in \mathbb{N}$ (see exercise 36). Then by the Schauder theorem, $S^* : X^{**} \to l_1$ is also compact. If we show that $T = S^* \circ K_X$, the exercise will be solved, since T will be a compact operator, by the ideal property for compact operators. Let $x \in X$. Then for each $n \in \mathbb{N}$ we have

$$[S^*(K_X(x))](e_n) = (S^*(\hat{x}))(e_n) = \hat{x}(Se_n) = \hat{x}(T^* p_n) = (T^* p_n)(x)$$
$$= p_n(Tx) = (K_{c_0}(e_n))(Tx) = (Tx)(e_n),$$

and then it follows that $T = S^* \circ K_X$.

48. i) Indeed, if the first condition from the exercise is satisfied, then in particular for the constant function $f = 1$ we have $(a_n/n)_{n \in \mathbb{N}} \in l_1$. Conversely, if $(a_n/n)_{n \in \mathbb{N}} \in l_1$ and $f \in C[0,1]$, then since $\left| a_n \int_0^{1/n} f(x) dx \right| \le \|f\| |a_n/n|$ $\forall n \in \mathbb{N}$, by the comparison test for series it follows that $\sum_{n=1}^{\infty} a_n \int_0^{1/n} f(x) dx$ is absolutely convergent. Moreover, $\|U(f)\| \le \|f\| \sum_{n=1}^{\infty} |a_n/n|$ $\forall f \in C[0,1]$ so since the linearity of U is trivial we deduce that U is continuous.

For the compactness consider for any $n \in \mathbb{N}$ the operator $U_n : C[0,1] \to l_1$,

$$U_n(f) = \left(a_1 \int_0^1 f(x) dx, ..., a_n \int_0^{1/n} f(x) dx, 0, 0, ... \right),$$

which is clearly a finite rank operator, hence compact. We have

$$\|U - U_n\| \le \sum_{k=n+1}^{\infty} |a_k/k| \to 0,$$

since the series $\sum_{n=1}^{\infty} a_n/n$ is absolutely convergent. From here we deduce that U is compact.

ii) Suppose that the first condition from the statement is true. For any $n \in \mathbb{N}$, consider the linear and continuous operator $U_n : H \to l_1$, $U_n(x) = (a_1 \langle x, x_1 \rangle, ..., a_n \langle x, x_n \rangle, 0, ...)$. The hypothesis assures that the operator $U : H \to l_1$, $U(x) = (a_n \langle x, x_n \rangle)_{n \in \mathbb{N}}$ is well defined. For any $x \in H$ we have $\|(U - U_n)(x)\| = \sum_{k=n+1}^{\infty} |a_k \langle x, x_k \rangle| \to 0$, since by hypothesis the series $\sum_{n=1}^{\infty} a_n \langle x, x_n \rangle$ is absolutely convergent. From the Banach–Steinhaus theorem it follows that U is also continuous. Then for any $n \in \mathbb{N}$ we have $\|U(\overline{a_1}x_1 + \cdots + \overline{a_n}x_n)\| \leq \|U\| \|\overline{a_1}x_1 + \cdots + \overline{a_n}x_n\|$. Using the orthogonality of the system $(x_n)_{n \in \mathbb{N}}$ and the Pythagoras theorem we obtain that

$$\sum_{k=1}^{n} |a_k|^2 \|x_k\|^2 \leq \|U\| \sqrt{\sum_{k=1}^{n} |a_k|^2 \|x_k\|^2},$$

i.e., $\sum_{k=1}^{n} |a_k|^2 \|x_k\|^2 \leq \|U\|^2$, so the series $\sum_{n=1}^{\infty} |a_n|^2 \|x_n\|^2$ is convergent. Conversely, by the Cauchy–Buniakowski–Schwarz inequality and the Bessel inequality we have that for any $x \in H$,

$$\sum_{n=1}^{\infty} |a_n \langle x, x_n \rangle| = \sum_{n=1}^{\infty} \left| a_n \|x_n\| \left\langle x, \frac{x_n}{\|x_n\|} \right\rangle \right| \leq \sqrt{\sum_{n=1}^{\infty} |a_n|^2 \|x_n\|^2} \sqrt{\sum_{n=1}^{\infty} \left| \left\langle x, \frac{x_n}{\|x_n\|} \right\rangle \right|^2}$$

$$\leq \|x\| \sqrt{\sum_{n=1}^{\infty} |a_n|^2 \|x_n\|^2} < \infty.$$

For the compactness we have that

$$\|(U - U_n)(x)\| = \sum_{k=n+1}^{\infty} |a_k \langle x, x_k \rangle| \leq \|x\| \sqrt{\sum_{k=n+1}^{\infty} |a_k|^2 \|x_k\|^2} \quad \forall x \in H,$$

i.e., $\|U - U_n\|^2 \leq \sum_{k=n+1}^{\infty} |a_k|^2 \|x_k\|^2 \to 0$, since the series $\sum_{k=1}^{\infty} |a_k|^2 \|x_k\|^2$ is convergent. Observing that $\{U_n\}_{n \in \mathbb{N}}$ are finite rank operators, and since the space of all compact operators is closed we deduce that U is compact.

49. See [19, chapter 2].

Let $x = (x_n)_{n \in \mathbb{N}}$ such that $(x^*(x_n))_{n \in \mathbb{N}} \in l_p \; \forall x^* \in X^*$. Then we can define $T : X^* \to l_p$, $T(x^*) = (x^*(x_n))_{n \in \mathbb{N}} \; \forall x^* \in X^*$. The operator T is well defined, linear, and using the closed graph theorem (see, for example, the solution for exercise 42) it follows that T is continuous. Then $\sup_{\|x^*\| \leq 1} \left(\sum_{n=1}^{\infty} |x^*(x_n)|^p \right)^{1/p} = \|T\| < \infty$, and so, for each $x \in w_p(X)$, $\|x\|_p^{\text{weak}} < \infty$. Obviously, if $x, y \in w_p(X)$, $\alpha, \beta \in \mathbb{K}$, then $\alpha x + \beta y \in w_p(X)$. It is easy to see that $\|\cdot\|_p^{\text{weak}}$ is indeed a norm on the linear space $w_p(X)$. Let us prove now that this norm is a complete one. Let therefore $x^{(k)} = \left(x_n^{(k)} \right)_{n \in \mathbb{N}} \in w_p(X) \; \forall k \in \mathbb{N}$ be a Cauchy sequence with respect to the norm $\|\cdot\|_p^{\text{weak}}$. For each $\varepsilon > 0$ there is an $n_\varepsilon \in \mathbb{N}$ such that

5.2 Solutions   145

$\forall m, q \geq n_\varepsilon$,

$$\sum_{n=1}^{\infty} \left|x^* \left(x_n^{(m)}\right) - x^* \left(x_n^{(q)}\right)\right|^p \leq \varepsilon^p \quad \forall x^* \in B_{X^*}.$$

Then

$$\left\|x_n^{(m)} - x_n^{(p)}\right\| = \sup_{\|x^*\| \leq 1} \left|x^* \left(x_n^{(m)}\right) - x^* \left(x_n^{(p)}\right)\right| \leq \varepsilon \quad \forall m, q \geq n_\varepsilon,$$

and hence the sequence $\left(x_n^{(k)}\right)_{k \in \mathbb{N}}$ is Cauchy $\forall n \in \mathbb{N}$. Since X is a Banach space, there are $x_n \in X$ $\forall n \in \mathbb{N}$ such that $\lim_{k \to \infty} x_n^{(k)} = x_n$ $\forall n \in \mathbb{N}$. We must prove that $x = (x_n)_{n \in \mathbb{N}} \in w_p(X)$ and that $x^{(k)} \to x$ in the norm $\|\cdot\|_p^{\text{weak}}$. For each $N \in \mathbb{N}$ we have

$$\sum_{n=1}^{N} \left|x^* \left(x_n^{(m)}\right) - x^* \left(x_n^{(q)}\right)\right|^p \leq \varepsilon^p \quad \forall x^* \in B_{X^*}, \ \forall m, q \geq n_\varepsilon.$$

For $q \to \infty$ it follows that

$$\sum_{n=1}^{N} \left|x^* \left(x_n^{(m)}\right) - x^* (x_n)\right|^p \leq \varepsilon^p \quad \forall x^* \in B_{X^*}, \ \forall m \geq n_\varepsilon.$$

Since $N \in \mathbb{N}$ is arbitrary, it follows that

$$\left(\sum_{n=1}^{\infty} \left|x^* \left(x_n^{(m)}\right) - x^* (x_n)\right|^p\right)^{1/p} \leq \varepsilon \quad \forall x^* \in B_{X^*}, \ \forall m \geq n_\varepsilon,$$

hence $x^{(n_\varepsilon)} - x \in w_p(X)$, and then, since $w_p(X)$ is a linear space it follows that $x \in w_p(X)$. We also have $\forall \varepsilon > 0$ $\exists n_\varepsilon \in \mathbb{N}$ such that $\left\|x^{(m)} - x\right\|_p^{\text{weak}} \leq \varepsilon$ $\forall m \geq n_\varepsilon$, hence $x^{(m)} \to x$ in the norm of $w_p(X)$.

50. Let $T : l_p \to X$ be a linear and continuous operator. Denoting $x_n = Te_n$ $\forall n \in \mathbb{N}$, then $x^*(x_n) = (T^*x^*)(e_n)$ $\forall x^* \in X^*$, i.e., $x^*(x_n) = (T^*x^*)_n$ $\forall n \in \mathbb{N}$ (the nth component of the sequence $T^*x^* \in l_q$). Then

$$\left(\sum_{n=1}^{\infty} |x^*(x_n)|^q\right)^{1/q} = \|T^*x^*\|_q \quad \forall x^* \in X^*,$$

and hence $(x_n)_{n \in \mathbb{N}} \in w_q(X)$ and $\left\|(x_n)_{n \in \mathbb{N}}\right\|_q^{\text{weak}} = \|T^*\| = \|T\|$.

Consider now a sequence $(x_n)_{n \in \mathbb{N}} \in w_q(X)$. Then $(x^*(x_n))_{n \in \mathbb{N}} \in l_q$ $\forall x^* \in X^*$ and we define the linear and continuous operator $S : X^* \to l_q$, $S(x^*) = (x^*(x_n))_{n \in \mathbb{N}}$. Then $\|S\| = \left\|(x_n)_{n \in \mathbb{N}}\right\|_q^{\text{weak}}$ (see exercise 49). We prove now that for every $\lambda = (\lambda_n)_{n \in \mathbb{N}} \in l_p$ the series $\sum_{n=1}^{\infty} \lambda_n x_n$ converges. Let $1 \leq m \leq n < \infty$. Then, using the Hölder inequality, we have

$$\left|x^* \left(\sum_{i=m}^{n} \lambda_i x_i\right)\right| \leq \sum_{i=m}^{n} |\lambda_i| |x^*(x_i)| \leq \left(\sum_{i=m}^{n} |\lambda_i|^p\right)^{1/p} \left(\sum_{i=m}^{n} |x^*(x_i)|^q\right)^{1/q} \quad \forall x^* \in X^*,$$

and then

$$\left\|\sum_{i=m}^{n}\lambda_i x_i\right\| = \sup_{\|x^*\|\leq 1}\left|x^*\left(\sum_{i=m}^{n}\lambda_i x_i\right)\right| \leq \left(\sum_{i=m}^{n}|\lambda_i|^p\right)^{1/p}\sup_{\|x^*\|\leq 1}\left(\sum_{i=m}^{n}|x^*(x_i)|^q\right)^{1/q}$$

$$\leq \left(\sum_{i=m}^{n}|\lambda_i|^p\right)^{1/p}\left\|(x_n)_{n\in\mathbb{N}}\right\|_q^{\text{weak}}.$$

Since $\sum_{i=1}^{\infty}|\lambda_i|^p < \infty$ and since X is complete it follows that the series $\sum_{n=1}^{\infty}\lambda_n x_n$ converges.

Let us define $T: l_p \to X$, $T\left((\lambda_n)_{n\in\mathbb{N}}\right) = \sum_{n=1}^{\infty}\lambda_n x_n$. Then T is linear and $x_n = Te_n$ $\forall n \in \mathbb{N}$. From what we have proved above we obtain that $\|T(\lambda)\| \leq \|\lambda\|\left\|(x_n)_{n\in\mathbb{N}}\right\|_q^{\text{weak}}$ $\forall \lambda \in l_p$, hence T is continuous. We also have $T^* = S$, and then $\|T\| = \|T^*\| = \|S\| = \left\|(x_n)_{n\in\mathbb{N}}\right\|_q^{\text{weak}}$.

It is easy now to see that we have an isometric isomorphism from $L(l_p, X)$ onto $w_q(X)$.

51. The solution is analogous to the one given for exercise 50.

Chapter 6

The Uniform Boundedness Principle

Theorem (The Uniform Boundedness Principle). Let X be a Banach space, Y a normed space, and $U_i : X \to Y$, $i \in I$, a family of linear and continuous operators which is bounded pointwise, i.e., $\forall x \in X$ the set $\{U_i(x) \mid i \in I\}$ is norm bounded. Then the family $(U_i)_{i \in I}$ is uniformly bounded, i.e., $\sup_{i \in I} \|U_i\| < \infty$.

The Banach–Steinhaus Theorem. Let X be a Banach space, Y a normed space, and $U_n : X \to Y$, $n \in \mathbb{N}$, a sequence of linear and continuous operators such that: $\forall x \in X$, there exists $\lim_{n \to \infty} U_n x =: Ux \in Y$. Then the family $(U_n)_{n \in \mathbb{N}}$ is uniformly bounded, i.e., $\sup_{n \in \mathbb{N}} \|U_n\| < \infty$, and the operator U is linear and continuous.

6.1 Exercises

Sequences of scalars which map null convergent or bounded sequences into null convergent sequences

1. i) Let $(a_n)_{n \in \mathbb{N}}$ be a sequence of scalars with the property $\forall (x_n)_{n \in \mathbb{N}} \in c_0$ it follows that $(a_n x_n)_{n \in \mathbb{N}} \in c_0$. Prove that $(a_n)_{n \in \mathbb{N}} \in l_\infty$.
   ii) Let $(a_n)_{n \in \mathbb{N}}$ be a sequence of scalars with the property $\forall (x_n)_{n \in \mathbb{N}} \in l_\infty$ it follows that $(a_n x_n)_{n \in \mathbb{N}} \in c_0$. Prove that $(a_n)_{n \in \mathbb{N}} \in c_0$.

Sequences of operators which map norm null convergent sequences into norm null convergent sequences

2. i) Let Y be a Banach space, Z a normed space, and $B_n : Y \to Z$, $n \in \mathbb{N}$, a sequence of linear and continuous operators with the property $\forall (y_n)_{n \in \mathbb{N}} \subseteq Y$ with $\|y_n\| \to 0$, it follows that $\|B_n(y_n)\| \to 0$. Prove that $\sup_{n \in \mathbb{N}} \|B_n\| < \infty$.

ii) Let $\varphi_n : [0,1] \to \mathbb{R}$, $n \in \mathbb{N}$, be a sequence of continuous functions. Using (i) prove that $\sup_{n \in \mathbb{N}} \sup_{x \in [0,1]} |\varphi_n(x)| < \infty$ is the necessary and sufficient condition for the property that for each sequence of continuous functions $f_n : [0,1] \to \mathbb{R}$, $n \in \mathbb{N}$, with the property that $f_n \to 0$ uniformly on $[0,1]$ it follows that $\varphi_n f_n \to 0$ uniformly on $[0,1]$.

iii) Let $\varphi_n : [0,1] \to \mathbb{R}$, $n \in \mathbb{N}$, be a sequence of bounded Lebesgue measurable functions. Using (i) prove that $\sup_{n \in \mathbb{N}} \|\varphi_n\|_{L_\infty} < \infty$ is the necessary and sufficient condition for the property that for each sequence of Lebesgue measurable functions $f_n : [0,1] \to \mathbb{R}$, $n \in \mathbb{N}$, with the property that $\int_0^1 |f_n(x)|\, dx \to 0$ it follows that $\int_0^1 |\varphi_n(x) f_n(x)|\, dx \to 0$.

iv) Let $(a_n)_{n \in \mathbb{N}} \subseteq [0,1]$ and $(b_n)_{n \in \mathbb{N}} \subseteq \mathbb{R}$. Using (i) prove that the sequence $(a_n b_n)_{n \in \mathbb{N}}$ being bounded is the necessary and sufficient condition for the fact that for each sequence of continuous functions $f_n : [0,1] \to \mathbb{R}$, $n \in \mathbb{N}$, with the property that $f_n \to 0$ uniformly on $[0,1]$ it follows that $b_n \int_0^{a_n} f_n(x)\, dx \to 0$.

v) Let Y and Z be two normed spaces, and $B_n : Y \to Z$, $n \in \mathbb{N}$, a sequence of linear and continuous operators with the property that: $\forall (y_n)_{n \in \mathbb{N}} \subseteq Y$ a bounded sequence, it follows that $\|B_n(y_n)\| \to 0$. Prove that $\|B_n\| \to 0$.

3. Is the part (i) from exercise 2 still true without the hypothesis that Y is a Banach space? I.e., if Y is a normed space, Z is a normed space and $T_n : Y \to Z$, $n \in \mathbb{N}$, is a sequence of linear and continuous operators with the property that $\forall (x_n)_{n \in \mathbb{N}} \subseteq Y$ with $\|x_n\| \to 0$ we have that $\|T_n(x_n)\| \to 0$, does it follow that $\sup_{n \in \mathbb{N}} \|T_n\| < \infty$?

Sequences of operators which map norm convergent series into norm null convergent sequences

4. Let X be a Banach space, Y a normed space, and $T_n : X \to Y$, $n \in \mathbb{N}$, a sequence of linear and continuous operators. Prove that the following assertions are equivalent:

i) For each norm convergent series $\sum_{n=1}^\infty x_n$ it follows that $T_n(x_n) \to 0$ in norm;

ii) $\sup_{n \in \mathbb{N}} \|T_n\| < \infty$.

Sequences of functionals which map norm null convergent sequences into convergent series

5. i) Let X be a Banach space, and $(x_n^*)_{n \geq 0} \subseteq X^*$. Prove that the following assertions are equivalent:

a) If $x_n \to 0$ in norm it follows that the series $\sum_{n=0}^\infty x_n^*(x_n)$ converges;

b) The series $\sum_{n=0}^\infty x_n^*$ is absolutely convergent;

ii) Let X be a Banach space, $A \subseteq X$ a dense subset, and $(x_n^*)_{n \geq 0} \subseteq X^*$ a uniformly bounded sequence of linear and continuous functionals. Using (i) prove that the following assertions are equivalent:

a) For any sequence $(x_n)_{n \geq 0} \subseteq A$ with $x_n \to 0$ in norm it follows that the series $\sum_{n=0}^\infty x_n^*(x_n)$ converges;

6.1 Exercises

b) The series $\sum_{n=0}^{\infty} x_n^*$ is absolutely convergent;

iii) Using (i) prove that if $(x_n)_{n\geq 0}$ is a sequence of scalars then the following assertions are equivalent:

a) For any null convergent scalar sequence $(a_n)_{n\geq 0}$ it follows that the series $\sum_{n=0}^{\infty} a_n x_n$ converges;

b) The series $\sum_{n=0}^{\infty} x_n$ is absolutely convergent;

iv) Can we find a sequence of scalars $(c_n)_{n\geq 0}$ with the property that a scalar series $\sum_{n=0}^{\infty} a_n$ converges if and only if $a_n c_n \to 0$?

v) Let $(a_n)_{n\geq 0} \subseteq \mathbb{R}$. Using (i) prove that the necessary and sufficient condition for the property that for any sequence of continuous functions $f_n : [0,1] \to \mathbb{R}, n \geq 0$, with the property that $f_n \to 0$ uniformly on $[0,1]$, it follows that the series $\sum_{n=0}^{\infty} a_n \int_0^1 x^n f_n(x^n) dx$ is convergent, is that the series $\sum_{n=0}^{\infty} a_n/(n+1)$ is absolutely convergent.

vi) Let $(a_n)_{n\geq 0} \subseteq \mathbb{R}$ be a bounded sequence. Using (ii) prove that the necessary and sufficient condition for the property that for any sequence of Lipschitz functions $f_n : [0,1] \to \mathbb{R}, n \geq 0$, with the property that $f_n \to 0$ uniformly on $[0,1]$, it follows that the series $\sum_{n=0}^{\infty} a_n \int_0^1 f_n(x^n) dx$ converges, is that the series $\sum_{n=0}^{\infty} a_n$ is absolutely convergent.

vii) Let $(a_n)_{n\geq 0} \subseteq \mathbb{R}$ be a sequence with the property that $\sup_{n\geq 0} \left(|a_n|/\sqrt{2n+1}\right) < \infty$. Using (ii) prove that the necessary and sufficient condition for the property that for any sequence of continuous functions $f_n : [0,1] \to \mathbb{R}, n \geq 0$, with the property that $\int_0^1 |f_n(x)|^2 dx \to 0$, it follows that the series $\sum_{n=0}^{\infty} a_n \int_0^1 x^n f_n(x) dx$ converges, is that the series $\sum_{n=0}^{\infty} \left(a_n/\sqrt{2n+1}\right)$ is absolutely convergent.

viii) Let $(b_n)_{n\geq 0} \subseteq \mathbb{R}$. Using (i) prove that the necessary and sufficient condition for the property that for any double-indexed sequence of scalars $(a_{nk})_{n,k\geq 0}$ with the properties that $\lim_{k\to\infty} a_{nk} = 0$ for any $n \geq 0$ and $\lim_{n\to\infty} a_{nk} = 0$ uniformly with respect to $k \geq 0$, it follows that the series $\sum_{n=0}^{\infty} a_{nn} b_n$ converges, is that the series $\sum_{n=0}^{\infty} b_n$ is absolutely convergent.

Sequences of operators which map norm null convergent sequences into convergent series

6. Does the part (i) from exercise 5 remain true if the linear and continuous functionals are replaced with linear and continuous operators? I.e., if X, Y are Banach spaces and $V_n : X \to Y, n \geq 0$, is a sequence of linear and continuous operators with the property that for each $x_n \to 0$ in norm it follows that the series $\sum_{n=0}^{\infty} V_n(x_n)$ is norm convergent, does it follow that the series $\sum_{n=0}^{\infty} V_n$ is absolutely convergent in the space $L(X,Y)$?

Sequences of functionals which map convergent series into convergent series

7. i) Let X, Y be two Banach spaces, and $T_n : X \to Y$, $n \geq 0$, a sequence of linear and continuous operators. Prove that the following assertions are equivalent:

a) For each norm convergent series $\sum_{n=0}^{\infty} x_n$ it follows that the series $\sum_{n=0}^{\infty} T_n(x_n)$ is norm convergent;

b) $\sup_{n \geq 0} \|T_n\| < \infty$, and for all $(x_n)_{n \geq 0} \subseteq X$ with $\|x_n\| \to 0$ it follows that the series $\sum_{n=0}^{\infty} (T_n - T_{n+1})(x_n)$ is norm convergent;

ii) Using (i) prove that if X is a Banach spaces and $(x_n^*)_{n \geq 0} \subseteq X^*$, then the following assertions are equivalent:

a) For each norm convergent series $\sum_{n=0}^{\infty} x_n$ it follows that the scalar series $\sum_{n=0}^{\infty} x_n^*(x_n)$ converges;

b) $\sum_{n=0}^{\infty} \|x_n^* - x_{n+1}^*\| < \infty$;

iii) Using (ii) prove that if X is a Banach spaces, $A \subseteq X$ is a dense subset and $(x_n^*)_{n \geq 0} \subseteq X^*$ is a uniformly bounded sequence of linear and continuous functionals, then the following assertions are equivalent:

a) For any sequence $(x_n)_{n \geq 0} \subseteq A$ for which the series $\sum_{n=0}^{\infty} x_n$ is norm convergent it follows that the scalar series $\sum_{n=0}^{\infty} x_n^*(x_n)$ converges;

b) $\sum_{n=0}^{\infty} \|x_n^* - x_{n+1}^*\| < \infty$;

iv) Using (ii) prove that if $(x_n)_{n \geq 0}$ is a sequence of scalars then the following assertions are equivalent:

a) For each scalar convergent series $\sum_{n=0}^{\infty} y_n$ it follows that the series $\sum_{n=0}^{\infty} x_n y_n$ converges;

b) $\sum_{n=0}^{\infty} |x_n - x_{n+1}| < \infty$;

v) Find all sequences of scalars $(c_n)_{n \geq 0}$ with the property that a scalar series $\sum_{n=0}^{\infty} a_n$ converges if and only if the series $\sum_{n=0}^{\infty} a_n c_n$ converges.

vi) Find the values for the real number α such that a scalar series $\sum_{n=0}^{\infty} a_n$ converges if and only if the series

$$\sum_{n=0}^{\infty} \left(1 - \frac{1}{2^\alpha} + \cdots + \frac{(-1)^n}{(n+1)^\alpha} \right) a_n$$

converges.

Sequences of operators which map convergent series into convergent series

8. Does the part (ii) from exercise 7 remain true if the linear and continuous functionals are replaced with linear and continuous operators? I.e., if X, Y are Banach spaces and

6.1 Exercises

$T_n : X \to Y$, $n \geq 0$, is a sequence of linear and continuous operators with the property that for each norm convergent series $\sum_{n=0}^{\infty} x_n$ it follows that the series $\sum_{n=0}^{\infty} T_n(x_n)$ is norm convergent, does it follow that the series $\sum_{n=0}^{\infty} (T_n - T_{n+1})$ is absolutely convergent in the space $L(X, Y)$?

Sequences of functionals which map absolutely convergent series into series with bounded partial sums

9. i) Let X be a Banach space and $(x_n^*)_{n \in \mathbb{N}} \subseteq X^*$ a sequence of linear and continuous functionals. Prove that the following assertions are equivalent:

a) For each absolutely convergent series $\sum_{n=1}^{\infty} x_n$ in X it follows that the sequence $\left(\sum_{k=1}^{n} x_k^*(x_k) \right)_{n \in \mathbb{N}}$ is bounded;

b) The sequence $(x_n^*)_{n \in \mathbb{N}} \subseteq X^*$ is norm bounded;

ii) Let $(a_n)_{n \in \mathbb{N}} \subseteq \mathbb{K}$. Using (i) prove that the following assertions are equivalent:

a) For each sequence $(x_n)_{n \in \mathbb{N}} \in l_1$ it follows that $\left(\sum_{k=1}^{n} a_k x_k \right)_{n \in \mathbb{N}} \in l_\infty$;

b) $(a_n)_{n \in \mathbb{N}} \in l_\infty$;

In the case when $(a_n)_{n \in \mathbb{N}} \in l_\infty$, prove that the operator $U : l_1 \to l_\infty$,

$$U(x_1, x_2, ...) = \left(\sum_{k=1}^{n} a_k x_k \right)_{n \in \mathbb{N}}$$

is linear and continuous, and calculate $\|U\|$.

Sequences of operators which map absolutely convergent series into series with bounded partial sums

10. Does the part (i) from exercise 9 remain true if the linear and continuous functionals are replaced with linear and continuous operators? I.e., if X, Y are Banach spaces and $T_n : X \to Y$, $n \in \mathbb{N}$, is a sequence of linear and continuous operators with the property that for each absolutely convergent series $\sum_{n=1}^{\infty} x_n$ it follows that the sequence $\left(\sum_{k=1}^{n} T_k(x_k) \right)_{n \in \mathbb{N}}$ is norm bounded, does it follow that the sequence $(T_n)_{n \in \mathbb{N}}$ is norm bounded in the space $L(X, Y)$?

Weak Cauchy series

11. Let X be a normed space, and $(x_n)_{n \in \mathbb{N}} \subseteq X$ with the property that $\sum_{n=1}^{\infty} |x^*(x_n)| < \infty$ $\forall x^* \in X^*$. Prove that

$$\sup_{\|x^*\| \leq 1} \sum_{n=1}^{\infty} |x^*(x_n)| < \infty.$$

Sequences of functionals which map p-absolutely convergent series into convergent series

12. Let $1 < p < \infty$, and q the conjugate of p, i.e., $1/p + 1/q = 1$.

i) Let X be a Banach spaces, and $(x_n^*)_{n \in \mathbb{N}} \subseteq X^*$ a sequence of linear and continuous functionals. Prove that the following assertions are equivalent:

a) For each p-absolutely norm convergent series $\sum\limits_{n=1}^{\infty} x_n$ (that is, $\sum\limits_{n=1}^{\infty} \|x_n\|^p < \infty$) it follows that the scalar series $\sum\limits_{n=1}^{\infty} x_n^*(x_n)$ converges;

b) The series $\sum\limits_{n=1}^{\infty} x_n^*$ is q-absolutely norm convergent, i.e., $\sum\limits_{n=1}^{\infty} \|x_n^*\|^q < \infty$;

ii) Using (i) prove that if $(x_n)_{n \in \mathbb{N}}$ is a sequence of scalars then the following assertions are equivalent:

a) $(x_n)_{n \in \mathbb{N}} \in l_q$;

b) For each $(y_n)_{n \in \mathbb{N}} \in l_p$, the series $\sum\limits_{n=1}^{\infty} x_n y_n$ converges.

Sequences of operators which map p-absolutely convergent series into convergent series

13. Does the part (i) from exercise 12 remain true if the linear and continuous functionals are replaced with linear and continuous operators? I.e., if $1 < p < \infty$, q is the conjugate of p, X, Y are Banach spaces, and $T_n : X \to Y$, $n \in \mathbb{N}$, is a sequence of linear and continuous operators with the property that for each p-absolutely norm convergent series $\sum\limits_{n=1}^{\infty} x_n$ it follows that the series $\sum\limits_{n=1}^{\infty} T_n(x_n)$ is norm convergent, does it follow that the series $\sum\limits_{n=1}^{\infty} \|T_n\|^q$ is convergent?

The 'Cauchy product' of a sequence of functionals with a norm null convergent sequence

14. i) Let X be a Banach space, and let $(x_n^*)_{n \geq 0} \subseteq X^*$. Prove that the following assertions are equivalent:

a) $\forall (x_n)_{n \geq 0} \subseteq X$ with $\|x_n\| \to 0$, it follows that $x_n^*(x_0) + x_{n-1}^*(x_1) + \cdots + x_0^*(x_n) \to 0$;

b) $\sum\limits_{n=0}^{\infty} \|x_n^*\| < \infty$, i.e., the series $\sum\limits_{n=0}^{\infty} x_n^*$ is absolutely convergent in the space X^*;

ii) Let X be a Banach space, $A \subseteq X$ a dense subset, and $(x_n^*)_{n \geq 0} \subseteq X^*$ a sequence of linear and continuous functionals with $\|x_n^*\| \to 0$. Using (i) prove that the following assertions are equivalent:

a) $\forall (x_n)_{n \geq 0} \subseteq A$ with $\|x_n\| \to 0$, it follows that $x_n^*(x_0) + x_{n-1}^*(x_1) + \cdots + x_0^*(x_n) \to 0$;

b) The series $\sum\limits_{n=0}^{\infty} \|x_n^*\|$ converges;

iii) Let $(a_n)_{n \geq 0} \subseteq [0, 1]$ and $(b_n)_{n \geq 0} \subseteq \mathbb{K}$. Using (i) prove that the series $\sum\limits_{n=0}^{\infty} a_n b_n$ being absolutely convergent is the necessary and sufficient condition for the fact that for each sequence of continuous functions $f_n : [0, 1] \to \mathbb{K}$, $n \geq 0$, with the property that

$f_n \to 0$ uniformly on $[0,1]$, it follows that $b_n \int_0^{a_n} f_0(x)dx + b_{n-1} \int_0^{a_{n-1}} f_1(x)dx + \cdots + b_0 \int_0^{a_0} f_n(x)dx \to 0$.

iv) Let $(b_n)_{n\geq 0} \subseteq \mathbb{K}$ be such that $b_n/\sqrt{2n+1} \to 0$. Using (ii) prove that the series $\sum_{n=0}^{\infty} \left(b_n/\sqrt{2n+1}\right)$ being absolutely convergent is the necessary and sufficient condition for the fact that for any sequence of continuous functions $f_n : [0,1] \to \mathbb{R}, n \geq 0$, with the property that $\int_0^1 |f_n(x)|^2 dx \to 0$, it follows that $b_n \int_0^1 x^n f_0(x)dx + b_{n-1} \int_0^1 x^{n-1} f_1(x)dx + \cdots + b_0 \int_0^1 f_n(x)dx \to 0$.

v) Let $(b_n)_{n\geq 0} \subseteq \mathbb{K}$. Using (i) prove that the series $\sum_{n=0}^{\infty} b_n$ being absolutely convergent is the necessary and sufficient condition for the fact that for any double-indexed sequence $(a_{nk})_{n,k\geq 0} \subseteq \mathbb{K}$ with the properties that $\lim_{k\to\infty} a_{nk} = 0$ for any $n \geq 0$ and $\lim_{n\to\infty} a_{nk} = 0$ uniformly with respect to $k \geq 0$, it follows that $b_n a_{0n} + b_{n-1} a_{1,n-1} + \cdots + b_0 a_{n0} \to 0$.

The 'Cauchy product' of a sequence of operators with a norm null convergent sequence

15. Does part (i) of exercise 14 remain true if the linear and continuous functionals are replaced with linear and continuous operators? I.e., if a X is a Banach space, Y a normed space, and $T_n : X \to Y$, $n \geq 0$, is a sequence of linear and continuous operators with the property that $\forall (x_n)_{n\geq 0} \subseteq X$ with $\|x_n\| \to 0$ we have $\|T_n(x_0) + T_{n-1}(x_1) + \cdots + T_0(x_n)\| \to 0$, does it follow that $\sum_{n=0}^{\infty} \|T_n\| < \infty$?

The conditions when the Uniform Boundedness Principle can be applied

16. Let $1 < p < \infty$, and $1 < q < \infty$ be the conjugate of p, i.e., $1/p + 1/q = 1$. Let $X = \left(C[0,1], \|\cdot\|_p\right)$, where $\|f\|_p = \left(\int_0^1 |f(x)|^p dx\right)^{1/p} \forall f \in C[0,1]$. Let $(b_n)_{n\in\mathbb{N}} \subseteq [0,1]$, $(c_n)_{n\in\mathbb{N}} \subseteq [0,1]$ with $b_n \leq c_n \forall n \in \mathbb{N}$ and let $(a_n)_{n\in\mathbb{N}} \subseteq \mathbb{K}$. For any $n \in \mathbb{N}$ define $x_n^* : X \to \mathbb{K}$ by $x_n^*(f) = a_n \int_{b_n}^{c_n} f(x) dx \; \forall f \in C[0,1]$.

i) Prove that $x_n^* \in X^* \; \forall n \in \mathbb{N}$.

ii) Prove that the sequence $(x_n^*)_{n\in\mathbb{N}} \subseteq X^*$ is pointwise bounded if and only if the sequence $(a_n(c_n - b_n))_{n\in\mathbb{N}}$ is bounded.

iii) Prove that the sequence $(x_n^*)_{n\in\mathbb{N}} \subseteq X^*$ is uniformly bounded if and only if the sequence $\left(a_n(c_n - b_n)^{1/q}\right)_{n\in\mathbb{N}}$ is bounded.

iv) Prove then that we can find $(x_n^*)_{n\in\mathbb{N}} \subseteq X^*$ pointwise bounded which is not uniformly bounded.

What can we deduce?

17. Let $\mathcal{R}[0,1] = \{f : [0,1] \to \mathbb{R} \mid f \text{ is Riemann integrable}\}$ which is a real linear space with respect to the usual operations for addition and scalar multiplication. Let $\|\cdot\|_1 : \mathcal{R}[0,1] \to \mathbb{R}$ defined by $\|f\|_1 = \int_0^1 |f(x)| dx$ and let

$$\mathcal{N} = \{f \in \mathcal{R}[0,1] \mid \|f\|_1 = 0\}.$$

Then on the quotient space $\widetilde{\mathcal{R}}[0,1] = \mathcal{R}[0,1]/\mathcal{N}$, the map $\|\widetilde{f}\|_1 = \int_0^1 |f(x)| dx$ gives a structure of a normed space (for $f \in \mathcal{R}[0,1]$ we denote by $\widetilde{f} \in \widetilde{\mathcal{R}}[0,1]$ the equivalence class for f).

i) Let $g_n : [0,1] \to [0,\infty)$, $n \in \mathbb{N}$, be a sequence of continuous functions, $(b_n)_{n\in\mathbb{N}} \subseteq \mathbb{R}$, and $x_n^* : \widetilde{\mathcal{R}}[0,1] \to \mathbb{R}$ defined by $x_n^*(\widetilde{f}) = b_n \int_0^1 g_n(x) f(x)\, dx$. Prove that

a) $(x_n^*)_{n\in\mathbb{N}}$ is pointwise bounded if and only if the sequence $\left(b_n \int_0^1 g_n(x)\, dx\right)_{n\in\mathbb{N}}$ is bounded;

b) $(x_n^*)_{n\in\mathbb{N}}$ is uniformly bounded if and only if the sequence $\left(b_n \sup_{x\in[0,1]} g_n(x)\right)_{n\in\mathbb{N}}$ is bounded;

ii) Let $(a_n)_{n\in\mathbb{N}} \subseteq [0,\infty)$ and $x_n^* : \widetilde{\mathcal{R}}[0,1] \to \mathbb{R}$ defined by

$$x_n^*(\widetilde{f}) = \sum_{k=1}^{n} a_k \int_0^1 x^k f(x)\, dx.$$

Prove that:

a) $(x_n^*)_{n\in\mathbb{N}}$ is pointwise bounded if and only if the series $\sum_{n=1}^{\infty} \dfrac{a_n}{n+1}$ converges;

b) $(x_n^*)_{n\in\mathbb{N}}$ is uniformly bounded if and only if the series $\sum_{n=1}^{\infty} a_n$ converges;

iii) Let $(a_n)_{n\in\mathbb{N}} \subseteq \mathbb{R}$ and $x_n^* : \widetilde{\mathcal{R}}[0,1] \to \mathbb{R}$ defined by $x_n^*(\widetilde{f}) = a_n \int_0^1 x^n f(x)\, dx$. Prove that:

a) $(x_n^*)_{n\in\mathbb{N}}$ is pointwise bounded if and only if the sequence $(a_n/(n+1))_{n\in\mathbb{N}}$ is bounded;

b) $(x_n^*)_{n\in\mathbb{N}}$ is uniformly bounded if and only if the sequence $(a_n)_{n\in\mathbb{N}}$ is bounded;

iv) Prove then that we can find $(x_n^*)_{n\in\mathbb{N}} \subseteq \left(\widetilde{\mathcal{R}}[0,1]\right)^*$ pointwise bounded, which is not uniformly bounded.

What can we deduce?

18. Let X be a normed space, and $(x_n)_{n\in\mathbb{N}} \subseteq X$. For every $n \in \mathbb{N}$ the operator $T_n : c_{00} \to X$ is defined by $T_n\left((a_k)_{k\in\mathbb{N}}\right) = \sum_{k=1}^{n} a_k x_k$. Prove that:

i) The sequence $(T_n)_{n\in\mathbb{N}}$ converges pointwise;

ii) If on c_{00} we consider the norm $\|(a_n)_{n\in\mathbb{N}}\| = \sup_{n\in\mathbb{N}} |a_n|$, then the sequence $(T_n)_{n\in\mathbb{N}}$ is uniformly bounded if and only if $\sup_{\|x^*\|\le 1} \sum_{n=1}^{\infty} |x^*(x_n)| < \infty$;

iii) If on c_{00} we consider the norm $\|(a_n)_{n\in\mathbb{N}}\| = \sum_{n=1}^{\infty} |a_n|$, then the sequence $(T_n)_{n\in\mathbb{N}}$ is uniformly bounded if and only if $\sup_{n\in\mathbb{N}} \|x_n\| < \infty$;

iv) If on c_{00} we consider the norm $\|(a_n)_{n\in\mathbb{N}}\| = \left(\sum_{n=1}^{\infty} |a_n|^p\right)^{1/p}$, then the sequence $(T_n)_{n\in\mathbb{N}}$ is uniformly bounded if and only if $\sup_{\|x^*\|\le 1} \left(\sum_{n=1}^{\infty} |x^*(x_n)|^q\right)^{1/q} < \infty$, where q is the conjugate of p ($1 < p < \infty$);

v) If on c_{00} we consider one of the above norms, prove that we can find a pointwise bounded sequence $(x_n^*)_{n\in\mathbb{N}} \subseteq c_{00}^*$, which is not uniformly bounded.

What can we deduce?

The Cantor Lemma in Banach spaces

19. Let X be a Banach space and $(x_n)_{n \in \mathbb{N}}$, $(y_n)_{n \in \mathbb{N}}$ two sequences of elements from X. If $x_n \cos nt + y_n \sin nt \to 0$ in norm on a non-degenerate interval, then $x_n \to 0$ and $y_n \to 0$ in norm.

A Banach–Steinhaus type of theorem for bilinear operators

20. i) Let X and Y be two normed spaces, one of them being a Banach space, and Z be a normed space. Consider a bilinear operator $B: X \times Y \to Z$ which is separately continuous. Prove that B is continuous.

ii) Let X and Y be Banach spaces, Z a normed space, and consider a bilinear operator $B: X \times Y \to Z$. If $B_n: X \times Y \to Z$, $n \in \mathbb{N}$, is a sequence of bilinear and continuous operators and $B_n(x, y) \to B(x, y) \ \forall (x, y) \in X \times Y$, prove that B is also continuous.

6.2 Solutions

1. i) Let $U: c_0 \to c_0$, $U((x_n)_{n \in \mathbb{N}}) = (a_n x_n)_{n \in \mathbb{N}}$. By our hypothesis, U is well defined. For $n \in \mathbb{N}$, let $U_n: c_0 \to c_0$, $U_n(x_1, x_2, ...) = (a_1 x_1, a_2 x_2, ..., a_n x_n, 0, ...)$. Then U_n is linear, and

$$\begin{aligned} \|U_n(x_1, x_2, ...)\| &= \|(a_1 x_1, a_2 x_2, ..., a_n x_n, 0, ...)\| = \max(|a_1 x_1|, ..., |a_n x_n|) \\ &\leq \|x\| \max(|a_1|, ..., |a_n|) \ \forall x \in c_0, \end{aligned}$$

i.e., U_n is continuous, $\|U_n\| \leq \max(|a_1|, ..., |a_n|)$. For $1 \leq k \leq n$ we have $\|U_n\| \geq \|U_n(e_k)\| = \|(0, ..., 0, a_k, 0, ...)\| = |a_k|$, and therefore $\max(|a_1|, ..., |a_n|) \leq \|U_n\|$. Hence $\|U_n\| = \max(|a_1|, ..., |a_n|) \ \forall n \in \mathbb{N}$. We also have $U_n(x) \to U(x) \ \forall x \in c_0$, because $\|U_n(x) - U(x)\| = \sup_{k \geq n+1} |a_k x_k| \to 0$. Now from the Uniform Boundedness Principle it follows that $\sup_{n \in \mathbb{N}} \|U_n\| < \infty$, i.e., $\sup_{n \in \mathbb{N}} |a_n| < \infty$.

ii) This follows easily, without using the Uniform Boundedness Principle. Indeed, since the sequence $(\text{sgn}(a_n))_{n \in \mathbb{N}}$ is in l_∞, from the hypothesis it follows that $a_n \text{sgn}(a_n) \to 0$, i.e., $|a_n| \to 0$.

2. i) For a normed space X we will denote $c_0(X) = \{(x_n)_{n \in \mathbb{N}} \subseteq X \mid \|x_n\| \to 0\}$, which is a linear space with the usual operations, and a normed space with respect to the norm $\|x\| = \sup_{n \in \mathbb{N}} \|x_n\| \ \forall x = (x_n)_{n \in \mathbb{N}} \in c_0(X)$. In a standard way one can prove that if X is in addition a Banach space then $c_0(X)$ is a Banach space.

For any $n \in \mathbb{N}$ let us consider the operator $h_n: c_0(Y) \to c_0(Z)$ defined by $h_n(y_1, y_2, ...) = (B_1(y_1), B_2(y_2), ..., B_n(y_n), 0, 0, ...)$, and let $h: c_0(Y) \to c_0(Z)$, $h(y_1, y_2, ...) = (B_1(y_1), B_2(y_2), ..., B_n(y_n), ...)$. By hypothesis, h is well defined. It is easy to see that for each $n \in \mathbb{N}$, h_n is a linear and continuous operator. Moreover, if $y = (y_n)_{n \in \mathbb{N}} \in c_0(Y)$, then $h(y) - h_n(y) = (0, 0, ..., 0, B_{n+1}(y_{n+1}), ...)$, from whence

$$\|h(y) - h_n(y)\| = \sup_{k \geq n+1} \|B_k(y_k)\| \to \limsup_{n \to \infty} \|B_n(y_n)\| = \lim_{n \to \infty} \|B_n(y_n)\| = 0,$$

by our hypothesis. Hence $h_n \to h$ pointwise, and h_n is linear and continuous $\forall n \in \mathbb{N}$. Since Y is a Banach space, $c_0(Y)$ is a Banach space. From the Banach–Steinhaus theorem it follows that h is linear and continuous, and therefore $\sup_{\|y\|\leq 1} \|h(y)\| = \|h\| < \infty$. Let now $n \in \mathbb{N}$ be fixed. For $y \in Y$, the element $(0, 0, ..., 0, y, 0, ...)$ is in $c_0(Y)$ (y on the nth position), and then $\|h(0, 0, ..., y, 0, ...)\| \leq \|h\| \|(0, 0, ..., y, 0, ...)\|$, i.e., by the definition for h and for the norm on $c_0(Y)$ we have $\|B_n(y)\| \leq \|h\| \|y\|$, from whence $\|B_n\| \leq \|h\|$. Since $n \in \mathbb{N}$ was arbitrary it follows that $\sup_{n \in \mathbb{N}} \|B_n\| \leq \|h\|$, i.e., the statement.

Let us observe that the converse is also true. Indeed, if $\sup_{n\in\mathbb{N}} \|B_n\| = M < \infty$, then $\forall (y_n)_{n\in\mathbb{N}} \subseteq Y$ with $\|y_n\| \to 0$ we have $\|B_n(y_n)\| \leq \|B_n\| \|y_n\| \leq M \|y_n\| \to 0$.

ii) Let $B_n : C[0, 1] \to C[0, 1]$ be the multiplication operator given by φ_n, i.e., $B_n f(x) = \varphi_n(x)f(x) \, \forall f \in C[0, 1]$. The condition from the hypothesis is that $\forall \|f_n\| \to 0$ it follows that $\|B_n(f_n)\| \to 0$. Using (i) this is equivalent to $\sup_{n\in\mathbb{N}} \|B_n\| < \infty$. But it is proved in exercise 14 of chapter 5 that $\|B_n\| = \|\varphi_n\| = \sup_{t \in [0,1]} |\varphi_n(t)|$, and therefore $\sup_{n\in\mathbb{N}} \|\varphi_n\| < \infty$.

iii) Let $B_n : L_1[0, 1] \to L_1[0, 1]$ be the multiplication operator given by φ_n, i.e., $B_n f(x) = \varphi_n(x)f(x) \, \forall f \in L_1[0, 1]$. The condition from the hypothesis is that $\forall \|f_n\| \to 0$ it follows that $\|B_n(f_n)\| \to 0$. Using (i) this is equivalent to $\sup_{n\in\mathbb{N}} \|B_n\| < \infty$. But we have $\|B_n\| = \|\varphi_n\|_{L^\infty} \, \forall n \in \mathbb{N}$, and we obtain the statement.

iv) Let $B_n : C[0, 1] \to \mathbb{R}$, $B_n(f) = b_n \int_0^{a_n} f(x)dx$. Then the condition from the statement becomes: $\forall (f_n)_{n\in\mathbb{N}} \subseteq C[0, 1]$ with $\|f_n\| \to 0$, it follows that $B_n(f_n) \to 0$. Using (i) it follows that the sequence $(\|B_n\|)_{n\in\mathbb{N}}$ is bounded. But as it is easy to see, $\|B_n\| = |a_n b_n| \, \forall n \in \mathbb{N}$.

v) Since Y is not a Banach space, we cannot apply the Banach–Steinhaus theorem. But for each $n \in \mathbb{N}$ from the definition for the operator norm there is $\|y_n\| \leq 1$ such that $\|B_n\| - 1/n < \|B_n(y_n)\|$. Now from the hypothesis it follows that $\|B_n(y_n)\| \to 0$, and from here $\|B_n\| \to 0$.

3. The answer is yes. We prove first that if $(a_n)_{n\in\mathbb{N}} \subseteq [0, \infty)$ is an unbounded sequence then there is a subsequence $(a_{k_n})_{n\in\mathbb{N}}$ such that $a_{k_n} > n \, \forall n \in \mathbb{N}$. Indeed, since the sequence $(a_n)_{n\in\mathbb{N}} \subseteq [0, \infty)$ is unbounded, there is $k_1 \geq 1$ such that $a_{k_1} > 1$. Then the sequence $(a_n)_{n \geq k_1+1} \subseteq [0, \infty)$ is unbounded, thus there is $k_2 \geq k_1 + 1 > k_1$ such that $a_{k_2} > 2$. The inductive construction is now clear.

Return now to our exercise. If, for a contradiction, $\sup_{n\in\mathbb{N}} \|T_n\| = \infty$ then by the above observation there is a subsequence $(T_{k_n})_{n\in\mathbb{N}}$ such that $\|T_{k_n}\| > n \, \forall n \in \mathbb{N}$. Since $n < \|T_{k_n}\| = \sup_{\|x\|\leq 1} \|T_{k_n}(x)\|$ we deduce that we can find $\|a_n\| \leq 1$ such that $\|T_{k_n}(a_n)\| > n$ $\forall n \in \mathbb{N}$. Consider now the sequence $(x_n)_{n\in\mathbb{N}}$, $x_n = 0$ if $n \neq k_j$, $j \in \mathbb{N}$, and $x_{k_j} = a_j/j$ $\forall j \in \mathbb{N}$. Clearly, this sequence is norm convergent to 0, and then by hypothesis we must have $\|T_n(x_n)\| \to 0$, that is, $\|T_{k_n}(a_n/n)\| \to 0$, which is false, since by our construction $\|T_{k_n}(a_n/n)\| > 1 \, \forall n \in \mathbb{N}$.

4. We will use the same notations as in the solution for exercise 2(i). We prove first that if $T_n : X \to Y$, $n \in \mathbb{N}$, is a sequence of linear and continuous operators, then for

6.2 Solutions

any n the operator $h_n : c_0(X) \to Y$ defined by $h_n(x_1, x_2, ...) = T_n(x_{n+1} - x_n)$ is linear and continuous, with $\|h_n\| = 2\|T_n\|$. Indeed, we have by the definition of the norm in the space $c_0(X)$ that

$$\|h_n(x)\| = \|T_n(x_{n+1} - x_n)\| \leq 2\|T_n\| \sup_{k \in \mathbb{N}} \|x_k\| = 2\|T_n\| \|x\| \quad \forall x \in c_0(X),$$

i.e., $\|h_n\| \leq 2\|T_n\|$. For $x \in X$, the element $(0, ..., 0, -x, x, 0, ...)$ is in $c_0(X)$ ($-x$ on the nth position), hence $\|h_n(0, ..., 0, -x, x, 0, ...)\| \leq \|h_n\| \|(0, ..., 0, -x, x, 0, ...)\|$, i.e., by the definition of the operator h_n and of the norm in the space $c_0(X)$, $2\|T_n(x)\| \leq \|h_n\| \|x\|$ $\forall x \in X$, thus $2\|T_n\| \leq \|h_n\|$.

'(i) $\Rightarrow$ (ii)' **First solution.** Let $x_n \to 0$ in norm. Then the series $\sum_{n=1}^{\infty}(x_{n+1} - x_n)$ is norm convergent, hence by hypothesis (i) it follows that $T_n(x_{n+1} - x_n) \to 0$ in norm. This shows that the sequence of linear and continuous operators $h_n : c_0(X) \to Y$ defined by $h_n(x) = T_n(x_{n+1} - x_n)$ $\forall x = (x_1, x_2, ...) \in c_0(X)$ has the property that $\forall x \in c_0(X)$, $\lim_{n \to \infty} h_n(x) = 0$. Since X is a Banach space then $c_0(X)$ is a Banach space. By the Uniform Boundedness Principle it follows that $\sup_{n \in \mathbb{N}} \|h_n\| < \infty$, and therefore $\sup_{n \in \mathbb{N}} \|T_n\| < \infty$, i.e., (ii).

Second solution. If $\sup_{n \in \mathbb{N}} \|T_n\| = \infty$, then $\sup_{n \in \mathbb{N}} \|h_n\| = \infty$. There is then a subsequence $(h_{k_n})_{n \in \mathbb{N}}$ such that $\|h_{k_n}\| > 2^n$ $\forall n \in \mathbb{N}$ (see the solution for exercise 3). Since $2^n < \|h_{k_n}\| = \sup_{\|x\| \leq 1} \|h_{k_n}(x)\|$ we deduce that there is $a_n = (x_n^1, x_n^2, ...) \in c_0(X)$, $\|a_n\| \leq 1$ (i.e., by the definition of the norm in the space $c_0(X)$, $\|x_n^i\| \leq 1$ $\forall i \in \mathbb{N}$), such that $\|h_{k_n}(a_n)\| > 2^n$. That is, $\|T_{k_n}(x_n^{k_n+1} - x_n^{k_n})\| > 2^n$. We consider now the sequence $(x_n)_{n \in \mathbb{N}} \subseteq X$ such that $x_n = 0$ if $n \neq k_j$, $j \in \mathbb{N}$, and $x_{k_n} = (x_n^{k_n+1} - x_n^{k_n})/2^n$ $\forall n \in \mathbb{N}$. Clearly for this sequence the series $\sum_{n=1}^{\infty} x_n$ is absolutely convergent, since

$$\left\|\frac{x_n^{k_n+1} - x_n^{k_n}}{2^n}\right\| \leq \frac{\|x_n^{k_n+1}\| + \|x_n^{k_n}\|}{2^n} \leq \frac{1}{2^{n-1}} \quad \forall n \in \mathbb{N}.$$

Since X is a Banach space the series $\sum_{n=1}^{\infty} x_n$ is norm convergent and then by hypothesis (i) we must have $\|T_n(x_n)\| \to 0$, that is, $\left\|T_{k_n}\left((x_n^{k_n+1} - x_n^{k_n})/2^n\right)\right\| \to 0$, which is false, since by our construction $\left\|T_{k_n}\left((x_n^{k_n+1} - x_n^{k_n})/2^n\right)\right\| > 1$ $\forall n \in \mathbb{N}$.

'(ii) $\Rightarrow$ (i)' It is obvious, since if $\sum_{n=1}^{\infty} x_n$ is a norm convergent series then $x_n \to 0$ in norm, and hence by hypothesis (ii) it follows that $T_n(x_n) \to 0$ in norm.

5. i) '(a) $\Rightarrow$ (b)' Let $V : c_0(X) \to \mathbb{K}$, $V((x_k)_{k \geq 0}) = \sum_{k=0}^{\infty} x_k^*(x_k)$, and, for $n \geq 0$, let $V_n : c_0(X) \to \mathbb{K}$, $V_n((x_k)_{k \geq 0}) = x_0^*(x_0) + \cdots + x_n^*(x_n)$. The hypothesis assures that if $(x_k)_{k \geq 0} \in c_0(X)$ then the series $\sum_{n=0}^{\infty} x_n^*(x_n)$ is convergent, i.e., V is well defined. In addition,

$$|V(x) - V_n(x)| = \left|\sum_{k=n+1}^{\infty} x_k^*(x_k)\right| \to 0 \quad \forall x \in c_0(X),$$

i.e., $V_n(x) \to V(x)\ \forall x \in c_0(X)$. We have $\|V_n\| = \sum_{k=0}^{n}\|x_k^*\|\ \forall n \geq 0$. Indeed,

$$|V_n(x)| \leq \sum_{k=0}^{n}\|x_k^*\|\,\|x_k\| \leq \|x\|\left(\sum_{k=0}^{n}\|x_k^*\|\right)\ \forall x = (x_k)_{k\geq 0} \in c_0(X),$$

i.e., $\|V_n\| \leq \sum_{k=0}^{n}\|x_k^*\|$. For $x_0, ..., x_n \in X$ with $\|x_0\| \leq 1, ..., \|x_n\| \leq 1$ and $\lambda_0 = \operatorname{sgn}(x_0^*(x_0)), ..., \lambda_n = \operatorname{sgn}(x_n^*(x_n))$, the element $(\lambda_0 x_0, ..., \lambda_n x_n, 0, 0, ...)$ is in $c_0(X)$, and then

$$\sum_{k=0}^{n}|x_k^*(x_k)| = |V_n(\lambda_0 x_0, ..., \lambda_n x_n, 0, ...)| \leq \|V_n\|\,\|(\lambda_0 x_0, ..., \lambda_n x_n, 0, ...)\| \leq \|V_n\|,$$

i.e., $\sum_{k=0}^{n}|x_k^*(x_k)| \leq \|V_n\|$. Passing to the supremum over $\|x_0\| \leq 1, ..., \|x_n\| \leq 1$ we obtain that $\sum_{k=0}^{n}\|x_k^*\| \leq \|V_n\|$.

We have $V_n \to V$ pointwise and V_n is linear and continuous $\forall n \geq 0$. Since X is a Banach space $c_0(X)$ is a Banach space. By the Uniform Boundedness Principle it follows that the sequence $(V_n)_{n\geq 0}$ is norm bounded, and therefore $\sum_{n=0}^{\infty}\|x_n^*\| < \infty$.

'(b) $\Rightarrow$ (a)' It is obvious since $x_n \to 0$ implies that $(x_n)_{n\geq 0}$ is norm bounded.

ii) '(a) $\Rightarrow$ (b)' Let $(x_n)_{n\geq 0} \subseteq X$ with $x_n \to 0$ in norm. Since A is dense, it follows that for any $n \geq 0$ there is $y_n \in A$ such that $\|x_n - y_n\| \leq 1/2^n$. Hence $x_n - y_n \to 0$ and from $x_n \to 0$ it follows that $y_n \to 0$. Since $(y_n)_{n\geq 0} \subseteq A$, by hypothesis (a) it follows that the series $\sum_{n=0}^{\infty} x_n^*(y_n)$ converges. Since the sequence $(x_n^*)_{n\geq 0} \subseteq X^*$ is uniformly bounded it follows that $|x_n^*(x_n - y_n)| \leq M\|x_n - y_n\| \leq M/2^n\ \forall n \geq 0$, where $M = \sup_{n\geq 0}\|x_n^*\| < \infty$, hence the scalar series $\sum_{n=0}^{\infty} x_n^*(x_n - y_n)$ is absolutely convergent, hence convergent. Since the series $\sum_{n=0}^{\infty} x_n^*(y_n)$ converges, it follows that the series $\sum_{n=0}^{\infty} x_n^*(x_n)$ converges. Using (i) it follows that (b) is true.

'(b) $\Rightarrow$ (a)' It is trivial.

iii) Let $x_n^* : \mathbb{R} \to \mathbb{R}$, $x_n^*(x) = a_n x\ \forall x \in \mathbb{R}$. We apply (i) observing that $\|x_n^*\| = |a_n|\ \forall n \geq 0$.

iv) Suppose, for a contradiction, that there is such a sequence. We will prove first that: $\exists k \in \mathbb{N}$ such that $c_n \neq 0\ \forall n \geq k$. Indeed, if this is not true, i.e., $\forall k \in \mathbb{N}\ \exists n \geq k$ such that $c_n = 0$, then we can construct a subsequence $(k_n)_{n\in\mathbb{N}}$ of $\mathbb{N}$ such that $c_{k_n} = 0\ \forall n \in \mathbb{N}$. Then for the sequence $(a_n)_{n\geq 0}$ whose components are all equal to 0, with the exception of those on the k_n^{th} positions which are equal to $(-1)^n$, we have $c_n a_n = 0\ \forall n \geq 0$. Therefore by hypothesis the series $\sum_{n=0}^{\infty} a_n$ converges, i.e., the series $\sum_{n=1}^{\infty}(-1)^n$ converges, which is false. Hence there is a $k \in \mathbb{N}$ such that $c_n \neq 0\ \forall n \geq k$, and therefore we may suppose that $c_n \neq 0\ \forall n \geq 0$. We know that for any convergent series $\sum_{n=0}^{\infty} a_n$ it follows that $a_n c_n \to 0$.

6.2 Solutions

Using exercise 4 it follows that the sequence $(c_n)_{n\geq 0}$ is bounded, i.e., $\exists M > 0$ such that $|c_n| \leq M \; \forall n \geq 0$. If $a_n \to 0$ then $c_n(a_n/c_n) \to 0$, and it follows by hypothesis that the series $\sum_{n=0}^{\infty} a_n/c_n$ converges. Using (iii) it follows that the series $\sum_{n=0}^{\infty} 1/c_n$ is absolutely convergent, in particular $1/c_n \to 0$, which is false since $1/|c_n| \geq 1/M \; \forall n \geq 0$.

v) Let $x^* : C[0,1] \to \mathbb{K}$, $x^*(f) = \int_0^1 x^n f(x^n) dx \; \forall f \in C[0,1]$. We have

$$|x^*(f)| \leq \int_0^1 x^n |f(x^n)| \, dx \leq \|f\| \int_0^1 x^n dx = \frac{1}{n+1} \|f\|,$$

and therefore $\|x^*\| \leq 1/(n+1)$. Taking $f_0 = 1$, the constant function, we obtain that $1/(n+1) = |x^*(f_0)| \leq \|x^*\|$, and therefore $\|x^*\| = 1/(n+1)$. For $n \geq 0$, let $x_n^*(f) = a_n \int_0^1 x^n f(x^n) dx \; \forall f \in C[0,1]$. Then the condition from the exercise becomes: $\forall (f_n)_{n\geq 0} \subseteq C[0,1]$ with $\|f_n\| \to 0$ it follows that the series $\sum_{n=0}^{\infty} x_n^*(f_n)$ converges. From (i) this is equivalent with the fact that the series $\sum_{n=0}^{\infty} \|x_n^*\|$ converges. Since $\|x_n^*\| = |a_n|/(n+1)$ $\forall n \geq 0$, we obtain the statement.

vi) Let $x_n^* : C[0,1] \to \mathbb{K}$ be defined by $x_n^*(f) = a_n \int_0^1 f(x^n) dx$. Then $\|x_n^*\| = |a_n|$ $\forall n \geq 0$. Since the set of all the Lipschitz functions on $[0,1]$ is dense in $C[0,1]$ (see exercise 4(ii) of chapter 9) the statement follows if we use (ii).

vii) Let $x_n^* : L_2[0,1] \to \mathbb{R}$ defined by $x_n^*(f) = a_n \int_0^1 x^n f(x) dx = \langle f, a_n x^n \rangle$, where $\langle \cdot, \cdot \rangle$ denotes the inner product on $L_2[0,1]$. Then

$$\|x_n^*\| = \|a_n x^n\|_2 = |a_n| \sqrt{\int_0^1 x^{2n} dx} = \frac{|a_n|}{\sqrt{2n+1}}.$$

The condition from the exercise is that $\forall (f_n)_{n\geq 0} \subseteq C[0,1]$ with $\|f_n\|_2 \to 0$ it follows that the series $\sum_{n=0}^{\infty} x_n^*(f_n)$ converges. Since

$$\|x_n^*\| = \frac{|a_n|}{\sqrt{2n+1}} \; \forall n \geq 0,$$

using the hypothesis it follows that the sequence $(x_n^*)_{n\geq 0}$ is uniformly bounded. Using the fact that $C[0,1] \subseteq (L_2[0,1], \|\cdot\|_2)$ is dense, from the part (ii) we obtain (vii).

viii) Consider $(a_{nk})_{n,k\geq 0}$ as in the statement. Let $x_n = (a_{n0}, a_{n1}, \ldots) \; \forall n \geq 0$. Then $x_n \in c_0 \; \forall n \geq 0$. Moreover $\|x_n\| = \sup_{k\geq 0} |a_{nk}| \; \forall n \geq 0$, hence $x_n \to 0$ in the space c_0. Let now $x_n^* : c_0 \to \mathbb{K}$ defined by $x_n^*(y_0, y_1, \ldots) = b_n y_n$, which is a linear and continuous functional, with $\|x_n^*\| = |b_n|$. The first condition from the exercise is that: $\forall x_n \to 0$ in the space c_0 it follows that the series $\sum_{n=0}^{\infty} x_n^*(x_n)$ converges. Using (i) this is equivalent with the fact that the series $\sum_{n=0}^{\infty} \|x_n^*\|$ converges, i.e., the series $\sum_{n=0}^{\infty} b_n$ is absolutely convergent.

6. For general Banach spaces X and Y the answer is: not necessarily. Let H be an infinite-dimensional Hilbert space, $(a_n)_{n\geq 0} \subseteq H \setminus \{0\}$ is an orthogonal system with

$\sum_{n=0}^{\infty} \|a_n\|^2 < \infty$, and consider a bounded sequence $(b_n)_{n\geq 0} \subseteq H$. We consider the rank 1 operators $V_n : H \to H$ defined by $V_n(x) = a_n \langle x, b_n \rangle$ $\forall x \in H$. We will show that for any bounded sequence $(x_n)_{n\geq 0} \subseteq H$ it follows that the series $\sum_{n=0}^{\infty} V_n(x_n)$ is norm convergent. Indeed for $n, p \geq 0$ from the Pythagoras theorem and the Cauchy–Buniakowski–Schwarz inequality it follows that

$$\|V_n(x_n) + V_{n+1}(x_{n+1}) + \cdots + V_{n+p}(x_{n+p})\|^2$$
$$= \|a_n \langle x_n, b_n \rangle + a_{n+1} \langle x_{n+1}, b_{n+1} \rangle + \cdots + a_{n+p} \langle x_{n+p}, b_{n+p} \rangle\|^2$$
$$= \|a_n\|^2 |\langle x_n, b_n \rangle|^2 + \|a_{n+1}\|^2 |\langle x_{n+1}, b_{n+1} \rangle|^2 + \cdots$$
$$+ \|a_{n+p}\|^2 |\langle x_{n+p}, b_{n+p} \rangle|^2$$
$$\leq M^2 L^2 \left(\|a_n\|^2 + \|a_{n+1}\|^2 + \cdots + \|a_{n+p}\|^2 \right),$$

where $M = \sup_{n \geq 0} \|b_n\| < \infty$ and $L = \sup_{n \geq 0} \|x_n\| < \infty$. Since the series $\sum_{n=0}^{\infty} \|a_n\|^2$ converges, using the Cauchy test and that H is a Hilbert space (therefore a complete space) we deduce the norm convergence of the series $\sum_{n=0}^{\infty} V_n(x_n)$.

If the assertion from exercise is true, then the series $\sum_{n=0}^{\infty} \|V_n\|$ must be convergent and using that $\|V_n\| = \|a_n\| \|b_n\|$ $\forall n \geq 0$, the series $\sum_{n=0}^{\infty} \|a_n\| \|b_n\|$ must be convergent. But a such situation is false in general, as the following example shows: take $V_n : l_2 \to l_2$ defined by $V_n(x) = (1/(n+1)) e_{n+1} \langle x, e_{n+1} \rangle$.

7. i) '(a) $\Rightarrow$ (b)' Let $x_n \to 0$ in norm. Then the sequence $s_n = x_n - x_{n+1}$ $\forall n \geq 0$ has the property that $\sum_{n=0}^{\infty} s_n$ is norm convergent, since $\sum_{k=0}^{n} s_k = x_0 - x_{n+1} \to x_0$. It follows, by hypothesis, that the series $\sum_{n=0}^{\infty} T_n(s_n) = \sum_{n=0}^{\infty} T_n(x_n - x_{n+1})$ converges, hence the operator $U : c_0(X) \to Y$, $U(x) = \sum_{n=0}^{\infty} T_n(x_n - x_{n+1})$ $\forall x = (x_n)_{n\geq 0} \in c_0(X)$ is well defined. For any $n \geq 0$, let $U_n : c_0(X) \to Y$ be defined by $U_n(x) = \sum_{k=0}^{n} T_k(x_k - x_{k+1})$ $\forall x = (x_k)_{k\geq 0} \in c_0(X)$. We have

$$\|U_n(x) - U(x)\| = \left\| \sum_{k=0}^{n} T_k(x_k - x_{k+1}) - s \right\| \to 0, \quad s = \sum_{k=0}^{\infty} T_k(x_k - x_{k+1}) \in Y.$$

From the Banach–Steinhaus theorem it follows that $\sup_{n \geq 0} \|U_n\| = M < \infty$, i.e.,

$$\left\| \sum_{k=0}^{n} T_k(x_k - x_{k+1}) \right\| \leq M \quad \forall n \geq 0, \ \forall (x_k)_{k\geq 0} \in c_0(X) \text{ with } \|(x_k)_{k\geq 0}\| \leq 1.$$

For $n \geq 0$ and $x \in X$ with $\|x\| \leq 1$, the element $(x_k)_{k \in \mathbb{N}} = (0, ..., 0, x, 0, ...)$ is in $c_0(X)$ (n times 0 before x),

$$\sum_{k=0}^{n} T_k(x_k - x_{k+1}) = T_0(x_0 - x_1) + \cdots + T_{n-1}(x_{n-1} - x_n) + T_n(x_n - x_{n+1}) = -T_n(x),$$

6.2 Solutions

and then $\|T_n(x)\| \leq M$, so $\|T_n\| \leq M$, and therefore $\sup_{n\geq 0} \|T_n\| \leq M$.

Using the linearity for the operators T_n we have
$$T_0(x_0 - x_1) + T_1(x_1 - x_2) + \cdots + T_{n-1}(x_{n-1} - x_n) + T_n(x_n - x_{n+1})$$
$$= T_0(x_0) + (T_1 - T_0)(x_1) + \cdots + (T_n - T_{n-1})(x_n) - T_n(x_{n+1}).$$
For every $n \in \mathbb{N}$ consider the operator $V_n : c_0(X) \to Y$,

$$V_n(x) = T_0(x_0) - \sum_{k=0}^{n-1}(T_k - T_{k+1})(x_{k+1}) \quad \forall x = (x_k)_{k\geq 0} \in c_0(X).$$

Using the above summation formula we have $U_n(x) - V_n(x) = -T_n(x_{n+1})$ $\forall x = (x_k)_{k\geq 0} \in c_0(X)$, so $\|U_n(x) - V_n(x)\| \leq 2M\|x_{n+1}\| \to 0$ $\forall x \in c_0(X)$. Hence we have $U_n(x) - V_n(x) \to 0$ in norm, $U_n(x) - U(x) \to 0$ in norm, from whence $V_n(x) \to U(x)$ in norm $\forall x \in c_0(X)$, i.e., $\forall (x_n)_{n\geq 0} \subseteq X$ with $\|x_n\| \to 0$ it follows that the series $\sum_{n=0}^{\infty}(T_n - T_{n+1})(x_{n+1})$ is norm convergent, and from here if $\|y_n\| \to 0$ then $\sum_{n=0}^{\infty}(T_n - T_{n+1})(y_n)$ converges (we consider the sequence $(x_n)_{n\geq 0} = (0, y_0, y_1, y_2, \ldots)$).

'(b) $\Rightarrow$ (a)' Let $\sum_{n=0}^{\infty} x_n$ be a norm convergent series. Let $s = \sum_{n=0}^{\infty} x_n \in X$. We denote $a_n = \sum_{k=0}^{n} x_k - s$, $a_n \to 0$ in norm. Let us observe that $a_n - a_{n-1} = x_n$ $\forall n \in \mathbb{N}$. By our hypothesis, $\sum_{n=0}^{\infty}(T_n - T_{n+1})(a_n)$ is norm convergent. We now prove that the series $\sum_{n=0}^{\infty} T_n(x_n)$ is norm convergent, i.e., since Y is a Banach space, it must be shown that $\forall \varepsilon > 0$ $\exists n_\varepsilon \in \mathbb{N}$ such that $\left\|\sum_{k=n}^{n+p} T_k(x_k)\right\| < \varepsilon$ $\forall n \geq n_\varepsilon$, $\forall p \geq 0$. To show this, we will use again the Abel summation formula to write
$$T_n(x_n) + T_{n+1}(x_{n+1}) + \cdots + T_{n+p}(x_{n+p})$$
$$= T_n(a_n - a_{n-1}) + T_{n+1}(a_{n+1} - a_n) + \cdots$$
$$+ T_{n+p-1}(a_{n+p-1} - a_{n+p-2}) + T_{n+p}(a_{n+p} - a_{n+p-1})$$
$$= -T_n(a_{n-1}) + (T_n - T_{n+1})(a_n) + \cdots$$
$$+ (T_{n+p-1} - T_{n+p})(a_{n+p-1}) + T_{n+p}(a_{n+p}).$$
Hence
$$\|T_n(x_n) + T_{n+1}(x_{n+1}) + \cdots + T_{n+p}(x_{n+p})\|$$
$$\leq \|T_n(a_{n-1})\| + \|(T_n - T_{n+1})(a_n) + \cdots + (T_{n+p-1} - T_{n+p})(a_{n+p-1})\|$$
$$+ \|T_{n+p}(a_{n+p})\|$$
$$\leq M\|a_{n-1}\| + M\|a_{n+p}\|$$
$$+ \|(T_n - T_{n+1})(a_n) + \cdots + (T_{n+p-1} - T_{n+p})(a_{n+p-1})\|,$$
where $M = \sup_{k\geq 0} \|T_k\| < \infty$ by hypothesis. Since the series $\sum_{k=0}^{\infty}(T_k - T_{k+1})(a_k)$ is norm convergent, $\forall \varepsilon > 0$ $\exists m_\varepsilon \in \mathbb{N}$ such that

$$\left\|\sum_{k=n}^{n+p-1}(T_k - T_{k+1})(a_k)\right\| < \varepsilon/2 \quad \forall n \geq m_\varepsilon, \forall p \geq 1.$$

Also, since $a_n \to 0$ in norm, there is $k_\varepsilon \in \mathbb{N}$ such that $\|a_n\| < \varepsilon/(4M)$ $\forall n \geq k_\varepsilon$. Let $n_\varepsilon = 1 + \max(m_\varepsilon, k_\varepsilon)$ and consider $n \geq n_\varepsilon$ and $p \geq 0$. If $p \geq 1$ we deduce that $\|T_n(x_n) + T_{n+1}(x_{n+1}) + \cdots + T_{n+p}(x_{n+p})\| < \varepsilon$. If $p = 0$, then $\|T_n(x_n)\| \leq M\|a_{n-1}\| + M\|a_n\| < \varepsilon$.

ii) '(a) $\Rightarrow$ (b)' Using (i) we have $\sup_{n \geq 0} \|x_n^*\| < \infty$ and $\forall (x_n)_{n \geq 0} \subseteq X$ with $\|x_n\| \to 0$ it follows that the series $\sum_{n=0}^{\infty}(x_n^* - x_{n+1}^*)(x_n)$ converges. From exercise 5(i) it follows that the series $\sum_{n=0}^{\infty} \|x_n^* - x_{n+1}^*\|$ is convergent.

'(b) $\Rightarrow$ (a)' Let us suppose now that the series $\sum_{n=0}^{\infty}(x_n^* - x_{n+1}^*)$ is absolutely convergent, hence convergent since X^* is a Banach space. Then the sequence $(x_n^*)_{n \geq 0}$ is convergent, hence bounded. If $(x_n)_{n \geq 0} \subseteq X$ with $\|x_n\| \to 0$, then there is $L > 0$ such that $\|x_n\| \leq L$ $\forall n \geq 0$. We have $|(x_n^* - x_{n+1}^*)(x_n)| \leq L\|x_n^* - x_{n+1}^*\|$ $\forall n \geq 0$ and therefore by our hypothesis (b) the series $\sum_{n=0}^{\infty}(x_n^* - x_{n+1}^*)(x_n)$ converges absolutely. Now we apply (i).

iii) '(a) $\Rightarrow$ (b)' Let $\sum_{n=0}^{\infty} x_n$ be a norm convergent series with elements from X. Since $A \subseteq X$ is dense for any $n \geq 0$ there is $y_n \in A$ such that $\|x_n - y_n\| \leq 1/2^n$. Then it follows that the series $\sum_{n=0}^{\infty}(x_n - y_n)$ is absolutely convergent and since X is a Banach space the series is norm convergent. It then follows that the series $\sum_{n=0}^{\infty} y_n$ is norm convergent. By our hypothesis (a) the scalar series $\sum_{n=0}^{\infty} x_n^*(y_n)$ converges. Since the sequence $(x_n^*)_{n \geq 0}$ is uniformly bounded we have $|x_n^*(x_n - y_n)| \leq M\|x_n - y_n\|$ $\forall n \geq 0$, so the scalar series $\sum_{n=0}^{\infty} x_n^*(x_n - y_n)$ is absolutely convergent, hence convergent. Then the series $\sum_{n=0}^{\infty} x_n^*(x_n)$ converges, and we use (ii).

'(b) $\Rightarrow$ (a)' It is trivial.

iv) See [22, exercises 7.1, 7.2, chapter 7].

Let $x_n^* : \mathbb{K} \to \mathbb{K}$, $x_n^*(x) = x_n x$. We apply (ii), observing that $\|x_n^* - x_{n+1}^*\| = |x_n - x_{n+1}|$.

v) Let $(c_n)_{n \geq 0} \subseteq \mathbb{K}$ be a sequence which satisfies the condition from the exercise. We have: $\exists k \geq 0$ such that $c_n \neq 0$ $\forall n \geq k$ (see the solution for exercise 5(iv)). Suppose that $c_n \neq 0$ $\forall n \geq 0$. From our hypothesis, for any convergent series $\sum_{n=0}^{\infty} a_n$ it follows that the series $\sum_{n=0}^{\infty} a_n c_n$ converges. Using (iv) it follows that the series $\sum_{n=0}^{\infty} |c_n - c_{n+1}|$ converges. Also, for any convergent series $\sum_{n=0}^{\infty} a_n$ the series $\sum_{n=0}^{\infty} a_n/c_n$ converges, and then by (iv) it follows that the series $\sum_{n=0}^{\infty} |1/c_n - 1/c_{n+1}|$ converges. Therefore the sequences $(c_n)_{n \geq 0}$ and $(1/c_n)_{n \geq 0}$ are convergent and then $\lim_{n \to \infty} c_n = x \neq 0$. We obtain that $\lim_{n \to \infty} c_n = x \neq 0$ and the series $\sum_{n=0}^{\infty} |c_n - c_{n+1}|$ converges.

6.2 Solutions

One can easily see now that the solutions for our exercise are the sequences $(c_n)_{n\geq 0}$ for which $\lim_{n\to\infty} c_n = x \neq 0$ and the series $\sum_{n=0}^{\infty} |c_n - c_{n+1}|$ converges.

vi) We use (v) with
$$c_n = 1 - \frac{1}{2^\alpha} + \cdots + \frac{(-1)^n}{(n+1)^\alpha}.$$

We have $|c_n - c_{n+1}| = 1/(n+2)^\alpha \ \forall n \geq 0$, thus the series $\sum_{n=0}^{\infty} |c_n - c_{n+1}| = \sum_{n=0}^{\infty} \frac{1}{(n+2)^\alpha}$ is convergent if and only if $\alpha > 1$. In this case we have

$$\lim_{n\to\infty} c_n = \sum_{n=0}^{\infty} \frac{(-1)^n}{(n+1)^\alpha} = \sum_{n=1}^{\infty} \frac{1}{(2n-1)^\alpha} - \sum_{n=1}^{\infty} \frac{1}{(2n)^\alpha} = \sum_{n=1}^{\infty} \frac{1}{n^\alpha} - 2\sum_{n=1}^{\infty} \frac{1}{(2n)^\alpha}$$

$$= \left(1 - 2^{1-\alpha}\right) \sum_{n=1}^{\infty} \frac{1}{n^\alpha} \neq 0,$$

for $\alpha > 1$. Hence the values of $\alpha \in \mathbb{R}$ which satisfy the condition from the exercise are: $\alpha \in (1, \infty)$.

8. For general Banach spaces X and Y the answer is: not necessarily. Let H be an infinite-dimensional Hilbert space, $(a_n)_{n\geq 0} \subseteq H\setminus\{0\}$ is an orthogonal system with the property that the series $\sum_{n=0}^{\infty} \|a_n\|^2$ is convergent, and consider a bounded sequence $(b_n)_{n\geq 0} \subseteq H$. We consider the rank 1 operators $V_n : H \to H$, $n \geq 0$, defined by $V_n(x) = a_n \langle x, b_n \rangle \ \forall x \in H$, and take $T_n = V_0 + V_1 + \cdots + V_n$. We have $T_n - T_{n+1} = -V_{n+1}$, and as in the solution for exercise 6 it follows that for any bounded sequence $(x_n)_{n\geq 0} \subseteq H$, the series $\sum_{n=0}^{\infty} (T_n - T_{n+1})(x_n)$ is norm convergent. Also, for $x \in X$ and $n \geq 0$ we have

$$\|T_n x\|^2 = \sum_{k=0}^{n} \|a_k \langle x, b_k \rangle\|^2 = \sum_{k=0}^{n} \|a_k\|^2 |\langle x, b_k \rangle|^2 \leq \left(\sum_{k=0}^{\infty} \|a_k\|^2 \sup_{k\geq 0} \|b_k\|^2\right) \|x\|^2,$$

and then $\sup_{n\geq 0} \|T_n\| < \infty$. We apply the part (i) from exercise 7 and we obtain that if $\sum_{n=0}^{\infty} x_n$ converges, then $\sum_{n=0}^{\infty} T_n(x_n)$ converges. If the assertion from the exercise is true then the series $\sum_{n=0}^{\infty} \|T_n - T_{n+1}\|$ must be convergent, i.e., $\sum_{n=1}^{\infty} \|V_n\|$ is convergent. Using that $\|V_n\| = \|a_n\| \|b_n\| \ \forall n \in \mathbb{N}$, it follows that the series $\sum_{n=1}^{\infty} \|a_n\| \|b_n\|$ is convergent.

A such situation is false in general, as the following example shows: take $V_n : l_2 \to l_2$ defined by $V_n(x) = (1/(n+1)) e_{n+1} \langle x, e_{n+1} \rangle$, thus $T_n : l_2 \to l_2$ are given by

$$T_n(x) = e_1 \langle x, e_1 \rangle + \frac{1}{2} e_2 \langle x, e_2 \rangle + \cdots + \frac{1}{n+1} e_{n+1} \langle x, e_{n+1} \rangle$$

$$= \left(x_1, \frac{x_2}{2}, \ldots, \frac{x_{n+1}}{n+1}, 0, \ldots\right) \ \forall x = (x_k)_{k\in\mathbb{N}} \in l_2.$$

9. i) We consider the space $l_1(X) = \left\{ (x_n)_{n \in \mathbb{N}} \subseteq X \,\Big|\, \sum_{n=1}^{\infty} \|x_n\| < \infty \right\}$ which is a linear space with respect to the usual operations. In a standard way one can prove that if X is a Banach space then $l_1(X)$ is also a Banach space with respect to the norm

$$\|x\| = \sum_{n=1}^{\infty} \|x_n\| \quad \forall x = (x_n)_{n \in \mathbb{N}} \in l_1(X).$$

For $n \in \mathbb{N}$ let $h_n : l_1(X) \to \mathbb{K}$, $h_n\left((x_k)_{k \in \mathbb{N}}\right) = \sum_{k=1}^{n} x_k^*(x_k)$, which is clearly a linear and continuous functional. Let us prove that $\|h_n\| = \max_{1 \le k \le n} \|x_k^*\|$ $\forall n \in \mathbb{N}$. We have

$$\left| h_n\left((x_k)_{k \in \mathbb{N}}\right) \right| = \left| \sum_{k=1}^{n} x_k^*(x_k) \right| \le \sum_{k=1}^{n} \|x_k^*\| \|x_k\| \le \left(\max_{1 \le k \le n} \|x_k^*\| \right) \left(\sum_{k=1}^{n} \|x_k\| \right)$$

$$\le \left(\max_{1 \le k \le n} \|x_k^*\| \right) \|x\| \quad \forall x = (x_k)_{k \in \mathbb{N}} \in l_1(X),$$

i.e., $\|h_n\| \le \max_{1 \le k \le n} \|x_k^*\|$. For every $1 \le k \le n$ and $x \in X$ the element $(0,...,0,x,0,...)$ is in $l_1(X)$ ($k-1$ times 0), and therefore $|h_n(0,...,0,x,0,...)| \le \|h_n\| \|(0,...,0,x,0,...)\|$, i.e., $|x_k^*(x)| \le \|h_n\| \|x\|$, and therefore $\|x_k^*\| \le \|h_n\|$. We obtain that $\max_{1 \le k \le n} \|x_k^*\| \le \|h_n\|$, and therefore $\|h_n\| = \max_{1 \le k \le n} \|x_k^*\|$. The hypothesis is that $\forall x \in l_1(X)$ the sequence $(h_n(x))_{n \in \mathbb{N}}$ is bounded. Since $l_1(X)$ is a Banach space by the Uniform Boundedness Principle it follows that $\sup_{n \in \mathbb{N}} \|h_n\| < \infty$, from whence $\sup_{n \in \mathbb{N}} \|x_n^*\| < \infty$.

The converse is clear.

ii) Take $x_n^* : \mathbb{K} \to \mathbb{K}$, $x_n^*(x) = a_n x$, and then apply (i).

We have

$$\left| \sum_{k=1}^{n} a_k x_k \right| \le \sum_{k=1}^{n} |a_k| |x_k| \le \left(\sup_{k \in \mathbb{N}} |a_k| \right) \|x\| \quad \forall n \in \mathbb{N},$$

from whence $\|U(x)\| \le \sup_{n \in \mathbb{N}} |a_n| \|x\|$ $\forall x \in l_1$, and therefore $\|U\| \le \sup_{n \in \mathbb{N}} |a_n|$. Since $\|U(e_n)\| \ge |a_n|$ $\forall n \in \mathbb{N}$, we have $\|U\| \ge |a_n|$ $\forall n \in \mathbb{N}$, and therefore $\|U\| = \sup_{n \in \mathbb{N}} |a_n| = \|a\|_\infty$.

10. The answer is yes! The same argument as in the solution for exercise 9 can be applied, working with the space Y instead of $\mathbb{K}$ (for $n \in \mathbb{N}$, $h_n : l_1(X) \to Y$, $h_n\left((x_k)_{k \in \mathbb{N}}\right) = \sum_{k=1}^{n} T_k(x_k)$, etc.).

11. Let $T : X^* \to l_1$, $T(x^*) = (x^*(x_n))_{n \in \mathbb{N}}$. The hypothesis assures that $T(x^*) \in l_1$ $\forall x^* \in X^*$. Let $T_n : X^* \to l_1$, $T_n(x^*) = (x^*(x_1),...,x^*(x_n), 0,...)$. Then for each $x^* \in X^*$, since the series $\sum_{n=1}^{\infty} |x^*(x_n)|$ converges it follows that $\sum_{k=n+1}^{\infty} |x^*(x_k)| \to 0$, i.e., $T_n(x^*) \to T(x^*)$ $\forall x^* \in X^*$. T_n is linear and continuous $\forall n \in \mathbb{N}$, $\left(\|T_n\| \le \sum_{k=1}^{n} \|x_k\| \right.$

6.2 Solutions

$\forall n \in \mathbb{N}$) and X^* is a Banach space. From the Banach–Steinhaus theorem it follows that T is linear and continuous. We have

$$\|T\| = \sup_{\|x^*\|\leq 1} \|T(x^*)\|_{l_1} = \sup_{\|x^*\|\leq 1} \sum_{n=1}^{\infty} |x^*(x_n)|,$$

and $\|T\| < \infty$.

12. i) '(a) $\Rightarrow$ (b)' For a Banach space X we will denote by

$$l_p(X) = \left\{ (x_n)_{n\in\mathbb{N}} \subseteq X \;\Big|\; \sum_{n=1}^{\infty} \|x_n\|^p < \infty \right\},$$

which is a linear space with respect to the usual operations. In a standard way, one can prove that $l_p(X)$ is a Banach space with respect to the norm

$$\|x\| = \left(\sum_{n=1}^{\infty} \|x_n\|^p \right)^{1/p} \qquad \forall x = (x_n)_{n\in\mathbb{N}} \in l_p(X).$$

Let $U : l_p(X) \to \mathbb{K}$, $U(x) = \sum_{n=1}^{\infty} x_n^*(x_n)$ $\forall x = (x_n)_{n\in\mathbb{N}} \in l_p(X)$. The hypothesis (a) assures that $U(x) \in \mathbb{K}$ $\forall x \in l_p(X)$. Let $U_n : l_p(X) \to \mathbb{K}$, $U_n(x) = \sum_{k=1}^{n} x_k^*(x_k)$. Then $U_n(x) \to U(x)$ $\forall x \in l_p(X)$ and from the Uniform Boundedness Principle it follows that $\sup_{n\in\mathbb{N}} \|U_n\| < \infty$. We have

$$|U_n(x)| \leq \sum_{k=1}^{n} \|x_k^*\| \|x_k\| \leq \left(\sum_{k=1}^{n} \|x_k^*\|^q\right)^{1/q} \left(\sum_{k=1}^{n} \|x_k\|^p\right)^{1/p}$$

$$\leq \|x\| \left(\sum_{k=1}^{n} \|x_k^*\|^q\right)^{1/q} \qquad \forall x \in l_p(X),$$

i.e., $\|U_n\| \leq \left(\sum_{k=1}^{n} \|x_k^*\|^q\right)^{1/q}$. Let $x_1, ..., x_n \in X$ with $\|x_1\| \leq 1, ..., \|x_n\| \leq 1$. Take $\lambda_1, ..., \lambda_n \in \mathbb{K}$ such that $\lambda_k x_k^*(x_k) = |x_k^*(x_k)|^q$ $\forall 1 \leq k \leq n$ (if $x_k^*(x_k) = 0$ then we take $\lambda_k = 0$). Then the element $x = (\lambda_1 x_1, ..., \lambda_n x_n, 0, ...)$ is in $l_p(X)$, and $U_n(x) = \sum_{k=1}^{n} |x_k^*(x_k)|^q$. We have $|U_n(x)| \leq \|U_n\| \|x\|$ and therefore

$$\sum_{k=1}^{n} |x_k^*(x_k)|^q \leq \|U_n\| \left(|\lambda_1|^p \|x_1\|^p + \cdots + |\lambda_n|^p \|x_n\|^p\right)^{1/p}$$

$$\leq \|U_n\| \left(|\lambda_1|^p + \cdots + |\lambda_n|^p\right)^{1/p},$$

and since $|\lambda_k| \leq |x_k^*(x_k)|^{q-1}$ it follows that

$$\sum_{k=1}^{n} |x_k^*(x_k)|^q \leq \|U_n\| \left(\sum_{k=1}^{n} |x_k^*(x_k)|^{(q-1)p}\right)^{1/p},$$

i.e., $\left(\sum_{k=1}^{n}|x_k^*(x_k)|^q\right)^{1/q} \leq \|U_n\|$. Passing to the supremum over $\|x_1\| \leq 1$, $\|x_2\| \leq 1$, ..., $\|x_n\| \leq 1$, we obtain that $\left(\sum_{k=1}^{n}\|x_k^*\|^q\right)^{1/q} \leq \|U_n\|$. Thus $\|U_n\| = \left(\sum_{k=1}^{n}\|x_k^*\|^q\right)^{1/q}$ $\forall n \in \mathbb{N}$, from whence $\sup_{n\in\mathbb{N}}\left(\sum_{k=1}^{n}\|x_k^*\|^q\right)^{1/q} = \sup_{n\in\mathbb{N}}\|U_n\| < \infty$, i.e., the series $\sum_{n=1}^{\infty}\|x_n^*\|^q$ converges.

'(b) $\Rightarrow$ (a)' We use the Hölder inequality and it follows easily from our hypothesis that the series $\sum_{n=1}^{\infty} x_n^*(x_n)$ is absolutely convergent, hence convergent.

ii) See [20, exercise 59, chapter 4].
Take $x_n^* : \mathbb{K} \to \mathbb{K}$, $x_n^*(y) = x_n y$ and then apply (i).

13. For general Banach spaces X and Y the answer is: not necessarily. Take $p = 2$, for example. Let H be an infinite-dimensional Hilbert space and $(a_n)_{n\in\mathbb{N}} \subseteq H\setminus\{0\}$ an orthogonal system. Consider also the sequence $(b_n)_{n\in\mathbb{N}} \subseteq H$ such that $\sup_{n\in\mathbb{N}}(\|a_n\|\|b_n\|) < \infty$. As in the solution for exercises 6 and 8, consider the rank 1 operators $V_n : H \to H$, $V_n(x) = a_n\langle x, y_n\rangle \; \forall x \in H$. We will show that for each 2-absolutely convergent series $\sum_{n=1}^{\infty} x_n$ it follows that the series $\sum_{n=1}^{\infty} V_n(x_n)$ is norm convergent. Indeed for $n \in \mathbb{N}$ and $p \geq 0$ from the Pythagoras theorem and the Cauchy–Buniakowski–Schwarz inequality it follows that
$$\|V_n(x_n) + V_{n+1}(x_{n+1}) + \cdots + V_{n+p}(x_{n+p})\|^2$$
$$= \|a_n\|^2|\langle x_n, b_n\rangle|^2 + \|a_{n+1}\|^2|\langle x_{n+1}, b_{n+1}\rangle| + \cdots$$
$$+ \|a_{n+p}\|^2|\langle x_{n+p}, b_{n+p}\rangle|^2$$
$$\leq M^2\left(\|x_n\|^2 + \|x_{n+1}\|^2 + \cdots + \|x_{n+p}\|^2\right),$$
where $M = \sup_{n\in\mathbb{N}}(\|a_n\|\|b_n\|)$. Since the series $\sum_{k=1}^{\infty}\|x_k\|^2$ converges, using the Cauchy test and the fact that H is a Hilbert (therefore complete) space we deduce the norm convergence for the series $\sum_{n=1}^{\infty} V_n(x_n)$.

If $\sum_{n=1}^{\infty}\|V_n\|^2 < \infty$, since $\|V_n\| = \|a_n\|\|b_n\| \; \forall n \in \mathbb{N}$, the series $\sum_{n=1}^{\infty}\|a_n\|^2\|b_n\|^2$ must be convergent. A such situation is false in general, as the following example shows: one can take $V_n : l_2 \to l_2$, $V_n(x) = e_n\langle x, e_n\rangle \; \forall x \in l_2$.

14. We recall first the following classical result: if $\sum_{n=0}^{\infty}\lambda_n$ is a scalar absolutely convergent series and $(\mu_n)_{n\geq 0}$ is a null convergent scalar sequence then
$$\lambda_n\mu_0 + \lambda_{n-1}\mu_1 + \cdots + \lambda_0\mu_n \to 0.$$

Indeed, let $s = \sum_{n=0}^{\infty}|\lambda_n|$. Since $\mu_n \to 0$, then $\forall \varepsilon > 0 \; \exists n_\varepsilon \in \mathbb{N}$ such that $|\mu_n| < \varepsilon/(1+2s)$, $\forall n \geq n_\varepsilon$ (1). Also, since the series $\sum_{n=0}^{\infty}\lambda_n$ is absolutely convergent it exists $k_\varepsilon \in \mathbb{N}$ such

6.2 Solutions

that $|\lambda_n| < \varepsilon/\left(2\left(|\mu_0| + |\mu_1| + \cdots + |\mu_{n_\varepsilon - 1}|\right)\right) \forall n \geq k_\varepsilon$ (2). Then for $n \geq n_\varepsilon + k_\varepsilon$ using (1) and (2) we have

$$
\begin{aligned}
|\lambda_n \mu_0 + \lambda_{n-1}\mu_1 + \cdots + \lambda_0 \mu_n| &\leq |\lambda_0|\,|\mu_n| + |\lambda_1|\,|\mu_{n-1}| + \cdots + |\lambda_{n-n_\varepsilon}|\,|\mu_{n_\varepsilon}| \\
&\quad + |\lambda_{n-n_\varepsilon+1}|\,|\mu_{n_\varepsilon - 1}| + \cdots + |\lambda_n|\,|\mu_0| \\
&\leq \frac{\varepsilon}{1+2s}\left(|\lambda_0| + |\lambda_1| + \cdots + |\lambda_{n-n_\varepsilon}|\right) + \\
&\quad \frac{\varepsilon}{2\left(|\mu_0| + |\mu_1| + \cdots + |\mu_{n_\varepsilon - 1}|\right)}\left(|\mu_{n_\varepsilon - 1}| + \cdots + |\mu_0|\right) \\
&\leq \frac{\varepsilon}{2} + \frac{\varepsilon}{2} = \varepsilon.
\end{aligned}
$$

Return now to the solution for our exercise. We shall use the notations from exercise 2.

i) '(a) $\Rightarrow$ (b)' For $n \geq 0$, let $h_n : c_0(X) \to c_0$ defined by

$$h_n(x_0, x_1, \ldots) = (x_0^*(x_0), x_1^*(x_0) + x_0^*(x_1), \ldots, x_n^*(x_0) + x_{n-1}^*(x_1) + \cdots + x_0^*(x_n), 0, 0, \ldots),$$

and let $h : c_0(X) \to c_0$ defined by

$$h(x_0, x_1, \ldots) = (x_0^*(x_0), x_1^*(x_0) + x_0^*(x_1), \ldots, x_n^*(x_0) + x_{n-1}^*(x_1) + \cdots + x_0^*(x_n), \ldots).$$

By our hypothesis (a) h is well defined. For each $n \geq 0$, h_n is a linear and continuous operator with $\|h_n\| = \|x_0^*\| + \cdots + \|x_n^*\|$ (see the final part for this solution). In addition, if $x = (x_k)_{k \geq 0} \in c_0(X)$, then $h(x) - h_n(x) = (0, 0, \ldots, 0, x_{n+1}^*(x_0) + x_n^*(x_1) + \cdots + x_0^*(x_{n+1}), \ldots)$, from whence

$$\|h(x) - h_n(x)\| = \sup_{k \geq n+1} \left|x_k^*(x_0) + x_{k-1}^*(x_1) + \cdots + x_0^*(x_k)\right| \to 0 \ (n \to \infty),$$

by our hypothesis (a). Hence $h_n \to h$ pointwise and h_n is a linear and continuous operator $\forall n \geq 0$. Since X is a Banach space then $c_0(X)$ is a Banach space. By the Uniform Boundedness Principle it follows that the sequence $(h_n)_{n \geq 0}$ is uniformly bounded, i.e., $\sup_{n \geq 0} \|h_n\| < \infty$.

Let now $n \geq 0$ be fixed. Let $x = (x_j)_{j \geq 0} \in c_0(X)$ with $\|x\| \leq 1$, i.e., from the definition of the norm on the space $c_0(X)$, $\|x_j\| \leq 1 \ \forall j \geq 0$. We have

$$
\begin{aligned}
\left|x_k^*(x_0) + x_{k-1}^*(x_1) + \cdots + x_0^*(x_k)\right| &\leq \left|x_k^*(x_0)\right| + \left|x_{k-1}^*(x_1)\right| + \cdots + \left|x_0^*(x_k)\right| \\
&\leq \|x_0^*\| + \cdots + \|x_k^*\| \\
&\leq \|x_0^*\| + \cdots + \|x_n^*\| \ \forall 0 \leq k \leq n,
\end{aligned}
$$

and therefore $\|h_n(x)\| \leq \|x_0^*\| + \cdots + \|x_n^*\| \ \forall x \in c_0(X)$ with $\|x\| \leq 1$, that is, h_n is linear and continuous, with $\|h_n\| \leq \|x_0^*\| + \cdots + \|x_n^*\|$. Conversely, let $x_0, x_1, \ldots, x_n \in X$ with $\|x_0\| \leq 1, \ldots, \|x_n\| \leq 1$, and let $\lambda_0 = \mathrm{sgn}\,(x_n^*(x_0)), \ldots, \lambda_n = \mathrm{sgn}\,(x_0^*(x_n))$. The element $(\lambda_0 x_0, \lambda_1 x_1, \ldots, \lambda_n x_n, 0, 0, \ldots)$ is in $c_0(X)$, from whence

$$\|h_n(\lambda_0 x_0, \lambda_1 x_1, \ldots, \lambda_n x_n, 0, 0, \ldots)\| \leq \|h_n\|\,\|(\lambda_0 x_0, \lambda_1 x_1, \ldots, \lambda_n x_n, 0, 0, \ldots)\|,$$

and then $|x_n^*(x_0)| + |x_{n-1}^*(x_1)| + \cdots + |x_0^*(x_n)| \leq \|h_n\|$. Passing to the supremum over $\|x_0\| \leq 1, \ldots, \|x_n\| \leq 1$ we obtain that $\|x_0^*\| + \cdots + \|x_n^*\| \leq \|h_n\|$, and then $\|h_n\| = \|x_0^*\| + \cdots + \|x_n^*\|$.

'(b) $\Rightarrow$ (a)' Let $(x_n)_{n\geq 0} \subseteq X$ with $\|x_n\| \to 0$. From (b) the series $\sum_{n=0}^{\infty} \|x_n^*\|$ is convergent. Now by the the classical result proved above we have $\|x_n^*\| \|x_0\| + \cdots + \|x_0^*\| \|x_n\| \to 0$. From here using the inequality

$$|x_n^*(x_0) + x_{n-1}^*(x_1) + \cdots + x_0^*(x_n)| \leq \|x_n^*\| \|x_0\| + \cdots + \|x_0^*\| \|x_n\|,$$

we obtain (a).

ii) '(a) $\Rightarrow$ (b)' Let $(x_n)_{n\geq 0} \subseteq X$ with $x_n \to 0$ in norm. Since $A \subseteq X$ is dense it follows that for any $n \geq 0$ there is $y_n \in A$ such that $\|x_n - y_n\| \leq 1/2^n$. Hence $x_n - y_n \to 0$ and from $x_n \to 0$ it follows that $y_n \to 0$. Since $(y_n)_{n\geq 0} \subseteq A$, by the hypothesis (a) it follows that $x_n^*(y_0) + x_{n-1}^*(y_1) + \cdots + x_0^*(y_n) \to 0$. Now since by hypothesis $\|x_n^*\| \to 0$, and $\sum_{n=0}^{\infty} \|x_n - y_n\| \leq \sum_{n=0}^{\infty} 1/2^n < \infty$, by the classical result proved above it follows that $\|x_n^*\| \|x_0 - y_0\| + \cdots + \|x_0^*\| \|x_n - y_n\| \to 0$, hence by the inequality

$$|x_n^*(x_0 - y_0) + x_{n-1}^*(x_1 - y_1) + \cdots + x_0^*(x_n - y_n)| \leq \|x_n^*\| \|x_0 - y_0\| \\ + \cdots + \|x_0^*\| \|x_n - y_n\|$$

we obtain that $x_n^*(x_0 - y_0) + x_{n-1}^*(x_1 - y_1) + \cdots + x_0^*(x_n - y_n) \to 0$ and hence $x_n^*(x_0) + x_{n-1}^*(x_1) + \cdots + x_0^*(x_n) \to 0$, since $x_n^*(y_0) + x_{n-1}^*(y_1) + \cdots + x_0^*(y_n) \to 0$. Now we can apply (i) to obtain (b).

'(b) $\Rightarrow$ (a)' It is proved in (i).

iii) Let $x_n^* : C[0,1] \to \mathbb{K}$, $x_n^*(f) = b_n \int_0^{a_n} f(x)dx$. Then the first condition from the statement becomes: $\forall (f_n)_{n\geq 0} \subseteq C[0,1]$ with $\|f_n\| \to 0$ it follows that $x_n^*(f_0) + x_{n-1}^*(f_1) + \cdots + x_0^*(f_n) \to 0$. Using (i) this is equivalent with the fact that the series $\sum_{n=0}^{\infty} \|x_n^*\|$ converges. Since $\|x_n^*\| = |a_n b_n| \ \forall n \geq 0$, we obtain the statement.

iv) Let $x_n^* : L_2[0,1] \to \mathbb{K}$ be defined by $x_n^*(f) = b_n \int_0^1 x^n f(x)dx = \langle f, \overline{b_n} x^n \rangle$. Then

$$\|x_n^*\| = \|\overline{b_n} x^n\|_2 = |b_n| \left(\int_0^1 x^{2n} dx \right)^{1/2} = \frac{|b_n|}{\sqrt{2n+1}}.$$

By hypothesis we have $\|x_n^*\| \to 0$. Also, the first condition from the exercise is: $\forall (f_n)_{n\geq 0} \subseteq C[0,1]$ with $\|f_n\|_2 \to 0$, it follows that $x_n^*(f_0) + x_{n-1}^*(f_1) + \cdots + x_0^*(f_n) \to 0$. Since $C[0,1] \subseteq L_2[0,1]$ is dense, we can use the results from (ii) to obtain that this is equivalent with the fact that the series $\sum_{n=0}^{\infty} \|x_n^*\|$ converges. Since

$$\|x_n^*\| = \frac{|b_n|}{\sqrt{2n+1}} \ \forall n \geq 0,$$

we obtain the statement.

v) Let $x_n = (a_{n0}, a_{n1}, \ldots) \ \forall n \geq 0$. Then the fact that $x_n \in c_0$ is equivalent with the fact that $\lim_{k\to\infty} a_{nk} = 0$, and since $\|x_n\| = \sup_{k\geq 0} |a_{nk}|$, the condition $x_n \to 0$ in the space c_0 is $\lim_{n\to\infty} a_{nk} = 0$ uniformly with respect to $k \geq 0$. Let now $x_n^* : c_0 \to \mathbb{K}$, $x_n^*(y_0, y_1, \ldots) = b_n y_n$, which are linear and continuous functionals with $\|x_n^*\| = |b_n|$. The first condition from the exercise is: $\forall x_n \to 0$ in the space c_0 it follows that $x_n^*(x_0) + x_{n-1}^*(x_1) + \cdots + x_0^*(x_n) \to 0$. We apply (i) and we obtain that this is equivalent with the fact that the series $\sum_{n=0}^{\infty} \|x_n^*\|$ converges, i.e., the series $\sum_{n=0}^{\infty} b_n$ is absolutely convergent.

15. For general Banach spaces X and Y the answer is: not necessarily. Let H be an infinite-dimensional Hilbert space, and $(a_n)_{n \geq 0} \subseteq H \setminus \{0\}$ an orthogonal system such that $\sum_{n=0}^{\infty} \|a_n\|^2 < \infty$. Let $(b_n)_{n \geq 0} \subseteq H$ be a bounded sequence. We consider the rank 1 operators, $T_n : H \to H$, $T_n(x) = a_n \langle x, b_n \rangle \ \forall x \in H$. We will show that $\forall (x_n)_{n \geq 0} \subseteq H$ with $\|x_n\| \to 0$, it follows that $\|T_n(x_0) + T_{n-1}(x_1) + \cdots + T_0(x_n)\| \to 0$. Indeed, from the Pythagoras theorem and the Cauchy–Buniakowski–Schwarz inequality it follows that
$$\|T_n(x_0) + T_{n-1}(x_1) + \cdots + T_0(x_n)\|^2$$
$$= \|a_n\|^2 |\langle x_0, b_n \rangle|^2 + \|a_{n-1}\|^2 |\langle x_1, b_{n-1}\rangle|^2 + \cdots + \|a_0\|^2 |\langle x_n, b_0\rangle|^2$$
$$\leq M^2 \left(\|a_n\|^2 \|x_0\|^2 + \|a_{n-1}\|^2 \|x_1\|^2 + \cdots + \|a_0\|^2 \|x_n\|^2 \right),$$
where $M = \sup_{n \geq 0} \|b_n\| < \infty$. Since the series $\sum_{n=0}^{\infty} \|a_n\|^2$ converges and $\|x_n\|^2 \to 0$, by the classical result proved in the beginning of the solution for exercise 14 it follows that $\|T_n(x_0) + T_{n-1}(x_1) + \cdots + T_0(x_n)\| \to 0$.

Suppose that the series $\sum_{n=0}^{\infty} \|T_n\|$ converges. Using that $\|T_n\| = \|a_n\| \|b_n\| \ \forall n \geq 0$, we obtain that the series $\sum_{n=0}^{\infty} \|a_n\| \|b_n\|$ converges. This is not always true: one can consider $T_n : l_2 \to l_2$, $T_n(x) = (1/(n+1)) e_{n+1} \langle x, e_{n+1}\rangle \ \forall x \in l_2$.

16. We will prove first that if $b, c \in [0, 1]$, $b < c$, and $x^* : X \to \mathbb{K}$ is defined by $x^*(f) = \int_b^c f(x) dx$, then $\|x^*\| = (c - b)^{1/q}$. Indeed, from the Hölder inequality we have
$$|x^*(f)| \leq \int_b^c |f(x)| dx \leq \left(\int_b^c |f(x)|^p dx \right)^{1/p} \left(\int_b^c 1 dx \right)^{1/q}$$
$$\leq (c - b)^{1/q} \|f\|_p \quad \forall f \in C[0, 1],$$
i.e., $\|x^*\| \leq (c-b)^{1/q}$. To prove the reverse inequality, we want to take $f = \chi_{(b,c)}$, in which case we obtain that $c - b = |x^*(f)| \leq \|x^*\| \|f\|_p = \|x^*\| (c - b)^{1/p}$ and hence $\|x^*\| \geq (c - b)^{1/q}$. But the function $f = \chi_{(b,c)}$ is not continuous. However, we can approximate this characteristic function by a sequence of continuous functions. Let $f_n : [0, 1] \to \mathbb{K}$ be defined by
$$f_n(x) = \begin{cases} 0 & \text{if } 0 \leq x \leq b - 1/n, \\ n(x - b + 1/n) & \text{if } b - 1/n < x < b, \\ 1 & \text{if } b \leq x \leq c, \\ n(c + 1/n - x) & \text{if } c < x \leq c + 1/n, \\ 0 & \text{if } c + 1/n < x \leq 1, \end{cases} \quad \forall n \geq n_0,$$
where n_0 is such that $b > 1/n_0$ and $c < 1 - 1/n_0$. We have $|x^*(f_n)| \leq \|x^*\| \|f_n\|_p$ $\forall n \geq n_0$. Now a simple calculation shows that $x^*(f_n) = c - b$ and $\|f_n\|_p =$

$(c - b + 2/((p+1)n))^{1/p}$. Thus $c - b \le \|x^*\| (c - b + 2/((p+1)n))^{1/p}$ $\forall n \ge n_0$. For $n \to \infty$ we obtain that $c - b \le \|x^*\| (c - b)^{1/p}$, i.e., $\|x^*\| \ge (c-b)^{1/q}$.

Return now to the solution for our exercise. Using the above relations we have $\|x_n^*\| = |a_n| (c_n - b_n)^{1/q}$ $\forall n \in \mathbb{N}$, and this proves (iii). For (ii), if the sequence $(x_n^*)_{n \in \mathbb{N}} \subseteq X^*$ is pointwise bounded then for the constant function $f = 1$ the sequence $(x_n^*(1))_{n \in \mathbb{N}} \subseteq \mathbb{K}$ is bounded, i.e., $(a_n(c_n - b_n))_{n \in \mathbb{N}}$ is bounded. Conversely, if $(a_n(c_n - b_n))_{n \in \mathbb{N}}$ is bounded, then for any $f \in X$ we have

$$|x_n^*(f)| \le |a_n| \int_{b_n}^{c_n} |f(x)| \, dx \le |a_n| (c_n - b_n) \sup_{x \in [0,1]} |f(x)| \le M \sup_{x \in [0,1]} |f(x)| \ \forall n \in \mathbb{N},$$

where $M = \sup_{n \in \mathbb{N}} |a_n| (c_n - b_n) < \infty$, i.e., the sequence $(x_n^*)_{n \in \mathbb{N}} \subseteq X^*$ is pointwise bounded. For (iv), taking $1/q < \alpha \le 1$, by (ii) and (iii) we have that the sequence $x_n^* : X \to \mathbb{K}$ defined by $x_n^*(f) = n^\alpha \int_0^{1/n} f(x) dx$ $\forall f \in C[0,1]$ is pointwise bounded but not uniformly bounded.

Remark. The norm $\|\cdot\|_p$ is not a complete one on $C[0,1]$ and therefore the Uniform Boundedness Principle cannot be applied.

17. It is clear the fact that $\|\cdot\|_1$ is a seminorm on $\mathcal{R}[0,1]$, hence the quotient space $\widetilde{\mathcal{R}}[0,1]$ makes sense, and $\|\cdot\|_1$ is a norm $\widetilde{\mathcal{R}}[0,1]$.

i) a) The fact that x_n^* is well defined and linear is obvious. We observe that $x_n^*(\widetilde{1}) = b_n \int_0^1 g_n(x) \, dx$. If $(x_n^*)_{n \in \mathbb{N}}$ is pointwise bounded, then in particular $(x_n^*(\widetilde{1}))_{n \in \mathbb{N}}$ is bounded, that is, $\left(b_n \int_0^1 g_n(x) \, dx\right)_{n \in \mathbb{N}}$ is bounded. Conversely, for any $f \in \mathcal{R}[0,1]$ and $n \in \mathbb{N}$ since the Riemann integrable functions are bounded we have

$$\left|x_n^*\left(\widetilde{f}\right)\right| \le |b_n| \int_0^1 |g_n(x) f(x)| \, dx \le \left(\sup_{x \in [0,1]} |f(x)|\right) |b_n| \int_0^1 g_n(x) \, dx$$
$$= \left(\sup_{x \in [0,1]} |f(x)|\right) |x_n^*(\widetilde{1})|.$$

Since $\left(x_n^*(\widetilde{1})\right)_{n \in \mathbb{N}}$ is bounded, from here it follows that the sequence $(x_n^*)_{n \in \mathbb{N}}$ is pointwise bounded.

b) We prove that if $g : [0,1] \to [0, \infty)$ is a continuous function and $x^* : \widetilde{\mathcal{R}}[0,1] \to \mathbb{R}$ is defined by $x^*(\widetilde{f}) = \int_0^1 g(x) f(x) \, dx$, then $\|x^*\| = \sup_{x \in [0,1]} g(x)$. Indeed, we have

$$\left|x^*(\widetilde{f})\right| \le \int_0^1 |g(x) f(x)| \, dx \le \left(\sup_{x \in [0,1]} g(x)\right) \int_0^1 |f(x)| \, dx,$$

hence $\|x^*\| \le \sup_{x \in [0,1]} g(x)$. For any $n \in \mathbb{N}$ we have $\left|x^*(\widetilde{g^n})\right| \le \|x^*\| \int_0^1 g^n(x) \, dx$, i.e., denoting by $\alpha_n = \int_0^1 g^n(x) \, dx$ we obtain that $\alpha_{n+1} \le \|x^*\| \alpha_n$, from whence $\alpha_n \le \alpha_1 \|x^*\|^{n-1}$, and hence $\sqrt[n]{\int_0^1 g^n(x) \, dx} \le \sqrt[n]{\alpha_1} \|x^*\|^{(n-1)/n}$. Now passing to the

6.2 Solutions

limit for $n \to \infty$ and using the well known property $\lim_{n\to\infty} \sqrt[n]{\int_0^1 g^n(x)\,dx} = \sup_{x\in[0,1]} g(x)$, we deduce that $\sup_{x\in[0,1]} g(x) \leq \|x^*\|$.

We have therefore $\|x_n^*\| = |b_n| \sup_{x\in[0,1]} g(x) \ \forall n \in \mathbb{N}$.

ii) We take $g_n(x) = \sum_{k=1}^{n} a_k x^k$, $b_n = 1 \ \forall n \in \mathbb{N}$ and we apply (i).

iii) We take $g_n(x) = x^n$, $b_n = a_n \ \forall n \in \mathbb{N}$ and we apply (i).

iv) It is easy to construct, using (i)–(iii), a sequence $(x_n^*)_{n\in\mathbb{N}} \subseteq (\widetilde{\mathcal{R}}[0,1])^*$, pointwise bounded, which is not uniformly bounded.

Remark. The norm $\|\cdot\|_1$ is not a complete one on $\widetilde{\mathcal{R}}[0,1]$ and therefore the Uniform Boundedness Principle cannot be applied.

18. i) Let $a = (a_n)_{n\in\mathbb{N}} \in c_{00}$. Then there is $k \in \mathbb{N}$ such that $a_n = 0 \ \forall n \geq k$ and therefore $T_n(a) = T_m(a) \ \forall m, n \geq k$.

ii) For every $x^* \in X^*$ and $a = (a_k)_{k\in\mathbb{N}} \in c_{00}$ we have

$$|x^*(T_n(a))| \leq \sum_{k=1}^{n} |a_k|\, |x^*(x_k)| \leq \|a\| \sum_{k=1}^{n} |x^*(x_k)|,$$

and then $\|T_n(a)\| \leq \|a\| \sup_{\|x^*\|\leq 1} \sum_{k=1}^{n} |x^*(x_k)|$, i.e., $\|T_n\| \leq \sup_{\|x^*\|\leq 1} \sum_{k=1}^{n} |x^*(x_k)|$. Let $x^* \in X^*$ with $\|x^*\| \leq 1$. There are $\lambda_k \in \mathbb{K}$ with $|\lambda_k| \leq 1$ such that $|x^*(x_k)| = \lambda_k x^*(x_k)$ for $1 \leq k \leq n$. Then taking $a = (\lambda_1, \lambda_2, ..., \lambda_n, 0, ...) \in c_{00}$ we have

$$\sum_{k=1}^{n} |x^*(x_k)| = x^*\left(\sum_{k=1}^{n} \lambda_k x_k\right) = x^*(T_n(a)) \leq \|x^*\|\, \|T_n\|\, \|a\| \leq \|T_n\|.$$

Hence $\|T_n\| = \sup_{\|x^*\|\leq 1} \sum_{k=1}^{n} |x^*(x_k)|$. Now the sequence $(T_n)_{n\in\mathbb{N}}$ is uniformly bounded if and only if $\sup_{n\in\mathbb{N}} \|T_n\| < \infty$, that is, $\sup_{n\in\mathbb{N}} \sup_{\|x^*\|\leq 1} \sum_{k=1}^{n} |x^*(x_k)| < \infty$, i.e., $\sup_{\|x^*\|\leq 1} \sum_{n=1}^{\infty} |x^*(x_n)| < \infty$.

iii) For any $a = (a_k)_{k\in\mathbb{N}} \in c_{00}$ we have

$$\|T_n(a)\| \leq \sum_{k=1}^{n} |a_k|\, \|x_k\| \leq \|a\| \sup_{1\leq k\leq n} \|x_k\|,$$

i.e., $\|T_n\| \leq \sup_{1\leq k\leq n} \|x_k\|$. Let $1 \leq k \leq n$. Then taking $e_k = (0, ..., 0, 1, 0, ...) \in c_{00}$ (1 on the kth position) we have

$$\|x_k\| = \|T_n(e_k)\| \leq \|T_n\|\, \|e_k\| = \|T_n\|,$$

and then $\sup_{1\leq k\leq n} \|x_k\| \leq \|T_n\|$. Therefore $\|T_n\| = \sup_{1\leq k\leq n} \|x_k\|$. The statement follows now in the same way as for (ii).

iv) For $x^* \in X^*$ and $a = (a_k)_{k\in\mathbb{N}} \in c_{00}$ we have

$$|x^*(T_n(a))| \leq \sum_{k=1}^{n} |a_k| |x^*(x_k)| \leq \left(\sum_{k=1}^{n} |a_k|^p\right)^{1/p} \left(\sum_{k=1}^{n} |x^*(x_k)|^q\right)^{1/q}$$

$$\leq \|a\| \left(\sum_{k=1}^{n} |x^*(x_k)|^q\right)^{1/q},$$

and then $\|T_n(a)\| \leq \|a\| \sup\limits_{\|x^*\|\leq 1} \left(\sum_{k=1}^{n} |x^*(x_k)|^q\right)^{1/q}$, i.e.,

$$\|T_n\| \leq \sup_{\|x^*\|\leq 1} \left(\sum_{k=1}^{n} |x^*(x_k)|^q\right)^{1/q}.$$

Let $x^* \in X^*$ with $\|x^*\| \leq 1$. There are $\lambda_k \in \mathbb{K}$ such that $|x^*(x_k)|^q = \lambda_k x^*(x_k)\ \forall 1 \leq k \leq n$ (we take $\lambda_k = 0$ if $x^*(x_k) = 0$). Observe that $|\lambda_k| \leq |x^*(x_k)|^{q-1}\ \forall 1 \leq k \leq n$. Then taking $a = (\lambda_1, \lambda_2, ..., \lambda_n, 0, ...) \in c_{00}$ we have

$$\sum_{k=1}^{n} |x^*(x_k)|^q = x^*\left(\sum_{k=1}^{n} \lambda_k x_k\right) = x^*(T_n(a)) \leq \|x^*\| \|T_n\| \|a\| \leq \|T_n\| \left(\sum_{k=1}^{n} |\lambda_k|^p\right)^{1/p}.$$

We have

$$\left(\sum_{k=1}^{n} |\lambda_k|^p\right)^{1/p} \leq \left(\sum_{k=1}^{n} |x^*(x_k)|^{p(q-1)}\right)^{1/p} = \left(\sum_{k=1}^{n} |x^*(x_k)|^q\right)^{1/p}$$

and from here we deduce that $\left(\sum_{k=1}^{n} |x^*(x_k)|^q\right)^{1/q} \leq \|T_n\|$ and therefore

$$\sup_{\|x^*\|\leq 1} \left(\sum_{k=1}^{n} |x^*(x_k)|^q\right)^{1/q} = \|T_n\|.$$

Now we continue as in (ii).

v) If we take $X = \mathbb{K}$, then using (i)–(iv), one can easily construct a sequence of linear and continuous functionals on c_{00} (on c_{00} we have, respectively, the norms from (i)–(iv)) which is pointwise bounded but not uniformly bounded. From the Uniform Boundedness Principle we deduce that the above norms are not complete on c_{00}.

Remark. It is proved at exercise 10(iv) of chapter 9 that there is no complete norm on c_{00}.

19. First, we give a proof for the Cantor Lemma in the scalar case (see [48, 13.8.1, chapter XIII]), i.e., we want to prove that if $(a_n)_{n\in\mathbb{N}}$, $(b_n)_{n\in\mathbb{N}}$ are two scalar sequences, and $a_n \cos nt + b_n \sin nt \to 0$ on a non-degenerate interval, then $a_n \to 0$ and $b_n \to 0$. Suppose that $|a_n|^2 + |b_n|^2$ does not converge towards 0. Then there are $\varepsilon > 0$ and a subsequence

6.2 Solutions

$(k_n)_{n \in \mathbb{N}}$ of $\mathbb{N}$ such that $|a_{k_n}|^2 + |b_{k_n}|^2 \geq \varepsilon \ \forall n \in \mathbb{N}$. Without loss of generality we will suppose that $k_n = n \ \forall n \in \mathbb{N}$, and therefore $|a_n|^2 + |b_n|^2 \geq \varepsilon \ \forall n \in \mathbb{N}$. Let $g_n : \mathbb{R} \to [0, \infty)$,

$$g_n(t) = \frac{|a_n \cos nt + b_n \sin nt|^2}{|a_n|^2 + |b_n|^2}.$$

Then $g_n(t) \to 0$ on a non-degenerate interval $(a, b) \subseteq \mathbb{R}$. Also $g_n(t) \leq 1 \ \forall n \in \mathbb{N}, \forall t \in \mathbb{R}$. Using now the Lebesgue dominated convergence theorem it follows that $\int_a^b g_n(t) dt \to 0$. But as a simple calculation shows, if λ, μ are complex numbers then

$$\int_a^b |\lambda \cos nt + \mu \sin nt|^2 \, dt = \frac{|\lambda|^2 + |\mu|^2}{2}(b-a) + \frac{|\lambda|^2 - |\mu|^2}{n} K_n + \frac{\Re(\lambda\mu)}{n} L_n,$$

where $|K_n| \leq 1/2$ and $|L_n| \leq 1$. Using this fact we obtain that

$$\int_a^b g_n(t) dt = \frac{b-a}{2} + \frac{|a_n|^2 - |b_n|^2}{(|a_n|^2 + |b_n|^2) n} K_n + \frac{\Re(a_n b_n)}{(|a_n|^2 + |b_n|^2) n} L_n,$$

where $|K_n| \leq 1/2$ and $|L_n| \leq 1$. It follows that $\int_a^b g_n(t) dt \to (b-a)/2 \neq 0$, and we obtain a contradiction!

Let now $x^* \in X^*$. Then $x^*(x_n) \cos nt + x^*(y_n) \sin nt \to 0$ on a non-degenerate interval. Since the Cantor Lemma is true for the scalar case, it follows that $x^*(x_n) \to 0$ and $x^*(y_n) \to 0$, that is, $x_n \to 0$ and $y_n \to 0$ *weak*. Using the Uniform Boundedness Principle, it follows that the sequences $(x_n)_{n \in \mathbb{N}}, (y_n)_{n \in \mathbb{N}}$ are bounded, i.e., there is an $M > 0$ such that $\|x_n\| \leq M$ and $\|y_n\| \leq M \ \forall n \in \mathbb{N}$. Let now $f_n : \mathbb{R} \to X$, $f_n(t) = x_n \cos nt + y_n \sin nt$. Then $\|f_n(t)\| \leq 2M \ \forall n \in \mathbb{N}, \forall t \in \mathbb{R}$, and by hypothesis, $\|f_n(t)\| \to 0$ on a non-degenerate interval $(a, b) \subseteq \mathbb{R}$. Again from the Lebesgue dominated convergence theorem it follows that $\int_a^b \|f_n(t)\|^2 dt \to 0$. But $|x^*(f_n(t))| \leq \|f_n(t)\| \ \forall x^* \in X^*, \|x^*\| \leq 1, \forall t \in \mathbb{R}$, and from here it follows that $\int_a^b |x^*(f_n(t))|^2 \, dt \to 0$ uniformly with respect to $\|x^*\| \leq 1$. We have

$$\int_a^b |x^*(f_n(t))|^2 \, dt = \frac{|x^*(x_n)|^2 + |x^*(y_n)|^2}{2}(b-a) + \frac{|x^*(x_n)|^2 - |x^*(y_n)|^2}{n} K_n$$
$$+ \frac{\Re(x^*(x_n) x^*(y_n))}{n} L_n,$$

where $|K_n| \leq 1/2$ and $|L_n| \leq 1$. Since $\|x_n\|, \|y_n\| \leq M \ \forall n \in \mathbb{N}$, it follows that

$$\left| \frac{|x^*(x_n)|^2 - |x^*(y_n)|^2}{n} K_n + \frac{\Re(x^*(x_n) x^*(y_n))}{n} L_n \right| \leq 2 \frac{M^2}{n} \to 0,$$

i.e,

$$\frac{|x^*(x_n)|^2 - |x^*(y_n)|^2}{n} K_n + \frac{\Re(x^*(x_n) x^*(y_n))}{n} L_n \to 0,$$

uniformly with respect to $\|x^*\| \leq 1$. It follows that

$$\frac{|x^*(x_n)|^2 + |x^*(y_n)|^2}{2}(b-a) \to 0,$$

uniformly with respect to $\|x^*\| \leq 1$, therefore $x^*(x_n) \to 0$, $x^*(y_n) \to 0$ uniformly with respect to $\|x^*\| \leq 1$, i.e., $\sup\limits_{\|x^*\|\leq 1} |x^*(x_n)| \to 0$ and $\sup\limits_{\|x^*\|\leq 1} |x^*(y_n)| \to 0$. Since $\sup\limits_{\|x^*\|\leq 1} |x^*(x)| = \|x\| \; \forall x \in X$ it follows that $\|x_n\| \to 0$ and $\|y_n\| \to 0$.

20. i) See [16, theorem 1.2, chapter I].

Suppose, for example, that Y is a Banach space. For any $x \in X$, the operator $B_x : Y \to Z$ defined by $B_x(y) = B(x,y)$ is linear and continuous, hence $A = \{z^* \circ B_x \mid \|z^*\| \leq 1, \|x\| \leq 1\} \subseteq Y^*$. We show that A is $weak^*$ bounded, i.e., $\forall y \in Y$ the set $\{(z^* \circ B_x)(y) \mid \|z^*\| \leq 1, \|x\| \leq 1\} \subseteq \mathbb{K}$ is bounded. Indeed, for $\|z^*\| \leq 1, \|x\| \leq 1$ since $B_y : X \to Z$ is linear and continuous we have

$$|z^* \circ B_x(y)| = |z^*(B(x,y))| = |z^* \circ B_y(x)| \leq \|z^*\| \, \|B_y\| \, \|x\| \leq \|B_y\|.$$

Since Y is a Banach space, from the Uniform Boundedness Principle it follows that $\sup\limits_{\|z^*\|\leq 1, \|x\|\leq 1} \|z^* \circ B_x\| = M < \infty$. Then $|z^* \circ B(x,y)| \leq M \|y\| \; \forall \|z^*\| \leq 1, \forall \|x\| \leq 1$, $\forall y \in Y$. Passing to the supremum over $\|z^*\| \leq 1$ it follows that $\|B(x,y)\| \leq M \|y\|$ $\forall \|x\| \leq 1, \forall y \in Y$, from whence using the homogeneity of B in the first variable, $\|B(x,y)\| \leq M \|x\| \|y\| \; \forall x \in X, \forall y \in Y$, i.e., B is continuous.

ii) See [16, exercise 1.11 (b), chapter I].

We will show that B is separately continuous, and then, using (i) it follows that B is continuous as a bilinear operator. With the usual notations, for any $y \in Y$ we obtain that $(B_n)_y (x) \to B_y(x) \; \forall x \in X$. Since $(B_n)_y \in L(X,Z) \; \forall n \in \mathbb{N}$, by the Banach–Steinhaus theorem (X is a Banach space) it follows that $B_y : X \to Z$ is linear and continuous for any $y \in Y$. Similarly, $B_x : Y \to Z$ defined by $B_x(y) = B(x,y) \; \forall y \in Y$ is linear and continuous for any $x \in X$ (Y is a Banach space). Then we apply (i).

Chapter 7

The Hahn–Banach theorem

Theorem (Hahn–Banach, the case $\mathbb{K} = \mathbb{R}$). Let X be a real linear space, $G \subseteq X$ a linear subspace, $p : X \to \mathbb{R}$ a sublinear functional and $f : G \to \mathbb{R}$ a linear functional such that $f(x) \leq p(x) \ \forall x \in G$. Then there is a linear functional $\overline{f} : X \to \mathbb{R}$ which extends f (that is, $\overline{f}(x) = f(x) \ \forall x \in G$) such that $\overline{f}(x) \leq p(x) \ \forall x \in X$.

Theorem (Hahn–Banach, the case $\mathbb{K} = \mathbb{R}$ or $\mathbb{C}$). Let X be a linear space, $G \subseteq X$ a linear subspace, $p : X \to \mathbb{R}$ a seminorm and $f : G \to \mathbb{K}$ a linear functional such that $|f(x)| \leq p(x) \ \forall x \in G$. Then there is $\overline{f} : X \to \mathbb{K}$, a linear functional which extends f such that $|\overline{f}(x)| \leq p(x) \ \forall x \in X$.

Theorem (Hahn–Banach, the normed space case). Let X be a normed space, $G \subseteq X$ a linear subspace and $f : G \to \mathbb{K}$ a linear and continuous functional. Then there is $\overline{f} : X \to \mathbb{K}$, a linear and continuous functional which extends f, such that $\|\overline{f}\| = \|f\|$. Such an $\overline{f}$ is called a *Hahn–Banach extension* for f.

Theorem. Let X be a normed space, $x \in X$, and $G \subseteq X$ a linear subspace such that $\delta = d(x, G) > 0$. Then there is $f : X \to \mathbb{K}$, a linear and continuous functional, such that $f = 0$ on G, $f(x) = 1$ and $\|f\| = 1/\delta$.

7.1 Exercises

Extensions for a linear functional

1. Let X be a linear space and $Y \subseteq X$ a linear subspace. Prove that each linear functional $f : Y \to \mathbb{K}$ has a linear extension $\tilde{f} : X \to \mathbb{K}$.

A surjectivity result

2. Let X be a normed space, $n \in \mathbb{N}$ and $\{x_1, ..., x_n\} \subseteq X$ a linearly independent system. Prove that for any $\alpha_1, ..., \alpha_n \in \mathbb{K}$ there is $x^* \in X^*$ such that $x^*(x_i) = \alpha_i \; \forall 1 \leq i \leq n$.

Riesz decomposition for a linear functional

3. i) Let X be a linear space, $n \in \mathbb{N}$, $\varphi : X \to \mathbb{K}$ a linear functional and $p_1, ..., p_n : X \to \mathbb{R}_+$ seminorms such that $|\varphi(x)| \leq \sum_{k=1}^{n} p_k(x) \; \forall x \in X$. Prove that there are $\varphi_1, ..., \varphi_n : X \to \mathbb{K}$, linear functionals, such that $\varphi = \sum_{k=1}^{n} \varphi_k$, $|\varphi_k(x)| \leq p_k(x) \; \forall x \in X, \forall 1 \leq k \leq n$.

ii) Let X be a real linear space, $p : X \to \mathbb{R}$ is a sublinear functional, $A \subseteq X$ is a convex cone, and $q : A \to \mathbb{R}$ is a supra-linear functional such that $q(x) \leq p(x) \; \forall x \in A$. Prove that there is a linear functional $f : X \to \mathbb{R}$ such that $f(x) \leq p(x) \; \forall x \in X$ and $q(x) \leq f(x) \; \forall x \in A$.

The structure for the closure of a linear subspace in a normed space

4. Let X be a normed space and $Y \subseteq X$ a linear subspace.

i) For $x_0 \in X \setminus Y$, we define $f : \text{Sp}\{Y, x_0\} \to \mathbb{K}$, $f(y + \lambda x_0) = \lambda \; \forall y \in Y, \forall \lambda \in \mathbb{K}$. Prove that f is well defined and linear.

ii) Prove that f is continuous if and only if $x_0 \notin \overline{Y}$.

iii) Prove that $\overline{Y} = \bigcap \{\ker x^* \mid x^* \in X^*, Y \subseteq \ker x^*\}$.

The Hahn–Banach theorem in separable normed spaces

5. i) Let X be a separable normed space, $(x_n)_{n \in \mathbb{N}} \subseteq X$ is a dense sequence, and $E \subseteq X$ is a closed linear subspace.

a) Prove that $\forall n \in \mathbb{N} \; \exists x_n^* \in X^*$ such that $x_n^* = 0$ on E, $\|x_n^*\| \leq 1$, and $x_n^*(x_n) = d(x_n, E)$.

b) Prove that $d(x, E) = \sup_{n \in \mathbb{N}} |x_n^*(x)| \; \forall x \in X$, and then deduce that $E = \bigcap_{n=1}^{\infty} \ker x_n^*$.

$((x_n^*)_{n \in \mathbb{N}}$ are given by (a).)

ii) Using (i) prove that there is a sequence $(x_n^*)_{n \in \mathbb{N}} \subseteq B_{X^*}$ such that $\|x\| = \sup_{n \in \mathbb{N}} |x_n^*(x)|$ $\forall x \in X$ and $\bigcap_{n=1}^{\infty} \ker x_n^* = \{0\}$.

Riesz representation for the differentiation operator

6. i) Let $n \in \mathbb{N}$. Prove that there is a finite Borel regular measure μ on $[0,1]$ such that $P'(0) = \int_0^1 P d\mu$ for any real polynomial P of degree at most n.

ii) Is there a finite Borel regular measure μ on $[0,1]$ such that $P'(0) = \int_0^1 P d\mu$ for any real polynomial P?

7.1 Exercises

Effective calculation for the set of all the Hahn–Banach extensions

7. In the space $\mathbb{R}^2$ let us consider the linear subspace $G = \{(x,y) \in \mathbb{R}^2 \mid 2x - y = 0\}$ and the functional $f : G \to \mathbb{R}$ defined by $f(x,y) = x$. Prove that $g : \mathbb{R}^2 \to \mathbb{R}$ defined by $g(x,y) = x/5 + 2y/5 \ \forall (x,y) \in \mathbb{R}^2$ is the unique Hahn–Banach extension of f. (On $\mathbb{R}^2$ we consider the Euclidean norm.)

8. Let $1 \leq p \leq \infty$, $\alpha \in \mathbb{K}$ be fixed and $G = \{(x_1, 0) \mid x_1 \in \mathbb{K}\} \subseteq \mathbb{K}^2$. Let us consider the linear and continuous functional $e : \left(G, \|\cdot\|_p\right) \to \mathbb{K}$ defined by $e(x) = \alpha x_1$, the norm $\|\cdot\|_p$ being the usual one. Prove that the Hahn–Banach extensions for $e \in G^*$ are the following:

i) For $p = 1$, $\varphi(x_1, x_2) = \alpha x_1 + v x_2 \ \forall \ (x_1, x_2) \in \mathbb{K}^2$, where $|v| \leq |\alpha|$, and so we have an infinity of such extensions if α is not zero.

ii) For $1 < p \leq \infty$, $\varphi(x_1, x_2) = \alpha x_1, \ \forall \ (x_1, x_2) \in \mathbb{K}^2$.

9. Let $G = \{(x_n)_{n \in \mathbb{N}} \in l_1 \mid x_1 - 3x_2 = 0\}$ and $f : G \to \mathbb{R}$ be defined by $f((x_n)_{n \in \mathbb{N}}) = x_1$. Prove that $g : l_1 \to \mathbb{R}$ defined by

$$g_1(x) = \frac{3}{4}x_1 + \frac{3}{4}x_2 \ \ \forall x = (x_n)_{n \in \mathbb{N}} \in l_1$$

is the unique Hahn-Banach extension for f.

Examples when the Hahn–Banach extension is unique

10. Prove that any linear and continuous functional on c_0 has a unique Hahn–Banach extension to l_∞.

11. Let X be a normed space. Prove that any linear and continuous functional on $c_0(X)$ has a unique Hahn–Banach extension to $l_\infty(X)$.

The Hahn–Banach extension in the Hilbert space case

12. Let H be a Hilbert space, and $G \subseteq H$ a closed linear subspace. Prove that any linear and continuous functional on G has a unique Hahn–Banach extension on H. More precisely:

i) If $f : G \to \mathbb{K}$ is a linear and continuous functional, then $\overline{f} : H \to \mathbb{K}$ defined by $\overline{f}(x) = f(\Pr_G(x)) \ \forall x \in H$ is the unique Hahn–Banach extension for f. (Here $\Pr_G(x)$ is the orthogonal projection of x onto G.)

ii) If $b \in H$ and $f : G \to \mathbb{K}$ is defined by $f(x) = \langle x, b \rangle \ \forall x \in G$, then the unique Hahn–Banach extension for f is $\overline{f} : H \to \mathbb{K}$ defined by $\overline{f}(x) = \langle x, \Pr_G(b) \rangle \ \forall x \in X$, and $\|f\| = \|\overline{f}\| = \|\Pr_G(b)\|$.

13. i) Let H be a Hilbert space, $a \in H$, $a \neq 0$, $G = \{x \in H \mid \langle x, a \rangle = 0\}$, and $b \in H$. Let $f : G \to \mathbb{K}$ be defined by $f(x) = \langle x, b \rangle \ \forall x \in G$. Prove that the unique Hahn–Banach extension for f is $\overline{f} : H \to \mathbb{K}$ defined by

$$\overline{f}(x) = \langle x, b \rangle - \frac{\langle a, b \rangle}{\|a\|^2} \langle x, a \rangle \ \ \forall x \in H.$$

ii) Let $G = \left\{ f \in L_2[0,1] \mid \int_0^1 x f(x) dx = 0 \right\}$ and $L : G \to \mathbb{K}$ be the linear and continuous functional defined by $L(f) = \int_0^1 x^2 f(x) dx$. Prove that the unique Hahn–Banach extension for L is $x^* : L_2[0,1] \to \mathbb{K}$ defined by $x^*(f) = \int_0^1 (x^2 - 3x/4) f(x) dx \ \forall f \in L_2[0,1]$.

iii) Let $\mathcal{M}_n(\mathbb{C})$ be the set of all the $n \times n$ complex matrices, which is a Hilbert space with respect to the scalar product $\langle A, B \rangle = \text{tr}(AB^*)$, where B^* is the adjoint of the matrix B. Let $G = \{A \in \mathcal{M}_n(\mathbb{C}) \mid \text{tr}(A) = 0\}$, $B \in \mathcal{M}_n(\mathbb{C})$ and $f : G \to \mathbb{C}$ be the linear and continuous functional defined by $f(A) = \text{tr}(AB^*)\ \forall A \in G$. Prove that the unique Hahn–Banach extension for f is $\overline{f} : \mathcal{M}_n(\mathbb{C}) \to \mathbb{C}$ defined by

$$\overline{f}(A) = \text{tr}(AB^*) - \frac{\overline{\text{tr}(B)}\text{tr}(A)}{n}\ \forall A \in \mathcal{M}_n(\mathbb{C}).$$

iv) Let $\mathcal{M}_n(\mathbb{C})$ be the Hilbert space given at (iii) and $\mathcal{M}_n(\mathbb{C}) \times \mathcal{M}_n(\mathbb{C})$ be the Hilbertian product. Let $G = \{A \in \mathcal{M}_n(\mathbb{C}) \times \mathcal{M}_n(\mathbb{C}) \mid \text{tr}(A) = \text{tr}(B)\}$, $C, D \in \mathcal{M}_n(\mathbb{C})$ and $f : G \to \mathbb{C}$ be the linear and continuous functional defined by $f(A,B) = \text{tr}(AC^*) + \text{tr}(BD^*)\ \forall(A,B) \in G$. Prove that the unique Hahn–Banach extension for f is $\overline{f} : \mathcal{M}_n(\mathbb{C}) \times \mathcal{M}_n(\mathbb{C}) \to \mathbb{C}$ defined by

$$\overline{f}(A,B) = \text{tr}(AC^*) + \text{tr}(BD^*) + \frac{\left(\overline{\text{tr}(C)} - \overline{\text{tr}(D)}\right)(\text{tr}(B) - \text{tr}(A))}{2n}.$$

An example when we have an infinity of Hahn–Banach extensions

14. Let $G \subseteq l_1$, $G = \{(x_n)_{n \in \mathbb{N}} \in l_1 \mid x_1 = x_3 = x_5 = \cdots = 0\}$. Prove that any non-null linear and continuous functional on G has an infinity of Hahn–Banach extensions.

The structure for the set of all the Hahn–Banach extensions

15. Let X be a normed space, $G \subseteq X$ is a linear subspace, and $f : G \to \mathbb{K}$ is a linear and continuous functional for which there exist $g, h : X \to \mathbb{K}$, two distinct Hahn–Banach extensions. Prove that f has an infinity of Hahn–Banach extensions.

More precisely, one can prove that the set of all the Hahn–Banach extensions is convex in X^*.

An algebraic result

16. Let X be a normed space and X^* its dual, $n \in \mathbb{N}$, $x_1^*, \ldots, x_n^*, f \in X^*$, $G = \{x \in X \mid x_1^*(x) = 0, \ldots, x_n^*(x) = 0\}$, and $g : G \to \mathbb{K}$ defined by $g(x) = f(x)\ \forall x \in G$. If $h : X \to \mathbb{K}$ is a linear and continuous functional which extends g, prove that there are $\alpha_1, \ldots, \alpha_n \in \mathbb{K}$ such that $h = f + \alpha_1 x_1^* + \cdots + \alpha_n x_n^*$.

Other examples for the effective calculation of all the Hahn–Banach extensions

17. Let $G = \{U \in L(l_2) \mid \langle Ue_1, e_1 \rangle - 2\langle Ue_1, e_2 \rangle = 0\}$, and $f : G \to \mathbb{R}$ be the linear and continuous functional defined by $f(U) = \langle Ue_1, e_1 \rangle$. Prove that $\overline{f} : L(l_2) \to \mathbb{R}$ defined by

$$\overline{f}(U) = \left\langle Ue_1, \frac{4}{5}e_1 + \frac{2}{5}e_2 \right\rangle\ \forall U \in L(l_2)$$

is the unique Hahn–Banach extension for f.

7.1 Exercises

18. Let $(a_n)_{n\in\mathbb{N}} \in l_1$ with $a_1 \neq 0$, $G = \left\{(x_n)_{n\in\mathbb{N}} \in c_0 \mid \sum_{n=1}^{\infty} a_n x_n = 0\right\}$ and $f : G \to \mathbb{R}$ defined by $f((x_n)_{n\in\mathbb{N}}) = x_1$. Prove that the Hahn–Banach extensions h for f are the following:

i) If $|a_1| < \sum_{n=2}^{\infty} |a_n|$ then $h((x_n)_{n\in\mathbb{N}}) = x_1$.

ii) If $|a_1| = \sum_{n=2}^{\infty} |a_n|$ then $h((x_n)_{n\in\mathbb{N}}) = x_1 + \alpha \sum_{n=1}^{\infty} a_n x_n$, where $-1 \leq \alpha a_1 \leq 0$.

iii) If $|a_1| > \sum_{n=2}^{\infty} |a_n|$ then $h((x_n)_{n\in\mathbb{N}}) = -\sum_{n=2}^{\infty} (a_n/a_1) x_n$.

19. Let $a \in \mathbb{R}$, $|a| < 1$, $a \neq 0$, $G = \left\{(x_n)_{n\in\mathbb{N}} \in c_0 \mid \sum_{n=1}^{\infty} a^n x_n = 0\right\}$ and $f : G \to \mathbb{R}$ be the functional defined by $f((x_n)_{n\in\mathbb{N}}) = x_1$. Prove that the Hahn–Banach extensions h for f are the following:

i) If $1/2 < |a| < 1$ then $h(x_1, x_2, ...) = x_1 \; \forall (x_1, x_2, ...) \in c_0$.

ii) If $a = 1/2$ then $h(x_1, x_2, ...) = x_1 + \alpha \sum_{n=1}^{\infty} a^n x_n \; \forall (x_1, x_2, ...) \in c_0$, where $-2 \leq \alpha \leq 0$.

iii) If $a = -1/2$ then $h(x_1, x_2, ...) = x_1 + \alpha \sum_{n=1}^{\infty} a^n x_n \; \forall (x_1, x_2, ...) \in c_0$, where $0 \leq \alpha \leq 2$.

iv) If $|a| < 1/2$ then $h(x_1, x_2, ...) = -\sum_{n=1}^{\infty} a^n x_{n+1} \; \forall (x_1, x_2, ...) \in c_0$.

20. Let $(a_n)_{n\in\mathbb{N}} \in l_\infty$ with $a_1 \neq 0$, $G = \left\{(x_n)_{n\in\mathbb{N}} \in l_1 \mid \sum_{n=1}^{\infty} a_n x_n = 0\right\}$ and $f : G \to \mathbb{R}$ be the functional defined by $f((x_n)_{n\in\mathbb{N}}) = x_1$. Prove that f has a unique Hahn–Banach extension, namely

$$h(x_1, x_2, ...) = x_1 - \sum_{n=1}^{\infty} \frac{\mathrm{sgn}(a_1)}{\lambda + |a_1|} a_n x_n \; \forall (x_1, x_2, ...) \in l_1,$$

where $\lambda = \sup_{n\geq 2} |a_n|$.

21. Let $1 < p < \infty$, q the conjugate of p, i.e., $1/p + 1/q = 1$, $(a_n)_{n\in\mathbb{N}} \in l_q$ with $a_1 \neq 0$, $G = \left\{(x_n)_{n\in\mathbb{N}} \in l_p \mid \sum_{n=1}^{\infty} a_n x_n = 0\right\}$ and $f : G \to \mathbb{R}$ be the functional defined by $f((x_n)_{n\in\mathbb{N}}) = x_1$. Prove that the Hahn–Banach extensions for f are of the form

$$h((x_n)_{n\in\mathbb{N}}) = x_1 + \alpha \sum_{n=1}^{\infty} a_n x_n \; \forall (x_n)_{n\in\mathbb{N}} \in l_p,$$

where $\alpha \in \mathbb{R}$ is a solution for the equation

$$(|1 + \alpha a_1|^q + \lambda^q |\alpha a_1|^q)^{1/q} = \left(\frac{\lambda^p}{\lambda^p + 1}\right)^{1/p}, \quad \lambda = \frac{\left(\sum_{n=2}^{\infty} |a_n|^q\right)^{1/q}}{|a_1|}.$$

22. Let $0 < b < 1$, $G = \left\{(x_n)_{n\in\mathbb{N}} \in c_0 \mid \sum_{n=1}^{\infty} \frac{x_n}{2^n} = 0\right\}$ and $f : G \to \mathbb{R}$ be the functional defined by $f((x_n)_{n\in\mathbb{N}}) = \sum_{n=1}^{\infty} b^n x_n$. Prove that the Hahn–Banach extensions for

f are of the form

$$h((x_n)_{n\in\mathbb{N}}) = \sum_{n=1}^{\infty} b^n x_n + \alpha \sum_{n=1}^{\infty} \frac{x_n}{2^n} \ \forall (x_n)_{n\in\mathbb{N}} \in c_0,$$

where $\alpha \in \mathbb{R}$ is a solution for the equation

$$\sum_{n=1}^{\infty} \left| b^n + \frac{\alpha}{2^n} \right| = \frac{b|2b-1|}{1-b}.$$

The Banach Limit

23. i) Let $M = \{(y_n)_{n\in\mathbb{N}} \mid \exists (x_n)_{n\in\mathbb{N}} \in l_\infty, y_n = x_n - x_{n+1} \ \forall n \in \mathbb{N}\}$ and consider $e = (1, 1, 1, ...) \in l_\infty$. (The space l_∞ is over $\mathbb{R}$.)
   a) Prove that $M \subseteq l_\infty$ is a linear subspace and that $d(e, M) = 1$.
   b) Deduce that there is $x^* \in l_\infty^*$ such that $\|x^*\| = x^*(e) = 1$, $x^*|_M = 0$.
   c) Prove that $c_0 \subseteq \ker x^*$ and $x^*\left((x_n)_{n\in\mathbb{N}}\right) = \lim_{n\to\infty} x_n$ if $(x_n)_{n\in\mathbb{N}}$ is convergent. Deduce from here that $x^* \in l_\infty^* \setminus K_{l_1}(l_1)$ and therefore that l_1 is not a reflexive space.
   d) Prove that $x^*(x_1, x_2, ...) = x^*(x_2, x_3, ...) \ \forall (x_1, x_2, ...) \in l_\infty$
   ii) Let $z_n = (0, ..., 0, 1, 1, ...) \in l_\infty$ (n times 0) $\forall n \in \mathbb{N}$. Prove that $z_n \to 0$ *weak** in l_∞, and using (i) prove that $(z_n)_{n\in\mathbb{N}}$ does not converge *weak* towards 0.
   iii) For $(x_1, x_2, ...) \in l_\infty$, we will denote by $LIM\{x_1, x_2, ...\} := x^*(x_1, x_2, ...)$, which usually is called a Banach Limit for the bounded sequence $(x_n)_{n\in\mathbb{N}}$. The part (c) from (i) affirms that the Banach Limit for a convergent sequence coincides with the usual limit.
   What is a Banach Limit for the sequence $a, b, a, b, ...$? I.e., calculate $LIM\{a, b, a, b, ...\}$. What about the Banach Limit for a periodic sequence?
   iv) Prove that we cannot find an additive, multiplicative, translation invariant function $f : l_\infty \to \mathbb{R}$, which coincides with the usual limit for convergent sequences, i.e., we cannot find f with the properties $f(x+y) = f(x) + f(y) \ \forall x, y \in l_\infty$, $f(xy) = f(x)f(y) \ \forall x, y \in l_\infty$, $f(x_1, x_2, ...) = f(x_2, x_3, ...) \ \forall (x_1, x_2, ...) \in l_\infty$, and $f\left((x_n)_{n\in\mathbb{N}}\right) = \lim_{n\to\infty} x_n$ if $(x_n)_{n\in\mathbb{N}}$ is convergent. (For $x = (x_n)_{n\in\mathbb{N}}, y = (y_n)_{n\in\mathbb{N}} \in l_\infty$ we write $xy = (x_n y_n)_{n\in\mathbb{N}}$.)

7.2 Solutions

1. Indeed, let $\{e_i\}_{i\in I} \subseteq Y$ be an algebraic basis of Y over $\mathbb{K}$. Then the system $\{e_i\}_{i\in I} \subseteq X$ is linearly independent, and hence we can extend it to a basis $\mathcal{B} = \{e_i\}_{i\in I} \cup \{f_j\}_{j\in J}$ of X. Define then $\widetilde{f} : \mathcal{B} \to \mathbb{K}$ by $\widetilde{f}(e_i) = f(e_i)$ for $i \in I$ and $\widetilde{f}(f_j) = \lambda_j \in \mathbb{K}$ for $j \in J$. Since $\mathcal{B} \subseteq X$ is an algebraic basis we can extend $\widetilde{f}$ to a linear functional $\widetilde{f} : X \to \mathbb{K}$. Obviously $\widetilde{f}\big|_Y = f$.

2. See [14, corollary 6.6, chapter III].
Let $Y = \mathrm{Sp}\{x_1, ..., x_n\}$. Since the system $\{x_1, ..., x_n\} \subseteq X$ is linearly independent, it follows that $\{x_1, ..., x_n\} \subseteq Y$ is an algebraic basis. Define then $f : Y \to \mathbb{K}$, linear, with $f(x_i) = \alpha_i \ \forall i = \overline{1, n}$. Since Y is of finite dimension, f is continuous (on Y we have the

7.2 Solutions

norm from X). By the Hahn–Banach theorem there is an extension $x^* \in X^*$ for f. Then $x^*(x_i) = f(x_i) = \alpha_i \ \forall i = \overline{1,n}$.

3. i) See [43, p. 313–358].

Let us consider the linear space X^n, on which we define $p : X^n \to \mathbb{R}_+$ by $p(x_1, ..., x_n) = \sum_{k=1}^{n} p_k(x_k)$. Then

$$p(\lambda(x_1, ..., x_n)) = \sum_{k=1}^{n} p_k(\lambda x_k) = |\lambda| \sum_{k=1}^{n} p_k(x_k) = |\lambda| p(x_1, ..., x_n)$$

and

$$p((x_1, ..., x_n) + (y_1, ..., y_n)) = \sum_{k=1}^{n} p_k(x_k + y_k) \leq \sum_{k=1}^{n} p_k(x_k) + \sum_{k=1}^{n} p_k(y_k)$$
$$= p(x_1, ..., x_n) + p(y_1, ..., y_n),$$

i.e., $p : X^n \to \mathbb{R}_+$ is a seminorm. Let $D_n = \{(x, ..., x) \in X^n \mid x \in X\}$ be the diagonal of the space X^n. Then $D_n \subseteq X^n$ is a linear subspace. We define $\varphi_0 : D_n \to \mathbb{K}$, $\varphi_0(x, ..., x) = \varphi(x)$. Then φ_0 is linear and by hypothesis and the definition of φ_0 we also have $|\varphi_0(x)| \leq p(x)\ \forall x \in D_n$.

From the Hahn–Banach theorem, there is a linear functional $\tilde{\varphi} : X^n \to \mathbb{K}$ with $\tilde{\varphi}|_{D_n} = \varphi_0$ and $|\tilde{\varphi}(x_1, ..., x_n)| \leq p(x_1, ..., x_n)\ \forall (x_1, ..., x_n) \in X^n$. For any $1 \leq k \leq n$, define $\varphi_k : X \to \mathbb{K}$, $\varphi_k(x) = \tilde{\varphi}(0, ..., 0, x, 0, ..., 0)\ \forall x \in X$ (x on the kth position). These linear functionals have all the required properties.

ii) See [7, p. 6–11].

Let $r : X \to \mathbb{R}$ defined by $r(x) = \inf\{p(x+a) - q(a) \mid a \in A\}$. We observe that for any $x \in X$ and any $a \in A$ we have $p(a) \leq p(x+a) + p(-x)$ and then $p(x+a) - q(a) \geq -p(-x) + [p(a) - q(a)] \geq -p(-x)$, i.e., the set on which we take the infimum is lower bounded. We will prove first that r is a sublinear functional. Indeed, let $x, y \in X$. Then for any $a, b \in A$ we have $a + b \in A$, since A is a convex cone. We have $r(x+y) \leq p(x+y+a+b) - q(a+b)$. Since p is subadditive and q supra-additive we have $p(x+y+a+b) \leq p(x+a) + p(y+b)$ and $-q(a+b) \leq -q(a) - q(b)$. This implies $p(x+y+a+b) - q(a+b) \leq [p(x+a) - q(a)] + [p(y+b) - q(b)]$, hence $r(x+y) \leq [p(x+a) - q(a)] + [p(y+b) - q(b)]$. Passing to the infimum over $a \in A$ and $b \in A$ we obtain that $r(x+y) \leq r(x) + r(y)$, i.e., r is subadditive. Let $x \in X$. Since $p(a) \geq q(a)\ \forall a \in A$ it follows that $r(0) = \inf\{p(a) - q(a) \mid a \in A\} \geq 0$. Also, since $0 \in A$ we deduce that $r(0) = \inf\{p(a) - q(a) \mid a \in A\} \leq p(0) - q(0) = 0$, i.e., $r(0) = 0$, and hence $r(0 \cdot x) = 0 \cdot r(x) = 0\ \forall x \in X$. If $\lambda > 0$ then since A is a convex cone for any $a \in A$ we have $\lambda a \in A$, hence $r(\lambda x) \leq p(\lambda x + \lambda a) - q(\lambda a) = \lambda[p(x+a) - q(a)]$. Passing to the infimum and using that $\lambda > 0$ we obtain that $r(\lambda x) \leq \lambda r(x)$. Replacing in this reasoning $x \to \lambda x$ and $\lambda \to 1/\lambda$ we obtain that $r(x) \leq r(\lambda x)/\lambda$, i.e., $\lambda r(x) \leq r(\lambda x)$. Since $0 \in A$ we obtain that $r(x) = \inf\{p(x+a) - q(a) \mid a \in A\} \leq p(x) - q(0) = p(x)$ $\forall x \in X$. Also if $a \in A$, then $r(-a) \leq p(-a+a) - q(a) = -q(a)$.

Taking $G = \{0\}$ and $g : G \to \mathbb{R}$ the zero functional, we have that $g(x) \leq r(x)$ $\forall x \in G$. From the Hahn–Banach theorem, there is a linear functional $f : X \to \mathbb{R}$ such that

$f(x) \le r(x) \ \forall x \in X$. Then $f(x) \le p(x) \ \forall x \in X$. If $a \in A$ then $f(-a) \le r(-a)$, i.e., since f is linear, $-r(-a) \le f(a)$, from whence $q(a) \le f(a)$.

4. i) Let us observe first that $\text{Sp}\{Y, x_0\} = \{y + \lambda x_0 \mid y \in Y, \lambda \in \mathbb{K}\}$. We show that if $y, y_0 \in Y$, $\lambda, \lambda' \in \mathbb{K}$ with $y + \lambda x_0 = y' + \lambda' x_0$ then $\lambda = \lambda'$, $y' = y$. Indeed, if $\lambda \ne \lambda'$ then $x_0 = (y' - y)/(\lambda - \lambda')$ would belong to Y, because Y is a linear subspace, and this contradicts our hypothesis. Thus $\lambda = \lambda'$, from whence $y = y'$. Then $f : \text{Sp}\{Y, x_0\} \to \mathbb{K}$ given by $f(y + \lambda x_0) = \lambda \ \forall y \in Y, \lambda \in \mathbb{K}$ is well defined, and by the uniqueness of the decomposition for the elements in $\text{Sp}\{Y, x_0\}$ it follows that f is also linear.

ii) Suppose that $f : \text{Sp}\{Y, x_0\} \to \mathbb{K}$ is continuous. We will prove that $x_0 \notin \overline{Y}$. Suppose, for a contradiction, that $x_0 \in \overline{Y}$. From the well known characterization of the closure in metric spaces it follows that there is $(y_n)_{n \in \mathbb{N}} \subseteq Y$ such that $y_n \to x_0$. We have $(y_n)_{n \in \mathbb{N}} \subseteq \text{Sp}\{Y, x_0\}$, $x_0 \in \text{Sp}\{Y, x_0\}$, and so by the continuity of f it follows that $f(y_n) \to f(x_0)$. Since $f(y_n) = 0 \ \forall n \in \mathbb{N}$ it follows that $f(y_n) \to 0$, from whence $f(x_0) = 0$, which is false, since $f(x_0) = 1$.

Suppose now that $x_0 \notin \overline{Y}$, i.e., $x_0 \in C_{\overline{Y}}$, which is an open set. By the definition for an open set in a normed space it follows that there is $\varepsilon > 0$ such that $B(x_0, \varepsilon) \subseteq C_{\overline{Y}}$, i.e., $\overline{Y} \cap B(x_0, \varepsilon) = \emptyset$. Let $x \in \text{Sp}\{Y, x_0\}$, $x = y + \lambda x_0$ with $y \in Y$ and $\lambda \in \mathbb{K}$. If $\lambda = 0$, then $|f(x)| = 0 \le \|x\|/\varepsilon$. If $\lambda \ne 0$, then $y/(-\lambda) \in Y$, from whence $y/(-\lambda) \notin B(x_0, \varepsilon)$, i.e., $\|-y/\lambda - x_0\| \ge \varepsilon$. Then $\|x\| = |\lambda| \|-y/\lambda - x_0\| \ge |\lambda|\varepsilon = \varepsilon |f(x)|$, and so $|f(x)| \le \|x\|/\varepsilon \ \forall x \in \text{Sp}\{Y, x_0\}$, i.e., $f : \text{Sp}\{Y, x_0\} \to \mathbb{K}$ is continuous.

iii) See [6, exercise 13, chapter 3].

If $x^* \in X^*$ and $Y \subseteq \ker x^*$ then it follows that $\overline{Y} \subseteq \ker x^*$ and therefore $\overline{Y} \subseteq \bigcap\{\ker x^* \mid x^* \in X^*, Y \subseteq \ker x^*\}$. Conversely let $x_0 \in \bigcap\{\ker x^* \mid x^* \in X^*, Y \subseteq \ker x^*\}$ and suppose that $x_0 \notin \overline{Y}$. Using (ii) there is $f : \text{Sp}\{Y, x_0\} \to \mathbb{K}$ linear and continuous such that $f|_Y = 0$, $f(x_0) = 1$. Let $x^* \in X^*$ be a Hahn–Banach extension for f given by the Hahn–Banach theorem. Then $x^* = f$ on $\text{Sp}\{Y, x_0\}$. In particular, $x^* = 0$ on Y, i.e., $Y \subseteq \ker x^*$ and $x^*(x_0) = f(x_0) = 1$, i.e., $x_0 \notin \ker x^*$, and so $x_0 \notin \bigcap\{\ker x^* \mid x^* \in X^*, Y \subseteq \ker x^*\}$, which contradicts our supposition.

5. i) a) We prove first that $\forall x \in X \ \exists x^* \in X^*$ such that $x^* = 0$ on E, $x^*(x) = d(x, E)$ and $\|x^*\| \le 1$ (1). Indeed, if $x \notin E$, then since E is a closed linear subspace, there is $f \in X^*$ such that $f = 0$ on E, $f(x) = 1$, and $\|f\| = 1/\delta$ where $\delta = d(x, E) > 0$. Then $x^* = \delta f$ has the properties: $x^* = 0$ on E, $x^*(x) = \delta = d(x, E)$ and $\|x^*\| = 1$. If $x \in E$, then the zero functional has the required properties.

Applying (1) to the elements $(x_n)_{n \in \mathbb{N}}$ it follows that there is $(x_n^*)_{n \in \mathbb{N}} \subseteq X^*$ such that $x_n^* = 0$ on E, $x_n^*(x_n) = d(x_n, E)$, and $\|x_n^*\| \le 1 \ \forall n \in \mathbb{N}$.

b) Let $x \in X$. Then for $y \in E$ using that for any $n \in \mathbb{N}$ we have $x_n^* = 0$ on E and $\|x_n^*\| \le 1$, it follows that

$$\|x - y\| = \sup_{\|x^*\| \le 1} |x^*(x - y)| \ge \sup_{n \in \mathbb{N}} |x_n^*(x - y)| = \sup_{n \in \mathbb{N}} |x_n^*(x)|,$$

from whence $d(x, E) = \inf_{y \in E} \|x - y\| \ge \sup_{n \in \mathbb{N}} |x_n^*(x)|$. Now let $\varepsilon > 0$. Since $(x_n)_{n \in \mathbb{N}} \subseteq X$ is dense, there is $n \in \mathbb{N}$ such that $\|x - x_n\| \le \varepsilon/2$. Using the relation $|d(x, E) - d(x_n, E)| \le \|x - x_n\|$ (see exercise 1(ii) of chapter 4) it follows that $|d(x, E) - d(x_n, E)| \le \varepsilon/2$, hence by (a) $|d(x, E) - x_n^*(x_n)| \leq \varepsilon/2$. Since

$$|x_n^*(x) - x_n^*(x_n)| = |x_n^*(x - x_n)| \leq \|x_n^*\| \|x - x_n\| \leq \varepsilon/2,$$

then

$$|d(x, E) - x_n^*(x)| \leq |d(x, E) - x_n^*(x_n)| + |x_n^*(x_n) - x_n^*(x)| \leq \varepsilon/2 + \varepsilon/2 = \varepsilon,$$

and therefore $d(x, E) \leq |x_n^*(x)| + \varepsilon$. Then $d(x, E) \leq \sup_{n \in \mathbb{N}} |x_n^*(x)| + \varepsilon$. Since $\varepsilon > 0$ is arbitrary we obtain that $d(x, E) \leq \sup_{n \in \mathbb{N}} |x_n^*(x)|$. Hence $d(x, E) = \sup_{n \in \mathbb{N}} |x_n^*(x)|$.

If $x \in E$ then $d(x, E) = 0$, and therefore $\sup_{n \in \mathbb{N}} |x_n^*(x)| = 0$, i.e., $x_n^*(x) = 0 \ \forall n \in \mathbb{N}$, $x \in \bigcap_{n=1}^{\infty} \ker x_n^*$, i.e., $E \subseteq \bigcap_{n=1}^{\infty} \ker x_n^*$. If $x \in \bigcap_{n=1}^{\infty} \ker x_n^*$, i.e., $x_n^*(x) = 0 \ \forall n \in \mathbb{N}$, then $\sup_{n \in \mathbb{N}} |x_n^*(x)| = 0$, i.e., $d(x, E) = 0$, hence by exercise 1(i) of chapter 4 we have $x \in \overline{E} = E$.

ii) Take $E = \{0\}$ in (i).

6. i) See [14, exercise 6, chapter III, section 6].

Let $\mathcal{P}_n$ be the linear space of all the real polynomials of degree $\leq n$, considered as a linear subspace of $C[0, 1]$. Since $\mathcal{P}_n$ has finite dimension, it follows that any linear functional from $\mathcal{P}_n$ to $\mathbb{R}$ is continuous. Let $f : \mathcal{P}_n \to \mathbb{R}$ be defined by $f(P) = P'(0)$ $\forall P \in \mathcal{P}_n$. Since f is continuous, it follows that there is $M \geq 0$ such that $|f(P)| \leq M \|P\|_\infty \forall P \in \mathcal{P}_n$. By the Hahn–Banach theorem there is $\widetilde{f} \in (C[0, 1])^*$ such that $\|\widetilde{f}\| \leq M$, and $\widetilde{f}|_{\mathcal{P}_n} = f$. Now from the Riesz representation theorem, there is a regular finite Borel measure μ on $[0, 1]$ such that $\widetilde{f}(g) = \int_0^1 g\, d\mu$ for any $g \in C[0, 1]$, and therefore $P'(0) = \int_0^1 P\, d\mu \ \forall P \in \mathcal{P}_n$.

ii) See [14, exercise 7, chapter III, section 6].

Let us suppose that there is a regular finite Borel measure μ on $[0, 1]$ such that $P'(0) = \int_0^1 P\, d\mu$, for any real polynomial P. Let $x^* \in (C[0, 1])^*$ given by μ. Then $x^*(P) = P'(0)$, $\forall P \in \mathcal{P}$, and since x^* is continuous, there is an $M \geq 0$ such that $|P'(0)| \leq M \|P\|_\infty$ $\forall P \in \mathcal{P}$. For any $k \in \mathbb{N}$, let P_k be the polynomial $P_k(x) = (x - 1)^k$. Then $|P_k'(0)| = k$, $\|P_k\|_\infty = 1 \ \forall k \in \mathbb{N}$, from whence $M \geq k \ \forall k \in \mathbb{N}$, which is false!

7. If $g : \mathbb{R}^2 \to \mathbb{R}$ is a linear functional, then $g(x, y) = xg(1, 0) + yg(0, 1) = ax + by$ $\forall (x, y) \in \mathbb{R}^2$, and as it is easy to see, $\|g\| = \sqrt{a^2 + b^2}$. Let us observe that $G = \{(x, 2x) \mid x \in \mathbb{R}\} = \mathrm{Sp}\{(1, 2)\}$. In addition, $|f(x, y)| = |x| = (1/\sqrt{5}) \|(x, y)\| \ \forall (x, y) \in G$, i.e., $\|f\| = 1/\sqrt{5}$. Let $g : \mathbb{R}^2 \to \mathbb{R}$ be a Hahn–Banach extension for f. A such extension exists, by the Hahn–Banach theorem. Then there are $a, b \in \mathbb{R}$ such that $g(x, y) = ax + by$ $\forall (x, y) \in \mathbb{R}^2$ and $\|g\| = \sqrt{a^2 + b^2}$. Since $g|_G = f$ it follows that $g(x, y) = x \ \forall (x, y) \in G$, i.e., $ax + by = x \ \forall (x, y) \in G$, that is, $ax + 2bx = x \ \forall x \in \mathbb{R}$, $a + 2b = 1$. Also g has the same norm as f, i.e., $\|g\| = \|f\|$, that is, $\sqrt{a^2 + b^2} = 1/\sqrt{5}$, and therefore $\begin{cases} a + 2b = 1, \\ a^2 + b^2 = 1/5. \end{cases}$

Solving this system we obtain that $b = 2/5$ and $a = 1/5$, i.e., the Hahn–Banach extension is unique and is given by the formula

$$g(x, y) = \frac{x}{5} + \frac{2y}{5} \ \forall (x, y) \in \mathbb{R}^2.$$

8. It is easy to see that $\|e\| = |\alpha|$ for any $1 \le p \le \infty$. Let now $\varphi : \mathbb{K}^2 \to \mathbb{K}$ be a Hahn–Banach extension for e. Then there are $u, v \in \mathbb{K}$ such that $\varphi(x_1, x_2) = ux_1 + vx_2$ $\forall (x_1, x_2) \in \mathbb{K}^2$. If on $\mathbb{K}^2$ we will consider the norm $\|\cdot\|_p$, $1 \le p \le \infty$, then from the duality we obtain that $\|\varphi\| = \begin{cases} \max\{|u|, |v|\} & \text{if } p = 1, \\ (|u|^q + |v|^q)^{1/q} & \text{if } 1 < p < \infty, \\ |u| + |v| & \text{if } p = \infty, \end{cases}$ where $1/p + 1/q = 1$.

Since $\varphi|_G = e$ then $u = \alpha$. We have the following cases:

i) If $p = 1$ then
$$\varphi(x_1, x_2) = \alpha x_1 + v x_2 \ \forall (x_1, x_2) \in \mathbb{K}^2,$$
where $|v| \le |\alpha|$.

ii) If $1 < p < \infty$ then $v = 0$, so $\varphi(x_1, x_2) = \alpha x_1 \ \forall (x_1, x_2) \in \mathbb{K}^2$.

iii) If $p = \infty$ then again $v = 0$, so $\varphi(x_1, x_2) = \alpha x_1 \ \forall (x_1, x_2) \in \mathbb{K}^2$.

9. We have
$$|f(x)| = |x_1| = \frac{3}{4}(|x_1| + |x_2|) \le \frac{3}{4}\|x\| \ \forall x \in G,$$
so $\|f\| \le 3/4$. Also, $|f(3, 1, 0, \ldots)| = 3 \le \|f\| \|(3, 1, 0, \ldots)\| = 4\|f\|$, and therefore $\|f\| = 3/4$.

Let $g : l_1 \to \mathbb{R}$ be a Hahn–Banach extension for f. Since $l_1^* = l_\infty$, there is $\xi = (\xi_n)_{n \in \mathbb{N}} \in l_\infty$ such that $g(x) = \sum_{n=1}^\infty \xi_n x_n \ \forall x = (x_n)_{n \in \mathbb{N}} \in l_1$ and $\|g\| = \|\xi\| = \sup_{n \in \mathbb{N}} |\xi_n|$. We have $g|_G = f$, hence $g(x) = x_1 \ \forall x = (x_n)_{n \in \mathbb{N}} \in l_1$ with $x_1 - 3x_2 = 0$, that is,
$$\xi_1 x_1 + \frac{\xi_2}{3} x_1 + \xi_3 x_3 + \cdots = x_1 \ \forall |x_1| + |x_3| + |x_4| + \cdots < \infty.$$

From this we deduce that $\xi_1 + \xi_2/3 = 1$ and $\xi_n = 0 \ \forall n \ge 3$, i.e., $3\xi_1 + \xi_2 = 3$, $\xi_n = 0$ $\forall n \ge 3$. So the Hahn–Banach extension g is given by
$$g(x) = \xi_1 x_1 + \xi_2 x_2 = \xi_1 x_1 + (3 - 3\xi_1) x_2 \ \forall x = (x_n)_{n \in \mathbb{N}} \in l_1.$$

Since $\|g\| = \|f\| = 3/4$, i.e., $\max(|\xi_1|, |3 - 3\xi_1|) = 3/4$ we have $\xi_1 = 3/4$, i.e., f has a unique Hahn–Banach extension to l_1, namely
$$g(x) = 3x_1/4 + 3x_2/4 \ \forall x = (x_n)_{n \in \mathbb{N}} \in l_1.$$

10. See [40, exercise 12.20].

Let $f \in c_0^*$. Since $c_0^* = l_1$, there is $(a_n)_{n \in \mathbb{N}} \in l_1$ such that $f(x) = \sum_{n=1}^\infty a_n x_n \ \forall x = (x_n)_{n \in \mathbb{N}} \in c_0$, and $\|f\| = \sum_{n=1}^\infty |a_n|$. Let $g : l_\infty \to \mathbb{K}$ be defined by $g(x) = \sum_{n=1}^\infty a_n x_n$ $\forall x = (x_n)_{n \in \mathbb{N}} \in l_\infty$. Then g is linear and
$$|g(x)| \le \sum_{n=1}^\infty |a_n||x_n| \le \|x\| \sum_{n=1}^\infty |a_n| \ \forall x \in l_\infty,$$

7.2 Solutions

i.e., g is continuous. It is easy to see that $\|g\| = \sum_{n=1}^{\infty} |a_n| = \|f\|$, and therefore g is a Hahn–Banach extension for f.

Now let $\overline{f} : l_\infty \to \mathbb{K}$ be another Hahn–Banach extension for f, i.e., $\overline{f}\,|_{c_0} = f$ and $\|\overline{f}\| = \|f\|$. We prove that $\overline{f} = g$. Let $x \in l_\infty$, $x = (x_n)_{n \in \mathbb{N}}$ with $\|x\| \leq 1$. For $n \in \mathbb{N}$ we denote by $y_n = (\text{sgn}(a_1), ..., \text{sgn}(a_n), x_{n+1}, ...) \in l_\infty$. Obviously, $\|y_n\| \leq 1$. Since $y_n - x = (\text{sgn}(a_1) - x_1, ..., \text{sgn}(a_n) - x_n, 0, ...) \in c_0$ and $\overline{f} = g = f$ on c_0 it follows that $\overline{f}(y_n - x) = g(y_n - x)$. Let us write $h = \overline{f} - g$. Then the above relation shows that $h(y_n - x) = 0$, that is, $h(y_n) = h(x)$. We have

$$\overline{f}(y_n) = h(y_n) + g(y_n) = h(x) + g(y_n) = h(x) + \sum_{k=1}^{n} |a_k| + \sum_{k=n+1}^{\infty} a_k x_k,$$

from whence

$$\left| \overline{f}(y_n) - h(x) - \sum_{k=1}^{n} |a_k| \right| = \left| \sum_{k=n+1}^{\infty} a_k x_k \right| \leq \sum_{k=n+1}^{\infty} |a_k| |x_k| \leq \sum_{k=n+1}^{\infty} |a_k| \to 0$$

for $n \to \infty$, since the series $\sum_{k=1}^{\infty} |a_k|$ converges. Then $\overline{f}(y_n) - h(x) - \sum_{k=1}^{n} |a_k| \to 0$, and since $\sum_{k=1}^{n} |a_k| \to \sum_{k=1}^{\infty} |a_n| = \|f\|$, we obtain that $\overline{f}(y_n) \to h(x) + \|f\|$.

We have proved that $\forall x \in l_\infty$ with $\|x\| \leq 1$ $\exists (y_n)_{n \in \mathbb{N}} \subseteq l_\infty$ with $\|y_n\| \leq 1$ $\forall n \in \mathbb{N}$ such that $\overline{f}(y_n) \to h(x) + \|f\|$ (1). Let now $x \in l_\infty$ with $\|x\| \leq 1$. There is $\lambda \in \mathbb{K}$, $|\lambda| \leq 1$ such that $|h(x)| = \lambda h(x) = h(\lambda x)$. Since $\|\lambda x\| = |\lambda| \|x\| \leq 1$ from the relation (1) it follows that there is $(y_n)_{n \in \mathbb{N}} \subseteq l_\infty$ with $\|y_n\| \leq 1$ $\forall n \in \mathbb{N}$ such that $\overline{f}(y_n) \to h(\lambda x) + \|f\|$, that is, $\overline{f}(y_n) \to |h(x)| + \|f\|$, and therefore $|\overline{f}(y_n)| \to |h(x)| + \|f\|$ (2). We have

$$|\overline{f}(y_n)| \leq \|\overline{f}\| \|y_n\| \leq \|\overline{f}\| = \|f\| \quad \forall n \in \mathbb{N}$$

from whence passing to the limit for $n \to \infty$, from (2) we obtain $|h(x)| + \|f\| \leq \|f\|$, $|h(x)| \leq 0$, $h(x) = 0$ $\forall x \in l_\infty$ with $\|x\| \leq 1$, and therefore, by homogeneity, $h = 0$, $\overline{f} - g = 0$, $\overline{f} = g$.

11. Let $f \in (c_0(X))^*$. Then there is $(x_n^*)_{n \in \mathbb{N}} \in l_1(X^*)$ such that $f(x) = \sum_{n=1}^{\infty} x_n^*(x_n)$ $\forall x = (x_n)_{n \in \mathbb{N}} \in c_0(X)$ and $\|f\| = \sum_{n=1}^{\infty} \|x_n^*\|$. Define $g : l_\infty(X) \to \mathbb{K}$, $g(x) = \sum_{n=1}^{\infty} x_n^*(x_n)$ $\forall x = (x_n)_{n \in \mathbb{N}} \in l_\infty(X)$. Obviously g is linear and

$$|g(x)| \leq \sum_{n=1}^{\infty} |x_n^*(x_n)| \leq \sum_{n=1}^{\infty} \|x_n^*\| \|x_n\| \leq \left(\sup_{n \in \mathbb{N}} \|x_n\| \right) \sum_{n=1}^{\infty} \|x_n^*\|$$
$$= \|x\| \sum_{n=1}^{\infty} \|x_n^*\| = \|x\| \|f\| \quad \forall x \in l_\infty(X),$$

so $\|g\| \leq \|f\|$. Let $n \in \mathbb{N}$ and $\|x_1\| \leq 1, ..., \|x_n\| \leq 1$. The sequence $x = (a_1 x_1, ..., a_n x_n, 0, ...)$ is in $l_\infty(X)$, where $a_k = \operatorname{sgn}(x_k^*(x_k))$, and

$$g(x) = \sum_{k=1}^n a_k x_k^*(x_k) = \sum_{k=1}^n |x_k^*(x_k)| \leq \|g\| \|x\| \leq \|g\|,$$

i.e., $\sum_{k=1}^n |x_k^*(x_k)| \leq \|g\|$, and taking the supremum over $\|x_1\| \leq 1, ..., \|x_n\| \leq 1$ it follows that $\sum_{k=1}^n \|x_k^*\| \leq \|g\|$ $\forall n \in \mathbb{N}$, and passing to the limit for $n \to \infty$, $\sum_{n=1}^\infty \|x_n^*\| \leq \|g\|$, i.e., $\|f\| \leq \|g\|$. Therefore g is an extension of f, with $\|g\| = \|f\| = \sum_{n=1}^\infty \|x^*\|$.

Now let $\overline{f} : l_\infty(X) \to \mathbb{K}$ be a linear and continuous functional which extends f and having the same norm as f, i.e., $\overline{f}|_{c_0(X)} = f$ and $\|\overline{f}\| = \|f\|$. We shall prove that $\overline{f} = g$. Denote $h = \overline{f} - g$ and let us consider $x \in l_\infty(X)$ with $\|x\| \leq 1$. There is $\lambda \in \mathbb{K}$ such that $|\lambda| \leq 1$ and $|h(x)| = h(\lambda x) = h(t)$, where $t = \lambda x \in l_\infty(X)$, $\|t\| = |\lambda| \|x\| \leq 1$. Let $t = (t_k)_{k \in \mathbb{N}} \in l_\infty(X)$, and consider $a_1, ..., a_n \in X$ with $\|a_k\| \leq 1$, $k = \overline{1, n}$. We consider $y = (\lambda_1 a_1, ..., \lambda_n a_n, t_{n+1}, ...) \in l_\infty(X)$, where $\lambda_k = \operatorname{sgn}(x_k^*(a_k))$, $k = \overline{1, n}$, and then $y - t = (\lambda_1 a_1 - t_1, ..., \lambda_n a_n - t_n, 0, ...) \in c_0(X)$. Since $h|_{c_0(X)} = 0$ it follows that $h(y - t) = 0$, hence by the linearity of h on $l_\infty(X)$ it follows that $h(y) = h(t)$, i.e.,

$$|h(x)| = h(y) = \overline{f}(y) - g(y) = \overline{f}(y) - \sum_{k=1}^n \lambda_k x_k^*(a_k) - \sum_{k=n+1}^\infty x_k^*(t_k),$$

therefore,

$$|h(x)| + \sum_{k=1}^n |x_k^*(a_k)| = \overline{f}(y) - \sum_{k=n+1}^\infty x_k^*(t_k) = \left|\overline{f}(y) - \sum_{k=n+1}^\infty x_k^*(t_k)\right|$$
$$\leq |\overline{f}(y)| + \sum_{k=n+1}^\infty |x_k^*(t_k)|$$
$$\leq \|\overline{f}\| \|y\| + \sum_{k=n+1}^\infty \|x_k^*\| \leq \|f\| + \sum_{k=n+1}^\infty \|x_k^*\|.$$

Hence

$$|h(x)| + \sum_{k=1}^n |x_k^*(a_k)| \leq \|f\| + \sum_{k=n+1}^\infty \|x_k^*\| \quad \forall \|a_1\| \leq 1, ..., \forall \|a_n\| \leq 1,$$

and passing to the supremum over $a_1, ..., a_n$ we obtain that

$$|h(x)| + \sum_{k=1}^n \|x_k^*\| \leq \|f\| + \sum_{k=n+1}^\infty \|x_k^*\| \quad \forall n \in \mathbb{N}.$$

For $n \to \infty$ this gives $|h(x)| + \sum_{n=1}^\infty \|x_n^*\| \leq \|f\|$, i.e., $|h(x)| + \|f\| \leq \|f\|$, $h(x) = 0$ $\forall x \in l_\infty(X)$ with $\|x\| \leq 1$, and then by homogeneity, $h(x) = 0$ $\forall x \in l_\infty(X)$, i.e., $h = 0$, that is, $\overline{f} = g$.

7.2 Solutions

12. See [40, exercise 12.18] or [45, exercise 6, chapter 5].

Existence. Let $f : G \to \mathbb{K}$ be linear and continuous. Since H is a Hilbert space by the orthogonal projection theorem (see chapter 11) we have the decomposition $H = G \oplus G^\perp$. Define now $\overline{f} : H \to \mathbb{K}$ by $\overline{f}(x) = f(\text{Pr}_G(x))$ $\forall x \in H$, where $x = \text{Pr}_G(x) + \text{Pr}_{G^\perp}(x)$. Then $\overline{f}$ is linear and $\overline{f}|_G = f$, so $\|\overline{f}\| \geq \|f\|$. However,

$$|\overline{f}(x)| = |f(\text{Pr}_G(x))| \leq \|f\| \|\text{Pr}_G(x)\| \leq \|f\| \|x\| \quad \forall x \in H$$

(we use the property $\|x\|^2 = \|\text{Pr}_G(x)\|^2 + \|\text{Pr}_{G^\perp}(x)\| \geq \|\text{Pr}_G(x)\|^2$). From here it follows that $\|\overline{f}\| \leq \|f\|$, and therefore $\|\overline{f}\| = \|f\|$.

Uniqueness. Let $g : H \to \mathbb{K}$ be a linear and continuous functional with $g|_G = f$ and $\|g\| = \|f\|$. By the Riesz representation theorem, there is $b \in H$ such that $g(x) = \langle x, b \rangle$ $\forall x \in H$. Since $b = \text{Pr}_G(b) + \text{Pr}_{G^\perp}(b)$, then for any $y \in G$ we have $f(y) = g(y) = \langle y, \text{Pr}_G(b) \rangle + \langle y, \text{Pr}_{G^\perp}(b) \rangle = \langle y, \text{Pr}_G(b) \rangle$, from whence $\|f\| = \|\text{Pr}_G(b)\|$. We have $\|f\| = \|g\|$, hence $\|g\| = \|\text{Pr}_G(b)\|$, from whence $\|b\|^2 = \|\text{Pr}_G(b)\|^2$. Since $\|b\|^2 = \|\text{Pr}_G(b)\|^2 + \|\text{Pr}_{G^\perp}(b)\|^2$, then $\text{Pr}_{G^\perp}(b) = 0$, and then $g(x) = \langle x, b \rangle = \langle x, \text{Pr}_G(b) \rangle = \langle \text{Pr}_G(x), b \rangle = \overline{f}(x)$ $\forall x \in X$, i.e., $g = \overline{f}$.

We have proved (i), and (ii) follows from (i).

Remark. This exercise shows that in the Hilbert space case, the problem of the effective calculation for the Hahn–Banach extensions requires the calculation of an orthogonal projection onto a closed linear subspace (see chapter 11).

13. i) We have

$$\text{Pr}_G(b) = b - \frac{\langle b, a \rangle}{\|a\|^2} a,$$

and we can use exercise 12.

ii) We have

$$\text{Pr}_G(x^2) = x^2 - \frac{\langle x^2, x \rangle}{\|x\|^2} x = x^2 - \frac{\int_0^1 x^3 dx}{\int_0^1 x^2 dx} x = x^2 - \frac{3}{4}x,$$

and we apply (i).

iii) For the fact that $(\mathcal{M}_n(\mathbb{C}), \langle \cdot, \cdot \rangle)$ is indeed a Hilbert space, see exercise 2 of chapter 10. Let us observe that if $A = (a_{ij})_{i,j} \in \mathcal{M}_n(\mathbb{C})$, then

$$\|A\| = \sqrt{\text{tr}(AA^*)} = \sqrt{\sum_{i,j=1}^n |a_{ij}|^2},$$

which is usually called the Frobenius norm for the matrix A.

Observe that $G = \{A \in \mathcal{M}_n(\mathbb{C}) \mid \text{tr}(AI_n^*) = 0\} = \{A \in \mathcal{M}_n(\mathbb{C}) \mid \langle A, I_n \rangle = 0\}$, where I_n is the unit matrix. Since

$$\text{Pr}_G(B) = B - \frac{\langle B, I_n \rangle}{\|I_n\|^2} I_n = B - \frac{\text{tr}(B)}{\text{tr}(I_n)} I_n = B - \frac{\text{tr}(B)}{n} I_n,$$

and $f(A) = \langle A, B \rangle$ $\forall A \in G$, we obtain that $\overline{f} : \mathcal{M}_n(\mathbb{C}) \to \mathbb{C}$ is defined by

$$\overline{f}(A) = \langle A, \text{Pr}_G(B) \rangle = \text{tr}(AB^*) - \frac{\overline{\text{tr}(B)}\text{tr}(A)}{n} \quad \forall A \in \mathcal{M}_n(\mathbb{C}).$$

iv) Recall that if H_1, H_2 are two Hilbert space then the Cartesian product $H_1 \times H_2$ is a Hilbert space with respect to the inner product $\langle (x_1, y_1), (x_2, y_2) \rangle = \langle x_1, x_2 \rangle + \langle y_1, y_2 \rangle$. Let us observe that

$$G = \{A \in \mathcal{M}_n(\mathbb{C}) \times \mathcal{M}_n(\mathbb{C}) \mid \langle (A, B), (I_n, -I_n) \rangle = 0\},$$

hence

$$\begin{aligned} \Pr_G(C, D) &= (C, D) - \frac{\langle (C, D), (I_n, -I_n) \rangle}{\|(I_n, -I_n)\|^2}(I_n, -I_n) \\ &= (C, D) - \frac{\operatorname{tr}(C) - \operatorname{tr}(D)}{2n}(I_n, -I_n) = (C - \lambda I_n, D + \lambda I_n), \end{aligned}$$

where

$$\lambda = (\operatorname{tr}(C) - \operatorname{tr}(D))/(2n).$$

Also, $f(A, B) = \operatorname{tr}(AC^*) + \operatorname{tr}(BD^*) = \langle (A, B), (C, D) \rangle \ \forall (A, B) \in G$. Therefore $\overline{f}(A, B) = \langle (A, B), \Pr_G(C, D) \rangle \ \forall (A, B) \in \mathcal{M}_n(\mathbb{C}) \times \mathcal{M}_n(\mathbb{C})$, i.e., the statement.

14. Let $f : G \to \mathbb{K}$ be a linear and continuous functional with $\|f\| \neq 0$. Using the Hahn–Banach theorem there is a Hahn–Banach extension $g : l_1 \to \mathbb{K}$ for f. Since $l_1^* = l_\infty$, there is $\xi = (\xi_n)_{n \in \mathbb{N}} \in l_\infty$ such that $g(x) = \sum_{n=1}^{\infty} x_n \xi_n \ \forall x = (x_n)_{n \in \mathbb{N}} \in l_1$, $\|g\| = \sup_{n \in \mathbb{N}} |\xi_n| = \|f\|$ and $g\mid_G = f$. So $f(x) = \sum_{k=1}^{\infty} x_{2k} \xi_{2k} \ \forall x \in G$ and $\|f\| = \|g\| \geq \sup_{k \in \mathbb{N}} |\xi_{2k}|$. However,

$$|f(x)| \leq \left(\sup_{k \in \mathbb{N}} |\xi_{2k}|\right) \sum_{k=1}^{\infty} |x_{2k}| \leq \left(\sup_{k \in \mathbb{N}} |\xi_{2k}|\right) \|x\|_1 \ \forall x \in G,$$

from whence $\|f\| \leq \sup_{n \in \mathbb{N}} |\xi_{2k}|$. We obtain that $f(x) = \sum_{k=1}^{\infty} x_{2k} \xi_{2k} \ \forall x \in G$ and $\|f\| = \sup_{k \in \mathbb{N}} |\xi_{2k}|$. Since $\|f\| \neq 0$, it follows that we can find an infinity of sequences $\tau = (\tau_{2k+1})_{k \geq 0} \in l_\infty$ such that $\sup_{k \geq 0} |\tau_{2k+1}| = \sup_{k \geq 1} |\xi_{2k}|$, and let us consider the sequence $a(\tau) = (\tau_1, \xi_2, \tau_3, \xi_4, ...) \in l_\infty$. We have $\|a(\tau)\|_\infty = \|f\|$. Using $a(\tau)$ we construct $g_\tau : l_1 \to \mathbb{K}$ defined by

$$g_\tau(x) = \sum_{n=0}^{\infty} \tau_{2n+1} x_{2n+1} + \sum_{n=1}^{\infty} \xi_{2n} x_{2n}.$$

Then $g_\tau \in l_1^*$ and $\|g_\tau\| = \|a(\tau)\| = \|f\|$, and for any $x \in G$ we have $g_\tau(x) = \sum_{n=1}^{\infty} x_{2n} \xi_{2n} = f(x)$, i.e., $g_\tau \mid_G = f$. Since $g_{\tau_1} \neq g_{\tau_2}$ for $\tau_1 \neq \tau_2$, the statement follows.

15. See [45, exercise 7, chapter 5].

For any $t \in [0, 1]$ we define $f_t : X \to \mathbb{K}$ by $f_t(x) = tg(x) + (1 - t)h(x)$. Clearly, f_t are linear and continuous functionals. Also, for $x \in G$ we have $f_t(x) = tg(x) + (1 - t)h(x) = tf(x) + (1 - t)f(x) = f(x)$, i.e., f_t extends f. If we prove that $\|f_t\| =$

7.2 Solutions

$\|f\|\ \forall t\in[0,1]$ then the exercise will be solved, since $f_{t_1}\neq f_{t_2}$ for $t_1\neq t_2$ (we have $g\neq h$). For any $x\in X$ we have

$$|f_t(x)|\leq t\,|g(x)|+(1-t)\,|h(x)|\leq (t\,\|g\|+(1-t)\,\|h\|)\,\|x\|=\|f\|\,\|x\|,$$

since $\|g\|=\|h\|=\|f\|$, i.e., $\|f_t\|\leq\|f\|\ \forall t\in[0,1]$. Since f_t extends f we also have $\|f_t\|\geq\|f\|$, and therefore $\|f_t\|=\|f\|\ \forall t\in[0,1]$.

16. Since h extends g we have $h(x)=f(x)\ \forall x\in G$, i.e., $G=\bigcap_{i=1}^{n}\ker x_i^*\subseteq\ker(h-f)$. Using now exercise 16(i) of chapter 14 we obtain that there are $\alpha_1,...,\alpha_n\in\mathbb{K}$ such that $h-f=\alpha_1 x_1^*+\cdots+\alpha_n x_n^*$.

Remark. This exercise shows that in reasonable situations, the problem of finding the explicit form for the extensions for a linear and continuous functional is not so difficult. The problem is much more complicated when we want to find the norm-preserving extensions.

17. For $U\in G$ we have

$$\frac{5}{4}|\langle Ue_1,e_1\rangle|^2=|\langle Ue_1,e_1\rangle|^2+|\langle Ue_1,e_2\rangle|^2\leq\|Ue_1\|^2,$$

by the Bessel inequality. Hence,

$$|f(U)|\leq\sqrt{\frac{4}{5}}\|Ue_1\|\leq\sqrt{\frac{4}{5}}\|U\|\quad\forall U\in G,$$

i.e. $\|f\|\leq\sqrt{4/5}$. Let $U\in L(l_2)$ be defined by $U(x_1,x_2,...)=(2x_1,x_1,0,...)=x_1(2e_1+e_2)$. Then $U\in G$, from whence $|f(U)|\leq\|f\|\,\|U\|$. We have $f(U)=\langle Ue_1,e_1\rangle=\langle 2e_1+e_2,e_1\rangle=2$, and $\|U(x_1,x_2,...)\|=\sqrt{5}|x_1|\leq\sqrt{5}\|x\|\ \forall x\in l_2$, and then $\|U\|\leq\sqrt{5}$. We obtain that $2\leq\sqrt{5}\|f\|$, and therefore $\|f\|=\sqrt{4/5}$. Let now $\overline{f}:L(l_2)\to\mathbb{R}$ be a Hahn–Banach extension for f. Using exercise 16 it follows that there is $\alpha\in\mathbb{R}$ such that

$$h(U)=\langle Ue_1,e_1\rangle+\alpha(\langle Ue_1,e_1\rangle-2\langle Ue_1,e_2\rangle)=\langle Ue_1,(1+\alpha)e_1-2\alpha e_2\rangle.$$

It is easy to see that if $x,y\in l_2$ and $g:L(l_2)\to\mathbb{R}$ is defined by $g(U)=\langle Ux,y\rangle$, then $\|g\|=\|x\|\,\|y\|$. Hence

$$\|h\|=\|(1+\alpha)e_1-2\alpha e_2\|=\sqrt{(1+\alpha)^2+4\alpha^2}.$$

Using the property $\|h\|=\|f\|=\sqrt{4/5}$, it follows that $(1+\alpha)^2+4\alpha^2=4/5$, $\alpha=-1/5$, i.e., the statement.

Remark. This exercise cannot be solved using with the same technique as the one used for exercises 7–14 since the structure for the dual space of $L(l_2)$ is not known.

18. If $\sum_{n=2}^{\infty}|a_n|=0$ then the assertion from the statement is clearly true. Suppose now that $\sum_{n=2}^{\infty}|a_n|>0$. We shall prove that $\|f\|=\min(1,\lambda)$, where $\lambda=\left(\sum_{n=2}^{\infty}|a_n|\right)/|a_1|$. We have $|f(x)| \leq \|x\|$ $\forall x \in G$, i.e., $\|f\| \leq 1$. For $x \in G$ we have $-a_1 x_1 = \sum_{n=2}^{\infty} a_n x_n$, from whence

$$|a_1| |x_1| \leq \sum_{n=2}^{\infty} |a_n| |x_n| \leq \sum_{n=2}^{\infty} |a_n| \left(\sup_{n \geq 2} |x_n|\right) \leq \|x\| \sum_{n=2}^{\infty} |a_n|,$$

i.e., $|x_1| \leq \lambda \|x\|$, and then $|f(x)| \leq \lambda \|x\|$ $\forall x \in G$, hence $\|f\| \leq \min(1, \lambda)$. Let $n \geq 2$ such that $\sum_{k=2}^{n} |a_k| > 0$. Choose $\alpha, \beta \in \mathbb{R} \backslash \{0\}$ such that

$$(\alpha, \beta \mathrm{sgn}(a_2), \beta \mathrm{sgn}(a_3), ..., \beta \mathrm{sgn}(a_n), 0, ...) \in G.$$

Then $\alpha a_1 + \beta(a_2 \mathrm{sgn}(a_2) + \cdots + a_n \mathrm{sgn}(a_n)) = 0$, and then $-\alpha a_1 = \beta(|a_2| + \cdots + |a_n|)$. We have

$$|\alpha| = |f(\alpha, \beta \mathrm{sgn}(a_2), \beta \mathrm{sgn}(a_3), ..., \beta \mathrm{sgn}(a_n), 0, ...)| \leq \|f\| \max(|\alpha|, |\beta|),$$

therefore $\min(1, |\alpha|/|\beta|) \leq \|f\|$, that is, $\|f\| \geq \min(1, (|a_2| + \cdots + |a_n|)/|a_1|)$. Passing to the limit for $n \to \infty$ we obtain that

$$\|f\| \geq \min\left(1, \frac{1}{|a_1|} \sum_{n=2}^{\infty} |a_n|\right) = \min(1, \lambda).$$

Let now $h : c_0 \to \mathbb{R}$ be a Hahn–Banach extension for f. Using exercise 16, there is $\alpha \in \mathbb{R}$ such that $h(x_1, ..., x_n, ...) = x_1 + \alpha \sum_{n=1}^{\infty} a_n x_n$ $\forall (x_1, ..., x_n, ...) \in c_0$. We have $\|h\| = |1 + \alpha a_1| + |\alpha| \sum_{n=2}^{\infty} |a_n|$. Since $\|h\| = \|f\|$ it follows that $|1 + \alpha a_1| + |\alpha| \sum_{n=2}^{\infty} |a_n| = \min(1, \lambda)$, that is, $|1 + \alpha a_1| + \lambda |\alpha a_1| = \min(1, \lambda)$, i.e., denoting by $x = \alpha a_1$, we have that $|1 + x| + \lambda |x| = \min(1, \lambda)$.

i) If $\lambda > 1$, we obtain the equation $|1 + x| + \lambda |x| = 1$, which has the real solution $x = 0$, that is, $\alpha a_1 = 0$, $\alpha = 0$.

ii) If $\lambda = 1$, we obtain the equation $|1 + x| + |x| = 1$, which has the real solutions $-1 \leq x \leq 0$, that is, $-1 \leq \alpha a_1 \leq 0$.

iii) If $\lambda < 1$, we obtain the equation $|1 + x| + \lambda |x| = \lambda$, which has the real solution $x = -1$, that is, $\alpha a_1 = -1$, $\alpha = -1/a_1$.

19. We have

$$\sum_{n=2}^{\infty} |a|^n = \frac{|a|^2}{1 - |a|},$$

and we use exercise 18.

20. We shall prove that $\|f\| = \lambda/(\lambda + |a_1|)$. If $\lambda = 0$, this is clearly true. Suppose now that $\lambda \neq 0$. For $x \in G$, we have $-a_1 x_1 = \sum_{n=2}^{\infty} a_n x_n$, so

$$|a_1| |x_1| \leq \lambda \sum_{n=2}^{\infty} |x_n| = \lambda(\|x\| - |x_1|),$$

i.e.,
$$|x_1| \leq \frac{\lambda}{\lambda + |a_1|} \|x\|,$$

hence
$$|f(x)| \leq \frac{\lambda}{\lambda + |a_1|} \|x\| \quad \forall x \in G,$$

i.e., $\|f\| \leq \lambda/(\lambda + |a_1|)$. Let $n \geq 2$ such that $a_n \neq 0$. Choose $\alpha \in \mathbb{R}$ such that $(1, 0, 0, ..., 0, \alpha \operatorname{sgn}(a_n), 0, ...) \in G$. Then $a_1 + \alpha |a_n| = 0$, i.e., $-a_1 = \alpha |a_n|$. We have $1 = |f(1, 0, 0, ..., 0, \alpha \operatorname{sgn}(a_n), 0, ...)| \leq \|f\|(1 + |\alpha|)$, that is, $|a_n| \leq \|f\|(|a_n| + |a_1|)$. The last inequality is also true for $a_n = 0$, and then $|a_n| \leq \|f\|(|a_n| + |a_1|) \; \forall n \geq 2$. We obtain that $\lambda \leq \|f\|(\lambda + |a_1|)$, i.e., $\|f\| \geq \lambda/(\lambda + |a_1|)$.

Let $h : l_1 \to \mathbb{R}$ be a Hahn–Banach extension for f. From exercise 16 we know that there is $\alpha \in \mathbb{R}$ such that $h(x_1, x_2, ...) = x_1 + \alpha \sum_{n=1}^{\infty} a_n x_n \; \forall (x_1, x_2, ...) \in l_1$. Since $\|h\| = \max(|1 + \alpha a_1|, |\alpha|\lambda)$ and $\|h\| = \|f\|$, it follows that

$$\max(|1 + \alpha a_1|, |\alpha|\lambda) = \frac{\lambda}{\lambda + |a_1|},$$

that is,
$$\max\left(|1 + \alpha a_1|, |\alpha a_1| \frac{\lambda}{|a_1|}\right) = \frac{\lambda}{\lambda + |a_1|},$$

i.e., denoting by $x = \alpha a_1$ and $M = \lambda/|a_1|$ we have $\max(|1 + x|, M|x|) = M/(M+1)$, equation which has a unique real solution $x = -1/(M+1)$, and then

$$\alpha = -\frac{1}{a_1(M+1)} = -\frac{|a_1|}{a_1(\lambda + |a_1|)} = -\frac{\operatorname{sgn}(a_1)}{\lambda + |a_1|},$$

i.e., the statement.

21. We shall prove that $\|f\| = (M^p/(M^p + |a_1|^p))^{1/p}$, where $M = \left(\sum_{n=2}^{\infty} |a_n|^q\right)^{1/q}$. If $M = 0$, this is clearly true. Suppose now that $M > 0$. For $x \in G$ we have $-a_1 x_1 = \sum_{n=2}^{\infty} a_n x_n$, hence

$$|a_1 x_1| \leq \sum_{n=2}^{\infty} |a_n| |x_n| \leq \left(\sum_{n=2}^{\infty} |a_n|^q\right)^{1/q} \left(\sum_{n=2}^{\infty} |x_n|^p\right)^{1/p} = M(\|x\|^p - |x_1|^p)^{1/p},$$

i.e.,
$$|f(x)| = |x_1| \leq \left(\frac{M^p}{M^p + |a_1|^p}\right)^{1/p} \|x\|,$$

and then
$$\|f\| \leq \left(\frac{M^p}{M^p + |a_1|^p}\right)^{1/p}.$$

Let $n \geq 2$ such that $|a_2| + \cdots + |a_n| > 0$. Choose $\alpha_n \in \mathbb{R}$ such that
$$(\alpha_n, |a_2|^{q-1} \operatorname{sgn}(a_2), |a_3|^{q-1} \operatorname{sgn}(a_3), ..., |a_n|^{q-1} \operatorname{sgn}(a_n), 0, ...) \in G.$$
Then
$$|\alpha_n| = \frac{|a_2|^q + \cdots + |a_n|^q}{|a_1|}.$$
We have
$$\begin{aligned}
|\alpha_n| &= \left| f(\alpha_n, |a_2|^{q-1} \operatorname{sgn}(a_2), |a_3|^{q-1} \operatorname{sgn}(a_3), ..., |a_n|^{q-1} \operatorname{sgn}(a_n), 0, ...) \right| \\
&\leq \|f\| \left(|\alpha_n|^p + \sum_{k=2}^{n} |a_k|^{(q-1)p} \right)^{1/p}.
\end{aligned}$$
Since $(q-1)p = q$, we have $\|f\| \left(1 + \left(\sum_{k=2}^{n} |a_k|^q \right) / |\alpha_n|^p \right)^{1/p} \geq 1$. Passing to the limit for $n \to \infty$ and using the fact that $|\alpha_n| \to M^q / |a_1|$ we obtain that
$$\|f\| \left(1 + \frac{M^q |a_1|^p}{M^{qp}} \right)^{1/p} \geq 1.$$
Since $q(p-1) = p$, we obtain that
$$\|f\| \geq \left(\frac{M^p}{M^p + |a_1|^p} \right)^{1/p}.$$

If $h : l_p \to \mathbb{R}$ is a Hahn–Banach extension for f, from exercise 16 we know that there is $\alpha \in \mathbb{R}$ such that $h(x_1, x_2, ...) = x_1 + \alpha \sum_{n=1}^{\infty} a_n x_n \ \forall (x_n)_{n \in \mathbb{N}} \in l_p$. Since
$$\|h\| = \left(|1 + \alpha a_1|^q + |\alpha|^q \sum_{n=2}^{\infty} |a_n|^q \right)^{1/q}$$
and $\|h\| = \|f\|$ we obtain that
$$(|1 + \alpha a_1|^q + |\alpha|^q M^q)^{1/q} = \left(\frac{M^p}{M^p + |a_1|^p} \right)^{1/p},$$
i.e., denoting by $\lambda = M / |a_1|$ it follows that $\alpha \in \mathbb{R}$ is a solution for the equation
$$(|1 + \alpha a_1|^q + \lambda^q |\alpha a_1|^q)^{1/q} = \left(\frac{\lambda^p}{\lambda^p + 1} \right)^{1/p}.$$

22. Put $1/2 = a$. For $x \in G$, $x_1 = -\sum_{n=2}^{\infty} a^{n-1} x_n$, from whence
$$\begin{aligned}
|f((x_n)_{n \in \mathbb{N}})| &= \left| \sum_{n=2}^{\infty} b^n x_n - b \sum_{n=2}^{\infty} a^{n-1} x_n \right| \leq b \sum_{n=2}^{\infty} |b^{n-1} - a^{n-1}| |x_n| \\
&\leq b \|x\| \sum_{n=2}^{\infty} |b^{n-1} - a^{n-1}| = b \|x\| \left| \sum_{n=2}^{\infty} (b^{n-1} - a^{n-1}) \right| \\
&= b \|x\| |b/(1-b) - a/(1-a)|,
\end{aligned}$$

i.e.,
$$\|f\| \le b\,|b/(1-b) - a/(1-a)|.$$

For a fixed $n \ge 2$, choose $x, y \in \mathbb{R} \setminus \{0\}$ such that $(x, y, y, ..., y, 0, ...) \in G$ ($n-1$ times y). Then $-x = y(a + \cdots + a^{n-1})$ and
$$|xb + y(b^2 + \cdots + b^n)| = |f(x, y, y, ..., y, 0, ...)| \le \|f\| \max(|x|, |y|),$$

that is,
$$|(x/y)b + (b^2 + \cdots + b^n)| \le \|f\| \max(1, |x|/|y|),$$
$$|-b(a + \cdots + a^{n-1}) + (b^2 + \cdots + b^n)| \le \|f\| \max(1, a + \cdots + a^{n-1}).$$

Passing to the limit for $n \to \infty$ we obtain that
$$|ba/(1-a) - b^2/(1-b)| \le \|f\| \max(1, a/(1-a)),$$

i.e.,
$$\|f\| \ge b\,|b/(1-b) - a/(1-a)| \min(1, (1-a)/a).$$

Since $a = 1/2$ we obtain that
$$\|f\| = b \left| \frac{b}{1-b} - 1 \right| = \frac{b\,|2b-1|}{1-b}.$$

Let $h : c_0 \to \mathbb{R}$ be a Hahn–Banach extension for f. Using exercise 16 we know that there is $\alpha \in \mathbb{R}$ such that $h((x_n)_{n \in \mathbb{N}}) = \sum_{n=1}^{\infty} b^n x_n + \alpha \sum_{n=1}^{\infty} x_n/2^n \; \forall (x_n)_{n \in \mathbb{N}} \in c_0$. Since $\|h\| = \sum_{n=1}^{\infty} |b^n + \alpha/2^n|$ it follows that $\alpha \in \mathbb{R}$ is a solution for the equation
$$\sum_{n=1}^{\infty} \left| b^n + \frac{\alpha}{2^n} \right| = \frac{b\,|2b-1|}{1-b}.$$

23. See [20, exercise 22, chapter II] or [5, chapter 2, section 2].

i) a) Let $S : l_\infty \to l_\infty$ be defined by $S((x_n)_{n \in \mathbb{N}}) = (x_2, x_3, ...) \; \forall (x_n)_{n \in \mathbb{N}} \in l_\infty$. Obviously $S \in L(l_\infty)$ and observing that $M = (I - S)(l_\infty)$ it follows that $M \subseteq l_\infty$ is a linear subspace. Since $0 \in M$, we obtain that $d(e, M) \le \|e\| = 1$. Let now $x \in l_\infty$. If there is $k \in \mathbb{N}$ such that $x_k - x_{k+1} \le 0$ then $|1 - (x_k - x_{k+1})| = 1 - (x_k - x_{k+1}) \ge 1$, and so by the definition for the norm on l_∞, $\|(1, 1, ...) - (I - S)((x_n)_{n \in \mathbb{N}})\| \ge |1 - (x_k - x_{k+1})| \ge 1$. If $x_k - x_{k+1} \ge 0 \; \forall k \in \mathbb{N}$ then the sequence $(x_n)_{n \in \mathbb{N}}$ is decreasing. Since the sequence is also bounded, there is $\lim_{n \to \infty} x_n \in \mathbb{R}$, so $\lim_{n \to \infty} (x_n - x_{n+1}) = 0$. Then, again by the definition for the norm in l_∞,

$$\|(1, 1, ...) - (I - S)((x_n)_{n \in \mathbb{N}})\| = \sup_{n \in \mathbb{N}} |1 - (x_n - x_{n+1})| \ge \lim_{n \to \infty} |1 - (x_n - x_{n+1})| = 1.$$

Hence $d(e, M) \ge 1$ and so $d(e, M) = 1$.

b) Using a corollary for the Hahn–Banach theorem, we obtain that there is $x^* \in l_\infty^*$ such that $\|x^*\| = 1/d(e, M) = 1$, $x^*(e) = 1$, and $x^*|_M = 0$.

c) Let $x \in c_0$. Define the sequence $(x^{(k)})_{k \in \mathbb{N}} \subseteq l_\infty$ by $x^{(1)} = S(x)$, $x^{(k+1)} = S(x^{(k)})$, i.e., $x^{(k)} = (x_{k+1}, x_{k+2}, ...)$ $\forall k \in \mathbb{N}$. Let us observe that $x - x^{(1)} = (I - S)(x) \in M$, $x^{(k-1)} - x^{(k)} = (I - S)(x^{(k-1)}) \in M$ $\forall k \in \mathbb{N}$, from whence since M is a linear subspace, $x^{(k)} - x = x^{(k)} - x^{(k-1)} + x^{(k-1)} - x^{(k-2)} + \cdots + x^{(1)} - x \in M$ $\forall k \in \mathbb{N}$. As $x^*|_M = 0$, it follows that $x^*(x^{(k)}) = x^*(x)$ $\forall k \in \mathbb{N}$. Let now $\varepsilon > 0$. Since $x \in c_0$, we can find $n_\varepsilon \in \mathbb{N}$ such that $|x_n| \le \varepsilon$ $\forall n \ge m_\varepsilon$, and so $\|x^{(n)}\| = \sup_{p \ge n+1} |x_n| \le \varepsilon$ $\forall n \ge m_\varepsilon$. Since $\|x^*\| = 1$ it follows that $|x^*(x^{(n)})| \le \|x^*\| \|x^{(n)}\| \le \varepsilon$ $\forall n \ge m_\varepsilon$. Since $x^*(x^{(k)}) = x^*(x)$ $\forall k \in \mathbb{N}$, it follows that $|x^*(x)| \le \varepsilon$ $\forall \varepsilon \ge 0$ and so $x^*(x) = 0$, i.e., $x \in \ker x^*$. Hence $c_0 \subseteq \ker x^*$. If $(x_n)_{n \in \mathbb{N}} \in c$, let $l = \lim_{n \to \infty} x_n$. Then $(x_n)_{n \in \mathbb{N}} - le = (x_n - l)_{n \in \mathbb{N}} \in c_0$, so $(x_n)_{n \in \mathbb{N}} - le \in \ker x^*$. Since x^* is linear, $x^*((x_n)_{n \in \mathbb{N}}) = lx^*(e) = l = \lim_{n \to \infty} x_n$.

Suppose now that there is $y = (y_n)_{n \in \mathbb{N}} \in l_1$ such that $x^* = \widehat{y}$. Then it follows that $x^*((x_n)_{n \in \mathbb{N}}) = \sum_{n=1}^{\infty} x_n y_n$ $\forall (x_n)_{n \in \mathbb{N}} \in l_\infty$. Since $x^*(e_n) = 0$ $\forall n \in \mathbb{N}$ we obtain that $y_n = 0$ $\forall n \in \mathbb{N}$, i.e., $y = 0$, so $x^* = 0$, which is false, since $x^*(e) = 1$.

d) Let $(x_n)_{n \in \mathbb{N}} \in l_\infty$. Then $(x_n - x_{n+1})_{n \in \mathbb{N}} \in M$, from whence $x^*((x_n - x_{n+1})_{n \in \mathbb{N}}) = 0$, and from the linearity of x^* we obtain the statement.

ii) (See also exercise 6(i) of chapter 14). If $y = (y_n)_{n \in \mathbb{N}} \in l_1$, then $\widehat{y}(z_n) = \sum_{k=n+1}^{\infty} y_n \to 0$, since the series $\sum_{k=1}^{\infty} y_n$ converges. Suppose that $z_n \to 0$ *weak* in l_∞. Then $x^*(z_n) \to 0$, where x^* is the element from l_∞^* given by (i). But for any $n \in \mathbb{N}$ the sequence $(0, ..., 0, 1, 1, 1, ...)$ converges towards 1 and therefore $x^*(z_n) = 1$ $\forall n \in \mathbb{N}$. We obtain a contradiction!

iii) Let us denote $l = LIM\{a, b, a, b, ...\}$. Then using the part (d) from (i) we have $l = LIM\{b, a, b, a, ...\}$, and therefore using the linearity of LIM, $2l = LIM\{a, b, a, b, ...\} + LIM\{b, a, b, a, ...\} = LIM\{a+b, a+b, ...\} = a+b$, because for a convergent sequence the Banach Limit coincides with the usual limit. Hence $l = (a+b)/2$. Using the same idea one can prove that the Banach Limit for a periodic sequence $\{x_1, ..., x_p, x_1, ..., x_p, ...\}$ is $(x_1 + \cdots + x_p)/p$.

iv) Using the additivity and the fact that f coincide with the usual limit for convergent sequences, we have $f(0, 1, 0, 1, ...) + f(1, 0, 1, 0, ...) = f(1, 1, 1, 1, ...) = 1$. Since f is multiplicative and coincide with the usual limit for convergent sequences we have $f(0, 1, 0, 1, ...) f(1, 0, 1, 0, ...) = f(0, 0, ...) = 0$. Since f is translation invariant, $f(0, 1, 0, 1, ...) = f(1, 0, 1, ...) = a$. Then $2a = 1$ and $a^2 = 0$, and we obtain a contradiction.

Chapter 8

Applications for the Hahn–Banach theorem

Definition. Let X be a normed space, or, more generally, a linear topological space. A subset $A \subseteq X$ is called *fundamental* if $\text{Sp}(A)$ is dense in X.

Theorem. Let X be a normed space, or, more generally, a Hausdorff locally convex space, and $A \subseteq X$. Then the following assertions are equivalent:
  i) A is fundamental;
  ii) For any $x^* \in X^*$ with the property that $x^* = 0$ on A it follows that $x^* = 0$ on X.

Theorem. Let X be a normed space, or, more generally, a linear topological space, and $A, B \subseteq X$ two non-empty convex subsets such that $\overset{\circ}{A} \neq \emptyset$. If $\overset{\circ}{A} \cap B = \emptyset$ then we can find $x^* \in X^* \setminus \{0\}$ and $t \in \mathbb{R}$ such that
$$\Re x^*(x) \leq t \leq \Re x^*(y) \quad \forall x \in A, \forall y \in B.$$

Theorem. i) Let X be a normed space, or, more generally, a linear topological space, and $A, B \subseteq X$ two non-empty convex subsets, such that A is open and $A \cap B = \emptyset$. Then we can find $x^* \in X^*$ and $t \in \mathbb{R}$ such that
$$\Re x^*(a) < t \leq \Re x^*(b) \quad \forall a \in A, \forall b \in B.$$

ii) Let X be a normed space, or, more generally, a linear topological space, and $A, B \subseteq X$ two non-empty open convex sets such that $A \cap B = \emptyset$. Then we can find $x^* \in X^*$ and $t \in \mathbb{R}$ such that
$$\Re x^*(a) < t < \Re x^*(b) \quad \forall a \in A, \forall b \in B.$$

Theorem. Let X be normed space, or, more generally, a Hausdorff locally convex space, and $A, B \subseteq X$ two non-empty convex subsets such that one is compact and the other is closed. If $A \cap B = \emptyset$ then A are B strictly separated, that is, we can find $x^* \in X^*$ and $t \in \mathbb{R}$ such that
$$\Re x^*(a) < t < \Re x^*(b) \quad \forall a \in A, \forall b \in B.$$

8.1 Exercises

Examples of dense linear subspaces in l_p and L_p

1. Let $1 < p < \infty$ and $G = \left\{ (x_n)_{n \in \mathbb{N}} \in l_p \,\middle|\, \sum_{n=1}^{\infty} x_n = 0 \right\}$. Prove that $G \subseteq l_p$ is a dense linear subspace.

2. Let $(a_n)_{n \in \mathbb{N}}$ be a sequence of scalars such that $\sum_{n=1}^{\infty} |a_n| \neq 0$, and let $1 < p < \infty$. Find a necessary and sufficient condition on the sequence $(a_n)_{n \in \mathbb{N}}$ for the linear subspace $G = \left\{ (x_n)_{n \in \mathbb{N}} \in l_p \,\middle|\, \sum_{n=1}^{\infty} a_n x_n = 0 \right\}$ be dense in l_p.

3. Let $1 < p < \infty$. Prove that $G = \{ x \in L_p(\mu) \mid \int_{\mathbb{R}} x(t) dt = 0 \}$ is a dense linear subspace in $L_p(\mu)$, where μ is the Lebesgue measure on $\mathbb{R}$.

4. Let $1 \leq p < \infty$. Consider an increasing function $\varphi : [0, \infty) \to [0, \infty)$ such that there are $M > 0, \alpha > 0$ with the properties:
$$\varphi(t + u) \leq M(\varphi(t) + \varphi(u)) \quad \forall t, u \geq 0,$$
and
$$\varphi(tu) \leq M t^\alpha \varphi(u) \quad \forall t > 0, \forall u \geq 0.$$
Prove that for any $(a_n)_{n \in \mathbb{N}} \subseteq (0, \infty)$ the set $G = \left\{ (x_n)_{n \in \mathbb{N}} \in l_p \,\middle|\, \sum_{n=1}^{\infty} a_n \varphi(|x_n|) < \infty \right\}$ is a dense linear subspace in l_p.

Deduce that if $\alpha \in \mathbb{R}$ then $G = \left\{ (x_n)_{n \in \mathbb{N}} \in l_2 \,\middle|\, \sum_{n=1}^{\infty} n^\alpha |x_n| < \infty \right\}$ is a dense linear subspace in l_2.

Fundamental sets in c_0, l_p and L_2

5. i) Let $f_n : [0, 1] \to \mathbb{R}$, $n \in \mathbb{N}$ be defined by $f_n(x) = \sum_{k=0}^{\infty} \frac{x^k}{2^{kn}}$. Prove that the set $\{ f_n \mid n \in \mathbb{N} \}$ is fundamental in $L_2[0, 1]$.

ii) Let $(f_k)_{k \geq 0} \subseteq C[0, 1]$ be a fundamental set in $L_2[0, 1]$ which is uniformly bounded, i.e., $\sup_{k \geq 0} \sup_{x \in [0,1]} |f_k(x)| < \infty$, and let $g_n : [0, 1] \to \mathbb{R}$, $n \in \mathbb{N}$, be defined by $g_n(x) = \sum_{k=0}^{\infty} \frac{f_k(x)}{2^{kn}}$. Prove that the set $\{ g_n \mid n \in \mathbb{N} \}$ is fundamental in $L_2[0, 1]$.

6. Let $1 \leq p < \infty$. For every $k \in \mathbb{N}$ let $x_k = (1, 1/2^k, 1/2^{2k}, ...) \in l_p$. Prove that the set $\{ x_k \mid k \in \mathbb{N} \}$ is fundamental in l_p.

7. Let $x_n = \left(e^{-(1+1/n)}, e^{-2(1+1/n)}, ... \right) \in l_1$ $\forall n \in \mathbb{N}$. Prove that $\{ x_n \mid n \in \mathbb{N} \}$ is a fundamental set in l_1.

8. Let $x_n = (e^{-1}, 2^{n-1} e^{-2}, 3^{n-1} e^{-3}, ..., k^{n-1} e^{-k}, ...) \in c_0$ $\forall n \in \mathbb{N}$. Prove that $\{ x_n \mid n \in \mathbb{N} \}$ is a fundamental set in c_0.

9. For every $z \in \mathbb{C}$ with $\Re z > 0$, let $x_z = (e^{-z}, e^{-2z}, e^{-3z}, ..., e^{-kz}, ...) \in c_0$. Prove that the set
$$A = \{ x_z \mid z \in \mathbb{C}, \Re z > 0 \}$$
is fundamental in c_0.

A way to construct a fundamental set from a fundamental one

10. i) Let $(a_{kn})_{(k,n) \in \mathbb{N} \times \mathbb{N}} \subseteq \mathbb{K} \setminus \{0\}$ be a double-indexed sequence of scalars such that

$$\lim_{n \to \infty} \frac{1}{|a_{pn}|} \sum_{k=p+1}^{n} |a_{kn}| = 0 \ \forall p \in \mathbb{N}.$$

Let X be a Banach space, $(x_n)_{n \in \mathbb{N}} \subseteq X$ a bounded sequence, and define $y_n = \sum_{k=1}^{n} a_{kn} x_k$ $\forall n \in \mathbb{N}$. Prove that if $x^* \in X^*$ has the property that $x^*(y_n) = 0 \ \forall n \in \mathbb{N}$ then $x^*(x_n) = 0$ $\forall n \in \mathbb{N}$, and then deduce that if $(x_n)_{n \in \mathbb{N}} \subseteq X$ is in addition a fundamental set then $(y_n)_{n \in \mathbb{N}} \subseteq X$ is also a fundamental set.

ii) Let X be a Banach space and $(x_n)_{n \in \mathbb{N}} \subseteq X$ a bounded sequence which is a fundamental set, and define $y_n = \sum_{k=1}^{n} x_k / k^n$ $\forall n \in \mathbb{N}$. Prove that the set $(y_n)_{n \in \mathbb{N}}$ is fundamental.

Non-reflexivity

11. For any $x \neq 0$ in a normed space X, by the Hahn–Banach theorem there is $x^* \in X^*$ of norm 1 such that $|x^*(x)| = \|x\|$. Give an example of a normed space X and an element $x^* \in X^*$ of norm 1 such that $|x^*(x)| < \|x\|$, for any $x \in X \setminus \{0\}$. (Obviously, X must be taken to be a non-reflexive space.)

Extension of equivalent norms

12. Let $(X, \|\cdot\|_1)$ be a normed space and $Y \subseteq X$ a linear subspace. Let $\|\cdot\|_2$ be a norm on X which is equivalent with the norm $\|\cdot\|_1$ on Y. Prove that there is a norm $\|\cdot\|$ on X which is equivalent with $\|\cdot\|_1$ on X and whose restriction to Y is $\|\cdot\|_2$.

Separation in finite-dimensional normed spaces

13. Let X be a finite-dimensional real normed space, and $A \subseteq X$ a convex subset. If $x_0 \in X \setminus A$, prove that there is a non-null linear and continuous functional which separates A and x_0.

14. Let X be a finite-dimensional real normed space, and $K \subseteq X_0$ be a convex subset with non-empty interior. Prove that for any $y \in \overline{K} \setminus \overset{\circ}{K}$ there is a hyperplane $H \subseteq X$ such that $y \in H$ and K is on one side of H.

Examples of non-separation

15. Consider the real Hilbert space l_2, and consider the set

$$A = \left\{ \sum_{i=1}^{n} \alpha_i e_i \ \middle| \ n \in \mathbb{N}, \alpha_i \in \mathbb{R}, i = \overline{1, n}, \ \alpha_n > 0 \right\} \subseteq l_2,$$

where $(e_n)_{n \in \mathbb{N}}$ is the standard basis from l_2. Let $B = -A$. Prove that A and B are convex disjoint sets, and that for any $x^* \in l_2^* \setminus \{0\}$ we have $x^*(A) = x^*(B) = \mathbb{R}$. What can we deduce?

16. In $\mathbb{R}^3$ we consider the convex cone C given by the equations: $x \geq 0$, $y \geq 0$, $z^2 \leq xy$. Let $d \subseteq \mathbb{R}^3$ be the line given by the equations: $x = 0$, $z = 1$. Prove that we cannot find a plane in $\mathbb{R}^3$ which contains d, and which has the cone strictly on one of its sides. However, the line and the cone are disjoint closed sets.

A characterization for the closed convex hull for a set in a normed space

17. Let X be a real normed space.
i) If $A \subseteq X$ is non-empty, prove that
$$\overline{\mathrm{co}}(A) = \left\{ x \in X \,\Big|\, x^*(x) \leq \sup_{a \in A} x^*(a) \ \forall x^* \in X^* \right\}.$$

ii) Let $K \subseteq X$ be a closed convex set with non-empty interior. Prove that for any $x_0 \in K \setminus \overset{\circ}{K}$ there is $x^* \in X^*$ such that $x^*(x_0) = \sup_{x \in K} x^*(x)$.

iii) In the case $X = \mathbb{R}^n$, $n \in \mathbb{N}$, with an arbitrary norm, prove that for any closed convex set $K \subseteq X$ and for any $x_0 \in K \setminus \overset{\circ}{K}$ there is an $x^* \in X^*$ such that $x^*(x_0) = \sup_{x \in K} x^*(x)$.

A set in $\mathbb{C}^n$ for which the convex cone generated by it is the entire $\mathbb{C}^n$

18. Prove that the convex cone generated by the set $\{(z, z^2, ..., z^n) \in \mathbb{C}^n \mid |z| \leq 1\}$ is exactly $\mathbb{C}^n$.

A concave function

19. Let X be a real normed space and $A \subseteq X$ a closed convex non-empty subset, different from X. Prove that the function $x \longmapsto d(x, \mathbf{C}_A)$ is concave on A.

Convexity inequalities for functions

20. Let X be a non-empty set, $\mathcal{P} = \{f : X \to \mathbb{R} \mid f(x) \geq 0 \ \forall x \in X\}$ and $M : \mathcal{P} \to \mathbb{R}_+$ be a positive homogeneous subadditive and increasing functional. Let $n \in \mathbb{N}$ and let $\varphi : [0, \infty)^n \to \mathbb{R}_+$ be a continuous function with the properties:
a) $t_1 > 0, ..., t_n > 0 \Rightarrow \varphi(t_1, ..., t_n) > 0$;
b) φ is positive homogeneous;
c) the set $A = \{(t_1, ..., t_n) \in [0, \infty)^n \mid \varphi(t_1, ..., t_n) \geq 1\}$ is convex.
Prove that $M(\varphi(f_1, f_2, ..., f_n)) \leq \varphi(M(f_1), ..., M(f_n))$, $\forall f_1, f_2, ..., f_n \in \mathcal{P}$.

The Phillips extension for an operator defined on a Hilbert space

21. i) Let H be a Hilbert space and $G \subseteq H$ a closed linear subspace. Prove that for any normed space X and for any linear and continuous operator $T : G \to X$ there is a linear and continuous operator $\overline{T} : H \to X$ such that $\overline{T}|_G = T$ and $\|T\| = \|\overline{T}\|$. The operator $\overline{T}$ with the above properties is called a Phillips extension of T.
More precisely, if $T : G \to X$ is a linear and continuous operator on G then $\overline{T} : H \to X$ defined by $\overline{T}(x) = T(\mathrm{Pr}_G(x)) \ \forall x \in H$ is a linear and continuous operator which extends T, and has the same norm as T. ($\mathrm{Pr}_G(x)$ is the orthogonal projection of x onto G.)
ii) Let $G = \{x \in l_2 \mid \langle x, e_1 \rangle = 0\}$ and $T : G \to c_0$ be the canonical inclusion, $T(x) = x \ \forall x \in G$. Prove that the Phillips extensions of T are of the form $\overline{T}_{a_1} : l_2 \to c_0$, $\overline{T}_{a_1}(x_1, x_2, x_3, ...) = (a_1 x_1, x_2, x_3, ...)$, where $|a_1| \leq 1$.

Extension for a seminorm

22. Let X be a linear space, $G \subseteq X$ a linear subspace and $p : X \to \mathbb{R}_+$ a seminorm. Let $q : G \to \mathbb{R}_+$ be a seminorm such that $q(x) \leq p(x)$ $\forall x \in G$.
Prove that there is a seminorm $\bar{q} : X \to \mathbb{R}_+$ such that $\bar{q}|_G = q$ and $\bar{q}(x) \leq p(x)$ $\forall x \in X$.

Strong separation in linear topological spaces

23. Let X be a linear topological space, and $A \subseteq X$ a non-empty convex and compact set. If x_0 is an element of X such that for each $x \in A$ there is a linear and continuous functional strongly separating x_0 and x, then there is a linear and continuous functional on X strongly separating x_0 and A.

Convex sets which cannot be separated

24. Let $H = L^2([-1,1]; \mathbb{R})$, where on $[-1,1]$ we consider the Lebesgue measure, and on H we consider the usual norm. For any $\alpha, \beta \in \mathbb{R}$, $\alpha \neq \beta$, we denote $E_\alpha = \{f \in H \mid f$ continuous, $f(0) = \alpha\}$, $E_\beta = \{f \in H \mid f$ continuous, $f(0) = \beta\}$. Prove that E_α and E_β are disjoint convex sets, which are dense in H, and that they cannot be separate by a linear and continuous functional on H.

8.2 Solutions

1. Let $f \in l_p^*$ be such that $f = 0$ on G. We shall show that $f = 0$, and then it will follow that G is dense in l_p. Since $f \in l_p^* = l_q$ there is $\xi = (\xi_n)_{n \in \mathbb{N}} \in l_q$ such that $f(x) = \sum\limits_{n=1}^{\infty} \xi_n x_n$ $\forall x = (x_n)_{n \in \mathbb{N}} \in l_p$ and $\|f\| = \|\xi\|_q$. Let $n \in \mathbb{N}$, $n \geq 2$. Then $x = (-1, 0, ..., 0, 1, 0, ...) \in G$ ($n - 2$ occurrences of 0) and by hypothesis $f(x) = 0$, i.e., $\xi_n - \xi_1 = 0$, $\xi_n = \xi_1$ $\forall n \geq 2$, i.e., $\xi = (\xi_1, \xi_1, ...) \in l_q$. We have $\sum\limits_{n=1}^{\infty} |\xi_1|^q < \infty$, from whence $\xi_1 = 0$, i.e., $\xi = 0$ and so $f(x) = 0$ $\forall x \in l_p$, $f = 0$.

Remarks.
i) Using the fact that $c_0^* = l_1$ the same reasoning as above assures that the linear subspace $G = \left\{ (x_n)_{n \in \mathbb{N}} \in c_0 \mid \sum\limits_{n=1}^{\infty} x_n = 0 \right\}$ is dense in c_0.
ii) In the case of l_2, which is a Hilbert space, the density can be also obtained using the following assertion: if G is a linear subspace in l_2, and from $x \perp G$ it follows that $x = 0$, then G is dense in l_2.

2. We prove that $G \subseteq l_p$ is dense if and only if $(a_n)_{n \in \mathbb{N}} \notin l_q$, $1/p + 1/q = 1$. Suppose that G is dense in l_p. Let us suppose, for a contradiction, that $(a_n)_{n \in \mathbb{N}} \in l_q$. Then $f : l_p \to \mathbb{K}$ defined by $f(x) = \sum\limits_{n=1}^{\infty} a_n x_n$ $\forall x = (x_n)_{n \in \mathbb{N}} \in l_p$ is a linear and continuous functional, and $G = \ker f$ is a closed linear subspace. Since G is dense it follows that $l_p = \overline{G} = \ker f$, i.e., $f(x) = 0$ $\forall x = (x_n)_{n \in \mathbb{N}} \in l_p$, i.e., $a_n = 0$ $\forall n \in \mathbb{N}$, a contradiction. Therefore $(a_n)_{n \in \mathbb{N}} \notin l_q$.

Conversely, suppose that $(a_n)_{n\in\mathbb{N}} \notin l_q$. Then $A = \{n \in \mathbb{N} \mid a_n \neq 0\}$ is infinite. Let us write A as a subsequence $(k_n)_{n\in\mathbb{N}}$ of $\mathbb{N}$. Let $f \in l_p^*$ such that $f = 0$ on G. Since $f \in l_p^* = l_q$, there is $\xi = (\xi_n)_{n\in\mathbb{N}} \in l_q$ such that $f(x) = \sum_{n=1}^{\infty} \xi_n x_n \ \forall x = (x_n)_{n\in\mathbb{N}} \in l_p$ and $\|f\| = \|\xi\|_q$. If $\mathbb{N} \neq A$, let $m \in \mathbb{N}\setminus A$. The element e_m belongs to G, and then $f(e_m) = 0$, i.e., $\xi_m = 0$. Hence $f(x) = \sum_{n=1}^{\infty} \xi_{k_n} x_{k_n} \ \forall x = (x_n)_{n\in\mathbb{N}} \in l_p$. For any $n \in \mathbb{N}, n \geq 2$, consider $x = (0, ..., 0, -1/a_{k_1}, 0, ..., 0, 1/a_{k_n}, 0, ...) \in G$ ($-1/a_{k_1}$ on the k_1^{th} position and $1/a_{k_n}$ on the k_n^{th} position). Then $f(x) = 0$, i.e.,

$$\frac{\xi_{k_n}}{a_{k_n}} - \frac{\xi_{k_1}}{a_{k_1}} = 0, \ \xi_{k_n} = \frac{a_{k_n}\xi_{k_1}}{a_{k_1}} \ \forall n \geq 2.$$

However, $\xi = (\xi_n)_{n\in\mathbb{N}} \in l_q$, i.e.,

$$\sum_{n=1}^{\infty} \frac{|a_n|^q |\xi_{k_1}|^q}{|a_{k_1}|^q} < \infty,$$

and since $(a_n)_{n\in\mathbb{N}} \notin l_q$, i.e., $\sum_{n=1}^{\infty} |a_n|^q = \infty$, it follows that $\xi_{k_1} = 0$, from whence $\xi_{k_n} = 0$ $\forall n \geq 2$. Then $f(x) = \sum_{n=1}^{\infty} \xi_{k_n} x_{k_n} = 0 \ \forall x = (x_n)_{n\in\mathbb{N}} \in l_p$, and therefore $G \subseteq l_p$ is dense.

3. Let $U \in (L_p(\mu))^*$ be such that $U = 0$ on G. We shall show that $U = 0$ and this will ensure that G is dense in $L_p(\mu)$. Since $U \in (L_p(\mu))^* = L_q(\mu)$, $1/p + 1/q = 1$, there is $g \in L_q(\mu)$ such that $U(\alpha) = \int_{\mathbb{R}} g(t)\alpha(t)dt \ \forall \alpha \in L_p(\mu)$ and $\|U\| = \|g\|_q$. Let $0 < h < \infty$ and $x, y \in \mathbb{R}$. The element $f = \chi_{[y,y+h]} - \chi_{[x,x+h]}$ belongs to G, and then by hypothesis $U(f) = 0$, i.e., $\int_y^{y+h} g(t)dt = \int_x^{x+h} g(t)dt$. Then

$$\frac{1}{h}\int_y^{y+h} (g(t) - g(y))\,dt = \frac{1}{h}\int_x^{x+h} (g(t) - g(x))\,dt + g(x) - g(y) \ \forall x, y \in \mathbb{R}.$$

From the Hölder inequality it follows that

$$\left|\frac{1}{h}\int_x^{x+h} (g(t) - g(x))\,dt\right| \leq \left(\frac{1}{h}\int_x^{x+h} |g(t) - g(x)|^q\,dt\right)^{1/q}$$

and since $g \in L_q(\mu)$, using the Lebesgue differentiation theorem it follows that

$$\lim_{h\to 0, h>0} \frac{1}{h}\int_x^{x+h} |g(t) - g(x)|^q\,dt = 0$$

μ-a.e. with respect to $x \in \mathbb{R}$. Then

$$\lim_{h\to 0, h>0} \frac{1}{h}\int_x^{x+h} (g(t) - g(x))\,dt = 0$$

8.2 Solutions

μ-a.e. with respect to $x \in \mathbb{R}$. Choosing y such that

$$\lim_{h\to 0, h>0} \frac{1}{h} \int_y^{y+h} (g(t) - g(y))\, dt = 0,$$

it follows that $g(x) - g(y) = 0$ μ-a.e. with respect to $x \in \mathbb{R}$, i.e., g is constant a.e.. But $g \in L_q(\mu)$, therefore $g = 0$, and then $U = 0$.

4. From $\varphi(tu) \leq Mt^\alpha \varphi(u)$ $\forall t > 0, \forall u \geq 0$ it follows that $\varphi(0) \leq Mt^\alpha \varphi(0)$ $\forall t > 0$, hence for $t \to 0$, $\varphi(0) \leq 0$, i.e., $\varphi(0) = 0$. From the inequality $\varphi(t + u) \leq M(\varphi(t) + \varphi(u))$ and that φ is increasing we obtain that $x + y \in G$ for $x, y \in G$. From the inequality $\varphi(tu) \leq Mt^\alpha \varphi(u)$ $\forall t > 0, \forall u \geq 0$ we obtain that $\forall x \in G, \forall \lambda \in \mathbb{K}$ it follows that $\lambda x \in G$, thus G is a linear subspace in l_p. Since $\varphi(0) = 0$ it follows that $\{e_n \mid n \in \mathbb{N}\} \subseteq G$ and since G is a linear subspace then $\mathrm{Sp}\{e_n \mid n \in \mathbb{N}\} \subseteq G$, and then $l_p = \overline{\mathrm{Sp}\{e_n \mid n \in \mathbb{N}\}} \subseteq \overline{G}$, i.e., G is dense in l_p.

5. i) Let $X = \mathrm{Sp}\{f_n \mid n \in \mathbb{N}\}$. To show that X is dense in $L_2[0,1]$, we will prove that if $g \in L_2[0,1]$ has the property that $g \perp X$, then it follows that $g = 0$. Let $g \in L_2[0,1]$ be such that $g \perp X$. Then $\langle g, f_n \rangle = 0$ $\forall n \in \mathbb{N}$, i.e., $\int_0^1 g(x) f_n(x)\, dx = 0$ $\forall n \in \mathbb{N}$. Let $p \in \mathbb{N}$ and let $s_p^n(x) = \sum_{k=0}^p \frac{x^k}{2^{kn}}$. Then

$$|s_p^n(x) - f_n(x)| = \sum_{k=p+1}^\infty \frac{x^k}{2^{kn}} \leq \sum_{k=p+1}^\infty \frac{1}{2^{kn}} = \frac{1}{2^{pn}(2^n - 1)} \quad \forall p, n \in \mathbb{N}, \forall x \in [0,1],$$

from whence $|s_p^n(x) g(x) - f_n(x) g(x)| \leq |g(x)|/2^{n-1}$ $\forall p, n \in \mathbb{N}, \forall x \in [0,1]$, and then $\left\|s_p^n g - f_n g\right\|^2 \leq |g|^2/2^{n-1}$ $\forall p, n \in \mathbb{N}$. Since $s_p^n g \to f_n g$ pointwise for $p \to \infty$, from the Lebesgue dominated convergence theorem it follows that $\lim_{p\to\infty} \left\|s_p^n g - f_n g\right\|_2 = 0$. But $\|\cdot\|_1 \leq \|\cdot\|_2$ on $L_2[0,1]$, by the Hölder inequality, from whence we deduce that $\lim_{p\to\infty} \left\|s_p^n g - f_n g\right\|_1 = 0$, and therefore

$$\lim_{p\to\infty} \left(\int_0^1 s_p^n(x) g(x)\, dx - \int_0^1 f_n(x) g(x)\, dx \right) = 0,$$

i.e.,

$$\sum_{k=0}^\infty \frac{1}{2^{kn}} \int_0^1 x^k g(x)\, dx = \int_0^1 f_n(x) g(x)\, dx = 0,$$

so $\sum_{k=0}^\infty (1/2^{kn}) \int_0^1 x^k g(x)\, dx = 0$ $\forall n \in \mathbb{N}$. Let $a_k = \int_0^1 x^k g(x)\, dx$. Then $\sum_{k=0}^\infty \frac{a_k}{2^{kn}} = 0$ $\forall n \in \mathbb{N}$ (1). Using now the Cauchy–Buniakowski–Schwarz inequality we obtain that

$$|a_k| \leq \left(\int_0^1 x^{2k}\, dx \right)^{1/2} \left(\int_0^1 |g(x)|^2\, dx \right)^{1/2} = \frac{1}{\sqrt{2k+1}} \left(\int_0^1 |g(x)|^2\, dx \right)^{1/2} \to 0,$$

i.e., $a_k \to 0$. In particular, the sequence $(a_k)_{k\in\mathbb{N}}$ is bounded. Let $\varphi : \mathbb{D} \to \mathbb{C}$ be the function defined by $\varphi(z) = \sum_{k=0}^\infty a_k z^k$, where $\mathbb{D} \subseteq \mathbb{C}$ is the open unit disc. Then we have

$$\sum_{k=0}^{\infty}|a_k||z|^k \leq \left(\sup_{k\geq 0}|a_k|\right)\sum_{k=0}^{\infty}|z|^k < \infty$$ on $\mathbb{D}$, and therefore φ is a well defined analytic function. The relation (1) implies that $\varphi(1/2^n) = 0 \ \forall n \in \mathbb{N}$ and since $1/2^n \to 0$ from the identity theorem for analytic functions it follows that $\varphi(z) = 0 \ \forall z \in \mathbb{D}$, i.e., $a_k = 0 \ \forall k \geq 0$, that is, $\int_0^1 x^k g(x)\,dx = 0 \ \forall k \geq 0$, i.e., $g \perp x^k \ \forall k \geq 0$, from whence $g \perp P$ for all polynomial P. Since the set of all the polynomials is dense in $L_2[0,1]$ it follows that $g = 0$.

ii) Exercise! One can use the same technique as above.

6. See [40, exercise 3.37].

Let $f \in l_p^*$ be such that $f = 0$ on G. We shall prove that $f = 0$ on l_p. Since $f \in l_p^* = l_q$ there is $\xi = (\xi_n)_{n\in\mathbb{N}} \in l_q$ such that $f(x) = \sum_{n=1}^{\infty}\xi_n x_n \ \forall x = (x_n)_{n\in\mathbb{N}} \in l_p$. Since $f = 0$ on G it follows that $f(x_k) = 0 \ \forall k \in \mathbb{N}$, i.e., $\sum_{n=1}^{\infty}\dfrac{\xi_n}{2^{k(n-1)}} = 0 \ \forall k \in \mathbb{N}$ (1). Consider $\varphi : \mathbb{D} \subseteq \mathbb{C} \to \mathbb{C}$ defined by $\varphi(z) = \sum_{n=1}^{\infty}\xi_n z^{n-1}$, where $\mathbb{D}$ is the open unit disc in the complex plane. Since $(\xi_n)_{n\in\mathbb{N}}$ is bounded $((\xi_n)_{n\in\mathbb{N}} \in l_q, \ 1 < q \leq \infty)$, we obtain that φ is a well defined analytic function on $\mathbb{D}$. The relation (1) implies that $\varphi(1/2^k) = 0 \ \forall k \in \mathbb{N}$. Since $1/2^k \to 0$, from the identity theorem for analytic functions it follows that $\varphi(z) = 0 \ \forall z \in \mathbb{D}$, i.e., $\sum_{n=1}^{\infty}\xi_n z^{n-1} = 0 \ \forall |z| < 1$, from whence it follows that $\xi_n = 0 \ \forall n \in \mathbb{N}$, and from here $f(x) = 0 \ \forall x \in l_p$.

7. See [47, p. 58].

Let $f \in l_1^*$ such that $f(x_n) = 0 \ \forall n \in \mathbb{N}$. Let $y = (y_n)_{n\in\mathbb{N}} \in l_\infty$ such that $f(\alpha_1, \alpha_2, ...) = \sum_{k=1}^{\infty}\alpha_k y_k \ \forall (\alpha_1, \alpha_2, ...) \in l_1$. Since $f(x_n) = 0 \ \forall n \in \mathbb{N}$, it follows that $\sum_{k=1}^{\infty}e^{-(1+1/n)k}y_k = 0 \ \forall n \in \mathbb{N}$. Let $\varphi(z) = \sum_{k=1}^{\infty}y_k e^{-zk}$, where $z \in \mathbb{C}, \ \Re z > 0$. We have

$$\sum_{k=1}^{\infty}|y_k|\left|e^{-zk}\right| = \sum_{k=1}^{\infty}|y_k|e^{-k\Re z} \leq \left(\sup_{k\in\mathbb{N}}|y_k|\right)\sum_{k=1}^{\infty}e^{-k\Re z} = \left(\sup_{k\in\mathbb{N}}|y_k|\right)\dfrac{e^{-\Re z}}{1-e^{-\Re z}}$$

and therefore $\varphi : H \subseteq \mathbb{C} \to \mathbb{C}$, where $H = \{z \in \mathbb{C} \mid \Re z > 0\}$, is a well defined and analytic function. But from our hypothesis $\varphi(1+1/n) = 0 \ \forall n \in \mathbb{N}$ and since $1+1/n \to 1$ from the identity theorem for analytic functions it follows that $\varphi(z) = 0 \ \forall z \in H$, therefore $\sum_{k=1}^{\infty}y_k e^{-zk} = 0 \ \forall z \in H$, and then $\sum_{k=1}^{\infty}y_k w^k = 0 \ \forall z \in \mathbb{D}\setminus\{0\}$, from whence $y_k = 0 \ \forall k \in \mathbb{N}$, i.e., $f = 0$.

8. See [21, exercise 2.1, chapter 2].

Let $\xi = (\xi_1, \xi_2, ..., \xi_k, ...) \in l_1$ such that $\sum_{k=1}^{\infty}\xi_k k^{n-1}e^{-k} = 0 \ \forall n \in \mathbb{N}$ (1). Then let $\varphi : \mathbb{D} \to \mathbb{C}$ be the function defined by $\varphi(z) = \sum_{k=1}^{\infty}\xi_k z^k$, where $\mathbb{D} \subseteq \mathbb{C}$ is the open unit disc in the complex plane. Clearly, $\varphi : \mathbb{D} \to \mathbb{C}$ is an analytic function. Define

8.2 Solutions

$\varphi_1(z) = z\varphi'(z)$ and $\varphi_{n+1}(z) = z\varphi'_n(z)$ if $n \geq 1$. We have $\varphi_2(z) = z(\varphi'(z) + z\varphi''(z)) = z\varphi'(z) + z^2\varphi''(z)$, and using induction one can easily deduce that $\varphi_n(z) = a_{1n}z\varphi'(z) + \cdots + a_{n-1,n}z^{n-1}\varphi^{(n-1)}(z) + z^n\varphi^{(n)}(z)$, $\forall n \in \mathbb{N}$ (2). From the relation (1) we have $\sum_{k=1}^{\infty} \xi_k k e^{-k} = 0$, i.e., $\varphi_1(1/e) = 0$, and by induction, using (1) we easily deduce that $\varphi_n(1/e) = 0 \ \forall n \in \mathbb{N}$. Using the relation (2), we obtain that $\varphi^{(n)}(1/e) = 0 \ \forall n \in \mathbb{N}$. Since φ is analytic and $\varphi^{(n)}(1/e) = 0 \ \forall n \geq 0$, it follows that $\varphi(z) = 0 \ \forall z \in \mathbb{D}$, i.e., $\xi_k = 0$ $\forall k \in \mathbb{N}$, and therefore if $x^* = \xi \in l_1$ has the property that $x^* = 0$ on $A = \{x_n \mid n \in \mathbb{N}\}$, then $x^* = 0$. This implies that $\mathrm{Sp}(A)$ is dense in c_0.

9. See [21, exercise 2.1].

Let $\xi = (\xi_1, \xi_2, \ldots, \xi_k, \ldots) \in l_1$ such that $\sum_{k=1}^{\infty} \xi_k e^{-kz} = 0 \ \forall z \in \mathbb{C}, \Re z > 0$. Consider the analytic function $\varphi : \mathbb{D} \subseteq \mathbb{C} \to \mathbb{C}$ defined by $\varphi(w) = \sum_{k=1}^{\infty} \xi_k w^k \ \forall w \in \mathbb{D}$. Then $\varphi(e^{-z}) = 0$ $\forall z \in H = \{z \in \mathbb{C} \mid \Re z > 0\}$, and then it follows that $\varphi = 0$. Then $\xi_k = 0 \ \forall k \geq 1$, that is, $\xi = 0$. Therefore if $x^* = \xi \in l_1$ has the property that $x^* = 0$ on A, then $x^* = 0$, and this implies that $\mathrm{Sp}(A)$ is dense in c_0.

10. i) From $x^*(y_n) = 0 \ \forall n \in \mathbb{N}$, using the formula for y_n and the linearity of x^* we obtain that $\sum_{k=1}^{n} a_{kn}x^*(x_k) = 0 \ \forall n \in \mathbb{N}$. Then $a_{1n}x^*(x_1) = -\sum_{k=2}^{n} a_{kn}x^*(x_k)$, $|a_{1n}||x^*(x_1)| \leq M \|x^*\| \sum_{k=2}^{n} |a_{kn}|$, from whence

$$|x^*(x_1)| \leq M \|x^*\| \frac{\sum_{k=2}^{n} |a_{kn}|}{|a_{1n}|} \to 0,$$

by our hypothesis (here $M = \sup_{n \in \mathbb{N}} \|x_n\| < \infty$). Hence $x^*(x_1) = 0$. Then $a_{2n}x^*(x_2) = -\sum_{k=3}^{n} a_{kn}x^*(x_k)$, from whence

$$|a_{2n}||x^*(x_2)| \leq M \|x^*\| \sum_{k=3}^{n} |a_{kn}|,$$

that is,

$$|x^*(x_2)| \leq M \|x^*\| \frac{\sum_{k=3}^{n} |a_{kn}|}{|a_{2n}|} \to 0,$$

by our hypothesis. Hence $x^*(x_2) = 0$. In this way, by induction, we obtain that $x^*(x_n) = 0$ $\forall n \in \mathbb{N}$.

If, in addition, $(x_n)_{n \in \mathbb{N}}$ is fundamental, then if $x^* \in X^*$, $x^*(y_n) = 0 \ \forall n \in \mathbb{N}$, we obtain that $x^*(x_n) = 0 \ \forall n \in \mathbb{N}$, and since $(x_n)_{n \in \mathbb{N}}$ is a fundamental set it follows that $x^* = 0$. Therefore the set $(y_n)_{n \in \mathbb{N}}$ is fundamental.

ii) See [47, p. 59].

For $a_{kn} = 1/k^n$ we have

$$\sum_{k=p+1}^{n} \frac{1}{k^n} \leq \sum_{k=p+1}^{n} \frac{1}{(p+1)^n} = \frac{n-p}{(p+1)^n},$$

from whence, denoting

$$\alpha_{pn} = p^n \left(\sum_{k=p+1}^{n} \frac{1}{k^n} \right),$$

we have

$$0 \leq \alpha_{pn} \leq \frac{(n-p)p^n}{(p+1)^n} = \frac{n-p}{(1+1/p)^n} \to 0,$$

since if $a > 1$ then $\lim_{x \to \infty} (x-p)/a^x = 0$. Now we apply (i).

11. For concrete examples see exercises 8–12 from chapter 3.

12. Since the norms $\|\cdot\|_1$ and $\|\cdot\|_2$ are equivalent on Y it follows that there are $m, M > 0$ such that $m\|y\|_1 \leq \|y\|_2 \leq M\|y\|_1$ $\forall y \in Y$. For any $f : (Y, \|\cdot\|_2) \to \mathbb{K}$ linear and continuous, it follows that $f : (Y, \|\cdot\|_1) \to \mathbb{K}$ is linear and continuous. Let $\widetilde{f} : (X, \|\cdot\|_1) \to \mathbb{K}$ be a Hahn–Banach extension for f (therefore, with the same norm as $f : (Y, \|\cdot\|_1) \to \mathbb{K}$). Since $\|y\|_2 \leq M\|y\|_1$ $\forall y \in Y$, we have $\|f\|_{(Y, \|\cdot\|_1)^*} \leq M\|f\|_{(Y, \|\cdot\|_2)^*}$, and then $\left\|\widetilde{f}\right\|_{(X, \|\cdot\|_1)^*} \leq M\|f\|_{(Y, \|\cdot\|_2)^*}$. For $x \in X$, define $\|x\|$ to be the maximum between $m\|x\|_1$ and $\sup\left\{|\widetilde{f}(x)|\right\}$, where the supremum is taken over all f and $\widetilde{f}$, where $\widetilde{f} : (X, \|\cdot\|_1) \to \mathbb{K}$ is a Hahn–Banach extension for $f|_{(Y, \|\cdot\|_1)}$, and $f \in (Y, \|\cdot\|_2)^*$, $\|f\| \leq 1$. Let us observe that $\|x\| \leq M\|x\|_1$ $\forall x \in X$. Indeed, if $\|f\|_{(Y, \|\cdot\|_2)^*} \leq 1$ then $\left\|\widetilde{f}\right\|_{(X, \|\cdot\|_1)^*} \leq M$, and then $|\widetilde{f}(x)| \leq M\|x\|_1$. So $\|x\| \in \mathbb{R}_+$ $\forall x \in X$.

If $\|x\| = 0$ then $\|x\|_1 = 0$, so $x = 0$. Obviously, $\|\lambda x\| = |\lambda| \|x\|$ $\forall \lambda \in \mathbb{K}$, $\forall x \in X$ and $\|x+y\| \leq \|x\| + \|y\|$ $\forall x, y \in X$. So, $(X, \|\cdot\|)$ is a normed space. For any $y \in Y$ we have

$$\|y\|_2 = \sup_{f \in (Y, \|\cdot\|_2)^*, \|f\| \leq 1} |f(y)| = \sup_{f \in (Y, \|\cdot\|_2)^*, \|f\| \leq 1} |\widetilde{f}(y)|,$$

and since $m\|y\|_1 \leq \|y\|_2$ $\forall y \in Y$, we obtain that $\|y\|_2 = \|y\|$ $\forall y \in Y$. Since $m\|x\|_1 \leq \|x\| \leq M\|x\|_1$ $\forall x \in X$ it follows that the norms $\|\cdot\|$ and $\|\cdot\|_1$ are equivalent on X.

13. See [20, exercise 24, chapter V, section 7].

Let $n = \dim_{\mathbb{R}} X$ and suppose that A does not contain n linearly independent elements. Let $k = \max\{m \mid \text{there are } m \text{ linearly independent elements in } A\}$. Then $k < n$ and if we consider $\{e_1, ..., e_k\} \subseteq A$ linearly independent, then for any $x \in A$ the system $\{e_1, ..., e_k, x\}$ will be linearly dependent, i.e., there are $\alpha_1, ..., \alpha_k, \alpha \in \mathbb{R}$ not all zero such that $\sum_{i=1}^{k} \alpha_i e_i + \alpha x = 0$. If $\alpha = 0$, since $\{e_1, ..., e_k\}$ are linearly independent it follows that $\alpha_1 = \cdots = \alpha_k = 0$, which is false! So $\alpha \neq 0$, and then $x = -\sum_{i=1}^{k} \alpha_i e_i / \alpha$, i.e., $A \subseteq \text{Sp}\{e_1, ..., e_k\}$. Complete now the linearly independent system $\{e_1, ..., e_k\}$ to a basis $\{e_1, ..., e_n\}$ of X. Consider then the dual basis $\{e_1^*, ..., e_n^*\}$ in X^*. Then $e_n^*(e_i) = 0$, $i = \overline{1,k}$, so $e_n^* = 0$ on $\text{Sp}\{e_1, ..., e_k\}$, from whence it follows that $e_n^*|_A = 0$. For $x_0 \in X \setminus A$, $e_n^*(x_0) \in \mathbb{R}$, and so $e_n^*(x_0) > 0$, or $e_n^*(x_0) = 0$, or $e_n^*(x_0) < 0$. Then

8.2 Solutions

$e_n^*(x) \geq 0 > e_n^*(x_0) \; \forall x \in A$, or $e_n^*(x) \leq 0 \leq e_n^*(x_0) \; \forall x \in A$, or $e_n^*(x) \leq 0 < e_n^*(x_0)$ $\forall x \in A$, so e_n^* separates A and x_0.

Suppose now that A contains n linearly independent elements. Using exercise 43(i) of chapter 1 it follows that $\overset{\circ}{A} \neq \varnothing$ and since A and $\{x_0\}$ are convex disjoint and non-empty sets we obtain that there is $x^* \in X^* \setminus \{0\}$ which separates A and x_0.

14. See [44, exercise 1, chapter 2].

Without loss of generality, we can suppose that $0 \in \overset{\circ}{K}$. Since $\overset{\circ}{K} \neq \varnothing$, using exercise 12(iii) of chapter 13 we obtain that $\overline{K} = \overline{\overset{\circ}{K}}$. Denoting by $A = \overset{\circ}{K}$, we have $A \neq \varnothing$, A is convex and open, and $y \in \overline{A}$, $y \notin A$. Since $0 \in A$, which is convex and open, and $y \notin A$, there is $x^* \in X^*$ such that $x^*(x) < 1 \; \forall x \in A$ and $x^*(y) = 1$. Let now $H = [x^* = 1]$. It follows that $y \in H$ and $A \subseteq [x^* < 1]$, from whence $\overline{A} \subseteq [x^* \leq 1]$. Since $\overline{A} = \overline{K}$ it follows that $K \subseteq [x^* \leq 1]$, i.e., K is on one side of $H = [x^* = 1]$.

15. Obviously, A and B are convex disjoint sets in l_2. Let $x^* \in (l_2)^* \setminus \{0\}$. We will show that $x^*(A) = \mathbb{R}$. Since x^* is not zero, there is $k \in \mathbb{N}$ such that $x^*(e_k) \neq 0$. But for $\lambda \in \mathbb{R}$ the element $\lambda e_k + e_{k+1} \in A$ (the coefficient for e_{k+1} is $1 > 0$) and then $x^*(\lambda e_k + e_{k+1}) \in x^*(A)$. So, $\lambda x^*(e_k) + x^*(e_{k+1}) \in x^*(A) \; \forall \lambda \in \mathbb{R}$. Since A is convex, it follows that $x^*(A) \subseteq \mathbb{R}$ is convex, i.e., an interval. Since $x^*(e_k) \neq 0$ for $\lambda \to \pm\infty$ it follows that $x^*(A)$ is unbounded on both sides of $\mathbb{R}$, i.e., $x^*(A) = \mathbb{R}$. We also have $x^*(B) = -x^*(A) = \mathbb{R}$.

We deduce that in order to have a separation theorem for two disjoint convex sets, we must also have a topological condition on one of the sets.

16. See [9, exercise 5, chapter II, section 3], or [35, section 3, chapter I].

We prove first that A is a cone. The conditions $x \geq 0$, $y \geq 0$, $z^2 \leq xy$ are equivalent with $xt^2 + 2zt + y \geq 0 \; \forall t \in \mathbb{R}$. Thus, if $(x_1, y_1, z_1), (x_2, y_2, z_2) \in A$ then $x_1 t^2 + 2z_1 t + y_1 \geq 0$ and $x_2 t^2 + 2z_2 t + y_2 \geq 0 \; \forall t \in \mathbb{R}$, from whence $(x_1 + x_2)t^2 + 2(z_1 + z_2)t + (y_1 + y_2) \geq 0$ $\forall t \in \mathbb{R}$, and then $x_1 + x_2 \geq 0$, $y_1 + y_2 \geq 0$, $(z_1 + z_2)^2 \leq (x_1 + x_2)(y_1 + y_2)$ so $(x_1 + x_2, y_1 + y_2, z_1 + z_2) \in A$. Clearly, if $\lambda \geq 0$ and $(x, y, z) \in A$, then $(\lambda x, \lambda y, \lambda z) \in A$. Suppose, for a contradiction, that there is a plane $\mathcal{P}$ in $\mathbb{R}^3$ with equation $ax + by + cz = \alpha$ such that $d \subseteq \mathcal{P}$ and $ax + by + cz < \alpha \; \forall (x, y, z) \in C$. From $d \subseteq \mathcal{P}$ it follows that $by + c = \alpha \; \forall y \in \mathbb{R}$, and so $b = 0$, $\alpha = c$. Hence the plane $\mathcal{P}$ has equation $ax + cz = c$. It follows that $ax + cz < c \; \forall (x, y, z) \in C$. For any $z \in \mathbb{R}$ the element $(1, z^2, z)$ belongs to C and so $a + cz < c \; \forall z \in \mathbb{R}$. For $z > 0$, $a/z + c < c/z$, from whence passing to the limit for $z \to \infty$, $c \leq 0$. Analogously, for $z < 0$, $a/z + c > c/z$, from whence passing to the limit for $z \to -\infty$, $c \geq 0$. Then $c = 0$ and the plane $\mathcal{P}$ has equation $x = 0$. But $(0, 1, 0) \in C$ and since C is strictly on one side of $\mathcal{P}$ we obtain a contradiction.

17. See [6, theorem 14, chapter 3].

i) Let $B = \overline{\mathrm{co}}(A)$ and $C = \left\{ x \in X \;\middle|\; x^*(x) \leq \sup_{a \in A} x^*(a) \; \forall x^* \in X^* \right\}$. Then $C = \bigcap_{x^* \in X^*} \left[x^* \leq \sup_{a \in A} x^*(a) \right]$, so C is an intersection of closed and convex sets (in the case when the supremum is infinite, the set $\left[x^* \leq \sup_{a \in A} x^*(a) \right]$ is the whole space X). It then follows that C is a closed convex set which contains A, and then $B \subseteq C$. Suppose, for a contradiction, that there is $x_0 \in C\setminus B$. Using a separation theorem it follows that there are $x^* \in X^*$ and $t \in \mathbb{R}$ such that $x^*(b) \leq t < x^*(x_0)$ $\forall b \in B$. Then from $x^*(b) \leq t$ $\forall b \in B$ and $A \subseteq B$, it follows that $x^*(a) \leq t$ $\forall a \in A$, i.e., $\sup_{a \in A} x^*(a) \leq t < x^*(x_0)$, that is, $\sup_{a \in A} x^*(a) < x^*(x_0)$, so, using the definition for the set C, $x_0 \notin C$, a contradiction.

ii) Since K is convex with non-empty interior using exercise 12 of chapter 13 it follows that $\overset{\circ}{K}$ is convex and that $\overline{\overset{\circ}{K}} = \overline{K} = K$, since K is closed. As $x_0 \in K \setminus \overset{\circ}{K}$, there are $x^* \in X^*$ and $c \in \mathbb{R}$ such that $x^*(x) < c \leq x^*(x_0)$ $\forall x \in \overset{\circ}{K}$ and so $\sup_{x \in \overset{\circ}{K}} x^*(x) \leq x^*(x_0)$. Since $\overline{\overset{\circ}{K}} = K$ from the continuity of $x^* \in X^*$ it follows that $\sup_{x \in \overset{\circ}{K}} x^*(x) = \sup_{x \in K} x^*(x)$. Then $\sup_{x \in K} x^*(x) \leq x^*(x_0)$. The reverse inequality follows from the fact that $x_0 \in K$.

iii) If the set K contains n linearly independent elements, then by exercise 43(i) of chapter 1 it follows that $\overset{\circ}{K} \neq \emptyset$ and we can apply (ii). If K contains at most $n-1$ linearly independent elements, then there is a linear subspace Y with $\dim Y \leq n-1$ such that $K \subseteq Y$. Since $Y \neq \mathbb{R}^n$, there is $x^* \in X^* \setminus \{0\}$ such that $x^* = 0$ on Y. Since $K \subseteq Y$ it follows that $x^*(x_0) = \sup_{x \in K} x^*(x) = 0$.

18. See [13, exercise 2, chapter II, section 8].

If X is a linear space and $A \subseteq X$ is a non-empty subset, then the convex cone generated by A is the set of all the linear combinations of the form $\sum_{i \in I} \lambda_i a_i$, where $(a_i)_{i \in I}$ is a finite non-empty family of elements from A and $\lambda_i > 0$ for every $i \in I$.

Let X be the convex cone from the statement. We will show first that X is dense in $\mathbb{C}^n$, i.e., $\overline{X} = \mathbb{C}^n$. Suppose, for a contradiction, that $\overline{X} \neq \mathbb{C}^n$, i.e., we can find $z_0 \in \mathbb{C}^n$, $z_0 \notin \overline{X}$ and $r > 0$ such that $\overline{X} \cap B(z_0, r) = \emptyset$, where $B(z_0, r) = \{w \in \mathbb{C}^n \mid \|w - z_0\|_2 < r\}$. Since $\overline{X}$ is convex and $B(z_0, r)$ is convex and open, we can find $x^* \in (\mathbb{C}^n)^* = \mathbb{C}^n$, not zero, $x^*(x_1, ..., x_n) = \sum_{i=1}^n c_i x_i$, and $t \in \mathbb{R}$ such that $\Re x^*(x) \leq t \leq \Re x^*(y)$ $\forall x \in B(z_0, r)$, $\forall y \in \overline{X}$. Since $\overline{X}$ is a convex cone we have that $\Re x^*(\lambda y) \geq t$ $\forall y \in \overline{X}$, $\forall \lambda \geq 0$, hence $\Re x^*(y) \geq t/\lambda$ $\forall y \in \overline{X}$, $\forall \lambda > 0$ and passing to the limit for $\lambda \to \infty$, it follows that $\Re x^*(y) \geq 0$ $\forall y \in \overline{X}$. Now $\Re x^*(y) \geq 0$ $\forall y = (z, z^2, ..., z^n) \in \mathbb{C}^n$, $|z| \leq 1$ implies that $\Re \left(\sum_{k=1}^n c_k z^k \right) \geq 0$ $\forall z \in \mathbb{C}$, $|z| \leq 1$. For any $\theta \in [0, 2\pi]$ we have $\left| e^{i\theta} \right| = 1$, and so $\Re \left(\sum_{k=1}^n c_k e^{ik\theta} \right) \geq 0$ for $\theta \in [0, 2\pi]$. But $\int_0^{2\pi} e^{ik\theta} d\theta = 0$ for every $k \in \mathbb{Z} \setminus \{0\}$ and so $\int_0^{2\pi} \left(\sum_{k=1}^n c_k e^{ik\theta} \right) d\theta = 0$, from whence it follows that $\Re \left(\int_0^{2\pi} \left(\sum_{k=1}^n c_k e^{ik\theta} \right) d\theta \right) = 0$, i.e., $\int_0^{2\pi} \Re \left(\sum_{k=1}^n c_k e^{ik\theta} \right) d\theta = 0$. Since $\Re \left(\sum_{k=1}^n c_k e^{ik\theta} \right) \geq 0$ $\forall \theta \in [0, 2\pi]$, from a well known property of the Riemann integral it follows that $\Re \left(\sum_{k=1}^n c_k e^{ik\theta} \right) = 0$ $\forall \theta \in [0, 2\pi]$. Let

8.2 Solutions

$P(z) = \sum_{k=1}^{n} c_k z^k$, $\forall z \in \mathbb{C}$ and $f(z) = e^{P(z)}$ $\forall z \in \mathbb{C}$. We have proved that $\Re P(z) = 0$ $\forall z \in \mathbb{C}$ with $|z| = 1$ and then $|f(z)| = e^{\Re P(z)} = 1$ $\forall z \in \mathbb{C}$ with $|z| = 1$. By the maximum modulus principle it follows that $|f(z)| \leq 1$ $\forall z \in \mathbb{C}$ with $|z| \leq 1$. Since $f(z) \neq 0$ on $\mathbb{D}$ applying the minimum modulus principle it follows that $|f(z)| \geq 1$ $\forall z \in \mathbb{C}$, $|z| \leq 1$ and then $|f(z)| = 1$ $\forall z \in \mathbb{C}$, $|z| \leq 1$. It follows that $\Re P(z) = 0$ $\forall z \in \mathbb{C}$, $|z| \leq 1$. Now from the Cauchy–Riemann equations it follows that $\Im P(z)$ is constant on the closed unit disc in the complex plane and since P is a polynomial it follows that P is constant on $\mathbb{C}$, which contradicts the form of P, $P(z) = \sum_{k=1}^{n} c_k z^k$, with $c_1, ..., c_n \in \mathbb{C}$ not all zero. So $\forall z_0 \in \mathbb{C}^n \backslash X$, $\forall r > 0$ we have $X \cap B(z_0, r) \neq \varnothing$, i.e., $X \subseteq \mathbb{C}^n$ is dense.

Let $T : \mathbb{C}^n \to \mathbb{R}^{2n}$, $T(z_1, ..., z_n) = (\Re z_1, \Im z_1, ..., \Re z_n, \Im z_n)$. If we consider $\mathbb{C}^n$ and $\mathbb{R}^{2n}$ as linear spaces over $\mathbb{R}$, then $T : \mathbb{C}^n \to \mathbb{R}^{2n}$ is $\mathbb{R}$-linear. Since $X \subseteq \mathbb{C}^n$ is convex, this implies that $T(X) \subseteq \mathbb{R}^{2n}$ is convex. Also, T is clearly a homeomorphism, and since $X \subseteq \mathbb{C}^n$ is dense then $T(X) \subseteq \mathbb{R}^{2n}$ is dense. We apply then exercise 43 (i) of chapter 1 to deduce that $T(X) = \mathbb{R}^{2n}$, and then $X = \mathbb{C}^n$.

19. See [13, exercise 18 (a), chapter II, section 6].

Since $A \neq X$ it follows that $\mathbf{C}_A \neq \varnothing$, hence $d(x, \mathbf{C}_A)$ has a meaning. Using exercise 17 we have that

$$A = \left\{ x \in X \,\middle|\, x^*(x) \leq \sup_{a \in A} x^*(a) \; \forall x^* \in E \right\},$$

where

$$E = \left\{ x^* \in X^* \,\middle|\, \sup_{a \in A} x^*(a) < \infty \right\}.$$

Let us denote $M_{x^*} = \sup_{a \in A} x^*(a)$ $\forall x^* \in E$. Then

$$A = \{ x \in X \mid x^*(x) \leq M_{x^*} \; \forall x^* \in E \} = \bigcap_{x^* \in E} [x^* \leq M_{x^*}].$$

We obtain that $\mathbf{C}_A = \bigcup_{x^* \in E} [x^* > M_{x^*}]$

Let us observe now that if X is a normed space, $(B_i)_{i \in I} \subseteq X$, and $Y = \bigcup_{i \in I} B_i$, then $\forall y \in X \backslash Y, d(y, Y) = \inf_{i \in I} d(y, B_i)$. Let us also observe that if $(f_i)_{i \in I}$ is family of concave functions on A, then the function $g : A \to \mathbb{R}$ defined by $g(x) = \inf_{i \in I} f_i(x)$ is also concave. Indeed, let $\varepsilon > 0$, $t \in [0, 1]$, $x, y \in A$. Then $g(tx + (1-t)y) = \inf_{i \in I} f_i(tx + (1-t)y)$, so there is $i_0 \in I$ such that $f_{i_0}(tx + (1-t)y) \leq g(tx + (1-t)y) + \varepsilon$, from whence, since f_{i_0} is concave,

$$g(tx + (1-t)y) + \varepsilon \geq t f_{i_0}(x) + (1-t) f_{i_0}(y) \geq t g(x) + (1-t) g(y).$$

Since $\varepsilon > 0$ is arbitrary we obtain that

$$g(tx + (1-t)y) \geq t g(x) + (1-t) g(y) \quad \forall x, y \in A, \; \forall t \in [0, 1],$$

i.e., the above claimed relation.

For $g : A \to \mathbb{R}$ defined by $g(x) = d(x, \mathbf{C}_A)$, we have

$$g(x) = d\left(x, \bigcup_{x^* \in E} [x^* > M_{x^*}]\right),$$

and then $g(x) = \inf_{x^* \in E} d(x, [x^* > M_{x^*}])$. But

$$d(x, [x^* > M_{x^*}]) = \frac{M_{x^*} - x^*(x)}{\|x^*\|} \quad \forall x \in A$$

(see, for example, exercises 5 and 6 of chapter 4), and then we obtain that $g(x) = \inf_{x^* \in X^*} (M_{x^*} - x^*(x)) / \|x^*\|$. Since the function $(M_{x^*} - x^*(\cdot)) / \|x^*\| : A \to \mathbb{R}$ is concave we obtain that $g : A \to \mathbb{R}$ is also concave.

20. See [10, proposition 1, chapter I].
Let us suppose first that $M(\varphi(f_1, f_2, ..., f_n)) = 1$. In this case we must prove that $\varphi(M(f_1), ..., M(f_n)) \geq 1$, i.e., $(M(f_1), ..., M(f_n)) \in A$. If $(M(f_1), ..., M(f_n)) \notin A$, since A is convex and closed, by a separation theorem it follows that there are $(\alpha_1, ..., \alpha_n) \in \mathbb{R}^n$ and $t \in \mathbb{R}$ such that

$$\alpha_1 M(f_1) + \cdots + \alpha_n M(f_n) < t \leq \alpha_1 t_1 + \cdots + \alpha_n t_n \quad \forall (t_1, ..., t_n) \in A \quad (1)$$

Let $(t_1, ..., t_n) \in (0, \infty)^n$. From (i) we have $\lambda_0 = 1/\varphi(t_1, ..., t_n) > 0$ and then for $\lambda \geq \lambda_0$, since φ is positive homogeneous,

$$\varphi(\lambda t_1, ..., \lambda t_n) = \lambda \varphi(t_1, ..., t_n) \geq \lambda_0 \varphi(t_1, ..., t_n) = 1,$$

i.e., $(\lambda t_1, ..., \lambda t_n) \in A$, and from (1) we have $t \leq \alpha_1 \lambda t_1 + \cdots + \alpha_n \lambda t_n$, that is, $t/\lambda \leq \alpha_1 t_1 + \cdots + \alpha_n t_n \ \forall \lambda \geq \lambda_0$. Passing to the limit for $\lambda \to \infty$ we obtain that $0 \leq \alpha_1 t_1 + \cdots + \alpha_n t_n$ $\forall (t_1, ..., t_n) \in (0, \infty)^n$ and for $t_2 \to 0, ..., t_n \to 0$, $\alpha_1 t_1 \geq 0 \ \forall t_1 > 0$, from whence $\alpha_1 \geq 0$. Analogously we obtain that $\alpha_2 \geq 0, ..., \alpha_n \geq 0$. Since $M(f_1) \geq 0, ..., M(f_n) \geq 0$, from (1) it follows that $t > 0$. Let $x \in X$. If $\lambda = \varphi(f_1(x), ..., f_n(x)) \neq 0$, i.e., $\lambda > 0$, then $\varphi(f_1(x)/\lambda, ..., f_n(x)/\lambda) = 1$, that is, $(f_1(x)/\lambda, ..., f_n(x)/\lambda) \in A$, from whence using (1),

$$t \leq \frac{\alpha_1 f_1(x)}{\lambda} + \cdots + \frac{\alpha_n f_n(x)}{\lambda},$$

that is, $\lambda t \leq \alpha_1 f_1(x) + \cdots + \alpha_n f_n(x)$, $t\varphi(f_1(x), ..., f_n(x)) \leq \alpha_1 f_1(x) + \cdots + \alpha_n f_n(x)$. If $\lambda = 0$ this inequality is clearly satisfied, since $\alpha_i \geq 0$, $f_i(x) \geq 0$, $i = \overline{1, n}$. Hence $t\varphi(f_1(x), ..., f_n(x)) \leq \alpha_1 f_1(x) + \cdots + \alpha_n f_n(x) \ \forall x \in X$, and therefore $t\varphi(f_1, ..., f_n) \leq \alpha_1 f_1 + \cdots + \alpha_n f_n$. Since M is increasing, positive homogeneous, and subadditive we obtain that

$$tM(\varphi(f_1, ..., f_n)) \leq \alpha_1 M(f_1) + \cdots + \alpha_n M(f_n),$$

that is,

$$t \leq \alpha_1 M(f_1) + \cdots + \alpha_n M(f_n),$$

which contradicts the relation (1). Hence the inequality from the statement is proved.

8.2 Solutions

For the general case, if $M(\varphi(f_1, ..., f_n)) = 0$ then obviously $M(\varphi(f_1, ..., f_n)) \leq \varphi(M(f_1), ..., M(f_n))$. If $M(\varphi(f_1, ..., f_n)) \neq 0$, then $\lambda = M(\varphi(f_1, ..., f_n)) > 0$. For $g_1 = f_1/\lambda, ..., g_n = f_n/\lambda$ we have $M(\varphi(g_1, ..., g_n)) = M(\varphi(f_1, ..., f_n))/\lambda = 1$, because M and φ are positive homogeneous. From the first case $1 \leq \varphi(M(g_1), ..., M(g_n))$, and using again the properties for M and φ we obtain that $\lambda \leq \varphi(M(f_1), ..., M(f_n))$, that is, $M(\varphi(f_1, ..., f_n)) \leq \varphi(M(f_1), ..., M(f_n))$.

21. i) Define $\overline{T} : H \to X$ by $\overline{T}(x) = T(\Pr_G(x))$. It is clear that $\overline{T}$ is linear, continuous, and that $\overline{T}$ extends T. We have

$$\|\overline{T}(x)\| = \|T(\Pr_G(x))\| \leq \|T\| \|\Pr_G(x)\| \leq \|T\| \|x\| \quad \forall x \in H,$$

from whence $\|\overline{T}\| \leq \|T\|$. Since $\overline{T}$ extends T, $\|\overline{T}\| \geq \|T\|$, and then $\|\overline{T}\| = \|T\|$.

ii) We have $\|T(x)\| \leq \|x\| \; \forall x \in l_2$, and then $\|T\| \leq 1$. From $\|T(e_2)\| \leq \|T\| \|e_2\|$ it follows that $\|T\| \geq 1$, and therefore $\|T\| = 1$.

Let $\overline{T} : l_2 \to c_0$ be a Phillips extension of T. Let $x \in l_2$. Then $\Pr_G(x) = x - \langle x, e_1 \rangle e_1 \in G$. Since $\overline{T}$ extends T it follows that $\overline{T}(\Pr_G(x)) = T(\Pr_G(x))$, that is, $\overline{T}(x - \langle x, e_1 \rangle e_1) = x - \langle x, e_1 \rangle e_1$ and since $\overline{T}$ is linear it follows that

$$\overline{T}(x) = x - \langle x, e_1 \rangle e_1 + \langle x, e_1 \rangle \overline{T}(e_1) \quad \forall x \in l_2.$$

Let us denote by $a = \overline{T}(e_1) \in l_2$. Then

$$\overline{T}(x) = x - \langle x, e_1 \rangle e_1 + \langle x, e_1 \rangle a \quad \forall x \in l_2$$

and we must find $a \in l_2$ such that $\|\overline{T}\| = \|T\|$.

Let $a = (a_1, a_2, ...)$. A simple calculation shows that

$$\overline{T}(x_1, x_2, ...) = x - \langle x, e_1 \rangle e_1 + \langle x, e_1 \rangle a = (x_1, x_2, ...) - (x_1, 0, ...) + x_1(a_1, a_2, ...)$$
$$= (a_1 x_1, x_2 + a_2 x_1, x_3 + a_3 x_1, ...).$$

To calculate the norm of $\overline{T}$, we have $|a_1 x_1| \leq |a_1| |x_1| \leq |a_1| \|x\|$, and for $n \geq 2$,

$$|x_n + a_n x_1| \leq \sqrt{|a_n|^2 + 1} \sqrt{|x_1|^2 + |x_n|^2} \leq \sqrt{|a_n|^2 + 1} \|x\|,$$

from whence, using the definition for the norm in the space c_0 we deduce that

$$\|\overline{T}(x)\| \leq \max\left(|a_1|, \sup_{n \geq 2} \sqrt{|a_n|^2 + 1}\right) \|x\| \quad \forall x \in l_2,$$

that is, $\|\overline{T}\| \leq \max\left(|a_1|, \sup_{n \geq 2} \sqrt{|a_n|^2 + 1}\right)$. Conversely, we have $\|\overline{T}(1, 0, 0, ..)\| \leq \|\overline{T}\| \|(1, 0, 0, ..)\|$, therefore $\sup_{n \geq 1} |a_n| \leq \|\overline{T}\|$, and then $|a_1| \leq \|\overline{T}\|$. For $n \geq 2$, we have

$$\|\overline{T}(\overline{a_n}, 0, 0, ..., 0, 1, 0, ...)\| \leq \|\overline{T}\| \|(\overline{a_n}, 0, 0, ..., 0, 1, 0, ...)\|$$

(1 on the nth position) and using again the definition for the norm in the space c_0 we have that $|a_n|^2 + 1 \leq \|\overline{T}\| \sqrt{|a_n|^2 + 1}$, i.e., $\sqrt{|a_n|^2 + 1} \leq \|\overline{T}\|$. Hence

$$\|\overline{T}\| = \max\left(|a_1|, \sup_{n \geq 2} \sqrt{|a_n|^2 + 1}\right).$$

Since $\|\overline{T}\| = \|T\|$ it follows that

$$\max\left(|a_1|, \sup_{n\geq 2}\sqrt{|a_n|^2 + 1}\right) = 1.$$

If $|a_1| > \sup_{n\geq 2}\sqrt{|a_n|^2 + 1}$ then $|a_1| = 1$, and then $\sup_{n\geq 2}\sqrt{|a_n|^2 + 1} < 1$, which is impossible. Then $|a_1| \leq \sup_{n\geq 2}\sqrt{|a_n|^2 + 1}$. It follows that $\sup_{n\geq 2}\sqrt{|a_n|^2 + 1} = 1$, and therefore $a_n = 0$ $\forall n \geq 2$. We also have $|a_1| \leq 1$, i.e. the statement.

22. See [9, exercise 15, chapter II, section 5].

Let $V = \{x \in G \mid q(x) \leq 1\}$ and $U = \{x \in X \mid p(x) \leq 1\}$. Since $q(x) \leq p(x)$ $\forall x \in G$ it follows that $U \cap G \subseteq V$. Indeed, if $x \in U \cap G$ then $x \in G$ and $p(x) \leq 1$. But $q(x) \leq p(x)$ on G, from whence $q(x) \leq 1$, $x \in G$, i.e., $x \in V$. Let $W = \text{co}(U \cup V)$. Since $U, V \subseteq X$ are convex (p and q are seminorms, and therefore $U \subseteq X$ and $V \subseteq G$ are convex and balanced), it follows that $\text{co}(U \cup V) = \{\lambda x + (1 - \lambda)y \mid x \in U, y \in V, 0 \leq \lambda \leq 1\}$. Since $U \subseteq W$ and U is absorbing (p is a seminorm on X), it follows that W is absorbing. Also, since U, V are balanced it follows that W is balanced (see exercise 7(i) of chapter 13). In addition, $W \cap G = V$. Indeed, we have $V \subseteq W$ and $V \subseteq G$, from whence $V \subseteq W \cap G$. Conversely, if $z \in W \cap G$ then there are $x \in U, y \in V, 0 \leq \lambda \leq 1$ such that $z = \lambda x + (1 - \lambda)y$. If $\lambda \neq 0$ then $z/\lambda \in G$. But

$$\frac{z}{\lambda} = x + \frac{1-\lambda}{\lambda}y$$

and since $y \in V$ it follows that $y \in G$, from whence since G is a linear subspace, it follows that $(1 - \lambda)y/\lambda \in G$ and then

$$x = \frac{z}{\lambda} - \frac{1-\lambda}{\lambda}y \in G.$$

But $x \in U$, i.e., $x \in G \cap U \subseteq V$, i.e., $x \in V$. Since $y \in V$ and V is a convex set, $z = \lambda x + (1 - \lambda)y \in V$. If $\lambda = 0$, then $z = \lambda x + (1 - \lambda)y = y \in V$. Hence in both cases $z \in V$, i.e., $W \cap G \subseteq V$.

Let now $\overline{q} = p_W : X \to \mathbb{R}_+$ be the Minkowski seminorm associated with W. We will prove that $\overline{q}$ has all the required properties. Let $x \in G$ and $\varepsilon > 0$. Then

$$\overline{q}\left(\frac{x}{\overline{q}(x) + \varepsilon}\right) = \frac{\overline{q}(x)}{\overline{q}(x) + \varepsilon} < 1,$$

so using the definition for the Minkowski seminorm, there is $0 < \mu < 1$ such that $x/(\overline{q}(x) + \varepsilon) \in \mu W$. Since $x \in G$ and G is a linear subspace, $x/(\mu(\overline{q}(x) + \varepsilon)) \in G$, i.e., $x/(\mu(\overline{q}(x) + \varepsilon)) \in W \cap G = V$, and then $q(x/(\mu(\overline{q}(x) + \varepsilon))) \leq 1$. Since q is a seminorm we obtain that $q(x) \leq \mu(\overline{q}(x) + \varepsilon)$ and since $\mu < 1$ we have $q(x) \leq \overline{q}(x) + \varepsilon$ $\forall \varepsilon > 0$ from whence for $\varepsilon \to 0$, $q(x) \leq \overline{q}(x)$. Let $x \in G$ and $\varepsilon > 0$. Then $x/(q(x) + \varepsilon) \in V = W \cap G$, i.e., $x/(q(x) + \varepsilon) \in W$, $x \in (q(x) + \varepsilon)W$, from whence using again the definition for the Minkowski seminorm, $\overline{q}(x) \leq q(x) + \varepsilon$ $\forall \varepsilon > 0$, i.e., $\overline{q}(x) \leq q(x)$. Hence $\overline{q}(x) = q(x)$ $\forall x \in G$.

8.2 Solutions

Let now $x \in X$ and $\varepsilon > 0$. Then $x/(p(x) + \varepsilon) \in U \subseteq W$, i.e., $x \in (p(x) + \varepsilon)W$, from whence $\overline{q}(x) \leq p(x) + \varepsilon \ \forall \varepsilon > 0$, that is, $\overline{q}(x) \leq p(x)$.

23. See [35, exercise A, chapter 4].

By hypothesis $\forall x \in A \ \exists x^* \in X^*$ such that $\Re x^*(x_0) < \Re x^*(x)$ (1). This suggests considering for each $x^* \in X^*$ the set $E_{x^*} = \{x \in X \mid \Re x^*(x_0) < \Re x^*(x)\}$, which is open, since x^* is continuous. With this notation, the relation (1) is: $A \subseteq \bigcup_{x^* \in X^*} E_{x^*}$. Since A is compact there are $x_1^*, ..., x_n^* \in X^*$ such that $A \subseteq \bigcup_{i=1}^n E_{x_i^*}$, i.e., $\forall x \in A \ \exists 1 \leq i \leq n$ such that $\Re x_i^*(x_0 - x) < 0$ (2). This suggests considering the closed convex cone $P = \{(\lambda_1, ..., \lambda_n) \in \mathbb{R}^n \mid \lambda_1 \geq 0, ..., \lambda_n \geq 0\}$ and the linear and continuous operator $h : X \to \mathbb{R}^n$, $h(x) = (\Re x_1^*(x), ..., \Re x_n^*(x))$. Then (2) implies that $h(x_0 - A) \cap P = \emptyset$ (3). Since A is a compact convex and non-empty set and h is linear and continuous, then $h(x_0 - A)$ is a compact convex and non-empty set, and then, by (3), it follows that we can strongly separate the sets $h(x_0 - A)$ and P in $\mathbb{R}^n$, i.e., we can find $(\alpha_1, ..., \alpha_n) \in \mathbb{R}^n \setminus \{0\}$ and $t \in \mathbb{R}$ such that $\sum_{i=1}^n \alpha_i \xi_i < t \ \forall (\xi_1, ..., \xi_n) \in P$ (4), and $t < \sum_{i=1}^n \alpha_i \Re x_i^*(x_0 - x) \ \forall x \in A$ (5). Taking in (4) $\xi_1 = 0, ..., \xi_n = 0$ we obtain that $t > 0$. From (5), $\sum_{i=1}^n \alpha_i \Re x_i^*(x_0) - t > \sum_{i=1}^n \alpha_i \Re x_i^*(x) \ \forall x \in A$, i.e., if we denote by $x^* = \sum_{i=1}^n \alpha_i x_i^* \in X^*$ we have, since $t > 0$, that $\Re x^*(x_0) > \Re x^*(x_0) - t > \Re x^*(x) \ \forall x \in A$, i.e., $\Re x^*$ strongly separates x_0 and A.

24. See [44, exercise 2, chapter 3].

If $f, g \in E_\alpha$, then $f, g \in C[-1, 1]$, $f(0) = g(0) = \alpha$. If $\lambda \in [0, 1]$ we obtain that $(\lambda f + (1 - \lambda)g)(0) = \alpha$, thus $E_\alpha \subseteq H$ is a convex set. Analogously $E_\beta \subseteq H$ is a convex set. If $f \in H$ then we can find $g \in C[-1, 1]$ such that $\|f - g\|_{L^2} < \varepsilon/2$. For $g \in C[-1, 1]$ we can find $h \in C[-1, 1]$, with $h(0) = \alpha$, such that $\|g - h\|_{L^2} < \varepsilon/2$. Indeed, let $n \in \mathbb{N}$, $n \geq 2$, such that

$$\sqrt{\frac{2}{n}} \left(2 \sup_{x \in [-1,1]} |g(x)| + |\alpha|\right) < \frac{\varepsilon}{2}.$$

Consider $h : [-1, 1] \to \mathbb{R}$ given by

$$h(x) = \begin{cases} g(x) & x \in [-1, -1/n], \\ -nxg(-1/n) + (1 + nx)\alpha & x \in (-1/n, 0], \\ nxg(1/n) + (1 - nx)\alpha & x \in (0, 1/n], \\ g(x) & x \in (1/n, 1]. \end{cases}$$

Then

$$\|g - h\|_{L^2}^2 = \int_{-1/n}^{1/n} |g(x) - h(x)|^2 \, dx \leq \int_{-1/n}^{1/n} (|g(x)| + |h(x)|)^2 \, dx$$

$$\leq \int_{-1/n}^{1/n} (2\|g\|_\infty + |\alpha|)^2 \, dx \leq \frac{2}{n} (2\|g\|_\infty + |\alpha|)^2,$$

thus $\|g - h\|_{L^2} < \varepsilon/2$. Then $h \in E_\alpha$, and $\|f - h\|_{L^2} \leq \|f - g\|_{L^2} + \|g - h\|_{L^2} < \varepsilon$, thus $E_\alpha \subseteq H$ is dense. Analogously $E_\beta \subseteq H$ is a dense set. Evidently, $E_\alpha \cap E_\beta = \emptyset$. Suppose now that we can find $x^* \in H^*\setminus\{0\}$ and $t_0 \in \mathbb{R}$ such that $x^*(x) \leq t_0 \leq x^*(v)$ $\forall x \in E_\alpha$, $\forall v \in E_\beta$. Since $E_\alpha \subseteq H$ is dense we obtain that $x^*(E_\alpha) \subseteq x^*(H) = \mathbb{R}$ is dense. But $E_\alpha \subseteq H$ is convex, and since x^* is linear we obtain that $x^*(E_\alpha) \subseteq \mathbb{R}$ is a convex set, thus an interval. We obtain that $x^*(E_\alpha) = \mathbb{R}$, and then $t_0 \geq \sup\limits_{x \in E_\alpha} x^*(x) = \infty$, a contradiction.

Let us observe that in the solution for exercise 5 of chapter 3 the density is proved in a different way.

Chapter 9

Baire's category. The open mapping and closed graph theorems

Definition. Let X be a topological space. A subset $A \subseteq X$ is called:
i) *nowhere dense* if $\overset{\circ}{\overline{A}} = \emptyset$;
ii) of the *first Baire category* if there is a sequence of nowhere dense sets $(A_n)_{n \in \mathbb{N}}$ such that $A = \bigcup_{n \in \mathbb{N}} A_n$;
iii) of the *second Baire category* if is not of the first Baire category.

The Baire category theorem. Any complete metric space is of the second Baire category. In particular, any Banach space is of the second Baire category.

The open mapping theorem. Let X be a Banach space, Y a normed space and $U : X \to Y$ a linear and continuous operator. Then either $U(X)$ is of the first Baire category in Y, or U is surjective, and in this second case U is an open map and Y is a Banach space.

The inverse mapping theorem. If X and Y are Banach spaces, and $U : X \to Y$ is a bijective linear and continuous operator, then $U^{-1} : Y \to X$ is also a linear and continuous operator.

The closed graph theorem. Let X and Y be two Banach spaces, and $U : X \to Y$ a linear operator with its graph closed. Then U is continuous.

Theorem. Let X and Y be two normed space, and $U : X \to Y$ a linear operator. Then the following assertions are equivalent:
i) U has its graph closed;
ii) If $x_n \to x$ in X, and $U(x_n) \to y$ in Y, then $U(x) = y$.

9.1 Exercises

Nowhere dense sets

1. i) Prove that $c_0 \subseteq c$ is a closed linear subspace, and also a nowhere dense set.

ii) Prove that $c \subseteq l_\infty$ is a closed linear subspace, and also a nowhere dense set.

iii) Let X be a normed space and $A \subseteq X$ a convex set with non-empty interior. Prove that $\operatorname{Fr}(A) = \overline{A} \setminus \overset{\circ}{A}$ is nowhere dense in X.

2. Let $X \neq \{0\}$ be a normed space, and $A \subseteq X$ a set which is not nowhere dense. Prove that there are $x \in X$ and $\varepsilon > 0$ such that $B(x, \varepsilon) \subseteq A'$, where A' is the derivative set of A, i.e., the set of all the accumulation points of A.

Examples of first Baire category sets which are not nowhere dense

3. Prove that the set $\left\{ \sum_{n=0}^\infty a_n x^n \mid \sum_{n=0}^\infty |a_n| < \infty \right\} \subseteq C[0,1]$ is of the first Baire category, that it is not nowhere dense, and it is not open.

4. i) Let X be a Banach space, Y a normed space, and $U \in L(X,Y)$ such that U is not surjective, but $U(X)$ is dense in Y. Prove that $U(X)$ is a first Baire category set in Y, and that it is not nowhere dense in Y.

ii) Using (i) prove that the following sets are of the first Baire category, but not nowhere dense sets in the indicated spaces:

a) for $0 < \alpha \leq 1$, $\operatorname{Lip}_\alpha[0,1]$ in the space $C[0,1]$;

b) for $1 \leq p < \infty$, l_p in the space c_0;

c) for $1 \leq p < q < \infty$, $L_q[0,1]$ in the space $L_p[0,1]$;

d) for $k \in \mathbb{N} \cup \{\infty\}$, $C^k[0,1]$ in the space $C[0,1]$;

e) for $1 \leq p < q < \infty$, l_p in the space l_q;

f) l_1 in

$$S_c = \left\{ (x_n)_{n \in \mathbb{N}} \subseteq \mathbb{K} \,\bigg|\, \left(\sum_{k=1}^n x_k \right)_{n \in \mathbb{N}} \text{ converges} \right\}$$

endowed with the norm $\|(x_n)_{n \in \mathbb{N}}\| = \sup_{n \in \mathbb{N}} \left| \sum_{k=1}^n x_k \right|$;

5. Let us consider the spaces $L_1[0,1]$ and $L_2[0,1]$, given by the Lebesgue measure on the unit interval $[0,1]$. Prove that $L_2[0,1] \subseteq L_1[0,1]$ is of the first Baire category, in the following three ways:

i) Proving that $\left\{ f \in L_1[0,1] \mid \int_0^1 |f(x)|^2 \, dx \leq 1 \right\}$ is closed in $L_1[0,1]$, but with empty interior;

ii) Considering the functions $g_n : [0,1] \to \mathbb{K}$ defined by $g_n = n\chi_{[0,1/n^3]}$, and proving that $\int_0^1 f(x) g_n(x) dx \to 0$ for any $f \in L_2[0,1]$, but not for any $f \in L_1[0,1]$;

iii) Proving that the inclusion operator from $L_2[0,1]$ into $L_1[0,1]$ is continuous, but not surjective.

6. Let $\varphi \in C[0,1]$ and $A = \{f\varphi \mid f \in C[0,1]\}$. Give a necessary and sufficient condition on the function φ for A to be a first Baire category set (respectively, a second Baire category set in $C[0,1]$).

9.1 Exercises

The algebraic sum of two first Baire category sets

7. i) Prove that the sets $A = \{f \in C[-1,1] \mid f \text{ is an even function}\}$ and $B = \{f \in C[-1,1] \mid f \text{ is an odd function}\}$ are nowhere dense in the space $C[-1,1]$. Then deduce that if A and B are two first Baire category sets in a Banach space then their algebraic sum $A+B$ can be a second Baire category set.

ii) Let X be a normed space. $P : X \to X$ is a linear and continuous projector, that is, $P^2 = P$, and suppose that $P \neq 0$ and $P \neq I$. Prove that $\ker P$ and $P(X)$ are nowhere dense sets in the space X and that their algebraic sum $\ker P + P(X)$ is of the second Baire category in X if and only if X is of the second Baire category.

Subsets which contain a ball

8. Let X be a Banach space. $F \subseteq X$ is a subset with the property that for any $x \in X$ there is $\varepsilon > 0$ such that $tx \in F \; \forall t \in [0, \varepsilon]$. Prove that we can find $x_0 \in X$ and $\varepsilon > 0$ such that $B(x_0, \varepsilon) \subseteq F'$. If, in addition, F is closed then F contains an open ball.

A characterization for infinite-dimensional Banach spaces

9. Let X be a Banach space. Prove that X is infinite-dimensional if and only if any totally bounded set is nowhere dense.

The algebraic dimension for an infinite-dimensional Banach space

10. i) If X is a normed space prove that any proper closed linear subspace of X is a nowhere dense set.

ii) If X is an infinite-dimensional normed space which can be written as a countable union of finite-dimensional linear subspaces, prove that X is of the first Baire category.

iii) Using (ii) prove that an infinite-dimensional Banach space cannot have a countable algebraic basis.

iv) Using (iii) prove that on c_{00} there is no complete norm.

Pointwise convergent sequences of continuous functions

11. Let X be a complete metric space, and $f_n : X \to \mathbb{K}$, $n \in \mathbb{N}$, be a sequence of continuous functions for which there is a function $f : X \to \mathbb{K}$ such that $\lim_{n \to \infty} f_n(x) = f(x)$ $\forall x \in X$, i.e., $f_n \to f$ pointwise on X.

i) Prove that there are $V \subseteq X$ open and non-empty and $M > 0$ such that $|f_n(x)| \leq M$ $\forall x \in V, \forall n \in \mathbb{N}$.

ii) Prove that for any $\varepsilon > 0$ there are $V \subseteq X$ open and non empty and $p \in \mathbb{N}$ such that $|f(x) - f_n(x)| \leq \varepsilon$ $\forall x \in V, \forall n \geq p$.

Continuous functions with null limit at infinity

12. Let $f : [1, \infty) \to \mathbb{R}$ be a continuous function. $(a_n)_{n \in \mathbb{N}} \subseteq [1, \infty)$ is a strictly increasing sequence with $\lim_{n \to \infty}(a_{n+1}/a_n) = 1$, $\lim_{n \to \infty} a_n = \infty$, and such that for any $x \in [1, \infty)$, $\lim_{n \to \infty} f(a_n x) = 0$. Prove that $\lim_{x \to \infty} f(x) = 0$.

Nowhere differentiable functions

13. i) For $n \in \mathbb{N}$ let $X_n = \{f \in C[0,1] \mid \exists t \in [0,1]$ such that $|f(s) - f(t)| \leq n|s-t|$ $\forall s \in [0,1]\}$. Prove that for each $n \in \mathbb{N}$, $X_n \subseteq C[0,1]$ is a closed and nowhere dense set.

ii) Prove that from (i) it follows the existence of a dense G_δ subset in $C[0,1]$, which contains only functions which are not differentiable at any point of $[0,1]$.

A common root for a sequence of functions

14. Let $f_n : \mathbb{R} \to \mathbb{R}$, $n \in \mathbb{N}$, be a sequence of functions with the property that $\forall x \in \mathbb{R}$, $\forall n \in \mathbb{N}$, $\lim_{t \to x} f_n(t) = 0$, and let $A \subseteq \mathbb{R}$ be a set of the first Baire category. Prove that there is $x \in \mathbb{R} \setminus A$ such that $f_n(x) = 0$ $\forall n \in \mathbb{N}$.

Dense sets and vertical asymptotes

15. i) Let X be a Banach space and $A \subseteq X$ a dense set. Can we find a function $f : X \to \mathbb{R}$ such that $\forall x \in A$, $\lim_{t \to x} |f(t)| = \infty$?

ii) Establish if there is a function $f : \mathbb{R} \to \mathbb{R}$ which has every rational line $x = q$ as a vertical asymptote, i.e., $\forall q \in \mathbb{Q}$, $\lim_{t \to q} |f(t)| = \infty$.

Pointwise unbounded sequences of continuous functions

16. i) Let X be a complete metric space and $A \subseteq X$ be a first Baire category set in X which is also dense in X. Does there exist a sequence of continuous functions $f_n : X \to \mathbb{R}$ such that the sequence $(f_n(x))_{n \in \mathbb{N}}$ is unbounded if and only if $x \in A$?

ii) Establish if there is a sequence of continuous functions $f_n : \mathbb{R} \to \mathbb{R}$ such that the sequence $(f_n(x))_{n \in \mathbb{N}}$ is unbounded if and only if x is rational.

iii) Establish if there is a sequence of continuous functions $f_n : c_0 \to \mathbb{R}$ such that the sequence $(f_n(x_1, x_2, \ldots))_{n \in \mathbb{N}}$ is unbounded if and only if the series $\sum_{k=1}^{\infty} x_k$ is absolutely convergent.

iv) Establish if there is a sequence of continuous functions $U_n : C[0,1] \to \mathbb{R}$ such that the sequence $(U_n(f))_{n \in \mathbb{N}}$ is unbounded if and only if f is a Lipschitz function.

A property for sequences, with respect to a first Baire category set

17. i) Let X be a complete metric space and $f : X \times X \to X$ be a function with the property that for any $y \in X$, the function $f(\cdot, y) : X \to X$ is a homeomorphism. Let $A \subseteq X$ be a first Baire category set in X and $(x_n)_{n \in \mathbb{N}} \subseteq X$. Prove that there is $x \in X$ such that $\forall n \in \mathbb{N}$, $f(x, x_n) \notin A$.

ii) Using (i) prove that if H is a Hilbert space and $(T_n)_{n \in \mathbb{N}} \subseteq L(H)$ then there is $T \in L(H)$ such that for any $n \in \mathbb{N}$, the operator $T + T_n$ is not self-adjoint.

iii) Using (i) prove that for any sequence of continuous functions $f_n : [0,1] \to \mathbb{R}$, there is a continuous function $f : [0,1] \to \mathbb{R}$ such that for any $n \in \mathbb{N}$ the function $f + f_n$ is not Lipschitz.

iv) Using (i) prove that for any sequence of continuous functions $f_n : [-1,1] \to \mathbb{R}$ there is a continuous function $f : [-1,1] \to \mathbb{R}$ such that for any $n \in \mathbb{N}$, $f + f_n$ is not an odd function.

9.1 Exercises

v) Using (i) prove that for any sequence of Lebesgue integrable functions $f_n : [0,1] \to \mathbb{R}$ there is a Lebesgue measurable function $f : [0,1] \to \mathbb{R}$ such that $\int_0^1 |f(x)| \, dx < \infty$, but $\int_0^1 |f(x) + f_n(x)|^2 \, dx = \infty$ for any $n \in \mathbb{N}$.

vi) Using (i) prove that for any sequence of continuous functions $f_n : [0,1] \to \mathbb{R}$ there is a continuous function $f : [0,1] \to \mathbb{R}$ such that for any $n \in \mathbb{N}$ the function $f + f_n$ is not differentiable at any point of the interval $[0,1]$.

A non-integrable function

18. Let $(S, \sum, \mu)$ be a measure space and let $f_n : S \to \mathbb{R}$, $n \in \mathbb{N}$, be a sequence of μ-integrable functions such that: $\sup_{n \in \mathbb{N}} |f_n(s)| < \infty$, μ-a.e., and $\sup_{n \in \mathbb{N}} \int_S |f_n(s)| \, d\mu(s) = \infty$. Prove that there is a scalar sequence $(a_n)_{n \in \mathbb{N}}$ such that $\sum_{n=1}^\infty |a_n| < \infty$ and the function $\sum_{n=1}^\infty a_n f_n$ is not μ-integrable.

The Baire category for the range of a linear and continuous operator

19. i) Let X be a Banach space, Y is a normed space, $U : X \to Y$ is a linear and continuous operator, and $G \subseteq X$ a closed linear subspace. Prove that $U(G)$ is either of the first Baire category, or $U(G) = Y$.

ii) Let X, Y be Banach spaces, Z a normed space, and $T : X \to Z$, $U : Y \to Z$ two linear and continuous operators. Prove that if $T^{-1}(U(Y))$ is of the second Baire category and $U^{-1}(T(X))$ is of the second Baire category, then $U(Y) = T(X)$.

20. i) Let $(X_n)_{n \in \mathbb{N}}$ be a sequence of Banach spaces, X a Banach space, and $U_n : X_n \to X$ a sequence of linear and continuous operators such that $X = \bigcup_{n \in \mathbb{N}} U_n(X_n)$. Prove that there is $n \in \mathbb{N}$ such that $U_n(X_n) = X$, i.e., there is $n \in \mathbb{N}$ such that U_n is surjective.

ii) Let $(X_n)_{n \in \mathbb{N}}$ be a sequence of Banach spaces, X is a Banach space, Y is a normed space, and $U_n : X_n \to Y$ is a sequence of linear and continuous operators. Let $U : X \to Y$ be a linear and continuous operator. Using (i) prove that if $U(X) \subseteq \bigcup_{n \in \mathbb{N}} U_n(X_n)$ then there is $n \in \mathbb{N}$ such that $U(X) \subseteq U_n(X_n)$.

iii) Prove that if $S \subseteq L_1[0,1]$ is a closed linear subspace with the property that $\forall f \in S$, there is $p > 1$ such that $f \in L_p[0,1]$ then there is $p > 1$ such that $S \subseteq L_p[0,1]$.

21. Let X be a Banach space, and let $(X_n)_{n \in \mathbb{N}}$ be a sequence of linear subspaces of X such that $X = \bigcup_{n \in \mathbb{N}} X_n$. Prove that there is $n \in \mathbb{N}$ such that $\overline{X_n} = X$.

In particular, if X is a Banach space and $(X_n)_{n \in \mathbb{N}}$ is a sequence of closed linear subspaces such that $X = \bigcup_{n \in \mathbb{N}} X_n$, prove that there is $n \in \mathbb{N}$ such that $X_n = X$.

The Baire category for the space of all finite rank operators

22. Let X be an infinite-dimensional Banach space.

i) For $n \in \mathbb{N}$ we denote by $\mathcal{F}_n$ the set of all linear operators from X into X of rank at most n. Prove that $\mathcal{F}_n \subseteq L(X)$ is a closed linear subspace. (The rank for $T \in L(X)$ is the dimension for the range of T.)

ii) Prove that there is $T \in L(X)$ with $\dim_{\mathbb{K}}(T(X)) = \infty$ such that T can be uniformly approximated (that is, with respect to the operator norm) by a sequence of finite rank operators.

In particular, deduce that if X is an infinite-dimensional Banach space, then the space of all finite rank operators is strictly contained in the space $K(X)$ of all compact operators from X into X.

A characterization for one-dimensional normed spaces

23. Let X, Y be two normed spaces, $X \neq \{0\}, Y \neq \{0\}$. Prove that the following assertions are equivalent:

i) Any non-null linear and continuous operator $U : X \to Y$ is open;
ii) Any non-null linear and continuous operator $U : X \to Y$ is surjective;
iii) $\dim_{\mathbb{K}} Y = 1$.

Multiplication operators between l_p-spaces

24. i) Let $1 \leq p < \infty$, $1 \leq q < \infty$, and let $a = (a_n)_{n \in \mathbb{N}}$ be a sequence of scalars with the property that: $\forall (x_n)_{n \in \mathbb{N}} \in l_p$ it follows that $(a_n x_n)_{n \in \mathbb{N}} \in l_q$. Prove that the multiplication operator $M_a : l_p \to l_q$ defined by $M_a((x_n)_{n \in \mathbb{N}}) = (a_n x_n)_{n \in \mathbb{N}}$ is linear and continuous.

ii) Let $1 \leq q < p < \infty$ and let $a = (a_n)_{n \in \mathbb{N}}$ be a sequence of scalars with the property that: $\forall (x_n)_{n \in \mathbb{N}} \in l_p$ it follows that $(a_n x_n)_{n \in \mathbb{N}} \in l_q$. Let $M_a : l_p \to l_q$ be the multiplication operator defined at (i). Prove that M_a is linear and continuous, that $a \in l_r$, and that $\|M_a\| = \|a\|_r$, where $1/p + 1/r = 1/q$.

iii) Let $1 \leq p \leq q < \infty$, and let $a = (a_n)_{n \in \mathbb{N}}$ be a sequence of scalars with the property that: $\forall x = (x_n)_{n \in \mathbb{N}} \in l_p$ it follows that $(a_n x_n)_{n \in \mathbb{N}} \in l_q$. Let $M_a : l_p \to l_q$ be the multiplication operator given at (i). Prove that M_a is linear and continuous, that $a \in l_\infty$, and that $\|M_a\| = \|a\|_\infty = \sup_{n \in \mathbb{N}} |a_n|$.

25. Prove that we cannot find a scalar sequence $(c_n)_{n \in \mathbb{N}}$ with the property that a scalar series $\sum_{n=1}^{\infty} a_n$ converges absolutely if and only if the scalar sequence $(a_n c_n)_{n \in \mathbb{N}}$ is bounded.

26. i) Find the scalar sequences $(c_n)_{n \in \mathbb{N}}$ with the property that a scalar series $\sum_{n=1}^{\infty} a_n$ is absolutely convergent if and only if the scalar series $\sum_{n=1}^{\infty} a_n c_n$ is absolutely convergent.

ii) Let $1 \leq p, q < \infty$ with $p \neq q$. Does there exist a scalar sequence $(c_n)_{n \in \mathbb{N}}$ with the property that a scalar series $\sum_{n=1}^{\infty} a_n$ is p-absolutely convergent if and only if the scalar series $\sum_{n=1}^{\infty} a_n c_n$ is q-absolutely convergent?

iii) Let $1 \leq p < \infty$. Find all the scalar sequences $(c_n)_{n \in \mathbb{N}}$ with the property that a scalar series $\sum_{n=1}^{\infty} a_n$ is p-absolutely convergent if and only if the scalar series $\sum_{n=1}^{\infty} a_n c_n$ is p-absolutely convergent.

9.2 Solutions

Indeed, if $z \in B(y, \delta)$ then

$$\|z - x\| \leq \|z - y\| + \|y - x\| < \delta + \|y - x\| \leq \frac{\varepsilon - \|y - x\|}{2} + \|y - x\|$$
$$= \frac{\varepsilon + \|y - x\|}{2} < \varepsilon,$$

i.e., $z \in B(x, \varepsilon)$. Obviously, $B(y, \delta) \subseteq B(y, r)$. So $B(y, \delta) \cap (A \setminus \{y\}) \subseteq B(y, r) \cap (A \setminus \{y\}) = \emptyset$, i.e., $B(y, \delta) \cap A \subseteq \{y\}$. Since $y \in B(x, \varepsilon) \subseteq \overline{A}$, $y \in \overline{A}$, from whence $B(y, \delta) \cap A \neq \emptyset$, and therefore $B(y, \delta) \cap A = \{y\}$. A well known topological result, which will be proved at the end of our solution, says that $D \cap \overline{A} \subseteq \overline{D \cap A}$, $\forall D$ open. Using this result we obtain that $B(y, \delta) \cap \overline{A} \subseteq \overline{B(y, \delta) \cap A}$. But $B(y, \delta) \subseteq B(x, \varepsilon) \subseteq \overline{A}$, i.e., $B(y, \delta) \cap \overline{A} = B(y, \delta)$, hence $B(y, \delta) = \{y\}$, which is false, since $X \neq \{0\}$ (there is $a \in X$, $a \neq 0$ and $b = a/\|a\| \in X$ with $\|b\| = 1$, and then $y + (\delta/2)b \in B(y, \delta) = \{y\}$, i.e., $y + (\delta/2)b = y$, $b = 0$, contradiction).

Let us justify the remaining claim. Let $x \in D \cap \overline{A}$. Let also V be a neighborhood of x. Since $x \in D$ and D is open, D is a neighborhood of x. It follows that $V \cap D$ is a neighborhood of x and as $x \in \overline{A}$ it follows that $D \cap V \cap A \neq \emptyset$, and therefore $x \in \overline{D \cap A}$.

3. In exercise 7 of chapter 2 it is proved that the operator $U : l_1 \to C[0, 1]$ defined by $(Ua)(x) = \sum_{n=0}^{\infty} a_n x^n$ $\forall a = (a_n)_{n \geq 0} \in l_1$ is linear and continuous. The set from the exercise is $U(l_1)$. If we suppose that U is surjective, then for the function $g(x) = 1/(x+1)$ we can find $a = (a_n)_{n \geq 0} \in l_1$ such that $Ua = g$, i.e., $\sum_{n=0}^{\infty} a_n x^n = 1/(x+1)$, that is, $a_0 + (a_0 + a_1)x + (a_1 + a_2)x^2 + \cdots = 1$ $\forall x \in [0, 1]$, and then $a_0 = 1$, $a_0 + a_1 = 0$, $a_1 + a_2 = 0$, ..., i.e., $a = (1, -1, 1, -1, \ldots) \notin l_1$. Thus U is not surjective, and since l_1 is a Banach space, from the open mapping theorem it follows that $U(l_1)$ is of the first Baire category. Since the set of all polynomial functions on the interval $[0, 1]$ is contained in A, by the well known Bernstein theorem it follows that A is dense in $C[0, 1]$ and hence A is not nowhere dense. If A were open, then since A is a linear subspace, by exercise 15 chapter 1 we must have $A = C[0, 1]$, and this is not true. Hence $U(l_1)$ is not an open set in $C[0, 1]$.

4. i) See [9, exercise 3, chapter I, section 3].

Since U is not surjective, from the open mapping theorem it follows that $U(X)$ is of the first Baire category. Since $\overline{U(X)} = Y$, then $\overset{\circ}{\overline{U(X)}} = Y \neq \emptyset$, i.e., $U(X)$ is not nowhere dense.

ii) a) Recall that, given $0 < \alpha \leq 1$, a function $f : [0, 1] \to \mathbb{R}$ (or $\mathbb{C}$) is called α-Lipschitz if there is $L > 0$, such that $|f(x) - f(y)| \leq L|x - y|^\alpha$ $\forall 0 \leq x, y \leq 1$. Let

$$\|f\|_\alpha := |f(0)| + \sup_{x \neq y} \frac{|f(x) - f(y)|}{|x - y|^\alpha},$$

and then the space $\text{Lip}_\alpha[0, 1] = \{f : [0, 1] \to \mathbb{R}\,(\mathbb{C}) \mid f \text{ is } \alpha\text{-Lipschitz}\}$ endowed with the norm $\|\cdot\|_\alpha$ is a Banach space. Let now $J : \text{Lip}_\alpha[0, 1] \to C[0, 1]$ be the inclusion operator, i.e., $J(f) = f \;\forall f \in \text{Lip}_\alpha[0,1]$. For $f \in \text{Lip}_\alpha[0,1]$, we have

$$|f(x) - f(0)| \leq (\|f\|_\alpha - |f(0)|)|x|^\alpha \leq \|f\|_\alpha - |f(0)|,$$
$$|f(x)| \leq |f(x) - f(0)| + |f(0)| \leq \|f\|_\alpha \quad \forall 0 \leq x \leq 1,$$

so $\sup_{x \in [0,1]} |f(x)| \leq \|f\|_\alpha$, i.e., $\|f\| \leq \|f\|_\alpha \;\forall f \in C[0,1]$, i.e., J is continuous. Also, J is not surjective. For example, the function $f : [0,1] \to \mathbb{R}$ defined by
$f(x) = \begin{cases} 1/\log x & 0 < x \leq 1, \\ 0 & x = 0, \end{cases}$ is continuous, but is not α-Lipschitz, for $0 < \alpha \leq 1$.

Now for $0 < \alpha \leq 1$, $n \in \mathbb{N}$, the functions $f_n : [0,1] \to \mathbb{R}$ defined by $f_n(x) = x^{n\alpha}$ are α-Lipschitz, since, as is easy to see, for $a, b \in [0,1]$, $(a+b)^\alpha \leq a^\alpha + b^\alpha$, and hence for $0 \leq x < y \leq 1$,

$$y^{n\alpha} \leq (y^n - x^n)^\alpha + x^{n\alpha} \leq n^\alpha (y-x)^\alpha + x^{n\alpha},$$

i.e, $y^{n\alpha} - x^{n\alpha} \leq n^\alpha (y-x)^\alpha$, that is, $|y^{n\alpha} - x^{n\alpha}| \leq n^\alpha |y-x|^\alpha$. Let $\mathcal{P}_\alpha = \text{Sp}\{1, x^\alpha, x^{2\alpha}, \ldots\} \subseteq \text{Lip}_\alpha[0,1]$. From the Bernstein theorem, if $g \in C[0,1]$, then for the Bernstein polynomials

$$B_n(x) = \sum_{k=0}^n \binom{n}{k} g\left(\frac{k}{n}\right) x^k (1-x)^{n-k}$$

we have $B_n \to g$ uniformly. Let $f \in C[0,1]$, and consider $g(x) = f(x^{1/\alpha})$. Then $P_n(x) := B_n(x^\alpha) \to g(x^\alpha) = f(x)$ uniformly on $[0,1]$. Since

$$P_n(x) = \sum_{k=0}^n \binom{n}{k} g\left(\frac{k}{n}\right) x^{\alpha k} (1 - x^\alpha)^{n-k} \in \mathcal{P}_\alpha$$

it follows that $\mathcal{P}_\alpha$ is dense in $C[0,1]$. Since $\mathcal{P}_\alpha \subseteq \text{Lip}_\alpha[0,1] \subseteq C[0,1]$, it follows that $\text{Lip}_\alpha[0,1]$ is dense in the space $C[0,1]$. From (i) it follows that $\text{Lip}_\alpha[0,1]$ is of the first Baire category in $C[0,1]$, and not nowhere dense.

b) Let $j : l_p \to c_0$ be again the inclusion operator, which is linear and continuous, and $j(l_p) = l_p \subseteq c_0$ is dense (see exercise 23 of chapter 1). Also, $l_p \neq c_0 \;\forall 1 \leq p < \infty$, i.e., j is not surjective. From (i) we obtain that l_p is of the first Baire category in c_0, and not nowhere dense.

c) From the Hölder inequality it follows that if $p < q$ we have $\left(\int_0^1 |f|^p d\mu\right)^{1/p} \leq \left(\int_0^1 |f|^q d\mu\right)^{1/q}$, i.e., $\|f\|_p \leq \|f\|_q$, and therefore the canonical inclusion $J : L_q[0,1] \to L_p[0,1]$ is continuous. Let us denote by $\mathcal{E}$ the linear subspace of all measurable step functions. We have $\mathcal{E} \subseteq J(L_q[0,1]) \subseteq L_p[0,1]$ and $\mathcal{E}$ is dense in $L_p[0,1]$, so it follows that $J(L_q[0,1])$ is dense in $L_p[0,1]$. Now if $\alpha \in \mathbb{R}$ then $f = \sum_{n=1}^\infty n^\alpha \chi_{(1/(n+1), 1/n)} \in L_p[0,1]$ if and only if

$$\int_0^1 |f|^p d\mu = \int_0^1 \left(\sum_{n=1}^\infty n^{\alpha p} \chi_{\left(\frac{1}{n+1}, \frac{1}{n}\right)}\right) d\mu = \sum_{n=1}^\infty \frac{n^{\alpha p}}{n(n+1)} < \infty,$$

9.2 Solutions

that is, if and only if $\alpha < 1/p$. (We have

$$\frac{n^{\alpha p}}{n(n+1)} \sim \frac{1}{n^{2-\alpha p}}$$

and we use the comparison test with the generalized harmonic series). Hence for $1/q \leq \alpha < 1/p$, the function f is in $L_p[0,1]$, but not in $L_q[0,1]$, i.e., J is not surjective. Using (i) it follows that $L_q[0,1]$ is of the first Baire category and not nowhere dense in $L_p[0,1]$.

d) Let $J : C^k[0,1] \to C[0,1]$ be the inclusion operator. As above, one can prove that $J(C^k[0,1]) = C^k[0,1]$ is of the first Baire category and not nowhere dense in $C[0,1]$.

e) Let $j : l_p \to l_q$ be the inclusion operator, which is well defined, since $p < q$ implies $\|x\|_q \leq \|x\|_p \, \forall x \in l_p$. Then J is continuous. Since $c_{00} \subseteq l_p \subseteq l_q$ and c_{00} is dense in l_q it follows that l_p is dense in l_q. We also have

$$x = \left(\frac{1}{n^\alpha}\right)_{n \in \mathbb{N}} \in l_p \Leftrightarrow \sum_{n=1}^{\infty} \frac{1}{n^{\alpha p}} < \infty \Leftrightarrow \alpha p > 1 \Leftrightarrow \alpha > \frac{1}{p},$$

and we obtain that $x \in l_q$ and $x \notin l_p$ if $1/q < \alpha \leq 1/p$, i.e., j is not surjective. Then (i) implies that l_p is of the first Baire category in l_q and not nowhere dense.

f) It is proved at exercise 27(ii) of chapter 1 that l_1 is dense in the space S_c. For the inclusion operator $j : l_1 \to S_c$, which is clearly linear and continuous, we have $j(l_1) = l_1$, and we apply (i).

5. See [44, exercise 4, chapter 2].

i) We shall prove that the set $B = \left\{ f \in L_1[0,1] \mid \int_0^1 |f(x)|^2 \, dx \leq 1 \right\}$ is closed in $(L_1[0,1], \|\cdot\|_1)$. Indeed, let $f \in L_1[0,1]$ for which there is $(f_n)_{n \in \mathbb{N}} \subseteq L_1[0,1]$ with $\int_0^1 |f_n(x)|^2 \, dx \leq 1 \, \forall n \in \mathbb{N}$ and $\|f_n - f\|_1 \to 0$ (we denote with $\|\cdot\|_p$ the usual norm in $L_p[0,1]$, for $1 \leq p \leq \infty$). Since $\|f_n - f\|_1 \to 0$, i.e., $\int_0^1 |f_n(x) - f(x)| \, dx \to 0$, then there is a subsequence $(n_k)_{k \in \mathbb{N}}$ such that $f_{n_k} \to f$, μ-a.e. on $[0,1]$. So $|f_{n_k}|^2 \to |f|^2$, μ-a.e.. Also $\int_0^1 |f_{n_k}(x)|^2 \, dx \leq 1 \, \forall k \in \mathbb{N}$. Now, from the well known Fatou lemma it follows that $\int_0^1 |f(x)|^2 \, dx \leq 1$, i.e., $f \in B$.

Let us suppose that $\overset{\circ}{B} \neq \emptyset$, i.e., $\overset{\circ}{B} \neq \emptyset$ in $(L_1[0,1], \|\cdot\|_1)$. Then there are $f_0 \in B, r > 0$ such that $B_{L_1[0,1]}(f_0, r) \subseteq B$. Let $f \in L_1[0,1], f \neq 0$. If we consider

$$g = \frac{f}{\|f\|_1} \frac{r}{2} + f_0,$$

it follows that $g \in B_{L_1[0,1]}(f_0, r)$, so $g \in B$, i.e., $g \in L_2[0,1]$, and then $f \in L_2[0,1]$. In this way we obtain that $L_2[0,1] = L_1[0,1]$, which is not true (see exercise 4(ii)).

So B is closed with empty interior, and therefore B is nowhere dense. Then nB is nowhere dense, $\forall n \in \mathbb{N}$. Since $L_2[0,1] = \bigcup_{n \in \mathbb{N}} nB$ it follows that $L_2[0,1]$ is of the first Baire category in $L_1[0,1]$.

ii) Let us observe that $g_n \in L_\infty[0,1] \, \forall n \in \mathbb{N}$. So the functional $\varphi_n : L_1[0,1] \to \mathbb{K}$, $\varphi_n(f) = \int_0^1 f(x) g_n(x) dx$ is linear and continuous, with $\|\varphi_n\| = \|g_n\|_\infty = n \, \forall n \in \mathbb{N}$. For

$f \in L_2[0,1]$ we have:

$$\left|\int_0^1 f(x)g_n(x)dx\right| \leq \left(\int_0^1 |f(x)|^2\, dx\right)^{1/2} \left(\int_0^1 |g_n(x)|^2\, dx\right)^{1/2} = \frac{\|f\|_2}{\sqrt{n}} \quad \forall n \in \mathbb{N},$$

and then $\lim_{n\to\infty} \varphi_n(f) = 0 \; \forall f \in L_2[0,1] \subseteq L_1[0,1]$.

If $L_2[0,1]$ were of the second Baire category in $L_1[0,1]$, since $\varphi_n(f) \to 0 \; \forall f \in L_2[0,1] \subseteq L_1[0,1]$ then the Uniform Boundedness Principle (see the proof for the Uniform Boundedness Principle) implies that $\varphi_n(f) \to 0 \; \forall f \in L_1[0,1]$. Then the same theorem implies that there is an $M > 0$ such that $\|\varphi_n\| \leq M \; \forall n \in \mathbb{N}$, i.e., $n = \|g_n\|_\infty \leq M \; \forall n \in \mathbb{N}$, which is impossible.

iii) See the part (c) from exercise 4(ii).

6. Let $U : C[0,1] \to C[0,1]$ be the multiplication operator given by $\varphi : Uf = f\varphi$ $\forall f \in C[0,1]$. Then U is linear and continuous. If U is surjective then for the constant function $g(x) = 1 \in C[0,1]$ there is $f \in C[0,1]$ such that $U(f) = g \Leftrightarrow f\varphi = g \Leftrightarrow f(x)\varphi(x) = g(x) = 1 \; \forall x \in [0,1] \Rightarrow \varphi(x) \neq 0 \; \forall x \in [0,1]$. Conversely, if $\varphi(x) \neq 0$ $\forall x \in [0,1]$, then $\forall f \in C[0,1] \; \exists f/\varphi \in C[0,1]$ such that $U(f/\varphi) = f$. Hence U is surjective $\Leftrightarrow \varphi(x) \neq 0 \; \forall x \in [0,1]$.

Since $C[0,1]$ is a Banach space and U linear and continuous, from the open mapping theorem it follows that $U(C[0,1])$ is either of the first Baire category, or U is surjective. As $A = U(C[0,1])$, it follows that:

1) A is of the first Baire category $\Leftrightarrow \exists x \in [0,1]$ such that $\varphi(x) = 0$.
2) A is of the second Baire category $\Leftrightarrow \varphi(x) \neq 0 \; \forall x \in [0,1]$.

Let us observe that in the first case, $A \subseteq \{f \in C[0,1] \mid f(x) = 0\}$, and the last set is a proper closed subspace in $C[0,1]$, hence by exercise 15 of chapter 1 or by exercise 10(i), A is nowhere dense.

7. i) Let $U : C[-1,1] \to C[-1,1]$ be defined by $(Uf)(x) = (f(x) - f(-x))/2$ $\forall f \in C[-1,1]$. At exercise 16 of chapter 5 is proved that U is a linear and continuous operator. We have $\ker(U) = A$. Then $A \subseteq C[-1,1]$ is a closed linear subspace. In fact, it is easy to see that $A \subseteq C[-1,1]$ is a proper closed linear subspace. Now, exercise 15 of chapter 1 shows that A is a nowhere dense set.

Let $V : C[-1,1] \to C[-1,1]$ be defined by $(Vf)(x) = (f(x) + f(-x))/2$. Then as above we obtain that B is a nowhere dense set.

Since for $f \in C[-1,1]$ we have $f = Uf + Vf \in A + B$, it follows that $A + B = C[-1,1]$, and $C[-1,1]$ being a Banach space, is of the second Baire category, i.e., $A + B$ is of the second Baire category. However, A and B are nowhere dense sets, in particular of the first Baire category.

ii) For $x \in X$ we have $x = Px + (x - Px)$ with $Px \in P(X)$. Since P is a linear projector we have $(x - Px) \in \ker P$, i.e., $X = \ker P + P(X)$. Also, if $x \in \ker P \cap P(X)$ then $Px = 0$ and there is $y \in X$ such that $Py = x$. Then $P^2y = Px$, $Py = 0$, so $x = 0$. Since P is linear and $P^2 = P$, it follows that $P(X) = \ker(I - P)$. By the continuity of P it follows that $\ker P$ and $P(X)$ are closed. We have $\ker P \neq X$, $P(X) \neq X$, and using exercise 15 of chapter 1 or exercise 10(i) below we obtain that $\ker P$ and $P(X)$ are nowhere dense sets.

9.2 Solutions

8. We will prove that $X = \bigcup_{n \in \mathbb{N}} nF$. Indeed, let $x \in X$. By our hypothesis, there is $\varepsilon > 0$ such that $tx \in F \ \forall t \in [0, \varepsilon]$. Since $1/n \to 0$ and $\varepsilon > 0$ we can find $n_\varepsilon \in \mathbb{N}$ such that $1/n_\varepsilon < \varepsilon$. Then $x/n_\varepsilon \in F$, i.e., $x \in n_\varepsilon F$, hence $x \in \bigcup_{n \in \mathbb{N}} nF$. Since X is Banach space from the Baire category theorem we obtain that there is $n_0 \in \mathbb{N}$ such that $\overset{\circ}{\overline{n_0 F}} \neq \varnothing$, from whence it follows that $\overset{\circ}{\overline{F}} \neq \varnothing$ (we use the fact that for $\lambda \neq 0$ the operator $x \to \lambda x$ from X to X is a homeomorphism). From exercise 2 there are $x_0 \in X$ and $r > 0$ such that $B(x_0, r) \subseteq F'$. In the case when F is closed we have $F' \subseteq F$, hence there is an open ball with centre at $x_0 \in X$ and radius $r > 0$ included in F.

9. Suppose that X is an infinite-dimensional Banach space. Let $A \subseteq X$ be a totally bounded set, i.e., $B := \overline{A} \subseteq X$ is compact. If A is not a nowhere dense set, then $\overset{\circ}{\overline{A}} \neq \varnothing$, i.e., $\overset{\circ}{B} \neq \varnothing$, from whence it follows that there are $x \in X$ and $\varepsilon > 0$ such that $\overline{B}(x, \varepsilon) \subseteq B$, and then $B_X \subseteq (1/\varepsilon)(-x + B)$. Since B is compact, $(1/\varepsilon)(-x + B)$ is also compact. Hence, the closed unit ball B_X is compact, which by the Riesz theorem (see exercise 23 of chapter 4) implies that X is finite-dimensional, and this contradicts our hypothesis.

Conversely, suppose that any totally bounded set is nowhere dense. If X is finite-dimensional, then the closed unit ball B_X is compact. Then by hypothesis B_X is nowhere dense, i.e., $\overset{\circ}{B_X} = \varnothing$. But $\overset{\circ}{B_X} = \{x \in X |\ \|x\| < 1\}$, which is a non-empty set, and we obtain a contradiction.

10. See [35, exercise F, chapter 3, section 9] or [20, exercise 30, chapter II].

i) Let $Y \subseteq X$ be a proper closed linear subspace. If Y is not nowhere dense, we have that $\overset{\circ}{\overline{Y}} \neq \varnothing$. Since Y is closed, $\overset{\circ}{Y} \neq \varnothing$, i.e., Y contains an open ball. Now exercise 15 of chapter 1 shows that $X = Y$, a contradiction.

ii) Let $X = \bigcup_{n \in \mathbb{N}} Y_n$, with Y_n finite-dimensional for any $n \in \mathbb{N}$. Every finite-dimensional space is closed. Since X is infinite-dimensional, $X \neq Y_n \ \forall n \in \mathbb{N}$. By (i) it follows that Y_n is nowhere dense $\forall n \in \mathbb{N}$, and hence X is of the first Baire category.

iii) Let X be a Banach space which has a countable algebraic basis, denoted by $\{e_n \mid n \in \mathbb{N}\}$. For $n \in \mathbb{N}$, $i_1, ..., i_n \in \mathbb{N}$ distinct, we denote by $Y_{i_1,...,i_n} = \mathrm{Sp}\{e_{i_1}, ..., e_{i_n}\}$. Then we have that $X = \bigcup_{n \in \mathbb{N}} \bigcup_{i_1,...,i_n \in \mathbb{N}} Y_{i_1, i_2,...,i_n}$. Now if we observe that the above union is countable we can apply (ii) and we obtain that X is of the first Baire category, which contradicts the fact that X is a Banach space.

iv) Let $e_n = (0, ..., 0, 1, 0, ...) \in c_{00}$ (1 on the nth position). If $x = (x_n)_{n \in \mathbb{N}} \in c_{00}$, then there is $k \in \mathbb{N}$ such that $x_n = 0 \ \forall n \geq k + 1$, and this shows that $x = \sum_{i=1}^{k} x_i e_i$. Also, it is obvious that the system $\{e_n \mid n \in \mathbb{N}\}$ is linearly independent and hence $\{e_n \mid n \in \mathbb{N}\}$ is an algebraic basis. Now we can apply (iii).

11. See [45, exercise 13, chapter 5].

i) For $n, k \in \mathbb{N}$ let $A_{nk} = \{x \in X \mid |f_n(x)| \leq k\}$. Let $A_k = \bigcap_{n \in \mathbb{N}} A_{nk} \ \forall k \in \mathbb{N}$ and $A = \bigcup_{k \in \mathbb{N}} A_k$. We observe that since the functions f_n are continuous, the sets A_{nk} are closed, and then the sets A_k are closed. Now, if $x \in X$ then the sequence $(f_n(x))_{n \in \mathbb{N}}$ is by hypothesis convergent, hence bounded, i.e., there is $k \in \mathbb{N}$ such that $|f_n(x)| \leq k \; \forall n \in \mathbb{N}$, i.e., $x \in A_k$, hence $x \in A$. It follows that $A = X$, that is, $X = \bigcup_{k \in \mathbb{N}} A_k$. Since X is complete and the sets A_k are closed, from the Baire category theorem it follows that there is $k \in \mathbb{N}$ such that $\overset{\circ}{A_k} \neq \emptyset$, i.e., there is an open and non-empty set $V \subseteq X$ such that $V \subseteq A_k$, i.e., $|f_n(x)| \leq k \; \forall x \in V, \forall n \in \mathbb{N}$.

ii) For $k \in \mathbb{N}$, let $A_k = \{x \in X \mid |f_n(x) - f_m(x)| \leq \varepsilon \; \forall m, n \geq k\}$, A_k being a closed set. Since the sequence $(f_n(x))_{n \in \mathbb{N}}$ converges it is Cauchy, and then it follows that $X = \bigcup_{k \in \mathbb{N}} A_k$. Applying again the Baire category theorem, we can find an open and non-empty set $V \subseteq X$ and $p \in \mathbb{N}$ such that $|f_n(x) - f_m(x)| \leq \varepsilon \; \forall x \in V, \forall n, m \geq p$. For $m \to \infty$ it follows that $|f(x) - f_n(x)| \leq \varepsilon \; \forall x \in V, \forall n \geq p$.

12. Let $\varepsilon \in (0, 1)$. For $n \in \mathbb{N}$ let $A_n \subseteq [1, \infty)$, $A_n = \bigcap_{m \geq n} F_m$, where $F_m = \{x \in [1, \infty) \mid |f(a_m x)| \leq \varepsilon\}$. Since f is continuous, F_m is closed $\forall m \in \mathbb{N}$, hence A_n is closed. Let $x \in [1, \infty)$. Since by our hypothesis $\lim_{n \to \infty} f(a_n x) = 0$, there is $k \in \mathbb{N}$ such that $|f(a_m x)| \leq \varepsilon \; \forall m \geq k$, i.e., $x \in A_k$. It follows that $[1, \infty) = \bigcup_{n \in \mathbb{N}} A_n$. From the Baire category theorem it follows that there is $n_\varepsilon \in \mathbb{N}$ such that $\overset{\circ}{A_{n_\varepsilon}} \neq \emptyset$, hence we can find $1 \leq a_\varepsilon < b_\varepsilon$ such that $[a_\varepsilon, b_\varepsilon] \subseteq A_{n_\varepsilon}$. Then $|f(a_m x)| \leq \varepsilon \; \forall m \geq n_\varepsilon, \forall x \in [a_\varepsilon, b_\varepsilon]$.

Since $\lim_{p \to \infty} a_{p+1}/a_p = 1$ and $a_\varepsilon < b_\varepsilon$, there is $p_\varepsilon \geq n_\varepsilon$ such that $a_{p+1}/a_p < b_\varepsilon/a_\varepsilon$ $\forall p \geq p_\varepsilon$, i.e., $a_{p+1} a_\varepsilon < a_p b_\varepsilon \; \forall p \geq p_\varepsilon$. We prove now that from this it follows that $(a_\varepsilon a_{p_\varepsilon}, \infty) \subseteq \bigcup_{p \geq p_\varepsilon} (a_p a_\varepsilon, a_p b_\varepsilon)$. Indeed, if $x > a_\varepsilon a_{p_\varepsilon}$ then $x/a_\varepsilon > a_{p_\varepsilon}$. Since the sequence $(a_p)_{p \in \mathbb{N}}$ is strictly increasing with limit ∞, there is $p \in \mathbb{N}$ such that $a_p < x/a_\varepsilon \leq a_{p+1}$. So $a_{p+1} > a_{p_\varepsilon}$, and since $(a_p)_{p \in \mathbb{N}}$ is strictly increasing we have $p + 1 > p_\varepsilon$, i.e., $p \geq p_\varepsilon$, and then $a_p < x/a_\varepsilon \leq a_{p+1} < a_p b_\varepsilon/a_\varepsilon$, i.e., $a_p a_\varepsilon < x < a_p b_\varepsilon$.

Let $M_\varepsilon = a_\varepsilon a_{p_\varepsilon}$. Then by the above proved relation, $\forall x > M_\varepsilon \; \exists k \geq p_\varepsilon$ such that $x = a_k y$, with $y \in (a_\varepsilon, b_\varepsilon)$. Hence $k \geq n_\varepsilon$ and $y \in (a_\varepsilon, b_\varepsilon)$. We have $|f(a_k y)| \leq \varepsilon$, i.e.. $|f(x)| \leq \varepsilon$. In conclusion $\forall \varepsilon > 0 \; \exists M_\varepsilon > 0$ such that $|f(x)| \leq \varepsilon \; \forall x > M_\varepsilon$, that is, $\lim_{x \to \infty} f(x) = 0$.

13. See [45, exercise 14, chapter 5].

i) We prove first that X_n is closed, for every n. Let $(f_k)_{k \in \mathbb{N}} \subseteq X_n$ such that $f_k \to f \in C[0, 1]$. We can find then $(t_k)_{k \in \mathbb{N}} \subseteq [0, 1]$ such that for every k, $|f_k(s) - f_k(t_k)| \leq n|s - t_k| \; \forall s \in [0, 1]$. Using Cesaro's lemma, without loss of generality we can suppose that $t_k \to t \in [0, 1]$. Then for $k \to \infty$ in the above inequality we obtain that $|f(s) - f(t)| \leq n|s - t| \; \forall s \in [0, 1]$, i.e., $f \in X_n$. (We use the fact that $f_k \to f$ uniformly on $[0, 1]$ to obtain that $f_k(t_k) \to f(t)$.)

Let $n \in \mathbb{N}$ and $D \subseteq C[0, 1]$ be an open and non-empty set. Let $f \in D$ and $r > 0$ be such that $B(f, 2r) \subseteq D$. The idea is that we can uniformly approximate the function f with zig-zag functions, the zig-zags being of arbitrarily large slope. Since f is continuous on the compact $[0, 1]$, it is uniformly continuous, hence for $r/8 > 0$ there is $\delta > 0$ such that: $|f(t + h) - f(t)| < r/8 \; \forall |h| < \delta, \; \forall t \in [0, 1]$ with $t + h \in [0, 1]$. Then for any

9.2 Solutions

$t \in [0,1]$ and any $|h| < \delta$ with $t + h \in [0,1]$ it follows that

$$\left| f(t+h) + \frac{r}{4} - f(t) \right| \geq \left| \frac{r}{4} - |f(t+h) - f(t)| \right| \geq \frac{r}{8}$$

and hence there is $\varepsilon \in (0, \delta)$ such that

$$\frac{|f(t+h) + r/4 - f(t)|}{\varepsilon} > n + 4 \quad \forall |h| < \delta, \ \forall t \in [0,1] \text{ with } t + h \in [0,1].$$

Then $\forall t \in [0,1]$ and $\forall h$ with $|h| \in (0, \varepsilon)$ and $t + h \in [0,1]$ we have that

$$\frac{|f(t+h) + r/4 - f(t)|}{|h|} > n + 4.$$

Let now $t \in [0, 1)$ and let $s = t + \varepsilon$. Let us suppose that $s \in [0,1]$. Define then $g : [t, s] \to \mathbb{R}$ by

$$g(x) = \begin{cases} f(t) + (x - t) \frac{f((t+s)/2) + r/4 - f(t)}{(s-t)/2} & x \in [t, (t+s)/2], \\ f\left(\frac{t+s}{2}\right) + \frac{r}{4} + \left(x - \frac{s+t}{2}\right) \frac{f(s) - r/4 - f((t+s)/2)}{(s-t)/2} & x \in ((t+s)/2, s]. \end{cases}$$

Then g is a continuous zig-zag function and $f(t) = g(t)$, $f(s) = g(s)$. The modulus for the slope of g on $[t, (t+s)/2]$ is

$$\left| \frac{f((t+s)/2) + r/4 - f(t)}{\varepsilon/2} \right| > n + 4$$

and, in the same way, the modulus for the slope of g on $((t+s)/2, s]$ is strictly bigger than $n + 4$. For any $x \in [t, (t+s)/2]$ we have

$$\begin{aligned} |g(x) - f(x)| &\leq |g(x) - f(t)| + |f(t) - f(x)| \\ &= 2|x - t| \left| \frac{f((t+s)/2) + r/4 - f(t)}{\varepsilon} \right| + \frac{r}{8} \\ &\leq 2\frac{\varepsilon}{2} \left| \frac{f((t+s)/2) + r/4 - f(t)}{\varepsilon} \right| + \frac{r}{8} \\ &\leq \left| f\left(\frac{t+s}{2}\right) - f(t) \right| + \frac{r}{4} + \frac{r}{8} \leq r. \end{aligned}$$

Analogously, $|g(x) - f(x)| \leq r \ \forall x \in ((t+s)/2, s]$.

The construction we will make is now clear. We begin with $t = 0$ and construct the function g from 0 to ε. After that we continue with $t = \varepsilon$ and construct the function g from ε to 2ε, etc. (we consider, if necessary, $f(t) = f(1)$ for $t \geq 1$). We obtain a function $g \in C[0,1]$ with $\|f - g\| \leq r$, g being a zig-zag function, and the modulus for the slopes of g being always strictly bigger than $n + 4$. Let $\tau > 0$ with $\tau < \min(r, \varepsilon/2)$ and let us consider $B(g, \tau) \subseteq C[0,1]$. Obviously $B(g, \tau) \subseteq B(f, 2r) \subseteq D$. We will prove that $B(g, \tau) \cap X_n = \emptyset$. Let us suppose that there is $h \in B(g, \tau) \cap X_n$. Since $h \in X_n$ there is $t \in [0,1]$ such that $|h(s) - h(t)| \leq n|s - t| \ \forall s \in [0,1]$. From $\|h - g\| \leq \varepsilon/2$ it follows that $|h(t) - g(t)| \leq \varepsilon/2$, $|h(s) - g(s)| \leq \varepsilon/2 \ \forall s \in [0,1]$ and so: $|g(t) - g(s)| \leq \varepsilon + n|s - t|$

$\forall s \in [0,1]$ (1). For $t \in [0,1]$ there is $k \geq 0$ such that $t \in [k\varepsilon, (k+1)\varepsilon)$. Dividing the interval $[k\varepsilon, (k+1)\varepsilon)$ into two parts let us suppose, for example, that $t \in [k\varepsilon, (k+1/2)\varepsilon]$. Then $|t - k\varepsilon| > \varepsilon/4$ or $|t - (k+1/2)\varepsilon| > \varepsilon/4$, and suppose, for example, that $|t - k\varepsilon| > \varepsilon/4$. For $s = k\varepsilon$ in the relation (1) we obtain that

$$|g(k\varepsilon) - g(t)| \leq \varepsilon + n|k\varepsilon - t| < 4|k\varepsilon - t| + n|k\varepsilon - t| = (n+4)|k\varepsilon - t|,$$

which contradicts the choice of g.

We have proved that $\forall D \subseteq C[0,1]$ open and $\forall n \in \mathbb{N}$, $\exists B(g,\tau) \subseteq D$ such that $B(g,\tau) \cap X_n = \emptyset$. Since X_n is closed $\forall n \in \mathbb{N}$ this easily implies that X_n is a nowhere dense set, $\forall n \in \mathbb{N}$.

ii) Since $C[0,1]$ is separable, we can find $(f_n)_{n\in\mathbb{N}}$, a dense sequence in $C[0,1]$. For $m, n, k \in \mathbb{N}$, applying to the set $B(f_n, 1/k)$ the results form (i), we obtain that there is an open set $D_{nk}^m \subseteq B(f_n, 1/k)$ with $D_{nk}^m \cap X_m = \emptyset$. Let $A_m = \bigcup_{n,k \geq 1} D_{nk}^m$ $\forall m \in \mathbb{N}$ and $G = \bigcap_{m \geq 1} A_m$. Obviously, G is a G_δ set. We will prove that $A_m \subseteq C[0,1]$ is dense $\forall m \in \mathbb{N}$, and then by Baire's category theorem it follows that $G \subseteq C[0,1]$ is dense. Let $f \in C[0,1]$, $m \in \mathbb{N}$ and $\varepsilon > 0$. There is $n \in \mathbb{N}$ such that $\|f - f_n\| < \varepsilon/2$. Let $k \in \mathbb{N}$ with $1/k < \varepsilon/2$. Since $D_{nk}^m \subseteq B(f_n, 1/k)$, choosing $g \in D_{nk}^m$, we have that $\|g - f_n\| < 1/k$. Then $\|f - g\| < \varepsilon$ and $g \in D_{nk}^m \subseteq A_m$.

Let us suppose that we can find $f \in G$ and $t_0 \in [0,1]$ such that f is differentiable at t_0. Then

$$\lim_{t \to t_0} \frac{f(t) - f(t_0)}{t - t_0} = f'(t_0),$$

so there are $\varepsilon > 0$ and $M > 0$ such that $|f(t) - f(t_0)| < M|t - t_0|$ $\forall t \in [0,1]$ with $|t - t_0| < \varepsilon$. If we consider

$$K = \sup_{t \in [0,1], |t - t_0| \geq \varepsilon} \left|\frac{f(t) - f(t_0)}{t - t_0}\right|$$

and $m \in \mathbb{N}$ with $m > \max(K, M)$, then $|f(t) - f(t_0)| \leq m|t - t_0|$ $\forall t \in [0,1]$, so $f \in X_m$. But $f \in G = \bigcap_{p \geq 1} A_p$, hence $f \in X_m \cap A_m$. It follows that there are n and k such that $f \in D_{nk}^m$, i.e., $D_{nk}^m \cap X_m \neq \emptyset$, which is not true, by our construction.

Remark. This exercise, known as the Banach theorem, shows that the set of all the continuous functions on $[0,1]$ which are not differentiable at any point of $[0,1]$ is a very large one. However, when we try to construct explicitly such a function, this is a difficult thing to do.

14. Let us suppose that the conclusion from the exercise is not true, i.e., $\forall x \in \mathbb{R} \backslash A$ $\exists n \in \mathbb{N}$ such that $f_n(x) \neq 0$. Let $E_n = \{x \in \mathbb{R}\backslash A \mid f_n(x) \neq 0\}$. Then $\mathbb{R}\backslash A = \bigcup_{n \in \mathbb{N}} E_n$. Since

$$E_n = \{x \in \mathbb{R}\backslash A \mid |f_n(x)| > 0\} = \bigcup_{k \in \mathbb{N}} \{x \in \mathbb{R}\backslash A \mid |f_n(x)| \geq 1/k\},$$

it follows that $\mathbb{R}\backslash A = \bigcup_{n,k \in \mathbb{N}} A_{n,k}$, where $A_{n,k} = \{x \in \mathbb{R}\backslash A \mid |f_n(x)| \geq 1/k\}$. Since $A \subseteq \mathbb{R}$ is a set of first Baire category, there is a sequence of nowhere dense sets $(B_p)_{p \in \mathbb{N}}$ such that

9.2 Solutions

$A = \bigcup_{p \in \mathbb{N}} B_p$. Then $\mathbb{R} = \bigcup_{p \in \mathbb{N}} B_p \cup \bigcup_{n,k \in \mathbb{N}} A_{n,k}$. Since $\mathbb{R}$ is of the second Baire category, it follows that there are $n_0, k_0 \in \mathbb{N}$ such that $\overset{\circ}{\overline{A_{n_0,k_0}}} \neq \emptyset$. Using now exercise 2, there are $x \in \mathbb{R}$ and $\varepsilon > 0$ such that $(x - \varepsilon, x + \varepsilon) \subseteq A'_{n_0,k_0}$. In particular, $x \in A'_{n_0,k_0}$, i.e., there is $(t_p)_{p \in \mathbb{N}} \subseteq A_{n_0,k_0}$ with $t_p \neq x \ \forall p \in \mathbb{N}$ and $t_p \to x$. However, by hypothesis, $\lim_{t \to x} f_{n_0}(t) = 0$, and it follows that $\lim_{p \to \infty} f_{n_0}(t_p) = 0$, which is contradictory, since $|f_{n_0}(t_p)| \geq 1/k_0 \ \forall p \in \mathbb{N}$.

15. i) Let us suppose that such a function exists. Since f takes finite values, $\forall x \in X$ $\exists k \in \mathbb{N}$ such that $|f(t)| \leq k$, i.e., $X = \bigcup_{k \in \mathbb{N}} A_k$, where $A_k = \{x \in \mathbb{R} \mid |f(t)| \leq k\}$. Since X is of the second Baire category, $\exists k \in \mathbb{N}$ such that $\overset{\circ}{\overline{A_k}} \neq \emptyset$, and using exercise 2 it follows that there are $x \in X$ and $\varepsilon > 0$ such that $B(x, \varepsilon) \subseteq A'_k$. Since A is dense, we have $B(x, \varepsilon) \cap A \neq \emptyset$, so $\emptyset \neq B(x, \varepsilon) \cap A \subseteq A'_k \cap A$, i.e., there is $a \in A$ such that $a \in A'_k$. Using now the characterization for the accumulation points in metric spaces, $\exists (x_n)_{n \in \mathbb{N}} \subseteq A_k$, $x_n \neq a \ \forall n \in \mathbb{N}$ such that $x_n \to a$. Since $a \in A$ and by hypothesis $\lim_{t \to a} |f(t)| = \infty$, we obtain that $\lim_{n \to \infty} |f(x_n)| = \infty$, which is not possible, since $(x_n)_{n \in \mathbb{N}} \subseteq A_k$, $|f(x_n)| \leq k$ $\forall n \in \mathbb{N}$, i.e., the sequence $(f(x_n))_{n \in \mathbb{N}}$ is bounded.

ii) See [45, exercise 21 (c), chapter 5].

Since $\mathbb{Q} \subseteq \mathbb{R}$ is dense, from (i) it follows that we cannot find such a function.

16. i) Let us suppose, for a contradiction, that such a sequence of functions exists. The hypothesis is that $A = \{x \in X \mid (f_n(x))_{n \in \mathbb{N}} \text{ is unbounded}\}$. Then $X \backslash A = \{x \in X \mid (f_n(x))_{n \in \mathbb{N}} \text{ is bounded}\} = \bigcup_{k \in \mathbb{N}} \{x \in X \mid |f_n(x)| \leq k \ \forall n \in \mathbb{N}\}$. Let $A_k = \{x \in X \mid |f_n(x)| \leq k \ \forall n \in \mathbb{N}\}$. Since the functions f_n are continuous, the set $\{x \in X \mid |f_n(x)| \leq k\}$ is closed, and then $A_k = \bigcap_{n \in \mathbb{N}} \{x \in X \mid |f_n(x)| \leq k\}$ is also closed. We have $X = A \cup (X \backslash A) = A \cup \bigcup_{k \in \mathbb{N}} A_k$. Since A is of the first Baire category, then $A = \bigcup_{k \in \mathbb{N}} B_k$, B_k nowhere dense $\forall k \in \mathbb{N}$, and so $X = \bigcup_{k \in \mathbb{N}} B_k \cup \bigcup_{k \in \mathbb{N}} A_k$. Since X is a complete metric space from the Baire category theorem we know that there is a $k \in \mathbb{N}$ such that $\overset{\circ}{\overline{A_k}} \neq \emptyset$. The set A_k being closed it follows that there are $x \in X$ and $\varepsilon > 0$ such that $B(x, \varepsilon) \subseteq A_k$. Since A is dense in X it follows that $B(x, \varepsilon) \cap A \neq \emptyset$. This is false, since $B(x, \varepsilon) \subseteq A_k \subseteq X \backslash A$.

ii) See [45, exercise 21 (a), chapter 5].

$\mathbb{Q}$ is dense and of the first Baire category in $\mathbb{R}$, hence by (i) we cannot find such a sequence.

iii) $l_1 \subseteq c_0$ is of the first Baire category (see exercise 4(ii)) and l_1 is dense in c_0. Hence, we cannot find such a sequence.

iv) $\text{Lip}[0, 1] \subseteq C[0, 1]$ is of the first Baire category (see exercise 4(ii)), and dense. Hence, by (i), we cannot find such a sequence.

17. i) Let us observe that if $h : X \to Y$ is a homeomorphism of topological spaces then $h(\overline{A}) = \overline{h(A)}$ and $h(\overset{\circ}{A}) = \overset{\circ}{\widetilde{h(A)}}$, so if A is nowhere dense (respectively, of the first Baire category), then $h(A)$ is nowhere dense (respectively, of the first Baire category).

Let $E_n = \{x \in X \mid f(x, x_n) \in A\} = h_n^{-1}(A)$, where $h_n : X \to X$ is defined by $h_n(x) = f(x, x_n)$, which by hypothesis is a homeomorphism of topological spaces. Hence, E_n is of the first Baire category. Let us suppose, for a contradiction, that the conclusion from the exercise is not true, i.e., $\forall x \in X \, \exists n \in \mathbb{N}$ such that $f(x, x_n) \in A$. Then $X = \bigcup_{n \in \mathbb{N}} E_n$. It follows that X is of the first Baire category, which contradicts the Baire theorem, X being a complete metric space.

If X is a normed space and $f : X \times X \to X$ is defined by $f(x, y) = x + y$, then for any $a \in X$, the translation $f(\cdot, a) = T_a : X \to X$ defined by $T_a(x) = x + a$ is a homeomorphism, so in this case we can apply (i) obtaining the following fact:

(i') Let X be a Banach space, $A \subseteq X$ a first Baire category set, and $(x_n)_{n \in \mathbb{N}} \subseteq X$. Then there is $x \in X$ such that $\forall n \in \mathbb{N}$, $x + x_n \notin A$.

ii) From exercise 9(iii) of chapter 12 the set $\mathcal{A}(H) = \{U \in L(H) \mid U \text{ self-adjoint}\}$ is nowhere dense in $L(H)$, so by (i') we obtain the statement.

iii) From exercise 4(ii), $\text{Lip}[0, 1] \subseteq C[0, 1]$ is of the first Baire category, and we can use again (i') to obtain the statement.

iv) From exercise 7(i), $A = \{f \in C[-1, 1] \mid f \text{ odd}\} \subseteq C[-1, 1]$ is nowhere dense, and we use (i').

v) From exercise 5, $L_2[0, 1] \subseteq L_1[0, 1]$ is of the first Baire category, hence by (i') it follows that there is $f \in L_1[0, 1]$ such that $f + f_n \notin L_2[0, 1] \, \forall n \in \mathbb{N}$, i.e.,
$$\int_0^1 |f(x)| \, dx < \infty,$$
but
$$\int_0^1 |f(x) + f_n(x)|^2 \, dx = \infty \, \forall n \in \mathbb{N}.$$

vi) It follows from exercise 13 and (i').

18. See [10, exercise 26, chapter VI, section 1].

Let us suppose, for a contradiction, that the conclusion is not true, i.e., $\forall (a_n)_{n \in \mathbb{N}} \subseteq \mathbb{K}$ such that $\sum_{n=1}^{\infty} |a_n| < \infty$, the function $\sum_{n=1}^{\infty} a_n f_n$ is μ-integrable (the series $\sum_{n=1}^{\infty} a_n f_n(s)$ converges for μ-almost every s since $\sum_{n=1}^{\infty} |a_n| < \infty$, and for μ-almost every $s \in S$ we have $\sup_{n \in \mathbb{N}} |f_n(s)| < \infty$, and then, if we denote $f(s) = \sum_{n=1}^{\infty} a_n f_n(s)$, we have that f is defined almost everywhere, is measurable, and $\int_S |f(s)| \, d\mu(s) < \infty$). Let us consider the operator $U : l_1 \to L_1(\mu)$ defined by $U((a_n)_{n \in \mathbb{N}}) = \sum_{n=1}^{\infty} a_n f_n$. From the above, U is well defined. Obviously, U is linear. We will prove that U has its graph closed. Let $(a_n^k)_{n \in \mathbb{N}} \to (a_n)_{n \in \mathbb{N}}$ in l_1 and $U((a_n^k)_{n \in \mathbb{N}}) \to g$ in $L_1(\mu)$. From measure theory we know that there is a subsequence $(k_p)_{p \in \mathbb{N}}$ of $\mathbb{N}$ such that $U((a_n^{k_p})_{n \in \mathbb{N}}) \to g$, μ-a.e., i.e., using the definition for U, $\sum_{n=1}^{\infty} a_n^{k_p} f_n(s) \to g(s)$, μ-a.e.. However, if $(\xi_n)_{n \in \mathbb{N}} \in l_{\infty}$ then

$$\left| \sum_{n=1}^{\infty} a_n^k \xi_n - \sum_{n=1}^{\infty} a_n \xi_n \right| \leq \sum_{n=1}^{\infty} |a_n^k - a_n| |\xi_n| \leq \left\| (a_n^k - a_n)_{n \in \mathbb{N}} \right\|_1 \left\| (\xi_n)_{n \in \mathbb{N}} \right\|_{\infty} \to 0,$$

9.2 Solutions

i.e., $\sum_{n=1}^{\infty} a_n^k \xi_n - \sum_{n=1}^{\infty} a_n \xi_n \to 0$. Since by hypothesis for μ-almost all $s \in S$, $(f_n(s))_{n \in \mathbb{N}} \in l_\infty$, it follows that $\sum_{n=1}^{\infty} a_n^k f_n(s) - \sum_{n=1}^{\infty} a_n f_n(s) \to 0$, μ-a.e.. But $\sum_{n=1}^{\infty} a_n^{k_p} f_n(s) \to g(s)$, μ-a.e., so $g(s) = \sum_{n=1}^{\infty} a_n f_n(s)$, μ-a.e., that is, $g = U((a_n)_{n \in \mathbb{N}})$. Now from the closed graph theorem it follows that U is continuous, i.e.,

$$\|U((a_n)_{n \in \mathbb{N}})\| \leq \|U\| \|(a_n)_{n \in \mathbb{N}}\| \quad \forall (a_n)_{n \in \mathbb{N}} \in l_1.$$

For the canonical basis in l_1, we obtain that $\|U(e_n)\| \leq \|U\| \ \forall n \in \mathbb{N}$. $U(e_n) = f_n$, and therefore $\int_S |f_n(s)| d\mu(s) \leq \|U\| \ \forall n \in \mathbb{N}$, i.e., $\sup_{n \in \mathbb{N}} \int_S |f_n(s)| d\mu(s) \leq \|U\| < \infty$, which contradicts our hypothesis.

19. i) Let $V : G \to Y$ be the restriction of U. Then V is linear and continuous. Since G is a closed linear subspace and X is a Banach space, it follows that G is a Banach space. From the open mapping theorem, it follows that $V(G) = U(G)$ is either of the first Baire category, or $U(G) = V(G) = Y$.

ii) Let us calculate $T^{-1}(U(Y))$. We have $x \in T^{-1}(U(Y)) \Leftrightarrow Tx \in U(Y) \Leftrightarrow \exists y \in Y$ such that $Tx = Uy$. Let $H = \{(x, y) \in X \times Y \mid Tx = Uy\} \subseteq X \times Y$. Since U and T are continuous operators, it follows that H is closed in $X \times Y$. Now the linearity of U and T implies that H is a linear subspace. Hence $H \subseteq X \times Y$ is a closed linear subspace. Since X and Y are Banach spaces, $X \times Y$ is a Banach space, and since H is a closed set it follows that H is complete, i.e., H is a Banach space. Let $p : X \times Y \to X$ be the canonical projection, $p(x, y) = x$, which is linear and continuous, hence the restriction $p : H \to X$ is also linear and continuous. From the open mapping theorem, or using (i), it follows that we have: either $p(H)$ is of the first Baire category, or $p(H) = X$ (i.e., p is surjective). Now, $x \in p(H) \Leftrightarrow \exists (a, y) \in H$ such that $p(a, y) = x \Leftrightarrow \exists y \in Y$ such that $(x, y) \in H \Leftrightarrow \exists y \in Y$ such that $Tx = Uy \Leftrightarrow x \in T^{-1}(U(Y))$, i.e., $p(H) = T^{-1}(U(Y))$. So, either $T^{-1}(U(Y))$ is of the first Baire category, or $T^{-1}(U(Y)) = X$. We have: $X = T^{-1}(U(Y)) \Leftrightarrow T(X) \subseteq U(Y)$. Indeed, suppose that $X = T^{-1}(U(Y))$, and let $y \in T(X)$. Then there is $x \in X$ such that $y = Tx$. But $X = T^{-1}(U(Y))$, from whence $x \in T^{-1}(U(Y))$, i.e., $Tx \in U(Y)$, $y \in U(Y)$. The converse can be proved in the same way.

Changing the role of U and T we obtain that either $U^{-1}(T(X))$ is of the first Baire category, or $U(Y) \subseteq T(X)$. Now if $T^{-1}(U(Y))$ is of the second Baire category, and $U^{-1}(T(X))$ is of the second Baire category, we must have $T(X) \subseteq U(Y)$ and $U(Y) \subseteq T(X)$, i.e., $U(Y) = T(X)$.

20. i) See [9, exercise 6, chapter I, section 3].

If for every $n \in \mathbb{N}$, $U_n(X_n)$ is of the first Baire category, in this case it follows that $\bigcup_{n \in \mathbb{N}} U_n(X_n)$ is of the first Baire category. Since by hypothesis $X = \bigcup_{n \in \mathbb{N}} U_n(X_n)$, we contradict the Baire category theorem, X being a Banach space. Therefore we can find $n_0 \in \mathbb{N}$ such that $U_{n_0}(X_{n_0})$ is of the second Baire category. Since $U_{n_0} : X_{n_0} \to X$ is a linear and continuous operator between two Banach spaces, from the open mapping theorem it follows that U_{n_0} is surjective.

ii) See [9, exercise 13 (a), chapter III, section 3].

From $U(X) \subseteq \bigcup_{n\in\mathbb{N}} U_n(X_n)$ it follows that $X \subseteq \bigcup_{n\in\mathbb{N}} U^{-1}(U_n(X_n))$, i.e., $X = \bigcup_{n\in\mathbb{N}} U^{-1}(U_n(X_n))$. But it is proved at exercise 19 that if $p_n : X \times X_n \to X$ is the canonical projection, $p_n(x,y) = x$, and $H_n = \{(x,y) \in X \times X_n \mid Ux = U_n y\}$, then $p_n(H_n) = U^{-1}(U_n(X_n))$. We have the following situation: $\{H_n\}_{n\in\mathbb{N}}$ are Banach spaces, and $p_n : H_n \to X$, $n \in \mathbb{N}$, are linear and continuous operators, with $X = \bigcup_{n\in\mathbb{N}} p_n(H_n)$. From (i), it follows that there is $n \in \mathbb{N}$ such that p_n is surjective, i.e., $p_n(H_n) = X$, $U^{-1}(U_n(X_n)) = X$, that is, $U(X) \subseteq U_n(X_n)$.

iii) See [42, exercise 32, chapter III].

First solution. The hypothesis is that $S \subseteq \bigcup_{p>1} L_p[0,1]$. Since $L_p[0,1] \subseteq L_q[0,1]$ if $q \leq p$, it follows that $S \subseteq \bigcup_{p>1,\, p\in\mathbb{Q}} L_p[0,1]$. Let $U : S \to L_1[0,1]$ and $U_p : L_p[0,1] \to L_1[0,1]$ be the inclusion maps. We have

$$U(S) = S \subseteq \bigcup_{p>1,\, p\in\mathbb{Q}} L_p[0,1] = \bigcup_{p>1,\, p\in\mathbb{Q}} U_p(L_p[0,1]).$$

From (ii) it follows that there is $p > 1$, $p \in \mathbb{Q}$, such that $S \subseteq L_p[0,1]$.

Second solution. From the hypothesis we have $\forall f \in S \; \exists p > 1$ such that $f \in L_p[0,1]$, so $\forall f \in S \; \exists p > 1$ and $q \in \mathbb{Q}$ with $p > q > 1$ such that $f \in L_p[0,1]$. Since $L_p[0,1] \subseteq L_q[0,1]$ we obtain that $\forall f \in S \; \exists q \in \mathbb{Q}, q > 1$, such that $f \in L_q[0,1]$. Since there is $n \in \mathbb{N}$ with $n \geq \|f\|_q$ we obtain that $\forall f \in S \; \exists q \in \mathbb{Q}, q > 1$, and $n \in \mathbb{N}$ such that $f = n(f/n) \in nB_{L_q[0,1]}$. Therefore

$$S \subseteq \bigcup_{p>1,\, p\in\mathbb{Q}} \bigcup_{n\in\mathbb{N}} nB_{L_p[0,1]},$$

so

$$S = \bigcup_{p>1,\, p\in\mathbb{Q}} \bigcup_{n\in\mathbb{N}} \left(S \cap nB_{L_p[0,1]}\right).$$

Since S is a closed subspace in $L_1[0,1]$, S is a Banach space. From the Baire category theorem there are $p \in \mathbb{Q}$ with $p > 1$ and $n \in \mathbb{N}$ such that $\overset{\circ}{\overline{S \cap nB_{L_p[0,1]}}} \neq \emptyset$, i.e., $\overset{\circ}{\overline{S \cap B_{L_p[0,1]}}} \neq \emptyset$. But using the Fatou lemma and standard results from measure theory it follows that $B_{L_p[0,1]}$ is closed in $L_1[0,1]$ (see also the solution for exercise 5). Hence $S \cap B_{L_p[0,1]}$ is closed in $L_1[0,1]$, and therefore we can find $g \in S \cap B_{L_p[0,1]}$ and $\varepsilon > 0$ such that $B_S(g,\varepsilon) \subseteq S \cap B_{L_p[0,1]}$. Let $f \in S$, $f \neq 0$. Then

$$\frac{\varepsilon}{2\|f\|_1} f + g \in B_S(g,\varepsilon) \subseteq S \cap B_{L_p[0,1]},$$

and therefore $(\varepsilon/(2\|f\|_1))f + g \in B_{L_p[0,1]}$. Since $g \in L_p[0,1]$ it follows that $f \in L_p$. Hence $S \subseteq L_p$.

21. Suppose first that X_n are all closed. From $X = \bigcup_{n\in\mathbb{N}} X_n$, since X is a Banach space, hence of the second Baire category, we obtain that there is $n \in \mathbb{N}$ such that $\overset{\circ}{\overline{X_n}} \neq \emptyset$, that is,

9.2 Solutions

$\overset{\circ}{X}_n \ne \emptyset$, i.e., X_n is a subspace and contains a ball. From exercise 15 of chapter 1 it follows that $X_n = X$.

For the general case, when the X_n are not supposed to be closed, we have $X = \bigcup_{n \in \mathbb{N}} X_n \subseteq \bigcup_{n \in \mathbb{N}} \overline{X_n} \subseteq X$, that is, $X = \bigcup_{n \in \mathbb{N}} \overline{X_n}$. We obtain that there is $n \in \mathbb{N}$ such that $\overline{X_n} = X$.

22. See [3, theorem 2.2.2].

i) Obviously, $\mathcal{F}_n \subseteq L(X)$ is a linear subspace. Let $(T_k)_{k \in \mathbb{N}} \subseteq \mathcal{F}_n$ be such that $T_k \to T$ in $L(X)$, but $T \notin \mathcal{F}_n$. Then there are $x_1, ..., x_{n+1} \in X$ such that $\{Tx_1, ..., Tx_{n+1}\} \subseteq X$ is linearly independent. Using now the stability property for finite linearly independent systems (see exercise 35 of chapter 1), there is $\varepsilon > 0$ such that if $y_1, ..., y_{n+1} \in X$ are such that $\|y_1\|, ..., \|y_{n+1}\| < \varepsilon$, then $\{Tx_1 + y_1, ..., Tx_{n+1} + y_{n+1}\} \subseteq X$ is linearly independent. Since $T_k \to T$ in $L(X)$, there is $k \in \mathbb{N}$ such that

$$\|T_k - T\| < \frac{\varepsilon}{1 + \max\{\|x_1\|, ..., \|x_{n+1}\|\}}.$$

Then $\|(T_k - T)x_j\| < \varepsilon$, $j = \overline{1, n+1}$, and hence for $y_j = T_k x_j - T x_j$ in the above relation it follows that the system $\{T_k x_1, ..., T_k x_{n+1}\} \subseteq X$ is linearly independent, so $T_k \notin \mathcal{F}_n$, a contradiction.

ii) Let $\mathcal{F}(X)$ be the set of all finite rank operators from X into X, and $\mathcal{A}(X)$ the closure of $\mathcal{F}(X)$ in the space $L(X)$. We have that $\mathcal{A}(X) \subseteq K(X)$. Since $\mathcal{A}(X)$ is a closed linear subspace in $L(X)$, $\mathcal{A}(X)$ is a Banach space. Now, by (i), for any $n \in \mathbb{N}$, $\mathcal{F}_n$ is a closed linear subspace in $\mathcal{A}(X)$. If $\mathcal{A}(X) = \bigcup_{n=1}^{\infty} \mathcal{F}_n$ then by exercise 21 there is $n \in \mathbb{N}$ such that $\mathcal{F}_n = \mathcal{A}(X)$. Then $\mathcal{F}(X) = \mathcal{F}_n$. Since $\dim_\mathbb{K} X = \infty$, we can find $n+1$ linearly independent elements $x_1, ..., x_{n+1} \in X$, and, considering $x_1^*, ..., x_{n+1}^* \in X^*$ such that $x_j^*(x_i) = \delta_{ij}$, $i, j = 1, ..., n+1$ (see exercise 2 of chapter 7) then for the operator $T \in L(X)$, $T(x) = \sum_{k=1}^{n+1} x_k^*(x) x_k$ $\forall x \in X$, we have that $\{x_1, ..., x_{n+1}\} \subseteq T(X)$, therefore $\dim_\mathbb{K}(T(X)) > n$, a contradiction, since $\mathcal{F}(X) = \mathcal{F}_n$. So $\mathcal{A}(X) \ne \bigcup_{n=1}^{\infty} \mathcal{F}_n$, i.e., there is $T \in \mathcal{A}(X)$ such that $\forall n \in \mathbb{N}$, $T \notin \mathcal{F}_n$, i.e., T can be uniformly approximated by a sequence of finite rank operators and $\dim_\mathbb{K}(T(X)) > n$ $\forall n \in \mathbb{N}$, i.e., $\dim_\mathbb{K}(T(X)) = \infty$.

The last assertion is now clear since we have $\bigcup_{n=1}^{\infty} \mathcal{F}_n \subseteq \mathcal{A}(X) \subseteq K(X)$, and also $\bigcup_{n=1}^{\infty} \mathcal{F}_n \ne \mathcal{A}(X)$.

23. '(i) $\Rightarrow$ (ii)' If a non-null linear and continuous operator $U : X \to Y$ is open, then the set $\{U(x) \mid \|x\| < 1\}$ is open, hence there is $\delta > 0$ such that $\{y \in Y \mid \|y\| < \delta\} \subseteq \{U(x) \mid \|x\| < 1\}$. Now if $y \in Y$ and $y \ne 0$, then $\|\delta y/(2\|y\|)\| = \delta/2 < \delta$, hence by the above there is $\|x\| < 1$ such that $U(x) = \delta y/(2\|y\|)$, so by the homogeneity of U, $U((2x\|y\|)/\delta) = y$, i.e., U is surjective.

'(ii) $\Rightarrow$ (iii)' Since $Y \ne \{0\}$, there is $y_0 \in Y$ with $y_0 \ne 0$. Also, $X \ne \{0\}$ and by the Hahn–Banach theorem there is $x^* \in X^*$, $x^* \ne 0$. By hypothesis any non-null linear and continuous operator is surjective, in particular the rank one operator $U : X \to Y$ defined by $U(x) = x^*(x)y_0$ is surjective, that is, $\forall y \in Y \; \exists x \in X$ such that $U(x) = y$, that is, $x^*(x)y_0 = y$, and therefore $\dim_{\mathbb{K}} Y = 1$.

'(iii) $\Rightarrow$ (i)' First we prove that any non-null linear and continuous functional $x^* : X \to \mathbb{K}$ is open (we cannot use the open mapping theorem since X is just normed, and not necessarily complete).

Since $x^* \neq 0$, there is $x_0 \in X$ such that $x^*(x_0) \neq 0$, and let $a = x_0/x^*(x_0) \in X$. Then $x^*(a) = 1$. Since x^* is linear it follows that $\forall \lambda \in \mathbb{K} \; \exists \lambda a \in X$ such that $x^*(\lambda a) = \lambda x^*(a) = \lambda$, i.e., x^* is surjective. Let us observe that $\forall x \in X$ we have $x - ax^*(x) \in \ker x^*$ and $x = (x - ax^*(x)) + ax^*(x) \in \ker x^* \oplus \mathbb{K}a$, i.e., $X = \ker x^* \oplus \mathbb{K}a$ (direct sum). Let $D \subseteq X$ be an open set such that $0 \in D$. Then D is a neighborhood of 0. Since the mapping $p : \mathbb{K} \times X \to X$ defined by $p(\lambda, x) = \lambda x$ is continuous at $(0, a)$, it follows that there are $\varepsilon > 0$ and V an open neighborhood for a such that $\forall (\lambda, x) \in \mathbb{K} \times X$ with $|\lambda| < \varepsilon$ and $x \in V$, it follows that $p(\lambda, x) = \lambda x \in D$. In particular, for $|\lambda| < \varepsilon$ it follows that $\lambda a \in D$, from whence $\lambda = \lambda x^*(a) = x^*(\lambda a) \in x^*(D)$, i.e., $\exists \varepsilon > 0$ such that $D(0, \varepsilon) \subseteq x^*(D)$, and therefore $x^*(D)$ is a neighborhood of 0. Let now $D \subseteq X$ be an arbitrary open set. Let also $x \in D$. Then $0 \in D - x$ and $D - x$ is open, hence $x^*(D - x) \subseteq \mathbb{K}$ is a neighborhood of 0 in $\mathbb{K}$, i.e., $\exists \varepsilon > 0$ such that $D(0, \varepsilon) \subseteq x^*(D - x)$, that is, $D(x^*(x), \varepsilon) \subseteq x^*(D)$. Hence $\forall \lambda \in x^*(D) \; \exists \varepsilon > 0$ such that $D(\lambda, \varepsilon) \subseteq x^*(D)$, i.e., $x^*(D) \subseteq \mathbb{K}$ is an open set in $\mathbb{K}$.

Consider now a non-null linear and continuous operator $U : X \to Y$. Since by hypothesis $\dim_{\mathbb{K}} Y = 1$, let $y_0 \in Y$ with $y_0 \neq 0$. Then for any $x \in X$ there is a unique $\lambda_x \in \mathbb{K}$ with the property that $U(x) = \lambda_x y_0$. From this it follows easily that the functional $x^* : X \to \mathbb{K}$ defined by $x^*(x) = \lambda_x$ is linear, continuous, and non-null, hence by the above an open operator. Now the mapping $h : \mathbb{K} \to Y$ defined by $h(\lambda) = \lambda y_0$ is a homeomorphism, so h is open. If we observe that $U = h \circ x^*$, U is open, since as is easy to see the composition of two open maps is an open map.

24. i) The hypothesis assures that M_a is well defined. Obviously, M_a is linear. We will prove that M_a has its graph closed, and since l_p and l_q are Banach spaces, we will obtain that M_a is continuous. Let $(x_k)_{k \in \mathbb{N}} \subseteq l_p$ be such that $x_k \to x \in l_p$ and $M_a(x_k) \to y \in l_q$. Let $x_k = (x_1^k, x_2^k, ...) \in l_p$, $x = (x_1, x_2, ...) \in l_p$, $y = (y_1, y_2, ...) \in l_q$. From $x_k \to x$ in l_p it follows that $x_i^k \to x_i \; \forall i \in \mathbb{N}$. Since $M_a(x_k) = (a_1 x_1^k, a_2 x_2^k, ...) \to y = (y_1, y_2, ...)$ in l_q it follows that $a_i x_i^k \to y_i \; \forall i \in \mathbb{N}$. Since $a_i x_i^k \to a_i x_i$ we obtain that $a_i x_i = y_i \; \forall i \in \mathbb{N}$, i.e., $y = M_a(x)$. Hence M_a is linear and continuous.

ii) By (i), M_a is linear and continuous. Let $x_n \in \mathbb{K}$ such that $|x_n| = |a_n|^{(r-q)/q}$. Then $|x_n|^q |a_n|^q = |a_n|^r$, and then

$$\|M_a(x_1, ..., x_k, 0, ...)\| = \left(\sum_{n=1}^{k} |a_n|^q |x_n|^q\right)^{1/q} = \left(\sum_{n=1}^{k} |a_n|^r\right)^{1/q} \quad \forall k \in \mathbb{N}.$$

We have

$$\|M_a(x_1, ..., x_k, 0, ...)\| \leq \|M_a\| \|(x_1, ..., x_k, 0, ...)\|_p,$$

and

$$\|(x_1, ..., x_k, 0, ...)\|_p = \left(\sum_{n=1}^{k} |a_n|^r\right)^{1/p}.$$

9.2 Solutions

Hence
$$\left(\sum_{n=1}^{k}|a_n|^r\right)^{1/q} \leq \|M_a\| \left(\sum_{n=1}^{k}|a_n|^r\right)^{1/p},$$

thus
$$\left(\sum_{n=1}^{k}|a_n|^r\right)^{1/r} \leq \|M_a\| \quad \forall k \in \mathbb{N},$$

so $\left(\sum_{n=1}^{\infty}|a_n|^r\right)^{1/r} \leq \|M_a\|$, i.e., $a \in l_r$ and $\|a\|_r \leq \|M_a\|$.

Conversely, if $a \in l_r$ then since $1/p + 1/r = 1/q$, from the Hölder inequality we have

$$\|(a_n x_n)_{n \in \mathbb{N}}\|_q \leq \|(x_n)_{n \in \mathbb{N}}\|_p \|(a_n)_{n \in \mathbb{N}}\|_r,$$

that is, $\|M_a(x)\|_q \leq \|x\|_p \|a\|_r \ \forall x \in l_p$, i.e., $\|M_a\| \leq \|a\|_r$.

iii) By (i), M_a is linear and continuous. Consider $e_n \in l_p$. Then $|a_n| = \|M_a(e_n)\| \leq \|M_a\| \|e_n\| = \|M_a\|$, i.e., $a = (a_n)_{n \in \mathbb{N}} \in l_\infty$ and $\|a\|_\infty = \sup_{n \in \mathbb{N}} |a_n| \leq \|M_a\|$. On the another hand, since $p \leq q$ we have

$$\|M_a(x)\| = \left(\sum_{n=1}^{\infty}|a_n|^q |x_n|^q\right)^{1/q} \leq \|a\|_\infty \left(\sum_{n=1}^{\infty}|x_n|^q\right)^{1/q}$$
$$\leq \|a\|_\infty \left(\sum_{n=1}^{\infty}|x_n|^p\right)^{1/p} = \|a\|_\infty \|x\|,$$

and then $\|M_a\| \leq \|a\|_\infty$.

25. See [29, exercise 14.28, chapter IV, section 14].

Suppose, for a contradiction, that we can find such a sequence $(c_n)_{n \in \mathbb{N}}$. This implies that the set $\{n \in \mathbb{N} \mid c_n = 0\}$ is finite, and then, without loss of generality, we may suppose that $c_n \neq 0 \ \forall n \in \mathbb{N}$. Consider then the operator $T : l_\infty \to l_1$, $T((a_n)_{n \in \mathbb{N}}) = (a_n/c_n)_{n \in \mathbb{N}}$. By our hypothesis T is a well defined operator. Indeed, if $(a_n)_{n \in \mathbb{N}} \in l_\infty$, i.e., $((a_n/c_n) c_n)_{n \in \mathbb{N}} \in l_\infty$, then by hypothesis $\sum_{n=1}^{\infty} |a_n/c_n| < \infty$. T is clearly linear, and we will use the closed graph theorem in order to prove that T is continuous. Suppose that $x_k = (x_n^k)_{n \in \mathbb{N}} \to 0$ in l_∞, and that $T(x_k) = (x_n^k/c_n)_{n \in \mathbb{N}} \to y$ in l_1. If $y = (y_n)_{n \in \mathbb{N}}$, then $x_n^k/c_n \to y_n \ \forall n \in \mathbb{N}$. But $x_n^k \to 0 \ \forall n \in \mathbb{N}$. Therefore $y_n = 0 \ \forall n \in \mathbb{N}$, and $y = 0$. We obtain that T is a bounded linear operator. Clearly, T is injective: if $T((a_n)_{n \in \mathbb{N}}) = 0$, then $a_n/c_n = 0 \ \forall n \in \mathbb{N}$, and then $a_n = 0 \ \forall n \in \mathbb{N}$. By our hypothesis, T is also surjective. Indeed, let $(b_n)_{n \in \mathbb{N}} \in l_1$. Then $\sum_{n=1}^{\infty} |b_n| < \infty$, and therefore $(b_n c_n)_{n \in \mathbb{N}} \in l_\infty$. Also $T((b_n c_n)_{n \in \mathbb{N}}) = (c_n)_{n \in \mathbb{N}}$. Thus, T is a bijective linear and bounded operator. Since l_∞ and l_1 are Banach spaces, we obtain that $T^{-1} : l_1 \to l_\infty$ is also a bounded linear operator. Therefore, we can find $\delta > 0$ such that $\|Tx\| \geq \delta \|x\| \ \forall x \in l_\infty$. If we consider, for each subset A of $\mathbb{N}$, $e_A(n) = 1$ if $n \in A$ and $e_A(n) = 0$ if $n \notin A$, we obtain $(e_A)_{A \in \mathcal{P}(\mathbb{N})} \subseteq l_\infty$ such that $\|e_A - e_B\| = 1 \ \forall A \neq B$.

Therefore $\|T(e_A) - T(e_B)\| \geq \delta$, $\forall A \neq B$. Using that $\mathcal{P}(\mathbb{N})$ is not countable, as in the solution for exercise 22 of chapter 2 we obtain that l_1 is not separable, a contradiction!

26. i) Let $A = \{n \in \mathbb{N} \mid c_n \neq 0\}$. If A is finite, then there is $k \in \mathbb{N}$ such that $c_n = 0$ $\forall n \geq k+1$. Then for the sequence $(-1)^n$ we have $(-1)^n c_n = 0$ $\forall n \geq k+1$, and therefore the series $\sum_{n=1}^{\infty}(-1)^n c_n$ converges absolutely. Hence, by our hypothesis, the series $\sum_{n=1}^{\infty}(-1)^n$ must be absolutely convergent, which is false. Hence A is infinite, and then suppose that $c_n \neq 0$ $\forall n \in \mathbb{N}$. Then our hypothesis are:

a) for each absolutely convergent series $\sum_{n=1}^{\infty} a_n$ it follows that the series $\sum_{n=1}^{\infty} a_n c_n$ is absolutely convergent. Using exercise 24(iii) this is equivalent with the fact that the sequence $(c_n)_{n \in \mathbb{N}}$ is bounded, i.e., $\sup_{n \in \mathbb{N}} |c_n| < \infty$

b) for each absolutely convergent series $\sum_{n=1}^{\infty} a_n = \sum_{n=1}^{\infty} c_n (a_n/c_n)$ it follows that the series $\sum_{n=1}^{\infty} a_n/c_n$ is absolutely convergent. Using the same exercise, this is equivalent with the fact that the sequence $(1/c_n)_{n \in \mathbb{N}}$ is bounded, i.e. $\sup_{n \in \mathbb{N}}(1/|c_n|) < \infty$, that is, $\inf_{n \in \mathbb{N}} |c_n| > 0$.

Hence the sequences which satisfy the conditions from the statement are those for which $B = \{n \in \mathbb{N} \mid c_n = 0\}$ is finite and $0 < \inf_{n \notin B} |c_n| \leq \sup_{n \in \mathbb{N}} |c_n| < \infty$.

ii) Suppose that we can find such a sequence. Let us say, for example, that $p < q$. Suppose (see (i)) that $c_n \neq 0$ $\forall n \in \mathbb{N}$. Let $M_c : l_p \to l_q$ be the multiplication operator $M_c((x_n)_{n \in \mathbb{N}}) = (c_n x_n)_{n \in \mathbb{N}}$. From our hypothesis it follows that M_c is well defined, and using exercise 24(iii) we obtain that $c = (c_n)_{n \in \mathbb{N}} \in l_\infty$.

Also, let us consider the multiplication operator associated with the sequence $1/c = (1/c_n)_{n \in \mathbb{N}}$, $M_{1/c} : l_q \to l_p$, $M_{1/c}((x_n)_{n \in \mathbb{N}}) = (x_n/c_n)_{n \in \mathbb{N}}$. From our hypothesis it follows that $M_{1/c}$ is well defined and then by exercise 24(ii) we have $1/c = (1/c_n)_{n \in \mathbb{N}} \in l_r$, where $1/q + 1/r = 1/p$. In particular, $(1/c_n)_{n \in \mathbb{N}} \in c_0$, which is false, since the sequence $(c_n)_{n \in \mathbb{N}}$ is bounded.

iii) We obtain as in (i) that the sequences which satisfy the conditions from the statement are those for which $B = \{n \in \mathbb{N} \mid c_n = 0\}$ is finite and $0 < \inf_{n \notin B} |c_n| \leq \sup_{n \in \mathbb{N}} |c_n| < \infty$.

27. See [35, exercise F, chapter 3].

Let us consider the differentiation operator $D : E \to C[0,1]$ defined by $D(f) = f'$ $\forall f \in E$. From our hypothesis on E, the operator D is well defined. The properties for the derivative show that D is linear. Since E is a closed linear subspace in $C[0,1]$, it follows that $(E, \|\cdot\|_\infty)$ is a Banach space. Since $C[0,1]$ is also a Banach space, if we prove that D has its graph closed it will follow from the closed graph theorem that D is continuous. Let $(f_n)_{n \in \mathbb{N}} \subseteq E$, $f_n \xrightarrow{\|\cdot\|_\infty} f$ and $Df_n \xrightarrow{\|\cdot\|_\infty} g \in C[0,1]$, i.e., $f_n \to f$ uniformly on $[0,1]$ and $f'_n \to g$ uniformly on $[0,1]$. But we know that if there is $x_0 \in [0,1]$ such that $(f_n(x_0))_{n \in \mathbb{N}}$ is Cauchy and $f'_n \to g$ uniformly on $[0,1]$, then there is $\widetilde{f} : [0,1] \to \mathbb{R}$ such that $f_n \to \widetilde{f}$ pointwise and $\widetilde{f}' = g$. Under our hypothesis, we have $f = \widetilde{f}$, and then $f' = g$, that is, $D(f) = g$, i.e., D has its graph closed. Then D is continuous and it follows that there is

$M > 0$ such that $\|D(f)\|_\infty \leq M \|f\|_\infty \ \forall f \in E$, i.e., $\|f'\|_\infty \leq M \ \forall f \in B_E$. Then $\forall \varepsilon > 0$ $\exists \delta_\varepsilon = \varepsilon/M$ such that $\forall x, y \in [0,1]$ with $|x - y| \leq \delta_\varepsilon$ it follows that $|f(x) - f(y)| \leq \varepsilon$ $\forall f \in B_E$ (we use the well known Lagrange formula for real functions). Since B_E is clearly uniformly bounded, the Arzela–Ascoli theorem assures that $B_E \subseteq (C[0,1], \|\cdot\|_\infty)$ is relatively compact. Since B_E is closed we obtain that B_E is compact in $(E, \|\cdot\|_\infty)$. The Riesz theorem (see exercise 23 of chapter 4) implies now that E is finite-dimensional.

28. See [44, theorem 5.2].
i) Since $X \subseteq L_p(\mu)$ is a closed linear space and $L_p(\mu)$ is a Banach space, it follows that $\left(X, \|\cdot\|_p\right)$ is a Banach space. Since also $(L_\infty(\mu), \|\cdot\|_\infty)$ is a Banach space and the inclusion operator $i : X \to L_\infty(\mu)$ is linear, in order to prove that i is continuous it is sufficient to prove that i has its graph closed. Let $(f_n)_{n \in \mathbb{N}} \subseteq X$ be such that $f_n \to f_0$ in $L_p(\mu)$ and $f_n \to g$ in $L_\infty(\mu)$. Then by the Hölder inequality and that the Lebesgue measure μ is finite it follows that $f_n \to g$ in $L_p(\mu)$, hence $f_0 = g$, μ-a.e., i.e., $f_0 = g$ in X.

ii) From (i) there is $k > 0$ such that $\|f\|_\infty \leq k \|f\|_p \ \forall f \in X$. If $1 \leq p \leq 2$, then $\|f\|_p \leq \|f\|_2$, and so $\|f\|_\infty \leq k \|f\|_2 \ \forall f \in X$. Suppose now that $p \in (2, \infty)$. Then we have

$$\|f\|_p^p = \int_0^1 |f(t)|^p \, dt = \int_0^1 |f(t)|^2 |f(t)|^{p-2} \, dt \leq \|f\|_\infty^{p-2} \int_0^1 |f(t)|^2 \, dt = \|f\|_\infty^{p-2} \|f\|_2^2.$$

Combining this with $\|f\|_\infty \leq k \|f\|_p$ we obtain that $\|f\|_p^p \leq k^{p-2} \|f\|_p^{p-2} \|f\|_2^2$, from whence $\|f\|_p \leq k^{(p-2)/2} \|f\|_2$ and $\|f\|_\infty \leq k \cdot k^{(p-2)/2} \|f\|_2$. In conclusion, $\|f\|_\infty \leq M \|f\|_2 \ \forall f \in X$, where $M = \max\left(k, k^{p/2}\right)$.

iii) Let $\{f_1, ..., f_n\} \subseteq X$ be an orthonormal set in X, X being considered as a linear subspace in $L_2(\mu)$. Let $\Gamma = \mathbb{Q}^n \cap B_{\mathbb{C}^n}$, where $B_{\mathbb{C}^n}$ is the closed unit ball of $\mathbb{C}^n$ in the Euclidean norm. For $z = (z_1, ..., z_n) \in B_{\mathbb{C}^n}$, let $f_z : [0,1] \to \mathbb{C}$ be defined by $f_z(\cdot) = \sum_{i=1}^n z_i f_i(\cdot)$. Then from the orthonormality we have $\|f_z\|_2^2 = \sum_{i=1}^n |z_i|^2$, hence $\|f_z\|_2 \leq 1$ $\forall z \in B_{\mathbb{C}^n}$.

From (ii) it follows that $\|f_z\|_\infty \leq M$, $\forall z \in B_{\mathbb{C}^n}$, i.e., $\forall z \in B_{\mathbb{C}^n}$ we have $|f_z| \leq M$ μ-a.e.. Since Γ is countable it follows that there is $A \subseteq [0,1]$, μ-measurable, such that $\mu(C_A) = 0$ and $|f_z(t)| \leq M \ \forall t \in A, \forall z \in \Gamma$. For a fixed $t \in A$, we have $|f_z(t)| \leq M$ $\forall z \in \Gamma$ and since $\Gamma \subseteq B_{\mathbb{C}^n}$ is dense, it follows that $|f_z(t)| \leq M \ \forall z \in B_{\mathbb{C}^n}$. Hence $\sup_{z \in B_{\mathbb{C}^n}} \left|\sum_{i=1}^n z_i f_i(t)\right| \leq M$ and then using the duality $(\mathbb{C}^n, \|\cdot\|_2)^* = (\mathbb{C}^n, \|\cdot\|_2)$ it follows that $\sum_{i=1}^n |f_i(t)|^2 \leq M^2 \ \forall t \in A$, i.e., $\sum_{i=1}^n |f_i|^2 \leq M^2$, μ-a.e.. Integrating and using again the orthonormality it follows that $n \leq M^2$.

Suppose that X is infinite-dimensional. Then we have an infinite linearly independent system in X. Then the Gram–Schmidt procedure shows that we have an infinite orthonormal system in X, which contradicts the facts proved above. Hence X is finite-dimensional, and the dimension of X is less than M^2.

29. i) Let us verify the axioms for a norm. If $\|x\|_p = 0$, then $|x_n|^p w_n = 0 \ \forall n \in \mathbb{N}$ and so $x_n = 0 \ \forall n \in \mathbb{N}$, since $w_n \neq 0 \ \forall n \in \mathbb{N}$. Also, if $\alpha \in \mathbb{C}$ and $x = (x_n)_{n \in \mathbb{N}} \in l_p(w)$, then

$$\|\alpha x\|_p = \left(\sum_{n=1}^{\infty} |\alpha x_n|^p w_n\right)^{1/p} = |\alpha|\, \|x\|_p.$$ If $x = (x_n)_{n\in\mathbb{N}} \in l_p(w)$ and $y = (y_n)_{n\in\mathbb{N}} \in l_p(w)$ then

$$\|x+y\|_p = \left(\sum_{n=1}^{\infty} |x_n + y_n|^p w_n\right)^{1/p} = \left(\sum_{n=1}^{\infty} |x_n w_n^{1/p} + y_n w_n^{1/p}|^p\right)^{1/p}$$

$$\leq \left(\sum_{n=1}^{\infty} |x_n w_n^{1/p}|^p\right)^{1/p} + \left(\sum_{n=1}^{\infty} |y_n w_n^{1/p}|^p\right)^{1/p} = \|x\|_p + \|y\|_p.$$

Hence $\left(l_p(w), \|\cdot\|_p\right)$ is a normed space. That it is a Banach space follows from l_p being a Banach space. Indeed, if $(x^k)_{k\in\mathbb{N}} \subseteq l_p(w)$ is a Cauchy sequence, $x^k = (x_n^k)_{n\in\mathbb{N}}$, then considering the sequence $(z^k)_{k\in\mathbb{N}} \subseteq l_p$, where $z^k = (x_n^k w_n^{1/p})_{n\in\mathbb{N}}$, we obtain a Cauchy sequence in l_p. Hence there is $z = (z_n)_{n\in\mathbb{N}} \in l_p$ such that $z^k \to z$ in l_p, and so $x^k \to x$ in $l_p(w)$, where $x = \left(z_n w_n^{-1/p}\right)_{n\in\mathbb{N}} \in l_p(w)$.

ii) Let us suppose that there are $r, s \in [1, \infty)$ with $r < s$ such that $L_r(\mu) = L_s(\mu)$ (equality of sets). Consider the inclusion operator $I : l_r(w) \to l_s(w)$. I is well defined, since if $x = (x_n)_{n\in\mathbb{N}} \in l_r(w)$ then for the function $f = \sum_{n=1}^{\infty} x_n \chi_{E_n}$ we have $f \in L_r(\mu)$, hence by hypothesis $f \in L_s(\mu)$, i.e., $\left(\sum_{n=1}^{\infty} |x_n|^s w_n\right)^{1/s} < \infty$, and so $x \in l_s(w)$. Obviously, I is linear. Since $(l_r(w), \|\cdot\|_r)$ and $(l_s(w), \|\cdot\|_s)$ are Banach spaces, in order to prove that I is continuous it is sufficient to show that I has its graph closed. Let $(x^k)_{k\in\mathbb{N}} \subseteq l_r(w)$ with $x^k \to x \in l_r(w)$ such that $I(x^k) \to y \in l_s(w)$. Hence $x^k \to y$ in $l_s(w)$. Then for all $n \in \mathbb{N}$, the nth component of x^k converges to the nth component of y. Since $x^k \to x$ in $l_r(w)$, then for any $n \in \mathbb{N}$, the nth component of x^k converges to the nth component of x. Hence $x = y$, i.e., $I(x) = y$. The closed graph theorem shows that I is continuous, hence there is $M > 0$ such that $\|x\|_s \leq M \|x\|_r$ $\forall x \in l_r(w)$. For $e_n = (0, 0, ..., 0, 1, 0, ...)$ (1 on the nth position), we obtain that: $w_n^{1/s} \leq M w_n^{1/r}$ $\forall n \in \mathbb{N}$, i.e., $w_n^{1/s - 1/r} \leq M$ $\forall n \in \mathbb{N}$, which is impossible since $w_n \to 0$ (the sets E_n are pairwise disjoint and μ is countably additive and finite).

Hence if $r, s \in [1, \infty)$ with $r \neq s$ and $L_r(\mu) = L_s(\mu)$, then we cannot find a sequence of pairwise disjoint sets $(E_n)_{n\in\mathbb{N}} \subseteq \Sigma$ such that $\mu(E_n) > 0$ $\forall n \in \mathbb{N}$.

iii) Let us suppose that $L_r(\mu)$ is not finite-dimensional. Let $\mathcal{E} \subseteq L_r(\mu)$ be the space of all measurable step functions. As is well known, $\mathcal{E} \subseteq L_r(\mu)$ is a dense linear subspace. If $\mathcal{E}$ is finite-dimensional then $\mathcal{E}$ is closed, and hence $L_r(\mu)$ will be finite-dimensional, contradiction. So $\mathcal{E}$ is not finite-dimensional, and in this case there are $A, B \subseteq X$ disjoint measurable sets with $A \cup B = X$, $\mu(A) > 0$, $\mu(B) > 0$, since in the contrary case we would have $\dim \mathcal{E} = 1$. Hence $\mathcal{E} = (\chi_A \mathcal{E}) \oplus (\chi_B \mathcal{E})$, direct sum. Since $\mathcal{E}$ is infinite-dimensional, it follows that either $\dim(\chi_A \mathcal{E}) = \infty$, or $\dim(\chi_B \mathcal{E}) = \infty$. If, for example, $\dim(\chi_B \mathcal{E}) = \infty$, then we can apply the same reasoning as above and construct $C, D \subseteq B$ disjoint measurable sets with $C \cup D = B$, $\mu(C) > 0$, $\mu(D) > 0$. We have $\mathcal{E} = (\chi_A \mathcal{E}) \oplus (\chi_C \mathcal{E}) \oplus (\chi_D \mathcal{E})$, with $A, C, D \subseteq X$ measurable and pairwise disjoint, $A \cup C \cup D = X$, $\mu(A) > 0$, $\mu(C) > 0$, $\mu(D) > 0$. In this way we can construct a sequence of disjoint sets

9.2 Solutions

$(E_n)_{n\in\mathbb{N}} \subseteq \Sigma$ such that $\mu(E_n) > 0, \forall n \in \mathbb{N}$. This contradicts what is proved at (ii).

30. See [20, exercise 3, chapter II].

i) Using that every element $x \in X$ has a unique representation as a summation of two elements $y \in Y$ and $z \in Z$, we can consider the operators $P : X \to Y$ and $Q : X \to Z$ defined by $P(x) = y$ and $Q(x) = z$, if $x = y + z$. Using the uniqueness for the representation and the fact that Y and Z are linear subspaces of X, it follows that $P : X \to Y$ and $Q : X \to Z$ are linear operators. Since Y and Z are closed linear subspaces and X is a Banach space, it follows that Y and Z are Banach spaces (with respect to the norm from X). In order to prove that P and Q are continuous we will use the closed graph theorem. Let $(x_n)_{n\in\mathbb{N}} \subseteq X$ be such that $x_n \to x \in X$ and $Px_n \to y \in Y$. For x_n we have a unique decomposition of the form $x_n = y_n + z_n$ with $y_n \in Y$ and $z_n \in Z$. Then $Px_n = y_n \; \forall n \in \mathbb{N}$ and passing to the limit for $n \to \infty$ in $x_n = y_n + z_n$ it follows that $z_n \to x - y$. Since $Z \subseteq X$ is closed it follows that $x - y = z \in Z$ and then $x = y + z$, $y \in Y, z \in Z$. Thus $Px = y$, i.e., P has its graph closed, so $P \in L(X, Y)$. In the same way, $Q \in L(X, Z)$. Let $k = \max\{\|P\|, \|Q\|\}$. For $x \in X, x = y + z, y \in Y, z \in Z$, we have $\|y\| = \|Px\| \le \|P\| \|x\| \le k\|x\|, \|z\| = \|Qx\| \le \|Q\| \|x\| \le k\|x\|$.

ii) It is proved in the solution of exercise 7(ii) that if T is a linear projector, then we have the decomposition $X = \ker T \oplus T(X)$. Hence any $x \in X$ has a unique representation of the form $x = a + b$ with $a \in T(X)$ and $b \in \ker(T)$. Since by our hypothesis $\ker T$ and $T(X) \subseteq X$ are closed linear subspaces, from (i) it follows that there is $k \ge 0$ such that $\|a\| \le k\|x\|, \|b\| \le k\|x\| \; \forall x \in X$ where $a = T(x) \in T(X)$ and $b = x - T(x) \in \ker T$, and therefore $\|Tx\| \le k\|x\| \; \forall x \in X$, i.e., T is continuous.

31. See [14, exercise 2, chapter III, section 12].

It is proved in the solution of exercise 27 that D has its graph closed. For any $n \in \mathbb{N}$ let $f_n : [0,1] \to \mathbb{R}$ be defined by $f_n(t) = t^n \; \forall t \in [0,1]$. We have $\|f_n\|_\infty = 1$ and $Df_n(t) = nt^{n-1} \; \forall t \in [0,1]$, and then $\|Df_n\|_\infty = n \; \forall n \in \mathbb{N}$, which shows that D is not continuous.

Since $(C[0,1], \|\cdot\|_\infty)$ is a Banach space from the closed graph theorem it follows that $(C^1[0,1], \|\cdot\|_\infty)$ is a normed space which is not a Banach space.

32. See [9, exercise 7, chapter I, section 3].

i) We observe that N is a closed set, being an intersection of closed sets. Let $y, z \in N$. Let V be a neighborhood of 0. Then there is U, a neighborhood of 0, such that $U + U \subseteq V$, hence, by the linearity of T, $T(U) + T(U) \subseteq T(V)$, and, by usual properties for the adherence, $\overline{T(U)} + \overline{T(U)} \subseteq \overline{T(U) + T(U)} \subseteq \overline{T(V)}$. Since $y, z \in N$ it follows that $y, z \in \overline{T(U)}$, so $y + z \in \overline{T(V)}$. Since V is an arbitrary neighborhood for 0, it follows that $y + z \in N$. It the same way, if $y \in N$ and $\lambda \in \mathbb{K}$ then $\lambda y \in N$.

ii) Suppose that $N = \{0\}$. In order to prove that T is continuous, since X and Y are Banach spaces, by the closed graph theorem it is sufficient to prove that T has its graph closed. Let $x_n \to x$ and $T(x_n) \to y$. Let also V be an arbitrary neighborhood of 0. Then there is k such that $x_n - x \in V \; \forall n \ge k$. Then $T(x_n) - T(x) \in T(V) \; \forall n \ge k$, hence passing to the limit for $n \to \infty$ we obtain that $y - T(x) \in \overline{T(V)}$, i.e., by the definition of N, $y - T(x) \in N = \{0\}$, i.e., $y = T(x)$.

Conversely, suppose that T is continuous. Let $y \in N$. Then, from the definition of N, for any $n \in \mathbb{N}$, we have $y \in \overline{T(B(0, 1/n))}$, hence there is $\|x_n\| < 1/n$ such that

$\|T(x_n) - y\| < 1/n$, i.e., $x_n \to 0$ and $T(x_n) \to y$. Since T is linear and continuous, it follows that $T(x_n) \to T(0) = 0$, and then $y = 0$, i.e., $N = \{0\}$.

iii) In order to prove that the composition $p \circ T : X \to Y/N$ is continuous, where $p : Y \to Y/N$ is the quotient map, since X and Y/N are Banach spaces, it is sufficient to prove that $p \circ T$ has its graph closed. We will denote by $p(y) = \widehat{y} \in Y/N$ the equivalence class for an element $y \in Y$.

Let $x_n \to 0$ and $\widehat{T(x_n)} \to \widehat{y}$. This means that $d(T(x_n) - y, N) = \left\|\widehat{T(x_n)} - \widehat{y}\right\| \to 0$. For any $n \in \mathbb{N}$ there is $a_n \in N$ such that $\|T(x_n) - y - a_n\| < 1/n + d(T(x_n) - y, N)$. We have $a_n \in N$, and, by the definition of N, $a_n \in \overline{T(B(0, 1/n))}$, hence by the definition of a closure point $B(a_n, 1/n) \cap T(B(0, 1/n)) \neq \emptyset$, therefore there is $b_n \in X$, $\|b_n\| < 1/n$, such that $\|T(b_n) - a_n\| < 1/n$. Then $\|T(x_n - b_n) - y\| \leq \|T(x_n) - a_n - y\| + \|a_n - T(b_n)\| \to 0$, and we have $T(x_n - b_n) \to y$, $x_n - b_n \to 0$. Then for any V, neighborhood of 0, it exists $n_0 \in \mathbb{N}$ such that $x_n - b_n \in V$ $\forall n \geq n_0$, from whence $T(x_n - b_n) \in T(V)$ $\forall n \geq n_0$, hence passing to the limit for $n \to \infty$ and using that $T(x_n - b_n) \to y$ it follows that $y \in \overline{T(V)}$, $\forall V$ a neighborhood of 0, i.e., by the definition of N, $y \in N$, i.e., $\widehat{y} = 0$.

iv) Let M be a closed linear subspace of Y with the property that the composition $q \circ T : X \to Y/M$ is continuous, where $q : Y \to Y/M$ is the quotient map. Let $y \in N$. From the definition of N, for any $n \in \mathbb{N}$ we have $y \in \overline{T(B(0, 1/n))}$. Since $B(y, 1/n)$ is a neighborhood of y, from the definition for the closure of a set we have that $T(B(0, 1/n)) \cap B(y, 1/n) \neq \emptyset$, i.e., there is $x_n \in X$ with $\|x_n\| < 1/n$ and $\|T(x_n) - y\| < 1/n$. Then $x_n \to 0$ and $T(x_n) \to y$. Since the quotient map $q : Y \to Y/M$ is continuous it follows that $q(T(x_n)) \to q(y)$. Also, by hypothesis, the composition $q \circ T : X \to Y/M$ is continuous, and then since $x_n \to 0$ it follows that $(q \circ T)(x_n) \to (q \circ T)(0) = \widehat{0}$. Hence $q(y) = \widehat{0}$, i.e., $y \in M$.

Part II

Hilbert spaces

Chapter 10

Hilbert spaces, general theory

Definition. Let H be a linear space. A map $\langle \cdot, \cdot \rangle : H \times H \to \mathbb{K}$ is called an *inner product* on H if and only if the following conditions are satisfied:
i) $\langle x, x \rangle \geq 0 \ \forall x \in H$;
ii) $\langle x, x \rangle = 0$ if and only if $x = 0$;
iii) $\langle \alpha x + \beta y, z \rangle = \alpha \langle x, z \rangle + \beta \langle y, z \rangle \ \forall \alpha, \beta \in \mathbb{K}, \forall x, y, z \in H$;
iv) $\overline{\langle y, x \rangle} = \langle x, y \rangle \ \forall x, y \in H$.
Then $(H, \langle \cdot, \cdot \rangle)$ is called an *inner product space*.

The Cauchy–Buniakowski–Schwarz inequality. Let $(H, \langle \cdot, \cdot \rangle)$ be an inner product space. Then
$$|\langle x, y \rangle|^2 \leq \langle x, x \rangle \langle y, y \rangle \quad \forall x, y \in H.$$

Theorem. Let $(H, \langle \cdot, \cdot \rangle)$ be an inner product space. The map $\|\cdot\| : H \to \mathbb{R}$ defined by $\|x\| = \sqrt{\langle x, x \rangle}$ is a norm called the *norm generated by the inner product (or associated to the inner product)*.

The von Neumann Theorem. Let $(X, \|\cdot\|)$ be a normed space. Then there is an inner product $\langle \cdot, \cdot \rangle : X \times X \to \mathbb{K}$ such that the norm associated to the inner product is $\|\cdot\|$ if and only if the parallelogram law is true:
$$\|x + y\|^2 + \|x - y\|^2 = 2 \left(\|x\|^2 + \|y\|^2 \right) \quad \forall x, y \in X.$$

Definition. An inner product space $(H, \langle \cdot, \cdot \rangle)$ is called a *Hilbert space* if and only if the norm generated by the inner product is complete, i.e., $(H, \|\cdot\|)$ is a Banach space.

Definition. Let $(H_1, \langle \cdot, \cdot \rangle)$ and $(H_2, \langle \cdot, \cdot \rangle)$ be two inner product spaces. On the Cartesian product $H_1 \times H_2$ the map $\langle (x_1, y_1), (x_2, y_2) \rangle = \langle x_1, x_2 \rangle + \langle y_1, y_2 \rangle$ is an inner product. If $(H_1, \langle \cdot, \cdot \rangle)$ and $(H_2, \langle \cdot, \cdot \rangle)$ are Hilbert spaces then $(H_1 \times H_2, \langle \cdot, \cdot \rangle)$ is a Hilbert space called the *Hilbertian product* of H_1 and H_2.

The Riesz Representation Theorem. Let $(H, \langle \cdot, \cdot \rangle)$ be a Hilbert space. Then for any linear and continuous functional $x^* : H \to \mathbb{K}$ there is a unique element $y \in H$ such that $x^*(x) = \langle x, y \rangle$ $\forall x \in X$, and conversely, if $y \in H$ then $x^* : H \to \mathbb{K}$, $x^*(x) = \langle x, y \rangle$ $\forall x \in H$, is linear and continuous, with $\|x^*\| = \|y\|$.

Definition. Let $(H, \langle \cdot, \cdot \rangle)$ be an inner product space, $A \subseteq H$ and $x \in H$. We say that $x \perp A$ if and only if $\langle x, a \rangle = 0$ $\forall a \in A$. The set $A^\perp = \{x \in H \mid x \perp A\}$ is called the *orthogonal complement* of the set A.

Definition. Let $(H, \langle \cdot, \cdot \rangle)$ be an inner product space. A subset $A \subseteq H$ is called:

i) an *orthogonal set* if and only if $\langle a, b \rangle = 0$ $\forall a, b \in A$ with $a \neq b$;

ii) an *orthonormal set* if and only if $\langle a, b \rangle = 0$ $\forall a, b \in A$ with $a \neq b$, and $\langle a, a \rangle = 1$ $\forall a \in A$;

iii) an *orthonormal basis* if and only if A is an orthonormal set and for any orthonormal set $B \subseteq H$ with the property that $A \subseteq B$ it follows that $A = B$.

The Pythagoras Theorem. i) Let $(H, \langle \cdot, \cdot \rangle)$ be an inner product space. If $n \in \mathbb{N}$ and $\{x_1, ..., x_n\} \subseteq H$ is an orthogonal system, then

$$\|x_1 + x_2 + \cdots + x_n\|^2 = \|x_1\|^2 + \|x_2\|^2 + \cdots + \|x_n\|^2.$$

ii) Let $(H, \langle \cdot, \cdot \rangle)$ be a Hilbert space. If $(x_n)_{n \in \mathbb{N}}$ is a countable orthogonal system, then the series $\sum_{n=1}^{\infty} x_n$ is convergent if and only if the series $\sum_{n=1}^{\infty} \|x_n\|^2$ is convergent, and in this case

$$\left\| \sum_{n=1}^{\infty} x_n \right\|^2 = \sum_{n=1}^{\infty} \|x_n\|^2.$$

Theorem. Let $(H, \langle \cdot, \cdot \rangle)$ be a Hilbert space, and $A \subseteq H$ an orthonormal set. Then A is an orthonormal basis if and only if is *total*, i.e., if $x \in H$ is an element with the property that $x \perp A$ then it follows that $x = 0$.

The Bessel inequality. Let $(H, \langle \cdot, \cdot \rangle)$ be an inner product space, and $(e_n)_{n \in \mathbb{N}} \subseteq H$ an orthonormal family. Then

$$\sum_{n=1}^{\infty} |\langle x, e_n \rangle|^2 \leq \|x\|^2 \quad \forall x \in H.$$

Theorem. Let $(H, \langle \cdot, \cdot \rangle)$ be a Hilbert space, and $(e_n)_{n \in \mathbb{N}} \subseteq H$ an orthonormal family. The following assertions are equivalent:

i) The family $(e_n)_{n \in \mathbb{N}}$ is an orthonormal basis in H;

ii) **The Fourier formula.** For any $x \in H$ we have the equality

$$x = \sum_{n=1}^{\infty} \langle x, e_n \rangle e_n;$$

iii) **The Parseval formula (for the norm).** For any $x \in H$ we have the equality

$$\|x\|^2 = \sum_{n=1}^{\infty} |\langle x, e_n \rangle|^2;$$

iv) The Parseval formula (for the inner product). For any $x, y \in H$ we have the equality
$$\langle x, y \rangle = \sum_{n=1}^{\infty} \langle x, e_n \rangle \overline{\langle y, e_n \rangle}.$$

Theorem. i) The map $\langle \cdot, \cdot \rangle : l_2 \times l_2 \to \mathbb{K}$,
$$\langle x, y \rangle = \sum_{n=1}^{\infty} x_n \overline{y_n} \quad \forall x = (x_n)_{n \in \mathbb{N}} \in l_2, \ y = (y_n)_{n \in \mathbb{N}} \in l_2$$

is an inner product and $(l_2, \langle \cdot, \cdot \rangle)$ is a Hilbert space.

ii) If (S, Σ, μ) is a measure space, then the map $\langle \cdot, \cdot \rangle : L_2(\mu) \times L_2(\mu) \to \mathbb{K}$,
$$\langle f, g \rangle = \int_S f(s) \overline{g(s)} d\mu(s) \quad \forall f, g \in L_2(\mu)$$

is an inner product and $(L_2(\mu), \langle \cdot, \cdot \rangle)$ is a Hilbert space.

10.1 Exercises

l_p (or L_p) is a Hilbert spaces if and only if $p = 2$

1. i) Let $1 \leq p < \infty$. Prove that $\left(l_p, \|\cdot\|_p \right)$ is a Hilbert space if and only if $p = 2$.

ii) Let $1 \leq p < \infty$ and $A \subseteq \mathbb{R}^n$ be a Lebesgue measurable set, $\lambda(A) > 0$ (λ is the Lebesgue measure on $\mathbb{R}^n$). Prove that $\left(L_p(A), \|\cdot\|_p \right)$ is a Hilbert space if and only if $p = 2$.

$\mathcal{M}_n(\mathbb{C})$ as a Hilbert space

2. Let $\mathcal{M}_n(\mathbb{C})$ be the set of all $n \times n$ complex matrices, and we define the map $\langle \cdot, \cdot \rangle : \mathcal{M}_n(\mathbb{C}) \times \mathcal{M}_n(\mathbb{C}) \to \mathbb{C}$ given by $\langle A, B \rangle = \text{tr}(AB^*)$, where B^* is the adjoint of the matrix B and tr is the trace. Prove that $(\mathcal{M}_n(\mathbb{C}), \langle \cdot, \cdot \rangle)$ is a Hilbert space, and deduce then that for two complex matrices A and B of order n we have
$$|\text{tr}(AB^*)|^2 \leq \text{tr}(AA^*) \text{tr}(BB^*).$$

A weighted inner product for the space of all polynomials of degree $\leq n$

3. Let $w : [0, 1] \to (0, \infty)$ be a continuous function. Let $n \in \mathbb{N}$ and consider P_n, the real linear space of all polynomials of degree $\leq n$, on which we consider the following inner product:
$$\langle p, q \rangle = \int_0^1 p(t) q(t) w(t) dt.$$

i) Prove that P_n has an orthonormal basis $\{p_0, p_1, ..., p_n\}$ such that $\deg p_k = k$, $k = 0, ..., n$.

ii) Prove that $\langle p_k, p'_k \rangle = 0$, $k = 0, ..., n$.

An inner product space which is not a Hilbert space

4. We denote by $l_0^2 = \{(x_n)_{n \in \mathbb{N}} \subseteq \mathbb{C} \mid x_n \neq 0 \text{ only for a finite number of } n\}$, and define $\langle \cdot, \cdot \rangle : l_0^2 \times l_0^2 \to \mathbb{C}$ by $\langle (x_n)_{n \in \mathbb{N}}, (y_n)_{n \in \mathbb{N}} \rangle = \sum_{n=1}^{\infty} x_n \overline{y_n}$. Prove that $(l_0^2, \langle \cdot, \cdot \rangle)$ is an inner product space but not a Hilbert space.

A non separable Hilbert space

5. Let X be the linear space of all functions of the form: $f : \mathbb{R} \to \mathbb{C}$, $f(t) = \sum_{k=1}^{n} c_k e^{i\alpha_k t}$, $n \in \mathbb{N}$, $c_k \in \mathbb{C}$, $\alpha_k \in \mathbb{R}$, $k = \overline{1, n}$.

i) Prove that the map $\langle \cdot, \cdot \rangle : X \times X \to \mathbb{C}$ defined by

$$\langle f, g \rangle = \lim_{A \to \infty} \frac{1}{2A} \int_{-A}^{A} f(t)\overline{g(t)} dt$$

is an inner product on X.

ii) Prove that if $\|\cdot\|$ is the norm generated by $\langle \cdot, \cdot \rangle$, then $\|f\| = \left(\sum_{k=1}^{n} |c_k|^2 \right)^{1/2}$ if $f \in X$, $f = \sum_{k=1}^{n} c_k e^{i\alpha_k t}$, with $\alpha_k \neq \alpha_j$ $\forall k \neq j$.

iii) If we consider the Hilbert space H obtained as the completion of X with respect to $\|\cdot\|$, prove that H is not separable.

The inversion and the Ptolemaic inequality in an inner product space

6. Let H be an inner product space and $T : H \setminus \{0\} \to H \setminus \{0\}$ be the inversion operator, defined by $T(x) = x/\|x\|^2$ $\forall x \in H \setminus \{0\}$.

i) Let $a \in H \setminus \{0\}$ and $V = \{x \in H \mid \Re \langle x, a \rangle = 1\}$. Prove that

$$T(V) = \{y \in H \setminus \{0\} \mid \|y - a/2\| = \|a\|/2\}.$$

ii) Let $a \in H \setminus \{0\}$ and $C = \{x \in H \mid \|x - a\| = \|a\|\}$. Prove that

$$T(C \setminus \{0\}) = \{y \in H \mid \Re \langle y, a \rangle = 1/2\}.$$

iii) Let $a \in H \setminus \{0\}$, $0 < r$, $r^2 \neq \|a\|^2$, and $C = \{x \in H \mid \|x - a\| = r\}$. Prove that

$$T(C) = \{y \in H \mid \|y - a/\lambda\| = r/|\lambda|\},$$

where $\lambda = \|a\|^2 - r^2$.

7. Let H be an inner product space.

i) Prove that if $a, b \in H \setminus \{0\}$, and if $a' = a/\|a\|^2$ and $b' = b/\|b\|^2$, then

$$\|a' - b'\| = \frac{\|a - b\|}{\|a\| \|b\|}.$$

ii) Using (i) prove the Ptolemaic inequality:

$$\|a - c\| \|b - d\| \leq \|a - b\| \|c - d\| + \|b - c\| \|a - d\| \quad \forall a, b, c, d \in H.$$

10.1 Exercises

Averages in inner product spaces

8. Let H be an inner product space, $n \in \mathbb{N}$, and $\{x_1, ..., x_n\} \subseteq H$. Prove that:

$$\sum_{(\varepsilon_1,...,\varepsilon_n) \in \{\pm 1\}^n} \left\| \sum_{i=1}^n \varepsilon_i x_i \right\|^2 = 2^n \sum_{i=1}^n \|x_i\|^2.$$

An orthonormal set in $L_2(\mathbb{R})$

9. Consider the sequence of functions $f_n : \mathbb{R} \to \mathbb{C}$ given by

$$f_n(x) = \pi^{-1/2} \frac{(x-i)^n}{(x+i)^{n+1}}.$$

Prove that the family $\{f_1, f_2, ...\}$ is orthonormal in $L_2(\mathbb{R})$, that is,

$$\int_{-\infty}^{\infty} f_m(x) \overline{f_n(x)} dx = \begin{cases} 1, & m = n, \\ 0, & m \neq n. \end{cases}$$

The topological structure for an orthonormal family

10. i) Let H be a Hilbert space and $A \subseteq H$ an infinite orthonormal family. Prove that $A \subseteq H$ is closed and bounded, but not totally bounded.

ii) If $(e_n)_{n \in \mathbb{N}} \subseteq H$ is an orthonormal system and $(a_n)_{n \in \mathbb{N}} \subseteq \mathbb{K}$, prove that the set $\{a_n e_n \mid n \in \mathbb{N}\} \subseteq H$ is totally bounded if and only if $a_n \to 0$. In particular, if $(x_n)_{n \in \mathbb{N}} \subseteq H \setminus \{0\}$ is an orthogonal system then the set $\{x_n \mid n \in \mathbb{N}\}$ is totally bounded if and only if $\|x_n\| \to 0$.

Weak convergence for an orthonormal sequence

11. i) Let H be a Hilbert space and $(x_n)_{n \in \mathbb{N}} \subseteq H$ an orthonormal system. Prove that $x_n \to 0$ weak.

ii) Let $A \subseteq [0, 2\pi]$ be a Lebesgue measurable set. Prove that

$$\lim_{n \to \infty} \int_A \sin(nt) \, dt = \lim_{n \to \infty} \int_A \cos(nt) \, dt = 0.$$

The explicit form for the element given by the Riesz representation theorem

12. Let H be a Hilbert space, $(e_n)_{n \in \mathbb{N}}$ is an orthonormal basis, and $x^* : H \to \mathbb{K}$ is a linear and continuous functional. Prove that $y = \sum_{n=1}^{\infty} \overline{x^*(e_n)} e_n$ is the unique element in H with the property that $x^*(x) = \langle x, y \rangle \; \forall x \in H$.

Hardy spaces on the open unit disc

13. Let $\mathbb{D} = \{z \in \mathbb{C} \mid |z| < 1\}$ be the open unit disc in the complex plane and $1 \leq p < \infty$. Define $H^p(\mathbb{D}) = \{f : \mathbb{D} \to \mathbb{C} \mid f \text{ analytic}, N_p(f) < \infty\}$, where

$$N_p(f) = \sup_{0 \leq r < 1} \left(\frac{1}{2\pi} \int_0^{2\pi} |f(re^{i\theta})|^p \, d\theta \right)^{1/p}.$$

Prove that:

i) $\forall z_0 \in \mathbb{D}, \forall f \in H^p(\mathbb{D})$, we have

$$|f(z_0)| \leq \frac{1}{(d(z_0, \partial \mathbb{D})^2)^{1/p}} N_p(f),$$

where $\partial \mathbb{D} = \{z \in \mathbb{C} \mid |z| = 1\}$;

ii) $(H^p(\mathbb{D}), N_p)$ is a normed space;

iii) $(H^p(\mathbb{D}), N_p)$ is a Banach space;

iv) $(H^p(\mathbb{D}), N_p)$ is a Hilbert space if and only if $p = 2$;

v) In the case of $H^2(\mathbb{D})$, prove that if $f(z) = \sum_{n=0}^{\infty} a_n z^n$, then $f \in H^2(\mathbb{D})$ if and only if $(a_n)_{n \geq 0} \in l_2$ and that the map $h : H^2(\mathbb{D}) \to l_2$ defined by $h(f) = (a_n)_{n \geq 0}$ is an isometric isomorphism of Hilbert spaces;

vi) Prove that the sequence $f_n(z) = z^n$, $n = 0, 1, 2, \ldots$, is an orthonormal basis for $H^2(\mathbb{D})$.

Riesz's representation theorem for the Hardy space

14. Let $x^* : H^2(\mathbb{D}) \to \mathbb{C}$ be a linear and continuous functional. $H^2(\mathbb{D})$ being a Hilbert space, using Riesz's theorem we obtain that there is a unique function $g \in H^2(\mathbb{D})$ such that $x^*(f) = \langle f, g \rangle_{H^2(\mathbb{D})} \; \forall f \in H^2(\mathbb{D})$. Calculate the expression of g in terms of x^*.

15. Let $\xi \in \mathbb{D}$ and $x^* : H^2(\mathbb{D}) \to \mathbb{C}$ be defined by $x^*(f) = f(\xi)$. Prove that x^* is a linear and continuous functional and find $g \in H^2(\mathbb{D})$ such that $x^*(f) = \langle f, g \rangle_{H^2(\mathbb{D})} \; \forall f \in H^2(\mathbb{D})$.

An extremal result in the Hardy space

16. Let $\xi \in \mathbb{D}$. How large can be $|f'(\xi)|$, if $f \in H^2(\mathbb{D})$ and $N_2(f) \leq 1$? Calculate the extremal functions.

The Bergman space

17. Let $\Omega \subseteq \mathbb{C}$ be an open non-empty set, $1 \leq p < \infty$, and let

$$\mathcal{A}^p(\Omega) = \left\{ f : \Omega \to \mathbb{C} \; \middle| \; f \text{ is holomorphic}, \iint_\Omega |f(x,y)|^p \, dxdy < \infty \right\}.$$

With respect to the usual operations for addition and scalar multiplication, the space $\mathcal{A}^p(\Omega)$ is a complex linear space, called the Bergman space.

10.1 Exercises

i) Prove that for any $z \in \Omega$ and for any $r > 0$ such that $\overline{D}(z,r) \subseteq \Omega$ we have
$$f(z) = \frac{1}{\pi r^2} \iint_{\overline{D}(z,r)} f(x,y)\,dxdy \quad \forall f \in \mathcal{A}^p(\Omega).$$

ii) Prove that $A_p : \mathcal{A}^p(\Omega) \to \mathbb{R}_+$ defined by
$$A_p(f) = \left(\iint_\Omega |f(x,y)|^p \, dxdy \right)^{1/p}$$
is a norm and that $(\mathcal{A}^p(\Omega), A_p)$ is a Banach space.

iii) Define $\langle \cdot, \cdot \rangle : \mathcal{A}^2(\Omega) \times \mathcal{A}^2(\Omega) \to \mathbb{C}$ by
$$\langle f, g \rangle = \iint_\Omega f(x,y)\,\overline{g(x,y)}\,dxdy.$$
Prove that $(\mathcal{A}^2(\Omega), \langle \cdot, \cdot \rangle)$ is a Hilbert space.

iv) Let $\Omega = \mathbb{D}$ be the open unit disc in the complex plane. Prove that the sequence $e_n(z) = \left(\sqrt{n+1}/\sqrt{\pi} \right) z^n$, $n = 0, 1, 2, \ldots$, is an orthonormal basis for $\mathcal{A}^2(\mathbb{D})$.

v) Let $h : \mathcal{A}^2(\mathbb{D}) \to l_2$ defined by $h(f) = \left(\left(\sqrt{\pi}/\sqrt{n+1} \right) a_n \right)_{n \geq 0} \Leftrightarrow f(z) = \sum_{n=0}^{\infty} a_n z^n$. Prove that h is an isometric isomorphism of Hilbert spaces.

Riesz's representation theorem for the Bergman space

18. Let $\Omega \subseteq \mathbb{C}$ be an open non-empty set, $\mathcal{A}^2(\Omega)$ is the Bergman space, and $(f_n)_{n \geq 0}$ is an orthonormal basis for $\mathcal{A}^2(\Omega)$.

i) Let $x^* : \mathcal{A}^2(\Omega) \to \mathbb{C}$ be a linear and continuous functional. $\mathcal{A}^2(\Omega)$ being a Hilbert space, using Riesz's theorem we obtain that there is a unique function $g \in \mathcal{A}^2(\Omega)$ such that $x^*(f) = \langle f, g \rangle_{\mathcal{A}^2(\Omega)} \ \forall f \in \mathcal{A}^2(\Omega)$. Calculate the expression of g in terms of x^* and the orthonormal basis $(f_n)_{n \geq 0}$.

ii) Let $\xi \in \Omega$ and $x_\xi^* : \mathcal{A}^2(\Omega) \to \mathbb{C}$ be defined by $x_\xi^*(f) = f(\xi)$. Prove that x_ξ^* is a linear continuous functional, and that there is a unique function $z \longmapsto K(z, \xi)$ in $\mathcal{A}^2(\Omega)$ such that $x_\xi^*(f) = \iint_\Omega f(z)\,\overline{K(z,\xi)}\,dxdy \ \forall f \in \mathcal{A}^2(\Omega)$. The function $(z, \xi) \longmapsto K(z, \xi)$ is called the Bergman kernel for Ω.

iii) Prove that the Bergman kernel is given by the formula
$$K(z,\xi) = \sum_{n=0}^{\infty} f_n(z)\overline{f_n(\xi)},$$
and that
$$\iint_\Omega |K(z,\xi)|^2 \, dxdy = \sum_{n=0}^{\infty} |f_n(\xi)|^2 = K(\xi,\xi) \ \ \forall \xi \in \Omega.$$

iv) Prove that the Bergman kernel for the open unit disc in the complex plane is given by $K(z, \xi) = 1/\left(\pi(1 - z\overline{\xi})^2 \right)$.

An extremal result in the Bergman space

19. Let $\xi \in \mathbb{D}$. How large can be $|f'(\xi)|$, if $f \in \mathcal{A}^2(\mathbb{D})$ and $\|f\| \leq 1$? Calculate the extremal functions.

Two functions associated to a weak convergent sequence in a Hilbert space

20. Let H be a Hilbert space, $(x_n)_{n \in \mathbb{N}} \subseteq H$ and $a \in H$ such that $x_n \to a$ weak. We define the functions $d, D : H \to \mathbb{R}_+$ by $d(x) = \liminf\limits_{n \to \infty} \|x_n - x\|$ and $D(x) = \limsup\limits_{n \to \infty} \|x_n - x\|$.

i) Prove that $d(x) = \sqrt{d^2(a) + \|x - a\|^2}$ and that $D(x) = \sqrt{D^2(a) + \|x - a\|^2}$ $\forall x \in H$.

ii) Let $0 \leq \alpha \leq \beta$. Give an example of a Hilbert space H, a sequence $(x_n)_{n \in \mathbb{N}} \subseteq H$, and an element $a \in H$ such that $d(a) = \alpha$ and $D(a) = \beta$.

10.2 Solutions

1. i) Evidently, if $p = 2$ then $(l_2, \|\cdot\|_2)$ is a Hilbert space. Consider now $1 \leq p < \infty$ such that $\left(l_p, \|\cdot\|_p\right)$ is a Hilbert space. For $k, t \in \mathbb{N}$ with $k \neq t$, we consider $e_k, e_t \in l_p$. Since $\left(l_p, \|\cdot\|_p\right)$ is a Hilbert space, by the parallelogram law we have

$$\|e_k + e_t\|_p^2 + \|e_k - e_t\|_p^2 = 2\left(\|e_k\|_p^2 + \|e_t\|_p^2\right),$$

that is, $2^{2/p} + 2^{2/p} = 2^2$, $2^{1+2/p} = 2^2$, i.e., $p = 2$.

ii) From the properties for the Lebesgue measure, we can find $B, C \subseteq A$, giving a partition for A with Lebesgue measurable sets, such that $\lambda(B) = \lambda(C) = \lambda(A)/2$. If $L_p(A)$ is a Hilbert space, then by the parallelogram law we have $\|\chi_B + \chi_C\|_p^2 + \|\chi_B - \chi_C\|_p^2 = 2\left(\|\chi_B\|_p^2 + \|\chi_C\|_p^2\right)$, that is,

$$(\lambda(A))^{2/p} + (\lambda(A))^{2/p} = 2\left((\lambda(A)/2)^{2/p} + (\lambda(A)/2)^{2/p}\right).$$

Since $\lambda(A) > 0$ we obtain that $2 = 2^{2/p}$, i.e., $p = 2$. Evidently, if $p = 2$ then $\left(L_p(A), \|\cdot\|_p\right)$ is a Hilbert space.

2. See [1, exercise 3, Spring 1982].

If $A = (a_{ij})$ then $A^* = (\overline{a_{ji}})$. Also, if $A = (a_{ij})$ and $B = (b_{ij})$ then $AB = (c_{ij})$, where $c_{ik} = \sum\limits_{j=1}^{n} a_{ij} b_{jk}$, and thus $\operatorname{tr}(AB) = \sum\limits_{i=1}^{n} c_{ii} = \sum\limits_{i,j=1}^{n} a_{ij} b_{ji}$. Then $\operatorname{tr}(AB^*) = \sum\limits_{i,j=1}^{n} a_{ij} \overline{b_{ij}}$. We have

$$\langle A, A \rangle = \operatorname{tr}(AA^*) = \sum_{i,j=1}^{n} a_{ij} \overline{a_{ij}} = \sum_{i,j=1}^{n} |a_{ij}|^2 \geq 0,$$

and

$$\langle A, A \rangle = 0 \Leftrightarrow |a_{ij}| = 0 \ \forall i, j \Leftrightarrow a_{ij} = 0 \Leftrightarrow A = O.$$

10.2 Solutions

Also,

$$\langle \lambda A_1 + \mu A_2, B \rangle = \operatorname{tr}\left((\lambda A_1 + \mu A_2) B^*\right) = \operatorname{tr}\left(\lambda A_1 B^* + \mu A_2 B^*\right)$$
$$= \lambda \operatorname{tr}(A_1 B^*) + \mu \operatorname{tr}(A_2 B^*) = \lambda \langle A_1, B \rangle + \mu \langle A_2, B \rangle,$$

and

$$\langle B, A \rangle = \operatorname{tr}(BA^*) = \sum_{i,j=1}^{n} b_{ij} \overline{a_{ij}} = \overline{\operatorname{tr}(AB^*)} = \overline{\langle A, B \rangle}.$$

Thus $\langle \cdot, \cdot \rangle$ is an inner product on $\mathcal{M}_n(\mathbb{C})$. The fact that $\mathcal{M}_n(\mathbb{C})$ is a Hilbert space follows from the fact that $\mathcal{M}_n(\mathbb{C})$ can be identified in a natural way with $\mathbb{C}^{n^2}$, and $\mathbb{C}^{n^2}$ is complete, if we consider an arbitrary norm on it.

The Cauchy–Buniakowski–Schwartz inequality implies $|\langle A, B \rangle|^2 \le \langle A, A \rangle \langle B, B \rangle$, i.e.,

$$|\operatorname{tr}(AB^*)| \le \operatorname{tr}(AA^*) \operatorname{tr}(BB^*) \quad \forall A, B \in \mathcal{M}_n(\mathbb{C}).$$

3. See [1, exercise 17, Fall 1993].

If $\langle p, p \rangle = 0$ then $\int_0^1 p^2(t) w(t) \, dt = 0$, so since the function under the integral is continuous and ≥ 0, it follows that $p^2(t) w(t) = 0 \; \forall t \in [0,1]$. Since $w(t) > 0 \; \forall t \in [0,1]$ it follows that $p(t) = 0 \; \forall t \in [0,1]$. The other axioms for an inner product are easy to verify.

i) Consider the polynomials $q_0(x) = 1$, $q_1(x) = x+1$, $q_2(x) = x^2+x+1$, ..., $q_n(x) = x^n + x^{n-1} + \cdots + x + 1$. We will prove that $\{q_0, q_1, ..., q_n\}$ is a basis for P_n. Suppose that $\sum_{i=0}^{n} a_i q_i = 0$. The coefficient for x^n in $\sum_{i=0}^{n} a_i q_i$ is a_n, thus $a_n = 0$, and then $\sum_{i=0}^{n-1} a_i q_i = 0$, and from here we obtain that $a_{n-1} = 0, ..., a_0 = 0$. Thus $\{q_0, q_1, ..., q_n\}$ is a linearly independent system. Let $f \in P_n$. Then

$$f(x) = \sum_{i=0}^{n} a_i x^i \Leftrightarrow f = a_n q_n + (a_{n-1} - a_n) q_{n-1} + \cdots + (a_2 - a_1) q_2 + (a_1 - a_0) q_1 + 2 a_0 q_0,$$

and therefore $\{q_0, q_1, ..., q_n\}$ generates P_n. We apply the Gram–Schmidt procedure to $\{q_0, q_1, ..., q_n\}$, and we obtain an orthonormal basis $\{p_0, p_1, ..., p_n\}$ in $(P_n, \langle \cdot, \cdot \rangle)$ such that $p_0 = q_0 / \|q_0\|_{\langle \cdot, \cdot \rangle}$ and

$$p_k = \frac{1}{\sqrt{D_k D_{k-1}}} \begin{vmatrix} q_0 & q_1 & \cdots & q_k \\ \langle q_0, q_0 \rangle & \langle q_1, q_0 \rangle & \cdots & \langle q_k, q_0 \rangle \\ \langle q_0, q_1 \rangle & \langle q_1, q_1 \rangle & \cdots & \langle q_k, q_1 \rangle \\ \cdots & \cdots & \cdots & \cdots \\ \langle q_0, q_{k-1} \rangle & \langle q_1, q_{k-1} \rangle & \cdots & \langle q_k, q_{k-1} \rangle \end{vmatrix} \quad \forall k = \overline{1, n},$$

where $D_k = \det\left(\langle q_i, q_j \rangle\right)_{0 \le i,j \le k} \; \forall k \ge 0$. Then, for $k \ge 0$, the polynomial p_k has degree k, since the coefficient of q_k in the expression for p_k as a linear combination of $\{q_0, q_1, ..., q_k\}$ is $D_{k-1} \ne 0$.

ii) For $k \ge 0$, we have $\left(\|p_k\|^2\right)' = 0$, $\left(\langle p_k, p_k \rangle\right)' = 0$, that is, $2 \langle p_k, p_k' \rangle = 0$, i.e., $\langle p_k, p_k' \rangle = 0$.

4. We observe that $\langle (x_n)_{n\in\mathbb{N}}, (y_n)_{n\in\mathbb{N}}\rangle$ is well defined for any $(x_n)_{n\in\mathbb{N}}, (y_n)_{n\in\mathbb{N}} \in l_0^2$, since the summation has only a finite number of terms. $\langle\cdot,\cdot\rangle$ being an inner product is analogous with the case l_2. We prove now that the norm generated by $\langle\cdot,\cdot\rangle$ is not a complete one. For any $(x_n)_{n\in\mathbb{N}} \in l_0^2$ we have $\|(x_n)_{n\in\mathbb{N}}\| = \left(\sum_{n=1}^{\infty}|x_n|^2\right)^{1/2}$. Let us consider the sequence $(y_n)_{n\in\mathbb{N}} \subseteq l_0^2$ defined by $y_n = (1, 1/2, ..., 1/n, 0, 0, ...)$ $\forall n \in \mathbb{N}$. Then for any $m > n$ we have

$$\|y_n - y_m\| = \left\|\left(0, ..., 0, \frac{1}{n+1}, ..., \frac{1}{m}, 0, 0, ...\right)\right\| = \left(\sum_{k=n+1}^{m}\frac{1}{k^2}\right)^{1/2}.$$

From this, using the convergence of the series $\sum_{k=1}^{\infty} 1/k^2$ it follows that the sequence $(y_n)_{n\in\mathbb{N}}$ is Cauchy. Suppose that there is $y_0 \in l_0^2$ such that $y_n \to y_0$ in norm. Then for any $k \in \mathbb{N}$ we have $y_n^k \to y_0^k$, with the usual notations. But $y_n^k = 1/k$ $\forall n \geq k$ and therefore $y_0^k = 1/k$ $\forall k \in \mathbb{N}$ and hence $y_0 = (1, 1/2, 1/3, ...)$, which contradicts the fact that $y_0 \in l_0^2$.

5. See [45, exercise 19, chapter 4].

We prove first that the limit in the definition of $\langle\cdot,\cdot\rangle$ exists. It is sufficient to prove that $\lim_{A\to\infty} \frac{1}{2A}\int_{-A}^{A} e^{i\alpha t}e^{-i\beta t}dt$ exists, for any $\alpha, \beta \in \mathbb{R}$. If $\alpha = \beta$ then $\int_{-A}^{A} e^{i\alpha t}e^{-i\beta t}dt = 2A$, and the limit is 1. If $\alpha \neq \beta$, then

$$\frac{1}{2A}\int_{-A}^{A} e^{i\alpha t}e^{-i\beta t}dt = \frac{1}{2iA(\alpha-\beta)}\left(e^{i(\alpha-\beta)A} - e^{-i(\alpha-\beta)A}\right) = \frac{\sin((\alpha-\beta)A)}{A(\alpha-\beta)},$$

and therefore the limit is 0. Thus the map $\langle\cdot,\cdot\rangle : X \times X \to \mathbb{C}$ is well defined. Evidently, $\langle af + bg, h\rangle = a\langle f, h\rangle + b\langle g, h\rangle$ $\forall f, g, h \in X$, $a, b \in \mathbb{C}$ and $\langle f, g\rangle = \overline{\langle g, f\rangle}$ $\forall f, g \in X$. For $f(t) = \sum_{k=1}^{n} c_k e^{i\alpha_k t}$ with the α_k different two by two, we have

$$\|f\|^2 = \langle f, f\rangle = \sum_{k=1}^{n} c_k \overline{c_k} \langle e^{i\alpha_k t}, e^{i\alpha_k t}\rangle = \sum_{k=1}^{n} |c_k|^2.$$

Thus $\langle f, f\rangle \geq 0$ $\forall f \in X$, and if $\langle f, f\rangle = 0$ then $\sum_{k=1}^{n} |c_k|^2 = 0$, hence $c_k = 0$ $\forall 1 \leq k \leq n$, i.e., $f = \sum_{k=1}^{n} c_k e^{i\alpha_k t} = 0$. We obtain that $(X, \langle\cdot,\cdot\rangle)$ is an inner product space.

It is not a complete space, since if $(c_k)_{k\in\mathbb{N}} \subseteq l_2$, $c_k \neq 0$ $\forall k \in \mathbb{N}$ then for the sequence $(f_n)_{n\in\mathbb{N}} \subseteq X$, $f_n(t) = \sum_{k=1}^{n} c_k e^{ikt}$, we have $\|f_n - f_m\|^2 = \sum_{k=n}^{m} |c_k|^2$, and from the convergence of the series $\sum_{k=1}^{\infty} |c_k|^2$ it follows that $(f_n)_{n\in\mathbb{N}} \subseteq X$ is a Cauchy sequence. Suppose that there is $f(t) = \sum_{k=1}^{m} b_k e^{i\alpha_k t}$ such that $\|f_n - f\| \to 0$. Let $M \in \mathbb{N}$ such that $\alpha_k < M$, $k = \overline{1, m}$. For $n \geq M$ we have

$$(f_n - f)(t) = \sum_{k \neq M} a_k e^{i\beta_k t} + c_M e^{iMt},$$

10.2 Solutions

with the β_k different two by two, and $\beta_k \neq M$ $\forall k \in \mathbb{N}$ (only a finite number of a_k are different from zero). Then $\|f_n - f\|^2 = \sum\limits_{k \neq M} |a_k|^2 + |c_M|^2 \geq |c_M|^2$, i.e., $\|f_n - f\| \geq |c_M|$ $\forall n \geq M$, and since $c_M \neq 0$ we obtain a contradiction. Thus $(X, \langle \cdot, \cdot \rangle)$ is an inner product space, which is not a Hilbert space.

iii) Let $(H, \langle \cdot, \cdot \rangle)$ be the Hilbert space obtained by completing X. For any $s \in \mathbb{R}$ we consider the function $f_s \in X$, $f_s(t) = e^{ist}$ $\forall t \in \mathbb{R}$. For $s \in \mathbb{R}$, $\|f_s\| = 1$, and for $s, u \in \mathbb{R}$, $s \neq u$, $\langle f_s, f_u \rangle = 0$. Thus $(f_s)_{s \in \mathbb{R}} \subseteq H$ is orthonormal and uncountable. Since $\|f_s - f_u\| = \sqrt{2}$ for $s \neq u$, using exercise 21 of chapter 2 it follows that H is not separable.

6. See [2, chapter II, section 11].

i) Let $x \in V$, i.e., $\Re \langle x, a \rangle = 1$. Then $y = Tx \in H \setminus \{0\}$ has the property that

$$\left\| y - \frac{a}{2} \right\|^2 = \|y\|^2 + \frac{\|a\|^2}{4} - \Re \langle y, a \rangle = \frac{1}{\|x\|^2} + \frac{\|a\|^2}{4} - \Re \langle Tx, a \rangle$$

$$= \frac{1}{\|x\|^2} + \frac{\|a\|^2}{4} - \frac{\Re \langle x, a \rangle}{\|x\|^2} = \frac{\|a\|^2}{4}.$$

Conversely, let $y \in H \setminus \{0\}$ with $\|y - a/2\| = \|a\|/2$, i.e., $\|y\|^2 - \Re \langle y, a \rangle = 0$. Then $x = Ty \in H \setminus \{0\}$ has the property that $\Re \langle x, a \rangle = \Re \langle Ty, a \rangle = \Re \langle y, a \rangle / \|y\|^2 = 1$. We have $x \in V$ and $Tx = T(Ty) = y$.

ii) Let $x \in C \setminus \{0\}$. Then $\|x - a\| = \|a\|$ and therefore $\|x\|^2 = 2\Re \langle x, a \rangle$. Then $\Re \langle Tx, a \rangle = \Re \langle x, a \rangle / \|x\|^2 = 1/2$. Conversely, let $y \in H$ with $\Re \langle y, a \rangle = 1/2$. Then $y \neq 0$ and let $x = Ty$. Then

$$\|x - a\|^2 = \|x\|^2 - 2\Re \langle x, a \rangle + \|a\|^2 = \frac{1}{\|y\|^2} - 2\frac{\Re \langle y, a \rangle}{\|y\|^2} + \|a\|^2 = \|a\|^2,$$

i.e., $\|x - a\| = \|a\|$, $x \in C \setminus \{0\}$. Then $y = Tx$ with $x \in C \setminus \{0\}$.

iii) Let $C(a, r) = \{x \in H \mid \|x - a\| = r\}$. We must prove that $T(C(a, r)) = \{y \in H \mid \|y - a/\lambda\| = r/|\lambda|\} = C(a/\lambda, r/|\lambda|)$, where $\lambda = \|a\|^2 - r^2$. Let $x \in C(a, r)$. Then $\|x - a\|^2 = r^2$, i.e., $\|x\|^2 - 2\Re \langle x, a \rangle = -\lambda$. Using this property we have

$$\left\| T(x) - \frac{a}{\lambda} \right\|^2 = \|T(x)\|^2 - 2\Re \left\langle T(x), \frac{a}{\lambda} \right\rangle + \frac{\|a\|^2}{\lambda^2} = \frac{1}{\|x\|^2} - \frac{2\Re \langle x, a \rangle}{\lambda \|x\|^2} + \frac{\|a\|^2}{\lambda^2}$$

$$= \frac{\lambda - 2\Re \langle x, a \rangle}{\lambda \|x\|^2} + \frac{\|a\|^2}{\lambda^2} = -\frac{1}{\lambda} + \frac{\|a\|^2}{\lambda^2} = \frac{r^2}{\lambda^2},$$

hence $T(x) \in C(a/\lambda, r/|\lambda|)$, i.e.,

$$T(C(a, r)) \subseteq C\left(a/\left(\|a\|^2 - r^2 \right), r/\left| \|a\|^2 - r^2 \right| \right).$$

For the reverse inclusion, we have

$$T((C(a/\lambda, r/|\lambda|)) \subseteq C\left(\frac{a/\lambda}{\|a/\lambda\|^2 - r^2/\lambda^2}, \frac{r/|\lambda|}{\left| \|a/\lambda\|^2 - (r/|\lambda|)^2 \right|} \right) = C(a, r).$$

Since $0 \notin C(a,r)$ and $T^2(x) = x \ \forall x \in H\setminus\{0\}$, it follows that $C(a/\lambda, r/|\lambda|) \subseteq T(C(a,r))$.

7. See [9, exercise 5, chapter V, section 1].
i) We have

$$\begin{aligned}\|a' - b'\|^2 &= \|a'\|^2 + \|b'\|^2 - 2\Re\langle a', b'\rangle = \frac{1}{\|a\|^2} + \frac{1}{\|b\|^2} - \frac{2\Re\langle a, b\rangle}{\|a\|^2 \|b\|^2} \\ &= \frac{\|a\|^2 + \|b\|^2 - 2\Re\langle a, b\rangle}{\|a\|^2 \|b\|^2} = \frac{\|a - b\|^2}{\|a\|^2 \|b\|^2}.\end{aligned}$$

ii) Let $a, b, c \in H\setminus\{0\}$ and $a', b', c' \in H$ as in (i). Then $\|a' - c'\| \leq \|a' - b'\| + \|b' - c'\|$. Using (i) we obtain that

$$\frac{\|a - c\|}{\|a\| \|c\|} \leq \frac{\|a - b\|}{\|a\| \|b\|} + \frac{\|b - c\|}{\|b\| \|c\|},$$

that is, $\|a - c\| \|b\| \leq \|a - b\| \|c\| + \|b - c\| \|a\|$ (1), i.e., the Ptolemaic theorem for $d = 0$. If one of a, b, c is zero, for example $a = 0$, then (1) becomes $\|c\| \|b\| \leq \|b\| \|c\|$. Therefore, (1) is true for any $a, b, c \in H$.

Let now $a, b, c, d \in H$. We apply (1) for $a - d, b - d, c - d$ and we obtain that

$$\|a - c\| \|b - d\| \leq \|a - b\| \|c - d\| + \|b - c\| \|a - d\|.$$

8. See [6, exercise 4, chapter 9].
We have

$$\sum_{(\varepsilon_1,\ldots,\varepsilon_n)\in\{\pm 1\}^n} \left\|\sum_{i=1}^n \varepsilon_i x_i\right\|^2 = \sum_{(\varepsilon_1,\ldots,\varepsilon_n)\in\{\pm 1\}^n} \sum_{i,j=1}^n \varepsilon_i\varepsilon_j \langle x_i, x_j\rangle$$

$$= \sum_{i,j=1}^n \langle x_i, x_j\rangle \sum_{(\varepsilon_1,\ldots,\varepsilon_n)\in\{\pm 1\}^n} \varepsilon_i\varepsilon_j.$$

For $i = j$, $\sum_{(\varepsilon_1,\ldots,\varepsilon_n)\in\{\pm 1\}^n} \varepsilon_i\varepsilon_j = 2^n$. For $i \neq j$, $\sum_{(\varepsilon_1,\ldots,\varepsilon_n)\in\{\pm 1\}^n} \varepsilon_i\varepsilon_j = 0$, since in our summation we have 2^{n-1} elements equal to 1 and 2^{n-1} elements equal to -1. We obtain that

$$\sum_{(\varepsilon_1,\ldots,\varepsilon_n)\in\{\pm 1\}^n} \left\|\sum_{i=1}^n \varepsilon_i x_i\right\|^2 = 2^n \sum_{i=1}^n \langle x_i, x_i\rangle = 2^n \sum_{i=1}^n \|x_i\|^2.$$

9. See [1, exercise 11, Spring 1986].
We have

$$\int_{-\infty}^{\infty} f_m(x)\overline{f_n(x)}dx = \frac{1}{\pi}\int_{-\infty}^{\infty} \frac{(x-i)^m}{(x+i)^{m+1}} \frac{(x+i)^n}{(x-i)^{n+1}}dx = \frac{1}{\pi}\int_{-\infty}^{\infty} \frac{(x-i)^{(m-n)-1}}{(x+i)^{(m-n)+1}}dx.$$

10.2 Solutions

For m and n we have the following cases:

a) $m - n \geq 1$. We consider then the function $f : \mathbb{C}\setminus\{-i\} \to \mathbb{C}$ given by $f(z) = (z-i)^{(m-n)-1}/(z+i)^{(m-n)+1}$ and the domain in the complex plane delimited by Ox and the upper half of the circle of center 0 and radius R (denoted by $\Gamma(R)$). On this domain f is a holomorphic function, and therefore

$$\frac{1}{\pi}\int_{-R}^{R}\frac{(x-i)^{(m-n)-1}}{(x+i)^{(m-n)+1}}dx + \int_{\Gamma(R)} f(z)\,dz = 0,$$

for any $R > 0$. But $\lim_{|z|\to\infty}|zf(z)| = 0$ and therefore $\lim_{R\to\infty}\int_{\Gamma(R)} f(z)\,dz = 0$. We obtain that

$$\int_{-\infty}^{\infty}\frac{(x-i)^{(m-n)-1}}{(x+i)^{(m-n)+1}}dx = 0.$$

b) $m - n \leq -1$. We obtain that

$$\int_{-\infty}^{\infty}\frac{(x-i)^{(m-n)-1}}{(x+i)^{(m-n)+1}}dx = \int_{-\infty}^{\infty}\frac{(x+i)^{(n-m)-1}}{(x-i)^{(n-m)+1}}dx = \int_{-\infty}^{\infty}\frac{(-t+i)^{(n-m)-1}}{(-t-i)^{(n-m)+1}}dt$$

$$= \int_{-\infty}^{\infty}\frac{(t-i)^{(n-m)-1}}{(t+i)^{(n-m)+1}}dt = 0,$$

by (a), since $n - m \geq 1$.

c) $m - n = 0$. Then

$$\int_{-\infty}^{\infty} f_m(x)\overline{f_m(x)}dx = \frac{1}{\pi}\int_{-\infty}^{\infty}\frac{1}{1+x^2}dx = \frac{1}{\pi}\arctan x\Big|_{-\infty}^{\infty} = 1.$$

10. i) Let $x \in \overline{A}$ and suppose that $x \notin A$. Then there is a sequence $(a_n)_{n\in\mathbb{N}} \subseteq A$ such that $a_n \to x$ in H. Since $x \notin A$ we have $a_n \neq x\ \forall n \in \mathbb{N}$. Let $n_1 = 1$. Then $\|a_{n_1} - x\| > 0$ and from $a_n \to x$, there is $n_2 > 1$ such that $\|a_{n_2} - x\| < \|a_{n_1} - x\|/2$. Then $a_{n_2} \neq a_{n_1}$. Let now $r_2 = \min\{\|a_{n_2} - x\|/2, \|x - a_{n_1}\|/2\}$. From $a_n \neq x\ \forall n \in \mathbb{N}$ it follows that $r_2 > 0$ and since $a_n \to x$ there is $n_3 > \max(n_1, n_2)$ such that $\|a_{n_3} - x\| < r_2$. From the definition of r_2 it follows that $a_{n_3} \neq a_{n_2}$ and $a_{n_3} \neq a_{n_1}$. In this way, by induction, we construct a subsequence $(n_k)_{k\in\mathbb{N}} \subseteq \mathbb{N}$ such that $a_{n_k} \neq a_{n_s}$ for $k \neq s$. Evidently $a_{n_k} \to x$ and therefore $(a_{n_k})_{k\in\mathbb{N}}$ is a Cauchy sequence, which is false, since by orthonormality, $\|a_{n_k} - a_{n_s}\| = \sqrt{2}\ \forall k \neq s$. Thus $x \in A$, that is, $\overline{A} \subseteq A$, and therefore $A \subseteq H$ is closed.

Since $\|b - a\| = \sqrt{2}\ \forall a, b \in A,\ a \neq b$, it is easy to see that A is not totally bounded (A cannot contain Cauchy sequences).

ii) Suppose that $(a_n)_{n\in\mathbb{N}}$ does not converge towards zero, and then we can find an $\varepsilon > 0$ and a subsequence $(n_k)_{k\in\mathbb{N}}$ of $\mathbb{N}$ such that $|a_{n_k}| \geq \varepsilon\ \forall k \in \mathbb{N}$. The set $\{a_{n_k}e_{n_k} \mid k \in \mathbb{N}\}$ is totally bounded, and then we can find a subsequence $(n_{k_p})_{p\in\mathbb{N}}$ of $(n_k)_{k\in\mathbb{N}}$ such that the sequence $\left(a_{n_{k_p}}e_{n_{k_p}}\right)_{p\in\mathbb{N}}$ converges. Then the sequence $\left(a_{n_{k_p}}e_{n_{k_p}}\right)_{p\in\mathbb{N}}$ is Cauchy, and this contradicts the fact that

$$\left\|a_{n_{k_p}}e_{n_{k_p}} - a_{n_{k_{p+s}}}e_{n_{k_{p+s}}}\right\|^2 = \left|a_{n_{k_p}}\right|^2 + \left|a_{n_{k_{p+s}}}\right|^2 \geq 2\varepsilon^2\ \forall p, s \in \mathbb{N}.$$

Conversely, we have $\|a_n e_n\| \to 0$, so the set $\{a_n e_n \mid n \in \mathbb{N}\} \cup \{0\}$ is compact, hence $\{a_n e_n \mid n \in \mathbb{N}\}$ is totally bounded.

For the second assertion we can apply what we have proved above, observing that $x_n = \|x_n\| e_n \ \forall n \in \mathbb{N}$, where $e_n = x_n / \|x_n\| \ \forall n \in \mathbb{N}$.

11. i) Consider $x^* \in H^*$. By Riesz's representation theorem there is a unique element $x \in H$ such that $x^*(y) = \langle y, x \rangle \ \forall y \in H$. For $x \in H$, using Bessel's inequality we have $\sum_{n=1}^{\infty} |\langle x_n, x \rangle|^2 \leq \|x\|^2$ and therefore $\lim_{n \to \infty} \langle x_n, x \rangle = 0$, that is, $\lim_{n \to \infty} x^*(x_n) = 0$.

ii) See [45, exercise 9, chapter 4].

Consider the space $L_2([0, 2\pi]; \mathbb{C})$ with the usual inner product. For any $n \in \mathbb{N}$, let $f_n : [0, 2\pi] \to \mathbb{C}$ defined by

$$f_n(t) = \frac{e^{int}}{\sqrt{2\pi}} = \frac{1}{\sqrt{2\pi}} (\cos nt + i \sin nt) \ \forall t \in [0, 2\pi].$$

Then the sequence $(f_n)_{n \in \mathbb{N}}$ has the properties from (i) and we obtain that

$$\lim_{n \to \infty} \int_{[0, 2\pi]} f_n(t) \overline{f(t)} dt = 0 \ \forall f \in L_2([0, 2\pi]; \mathbb{C}).$$

Since $A \subseteq [0, 2\pi]$ is a Lebesgue measurable set we obtain that $\chi_A \in L_2([0, 2\pi]; \mathbb{C})$, and therefore $\lim_{n \to \infty} \int_{[0, 2\pi]} f_n(t) \chi_A(t) dt = 0$, that is, $\lim_{n \to \infty} \int_A f_n(t) dt = 0$, and then $\lim_{n \to \infty} \int_A \cos(nt) dt = 0$, $\lim_{n \to \infty} \int_A \sin(nt) dt = 0$.

12. Indeed, by the Riesz representation theorem there is a unique element $y \in H$ such that $x^*(x) = \langle x, y \rangle \ \forall x \in H$. In particular, $x^*(e_n) = \langle e_n, y \rangle \ \forall n \in \mathbb{N}$. Since $(e_n)_{n \in \mathbb{N}}$ is an orthonormal basis we have $y = \sum_{n=1}^{\infty} \langle y, e_n \rangle e_n$, so by the above $y = \sum_{n=1}^{\infty} \overline{x^*(e_n)} e_n$, i.e., the statement. Since $\|x^*\| = \|y\|$ and $(e_n)_{n \in \mathbb{N}}$ is an orthonormal basis, we have $\|x^*\| = \sqrt{\sum_{n=1}^{\infty} |x^*(e_n)|^2}$.

13. We begin with the following remark. For any $f : \mathbb{D} \to \mathbb{C}$ and any $0 \leq r < 1$, let $f_r : [0, 2\pi] \to \mathbb{C}$ be defined by $f_r(\theta) = f(re^{i\theta})$. Also let μ be the measure defined by $\mu(E) = (1/(2\pi)) m(E) \ \forall E \subseteq [0, 2\pi]$ Lebesgue measurable, where m is the Lebesgue measure on $[0, 2\pi]$. Then

$$\|f_r\|_p = \left(\int_0^{2\pi} |f_r|^p d\mu \right)^{1/p} = \left(\frac{1}{2\pi} \int_0^{2\pi} |f(re^{i\theta})|^p d\theta \right)^{1/p}.$$

From here it follows that $f \in H^p(\mathbb{D})$ if and only if $N_p(f) = \sup_{0 \leq r < 1} \|f_r\|_p < \infty$.

i) We will prove that if $f : \mathbb{D} \to \mathbb{C}$ is an analytic function, then

$$\iint_{|z - z_0| \leq R} |f(z)|^p \, dx dy = \int_0^R r \left(\int_0^{2\pi} |f(z_0 + re^{i\theta})|^p d\theta \right) dr \qquad (1)$$

10.2 Solutions

for $z_0 \in \mathbb{D}$ and $0 < R < 1 - |z_0|$,

$$\iint_{|z| \leq \rho} |f(z)|^p \, dxdy \leq \pi \rho^2 [N_p(f)]^p \tag{2}$$

for $0 < \rho < 1$,

$$\iint_{|z| \leq \rho} |f(z_0 + z)|^p \, dxdy = \iint_{|z-z_0| \leq \rho} |f(z)|^p \, dxdy \tag{3}$$

for $z_0 \in \mathbb{D}$ and $0 < \rho < 1 - |z_0|$,

$$\frac{\rho^2}{2} |f(0)|^p \leq \frac{1}{2\pi} \iint_{|z| \leq \rho} |f(z)|^p \, dxdy \tag{4}$$

for $0 \leq \rho < 1$, and then we will obtain the relation from the statement.

For (1), let $z = x + iy$ and $z_0 = x_0 + iy_0$, $f = u + iv$, that is, $f(z) = u(x,y) + iv(x,y)$, $|f(z)|^2 = |u(x,y)|^2 + |v(x,y)|^2$, and then for the left side in (1) we have

$$\iint_{|z-z_0| \leq R} |f(z)|^p \, dxdy = \iint_{(x-x_0)^2 + (y-y_0)^2 \leq R^2} (u^2(x,y) + v^2(x,y))^{p/2} dxdy.$$

Using polar coordinates, $x = x_0 + r\cos\theta$ and $y = y_0 + r\cos\theta$ with $r \in [0, R]$ and $\theta \in [0, 2\pi]$, we obtain that

$$\iint_{|z-z_0| \leq R} |f(z)|^p \, dxdy = \iint_{[0,R] \times [0,2\pi]} \left((u^2 + v^2)(x_0 + r\cos\theta, y_0 + r\sin\theta) \right)^{p/2} r \, dr d\theta.$$

From the Fubini theorem the right side is equal to

$$\int_0^R r \left(\int_0^{2\pi} |f(x_0 + r\cos\theta + iy_0 + ri\sin\theta)|^p \, d\theta \right) dr,$$

and therefore equal to $\int_0^R r \left(\int_0^{2\pi} |f(z_0 + re^{i\theta})|^p \, d\theta \right) dr$, and (1) is proved.

For (2), we will use (1) for $z_0 = 0$ to deduce that

$$\iint_{|z| \leq \rho} |f(z)|^p \, dxdy = \int_0^\rho r \left(\int_0^{2\pi} |f(re^{i\theta})|^p \, d\theta \right) dr \leq 2\pi (N_p(f))^p \int_0^\rho r \, dr$$

$$= \pi \rho^2 (N_p(f))^p \quad \forall 0 < \rho < 1.$$

For (3) we have that

$$\iint_{|z| \leq \rho} |f(z_0 + z)|^p \, dxdy = \iint_{x^2 + y^2 \leq \rho^2} |f(x_0 + x, y_0 + y)|^p \, dxdy.$$

If $x = u - x_0$ and $y = v - y_0$, then

$$\iint_{|z|\leq \rho} |f(z_0 + z)|^p\, dxdy = \iint_{(u-x_0)^2+(v-y_0)^2\leq \rho^2} |f(u,v)|^p\, dudv = \iint_{|z-z_0|\leq \rho} |f(z)|^p\, dxdy.$$

To prove (4), let us consider $0 \leq r < 1$. Then $\overline{D}(0,r) = \{z \in \mathbb{C} \mid |z| \leq r\} \subseteq \mathbb{D}$. Using Cauchy's formula, we obtain that $f(0) = (1/(2\pi i)) \int_{|w|=r} (f(w)/w)\, dw$. If we write $w = re^{i\theta}$, $\theta \in [0, 2\pi]$, $dw = rie^{i\theta} d\theta$, we deduce that $f(0) = (1/(2\pi)) \int_0^{2\pi} f(re^{i\theta}) d\theta$, from whence $|f(0)| \leq (1/(2\pi)) \int_0^{2\pi} |f(re^{i\theta})|\, d\theta$. If $p > 1$, then

$$|f(0)| \leq \frac{1}{2\pi} \left(\int_0^{2\pi} |f(re^{i\theta})|^p\, d\theta \right)^{1/p} \left(\int_0^{2\pi} 1 d\theta \right)^{1/q} = \left(\frac{1}{2\pi} \int_0^{2\pi} |f(re^{i\theta})|^p\, d\theta \right)^{1/p},$$

by Hölder's inequality (here $1/p + 1/q = 1$). Hence for all $p \geq 1$ we have proved that $|f(0)|^p \leq (1/2\pi) \int_0^{2\pi} |f(re^{i\theta})|^p\, d\theta$, and therefore

$$|f(0)|^p \int_0^\rho r\, dr \leq \frac{1}{2\pi} \int_0^\rho r \left(\int_0^{2\pi} |f(re^{i\theta})|^p\, d\theta \right) dr.$$

Using (1) for the right side of the inequality we obtain that

$$\frac{\rho^2}{2} |f(0)|^p \leq \frac{1}{2\pi} \iint_{|z|\leq \rho} |f(z)|^p\, dxdy.$$

(We observe that for (1), (3) and (4) it is not necessary for f to be defined on all the unit disc.)

Now let $f \in H^p(\mathbb{D})$, $z_0 \in \mathbb{D}$ and $0 < r < d(z_0, \partial \mathbb{D})$. We consider $g : \overline{D}(0,r) \to \mathbb{C}$ defined by $g(z) = f(z + z_0)$. Using (4) we obtain that for any $0 \leq \rho < r$ we have

$$\frac{\rho^2}{2} |g(0)|^p \leq \frac{1}{2\pi} \iint_{|z|\leq \rho} |g(z)|^p\, dxdy,$$

that is,

$$\frac{\rho^2}{2} |f(z_0)|^p \leq \frac{1}{2\pi} \iint_{|z|\leq \rho} |f(z + z_0)|^p\, dxdy.$$

Using (3) we obtain that

$$\frac{\rho^2}{2} |f(z_0)|^p \leq \frac{1}{2\pi} \iint_{|z-z_0|\leq \rho} |f(z)|^p\, dxdy.$$

But

$$\frac{1}{2\pi} \iint_{|z-z_0|\leq \rho} |f(z)|^p\, dxdy \leq \frac{1}{2\pi} \iint_{|z|\leq \rho+|z_0|} |f(z)|^p\, dxdy.$$

10.2 Solutions

Since $\rho \leq r < d(z_0, \partial \mathbb{D})$ it follows that $\rho + |z_0| < 1$, and using (2) we obtain that

$$\frac{1}{2\pi} \iint\limits_{|z| \leq \rho + |z_0|} |f(z)|^p \, dx dy \leq \frac{\pi (N_p(f))^p}{2\pi}.$$

Then

$$\frac{\rho^2}{2} |f(z_0)|^p \leq \frac{\pi (N_p(f))^p}{2\pi},$$

and for $\rho \to r$ we obtain that $|f(z_0)|^p \leq (N_p(f))^p / r^2$, that is, $|f(z_0)| \leq N_p(f)/r^{2/p}$ for any $0 < r < d(z_0, \partial \mathbb{D})$. Now passing to the limit for $r \to d(z_0, \partial \mathbb{D})$ we obtain that $|f(z_0)| \leq N_p(f)/(d(z_0, \partial \mathbb{D}))^{2/p}$, and (i) is proved.

ii) If $f, g \in H^p(\mathbb{D})$, then obviously $(f+g)_r = f_r + g_r$ and since $f_r, g_r \in L_p(\mu)$ and $L_p(\mu)$ is a linear space, we obtain that $f_r + g_r \in L_p(\mu)$. Using Minkowski's inequality we have $\|(f+g)_r\|_p \leq \|f_r\|_p + \|g_r\|_p \leq N_p(f) + N_p(g) \; \forall 0 \leq r < 1$ and therefore $N_p(f+g) = \sup\limits_{0 \leq r < 1} \|(f+g)_r\|_p \leq N_p(f) + N_p(g) < \infty$, $f + g \in H^p(\mathbb{D})$. If $f \in H^p(\mathbb{D})$ and $\lambda \in \mathbb{C}$ then $(\lambda f)_r = \lambda f_r$, $\|(\lambda f)_r\|_p = |\lambda| \|f_r\|_p \; \forall 0 \leq r < 1$, and therefore $\lambda f \in H^p(\mathbb{D})$, $N_p(\lambda f) = |\lambda| N_p(f) < \infty$. If $N_p(f) = 0$, then using (i) we have $|f(z)| \leq 0 \; \forall z \in \mathbb{D}$, i.e., $f = 0$.

Thus $(H^p(\mathbb{D}), N_p)$ is a normed space.

iii) Consider a Cauchy sequence $(f_n)_{n \in \mathbb{N}} \subseteq H^p(\mathbb{D})$. Let $K \subseteq \mathbb{D}$ be a compact, nonempty set. Then $d(K, \partial \mathbb{D}) > 0$ and $d(z, \partial \mathbb{D}) \geq d(K, \partial \mathbb{D}) \; \forall z \in K$. Using (i) we obtain that

$$|f_n(z) - f_m(z)| \leq \frac{1}{(d(z, \partial \mathbb{D})^2)^{1/p}} N_p(f_n - f_m) \leq \frac{1}{(d(K, \partial \mathbb{D})^2)^{1/p}} N_p(f_n - f_m).$$

Since the sequence $(f_n)_{n \in \mathbb{N}} \subseteq H^p(\mathbb{D})$ is Cauchy from this last inequality we deduce that $(f_n)_{n \in \mathbb{N}}$ is uniformly Cauchy on K. Using the Weierstrass theorem we obtain that there is $f : \mathbb{D} \to \mathbb{C}$ analytic such that $f_n \to f$ uniformly on every compact set $K \subseteq \mathbb{D}$. Since $(f_n)_{n \in \mathbb{N}} \subseteq H^p(\mathbb{D})$ is Cauchy it is bounded, i.e., $\sup\limits_{n \in \mathbb{N}} N_p(f_n) = M < \infty$. Let $0 \leq r < 1$. Since $f_n \to f$ uniformly on every compact set in $\mathbb{D}$ we obtain that $f_n \to f$ uniformly on $\overline{D}(0, r)$, i.e., $f_n(z) \to f(z)$ uniformly with respect to $|z| \leq r$, hence $f_n(re^{i\theta}) \to f(re^{i\theta})$ uniformly with respect to $\theta \in [0, 2\pi]$. Then $(f_n)_r \to f_r$ uniformly on $[0, 2\pi]$, from whence $\|(f_n)_r - f_r\|_p \to 0$ and $\|(f_n)_r\|_p \to \|f_r\|_p$. Therefore $\|f_r\|_p = \lim\limits_{n \to \infty} \|(f_n)_r\|_p \leq \sup\limits_{n \in \mathbb{N}} N_p(f_n) = M \; \forall 0 \leq r < 1$, thus $N_p(f) = \sup\limits_{0 \leq r < 1} \|f_r\|_p \leq M < \infty$, i.e., $f \in H^p(\mathbb{D})$. The sequence $(f_n)_{n \in \mathbb{N}}$ is Cauchy, i.e., $\forall \varepsilon > 0 \; \exists n_\varepsilon \in \mathbb{N}$ such that $N_p(f_n - f_m) < \varepsilon \; \forall n, m \geq n_\varepsilon$, that is, $\|(f_n)_r - (f_m)_r\|_p < \varepsilon \; \forall n, m \geq n_\varepsilon, \; \forall 0 \leq r < 1$. Since we have proved above that $(f_m)_r \to f_r$ in $\|\cdot\|_p$, passing to the limit for $m \to \infty$ in the last inequality, we obtain that that $\|(f_n)_r - f_r\|_p \leq \varepsilon \; \forall n \geq n_\varepsilon, \; \forall 0 \leq r < 1$, from whence $N_p(f_n - f) \leq \varepsilon \; \forall n \geq n_\varepsilon$, i.e., $f_n \to f$ in $H^p(\mathbb{D})$.

iv) We observe first that if $f : [-1, 1] \to \mathbb{R}$ is a continuous function, then we have $\int_0^{2\pi} f(\cos \theta) d\theta = 2 \int_0^{\pi/2} [f(\cos \theta) + f(-\cos \theta)] d\theta$. For $f(t) = (1 - 2rt + r^2)^s$, $s > 0$, it follows that

$$\frac{1}{2\pi}\int_0^{2\pi}(1-2r\cos\theta+r^2)^s d\theta = \frac{1}{\pi}\int_0^{\pi/2}[(1-2r\cos\theta+r^2)^s+(1+2r\cos\theta+r^2)^s]d\theta$$
$$= \frac{1}{2\pi}\int_0^{2\pi}(1+2r\cos\theta+r^2)^s d\theta.$$

Suppose that $s>0$ is such that

$$\sup_{0\le r<1}\frac{1}{\pi}\int_0^{\pi/2}[(1-2r\cos\theta+r^2)^s+(1+2r\cos\theta+r^2)^s]d\theta = 2^s.$$

We will show that $s=1$. Let $h:[0,1]\to\mathbb{R}$ be defined by

$$h(r) = \frac{1}{\pi}\int_0^{\pi/2}[(1-2r\cos\theta+r^2)^s+(1+2r\cos\theta+r^2)^s]d\theta.$$

Using the Lebesgue dominated convergence theorem, it is easy to see that h is continuous. Our hypothesis is that $\sup_{0\le r<1} h(r)=2^s$. We can have two cases:

Case I: $0<s\le 1$. Since the function $x\longmapsto x^s$ is concave, we have $(a^s+b^s)/2 \le ((a+b)/2)^s$ $\forall a,b\ge 0$, hence

$$(1-2r\cos\theta+r^2)^s+(1+2r\cos\theta+r^2)^s \le 2(1+r^2)^s\ \forall 0\le\theta\le\pi/2,\ \forall 0\le r\le 1,$$

and by integration, $h(r)\le(1+r^2)^s$ $\forall 0\le r\le 1$. Since our hypothesis is that $\sup_{0\le r<1} h(r)=2^s$, there is a sequence $(r_k)_{k\in\mathbb{N}}\subseteq(0,1)$ such that $h(r_k)\to 2^s$. By the Cesaro lemma there is a convergent subsequence $r_{k_n}\to\rho\in[0,1]$. By the continuity of h on $[0,1]$ it follows that $h(\rho)=2^s$. Since $h(\rho)\le(1+\rho^2)^s$ we obtain that $2^s\le(1+\rho^2)^s$, hence $\rho=1$, thus $h(1)=2^s$, i.e.,

$$\frac{1}{\pi}\int_0^{\pi/2}[(2-2\cos\theta)^s+(2+2\cos\theta)^s]d\theta = 2^s.$$

Again, since $0<s\le 1$, we have $(2-2\cos\theta)^s+(2+2\cos\theta)^s\le 2\cdot 2^s$ $\forall 0\le\theta\le\pi/2$. Also,

$$\frac{1}{\pi}\int_0^{\pi/2}[(2-2\cos\theta)^s+(2+2\cos\theta)^s]d\theta = \frac{1}{\pi}\int_0^{\pi/2}2\cdot 2^s d\theta.$$

Since both functions under the integrals are continuous, we must have $(2-2\cos\theta)^s+(2+2\cos\theta)^s=2\cdot 2^s$ $\forall 0\le\theta\le\pi/2$. Denoting with $\varphi(\theta)=(2-2\cos\theta)^s+(2+2\cos\theta)^s$, we have $\varphi(\theta)=2^{s+1}$ $\forall 0\le\theta\le\pi/2$, from whence $\varphi'(\theta)=0$ $\forall 0\le\theta\le\pi/2$. But $\varphi'(\theta)=2s\sin\theta[(2-2\cos\theta)^{s-1}-(2+2\cos\theta)^{s-1}]$ and since $\varphi'(\pi/4)=0$ we obtain that $(2-\sqrt{2})^{s-1}-(2+\sqrt{2})^{s-1}=0$, and therefore $s=1$.

Case II: $1<s<\infty$. Our hypothesis is that $\sup_{0\le r<1} h(r)=2^s$, and therefore $h(r)\le 2^s$ $\forall 0\le r<1$. Passing to the limit and using the fact that h is continuous we obtain that $h(1)\le 2^s$, i.e., $(1/\pi)\int_0^{\pi/2}[(2-2\cos\theta)^s+(2+2\cos\theta)^s]d\theta\le 2^s$ (1). Since $s>$

10.2 Solutions

1, the function $x \longmapsto x^s$ is convex. Then $(a^s + b^s)/2 \geq ((a+b)/2)^s$ $\forall a, b \geq 0$, so $(2 - 2\cos\theta)^s + (2 + 2\cos\theta)^s \geq 2 \cdot 2^s$ $\forall 0 \leq \theta \leq \pi/2$, hence by integration

$$\frac{1}{\pi} \int_0^{\pi/2} [(2 - 2\cos\theta)^s + (2 + 2\cos\theta)^s] d\theta \geq \frac{1}{\pi} \int_0^{\pi/2} 2 \cdot 2^s d\theta = 2^s,$$

and combining this with (1), we have that

$$\frac{1}{\pi} \int_0^{\pi/2} [(2 - 2\cos\theta)^s + (2 + 2\cos\theta)^s] d\theta = \frac{1}{\pi} \int_0^{\pi/2} 2 \cdot 2^s d\theta.$$

Since both functions under the integrals are continuous, we must have $(2 - 2\cos\theta)^s + (2 + 2\cos\theta)^s = 2 \cdot 2^s$ $\forall 0 \leq \theta \leq \pi/2$. From here we obtain that $s = 1 \notin (1, \infty)$.

Return now to the solution for our exercise. If $H^p(\mathbb{D})$ is a Hilbert space then we must have $\|f + g\|^2 + \|f - g\|^2 = 2(\|f\|^2 + \|g\|^2)$ $\forall f, g \in H^p(\mathbb{D})$ (to simplify the notations, we write here $\|\cdot\|$ instead of N_p). Take $f(z) = 1$, $g(z) = z$ (these functions are the analogous for e_1 and e_2, elements which can be used in the solution for exercise 1 to show that if l_p is Hilbert then $p = 2$). We have $\|f\| = \|g\| = 1$,

$$\|f + g\| = \sup_{0 \leq r < 1} \left(\frac{1}{2\pi} \int_0^{2\pi} |(f+g)(re^{i\theta})|^p d\theta \right)^{1/p} = \sup_{0 \leq r < 1} \left(\frac{1}{2\pi} \int_0^{2\pi} |1 + re^{i\theta}|^p d\theta \right)^{1/p}$$
$$= \sup_{0 \leq r < 1} \left(\frac{1}{2\pi} \int_0^{2\pi} (1 + 2r\cos\theta + r^2)^{p/2} d\theta \right)^{1/p}$$

and

$$\|f - g\| = \sup_{0 \leq r < 1} \left(\frac{1}{2\pi} \int_0^{2\pi} |(f-g)(re^{i\theta})|^p d\theta \right)^{1/p} = \sup_{0 \leq r < 1} \left(\frac{1}{2\pi} \int_0^{2\pi} |1 - re^{i\theta}|^p d\theta \right)^{1/p}$$
$$= \sup_{0 \leq r < 1} \left(\frac{1}{2\pi} \int_0^{2\pi} (1 - 2r\cos\theta + r^2)^{p/2} d\theta \right)^{1/p}.$$

Then $\|f + g\| = \|f - g\|$, hence we must have $\|f + g\|^2 = 2$, i.e.,

$$\sup_{0 \leq r < 1} \left(\frac{1}{2\pi} \int_0^{2\pi} (1 - 2r\cos\theta + r^2)^{p/2} d\theta \right)^{2/p} = 2,$$

that is,

$$\sup_{0 \leq r < 1} \frac{1}{2\pi} \int_0^{2\pi} (1 - 2r\cos\theta + r^2)^{p/2} d\theta = 2^{p/2},$$

i.e., if we denote $s = p/2 > 0$, we have

$$\sup_{0 \leq r < 1} \frac{1}{2\pi} \int_0^{2\pi} (1 - 2r\cos\theta + r^2)^s d\theta = 2^s.$$

It is proved above that this implies $s = 1$, i.e., $p = 2$.

See (v) for the fact that $H^2(\mathbb{D})$ is indeed a Hilbert space.

v) See [45, theorem 17.10].

Let $f : \mathbb{D} \to \mathbb{C}$ be an analytic function, $f(z) = \sum_{n=0}^{\infty} a_n z^n$. Consider $0 \leq r < 1$. Then

$$\left| f_r(\theta) - \sum_{k=0}^{n} a_k r^k e^{i\theta k} \right| \leq \sum_{k=n+1}^{\infty} |a_k| r^k \to 0$$

uniformly with respect to $\theta \in [0, 2\pi]$, and therefore

$$\begin{aligned}
\frac{1}{2\pi} \int_0^{2\pi} |f_r(\theta)|^2 \, d\theta &= \lim_{n \to \infty} \frac{1}{2\pi} \int_0^{2\pi} \left| \sum_{k=0}^{n} a_k r^k e^{ik\theta} \right|^2 d\theta \\
&= \lim_{n \to \infty} \frac{1}{2\pi} \sum_{k,l=0}^{n} a_k \overline{a_l} r^{k+l} \int_0^{2\pi} e^{i(k-l)\theta} d\theta \\
&= \lim_{n \to \infty} \sum_{k,l=0}^{n} a_k \overline{a_l} r^{k+l} \delta_{kl} = \lim_{n \to \infty} \sum_{k=0}^{n} a_k \overline{a_k} r^{2k} \\
&= \sum_{n=0}^{\infty} |a_n|^2 r^{2n},
\end{aligned}$$

i.e., $\|f_r\|_2 = \sqrt{\sum_{n=0}^{\infty} |a_n|^2 r^{2n}}$ for $0 \leq r < 1$. Hence

$$\begin{aligned}
N_2(f) &= \sup_{0 \leq r < 1} \sqrt{\sum_{n=0}^{\infty} |a_n|^2 r^{2n}} = \sup_{0 \leq r < 1} \sup_{n \in \mathbb{N}} \sqrt{\sum_{k=0}^{n} |a_k|^2 r^{2k}} = \sup_{n \in \mathbb{N}} \sup_{0 \leq r < 1} \sqrt{\sum_{k=0}^{n} |a_k|^2 r^{2k}} \\
&= \sup_{n \in \mathbb{N}} \sqrt{\sum_{k=0}^{n} |a_k|^2} = \sqrt{\sum_{n=0}^{\infty} |a_n|^2}.
\end{aligned}$$

Thus, if $f : \mathbb{D} \to \mathbb{C}$ is defined by $f(z) = \sum_{n=0}^{\infty} a_n z^n \; \forall |z| < 1$ then $f \in H^2(\mathbb{D}) \Leftrightarrow (a_n)_{n \geq 0} \in l_2$, and $N_2(f) = \|(a_n)_{n \geq 0}\|_{l_2}$. From this we obtain that if $f, g \in H^2(\mathbb{D})$, $f(z) = \sum_{n=0}^{\infty} a_n z^n$ and $g(z) = \sum_{n=0}^{\infty} b_n z^n$, then

$$\begin{aligned}
(N_2(f+g))^2 + (N_2(f-g))^2 &= \|(a_n + b_n)_{n \geq 0}\|_{l_2}^2 + \|(a_n - b_n)_{n \geq 0}\|_{l_2}^2 \\
&= 2 \left(\|(a_n)_{n \geq 0}\|_{l_2}^2 + \|(b_n)_{n \geq 0}\|_{l_2}^2 \right) \\
&= 2 \left((N_2(f))^2 + (N_2(g))^2 \right),
\end{aligned}$$

i.e., $N_2 : H^2(\mathbb{D}) \to \mathbb{R}_+$ satisfies the parallelogram law. Since by (iii) $H^2(\mathbb{D})$ is a Banach space, using von Neumann's theorem we obtain that $H^2(\mathbb{D})$ is a Hilbert space.

The map $h : H^2(\mathbb{D}) \to l_2$ defined by $h(f) = (a_n)_{n \geq 0} \Leftrightarrow f(z) = \sum_{n=0}^{\infty} a_n z^n$ is obviously linear. We have already proved that $\|h(f)\|_{l_2} = \|f\|_{H^2(\mathbb{D})} \; \forall f \in H^2(\mathbb{D})$, i.e., h is an

10.2 Solutions

isometry. If $(a_n)_{n\geq 0} \in l_2$, we consider $f : \mathbb{D} \to \mathbb{C}$ defined by $f(z) = \sum_{n=0}^{\infty} a_n z^n$. We have

$$\sum_{n=0}^{\infty} |a_n||z|^n \leq \sqrt{\sum_{n=0}^{\infty} |a_n|^2} \cdot \sqrt{\sum_{n=0}^{\infty} |z|^{2n}} = \frac{1}{\sqrt{1-|z|^2}} \sqrt{\sum_{n=0}^{\infty} |a_n|^2} < \infty \quad \forall |z| < 1,$$

i.e., the series $\sum_{n=0}^{\infty} a_n z^n$ is absolutely convergent on $\mathbb{D}$, thus f is a well defined analytic function. Since $N_2(f) = \|(a_n)_{n\geq 0}\|_{l_2} < \infty$ we obtain that $f \in H^2(\mathbb{D})$ and $h(f) = (a_n)_{n\geq 0}$, i.e., h is surjective.

vi) The fact that the sequence $f_n(z) = z^n$, $n = 0, 1, 2, ...$, is an orthonormal basis for $H^2(\mathbb{D})$ follows easily from the fact that $h(f_n) = e_n$ $\forall n \geq 0$, h is an isometric isomorphism of Hilbert spaces, and $(e_n)_{n\geq 0}$ is an orthonormal basis in l_2.

Remark. We have

$$\langle f, g \rangle_{H^2(\mathbb{D})} = \lim_{r \to 1, r<1} \frac{1}{2\pi} \int_0^{2\pi} f(re^{i\theta}) \overline{g(re^{i\theta})} d\theta.$$

Indeed, if $f(z) = \sum_{n=0}^{\infty} a_n z^n$, $g(z) = \sum_{n=0}^{\infty} b_n z^n$, then $\langle f, g \rangle_{H^2(\mathbb{D})} = \langle h(f), h(g) \rangle_{l_2} = \sum_{n=0}^{\infty} a_n \overline{b_n}$.

But $f(re^{i\theta}) \overline{g(re^{i\theta})} = \left(\sum_{n=0}^{\infty} a_n r^n e^{in\theta} \right) \overline{g(re^{i\theta})}$ uniformly with respect to $\theta \in [0, 2\pi]$. Thus

$$\frac{1}{2\pi} \int_0^{2\pi} f(re^{i\theta}) \overline{g(re^{i\theta})} d\theta = \frac{1}{2\pi} \left(\sum_{n=0}^{\infty} a_n r^n \right) \int_0^{2\pi} e^{in\theta} \overline{g(re^{i\theta})} d\theta.$$

Analogously,

$$\frac{1}{2\pi} \int_0^{2\pi} e^{in\theta} \overline{g(re^{i\theta})} d\theta = \sum_{m=0}^{\infty} \overline{b_m} r^m \frac{1}{2\pi} \int_0^{2\pi} e^{in\theta} e^{-im\theta} d\theta = \sum_{m=0}^{\infty} \overline{b_m} r^m \delta_{nm} = \overline{b_n} r^n,$$

and therefore $(1/2\pi) \int_0^{2\pi} f(re^{i\theta}) \overline{g(re^{i\theta})} d\theta = \sum_{n=0}^{\infty} a_n \overline{b_n} r^{2n}$. Since by the Cauchy–Buniakowski–Schwarz inequality the series $\sum_{n=0}^{\infty} a_n \overline{b_n}$ converges absolutely we deduce that

$$\lim_{r \to 1, r<1} \sum_{n=0}^{\infty} a_n \overline{b_n} r^{2n} = \sum_{n=0}^{\infty} a_n \overline{b_n}.$$

14. First solution. It is proved in exercise 13(vi) that the set of functions $f_n(z) = z^n$, $n \geq 0$, is an orthonormal basis in $H^2(\mathbb{D})$, hence by the result from exercise 12,

$$g(z) = \sum_{n=0}^{\infty} \overline{x^*(f_n)} f_n(z) = \sum_{n=0}^{\infty} \overline{x^*(f_n)} z^n$$

is the only function which has the property that $x^*(f) = \langle f, g \rangle_{H^2(\mathbb{D})}$ $\forall f \in H^2(\mathbb{D})$.

Second solution. Let $h : H^2(\mathbb{D}) \to l_2$ be the isometric isomorphism given by exercise 13(v): $h(f) = (a_n)_{n \geq 0} \in l_2 \Leftrightarrow f(z) = \sum_{n=0}^{\infty} a_n z^n \ \forall z \in \mathbb{D}$. We observe that $h(z^n) = e_n$ and then $h^{-1}(e_n) = z^n \ \forall n \geq 0$ (by z^n we understand the function $z \mapsto z^n$). Since $x^* : H^2(\mathbb{D}) \to \mathbb{C}$ is a linear and continuous functional, then $x^* h^{-1} : l_2 \to \mathbb{C}$ is a linear continuous functional. Since l_2 is a Hilbert space, using Riesz's representation theorem there is $a = (a_n)_{n \geq 0} \in l_2$ such that $x^* h^{-1}(b) = \langle b, a \rangle_{l_2} = \sum_{n=0}^{\infty} b_n \overline{a_n} \ \forall b \in l_2$ (1). We have $x^*(h^{-1}(e_n)) = \langle e_n, a \rangle_{l_2} = \overline{a_n}$. But $h^{-1}(e_n) = z^n$ and therefore $\overline{a_n} = x^*(z^n)$, i.e., $a_n = \overline{x^*(z^n)} \ \forall n \geq 0$. Let $g = h^{-1}(a)$, i.e., $g(z) = \sum_{n=0}^{\infty} a_n z^n = \sum_{n=0}^{\infty} \overline{x^*(z^n)} z^n \ \forall z \in \mathbb{D}$. Then for every $f \in H^2(\mathbb{D})$, denoting $b = h(f)$, from (1) we obtain that $x^*(f) = \langle b, a \rangle_{l_2} = \langle h(f), h(g) \rangle_{l_2} = \langle f, g \rangle_{H^2(\mathbb{D})}$. Thus the function $g(z) = \sum_{n=0}^{\infty} \overline{x^*(z^n)} z^n$ belongs to $H^2(\mathbb{D})$, and has the property that $x^*(f) = \langle f, g \rangle_{H^2(\mathbb{D})} \ \forall f \in H^2(\mathbb{D})$. Also $\|x^*\| = \|g\| = \sqrt{\sum_{n=0}^{\infty} |x^*(z^n)|^2}$.

15. See [45, exercise 10, chapter 17].

Obviously x^* is linear. Using exercise 13(i), we have
$$|x^*(f)| = |f(\xi)| \leq \frac{1}{d(\xi, \partial \mathbb{D})} N_2(f) \ \forall f \in H^2(\mathbb{D}),$$
i.e., x^* is also continuous. Then exercise 14 gives us a unique function $g \in H^2(\mathbb{D})$, $g(z) = \sum_{n=0}^{\infty} \overline{x^*(z^n)} z^n \ \forall z \in \mathbb{D}$, such that $x^*(f) = \langle f, g \rangle_{H^2(\mathbb{D})} \ \forall f \in H^2(\mathbb{D})$. But $x^*(z^n) = \xi^n$, and therefore $g(z) = \sum_{n=0}^{\infty} \overline{\xi}^n z^n = 1/(1 - \overline{\xi} z)$. Thus $g : \mathbb{D} \to \mathbb{C}$ defined by $g(z) = 1/(1 - \overline{\xi} z)$ is the only function $g \in H^2(\mathbb{D})$ such that $x^*(f) = \langle f, g \rangle_{H^2(\mathbb{D})} \ \forall f \in H^2(\mathbb{D})$.

16. See [45, exercise 11, chapter 17].

We prove first the following remark. Let H be a Hilbert space, $x^* : H \to \mathbb{K}$ is a non-null linear and continuous functional, and $y \in H$ the element given by the Riesz representation theorem, applied to x^*. Then $|x^*(x)|$, with $\|x\| \leq 1$, can be at most $\|y\|$, and if $z \in H$ has the property that $\|z\| \leq 1$ and $|x^*(z)| = \|y\|$ then there is $\lambda \in \mathbb{K}$ with $|\lambda| = 1/\|y\|$ such that $z = \lambda y$. Indeed, if $\|x\| \leq 1$, then $|x^*(x)| \leq \|x^*\| \|x\| \leq \|x^*\| = \|y\|$ where the last equality follows from the Riesz representation theorem. If $z \in H$, $\|z\| \leq 1$, has the property that $|x^*(z)| = \|y\|$ then we have $|\langle z, y \rangle| = \|y\|$. From the Cauchy–Buniakowski–Schwarz inequality we have $|\langle z, y \rangle| \leq \|y\| \|z\| \leq \|y\|$. Therefore $|\langle z, y \rangle| = \|y\| \|z\|$, i.e., we have equality in the Cauchy–Buniakowski–Schwarz inequality and therefore there is $\lambda \in \mathbb{K}$ such $z = \lambda y$. Since $|x^*(z)| = \|y\|$, then $|\lambda| \|y\|^2 = \|y\|$, i.e., $|\lambda| = 1/\|y\|$.

Return now to the solution for our exercise. Let $x^* : H^2(\mathbb{D}) \to \mathbb{C}$ defined by $x^*(f) = f'(\xi)$. If $f(z) = \sum_{n=0}^{\infty} a_n z^n$ then $x^*(f) = \sum_{n=1}^{\infty} n a_n \xi^{n-1}$ and

$$|x^*(f)| \leq \sum_{n=1}^{\infty} n |\xi|^{n-1} |a_n| \leq \sqrt{\sum_{n=1}^{\infty} n^2 |\xi|^{2(n-1)}} \sqrt{\sum_{n=1}^{\infty} |a_n|^2} \leq M N_2(f) \ \forall f \in H^2(\mathbb{D}),$$

10.2 Solutions

where $M = \sqrt{\sum_{n=1}^{\infty} n^2 |\xi|^{2(n-1)}}$. We obtain that $x^* : H^2(\mathbb{D}) \to \mathbb{C}$ is a linear and continuous functional. Using exercise 14, there is a unique function $g \in H^2(\mathbb{D})$, $g(z) = \sum_{n=0}^{\infty} \overline{x^*(z^n)} z^n$, such that $x^*(f) = \langle f, g \rangle_{H^2(\mathbb{D})} \; \forall f \in H^2(\mathbb{D})$. Since $x^*(1) = 0$, $x^*(z^n) = (z^n)'|_{z=\xi} = n\xi^{n-1}$, we have that
$$g(z) = \sum_{n=1}^{\infty} n\overline{\xi}^{n-1} z^n = z \sum_{n=1}^{\infty} n(\overline{\xi} z)^{n-1} = \frac{z}{(1-\overline{\xi}z)^2}.$$
From the above remark we deduce that $|f'(\xi)|$, if $N_2(f) \leq 1$, can be at most
$$\|g\|_{H^2(\mathbb{D})} = \sqrt{\sum_{n=0}^{\infty} |x^*(z^n)|^2} = \sqrt{\sum_{n=1}^{\infty} n^2 |\xi|^{2(n-1)}} = M.$$
But for $0 \leq r < 1$, $\sum_{n=1}^{\infty} nr^{n-1} = 1/(1-r)^2$, $\sum_{n=1}^{\infty} nr^n = r/(1-r)^2$, and by derivation, $\sum_{n=1}^{\infty} n^2 r^{n-1} = (1+r)/(1-r)^3$. Taking $r = |\xi|^2$, we obtain that $\sum_{n=1}^{\infty} n^2 |\xi|^{2(n-1)} = (1+|\xi|^2)/(1-|\xi|^2)^3$, i.e., $\|g\|_{H^2(\mathbb{D})} = \sqrt{(1+|\xi|^2)/(1-|\xi|^2)}/(1-|\xi|^2)$, and therefore $|f'(\xi)|$ can be at most $\sqrt{(1+|\xi|^2)/(1-|\xi|^2)}/(1-|\xi|^2)$. Now let $f_0 \in H^2(\mathbb{D})$ with $N_2(f_0) \leq 1$ such that $|f_0'(\xi)| = \|g\|_{H^2(\mathbb{D})}$. Then, again by the above remark, we deduce that there is $\lambda \in \mathbb{C}$ such that $f_0 = \lambda g$, and $|\lambda| = 1/\|g\|_{H^2(\mathbb{D})} = (1-|\xi|^2)\sqrt{(1-|\xi|^2)/(1+|\xi|^2)}$. Hence the extremal functions are $f_0(z) = \lambda g(z) = \lambda z/(1-\overline{\xi}z)^2$, with $|\lambda| = (1-|\xi|^2)\sqrt{(1-|\xi|^2)/(1+|\xi|^2)}$.

17. See [51, chapter 1] or [52, chapter 4, section 4.1].
    i) Let $f \in \mathcal{A}^p(\Omega)$. For $x = \Re z + s \cos t$ and $y = \Im z + s \sin t$ we have
    $$\int_{\overline{D}(z,r)} f(x,y) \, dxdy = \int_0^r \int_0^{2\pi} f(z + se^{it}) \, sdsdt = \int_0^r s\left(\int_0^{2\pi} f(z + se^{it}) \, dt\right) ds.$$
    Using the mean value property for holomorphic functions, we obtain that $\int_0^{2\pi} f(z + se^{it}) \, dt = 2\pi f(z) \; \forall s \in [0, r]$. Thus
    $$\int_{\overline{D}(z,r)} f(x,y) \, dxdy = 2\pi f(z) \int_0^r sds = \pi r^2 f(z).$$

    ii) Since $\Omega \subseteq \mathbb{C}$ is open, for any $z \in \Omega$ there is $r_z > 0$ such that $\overline{D}(z, r_z) \subseteq \Omega$. Then, for any $f \in \mathcal{A}^p(\Omega)$, by (i) and the Hölder inequality we have
    $$\begin{aligned}
    |f(z)| &= \left|\frac{1}{\pi r_z^2} \int_{\overline{D}(z,r_z)} f(x,y) \, dxdy\right| \\
    &\leq \frac{1}{\pi r_z^2} \left(\int_{\overline{D}(z,r_z)} |f(x,y)|^p \, dxdy\right)^{1/p} \left(\int_{\overline{D}(z,r_z)} 1 \, dxdy\right)^{1/q} \\
    &\leq \frac{1}{\pi r_z^2} (\pi r_z^2)^{1/q} \|f\|_{L_p(\Omega;\mathbb{C})} = \frac{\|f\|_{L_p(\Omega;\mathbb{C})}}{(\pi r_z^2)^{1/p}},
    \end{aligned}$$

thus $|f(z)| \leq \|f\|_{L_p(\Omega;\mathbb{C})}/(\pi r_z^2)^{1/p}\ \forall f \in \mathcal{A}^p(\Omega)$ (if $p = 1$, this relation follows directly from (i)).

It is easy to see that A_p is a norm on $\mathcal{A}^p(\Omega)$. Let us prove that this norm is a complete one. Let $K \subseteq \Omega$ be a compact set, and let $r = d(K, \mathbb{C}\setminus\Omega) > 0$. If $z \in \mathbb{C}$ with $d(z, K) \leq r$, then $z \in \Omega$. We deduce that for any $f \in \mathcal{A}^p(\Omega)$ and for any $z \in K$ we have

$$|f(z)| \leq \frac{\|f\|_{L_p(\Omega;\mathbb{C})}}{(\pi r_z^2)^{1/p}} \leq \frac{\|f\|_{L_p(\Omega;\mathbb{C})}}{(\pi r^2)^{1/p}}.$$

Let $(f_n)_{n\in\mathbb{N}} \subseteq \mathcal{A}^p(\Omega)$ be a Cauchy sequence, i.e., $A_p(f_n - f_m) = \|f_n - f_m\|_{L_p(\Omega;\mathbb{C})} \to 0$. Then we have

$$|f_n(z) - f_m(z)| \leq \frac{\|f_n - f_m\|_{L_p(\Omega;\mathbb{C})}}{(\pi r^2)^{1/p}}\ \forall z \in K,$$

and it follows that $(f_n)_{n\in\mathbb{N}}$ is a uniformly Cauchy sequence on any compact set $K \subseteq \Omega$. From the Weierstrass theorem we obtain that there is $f : \Omega \to \mathbb{C}$ holomorphic such that $f_n \to f$ uniformly on every compact set $K \subseteq \Omega$. Since $L_p(\Omega)$ is a Banach space we know that there is $g \in L_p(\Omega)$ such that $f_n \to g$ in $L_p(\Omega)$. Then $f = g$ a.e. on Ω, and therefore $f \in \mathcal{A}^p(\Omega)$, $A_p(f_n - f) \to 0$. Thus, $(\mathcal{A}^p(\Omega), A_p)$ is a Banach space.

(iii) The axioms for an inner product are easy to verify. The completeness is proved above.

iv) For any $n \geq 0$ we have

$$\|e_n\| = \left(\frac{n+1}{\pi}\int_0^1 r\left(\int_0^{2\pi}|re^{i\theta}|^{2n}d\theta\right)dr\right)^{1/2} = \left(\frac{n+1}{\pi}\int_0^1 2\pi r^{2n+1}dr\right)^{1/2} = 1.$$

Also, for $m, n \geq 0$, $m > n$, we have that

$$\langle e_n, e_m\rangle = \frac{\sqrt{n+1}\sqrt{m+1}}{\pi}\int_0^1 r^{2n+1}\left(\int_0^{2\pi}(re^{i\theta})^{m-n}d\theta\right)dr = 0,$$

i.e., the family $\{e_n \mid n \geq 0\} \subseteq \mathcal{A}^2(\mathbb{D})$ is an orthonormal one. We will prove that $\|f\|^2 = \sum_{n=0}^{\infty}|\langle f, e_n\rangle|^2\ \forall f \in \mathcal{A}^2(\mathbb{D})$, and then we deduce that $\{e_n \mid n \geq 0\} \subseteq \mathcal{A}^2(\mathbb{D})$ is an orthonormal basis.

We prove first that if $f : \mathbb{D} \to \mathbb{C}$ is holomorphic, $f(z) = \sum_{n=0}^{\infty}a_n z^n\ \forall z \in \mathbb{D}$, then

$$\sum_{n=0}^{\infty}\frac{|a_n|^2}{n+1} = \frac{1}{\pi}\int_{\mathbb{D}}|f(x,y)|^2\,dxdy.$$

In particular, if $f \in \mathcal{A}^2(\mathbb{D})$, $f(z) = \sum_{n=0}^{\infty}a_n z^n$, then

$$\sum_{n=0}^{\infty}\frac{\pi|a_n|^2}{n+1} = \int_{\mathbb{D}}|f(x,y)|^2\,dxdy = \|f\|^2.$$

10.2 Solutions

Indeed, using Parseval's identity, for any $0 < r < 1$ we have

$$\sum_{n=0}^{\infty} |a_n|^2 r^{2n} = \frac{1}{2\pi} \int_0^{2\pi} |f(re^{i\theta})|^2 \, d\theta.$$

If we consider $R \in (0,1)$ then the series $\sum_{n=0}^{\infty} |a_n|^2 R^{2n}$ converges, and therefore the series $\sum_{n=0}^{\infty} |a_n|^2 r^{2n}$ converges uniformly for $r \in [0, R]$. Then

$$\sum_{n=0}^{\infty} |a_n|^2 \frac{R^{2n+2}}{2n+2} = \frac{1}{2\pi} \int_0^R s \left(\int_0^{2\pi} |f(se^{i\theta})|^2 \, d\theta \right) ds = \frac{1}{2\pi} \int_{\overline{D}(0,R)} |f(x,y)|^2 \, dxdy.$$

If we take now $R_k \nearrow 1$, then $\chi_{\overline{D}(0,R_k)} |f|^2 \nearrow |f|^2$, hence by the Beppo–Levi theorem it follows that

$$\frac{1}{2\pi} \int_{\overline{D}(0,R_k)} |f(x,y)|^2 \, dxdy \nearrow \frac{1}{2\pi} \int_{\mathbb{D}} |f(x,y)|^2 \, dxdy,$$

and then

$$\lim_{R \to 1,\, R<1} \frac{1}{2\pi} \int_{\overline{D}(0,R)} |f(x,y)|^2 \, dxdy = \frac{1}{2\pi} \int_{\mathbb{D}} |f(x,y)|^2 \, dxdy.$$

Since the function $R \longmapsto \sum_{n=0}^{\infty} |a_n|^2 R^{2n+2}/(2n+2)$ is increasing, we have

$$\lim_{R \to 1,\, R<1} \sum_{n=0}^{\infty} |a_n|^2 \frac{R^{2n+2}}{2n+2} = \sup_{0<R<1} \sum_{n=0}^{\infty} |a_n|^2 \frac{R^{2n+2}}{2n+2} = \sup_{0<R<1} \sup_{n \in \mathbb{N}} \sum_{k=0}^{n} |a_k|^2 \frac{R^{2k+2}}{2k+2}$$

$$= \sup_{n \in \mathbb{N}} \sup_{0<R<1} \sum_{k=0}^{n} |a_k|^2 \frac{R^{2k+2}}{2k+2} = \sum_{n=0}^{\infty} \frac{|a_n|^2}{2n+2},$$

and therefore

$$\sum_{n=0}^{\infty} \frac{|a_n|^2}{n+1} = \frac{1}{\pi} \int_{\mathbb{D}} |f(x,y)|^2 \, dxdy.$$

Hence, if $f = \sum_{n=0}^{\infty} a_n z^n \in \mathcal{A}^2(\mathbb{D})$, then $\sum_{n=0}^{\infty} \frac{\pi |a_n|^2}{n+1} = \|f\|^2$. If we observe now that

$$\langle f, e_n \rangle = \frac{\sqrt{n+1}}{\sqrt{\pi}} \int_0^1 r^{n+1} \left(\int_0^{2\pi} f(re^{i\theta}) e^{-in\theta} \, d\theta \right) dr,$$

and that

$$\int_0^{2\pi} f(re^{i\theta}) e^{-in\theta} \, d\theta = \sum_{k=0}^{\infty} a_k r^k \int_0^{2\pi} e^{ik\theta} e^{-in\theta} \, d\theta = 2\pi a_n r^n,$$

we obtain that

$$\langle f, e_n \rangle = \frac{2\pi \sqrt{n+1}}{\sqrt{\pi}} a_n \int_0^1 r^{2n+1} \, dr = \frac{\sqrt{\pi}}{\sqrt{n+1}} a_n.$$

Hence, $\|f\|^2 = \sum_{n=0}^{\infty} |\langle f, e_n \rangle|^2$.

v) Let $h : \mathcal{A}^2(\mathbb{D}) \to l_2$ defined by $h(f) = \left(\sqrt{\pi} a_n / \sqrt{n+1}\right)_{n \geq 0} \Leftrightarrow f(z) = \sum_{n=0}^{\infty} a_n z^n$.
It is proved at (iv) that $\|h(f)\| = \|f\|$ $\forall f \in \mathcal{A}^2(\mathbb{D})$, i.e., h is an isometry. It remains to be proved that h is surjective. Let $\xi = (\xi_n)_{n \geq 0} \in l_2$. Then, for any $|z| < 1$, we have

$$\sum_{n=0}^{\infty} \frac{\sqrt{n+1}}{\sqrt{\pi}} |\xi_n| |z|^n \leq \sqrt{\sum_{n=0}^{\infty} \frac{n+1}{\pi} |z|^{2n}} \sqrt{\sum_{n=0}^{\infty} |\xi_n|^2} < \infty,$$

hence the series $\sum_{n=0}^{\infty} \left(\sqrt{n+1}/\sqrt{\pi}\right) \xi_n z^n$ converges absolutely, i.e., the function $f : \mathbb{D} \to \mathbb{C}$ defined by $f(z) = \sum_{n=0}^{\infty} a_n z^n$ where $a_n = \left(\sqrt{n+1}/\sqrt{\pi}\right) \xi_n$ is well defined and holomorphic. From (iv) we have

$$\int_{\mathbb{D}} |f(x,y)|^2 \, dxdy = \sum_{n=0}^{\infty} \frac{\pi |a_n|^2}{n+1} = \sum_{n=0}^{\infty} |\xi_n|^2 < \infty,$$

i.e., $f \in \mathcal{A}^2(\mathbb{D})$. Obviously, $h(f) = \xi$.

18. See [21, exercises 1.9 and 1.11, chapter 1] or [52, chapter 4, section 4.1].

i) Since $\mathcal{A}^2(\Omega)$ is a Hilbert space (see exercise 17(iii)) and $(f_n)_{n \geq 0}$ is an orthonormal basis we can apply exercise 12 to obtain that $g \in \mathcal{A}^2(\Omega)$ given by $g(z) = \sum_{n=0}^{\infty} a_n f_n(z)$ where $a_n = \overline{x^*(f_n)}$ is the only function which satisfies $x^*(f) = \langle f, g \rangle_{\mathcal{A}^2(\Omega)}$ $\forall f \in \mathcal{A}^2(\Omega)$. We also have $\|x^*\| = \sqrt{\sum_{n=0}^{\infty} |x^*(f_n)|^2}$.

ii) It is proved in the solution for exercise 17 that $x_\xi^* : \mathcal{A}^2(\Omega) \to \mathbb{C}$ is a linear and continuous functional. Applying now (i) we obtain that $g_\xi \in \mathcal{A}^2(\Omega)$ given by $g_\xi(z) = \sum_{n=0}^{\infty} \overline{x_\xi^*(f_n)} f_n(z)$ is the only function which satisfies $x_\xi^*(f) = \langle f, g_\xi \rangle_{\mathcal{A}^2(\Omega)}$ $\forall f \in \mathcal{A}^2(\Omega)$ and, moreover,

$$\|x_\xi^*\| = \|g_\xi\| = \sqrt{\sum_{n=0}^{\infty} |x_\xi^*(f_n)|^2}.$$

If we denote by $K(z, \xi) = g_\xi(z)$, then $f(\xi) = \iint_\Omega f(z) \overline{K(z, \xi)} dxdy$ $\forall f \in \mathcal{A}^2(\Omega)$.

iii) With the same notations as in (ii) we have $K(z, \xi) = g_\xi(z) = \sum_{n=0}^{\infty} f_n(z) \overline{f_n(\xi)}$ and, in addition, by the definition of the norm in the Hilbert space $\mathcal{A}^2(\Omega)$, $\|g_\xi\|^2 = \iint_\Omega |K(z, \xi)|^2 dxdy$. By the definition for x_ξ^*, $\sum_{n=0}^{\infty} |x_\xi^*(f_n)|^2 = \sum_{n=0}^{\infty} |f_n(\xi)|^2$, so $\iint_\Omega |K(z, \xi)|^2 dxdy = \sum_{n=0}^{\infty} |f_n(\xi)|^2$. Also, we observe that

$$K(\xi, \xi) = \sum_{n=0}^{\infty} f_n(\xi) \overline{f_n(\xi)} = \sum_{n=0}^{\infty} |f_n(\xi)|^2,$$

10.2 Solutions

and (iii) is proved.

iv) It is proved at exercise 17(iv) that $f_n(z) = \left(\sqrt{n+1}/\sqrt{\pi}\right) z^n$, $n \geq 0$, is an orthonormal basis for $\mathcal{A}^2(\mathbb{D})$, hence by (iii), the Bergman kernel for the open unit disc is

$$K(z,\xi) = \sum_{n=0}^{\infty} f_n(z)\overline{f_n(\xi)} = \frac{1}{\pi} \sum_{n=0}^{\infty} (n+1)(z\overline{\xi})^n = \frac{1}{\pi(1 - z\overline{\xi})^2},$$

since if $|z| < 1$ then $\sum_{n=0}^{\infty} z^{n+1} = z/(1-z)$, and by differentiation,

$$\sum_{n=0}^{\infty} (n+1)z^n = 1/(1-z)^2.$$

19. Let $x^* : \mathcal{A}^2(\mathbb{D}) \to \mathbb{C}$ defined by $x^*(f) = f'(\xi)$. If $f(z) = \sum_{n=0}^{\infty} a_n z^n$, then $x^*(f) = \sum_{n=1}^{\infty} n a_n \xi^{n-1}$, and

$$|x^*(f)| \leq \sqrt{\sum_{n=1}^{\infty} \frac{n^2(n+1)}{\pi} |\xi|^{2(n-1)}} \sqrt{\sum_{n=1}^{\infty} \frac{\pi}{n+1} |a_n|^2} \leq M \|f\| \quad \forall f \in \mathcal{A}^2(\mathbb{D}),$$

where

$$M = \sqrt{\sum_{n=1}^{\infty} \frac{n^2(n+1)}{\pi} |\xi|^{2(n-1)}}.$$

We obtain that $x^* : \mathcal{A}^2(\mathbb{D}) \to \mathbb{C}$ is a linear and continuous functional. Using exercise 18(i) there is a unique function $g \in \mathcal{A}^2(\mathbb{D})$,

$$g(z) = \sum_{n=0}^{\infty} \overline{x^*\left(\frac{\sqrt{n+1}}{\sqrt{\pi}} z^n\right)} \frac{\sqrt{n+1}}{\sqrt{\pi}} z^n,$$

such that $x^*(f) = \langle f, g \rangle_{\mathcal{A}^2(\mathbb{D})}$ $\forall f \in \mathcal{A}^2(\mathbb{D})$, and

$$\|g\| = \sqrt{\sum_{n=0}^{\infty} \left| x^*\left(\frac{\sqrt{n+1}}{\sqrt{\pi}} z^n\right) \right|^2}.$$

But $x^*\left(\left(\sqrt{n+1}/\sqrt{\pi}\right) z^n\right) = \left(\sqrt{n+1}/\sqrt{\pi}\right) (z^n)'\big|_{z=\xi} = \left(n\sqrt{n+1}/\sqrt{\pi}\right) \xi^{n-1}$ for $n \geq 1$ and $x^*(1/\sqrt{\pi}) = 0$, and therefore

$$g(z) = \frac{z}{\pi} \sum_{n=1}^{\infty} n(n+1)(z\overline{\xi})^{n-1} = \frac{2z}{\pi(1-z\overline{\xi})^3},$$

since $\sum_{n=1}^{\infty} nz^{n-1} = 1/(1-z)^2$, $\sum_{n=1}^{\infty} nz^{n+1} = z^2/(1-z)^2$, and by derivation, $\sum_{n=1}^{\infty} n(n+1)z^n = 2z/(1-z)^3$, and then $\sum_{n=1}^{\infty} n(n+1)z^{n-1} = 2/(1-z)^3$. We also have

$$\sum_{n=1}^{\infty} n^2(n+1)z^{n-1} = \frac{2(1-z)^3 + 6z(1-z)^2}{(1-z)^6} = \frac{2+4z}{(1-z)^4},$$

and then

$$\|g\| = M = \sqrt{\frac{2+4|\xi|^2}{\pi(1-|\xi|^2)^4}} = \sqrt{\frac{2}{\pi}} \frac{\sqrt{1+2|\xi|^2}}{(1-|\xi|^2)^2}.$$

Now, if we use the remark from the beginning of the solution for exercise 16, we obtain that $|f'(\xi)|$, for $\|f\| \leq 1$, can be at most $\|g\| = \sqrt{2/\pi}\sqrt{1+2|\xi|^2}\big/\left(1-|\xi|^2\right)^2$, and that if $f_0 \in \mathcal{A}^2(\mathbb{D})$ with $\|f_0\| \leq 1$ is such that $|f_0'(\xi)| = \|g\|_{\mathcal{A}^2(\mathbb{D})}$, then there is $\lambda \in \mathbb{C}$ with $|\lambda| = 1/\|g\|$ such that $f_0 = \lambda g$.

20. See [9, exercise 19, chapter IV, section 2].

i) We shall use the relation $\|x-y\|^2 = \|x\|^2 + \|y\|^2 - 2\Re\langle x,y\rangle$ $\forall x,y \in H$. We have

$$\|x_n - x\|^2 = \|(x_n - a) + (a - x)\|^2 = \|x_n - a\|^2 + \|a - x\|^2 - 2\Re\langle x_n - a, a - x\rangle.$$

That $x_n \to a$ *weak* means that $\langle x_n, y\rangle \to \langle a, y\rangle$ $\forall y \in H$, that is, $\langle x_n - a, y\rangle \to 0$ $\forall y \in H$, and therefore $\langle x_n - a, a - x\rangle \to 0$. Using that if $(b_n)_{n\in\mathbb{N}}$ converges then $\liminf_{n\to\infty}(a_n + b_n) = \liminf_{n\to\infty} a_n + \lim_{n\to\infty} b_n$ and $\limsup_{n\to\infty}(a_n + b_n) = \limsup_{n\to\infty} a_n + \lim_{n\to\infty} b_n$, we obtain that $\liminf_{n\to\infty} \|x_n - x\|^2 = \liminf_{n\to\infty} \|x_n - a\|^2 + \|x - a\|^2$, i.e., $d^2(x) = d^2(a) + \|x - a\|^2$, and, also, $D^2(x) = D^2(a) + \|x - a\|^2$.

ii) For $(a_n)_{n\in\mathbb{N}} \in l_\infty$ we consider $x_n = a_n e_n \in l_2$ $\forall n \in \mathbb{N}$. Then $\langle x_n, y\rangle = \langle a_n e_n, y\rangle = a_n y_n$ $\forall y = (y_1, y_2, \ldots) \in l_2$. Since $(a_n)_{n\in\mathbb{N}}$ is bounded, $|a_n y_n| \leq \left(\sup_{n\in\mathbb{N}} |a_n|\right) |y_n|$ $\forall n \in \mathbb{N}$, and since $|y_n| \to 0$, we obtain that $a_n y_n \to 0$, i.e., $x_n \to 0$ *weak*. Then

$$d(0) = \liminf_{n\to\infty} \|x_n\| = \liminf_{n\to\infty} |a_n|\|e_n\| = \liminf_{n\to\infty} |a_n|$$

and $D(0) = \limsup_{n\to\infty} |a_n|$. This shows that if we choose $(a_n)_{n\in\mathbb{N}} \in l_\infty$ such that $\liminf_{n\to\infty} |a_n| = \alpha$ and $\limsup_{n\to\infty} |a_n| = \beta$ then the sequence $x_n = a_n e_n \in l_2$ is a solution of our exercise. One can take $(a_n)_{n\in\mathbb{N}}$ as being the sequence $\alpha, \beta, \alpha, \beta, \ldots$, i.e., $a_{2n-1} = \alpha$, $a_{2n} = \beta$ $\forall n \in \mathbb{N}$.

Chapter 11

The projection in Hilbert spaces

Theorem (the projection onto a closed convex non-empty set). Let H be a Hilbert space, $A \subseteq H$ a closed convex non-empty set, and $x \in H$. Then there is a unique element $y \in A$ such that $d(x, A) = \|x - y\|$. The element $y \in A$ with this property is denoted by $\Pr_A(x)$ and is called the *projection* of x on A.

Equivalently, an element $y \in H$ is the projection of x on A if and only if $y \in A$ and $\Re \langle x - y, y - a \rangle \geq 0 \ \forall a \in A$.

Theorem (the projection onto closed linear subspaces). Let H be a Hilbert space, $G \subseteq H$ a closed linear subspace, and $x \in H$. Then there is a unique element $y \in G$ such that $x - y \in G^\perp$. The element $y \in G$ with the property that $x - y \in G^\perp$ is called the *orthogonal projection* of x onto G, and is denoted by $\Pr_G(x)$.

We have $H = G + G^\perp$ and $G \cap G^\perp = \{0\}$, i.e., $H = G \oplus G^\perp$. The operator $\Pr_G : H \to H$ defined by $x \longmapsto \Pr_G(x)$ is called the *orthogonal projection* onto G.

Definition. Let H be a Hilbert space. A linear and continuous operator $T : H \to H$ is called an *orthogonal projection* if and only if there is a closed linear subspace $G \subseteq H$ such that $T = \Pr_G$.

11.1 Exercises

The orthogonal projection is a non-expansive map

1. Let H be a Hilbert space, and $A \subseteq H$ a closed convex and non-empty set. Prove that $\Pr_A : H \to H$ is non-expansive, i.e., $\|\Pr_A(x) - \Pr_A(y)\| \leq \|x - y\| \ \forall x, y \in H$.

A geometrical result

2. Let H be a Hilbert space, and $A \subseteq H$ a closed, convex and non-empty set.

i) Prove that if $x \notin A$ then $\Pr_A(x) \in \operatorname{Fr}(A)$ and $d(x, A) = d(x, \operatorname{Fr}(A))$.

ii) Then deduce that if $x \notin A$ and, in addition, $\operatorname{Fr}(A)$ is convex and non-empty, then $\Pr_A(x) = \Pr_{\operatorname{Fr}(A)}(x)$.

(Here $\operatorname{Fr}(A)$ is the boundary of A, $\operatorname{Fr}(A) = \overline{A} \setminus \overset{\circ}{A}$.)

The orthogonal projection onto the subgraph of a convex function

3. i) Let $\varphi : \mathbb{R} \to \mathbb{R}$ be a C^1 convex function and $A = \{(x, y) \in \mathbb{R}^2 \mid \varphi(x) \leq y\}$. Prove that if $(a, b) \notin A$ then $\Pr_A(a, b) = (x, \varphi(x))$, where $x \in \mathbb{R}$ is a solution for the equation $(b - \varphi(x))\varphi'(x) + a - x = 0$.

ii) Let $A = \{(x, y) \in \mathbb{R}^2 \mid x^2 \leq y\}$ and $(a, b) \notin A$. Prove that $\Pr_A(a, b) = (x, x^2)$ where $x \in \mathbb{R}$ is a solution of the equation $2x^3 - (2b - 1)x - a = 0$. In particular, if $a \in (-\infty, -1/\sqrt{2}) \cup (1/\sqrt{2}, \infty)$ and $A = \{(x, y) \in \mathbb{R}^2 \mid x^2 \leq y\}$, then $\Pr_A(a, 1/2) = \left((a/2)^{1/3}, (a^2/4)^{1/3}\right)$.

4. i) Let $\varphi : \mathbb{R}^n \to \mathbb{R}$ be a C^1 convex function, $A = \{(x, y) \in \mathbb{R}^n \times \mathbb{R} \mid \varphi(x) \leq y\}$ and $a = (a_1, a_2, ..., a_n) \in \mathbb{R}^n$, $b \in \mathbb{R}$. Prove that if $(a, b) \notin A$ then $\Pr_A(a, b) = (x, \varphi(x))$ where $x = (x_1, x_2, ..., x_n) \in \mathbb{R}^n$ is a solution of the system

$$\begin{cases} a_1 - x_1 + (b - \varphi(x))\dfrac{\partial \varphi}{\partial x_1}(x) = 0 \\ a_2 - x_2 + (b - \varphi(x))\dfrac{\partial \varphi}{\partial x_2}(x) = 0 \\ \cdots\cdots\cdots\cdots\cdots\cdots\cdots\cdots\cdots\cdots \\ a_n - x_n + (b - \varphi(x))\dfrac{\partial \varphi}{\partial x_n}(x) = 0. \end{cases}$$

ii) Let $A = \{(x, y, z) \in \mathbb{R}^3 \mid x^2 + y^2 \leq z\}$. Prove that $\Pr_A(1, \sqrt{3}, 1/2) = (1/2, \sqrt{3}/2, 1)$.

Orthogonal projections in a Hilbertian product

5. i) Let $a, b \in \mathbb{R}$, $a \leq b$. Prove that

$$\Pr_{[a,\infty)}(x) = \begin{cases} a, & x < a, \\ x, & x \geq a, \end{cases} \quad \Pr_{(-\infty,a]}(x) = \begin{cases} x, & x \leq a, \\ a, & x > a, \end{cases}$$

$$\Pr_{[a,b]}(x) = \begin{cases} a, & x < a, \\ x, & a \leq x \leq b, \\ b, & x > b, \end{cases} \quad \forall x \in \mathbb{R}.$$

ii) Let $D = [a, b] \times [c, d] \subseteq \mathbb{R}^2$. Prove that $\Pr_D(x, y) = (\Pr_{[a,b]}(x), \Pr_{[c,d]}(y))$.

iii) Let H_1, H_2 be Hilbert spaces, and $H_1 \times H_2$ their Hilbertian product. Suppose that $A \subseteq H_1$ and $B \subseteq H_2$ are closed, convex, and non-empty sets. Prove that

$$\Pr_{A \times B}(x, y) = (\Pr_A(x), \Pr_B(y)) \quad \forall (x, y) \in H_1 \times H_2.$$

11.1 Exercises

Orthogonal projections in l_2

6. i) Let $A = \{(x_n)_{n\in\mathbb{N}} \in l_2 \mid 0 \leq x_n \leq 1/n \ \forall n \in \mathbb{N}\} \subseteq l_2$. Calculate $\Pr_A(x)$, $x \in l_2$. In particular, calculate $\Pr_A(e_k)$, $k \in \mathbb{N}$, and $\lim_{k\to\infty} d(e_k, A)$.

ii) Let $(a_n)_{n\in\mathbb{N}} \subseteq [0, \infty)$ and $A = \{(x_n)_{n\in\mathbb{N}} \in l_2 \mid |x_n| \leq a_n \ \forall n \in \mathbb{N}\} \subseteq l_2$. Calculate $\Pr_A(x)$, $x \in l_2$.

iii) Let $A = \{(x_n)_{n\in\mathbb{N}} \in l_2 \mid |x_n| \leq 1 \ \forall n \in \mathbb{N}\} \subseteq l_2$. Calculate $\Pr_A(ae_k)$, where $a \in \mathbb{R}$, and calculate $\lim_{k\to\infty} d(((k+1)/k) e_k, A)$.

iv) Let $m \in \mathbb{N}$, $a_1 \geq 0, ..., a_m \geq 0$ and $A = \{(x_n)_{n\in\mathbb{N}} \in l_2 \mid 0 \leq x_1 \leq a_1, 0 \leq x_2 \leq a_2, ..., 0 \leq x_m \leq a_m\} \subseteq l_2$. Calculate $\Pr_A(x)$, $x \in l_2$.

7. Let $(a_n)_{n\in\mathbb{N}}$ and $(b_n)_{n\in\mathbb{N}}$ be two sequences of real numbers with $a_n \leq b_n \ \forall n \in \mathbb{N}$ and $A = \{(x_n)_{n\in\mathbb{N}} \in l_2 \mid a_n \leq x_n \leq b_n \ \forall n \in \mathbb{N}\} \subseteq l_2$. Suppose that A is not empty. Calculate $\Pr_A(x)$, where $x \in l_2$.

8. Consider the following linear subspaces in l_2:

$$G_1 = \{(\xi_1, \xi_1, \xi_1, \xi_2, \xi_2, \xi_2, ...) \mid (\xi_n)_{n\in\mathbb{N}} \in l_2\},$$
$$G_2 = \{(\xi_1, \varepsilon\xi_1, \varepsilon^2\xi_1, \xi_2, \varepsilon\xi_2, \varepsilon^2\xi_2, ...) \mid (\xi_n)_{n\in\mathbb{N}} \in l_2\},$$
$$G_3 = \{(\xi_1, \varepsilon^2\xi_1, \varepsilon\xi_1, \xi_2, \varepsilon^2\xi_2, \varepsilon\xi_2, ...) \mid (\xi_n)_{n\in\mathbb{N}} \in l_2\},$$

where $\varepsilon = -1/2 + i\sqrt{3}/2$. Find the orthogonal projection $\Pr_{G_j}(x)$, $j = 1, 2, 3$, where $x \in l_2$.

Properties for projections

9. i) Let H_1, H_2 be Hilbert spaces and $T \in L(H_1, H_2)$ such that $TT^* = I$. Prove that if $B \subseteq H_2$ is a closed convex and non-empty set then $T^*(\Pr_B y) = \Pr_{T^*(B)}(T^*y) \ \forall y \in H_2$.

ii) Let H be a Hilbert space, $A \subseteq H$ is a closed convex and non-empty set, and $x \in H$. Prove that $\Pr_{x+A}(y) = x + \Pr_A(y - x) \ \forall y \in H$.

iii) Let H be a Hilbert space, $A \subseteq H$ is a closed convex and non-empty set, and $\lambda \in \mathbb{K}$. Prove that $\Pr_{\lambda A}(\lambda y) = \lambda \Pr_A(y) \ \forall y \in H$. Then deduce that if, in addition, A is symmetric, then $\Pr_A(-x) = -\Pr_A(x) \ \forall x \in H$.

A continuity property for a sequence of orthogonal projections

10. i) Let X be a normed space, $G_1 \subseteq G_2 \subseteq \cdots \subseteq G_n \subseteq G_{n+1} \subseteq \cdots$ is a sequence of closed linear subspaces of X, and $G = \overline{\mathrm{Sp}(\bigcup_{n\in\mathbb{N}} G_n)}$.

a) Prove that $d(x, G) = \lim_{n\to\infty} d(x, G_n) \ \forall x \in X$.

b) If X is a Hilbert space prove that $\Pr_G(x) = \lim_{n\to\infty} \Pr_{G_n}(x) \ \forall x \in X$.

ii) Let X be a Hilbert space, $G_1 \supseteq G_2 \supseteq \cdots \supseteq G_n \supseteq G_{n+1} \supseteq \cdots$ is a sequence of closed linear subspaces, and $G = \bigcap_{n\in\mathbb{N}} G_n$. Prove that $\Pr_G(x) = \lim_{n\to\infty} \Pr_{G_n}(x) \ \forall x \in X$.

The Gram determinant

11. i) Let H be an inner product space, $n \in \mathbb{N}$, and $\{x_1, ..., x_n\} \subseteq H \backslash \{0\}$. The determinant
$$\Gamma(x_1, ..., x_n) = \begin{vmatrix} \langle x_1, x_1 \rangle & \langle x_1, x_2 \rangle & ... & \langle x_1, x_n \rangle \\ \langle x_2, x_1 \rangle & \langle x_2, x_2 \rangle & ... & \langle x_2, x_n \rangle \\ ... & ... & ... & ... \\ \langle x_n, x_1 \rangle & \langle x_n, x_2 \rangle & ... & \langle x_n, x_n \rangle \end{vmatrix}$$
is called the Gram determinant of the system $\{x_1, ..., x_n\}$. Prove that:
  a) $\Gamma(x_1, ..., x_n) \geq 0$;
  b) $\Gamma(x_1, ..., x_n) = \Gamma(y_1, ..., y_n)$, where $\{y_1, ..., y_n\}$ is the system obtained by applying the Gram–Schmidt procedure to the system $\{x_1, ..., x_n\}$;
  c) $\Gamma(x_1, ..., x_n) = 0$ if and only if the system $\{x_1, ..., x_n\}$ is linearly dependent.

ii) Let $a, b, c \in \mathbb{R}$ with $|a| < 1$, $|b| < 1$, $|c| < 1$. Using (i) prove that
$$\begin{vmatrix} 1/(1-|a|^2) & 1/(1-ab) & 1/(1-ac) \\ 1/(1-ab) & 1/(1-|b|^2) & 1/(1-bc) \\ 1/(1-ac) & 1/(1-bc) & 1/(1-|c|^2) \end{vmatrix} \geq 0.$$

Orthogonal projections onto finite-dimensional linear subspaces and the Gram determinant

12. i) Let H be a Hilbert space, $n \in \mathbb{N}$, $x_1, ..., x_n \in H$ are n linearly independent elements, and $G = \text{Sp}\{x_1, ..., x_n\}$. Prove that for any $x \in H$,
$$\text{Pr}_G(x) = \frac{\overline{\Delta_1}}{\Delta} x_1 + \frac{\overline{\Delta_2}}{\Delta} x_2 + \cdots + \frac{\overline{\Delta_n}}{\Delta} x_n,$$
where $\Delta = \Gamma(x_1, x_2, ..., x_n)$ is the Gram determinant and Δ_i are obtained by replacing in Δ the column i with the column
$$\begin{pmatrix} \langle x_1, x \rangle \\ \langle x_2, x \rangle \\ ... \\ \langle x_n, x \rangle \end{pmatrix}.$$

ii) Using (i) calculate the orthogonal projection of $x = (4, -1, -3, 4)$ onto the linear subspace generated by $x_1 = (1, 1, 1, 1)$, $x_2 = (1, 2, 2, -1)$ and $x_3 = (1, 0, 0, 3)$.

Orthogonal projections onto cofinite-dimensional linear subspaces and the Gram determinant

13. i) Let H be a Hilbert space, $n \in \mathbb{N}$, $x_1, ..., x_n \in H$ are n linearly independent elements, and $L = \{x \in H \mid \langle x, x_1 \rangle = 0, ..., \langle x, x_n \rangle = 0\}$. Calculate $\text{Pr}_L(x)$, $x \in H$. More precisely, prove that
$$\text{Pr}_L(x) = x - \left(\frac{\overline{\Delta_1}}{\Delta} x_1 + \cdots + \frac{\overline{\Delta_n}}{\Delta} x_n \right),$$

where $\triangle$ and $\triangle_k$ are obtained as in exercise 12.

ii) Calculate the orthogonal projection of $x = (7, -4, -1, 2)$ onto the linear space L given by the system of equations

$$\begin{cases} 2y_1 + y_2 + y_3 + 3y_4 = 0 \\ 3y_1 + 2y_2 + 2y_3 + y_4 = 0 \\ y_1 + 2y_2 + 2y_3 - 9y_4 = 0 \end{cases}.$$

Orthogonal projections onto closed linear manifolds and the Gram determinant

14. i) Let H be a Hilbert space, $n \in \mathbb{N}$, $x_1, ..., x_n \in H$ are n linearly independent elements, $c_1, ..., c_n \in \mathbb{K}$, and $V = \{x \in H \mid \langle x, x_1 \rangle = c_1, ..., \langle x, x_n \rangle = c_n\}$ is a closed linear manifold. Prove that

$$\mathrm{Pr}_V(x) = x - \left(\overline{\frac{\Delta_1}{\Delta}} x_1 + \cdots + \overline{\frac{\Delta_n}{\Delta}} x_n \right),$$

where $\Delta = \Gamma(x_1, ..., x_n)$, and Δ_i is obtained by replacing in Δ the column i with the column

$$\begin{pmatrix} \langle x_1, x \rangle - \overline{c_1} \\ \cdots \\ \langle x_n, x \rangle - \overline{c_n} \end{pmatrix}.$$

ii) Using (i) find the orthogonal projection of $x = (4, 2, -5, 1)$ onto the linear manifold given by

$$\begin{cases} 2y_1 - y_2 + y_3 + 2y_4 = 9 \\ 2y_1 - 4y_2 + 2y_3 + 3y_4 = 12 \end{cases}.$$

The distance from a point to a finite-dimensional linear subspace and the Gram determinant

15. Let H be a Hilbert space, $n \in \mathbb{N}$, $\{x_1, ..., x_n\} \subseteq H$ is a linearly independent system, and $G = \mathrm{Sp}\{x_1, ..., x_n\}$. Prove that

$$d(x, G) = \sqrt{\frac{\Gamma(x_1, ..., x_n, x)}{\Gamma(x_1, ..., x_n)}}.$$

16. i) Let H be a Hilbert space and $n \in \mathbb{N}$. Prove that if $\{x_1, ..., x_n\} \subseteq H$ is a linearly independent system, and $V = x_0 + G$, $G = \mathrm{Sp}\{x_1, ..., x_n\}$, $x_0 \in H$, then

$$d(x, V) = \sqrt{\frac{\Gamma(x_1, ..., x_n, x - x_0)}{\Gamma(x_1, ..., x_n)}}.$$

ii) Using (i) find the distance from the point x to the linear manifold $x_0 + \lambda_1 x_1 + \lambda_2 x_2$, $\lambda_1, \lambda_2 \in \mathbb{K}$, where $x = (1, 2, -1, 1)$, $x_0 = (0, -1, 1, 1)$, $x_1 = (0, -3, -1, 5)$, and $x_2 = (4, -1, -3, 3)$.

iii) Using (i) find the distance from the origin to the linear manifold defined by the set of all polynomials of the form $t^2 + a_1 t + a_2$ in the Hilbert space $L_2[0, 1]$.

17. Let H be a Hilbert space. A system $(x_n)_{n\in\mathbb{N}} \subseteq H$ is called topologically linearly independent if and only if $\forall n \in \mathbb{N}$, $x_n \notin \overline{\mathrm{Sp}\{x_1, x_2, ..., x_{n-1}, x_{n+1}, ...\}}$. Prove that $(x_n)_{n\in\mathbb{N}}$ is a topologically linearly independent if and only if $\forall p \geq 1$ we have

$$\sup_{n \geq p+1} \frac{\Gamma(x_1, ..., x_{p-1}, x_{p+1}, ..., x_n)}{\Gamma(x_1, ..., x_n)} < \infty.$$

The distance between two closed linear manifolds and the Gram determinant

18. i) Let H be a Hilbert space and $V_1, V_2 \subseteq H$ two closed linear manifolds, $V_1 = x_1 + G_1$, $V_2 = x_2 + G_2$, where $G_1, G_2 \subseteq H$ are two closed linear subspaces and $x_1, x_2 \in H$. Prove that $d(V_1, V_2) = \|\mathrm{Pr}_{G_1^\perp \cap G_2^\perp}(x_1 - x_2)\|$.

ii) Let H be a Hilbert space, $\{a_1, ..., a_n, a_{n+1}, ..., a_{n+m}\} \subseteq H$ is a linearly independent system, $G_1 = \mathrm{Sp}\{a_1, ..., a_n\}$, $G_2 = \mathrm{Sp}\{a_{n+1}, ..., a_{n+m}\}$, and $V_1, V_2 \subseteq H$ are the closed linear manifolds defined by $V_1 = x_1 + G_1$ and $V_2 = x_2 + G_2$. Prove that

$$d(V_1, V_2) = \left\| x_1 - x_2 - \left(\frac{\overline{\Delta_1}}{\Delta} a_1 + \cdots + \frac{\overline{\Delta_{n+m}}}{\Delta} a_{n+m} \right) \right\|,$$

where $\Delta = \Gamma(a_1, ..., a_n, a_{n+1}, ..., a_{n+m})$ and Δ_i are obtained by replacing in Δ the column i with the column

$$\begin{pmatrix} \langle a_1, x_1 - x_2 \rangle \\ \langle a_2, x_1 - x_2 \rangle \\ ... \\ \langle a_{n+m}, x_1 - x_2 \rangle \end{pmatrix}.$$

iii) Using (ii) calculate the distance between the planes $x = a_1 t_1 + a_2 t_2 + x_1$ and $x = a_3 t_1 + a_4 t_2 + x_2$, where $a_1 = (1, 2, 2, 2)$, $a_2 = (2, -2, 1, 2)$, $a_3 = (2, 0, 2, 1)$, $a_4 = (1, -2, 0, -1)$, $x_1 = (4, 5, 3, 2)$ and $x_2 = (1, -2, 1, -3)$.

Orthogonal projections in L_2

19. i) Let (S, Σ, μ) be a measure space, $I \subseteq \mathbb{R}$ is a closed interval which contains 0, and $A = \{f \in L_2(\mu) \mid f(s) \in I \text{ a.e.}\}$. Find the orthogonal projection $\mathrm{Pr}_A(f)$, $f \in L_2(\mu)$.

ii) Let $A = \{f \in L_2[-1, 1] \mid f(x) \in [0, 1] \text{ a.e.}\}$. Using (i) find the orthogonal projection $\mathrm{Pr}_A(e^x)$ and the distance $d(e^x, A)$.

iii) Let (S, Σ, μ) be a finite measure space and $A = \{f \in L_2(\mu) \mid f(s) \in [a, b] \text{ a.e.}\}$. Find the orthogonal projection $\mathrm{Pr}_A(f)$, $f \in L_2(\mu)$.

iv) If (S, Σ, μ) is a measure space with $\mu(S) = \infty$ and $a > 0$, prove that the set $A = \{f \in L_2(\mu) \mid f(s) \in [a, \infty) \text{ a.e.}\}$ is empty.

v) Let (S, Σ, μ) be a finite measure space, $a > 0$ and $A = \{f \in L_2(\mu) \mid f(s) \in [a, \infty) \text{ a.e.}\}$. Find the orthogonal projection $\mathrm{Pr}_A(f)$, $f \in L_2(\mu)$.

vi) If (S, Σ, μ) is a measure space with $\mu(S) = \infty$ and $a < 0$, then the set $A = \{f \in L_2(\mu) \mid f(s) \in (-\infty, a] \text{ a.e.}\}$ is empty.

vii) Let (S, Σ, μ) be a finite measure space, $a < 0$ and $A = \{f \in L_2(\mu) \mid f(s) \in (-\infty, a] \text{ a.e.}\}$. Find the orthogonal projection $\mathrm{Pr}_A(f)$, $f \in L_2(\mu)$.

viii) Let $g : [0, 1] \to [0, \infty)$ be a Lebesgue integrable function and let $A = \{f \in L_2[0, 1] \mid |f(x)| \leq g(x) \text{ a.e.}\}$. Find the orthogonal projection $\mathrm{Pr}_A(f)$, $f \in L_2[0, 1]$.

11.1 Exercises

ix) Let $A = \{f \in L_2[0,1] \mid |f(x)| \leq x \text{ a.e.}\}$. Using (viii) find the orthogonal projection $\Pr_A(x - 1/2)$ and the distance $d(x - 1/2, A)$.

20. Let $L = \left\{f \in L_2[0,1] \mid \int_0^{1/2} f(x)dx = 0, \int_{1/2}^1 f(x)dx = 0\right\}$. Find the orthogonal projection $\Pr_L(f)$, where $f \in L_2[0,1]$.

21. Let $A = \left\{f \in L_2[0,1] \mid \int_0^1 |f(x)| dx \leq 1\right\} \subseteq L_2[0,1]$.
   i) Prove that A is closed, convex, and non-empty.
   ii) For $g \in L_2[0,1]$ find the orthogonal projection $\Pr_A(g)$.
   iii) Find the orthogonal projection $\Pr_A(8x)$.

22. Let $G = \left\{f \in L_2(0,\infty) \mid \int_0^n f(x) dx = 0 \,\forall n \in \mathbb{N}\right\}$. Find the orthogonal projection $\Pr_G(f)$, $f \in L_2(0,\infty)$.

Projections in Banach spaces

23. Let X be a Banach space and $L, M \subseteq X$ two linear subspaces such that $L \cap M = \{0\}$. Let the operator $P_{L\|M} : L + M \to L$ be defined by $P_{L\|M}(x + y) = x \, \forall x \in L, y \in M$. Prove that:
   i) $P_{L\|M}$ is linear, $P_{L\|M}^2 = P_{L\|M}$, $P_{L\|M}\big|_L = I$, $P_{L\|M}\big|_M = 0$.
   ii) Prove that $P_{L\|M}$ is continuous if and only if $\overline{L} \cap \overline{M} = \{0\}$ and $P_{\overline{L}\|\overline{M}}$ is continuous.
   iii) If L and M are closed linear subspaces, prove that $P_{L\|M}$ is continuous if and only if $L + M \subseteq X$ is a closed linear subspace.

In the case in which it is well defined the operator $P_{L\|M}$ is called the *oblique projection* on L, parallel with M.

Angles in Hilbert spaces

24. i) Let H be a real Hilbert space, and $x, y \in H$, $x \neq 0$, $y \neq 0$. Prove that there is a unique $\alpha \in [0, \pi]$ such that
$$\cos \alpha = \frac{\langle x, y \rangle}{\|x\| \|y\|}.$$
$\alpha \in [0, \pi]$ with this property is called the angle between x and y and we will denote $\alpha = \widehat{(x, y)}$.

   ii) Let H be a real or a complex Hilbert space, $G \subseteq H$ is a linear subspace, $G \neq \{0\}$, and $x \in H$, $x \neq 0$. Prove that
$$\sup \left\{ \frac{|\langle x, g \rangle|}{\|x\| \|g\|} \,\bigg|\, g \in G, g \neq 0 \right\} \in [0, 1]$$
and that there is a unique $\alpha \in [0, \pi/2]$ such that
$$\cos \alpha = \sup \left\{ \frac{|\langle x, g \rangle|}{\|x\| \|g\|} \,\bigg|\, g \in G, g \neq 0 \right\}$$
$\alpha \in [0, \pi/2]$ with this property is called the angle between x and the subspace G and we will denote $\alpha = \widehat{(x, G)}$. Prove also that $\widehat{(x, G)} = \widehat{(x, \overline{G})}$.

   iii) Let H be a real or a complex Hilbert space, $G \subseteq H$ is a closed linear subspace, $G \neq \{0\}$, and $x \in H$, $x \neq 0$. Prove that if $\alpha = \widehat{(x, G)}$, then it is true the well known formula from the analytic geometry, $\cos\alpha = \|\mathrm{Pr}_G(x)\| / \|x\|$, and therefore the supremum from (ii) is attained at $\mathrm{Pr}_G(x)$.

iv) Let H be a real Hilbert space, $G \subseteq H$ is a closed linear subspace, $G \neq \{0\}$, and $x \in H$, $x \neq 0$. Prove that
$$\sup\left\{\frac{|\langle x,g\rangle|}{\|x\|\,\|g\|}\,\bigg|\,g \in G, g \neq 0\right\} = \sup\left\{\frac{\langle x,g\rangle}{\|x\|\,\|g\|}\,\bigg|\,g \in G, g \neq 0\right\},$$
i.e., if $\alpha = \widehat{(x,G)}$ then
$$\cos\alpha = \sup\left\{\frac{\langle x,g\rangle}{\|x\|\,\|g\|}\,\bigg|\,g \in G, g \neq 0\right\}.$$

Hence, in the real case, we can eliminate the modulus in the definition for $\cos\alpha$.

v) Let H be a real Hilbert space, $x, y \in H$, $x \neq 0$, $y \neq 0$, and $G = \mathrm{Sp}\{y\} = \{\lambda y \mid \lambda \in \mathbb{R}\}$. With the notations from (i) and (ii) prove that
$$\widehat{(x,G)} = \begin{cases} \widehat{(x,y)} & \text{if } \widehat{(x,y)} \in [0, \pi/2], \\ \pi - \widehat{(x,y)} & \text{if } \widehat{(x,y)} \in (\pi/2, \pi]. \end{cases}$$

vi) Let H be a real or a complex Hilbert space, and L, M two linear subspaces of H. Prove that
$$\sup_{x\in L\setminus\{0\},\, y\in M\setminus\{0\}} \frac{|\langle x,y\rangle|}{\|x\|\,\|y\|} \in [0,1]$$
and that there is a unique $\alpha \in [0, \pi/2]$ such that
$$\cos\alpha = \sup_{x\in L\setminus\{0\},\, y\in M\setminus\{0\}} \frac{|\langle x,y\rangle|}{\|x\|\,\|y\|}.$$

$\alpha \in [0, \pi/2]$ with this property is called the angle between the linear subspaces L and M and we will denote $\alpha = \widehat{(L,M)}$.

vii) Let H be a real or a complex Hilbert space, and L, M two linear subspaces of H, with $L \cap M = \{0\}$. Prove that:
$$\cos\widehat{(L,M)} = \cos\widehat{(\overline{L},\overline{M})} = \|P_{\overline{M}} P_{\overline{L}}\|,$$
$$\sin\widehat{(L,M)} = \sin\widehat{(\overline{L},\overline{M})} = \frac{1}{\|P_{L\|M}\|},$$
where $P_{\overline{M}}$ and $P_{\overline{L}}$ are respectively the orthogonal projections on $\overline{M}$, $\overline{L}$ and $P_{L\|M}$ is the oblique projection on L, parallel with M.

25. i) Calculate the angle between x and the linear subspace generated by x_1 and x_2, where $x = (1, 3, -1, 3)$, $x_1 = (1, -1, 1, 1)$, $x_2 = (5, 1, -3, 3)$.

ii) Let $m, n \in \mathbb{N}$ with $m < n$, $x = (1, 1, ..., 1) \in \mathbb{R}^n$, and $G = \{(x_1, ..., x_m, 0, ..., 0) \mid x_1, ..., x_m \in \mathbb{R}\}$. Calculate the angle between x and G.

iii) Calculate the angle between the line $d : x_1 = x_2 = \cdots = x_n$ and the coordinate axis of the space $\mathbb{R}^n$, where $n \in \mathbb{N}$.

iv) Calculate the angle between the line d given by the equations $x_0 = 2x_1 = 2^2 x_2 = \cdots = 2^n x_n = \cdots$, and the coordinate axis of the space l_2, i.e., calculate $\widehat{(e_n, d)}$, $n \geq 0$.

11.1 Exercises

The orthogonal projection onto a closed ball, a hyperplane, or a half-plane

26. i) Let H be a Hilbert space, $a \in H$, $r > 0$, and $\overline{B}(a,r)$ be the closed ball with center a and radius r in H. Find the orthogonal projection $\Pr_{\overline{B}(a,r)}(x)$, where $x \in H$.

ii) Let H be a Hilbert space, $a \in H \setminus \{0\}$, $c \in \mathbb{K}$, and consider the closed hyperplane $V = \{x \in H \mid \langle x, a \rangle = c\}$. Find the orthogonal projection $\Pr_V(x)$, where $x \in H$. In particular, prove that if $a \in H$, $a \neq 0$ and $A = \{x \in H \mid \langle x, a \rangle = 1\}$, then $\Pr_A(0) = a/\|a\|^2$.

iii) Let H be a real Hilbert space, $a \in H \setminus \{0\}$, $c \in \mathbb{R}$, and let $A = \{x \in H \mid \langle x, a \rangle \leq c\}$. Find the orthogonal projection $\Pr_A(x)$, where $x \in H$.

iv) Let H be a real Hilbert space, $a \in H \setminus \{0\}$, $m, M \in \mathbb{R}$, $m < M$, and let $A = \{x \in H \mid m \leq \langle x, a \rangle \leq M\}$. Find the orthogonal projection $\Pr_A(x)$, where $x \in H$.

v) Let H be a Hilbert space, $a \in H \setminus \{0\}$, and let $A = \{x \in H \mid \Re\langle x, a \rangle = 1\}$, $B = \{x \in H \mid |\Re\langle x, a \rangle| \leq 1\}$. Find the orthogonal projections $\Pr_A(x)$, $\Pr_B(x)$, where $x \in H$.

vi) Consider the Hilbert space $\mathcal{M}_n(\mathbb{K})$ with the usual inner product $\langle A, B \rangle = \text{tr}(AB^*)$, and consider $E = \{A \in \mathcal{M}_n(\mathbb{K}) \mid \Re(\text{tr}(A)) = 1\}$, $F = \{A \subset \mathcal{M}_n(\mathbb{K}) \mid |\Re(\text{tr}(A))| \leq 1\}$. Using (v) find the orthogonal projections $\Pr_E(B)$, $\Pr_F(B)$, where $B \in \mathcal{M}_n(\mathbb{K})$. In particular, find the orthogonal projections $\Pr_E(aI_n)$, $\Pr_F(aI_n)$, where $I_n \in \mathcal{M}_n(\mathbb{K})$ is the unit matrix and $a \in \mathbb{K}$.

Projections in the Hardy space

27. Let $\xi \in \mathbb{D}$ and let $A = \{f \in H^2(\mathbb{D}) \mid f(\xi) = 1\}$. Find the minimal norm element and the norm for the minimal norm element for the set A.

28. Let $\xi \in \mathbb{D}$ and let $A = \{f \in H^2(\mathbb{D}) \mid f'(\xi) = 1\}$. Find the minimal norm element and the norm for the minimal norm element for the set A.

Projections in the Bergman space

29. Let $\Omega \subseteq \mathbb{C}$ be an open non-empty set. Let $\mathcal{A}^2(\Omega)$ be the Bergman space (see exercise 17 of chapter 10) and $(f_n)_{n \geq 0}$ be an orthonormal basis in $\mathcal{A}^2(\Omega)$.

i) Let $x^* : \mathcal{A}^2(\Omega) \to \mathbb{C}$ be a linear and continuous functional and $H = \{f \in \mathcal{A}^2(\Omega) \mid x^*(f) = 1\}$. Find the minimal norm element and the norm for the minimal norm element for the set H in terms of x^* and the orthonormal basis $(f_n)_{n \geq 0}$.

ii) Let $\xi \in \Omega$ and $H = \{f \in \mathcal{A}^2(\Omega) \mid f(\xi) = 1\}$. Prove that the minimal norm element of the set H is the function f_ξ, where $f_\xi(z) = K(z, \xi)/K(\xi, \xi)$, and that the norm for the minimal norm element is $1/\sqrt{K(\xi, \xi)}$, where K is the Bergman kernel for Ω.

iii) Let $\mathbb{D}$ be the open unit disc in the complex plane, $\xi \in \mathbb{D}$ and $H = \{f \in \mathcal{A}^2(\mathbb{D}) \mid f(\xi) = 1\}$. Prove that the minimal element for the set H is the function f_ξ, where $f_\xi(z) = \left((1 - |\xi|^2)/(1 - z\bar{\xi})\right)^2$, and that the norm for the minimal element is $\sqrt{\pi}(1 - |\xi|^2)$.

A minimum problem

30. i) Let H be a real Hilbert space and $f : H \to \mathbb{R}$ a linear and continuous functional on H. Prove that the function $\varphi : H \to \mathbb{R}$ defined by $\varphi(x) = \|x\|^2 - f(x)$ is lower bounded and attains its infimum in a unique point on any closed, convex, and non-empty set.

ii) Let $\varphi : L_2[0,1] \to \mathbb{R}$ be defined by $\varphi(f) = \int_0^1 |f(x)|^2 \, dx - 16 \int_0^1 xf(x) \, dx$, and $A = \left\{ f \in L_2[0,1] \mid \int_0^1 |f(x)| \, dx \leq 1 \right\}$. Using (i) find the function $f_0 \in A$ such that $\varphi(f_0) = \inf_{f \in A} \varphi(f)$.

A Cantor type of theorem in Hilbert spaces

31. Prove that in a Hilbert space any decreasing sequence of closed, convex, bounded, and non-empty sets has non-empty intersection.

The Riesz representation theorem and the projection theorem are not true in general inner product spaces

32. We consider l_2 with the usual inner product and $c_{00} \subseteq l_2$, the linear subspace of all finite rank sequences. We consider on c_{00} the inner product from l_2 and the norm generated by this inner product. Let

$$M = \left\{ x = (x_n)_{n \in \mathbb{N}} \in c_{00} \,\Big|\, \sum_{n=1}^{\infty} \frac{x_n}{n} = 0 \right\}.$$

i) Prove that there is a unique $x^* \in c_{00}^*$ such that $M = \ker x^*$ and deduce that $M \subseteq c_{00}$ is a closed linear subspace.
ii) Prove that we cannot find y in c_{00} such that $x^*(x) = \langle x, y \rangle \ \forall x \in c_{00}$.
iii) Calculate $M^\perp$. Prove that $c_{00} \neq M \oplus M^\perp$, and that $M \neq M^{\perp\perp}$.
What can we deduce?

A projector that is not an orthogonal projection

33. Let $H = (\mathbb{R}^2, \langle \cdot, \cdot \rangle_2)$ and $M = \{(x, 0) \mid x \in \mathbb{R}\}$, $N = \{(x, x \tan \theta) \mid x \in \mathbb{R}\}$, where $\theta \in (0, \pi/2)$. Find $T_\theta \in L(H)$ such that $T_\theta^2 = T_\theta$, $T_\theta(H) = M$, $\ker T_\theta = N$. Prove that $\|T_\theta\| = 1/\sin \theta$ and that T_θ is not an orthogonal projection.

Characterizations for orthogonal projections

34. Let H be a Hilbert space and $T \in L(H)$, $T \neq 0$ such that $T^2 = T$. Prove that the following assertions are equivalent:
 i) T is an orthogonal projection;
 ii) $\ker T = (T(H))^\perp$;
 iii) $\|T\| = 1$;
 iv) $T = T^*$;
 v) T is normal, i.e., $T^*T = TT^*$;
 iv) $\langle Tx, x \rangle \geq 0 \ \forall x \in H$.

35. Let H a Hilbert space and $A \subseteq H$ a non-empty set. Let $T : H \to H$ be a linear operator such that $T(H) \subseteq A$ and $(I - T)(H) \subseteq A^\perp$. Prove that $A \subseteq H$ is a closed linear subspace and that T is the orthogonal projection onto A.

Algebraic operations with orthogonal projections

36. Let H be a Hilbert space and P, Q two orthogonal projections onto two closed linear subspaces of H. Then:

i) $P + Q$ is an orthogonal projection if and only if $P(H) \perp Q(H)$. If $P + Q$ is an orthogonal projection then

$$\begin{aligned}(P+Q)(H) &= P(H) + Q(H), \\ \ker(P+Q) &= \ker P \cap \ker Q.\end{aligned}$$

ii) PQ is an orthogonal projection if and only if $PQ = QP$. If PQ is an orthogonal projection then

$$\begin{aligned}(PQ)(H) &= P(H) \cap Q(H), \\ \ker(PQ) &= \overline{\ker P + \ker Q}.\end{aligned}$$

iii) The following assertions are equivalent:
a) $P - Q$ is an orthogonal projection;
b) $Q(H) \subseteq P(H)$;
c) $PQ = Q$;
d) $QP = Q$.

If $P - Q$ is an orthogonal projection, then

$$\begin{aligned}(P-Q)(H) &= P(H) \ominus Q(H), \\ \ker(P-Q) &= Q(H) \oplus \ker P.\end{aligned}$$

(If L and M are two closed linear subspaces of a Hilbert space H and $L \subseteq M$, then $M \ominus L$ is the closed subspace of M such that $L \oplus (M \ominus L) = M$.)

iv) $PQ = QP$ if and only if $P + Q - PQ$ is an orthogonal projection.

The distance between two orthogonal projections

37. i) Let X be a normed space and $T, S \in L(X)$ such that $T^2 = T$, $S^2 = S$, and $TS = ST$. Prove that either $T = S$, or $\|T - S\| \geq 1$.

ii) Let us consider two closed linear subspaces L_1 and L_2 in a Hilbert H space and denote by P_1 and P_2 the orthogonal projections respectively onto L_1 and L_2. Prove that $\|P_1 - P_2\| \leq 1$.

iii) Let H be a Hilbert space. If P and Q are two orthogonal projections on H such that PQ is also an orthogonal projection prove that either $P = Q$, or $\|P - Q\| = 1$.

11.2 Solutions

1. From the projection theorem, we have $\Re \langle x - \Pr_A(x), \Pr_A(x) - a \rangle \geq 0 \ \forall a \in A$. In particular, replacing a with $\Pr_A(y)$ we obtain that

$$\Re \langle x - \Pr_A(x), \Pr_A(x) - \Pr_A(y) \rangle \geq 0. \tag{1}$$

Analogously, $\Re \langle y - \Pr_A(y), \Pr_A(y) - \Pr_A(x) \rangle \geq 0$, and therefore

$$\Re \langle \Pr_A(y) - y, \Pr_A(x) - \Pr_A(y) \rangle \geq 0. \tag{2}$$

Adding (1) and (2) we get $\Re \langle x - y - (\Pr_A(x) - \Pr_A(y)), \Pr_A(x) - \Pr_A(y) \rangle \geq 0$, i.e.,

$$\Re \langle x - y, \Pr_A(x) - \Pr_A(y) \rangle \geq \|\Pr_A(x) - \Pr_A(y)\|^2 \tag{3}$$

From the Cauchy–Buniakowski–Schwarz inequality, we have

$$\Re \langle x - y, \Pr_A(x) - \Pr_A(y) \rangle \leq |\langle x - y, \Pr_A(x) - \Pr_A(y) \rangle|$$
$$\leq \|x - y\| \|\Pr_A(x) - \Pr_A(y)\|$$

and from here, replacing in (3), we obtain that

$$\|\Pr_A(x) - \Pr_A(y)\|^2 \leq \|x - y\| \|\Pr_A(x) - \Pr_A(y)\|,$$

and therefore $\|\Pr_A(x) - \Pr_A(y)\| \leq \|x - y\|$.

2. i) Let $y = \Pr_A(x) \in A$. Let $h : \mathbb{R} \to H$ be defined by $h(t) = tx + (1 - t)y$. Then h is continuous and $h(0) = y$. If, for a contradiction, $y \in \overset{\circ}{A}$, then $h(0) \in \overset{\circ}{A}$ and since h is continuous and $\overset{\circ}{A}$ open, there is $0 < \delta < 1$ such that $\forall t \in [-\delta, \delta]$ it follows that $h(t) \in A$. In particular $h(\delta) \in A$, i.e., $\delta x + (1 - \delta)y \in A$. Then $d(x, A) \leq \|x - (\delta x + (1-\delta)y)\| = (1-\delta)\|x-y\|$. But $y = \Pr_A(x)$ so $d(x, A) = \|x-y\|$, from whence $\|x - y\| \leq (1 - \delta)\|x - y\|$, and since $\|x - y\| > 0$ we obtain that $1 \leq 1 - \delta$, i.e., $\delta \leq 0$, which is false. Hence $y \notin \overset{\circ}{A}$ and since $y \in A$ it follows that $y \in A \setminus \overset{\circ}{A} = \overline{A} \setminus \overset{\circ}{A} = \mathrm{Fr}(A)$.

ii) Since $\mathrm{Fr}(A) \neq \emptyset$ and $\mathrm{Fr}(A) \subseteq A$ it follows that $d(x, A) \leq d(x, \mathrm{Fr}(A))$. But, from (i), $y = \Pr_A(x) \in \mathrm{Fr}(A)$, so $d(x, \mathrm{Fr}(A)) \leq \|x - y\|$. Now from the projection theorem, $\|x - y\| = d(x, A)$, and then $d(x, A) = \|x - y\| = d(x, \mathrm{Fr}(A))$. Since by hypothesis $\mathrm{Fr}(A)$ is convex, non-empty, and always closed, there exists $\Pr_{\mathrm{Fr}(A)}(x) \in \mathrm{Fr}(A)$ and this is the unique element with the property that $d(x, \mathrm{Fr}(A)) = \|x - \Pr_{\mathrm{Fr}(A)}(x)\|$. Since $\|x - y\| = d(x, \mathrm{Fr}(A))$ and $y \in \mathrm{Fr}(A)$ it follows that $y = \Pr_{\mathrm{Fr}(A)}(x)$.

3. i) Using the convexity of φ, it is easy to see that A is a convex set. Also from the continuity of φ it follows that A is closed and that $\overset{\circ}{A} = \{(x, y) \in \mathbb{R}^2 \mid \varphi(x) < y\}$. From exercise 2, $\Pr_A(a, b) \in \mathrm{Fr}(A) = \{(x, y) \in \mathbb{R}^2 \mid \varphi(x) = y\}$, i.e., $\exists x \in \mathbb{R}$ such that $\Pr_A(a, b) = (x, \varphi(x))$. Now, the projection condition is $\langle (a, b) - (x, \varphi(x)), (x, \varphi(x)) - (t, y) \rangle \geq 0$ $\forall t, y \in \mathbb{R}$ with $\varphi(t) \leq y$. In particular, $(a - x)(x - t) + (b - \varphi(x))(\varphi(x) - \varphi(t)) \geq 0$ $\forall t < x$, and therefore

$$a - x + (b - \varphi(x))\frac{\varphi(x) - \varphi(t)}{x - t} \geq 0 \quad \forall t < x.$$

For $t \to x$, $t < x$ it follows that $a - x + (b - \varphi(x))\varphi'(x) \geq 0$. Analogously, for $t \to x$, $t > x$ it follows that $a - x + (b - \varphi(x))\varphi'(x) \leq 0$, and so $a - x + (b - \varphi(x))\varphi'(x) = 0$.

ii) We apply (i).

4. i) From exercise 2, $\Pr_A(a, b) \in \mathrm{Fr}(A)$, i.e., $\exists x \in \mathbb{R}^n$ such that $\Pr_A(a, b) = (x, \varphi(x))$. The projection condition implies $\langle a - x, x - t \rangle + (b - \varphi(x))(\varphi(x) - \varphi(t)) \geq 0$ $\forall t \in \mathbb{R}^n$.

11.2 Solutions

Let $1 \leq i \leq n$. For $t = x + \lambda e_i$, with $\lambda \in \mathbb{R}$, we have $\langle a - x, -\lambda e_i \rangle + (b - \varphi(x))(\varphi(x) - \varphi(x + \lambda e_i)) \geq 0 \ \forall \lambda \in \mathbb{R}$, that is, $\lambda \langle a - x, e_i \rangle + (b - \varphi(x))(\varphi(x + \lambda e_i) - \varphi(x)) \leq 0 \ \forall \lambda \in \mathbb{R}$. For $\lambda > 0$ we have

$$\langle a - x, e_i \rangle + (b - \varphi(x)) \frac{\varphi(x + \lambda e_i) - \varphi(x)}{\lambda} \leq 0,$$

hence passing to the limit for $\lambda \to 0$, $\lambda > 0$ it follows that $a_i - x_i + (b - \varphi(x))(\partial \varphi / \partial x_i)(x) \leq 0$. Analogously, working with $\lambda < 0$ we obtain that $a_i - x_i + (b - \varphi(x))(\partial \varphi / \partial x_i)(x) \geq 0$, i.e., the statement.

ii) Let $\varphi : \mathbb{R}^2 \to \mathbb{R}$ be defined by $\varphi(x, y) = x^2 + y^2$. Obviously φ is C^1, convex, and $\partial \varphi / \partial x = 2x$, $\partial \varphi / \partial y = 2y$. From (i), $\mathrm{Pr}_A(1, \sqrt{3}, 1/2) = (x, y, z)$, where $z = x^2 + y^2$ and (x, y) is a solution for the system

$$\begin{cases} 1 - x + 2x(1/2 - (x^2 + y^2)) = 0 \\ \sqrt{3} - y + 2y(1/2 - (x^2 + y^2)) = 0 \end{cases}.$$

We obtain that $x = 1/2$, $y = \sqrt{3}/2$, $z = 1$.

5. i) On $\mathbb{R}$ the inner product is given by $\langle x, y \rangle = xy$. Let $z = \mathrm{Pr}_{[a,b]}(x)$. Then $\langle x - z, z - y \rangle \geq 0 \ \forall y \in [a, b]$, that is, $(x - z)(z - y) \geq 0 \ \forall y \in [a, b]$. We have 3 cases:

1. $x < a$. Since $z \in [a, b]$ then $x - z < 0$, and it follows that $z - y \leq 0 \ \forall y \in [a, b]$, i.e., $z \leq y \ \forall y \in [a, b]$. In particular $z \leq a$. But $z \in [a, b]$, $z \geq a$, and then $z = a$.
2. $x \in [a, b]$. Then $\mathrm{Pr}_{[a,b]}(x) = x$.
3. $x > b$. Since $z \in [a, b]$ then $x - z > 0$, and we obtain that $z - y \geq 0 \ \forall y \in [a, b]$, so $z \geq y \ \forall y \in [a, b]$. In particular $z \geq b$. But $z \in [a, b]$, $z \leq b$, i.e., $z = b$.

Analogously for the other projections.

ii) Let $x_1 = \mathrm{Pr}_{[a,b]}(x)$ and $y_1 = \mathrm{Pr}_{[c,d]}(y)$. Then $(x_1, y_1) \in D$. In order to prove that $(x_1, y_1) = \mathrm{Pr}_D(x, y)$, we must show that

$$\langle (x, y) - (x_1, y_1), (x_1, y_1) - (u, v) \rangle \geq 0 \ \forall (u, v) \in D,$$

that is, $(x - x_1)(x_1 - u) + (y - y_1)(y_1 - v) \geq 0$, which is true, since by (i) $(x - x_1)(x_1 - u) \geq 0 \ \forall u \in [a, b]$ and $(y - y_1)(y_1 - v) \geq 0 \ \forall v \in [c, d]$.

iii) Recall that the inner product on $H_1 \times H_2$ is defined by the formula $\langle (x_1, x_2), (y_1, y_2) \rangle = \langle x_1, y_1 \rangle + \langle x_2, y_2 \rangle$. Let $x_1 = \mathrm{Pr}_A(x)$ and $y_1 = \mathrm{Pr}_B(y)$. Then $(x_1, y_1) \in D = A \times B$. To prove that $(x_1, y_1) = \mathrm{Pr}_D(x, y)$, we must prove that

$$\Re \langle (x, y) - (x_1, y_1), (x_1, y_1) - (u, v) \rangle \geq 0 \ \forall (u, v) \in D,$$

that is,

$$\Re \langle x - x_1, x_1 - u \rangle + \Re \langle y - y_1, y_1 - v \rangle \geq 0,$$

which is true, since $\Re \langle x - x_1, x_1 - u \rangle \geq 0 \ \forall u \in A$ and $\Re \langle y - y_1, y_1 - v \rangle \geq 0 \ \forall v \in B$.

6. i) Let $y_n = \mathrm{Pr}_{[0,1/n]}(x_n)$, that is, $y_n = \begin{cases} 0, & x_n < 0, \\ x_n, & 0 \leq x_n \leq 1/n, \\ 1/n, & x_n > 1/n, \end{cases}$ (see exercise 5(i)).

We will prove that $y = (y_n)_{n \in \mathbb{N}} = \mathrm{Pr}_A(x)$, i.e., equivalently, $y \in A$ and $\Re \langle x - y, y - z \rangle \geq$

$0 \, \forall z \in A$. But obviously $(y_n)_{n \in \mathbb{N}} \in l_2$, $y_n \in [0, 1/n] \, \forall n \in \mathbb{N}$, and $\langle x - y, y - z \rangle = \sum_{n=1}^{\infty} (x_n - y_n)(y_n - z_n) \geq 0$ (since $z = (z_n)_{n \in \mathbb{N}} \in A$, $z_n \in [0, 1/n]$, and from $y_n = \Pr_{[0,1/n]}(x_n)$ we have $(x_n - y_n)(y_n - z_n) \geq 0 \, \forall n \in \mathbb{N}$). Then

$$\Pr_A(e_k) = \left(\Pr_{[0,1]}(0), ..., \Pr_{[0,1/k]}(1), ...\right) = (0, ..., 0, 1/k, 0, ...) = e_k/k \quad \forall k \geq 2,$$

$\Pr_A(e_1) = e_1$, since $e_1 \in A$. Now from the projection theorem, $d(e_k, A) = \|e_k - \Pr_A(e_k)\| = (k-1)/k$, and so $\lim_{k \to \infty} d(e_k, A) = 1$.

ii) We have $A = \{(x_n)_{n \in \mathbb{N}} \in l_2 \mid x_n \in [-a_n, a_n] \, \forall n \in \mathbb{N}\}$. Let $y_n = \Pr_{[-a_n, a_n]}(x_n)$, i.e., $y_n = \begin{cases} -a_n, & x_n < -a_n, \\ x_n, & -a_n \leq x_n \leq a_n, \\ a_n, & x_n > a_n, \end{cases}$ and take $y = (y_n)_{n \in \mathbb{N}}$. The proof for the fact that $y = \Pr_A(x)$ is the same as above, and we will omit it.

iii) Using (ii) we obtain that $\Pr_A(ae_k) = \begin{cases} -e_k, & a < -1, \\ ae_k, & a \in [-1, 1], \\ e_k, & a > 1. \end{cases}$ Also,

$$d(((k+1)/k) e_k, A) = \|((k+1)/k) e_k - \Pr_A(((k+1)/k) e_k)\|.$$

Since $(k+1)/k > 1$ we have $\Pr_A(((k+1)/k) e_k) = e_k$, from whence

$$d(((k+1)/k) e_k, A) = \|((k+1)/k) e_k - e_k\| = 1/k,$$

and then $\lim_{k \to \infty} d(((k+1)/k) e_k, A) = 0$.

iv) Let $x = (x_1, x_2, ...) \in l_2$. Define $y_1 = \Pr_{[0,a_1]}(x_1)$, $y_2 = \Pr_{[0,a_2]}(x_2)$, ..., $y_m = \Pr_{[0,a_m]}(x_m)$, and consider $y = (y_1, y_2, ..., y_m, x_{m+1}, ...) \in A$. Then $y = \Pr_A(x)$. Indeed, the projection condition is $\langle x - y, y - z \rangle \geq 0 \, \forall z \in A$, that is, $(x_1 - y_1)(y_1 - z_1) + \cdots + (x_m - y_m)(y_m - z_m) \geq 0 \, \forall 0 \leq z_1 \leq a_1, ..., 0 \leq z_m \leq a_m$, which is clearly true.

7. Let $x = (x_n)_{n \in \mathbb{N}} \in l_2$. For any $n \in \mathbb{N}$ we denote $A_n = \{(z_k)_{k \in \mathbb{N}} \in l_2 \mid a_1 \leq z_1 \leq b_1, ..., a_n \leq z_n \leq b_n\}$, $y_n = \Pr_{[a_n, b_n]}(x_n)$, and let $y = (y_1, y_2, ...)$. Then $A = \bigcap_{n \in \mathbb{N}} A_n$, $A_{n+1} \subseteq A_n$, and obviously A_n is closed, convex, and non-empty, $\forall n \in \mathbb{N}$. Since by hypothesis $A \neq \emptyset$ then $d(x, A)$ has a meaning, and belongs to $\mathbb{R}_+$. Since $A \subseteq A_n \, \forall n \in \mathbb{N}$ it follows $d(x, A_n) \leq d(x, A) \, \forall n \in \mathbb{N}$, i.e., $\{d(x, A_n) \mid n \in \mathbb{N}\}$ is bounded. Thus $\exists M > 0$ such that $d(x, A_n) \leq M \, \forall n \in \mathbb{N}$, that is, $\|x - \Pr_{A_n}(x)\| \leq M \, \forall n \in \mathbb{N}$. From exercise 6(iv), for any $n \in \mathbb{N}$ we have $\Pr_{A_n}(x) = (y_1, ..., y_n, x_{n+1}, ...)$. Then

$$\|x - \Pr_{A_n}(x)\| = \sqrt{|x_1 - y_1|^2 + \cdots + |x_n - y_n|^2} \leq M \quad \forall n \in \mathbb{N},$$

from whence $\sum_{n=1}^{\infty} |x_n - y_n|^2 \leq M^2$, i.e., $(x_n - y_n)_{n \in \mathbb{N}} \in l_2$, and since $(x_n)_{n \in \mathbb{N}} \in l_2$ it follows that $y = (y_n)_{n \in \mathbb{N}} \in l_2$. We will prove now that $y = \Pr_A(x)$. Obviously, $a_n \leq y_n \leq b_n \, \forall n \in \mathbb{N}$, i.e., $y \in A$, and we must verify the projection condition, i.e., $\langle x - y, y - z \rangle \geq 0 \, \forall z \in A$, or $\sum_{n=1}^{\infty} (x_n - y_n)(y_n - z_n) \geq 0 \, \forall z = (z_n)_{n \in \mathbb{N}} \in A$. But this is an easy consequence of the fact that $(x_n - y_n)(y_n - z_n) \geq 0 \, \forall n \in \mathbb{N}$.

11.2 Solutions

8. See [26, exercise 49, chapter II].

Let $\alpha_1 = (1, 1, 1, 0, 0, 0, \ldots)$, $\alpha_2 = (0, 0, 0, 1, 1, 1, 0, \ldots)$, etc.. For $x = (x_n)_{n \in \mathbb{N}}$, $(\xi_1, \xi_1, \xi_1, \xi_2, \xi_2, \xi_2, \ldots) = \mathrm{Pr}_{G_1}(x)$ is given by the conditions: $\mathrm{Pr}_{G_1}(x) \in G_1$ and $x - \mathrm{Pr}_{G_1}(x) \bot G_1$. From $x - \mathrm{Pr}_{G_1}(x) \bot G_1$ it follows that $x - \mathrm{Pr}_{G_1}(x) \bot \alpha_1$, i.e., $x_1 - \xi_1 + x_2 - \xi_1 + x_3 - \xi_1 = 0$, $\xi_1 = (x_1 + x_2 + x_3)/3$. From $x - \mathrm{Pr}_{G_1}(x) \bot \alpha_2$ it follows that $\xi_2 = (x_4 + x_5 + x_6)/3$, etc.. Analogously one can calculate the projections onto G_2 and G_3.

9. i) Let us observe first that $T^*(B) \subseteq H_1$ is a closed convex set. The fact that T^* is linear implies $T^*(B)$ convex, and that T^* is an isometry implies that $T^*(B)$ is closed. Let $y_1 = \mathrm{Pr}_B(y) \in B$. What we must prove is that $T^*(y_1) = \mathrm{Pr}_{T^*(B)}(T^*y)$, that is, $T^*(y_1) \in T^*(B)$ (this is obviously true) and $\Re \langle T^*y - T^*y_1, T^*y_1 - a \rangle \geq 0 \, \forall a \in T^*(B)$. Let $a \in T^*(B)$. Then there is $b \in B$ such that $a = T^*(b)$. Since $y_1 = \mathrm{Pr}_B(y)$ we have $\Re \langle y - y_1, y_1 - b \rangle \geq 0$. From $TT^* = I$ it follows that $\Re \langle y - y_1, TT^*(y_1 - b) \rangle \geq 0$, or $\Re \langle T^*y - T^*y_1, T^*y_1 - T^*b \rangle \geq 0$, that is, $\Re \langle T^*y - T^*y_1, T^*y_1 - a \rangle \geq 0$, which is exactly what we wanted to prove.

ii) Let us observe that under the hypothesis from the statement all the projections have a meaning. We will use the existence and uniqueness property from the projection theorem, i.e., $\mathrm{Pr}_A(u)$ is the only element in A with the property that $d(u, A) = \|u - \mathrm{Pr}_A(u)\|$. We have $d(y, x + A) = \inf_{a \in A} \|y - x - a\| = d(y - x, A)$. Also there exists $\mathrm{Pr}_{x+A}(y) \in x + A$ and this is the only element in $x + A$ with the property that $d(y, x + A) = \|y - \mathrm{Pr}_{x+A}(y)\|$. Analogously $\mathrm{Pr}_A(y - x) \in A$ and this is the only element with the property that $d(y - x, A) = \|y - x - \mathrm{Pr}_A(y - x)\|$. The element $x + \mathrm{Pr}_A(y - x)$ is in $x + A$, and by the above it has the property that $d(y, x + A) = \|y - (x + \mathrm{Pr}_A(y - x))\|$. It follows that $\mathrm{Pr}_{x+A}(y) = x + \mathrm{Pr}_A(y - x)$.

iii) **First solution.** We remark first that under our hypothesis all the projections have a meaning. We will proceed as in (ii). We have $d(\lambda x, \lambda A) = |\lambda| d(x, A)$. $\mathrm{Pr}_A(x) \in A$ is the only element with the property that $d(x, A) = \|x - \mathrm{Pr}_A(x)\|$, so $d(\lambda x, \lambda A) = |\lambda| \|x - \mathrm{Pr}_A(x)\| = \|\lambda x - \lambda \mathrm{Pr}_A(x)\|$, i.e., the element $\lambda \mathrm{Pr}_A(x) \in \lambda A$ and has the property that $d(\lambda x, \lambda A) = \|\lambda x - \lambda \mathrm{Pr}_A(x)\|$. Therefore $\lambda \mathrm{Pr}_A(x) = \mathrm{Pr}_{\lambda A}(\lambda x)$.

Second solution. If $\lambda = 0$, the property $\lambda \mathrm{Pr}_A(x) = \mathrm{Pr}_{\lambda A}(\lambda x)$ is obvious. Suppose now that $\lambda \neq 0$. Let $x_1 = \mathrm{Pr}_{\lambda A}(\lambda x) \in \lambda A$. Then $\Re \langle \lambda x - x_1, x_1 - z \rangle \geq 0 \, \forall z \in \lambda A$, or $\Re \langle \lambda x - x_1, x_1 - \lambda a \rangle \geq 0 \, \forall a \in A$, or $\Re \left(\lambda \overline{\lambda} \langle x - x_1/\lambda, x_1/\lambda - a \rangle \right) \geq 0 \, \forall a \in A$. Since $\lambda \overline{\lambda} = |\lambda|^2 > 0$, it follows that $\Re \langle x - x_1/\lambda, x_1/\lambda - a \rangle \geq 0 \, \forall a \in A$. Since $x_1 \in \lambda A$ then $x_1/\lambda \in A$, so $x_1/\lambda = \mathrm{Pr}_A(x)$, i.e., $x_1 = \lambda \mathrm{Pr}_A(x)$.

A is symmetric if and only if $-A = A$, and we can apply what is proved above, taking $\lambda = -1$.

10. i) a) Let $A = \mathrm{Sp}\left(\bigcup_{n \in \mathbb{N}} G_n\right)$. Then using exercise 1(i) chapter 4, $d(x, G) = d(x, \overline{A}) = d(x, A)$. Let $\varepsilon > 0$. Then $d(x, G) + \varepsilon = d(x, A) + \varepsilon > d(x, A) = \inf_{a \in A} \|x - a\|$, from whence we deduce that there is $a_\varepsilon \in A$ such that $d(x, G) + \varepsilon > \|x - a_\varepsilon\|$. Since $a_\varepsilon \in \mathrm{Sp}\left(\bigcup_{n \in \mathbb{N}} G_n\right)$ we can find $\lambda_1, \ldots, \lambda_k \in \mathbb{K}$, $x_1, \ldots, x_k \in \bigcup_{n \in \mathbb{N}} G_n$ such that $a_\varepsilon = \lambda_1 x_1 + \cdots + \lambda_k x_k$. There are $n_1, \ldots, n_k \in \mathbb{N}$ such that $x_1 \in G_{n_1}, \ldots, x_k \in G_{n_k}$. Let $n = \max(n_1, \ldots, n_k)$. Then $G_{n_1}, G_{n_2}, \ldots, G_{n_k} \subseteq G_n$, from whence $x_1, \ldots, x_k \in G_n$, and since G_n is a linear subspace, $a_\varepsilon = \lambda_1 x_1 + \cdots + a_k x_k \in G_n$. Hence $\forall \varepsilon > 0$ $\exists n \in \mathbb{N}$, $a_\varepsilon \in G_n$ such that $d(x,G) + \varepsilon > \|x - a_\varepsilon\| \geq d(x, G_n)$. From $G_n \subseteq G$ it follows that $d(x,G) \leq d(x, G_n)$ $\forall n \in \mathbb{N}$. Hence $d(x,G) = \inf_{n \in \mathbb{N}} d(x, G_n)$. From $G_n \subseteq G_{n+1}$ it follows that $d(x, G_{n+1}) \leq d(x, G_n)$, so from the Weierstrass theorem in $\mathbb{R}$ we have $\inf_{n \in \mathbb{N}} d(x, G_n) = \lim_{n \to \infty} d(x, G_n) = d(x,G)$.

b) See [9, proposition 7, chapter V, section 1].

From (a) we have $d(x,G) = \lim_{n \to \infty} d(x, G_n)$. Let $a_n = \mathrm{Pr}_{G_n}(x)$ and $a = \mathrm{Pr}_G(x)$. Then we have $\lim_{n \to \infty} \|x - a_n\| = \lim_{n \to \infty} d(x, G_n) = d(x,G)$. Since $G_n \subseteq G$ $\forall n \in \mathbb{N}$, then $(a_n)_{n \in \mathbb{N}} \subseteq G$, and then from the proof for the projection theorem (or see the solution for the part (ii) below) it follows that $a_n \to \mathrm{Pr}_G(x)$ in norm, i.e., $\mathrm{Pr}_G(x) = \lim_{n \to \infty} \mathrm{Pr}_{G_n}(x)$.

ii) See [9, proposition 6, chapter V, section 1].

Since for any $n \in \mathbb{N}$ we have $G_n \supseteq G_{n+1}$ and $G \subseteq G_n$, then $d(x, G_n) \leq d(x, G_{n+1})$ and $d(x, G_n) \leq d(x, G)$, hence $\sup_{n \in \mathbb{N}} d(x, G_n) = \lim_{n \to \infty} d(x, G_n) \leq d(x,G)$. Let $a_n = \mathrm{Pr}_{G_n}(x)$. Then for any $n, p \in \mathbb{N}$ we have $a_n \in G_n$ and $a_{n+p} \in G_{n+p} \subseteq G_n$, i.e., $a_n, a_{n+p} \in G_n$, hence $(a_n + a_{n+p})/2 \in G_n$, so $\|x - (a_n + a_{n+p})/2\| \geq d(x, G_n)$. By the parallelogram law we have

$$\|(x - a_n) + (x - a_{n+p})\|^2 + \|(x - a_n) - (x - a_{n+p})\|^2 = 2\left(\|x - a_n\|^2 + \|x - a_{n+p}\|^2\right),$$

i.e.,

$$\begin{aligned} \|a_{n+p} - a_n\|^2 &= 2\left(\|x - a_n\|^2 + \|x - a_{n+p}\|^2\right) - 4\|x - (a_n + a_{n+p})/2\|^2 \\ &\leq 2(d(x, G_n)^2 + d(x, G_{n+p})^2) - 4d(x, G_n)^2. \end{aligned}$$

Since the sequence $(d(x, G_n))_{n \in \mathbb{N}}$ converges, from this inequality it follows that the sequence $(a_n)_{n \in \mathbb{N}} \subseteq X$ is Cauchy, hence convergent to an element $a \in X$, i.e., $a_n \to a$. Let $n \in \mathbb{N}$. Since $a_{n+p} \in G_{n+p} \subseteq G_n$, i.e., $a_{n+p} \in G_n$ $\forall p \in \mathbb{N}$, passing to the limit for $p \to \infty$ we obtain that $a \in \overline{G_n} = G_n$, since G_n is closed, hence $a \in \bigcap_{n \in \mathbb{N}} G_n = G$, and $d(x,G) \leq \|x - a\| = \lim_{n \to \infty} \|x - a_n\| = \lim_{n \to \infty} d(x, G_n) \leq d(x,G)$, i.e., $d(x,G) = \lim_{n \to \infty} d(x, G_n) = \|x - a\|$. It follows that $a = \mathrm{Pr}_G(x)$ and $\mathrm{Pr}_G(x) = \lim_{n \to \infty} \mathrm{Pr}_{G_n}(x)$.

11. See [9, exercise 5, chapter V, section 2].

i) Let $y_1 = x_1$ and $y_2 = \alpha_1 y_1 + x_2$, where $\alpha_1 \in \mathbb{K}$ is such that $y_2 \perp y_1$. Multiplying the first column by $\overline{\alpha}_1$ and adding it to the second column, we have that

$$\Gamma(x_1, ..., x_n) = \begin{vmatrix} \langle y_1, y_1 \rangle & \langle y_1, \alpha_1 y_1 + x_2 \rangle & \cdots & \langle y_1, x_n \rangle \\ \langle x_2, y_1 \rangle & \langle x_2, \alpha_1 y_1 + x_2 \rangle & \cdots & \langle x_2, x_n \rangle \\ \cdots & \cdots & \cdots & \cdots \\ \langle x_n, y_1 \rangle & \langle x_n, \alpha_1 y_1 + x_2 \rangle & \cdots & \langle x_n, x_n \rangle \end{vmatrix}$$

$$= \begin{vmatrix} \langle y_1, y_1 \rangle & \langle y_1, y_2 \rangle & \cdots & \langle y_1, x_n \rangle \\ \langle x_2, y_1 \rangle & \langle x_2, y_2 \rangle & \cdots & \langle x_2, x_n \rangle \\ \cdots & \cdots & \cdots & \cdots \\ \langle x_n, y_1 \rangle & \langle x_n, y_2 \rangle & \cdots & \langle x_n, x_n \rangle \end{vmatrix}.$$

11.2 Solutions

Now we multiply the first line by α_1 and we add it to the second line and we obtain that

$$\Gamma(x_1,...,x_n) = \begin{vmatrix} \langle y_1, y_1 \rangle & \langle y_1, y_2 \rangle & \cdots & \langle y_1, x_n \rangle \\ \langle \alpha_1 y_1 + x_2, y_1 \rangle & \langle \alpha_1 y_1 + x_2, y_2 \rangle & \cdots & \langle \alpha_1 y_1 + x_2, x_n \rangle \\ \cdots & \cdots & \cdots & \cdots \\ \langle x_n, y_1 \rangle & \langle x_n, y_2 \rangle & \cdots & \langle x_n, x_n \rangle \end{vmatrix}$$

$$= \begin{vmatrix} \langle y_1, y_1 \rangle & \langle y_1, y_2 \rangle & \cdots & \langle y_1, x_n \rangle \\ \langle y_2, y_1 \rangle & \langle y_2, y_2 \rangle & \cdots & \langle y_2, x_n \rangle \\ \cdots & \cdots & \cdots & \cdots \\ \langle x_n, y_1 \rangle & \langle x_n, y_2 \rangle & \cdots & \langle x_n, x_n \rangle \end{vmatrix}.$$

Hence $\Gamma(x_1, x_2, x_3, ..., x_n) = \Gamma(y_1, y_2, x_3, ..., x_n)$. Let $y_3 = \beta_1 y_1 + \beta_2 y_2 + x_3$, where $\beta_1, \beta_2 \in \mathbb{K}$ are such that $y_3 \perp y_1$ and $y_3 \perp y_2$. We add to the third column the first column multiplied by $\overline{\beta_1}$ and the second column multiplied by $\overline{\beta_2}$, and the same operations are made for the lines (we multiply with β_1 and β_2, respectively). We obtain a determinant in which x_3 is replaced by y_3, i.e., $\Gamma(y_1, y_2, x_3, ..., x_n) = \Gamma(y_1, y_2, y_3, x_4, ..., x_n)$. In this way we obtain that $\Gamma(x_1, x_2, x_3, ..., x_n) = \Gamma(y_1, y_2, y_3, ..., y_n)$. Since the system $\{y_1, y_2, y_3, ..., y_n\}$ is an orthogonal one, we obtain that

$$\Gamma(x_1, x_2, x_3, ..., x_n) = \Gamma(y_1, y_2, y_3, ..., y_n) = \|y_1\|^2 \|y_2\|^2 \cdots \|y_n\|^2 \geq 0,$$

and that $\Gamma(x_1, x_2, x_3, ..., x_n) = 0$ if and only if $\Gamma(y_1, y_2, y_3, ..., y_n) = 0$, that is, if and only if there is $1 \leq i \leq n$ such that $y_i = 0$, i.e., if and only if the system $\{x_1, x_2, x_3, ..., x_n\}$ is linearly dependent.

ii) Let $x_1 = (a^n)_{n \geq 0}$, $x_2 = (b^n)_{n \geq 0}$, $x_3 = (c^n)_{n \geq 0}$. Then $x_1, x_2, x_3 \in l_2$ since, for instance, $\sum\limits_{n=0}^{\infty} |a|^{2n} = 1/\left(1 - |a|^2\right)$. From (i) we have $\Gamma(x_1, x_2, x_3) \geq 0$. But $\langle x_1, x_1 \rangle = \|x_1\|^2 = 1/\left(1 - |a|^2\right)$, $\langle x_1, x_2 \rangle = \sum\limits_{n=0}^{\infty} (ab)^n = 1/\left(1 - ab\right)$, $\langle x_1, x_3 \rangle = 1/\left(1 - ac\right)$, etc..
Writing the Gram determinant we obtain the inequality from the statement.

12. i) Let $y = \Pr_G(x) \in G = \text{Sp}\{x_1, x_2, ..., x_n\}$, since G is closed. Then we can find n unique scalars $\lambda_1, ..., \lambda_n \in \mathbb{K}$ such that $y = \lambda_1 x_1 + \cdots + \lambda_n x_n$. But $y - x \in G^\perp$, i.e., $y - x \perp G$, $\langle x_i, y - x \rangle = 0$, $i = \overline{1,n}$, that is, $\overline{\lambda_1} \langle x_i, x_1 \rangle + \cdots + \overline{\lambda_n} \langle x_i, x_n \rangle = \langle x_i, x \rangle$, $i = \overline{1,n}$, i.e., $\overline{\lambda_1}, ..., \overline{\lambda_n}$ is a solution for the linear system

$$\begin{cases} \overline{\lambda_1} \langle x_1, x_1 \rangle + \cdots + \overline{\lambda_n} \langle x_1, x_n \rangle = \langle x_1, x \rangle \\ \cdots \\ \overline{\lambda_1} \langle x_n, x_1 \rangle + \cdots + \overline{\lambda_n} \langle x_n, x_n \rangle = \langle x_n, x \rangle \end{cases}.$$

The determinant for this system is $\triangle = \Gamma(x_1, x_2, ..., x_n)$, and $\triangle \neq 0$ by exercise 11, since $x_1, x_2, ..., x_n$ are linearly independent. Now the well known Cramer law shows that this system has a unique solution, and that $\lambda_i = \overline{\triangle_i}/\triangle$, where $\triangle_i$ is obtained as in the statement.

ii) Let $A = \begin{pmatrix} 1 & 1 & 1 & 1 \\ 1 & 2 & 2 & -1 \\ 1 & 0 & 0 & 3 \end{pmatrix}$. It is easy to see that rank $(A) = 2$ and that $\begin{vmatrix} 1 & 1 \\ 1 & 2 \end{vmatrix} \neq$

0, i.e., $x_3 \in \text{Sp}\{x_1, x_2\}$ and $\text{Sp}\{x_1, x_2, x_3\} = \text{Sp}\{x_1, x_2\}$. From (i),

$$\Gamma(x_1, x_2) = \begin{vmatrix} \langle x_1, x_1 \rangle & \langle x_1, x_2 \rangle \\ \langle x_2, x_1 \rangle & \langle x_2, x_2 \rangle \end{vmatrix} = 24,$$

$$\triangle_1 = \begin{vmatrix} \langle x_1, x \rangle & \langle x_1, x_2 \rangle \\ \langle x_2, x \rangle & \langle x_2, x_2 \rangle \end{vmatrix} = 72,$$

$$\triangle_2 = \begin{vmatrix} \langle x_1, x_1 \rangle & \langle x_1, x \rangle \\ \langle x_2, x_1 \rangle & \langle x_2, x \rangle \end{vmatrix} = -48,$$

so $\text{Pr}_G(x) = (\overline{\triangle_1}/\triangle) x_1 + (\overline{\triangle_2}/\triangle) x_2 = 3x_1 - 2x_2 = (1, -1, -1, 5)$.

13. i) Let us observe that $L = (\text{Sp}\{x_1, ..., x_n\})^\perp$ and that $L^\perp = \text{Sp}\{x_1, ..., x_n\} = G$. Since $x = \text{Pr}_L(x) + \text{Pr}_G(x)$, it follows that $\text{Pr}_L(x) = x - \text{Pr}_G(x)$ and from exercise 12 we obtain the statement.

ii) Let $x_1 = (2, 1, 1, 3)$, $x_2 = (3, 2, 2, 1)$, $x_3 = (1, 2, 2, -9)$. Then

$$L = \{y \in \mathbb{R}^4 \mid \langle y, x_1 \rangle = 0, \langle y, x_2 \rangle = 0, \langle y, x_3 \rangle = 0\}.$$

The matrix $A = \begin{pmatrix} 2 & 1 & 1 & 3 \\ 3 & 2 & 2 & 1 \\ 1 & 2 & 2 & -9 \end{pmatrix}$ has rank 2 and $\begin{vmatrix} 2 & 1 \\ 3 & 2 \end{vmatrix} \neq 0$, i.e., $x_3 \in \text{Sp}\{x_1, x_2\}$, that is, $L = \{y \mid \langle y, x_1 \rangle = 0, \langle y, x_2 \rangle = 0\}$. Let $G = \text{Sp}\{x_1, x_2\}$. We have $\Gamma(x_1, x_2) = 101$, $\triangle_1 = 101$ and $\triangle_2 = 0$. Then $\text{Pr}_L(x) = x - x_1 = (5, -5, -2, -1)$.

14. i) Since $x_1, ..., x_n \in H$ are linearly independent, then by exercise 2 of chapter 7 there is $x^* \in H^*$ such that $x^*(x_1) = \overline{c_1}, ..., x^*(x_n) = \overline{c_n}$. By Riesz representation theorem it follows that there is $x_0 \in H$ such that $x^*(x) = \langle x, x_0 \rangle \; \forall x \in H$. Then $\langle x_0, x_1 \rangle = c_1, ..., \langle x_0, x_n \rangle = c_n$, i.e., $x_0 \in V$. Then $V = x_0 + G$, where

$$G = \{x \in H \mid \langle x, x_1 \rangle = 0, ..., \langle x, x_n \rangle = 0\} = (\text{Sp}\{x_1, ..., x_n\})^\perp.$$

From exercise 9(ii) we have that $\text{Pr}_V(x) = x_0 + \text{Pr}_G(x - x_0)$, and since $G = L^\perp$ we have that $\text{Pr}_G(x - x_0) + \text{Pr}_L(x - x_0) = x - x_0$, from whence $\text{Pr}_V(x) = x - \text{Pr}_L(x - x_0)$. Since $L = \text{Sp}\{x_1, ..., x_n\}$, we can apply the results from exercise 12 to obtain the statement, i.e., $\triangle = \Gamma(x_1, ..., x_n)$ and $\triangle_i$ is obtained by replacing in $\triangle$ the column i with the column

$$\begin{pmatrix} \langle x_1, x - x_0 \rangle \\ ... \\ \langle x_n, x - x_0 \rangle \end{pmatrix} = \begin{pmatrix} \langle x_1, x \rangle - \overline{c}_1 \\ ... \\ \langle x_n, x \rangle - \overline{c}_n \end{pmatrix},$$

since $x_0 \in V$, i.e., $\langle x_0, x_i \rangle = c_i$, $1 \leq i \leq n$.

ii) We have $x = (4, 2, -5, 1)$, $x_1 = (2, -1, 1, 2)$, $x_2 = (2, -4, 2, 3)$. Then $\triangle = \Gamma(x_1, x_2) = 74$, $\triangle_1 = 106$, $\triangle_2 = -94$, so $\text{Pr}_V(x) = (136/37, -61/37, -144/37, 72/37)$.

15. See [9, exercise 5 (b), chapter V, section 2].

Let $y = \text{Pr}_G(x) = \lambda_1 x_1 + \cdots + \lambda_n x_n$. For $x \in H$ we have that

$$\Gamma(x_1, ..., x_n, x) = \begin{vmatrix} \langle x_1, x_1 \rangle & \langle x_1, x_2 \rangle & ... & \langle x_1, x_n \rangle & \langle x_1, x \rangle \\ ... & ... & ... & ... & ... \\ \langle x_n, x_1 \rangle & \langle x_n, x_2 \rangle & ... & \langle x_n, x_n \rangle & \langle x_n, x \rangle \\ \langle x, x_1 \rangle & \langle x, x_2 \rangle & ... & \langle x, x_n \rangle & \langle x, x \rangle \end{vmatrix}.$$

11.2 Solutions

We add to the last column the first column multiplied by $-\overline{\lambda}_1$, the second column multiplied by $-\overline{\lambda}_2$, ..., the nth column multiplied by $-\overline{\lambda}_n$. On the last column we have now $\langle x_j, x - \lambda_1 x_1 - \cdots - \lambda_n x_n \rangle = \langle x_j, x - y \rangle = 0$, since $x - y = \Pr_G(x) \in G^{\perp}$ and $x_j \in G$. Also $\langle x, x - \lambda_1 x_1 - \cdots - \lambda_n x_n \rangle = \langle x, x - y \rangle$. But $x - y \in G^{\perp}$, $y = \Pr_G(x) \in G$, hence $\langle y, x - y \rangle = 0$, from whence $\langle x - y, x - y \rangle = \langle x, x - y \rangle$, i.e.,

$$\langle x, x - \lambda_1 x_1 - \cdots - \lambda_n x_n \rangle = \|x - y\|^2 = \|x - \Pr_G(x)\|^2.$$

But from the orthogonal projection theorem, $\|x - \Pr_G(x)\| = d(x, G) = d$. Hence

$$\Gamma(x_1, ..., x_n, x) = \begin{vmatrix} \langle x_1, x_1 \rangle & \langle x_1, x_2 \rangle & \cdots & \langle x_1, x_n \rangle & 0 \\ \cdots & \cdots & \cdots & \cdots & \cdots \\ \langle x_n, x_1 \rangle & \langle x_n, x_2 \rangle & \cdots & \langle x_n, x_n \rangle & 0 \\ \langle x, x_1 \rangle & \langle x, x_2 \rangle & \cdots & \langle x, x_n \rangle & d^2 \end{vmatrix},$$

and then $\Gamma(x_1, ..., x_n, x) = d^2 \Gamma(x_1, ..., x_n)$.

16. i) We have

$$d(x, V) = \inf_{v \in V} \|x - v\| = \inf_{g \in G} \|x - x_0 - g\| = d(x - x_0, G)$$

and we can use the result from exercise 15.

ii) We have

$$d(x, V) = \sqrt{\frac{\Gamma(x_1, x_2, x - x_0)}{\Gamma(x_1, x_2)}},$$

where $\Gamma(x_1, x_2) = 7^2 \cdot 16$, $\Gamma(x_1, x_2, x - x_0) = 7^3 \cdot 16$, so $d(x, V) = \sqrt{7}$.

iii) We have $V = t^2 + \text{Sp}\{1, t\}$, hence

$$d = \sqrt{\frac{\Gamma(1, t, -t^2)}{\Gamma(1, t)}} = \frac{1}{6\sqrt{5}},$$

since $\Gamma(1, t) = 1/12$ and $\Gamma(1, t, -t^2) = 1/(240 \cdot 9)$.

17. See [9, exercise 5 (c), chapter V, section 2].
$(x_n)_{n \in \mathbb{N}}$ is a topologically linearly independent system if and only if

$$\forall p \geq 1, \; x_p \notin \overline{\text{Sp}\{x_1, x_2, ..., x_{p-1}, x_{p+1}, ...\}}.$$

Let $p \geq 1$. Let $G_n = \text{Sp}\{x_1, x_2, ..., x_{p-1}, x_{p+1}, ..., x_n\}$, $n \geq p + 1$. We have $G_n \subseteq G_{n+1}$, and

$$G^p := \overline{\text{Sp}\{x_1, x_2, ..., x_{p-1}, x_{p+1}, ...\}} = \overline{\bigcup_{n \geq p+1} G_n}.$$

From exercise 10(i),

$$d(x_p, G^p) = \lim_{n \to \infty} d(x_p, G_n) = \inf_{n \geq p+1} d(x_p, G_n).$$

Now using exercise 15,
$$d(x_p, G_n) = \left(\frac{\Gamma(x_1, ..., x_{p-1}, x_{p+1}, ..., x_n)}{\Gamma(x_1, ..., x_n)} \right)^{-1/2},$$
so
$$d(x_p, G^p) = \inf_{n \geq p+1} \left(\frac{\Gamma(x_1, ..., x_{p-1}, x_{p+1}, ..., x_n)}{\Gamma(x_1, ..., x_n)} \right)^{-1/2}.$$
Then $x_p \notin G^p$ if and only if
$$\sup_{n \geq p+1} \sqrt{\frac{\Gamma(x_1, ..., x_{p-1}, x_{p+1}, ..., x_n)}{\Gamma(x_1, ..., x_n)}} < \infty.$$

18. i) We have, by definition,
$$\begin{aligned} d(V_1, V_2) &= \inf_{x \in V_1, y \in V_2} \|x - y\| = \inf_{g_1 \in G_1, g_2 \in G_2} \|x_1 + g_1 - x_2 - g_2\| \\ &= \inf_{g \in G_1 + G_2} \|x_1 - x_2 - g\| = d(x_1 - x_2, G) = d\left(x_1 - x_2, \overline{G}\right), \end{aligned}$$
where $G = G_1 + G_2$. Now from the orthogonal projection theorem we have that $d\left(x, \overline{G}\right) = \|x - \Pr_{\overline{G}}(x)\| = \|\Pr_{\overline{G}^\perp}(x)\|$, and then $d(V_1, V_2) = \|\Pr_{(\overline{G_1 + G_2})^\perp}(x_1 - x_2)\|$. Since $\left(\overline{G_1 + G_2}\right)^\perp = G_1^\perp \cap G_2^\perp$ we obtain the statement.

ii) We have $G_1^\perp \cap G_2^\perp = \{x \in H \mid \langle x, a_1 \rangle = 0, ..., \langle x, a_{n+m} \rangle = 0\}$, and then by exercise 13(i),
$$\Pr_{G_1^\perp \cap G_2^\perp}(x_1 - x_2) = x_1 - x_2 - \left(\frac{\overline{\Delta}_1}{\Delta} a_1 + \cdots + \frac{\overline{\Delta}_{n+m}}{\Delta} a_{n+m} \right),$$
thus the statement follows from (i).

iii) Use (ii)!

19. See [4, example 1.4.2, chapter I, section 1.4].

i) The same proof as in the solution for exercise 20 of chapter 1 shows that A is closed convex and non-empty, hence the projection $\Pr_A(f) = g$ has a meaning. We will prove that $g(s) = \Pr_I(f(s)) \; \forall s \in S$, i.e., $g = \Pr_I \circ f$.

From exercise 1 we have that $\Pr_I$ is non-expansive, thus continuous, and then since f is measurable, $g(s) = \Pr_I(f(s))$ is measurable. We have $\Pr_I(f(s)) \in I \; \forall s \in S$ and since $\Pr_I$ is non-expansive we have $|\Pr_I(f(s)) - \Pr_I(0)| \leq |f(s) - 0|, \; \forall s \in S$. Using that $0 \in I$, we have $\Pr_I(0) = 0$, hence $|\Pr_I(f(s))| \leq |f(s)| \; \forall s \in I$, i.e., $|g(s)| \leq |f(s)| \; \forall s \in S$ and then $\int_S |g|^2 d\mu \leq \int_S |f|^2 d\mu < \infty$, and therefore $g \in L_2(\mu)$. In order to prove that $g(s) = \Pr_I(f(s))$ is the projection of f on A, we must verify that $\langle f - g, g - h \rangle \geq 0 \; \forall h \in A$, i.e., $\int_S (f(s) - g(s))(g(s) - h(s)) d\mu(s) \geq 0 \; \forall h \in A$. Since for any $s \in S$, $g(s) = \Pr_I(f(s))$, we have $(f(s) - \Pr_I(f(s)))(\Pr_I(f(s)) - h(s)) \geq 0$ a.e., on S (if $h \in A$, $h(s) \in I$ a.e.), hence by integration we obtain the relation which we wanted to prove.

ii) By (i),
$$g(x) = \Pr_{[0,1]}(e^x) = \begin{cases} e^x, & 0 \leq e^x \leq 1, \\ 1, & e^x > 1, \end{cases}$$

11.2 Solutions

i.e.,
$$g(x) = \begin{cases} e^x, & -1 \le x \le 0, \\ 1, & 0 < x \le 1. \end{cases}$$

Also,
$$d(e^x, A) = \|e^x - \Pr_A(e^x)\| = \sqrt{(e^2 - 4e + 5)/2}.$$

iii) We will prove that for $f \in L_2(\mu)$ we have $\Pr_A(f) = g$, where $g(s) = \Pr_{[a,b]}(f(s))$. Indeed, in the same way as in (i) we obtain that g is measurable. Also, $g(s) \in [a, b] \, \forall s \in S$, from whence $|g(s)| \le M = \max(|b|, |a|)$. Since $\mu(S) < \infty$ then $\int_S |g|^2 d\mu \le M\mu(S) < \infty$. It follows that $g \in L_2(\mu)$, and we continue as in (i). If we use exercise 5(i) we obtain that
$$g(s) = \begin{cases} a, & f(s) < a, \\ f(s), & a \le f(s) \le b, \\ b, & f(s) > b. \end{cases}$$

iv) If $f \in A$ then $f \ge a$ a.e., and then $|f|^2 = f^2 \ge a^2$ a.e., from whence $\int_S |f|^2 d\mu \ge a^2 \mu(S) = \infty$, which is false.

v) Let $I = [a, \infty)$ and $g(s) = \Pr_I(f(s)) \, \forall s \in S$. As in (i) it follows that g is measurable, and by exercise 1 we have $|\Pr_I(f(s)) - \Pr_I(a)| \le |f(s) - a|$, i.e., $|g(s) - a| \le |f(s) - a| \, \forall s \in S$. Since the measure space is finite it follows that the constant functions belong to $L_2(\mu)$, from whence $f - a \in L_2(\mu)$, hence by the above inequality $g - a \in L_2(\mu)$, so $g \in L_2(\mu)$. Now we continue as in (iii).

vi) If $f \in A$ then $f(s) \le a < 0$ a.e., from whence $|f(s)| = -f(s) \ge -a = |a|$ a.e., $\int_S |f|^2 d\mu \ge |a|^2 \mu(S) = \infty$, contradiction.

vii) As above we obtain that if $g = \Pr_A(f)$ then $g(s) = \Pr_{(-\infty, a]}(f(s))$ a.e..

viii) Again, we will prove that if $h = \Pr_A(f)$, then $h(x) = \Pr_{[-g(x), g(x)]}(f(x))$ $\forall x \in [0, 1]$. We first prove that h is measurable. For this, we will use that the projection map is non-expansive. Since f is measurable there is a sequence of step functions $(s_n)_{n \in \mathbb{N}}$ such that $s_n(x) \to f(x) \, \forall x \in [0, 1]$. We have $\left|\Pr_{[-g(x),g(x)]}(f(x)) - \Pr_{[-g(x),g(x)]}(s_n(x))\right| \le |f(x) - s_n(x)| \to 0$. Now it is easy to see that if s is a step function then $x \to \Pr_{[-g(x),g(x)]}(s(x))$ is measurable. Since the pointwise convergence preserves the measurability, it follows that h is measurable. From exercise 1 we also have $|h(x)| = \left|\Pr_{[-g(x),g(x)]}(f(x)) - \Pr_{[-g(x),g(x)]}(0)\right| \le |f(x)|$, that is, $|h(x)|^2 \le |f(x)|^2$, and since h is measurable and $|f|^2 \in L_1[0,1]$, it follows that $|h|^2 \in L_1[0,1]$, i.e., $h \in L_2[0,1]$. In order to prove that h is the projection of f on A we must show that $\langle f - h, h - u \rangle \ge 0 \, \forall u \in A$, i.e., $\int_0^1 (f(x) - h(x))(h(x) - u(x)) dx \ge 0 \, \forall u \in A$. Indeed, from $h(x) = \Pr_{[-g(x),g(x)]}(f(x))$ $\forall x \in [0,1]$ it follows that $(f - h)(h - u) \ge 0$ a.e., $\forall u \in A$ (if $u \in A$, $u(x) \in [-g(x), g(x)]$ a.e. $x \in [0, 1]$), so by integration we obtain the relation we wanted to prove.

ix) From (viii) it follows that $\Pr_A(x - 1/2) = g$, where $g(x) = \Pr_{[-x,x]}(x - 1/2)$, that is,
$$g(x) = \begin{cases} -x & 0 \le x < 1/4, \\ x - 1/2 & 1/4 \le x \le 1. \end{cases}$$

Also $d(x - 1/2, A) = \|x - 1/2 - g(x)\| = 1/(4\sqrt{3})$.

20. Let r_1 and r_2 be the first two Rademacher functions, i.e., $r_1, r_2 : [0, 1] \to \mathbb{R}$, $r_1(t) = 1$ for $t \in [0, 1]$, $r_2(t) = 1$ if $t \in [0, 1/2]$ and $r_2(t) = -1$ if $t \in (1/2, 1]$. Then $L =$

$\{f \in L_2[0,1] \mid \langle f, r_1 \rangle = 0, \langle f, r_2 \rangle = 0\}$, and so $\Pr_L(f) = f - \left(\overline{\Delta_1}/\Delta\right) r_1 - \left(\overline{\Delta_2}/\Delta\right) r_2$, where

$$\Delta = \Gamma(r_1, r_2) = \begin{vmatrix} \langle r_1, r_1 \rangle & \langle r_1, r_2 \rangle \\ \langle r_2, r_1 \rangle & \langle r_2, r_2 \rangle \end{vmatrix} = 1,$$

$$\Delta_1 = \begin{vmatrix} \langle r_1, f \rangle & \langle r_1, r_2 \rangle \\ \langle r_2, f \rangle & \langle r_2, r_2 \rangle \end{vmatrix} = \langle r_1, f \rangle,$$

$$\Delta_2 = \begin{vmatrix} \langle r_1, r_1 \rangle & \langle r_1, f \rangle \\ \langle r_2, r_1 \rangle & \langle r_2, f \rangle \end{vmatrix} = \langle r_2, f \rangle.$$

Hence $\Pr_L(f) = f - \overline{\langle r_1, f \rangle} r_1 - \overline{\langle r_2, f \rangle} r_2$.

21. See [4, example 1.4.3, chapter I, section 1.4].

i) If $f, g \in A$ and $0 \le \lambda \le 1$ then $\int_0^1 |f(x)| \, dx \le 1$, $\int_0^1 |g(x)| \, dx \le 1$ and

$$\int_0^1 |\lambda f(x) + (1-\lambda) g(x)| \, dx \le \lambda \int_0^1 |f(x)| \, dx + (1-\lambda) \int_0^1 |g(x)| \, dx \le 1,$$

thus $\lambda f + (1-\lambda) g \in A$, i.e., A is convex.

Let $(f_n)_{n \in \mathbb{N}} \subseteq A$ and $f \in L_2[0,1]$ such that $f_n \to f$ in $L_2[0,1]$, i.e., $\int_0^1 |f_n(x) - f(x)|^2 \, dx \to 0$. But

$$\int_0^1 |f(x)| \, dx \le \int_0^1 |f_n(x) - f(x)| \, dx + \int_0^1 |f_n(x)| \, dx$$
$$\le \|f_n - f\|_1 + 1 \le \|f_n - f\|_2 + 1 \quad \forall n \in \mathbb{N},$$

and passing to the limit for $n \to \infty$ we obtain that $\int_0^1 |f(x)| \, dx \le 1$, i.e., $f \in A$, and therefore A is closed.

ii) Let $g \in L_2[0,1]$ be such that $g \notin A$, i.e., $\int_0^1 |g(x)| \, dx > 1$, and let $\delta > 0$ be such that $\int_0^1 |g(x)| \, dx = 1 + \delta$. For any $t \ge 0$ we define $A_t = \{x \in [0,1] \mid |g(x)| > t\}$ and $\varphi(t) = \int_{A_t} |g(x)| \, dx - t\mu(A_t)$. Since g is Lebesgue measurable it follows that A_t is a Lebesgue measurable set (μ is the Lebesgue measure on $[0,1]$). We will prove that $\varphi : [0, \infty) \to \mathbb{R}$ is continuous, that $\varphi(0) = 1 + \delta$, and that $\lim_{t \to \infty} \varphi(t) = 0$.

Let $t_n \nearrow t$. Then $A_{t_{n+1}} \subseteq A_{t_n}$ and $\bigcap_{n \in \mathbb{N}} A_{t_n} = [|g| \ge t]$. Now since the Lebesgue measure is countably additive and the measure $A \longmapsto \int_A |g(x)| \, dx$ is countably additive, it follows that

$$\lim_{n \to \infty} \varphi(t_n) = \int_{[|g| \ge t]} |g| \, d\mu - t\mu(|g| \ge t)$$
$$= \int_{A_t} |g(x)| \, dx + \int_{[|g|=t]} |g(x)| \, dx - t\mu(A_t) - t\mu(|g| = t) = \varphi(t),$$

since $\int_{[|g|=t]} |g(x)| \, dx = \int_{[|g|=t]} t \, d\mu = t\mu(|g| = t)$. Thus φ is left continuous. Let now $t_n \searrow t$. Then $A_{t_n} \subseteq A_{t_{n+1}}$ and $A_t = \bigcup_{n \in \mathbb{N}} A_{t_n}$. Again, using the properties for the Lebesgue measure, $\lim_{n \to \infty} \varphi(t_n) = \varphi(t)$, i.e., φ is right continuous. Also

$$\varphi(0) = \int_{[|g|>0]} |g(x)| \, dx - 0 \cdot \mu([|g|>0]) = \int_{[|g|>0]} |g(x)| \, dx.$$

11.2 Solutions

Since $[0,1] = [|g| \geq 0] = [|g| > 0] \cup [|g| = 0]$ we obtain that

$$\varphi(0) = \int_0^1 |g(x)|\, dx - \int_{[|g|=0]} |g(x)|\, dx = \int_0^1 |g(x)|\, dx = 1 + \delta.$$

We have $|g|^2 \chi_{A_t} \geq t^2 \chi_{A_t}$, and by integration $\int_0^1 t^2 \chi_{A_t} d\mu \leq \int_0^1 |g(x)|^2\, dx$, $t^2 \mu(A_t) \leq \int_0^1 |g(x)|^2\, dx$, i.e., $0 \leq t\mu(A_t) \leq (1/t) \int_0^1 |g(x)|^2\, dx \to 0$ for $t \to \infty$. Also, if $t_n \nearrow \infty$ then $A_{t_n} \searrow \varnothing$, and therefore $\lim_{t \to \infty} \varphi(t) = 0$.

Using now the Darboux property for φ we deduce that there is $k \in [0, \infty)$ such that $\varphi(k) = 1$, i.e., $\int_{A_k} |g(x)|\, dx - k\mu(A_k) = 1$. We will prove now that $\Pr_A(g) = \widehat{g}$, where

$$\widehat{g}(x) = \begin{cases} g(x) - kg(x)/|g(x)|, & x \in A_k, \\ 0, & x \notin A_k. \end{cases}$$

For this, we will prove that $\widehat{g} \in A$ and that $\Re\langle g - \widehat{g}, \widehat{g}\rangle \geq \Re\langle g - \widehat{g}, f\rangle \ \forall f \in A$. We have

$$\int_0^1 |\widehat{g}(x)|\, dx = \int_{A_k} |g(x) - kg(x)/|g(x)||\, dx = \int_{A_k} ||g(x)| - k|\, dx$$
$$= \int_{A_k} (|g(x)| - k)\, dx = \int_{A_k} |g(x)|\, dx - k\mu(A_k) = \varphi(k) = 1.$$

Since

$$g(x) - \widehat{g}(x) = \begin{cases} kg(x)/|g(x)|, & x \in A_k, \\ g(x), & x \notin A_k, \end{cases}$$

it follows that $|g(x) - \widehat{g}(x)| \leq k \ \forall x \in [0,1]$. From this we obtain that

$$\Re\langle g - \widehat{g}, f\rangle = \Re\left(\int_0^1 (g(x) - \widehat{g}(x))\overline{f(x)}\, dx\right) \leq \left|\int_0^1 (g(x) - \widehat{g}(x))\overline{f(x)}\, dx\right|$$
$$\leq \int_0^1 |g(x) - \widehat{g}(x)|\left|\overline{f(x)}\right|\, dx \leq k \int_0^1 |f(x)|\, dx \leq k,$$

for $f \in A$. So $\Re\langle g - \widehat{g}, f\rangle \leq k$. If we will prove that $k = \Re\langle g - \widehat{g}, \widehat{g}\rangle$, we will obtain the statement. But this is equivalent with the fact that

$$k = \Re\left(\int_0^1 (g(x) - \widehat{g}(x))\overline{\widehat{g}(x)}\, dx\right)$$
$$= \Re\left(\int_{A_k} (g(x) - \widehat{g}(x))\overline{g(x)}\,(1 - k/|g(x)|)\, dx\right)$$
$$= \Re\left(\int_{A_k} \frac{kg(x)\overline{g(x)}}{|g(x)|^2}(|g(x)| - k)\, dx\right) = \Re\left(\int_{A_k} \frac{k|g(x)|^2}{|g(x)|^2}(|g(x)| - k)\, dx\right)$$
$$= k\int_{A_k} (|g(x)| - k)\, dx = k\int_{A_k} |g(x)|\, dx - k^2\mu(A_k) = k\varphi(k),$$

which is true.

iii) We shall use the results from (ii). We have

$$A_t = \{x \in [0,1] \mid 8x > t\} = \begin{cases} (t/8, 1], & 0 \leq t \leq 8, \\ \emptyset, & t > 8, \end{cases}$$

and

$$\varphi(t) = \int_{A_t} 8x\,dx - t\mu(A_t) = \begin{cases} \int_{t/8}^{1} 8x\,dx - t(1 - t/8), & 0 \leq t \leq 8 \\ 0, & t > 8 \end{cases}$$
$$= \begin{cases} (t^2 - 16t + 64)/16, & 0 \leq t \leq 8, \\ 0, & t > 8. \end{cases}$$

We calculate $k \geq 0$ such that $\varphi(k) = 1$ and we obtain that $k = 4$. If $g(x) = 8x$, then $\Pr_A(g) = \widehat{g}$, where

$$\widehat{g}(x) = \begin{cases} g(x) - kg(x)/|g(x)|, & x \in A_k \\ 0, & x \notin A_k \end{cases} = \begin{cases} 8x - 4, & x \in (1/2, 1], \\ 0, & x \in [0, 1/2]. \end{cases}$$

22. We observe first that if H is a Hilbert space and $(e_n)_{n \in \mathbb{N}} \subseteq H$ is a sequence, then $\{x \in H \mid \langle x, e_1 + e_2 + \cdots + e_n \rangle = 0 \ \forall n \in \mathbb{N}\} = \{x \in H \mid \langle x, e_n \rangle = 0 \ \forall n \in \mathbb{N}\}$. Applying this result to our exercise (take $e_n = \chi_{[n,n+1)}$), it follows that

$$G = \left\{ f \in L_2(0, \infty) \mid \int_n^{n+1} f(x)\,dx = 0 \ \forall n \geq 0 \right\}.$$

Let now

$$G_n = \left\{ f \in L_2(0, \infty) \mid \int_k^{k+1} f(x)\,dx = 0, \ k = 0, 1, ..., n \right\}.$$

Then

$$G_n = \{ f \in L_2(0, \infty) \mid \langle f, e_k \rangle = 0, \ k = 0, 1, ..., n \},$$

where $e_k = \chi_{[k,k+1)}$. We have $G_{n+1} \subseteq G_n$ and $G = \bigcap_{n \in \mathbb{N}} G_n$. Using exercise 10(ii), $\Pr_G(f) = \lim_{n \to \infty} \Pr_{G_n}(f) \ \forall f \in L_2(0, \infty)$. To calculate $\Pr_{G_n}(f)$ we use the result from exercise 13(i) to obtain that $\Pr_{G_n}(f) = f - \sum_{k=1}^{n} \langle f, e_k \rangle e_k$, because $\langle e_i, e_j \rangle = \delta_{ij} \ \forall i \neq j$. Hence

$$\Pr_G(f) = f - \sum_{n=1}^{\infty} \langle f, e_n \rangle e_n = f - \sum_{n=1}^{\infty} \chi_{[n,n+1)} \int_n^{n+1} f(x)\,dx.$$

23. i) The condition $L \cap M = \{0\}$ implies that the operator $P_{L\|M}$ is well defined. The other relations are obvious.

ii) Let us suppose that $P_{L\|M} : L + M \to L$ is continuous (the norm on $L + M$ and L is the norm from the space X). Then there is $k \geq 0$ such that $\|P_{L\|M}(x)\| \leq k\|x\|$ $\forall x \in L + M$. Let $x \in \overline{L} \cap \overline{M}$. It follows that there are two sequences $(x_n)_{n \in \mathbb{N}} \subseteq L$ and $(y_n)_{n \in \mathbb{N}} \subseteq M$ such that $x_n \to x$ and $y_n \to x$. Then $x_n - y_n \to 0$ and so $P_{L\|M}(x_n - y_n) \to$

11.2 Solutions

0. Hence $x_n \to 0$, so $x = 0$, i.e., $\overline{L} \cap \overline{M} = \{0\}$. Hence $P_{\overline{L}\|\overline{M}}$ is well defined. Let $z \in \overline{L} + \overline{M}$, $z = x + y$ with $x \in \overline{L}$ and $y \in \overline{M}$. Then there are $(x_n)_{n \in \mathbb{N}} \subseteq L$ and $(y_n)_{n \in \mathbb{N}} \subseteq M$ such that $x_n \to x$ and $y_n \to y$. We obtain

$$\left\| P_{\overline{L}\|\overline{M}}(z) \right\| = \|x\| = \lim_{n \to \infty} \|x_n\| = \lim_{n \to \infty} \left\| P_{L\|M}(x_n + y_n) \right\|$$
$$\leq \left\| P_{L\|M} \right\| \limsup_{n \to \infty} \|x_n + y_n\| = \left\| P_{L\|M} \right\| \|z\|,$$

hence $P_{\overline{L}\|\overline{M}}$ is continuous. Let us observe that $\left\| P_{\overline{L}\|\overline{M}} \right\| = \left\| P_{L\|M} \right\|$.

The converse is clear.

iii) Suppose now that L and M are closed and that $P_{L\|M} : L + M \to L$ is continuous. Let $(x_n)_{n \in \mathbb{N}} \subseteq L$ and $(y_n)_{n \in \mathbb{N}} \subseteq M$ be such that $x_n + y_n \to z \in X$. Then $(x_n + y_n)_{n \in \mathbb{N}} \subseteq L + M$ is a Cauchy sequence and so $\left(P_{L\|M}(x_n + y_n)\right)_{n \in \mathbb{N}} \subseteq L$ is a Cauchy sequence. Hence there is $x \in L$ such that $x_n \to x$. Then $y_n \to y = z - x$, $y \in M$, i.e., $z \in L + M$, and so $L + M \subseteq X$ is a closed linear subspace.

Conversely, suppose that L, M, $L + M \subseteq X$ are closed linear subspaces. Then $L + M$ and L are Banach spaces with respect to the norm from X. We will prove that $P_{L\|M} : L + M \to L$ is continuous using the closed graph theorem. Let $(x_n)_{n \in \mathbb{N}} \subseteq L$ and $(y_n)_{n \in \mathbb{N}} \subseteq M$ such that $x_n + y_n \to z \in L + M$ and $x_n \to x \in L$. Then $y_n \to z - x$, hence $z - x \in M$. We obtain that $z = x + (z - x)$ with $x \in L$, $z - x \in M$, hence $P_{L\|M}(z) = x$.

24. i) From the Cauchy–Buniakowski–Schwarz inequality, we have

$$\frac{|\langle x, y \rangle|}{\|x\| \|y\|} \leq 1,$$

and since the Hilbert space is over $\mathbb{R}$, we have $-1 \leq \langle x, y \rangle / (\|x\| \|y\|) \leq 1$. Now since the function $\cos : [0, \pi] \to [-1, 1]$ is bijective, the statement follows.

ii) Let $g \in G$, $g \neq 0$. Again from the Cauchy–Buniakowski–Schwarz inequality we have $|\langle x, g \rangle| / (\|x\| \|g\|) \leq 1$, i.e., the set for which we take the supremum is bounded, and

$$\sup \left\{ \frac{|\langle x, g \rangle|}{\|x\| \|g\|} \,\middle|\, g \in G,\ g \neq 0 \right\} \in [0, 1].$$

We use now the fact that the function $\cos : [0, \pi/2] \to [0, 1]$ is bijective to obtain the statement. The fact that $\widehat{(x, G)} = \widehat{(x, \overline{G})}$ is obvious.

iii) From the orthogonal projection theorem, $x = \mathrm{Pr}_G(x) + \mathrm{Pr}_{G^\perp}(x)$, $\mathrm{Pr}_{G^\perp}(x) \in G^\perp$, i.e., $\langle \mathrm{Pr}_{G^\perp}(x), g \rangle = 0 \ \forall g \in G$, and then $\langle x, g \rangle = \langle \mathrm{Pr}_G(x), g \rangle \ \forall g \in G$. For $g \in G$, $g \neq 0$ we have

$$\frac{|\langle x, g \rangle|}{\|x\| \|g\|} = \frac{|\langle \mathrm{Pr}_G(x), g \rangle|}{\|x\| \|g\|} \leq \frac{\|\mathrm{Pr}_G(x)\| \|g\|}{\|x\| \|g\|} = \frac{\|\mathrm{Pr}_G(x)\|}{\|x\|}.$$

Hence

$$\sup \left\{ \frac{|\langle x, g \rangle|}{\|x\| \|g\|} \,\middle|\, g \in G,\ g \neq 0 \right\} \leq \frac{\|\mathrm{Pr}_G(x)\|}{\|x\|}.$$

If $\mathrm{Pr}_G(x) = 0$ we have equality. If not, for $\mathrm{Pr}_G(x) \in G$ we obtain that

$$\sup \left\{ \frac{|\langle x, g \rangle|}{\|x\| \|g\|} \,\middle|\, g \in G,\ g \neq 0 \right\} \geq \frac{|\langle x, \mathrm{Pr}_G(x) \rangle|}{\|x\| \|\mathrm{Pr}_G(x)\|} = \frac{\langle \mathrm{Pr}_G(x), \mathrm{Pr}_G(x) \rangle}{\|x\| \|\mathrm{Pr}_G(x)\|} = \frac{\|\mathrm{Pr}_G(x)\|}{\|x\|},$$

and (iii) is proved.

iv) It follows easily from (iii).

v) We have $G = \text{Sp}\{y\} = \{\lambda y \mid \lambda \in \mathbb{R}\}$. Then from (iv) if $\alpha = \widehat{(x,y)}$, then

$$\cos\widehat{(x,G)} = \sup\left\{\frac{\langle x, \lambda y\rangle}{\|x\|\,\|\lambda y\|}\;\Big|\;\lambda\in\mathbb{R},\;\lambda\neq 0\right\} = \sup\left\{\frac{\lambda}{|\lambda|}\frac{\langle x,y\rangle}{\|x\|\,\|y\|}\;\Big|\;\lambda\in\mathbb{R},\;\lambda\neq 0\right\}$$

$$= \max\left\{\frac{\langle x,y\rangle}{\|x\|\,\|y\|},\;\frac{-\langle x,y\rangle}{\|x\|\,\|y\|}\right\} = \max(\cos\alpha,\,-\cos\alpha).$$

If $\alpha\in[0,\pi/2]$ then $\cos\alpha\geq 0$, thus $\cos\widehat{(x,G)} = \cos\alpha$, and since $\cos$ on the interval $[0,\pi/2]$ is injective it follows that $\widehat{(x,G)} = \alpha$. If $\alpha\in(\pi/2,\pi]$, then $\cos\alpha < 0$, from whence $\cos\widehat{(x,G)} = -\cos\alpha = \cos(\pi-\alpha)$, and since $\cos$ on the interval $[0,\pi/2]$ is injective it follows that $\widehat{(x,G)} = \pi - \alpha$, i.e., the statement.

vi) See the part (ii) above.

vii) We have

$$\cos\widehat{(L,M)} = \sup_{x\in L\setminus\{0\},\,y\in M\setminus\{0\}}\frac{|\langle x,y\rangle|}{\|x\|\,\|y\|} = \sup_{x\in\overline{L}\setminus\{0\},\,y\in\overline{M}\setminus\{0\}}\frac{|\langle x,y\rangle|}{\|x\|\,\|y\|}$$

$$= \sup_{x\in\overline{L}\setminus\{0\}}\sup_{y\in\overline{M}\setminus\{0\}}\frac{|\langle x,y\rangle|}{\|x\|\,\|y\|} = \sup_{x\in\overline{L}\setminus\{0\}}\frac{\|P_{\overline{M}}x\|}{\|x\|}$$

$$= \sup_{x\in\overline{L}\setminus\{0\}}\frac{\|P_{\overline{M}}P_{\overline{L}}x\|}{\|x\|} = \sup_{x\in H\setminus\{0\}}\frac{\|P_{\overline{M}}P_{\overline{L}}x\|}{\|x\|} = \|P_{\overline{M}}P_{\overline{L}}\|.$$

Also,

$$\sin^2\widehat{(L,M)} = 1 - \cos^2\widehat{(L,M)} = 1 - \sup_{x\in\overline{L}\setminus\{0\}}\frac{\|P_{\overline{M}}x\|^2}{\|x\|^2} = \inf_{x\in\overline{L}\setminus\{0\}}\frac{\|x\|^2 - \|P_{\overline{M}}x\|^2}{\|x\|^2}$$

$$= \inf_{x\in\overline{L}\setminus\{0\}}\frac{\|(I-P_{\overline{M}})x\|^2}{\|x\|^2} = \frac{1}{\sup_{x\in\overline{L}\setminus\{0\}}\frac{\|x\|^2}{\|(I-P_{\overline{M}})x\|^2}} = \frac{1}{\sup_{x\in\overline{L}\setminus\{0\}}\frac{\|x\|^2}{\|P_{\overline{M}^\perp}(x)\|^2}}$$

$$= \frac{1}{\sup_{x\in\overline{L}\setminus\{0\}}\sup_{y\in\overline{M}\setminus\{0\}}\frac{\|x\|^2}{\|x+y\|^2}} = \frac{1}{\|P_{L\|M}\|^2}.$$

Let us observe that the operator $P_{L\|M}$ is continuous if and only if the angle between L and M is strictly positive.

Remark. We observe that (i), (ii), (v) and (vi) are true in any inner product space.

25. It is proved at exercise 24(iii) that if $\alpha = \widehat{(x,G)}$ then $\cos\alpha = \|\text{Pr}_G(x)\|/\|x\|$.

i) We will use exercise 12(i). We have $\text{Pr}_G(x) = (\Delta_1/\Delta)\,x_1 + (\Delta_2/\Delta)\,x_2$, where $\Delta = \Gamma(x_1,x_2) = 160$, $\Delta_1 = -80$, $\Delta_2 = 80$. So $\text{Pr}_G(x) = 1/2(x_2 - x_1) = (2,1,-2,1)$, and then $\cos\alpha = 1/\sqrt{2}$, i.e., $\alpha = \pi/4$.

ii) We have $\text{Pr}_G(x) = \langle e_1,x\rangle e_1 + \cdots + \langle e_m,x\rangle e_m$ and we obtain that $\cos\alpha = (1/\sqrt{n})\left(\sum_{i=1}^m|\langle e_i,x\rangle|^2\right)^{1/2} = \sqrt{m/n}$.

11.2 Solutions

iii) Let us observe that

$$G = \{(x_1, x_2, ..., x_n) \mid x_1 = x_2 = \cdots = x_n\} = \{(x_1, x_1, ..., x_1) \mid x_1 \in \mathbb{R}\}$$
$$= \operatorname{Sp}\{e\},$$

where $e = (1, 1, ..., 1)$. Taking $e_i = (0, ..., 0, 1, 0, ...0)$ we have that $\cos \widehat{(e_i, G)} = 1/\sqrt{n}$.

iv) Let $q = 1/2$. Then $d = \{(x_0, qx_0, ..., q^n x_0, ...) \mid x_0 \in \mathbb{R}\} = \operatorname{Sp}\{e\}$, where $e = (1, q, ..., q^n, ...)$. We have $\|e\| = \sqrt{1/(1-q^2)}$. If $\alpha_n = \widehat{(e_n, e)}$ then

$$\cos \alpha_n = \frac{\langle e_n, e \rangle}{\|e_n\| \|e\|} = q^n \sqrt{1-q^2} \in [0, 1],$$

and so $\cos \widehat{(e_n, G)} = \sqrt{3}/2^{n+1}$.

26. i) We prove first that

$$\operatorname{Pr}_{B_H}(x) = \begin{cases} x & \text{if } x \in B_H, \\ x/\|x\| & \text{if } x \notin B_H. \end{cases}$$

If $x \notin B_H$, then $\|x\| > 1$ and $d(x, B_H) \leq \|x - x/\|x\|\| = \|x\| - 1$. Also if $y \in B_H$ then $\|y\| \leq 1$, and then $\|x\| > 1 \geq \|y\|$, so $\|x - y\| \geq \|\|x\| - \|y\|\| = \|x\| - \|y\| \geq \|x\| - 1$, i.e., $d(x, B_H) = \|x - x/\|x\|\| = \|x\| - 1$. From the projection theorem it follows that $\operatorname{Pr}_{B_H}(x) = x/\|x\|$ if $x \notin B_H$, and, obviously, $\operatorname{Pr}_{B_H}(x) = x$ if $x \in B_H$. Since $\overline{B}(a, r) = a + rB_H$, using exercise 9 (i) and (iii) we have successively

$$\operatorname{Pr}_{\overline{B}(a,r)}(x) = a + \operatorname{Pr}_{rB_H}(x - a) = a + r\operatorname{Pr}_{B_H}((x-a)/r)$$
$$= \begin{cases} x & \text{if } x \in \overline{B}(a, r) \\ a + r(x-a)/\|x-a\| & \text{if } x \notin \overline{B}(a, r). \end{cases}$$

ii) Let $x_0 = ca/\|a\|^2 \in V$. Then $V = x_0 + G$, where $G = \{x \in H \mid \langle x, a \rangle = 0\} = \{a\}^\perp$. From exercise 9(ii) we know that $\operatorname{Pr}_V(x) = x_0 + \operatorname{Pr}_G(x - x_0)$. But if $L = \operatorname{Sp}\{a\}$ then $\operatorname{Pr}_L(x) = a \langle x, a \rangle / \|a\|^2$, and since $L^\perp = G$ it follows that $\operatorname{Pr}_G(x) = x - \operatorname{Pr}_L(x) = x - a\langle x, a \rangle / \|a\|^2 \; \forall x \in H$. We deduce that

$$\operatorname{Pr}_V(x) = x_0 + (x - x_0) - \frac{\langle x - x_0, a \rangle}{\|a\|^2} a = x - \frac{\langle x - x_0, a \rangle}{\|a\|^2} a = x + \frac{c - \langle x, a \rangle}{\|a\|^2} a,$$

since $\langle x_0, a \rangle = c$. In particular, if $x = 0$ we obtain that $\operatorname{Pr}_V(0) = ca/\|a\|^2$.

iii) We have $\operatorname{Fr}(A) = \{x \in H \mid \langle x, a \rangle = c\} = V$. If $x \in A$ then $\operatorname{Pr}_A(x) = x$. If $x \notin A$ then using exercise 2(ii) we have $\operatorname{Pr}_A(x) = \operatorname{Pr}_{\operatorname{Fr}(A)}(x)$, and by (ii) $\operatorname{Pr}_V(x) = x + (c - \langle x, a \rangle) a/\|a\|^2$, and then $\operatorname{Pr}_A(x) = x + (c - \langle x, a \rangle) a/\|a\|^2$.

iv) If $x \notin A$ then from exercise 2(ii), $y = \operatorname{Pr}_A(x) \in \operatorname{Fr}(A) = E \cup F$, where $E = \{x \in H \mid \langle x, a \rangle = m\}$ and $F = \{x \in H \mid \langle x, a \rangle = M\}$, and the sets E, F are closed, convex, and non-empty. If $y \in E$ then since $E \subseteq \operatorname{Fr}(A) \subseteq A$ we have $d(x, A) \leq d(x, E) \leq \|x - y\| = d(x, A)$, i.e., $d(x, E) = \|x - y\|$, and since $y \in E$ from uniqueness reasons in the projection theorem it follows that $y = \operatorname{Pr}_E(x)$. Using (iii) we have $\operatorname{Pr}_E(x) = x + (m - \langle x, a \rangle) a/\|a\|^2$. In the same way, if $y \in F$ it follows that $y = \operatorname{Pr}_F(x) = $

$\Pr_A(x) = x + (M - \langle x, a \rangle) a / \|a\|^2$. Let us remark that if $x \notin A$ then $\Pr_A(x) = x + (m - \langle x, a \rangle) a / \|a\|^2 = y_1$ or $\Pr_A(x) = x + (M - \langle x, a \rangle) a / \|a\|^2 = y_2$. Since $x \notin A$ it follows that either $\langle x, a \rangle < m$, or $\langle x, a \rangle > M$. If $\langle x, a \rangle < m$ then

$$\|x - y_1\| = \frac{m - \langle x, a \rangle}{\|a\|} < \frac{M - \langle x, a \rangle}{\|a\|} = \|x - y_2\|.$$

It follows in this case that $y_1 = \Pr_A(x)$. If $\langle x, a \rangle > M$, then

$$\|x - y_2\| = \frac{\langle x, a \rangle - M}{\|a\|} < \frac{\langle x, a \rangle - m}{\|a\|} = \|x - y_1\|.$$

It follows in this case that $y_2 = \Pr_A(x)$. Therefore

$$\Pr_A(x) = \begin{cases} x + (m - \langle x, a \rangle) a / \|a\|^2 & \text{if } \langle x, a \rangle < m, \\ x + (M - \langle x, a \rangle) a / \|a\|^2 & \text{if } \langle x, a \rangle > M, \\ x & \text{if } m \leq \langle x, a \rangle \leq M. \end{cases}$$

v) Consider $(H, \langle \cdot, \cdot \rangle_1)$, where $\langle x, y \rangle_1 = \Re \langle x, y \rangle \ \forall x, y \in H$. Then $(H, \langle \cdot, \cdot \rangle_1)$ is a real Hilbert space and the set from the exercise is $A = \{x \in H \mid \langle x, a \rangle_1 = 1\}$. From (ii)

$$\Pr_A(x) = x + \frac{1 - \langle x, a \rangle_1}{\|a\|^2} a = x + \frac{1 - \Re \langle x, a \rangle}{\|a\|^2} a.$$

Analogously $B = \{x \in H \mid -1 \leq \langle x, a \rangle_1 \leq 1\}$. Using (iv) we obtain that

$$\Pr_A(x) = \begin{cases} x + (-1 - \Re \langle x, a \rangle) a / \|a\|^2 & \text{if } \Re \langle x, a \rangle < -1, \\ x + (1 - \Re \langle x, a \rangle) a / \|a\|^2 & \text{if } \Re \langle x, a \rangle > 1, \\ x & \text{if } -1 \leq \Re \langle x, a \rangle \leq 1. \end{cases}$$

The above projections are calculated in $(H, \langle \cdot, \cdot \rangle_1)$. But using the definition for the projection (as the point where an infimum is attained) and the fact that the norm associated to $\langle \cdot, \cdot \rangle_1$ is equal to the norm associated to $\langle \cdot, \cdot \rangle$, we obtain that $\Pr_A(x)$ and $\Pr_B(x)$ are the wanted projections.

vi) Let us observe that $E = \{A \in \mathcal{M}_n(\mathbb{K}) \mid \Re \langle A, I_n \rangle = 1\}$. From (v) we have

$$\Pr_E(B) = B + \frac{1 - \Re \langle B, I_n \rangle}{\|I_n\|^2} I_n = B + \frac{1 - \Re (\operatorname{tr}(B))}{n} I_n,$$

so

$$\Pr_E(aI_n) = aI_n + \frac{1 - \Re (\operatorname{tr}(aI_n))}{n} I_n = \frac{n \Im (a) i + 1}{n} I_n.$$

We have $F = \{A \in \mathcal{M}_n(\mathbb{K}) \mid |\Re (\langle A, I_n \rangle)| \leq 1\}$, hence from (v)

$$\Pr_F(B) = \begin{cases} B + ((-1 - \Re (\operatorname{tr}(B))) / n) I_n & \text{if } \Re (\operatorname{tr}(B)) < -1, \\ B + ((1 - \Re (\operatorname{tr}(B))) / n) I_n & \text{if } \Re (\operatorname{tr}(B)) > 1, \\ B & \text{if } -1 \leq \Re (\operatorname{tr}(B)) \leq 1, \end{cases}$$

and from here
$$\mathrm{Pr}_F(aI_n) = \begin{cases} ((n\Im(a)i - 1)/n)\,I_n & \text{if } \Re(a) < -1/n, \\ ((n\Im(a)i + 1)/n)\,I_n & \text{if } \Re(a) > 1/n, \\ aI_n & \text{if } -1/n \leq \Re(a) \leq 1/n. \end{cases}$$

27. Let $x^* : H^2(\mathbb{D}) \to \mathbb{C}$ be a linear and continuous functional. From exercise 15 of chapter 10, the function $g \in H^2(\mathbb{D})$ given by $g(z) = \sum_{n=0}^{\infty} \overline{x^*(z^n)} z^n$ has the properties that $x^*(f) = \langle f, g \rangle_{H^2(\mathbb{D})} \ \forall f \in H^2(\mathbb{D})$ and $\|x^*\| = \left(\sum_{n=0}^{\infty} |x^*(z^n)|^2\right)^{1/2} = \|g\|$. The set from the exercise is of the form $A = \{f \in H^2(\mathbb{D}) \mid \langle f, g \rangle = 1\}$, and then by exercise 26(ii) the minimal norm element is $\mathrm{Pr}_A(0) = g/\|g\|^2$, i.e., $\mathrm{Pr}_A(0)$ is the function

$$h(z) = \frac{g(z)}{\|g\|^2} = \frac{1}{\sum_{n=0}^{\infty} |x^*(z^n)|^2} \sum_{n=0}^{\infty} \overline{x^*(z^n)} z^n,$$

and $\|\mathrm{Pr}_A(0)\| = 1/\|g\|$.

For our exercise we have $x^*(f) = f(\xi)$. Applying what we have proved above we obtain that $\|\mathrm{Pr}_A(0)\| = \sqrt{1 - |\xi|^2}$ and $\mathrm{Pr}_A(0)$ is the function $h(z) = (1 - |\xi|^2)/(1 - \bar{\xi}z)$.

28. Let $x^* : H^2(\mathbb{D}) \to \mathbb{C}$ be defined by $x^*(f) = f'(\xi)$. Using exercise 16 of chapter 10, $x^*(f) = \langle f, g \rangle \ \forall f \in H^2(\mathbb{D})$, where $g(z) = z/(1 - \bar{\xi}z)^2$ and

$$\|g\| = \frac{1}{1 - |\xi|^2} \sqrt{\frac{1 + |\xi|^2}{1 - |\xi|^2}}.$$

Then $A = \{f \mid \langle f, g \rangle = 1\}$ and from exercise 26(ii), $\mathrm{Pr}_A(0) = g/\|g\|^2$ and $\|\mathrm{Pr}_A(0)\| = 1/\|g\|$, i.e., $\mathrm{Pr}_A(0)$ is the function

$$h(z) = \frac{g(z)}{\|g\|^2} = \frac{z(1 - |\xi|^2)^3}{(1 - \bar{\xi}z)^2(1 + |\xi|^2)}$$

and $\|\mathrm{Pr}_A(0)\| = (1 - |\xi|^2)\sqrt{(1 - |\xi|^2)/(1 + |\xi|^2)}$.

29. See [21, exercise 1.10, chapter 1].

We observe that the minimal norm element in a closed convex non-empty set in a Hilbert space is the projection of 0 on that set.

i) Since $x^* : \mathcal{A}^2(\Omega) \to \mathbb{C}$ is a linear and continuous functional then by exercise 18(i) of chapter 10 there is a unique function $g \in \mathcal{A}^2(\Omega)$, $g(z) = \sum_{n=0}^{\infty} a_n f_n(z)$, where $a_n = \overline{x^*(f_n)}$, which satisfies $x^*(f) = \langle f, g \rangle_{\mathcal{A}^2(\Omega)} \ \forall f \in \mathcal{A}^2(\Omega)$. We also have

$$\|x^*\| = \|g\|_{\mathcal{A}^2(\Omega)} = \sqrt{\sum_{n=0}^{\infty} |x^*(f_n)|^2}.$$

The set from the exercise is $H = \{f \in \mathcal{A}^2(\Omega) \mid \langle f, g \rangle_{\mathcal{A}^2(\Omega)} = 1\}$. Then by exercise 26(ii) we have
$$\Pr_H(0) = \frac{g}{\|g\|_{\mathcal{A}^2(\Omega)}^2} = \frac{\sum_{n=0}^{\infty} \overline{x^*(f_n)} f_n(z)}{\sum_{n=0}^{\infty} |x^*(f_n)|^2},$$
and $\|\Pr_H(0)\| = \left(\sum_{n=0}^{\infty} |x^*(f_n)|^2 \right)^{-1/2}$.

ii) Let $x^* : \mathcal{A}^2(\Omega) \to \mathbb{C}$ be defined by $x^*(f) = f(\xi)$. It is proved at exercise 18(ii) of chapter 10 that $x^*(f) = \langle f, K(\cdot, \xi) \rangle$, where K is the Bergman kernel for Ω. Then the set from the exercise is $H = \{f \in \mathcal{A}^2(\Omega) \mid \langle f, K(\cdot, \xi) \rangle = 1\}$. From exercise 26(ii) we know that $f_\xi = \Pr_H(0) = K(\cdot, \xi)/\|K(\cdot, \xi)\|_{\mathcal{A}^2(\Omega)}^2$, and since it is proved at exercise 18 of chapter 10 that $\|K(\cdot, \xi)\|_{\mathcal{A}^2(\Omega)}^2 = K(\xi, \xi)$ we obtain that $f_\xi(z) = K(z, \xi)/K(\xi, \xi)$ and also that $\|f_\xi\| = 1/\sqrt{K(\xi, \xi)}$, i.e., the statement.

iii) It is proved at exercise 18(iv) of chapter 10 that the Bergman kernel for the open unit disc is $K(z, \xi) = 1/(\pi(1 - z\bar{\xi})^2)$. Then we deduce that $\Pr_A(0) = f_\xi$ where $f_\xi(z) = \left((1 - |\xi|^2)/(1 - z\bar{\xi}) \right)^2$ and that the norm for the minimal norm element in H is $\sqrt{\pi}(1 - |\xi|^2)$.

30. i) See [9, exercise 18, chapter V, section 1].

From the Riesz representation theorem, there is a unique element $y \in H$ such that $f(x) = \langle x, y \rangle \ \forall x \in H$. Then $\varphi(x) = \|x\|^2 - \langle x, y \rangle = \langle x, x \rangle - \langle x, y \rangle = \langle x, x - y \rangle$. Since
$$\langle a, b \rangle = \frac{\|a + b\|^2 - \|a - b\|^2}{4} \quad \forall a, b \in H,$$
we obtain that $\varphi(x) = \|x - y/2\|^2 - \|y/2\|^2 \ \forall x \in H$. Then $\inf_{x \in A} \varphi(x) = \inf_{x \in A} \|x - y/2\|^2 - \|y/2\|^2 = d(y/2, A) - \|y/2\|^2$. Now using the projection theorem we obtain that $\inf_{x \in A} \varphi(x) = \|y/2 - \Pr_A(y/2)\|^2 - \|y/2\|^2 = \varphi(\Pr_A(y/2))$, i.e., φ attains its minimum on A at $\Pr_A(y/2)$.

ii) By (i) we have $f_0 = \Pr_A(g/2) = \Pr_A(8x)$, since g is the function $16x$. To calculate f_0, we will use the results from exercise 21. We have
$$f_0(x) = \begin{cases} 0, & x \in [0, 1/2), \\ 8x - 4, & x \in [1/2, 1], \end{cases}$$
and
$$\inf_{f \in A} \varphi(f) = \varphi(\Pr_A(g/2)) = \varphi(f_0) = \int_{1/2}^1 (8x - 4)^2 dx - 16 \int_{1/2}^1 x(8x - 4) dx = -32/3.$$

31. Let H be a Hilbert space, and $\{A_n\} \subseteq H$ a sequence of closed, convex, and bounded sets with $A_n \neq \emptyset$ and $A_{n+1} \subseteq A_n \ \forall n \in \mathbb{N}$. From the projection theorem it follows that there is an element of minimal norm $a_n \in A_n$, i.e., $\|a_n\| = \min\{\|a\| \mid a \in A_n\} = \lambda_n$, ($a_n = \Pr_{A_n}(0)$). From $A_{n+1} \subseteq A_n$ it follows that $\|a_n\| \leq \|a_{n+1}\| \ \forall n \in \mathbb{N}$. Now let $n, p \in \mathbb{N}$. Then $a_{n+p} \in A_{n+p} \subseteq A_n$, hence $a_n, a_{n+p} \in A_n$ and as A_n is convex it follows that

11.2 Solutions

$(a_n + a_{n+p})/2 \in A_n$ and $\|(a_n + a_{n+p})/2\| \geq \lambda_n$. Using now the parallelogram law we have

$$\left\|\frac{a_{n+p} - a_n}{2}\right\|^2 = \frac{\|a_n\|^2 + \|a_{n+p}\|^2}{2} - \left\|\frac{a_{n+p} + a_n}{2}\right\|^2$$

$$\leq \frac{\lambda_n^2 + \lambda_{n+p}^2}{2} - \lambda_n^2 = \frac{\lambda_{n+p}^2 - \lambda_n^2}{2},$$

that is, $\|a_{n+p} - a_n\|^2 \leq 2(\lambda_{n+p}^2 - \lambda_n^2) \; \forall n, p \in \mathbb{N}$. But $(a_n)_{n \geq 2} \subseteq A_1$ and as A_1 is bounded it follows that $(\lambda_n)_{n \in \mathbb{N}}$ is an increasing and bounded sequence of real numbers, hence convergent, so a Cauchy sequence. It follows that $(a_n)_{n \in \mathbb{N}} \subseteq H$ is a Cauchy sequence. Since H is a Hilbert space, there is $x \in H$ such that $a_n \to x$. Let $n \in \mathbb{N}$. Then $\lim\limits_{p \to \infty} a_{n+p} = x$. Since $a_{n+p} \in A_{n+p} \subseteq A_n \; \forall p \in \mathbb{N}$ we obtain that $x \in \overline{A_n} = A_n$, and therefore $\bigcap\limits_{n \in \mathbb{N}} A_n \neq \emptyset$.

32. i) Let $z = (1, 1/2, 1/3, ...) \in l_2$ and we define $z^* \in l_2^*$, $z^*(x) = \langle x, z \rangle \; \forall x \in l_2$. Let $x^* = z^* \mid_{c_{00}}$. Then $x^* \in c_{00}^*$ and for any $x = (x_n)_{n \in \mathbb{N}} \in c_{00}$ we have $x^*(x) = \sum\limits_{n=1}^{\infty} x_n/n$. Thus $M = \ker x^*$ and obviously $x^* \in c_{00}^*$ is unique with this property.

ii) Suppose that there is $y \in c_{00}$ such that $x^*(x) = \langle x, y \rangle \; \forall x \in c_{00}$. We consider now the element y as belonging to l_2 and we obtain that $\langle x, y \rangle = \langle x, z \rangle \; \forall x \in c_{00}$, that is, $(y - z) \perp c_{00}$. Since $e_n \in c_{00} \; \forall n \in \mathbb{N}$ it follows that $(y - z) \perp e_n \; \forall n \in \mathbb{N}$, hence $y = z$. However $z \notin c_{00}$, and we obtain a contradiction.

iii) Let $w = (w_n)_{n \in \mathbb{N}} \in M^\perp \subseteq c_{00}$. Then $\langle w, (1, 0, ..., 0, -k, 0, ...) \rangle = 0$, since for any $k \in \mathbb{N}$, $k \geq 2$ we have $(1, 0, ..., 0, -k, 0, ...) \in M$, and then $w_k = w_1/k \; \forall k \in \mathbb{N}$. Since $w \in c_{00}$ we obtain that $w_1 = 0$ and therefore $w = 0$. We obtain that $M^\perp = \{0\}$. Obviously $M \neq c_{00}$, hence $c_{00} \neq M \oplus M^\perp = M$. Since $M^\perp = \{0\}$ we obtain that $M^{\perp\perp} = c_{00}$, and therefore $M \neq M^{\perp\perp}$.

Remark. $(c_{00}, \langle \cdot, \cdot \rangle)$ is an inner product space which is not a complete one. We observe that $c_{00} \subseteq (l_2, \|\cdot\|_2)$ is not a closed linear subspace.

33. See [14, exercise 1, chapter II, section 3].

Choose $A = \begin{pmatrix} a & b \\ c & d \end{pmatrix} \in M_2(\mathbb{R})$ such that $cx + dy = 0 \; \forall x, y \in \mathbb{R}$ and $ax + by = 0$ if and only if $y = x \tan \theta$. It follows that $A = a \begin{pmatrix} 1 & -1/\tan \theta \\ 0 & 0 \end{pmatrix}$, with $a \in \mathbb{R} \setminus \{0\}$. Then $A^2 = a^2 \begin{pmatrix} 1 & -1/\tan \theta \\ 0 & 0 \end{pmatrix}$. If $A^2 = A$, then $a \in \{-1, 1\}$. Taking for example $a = 1$ we obtain that $A = \begin{pmatrix} 1 & -1/\tan \theta \\ 0 & 0 \end{pmatrix}$. The operator $T_\theta \in L(H)$ will be given by the formula $T_\theta(x, y) = (x - y/\tan \theta, 0) \; \forall (x, y) \in H$. We have

$$\|T_\theta\| = \sup_{|x|^2 + |y|^2 \leq 1} \frac{1}{\sin \theta} |x \sin \theta - y \cos \theta| = \frac{1}{\sin \theta} (\sin^2 \theta + \cos^2 \theta)^{1/2} = \frac{1}{\sin \theta}.$$

Since $\|T_\theta\| > 1$ it follows that T_θ is not an orthogonal projection.

34. See [14, proposition 3.3, chapter II].

'(i) $\Rightarrow$ (ii)' Let $M \subseteq H$ be a closed linear subspace and $T : H \to M \subseteq H$ be the orthogonal projection onto M. Then $T \in L(H)$, $T^2 = T$, $T(H) = M$, $\ker T = M^\perp$, so $\ker T = (T(H))^\perp$.

'(ii) $\Rightarrow$ (i)' Since T is idempotent we have $\ker(I - T) = T(H)$, i.e., $T(H)$ is a closed linear subspace. Let $M = T(H)$. Then by the orthogonal projection theorem, $M \oplus M^\perp = H$, so for any $x \in H$ there are two unique elements $y \in M$ and $z \in M^\perp = \ker T$ (by hypothesis (ii), $\ker T = (T(H))^\perp$) such that $x = y + z$. Then $Tx = Ty$. There is $w \in H$ such that $Tw = y$ so $T^2 w = Ty$ and then $Tw = Ty$, i.e., $Ty = y$. It follows that $Tx = y$ and therefore T is the orthogonal projection onto $T(H)$.

'(i) $\Rightarrow$ (iii)' It is obvious.

'(iii) $\Rightarrow$ (ii)' Let now T be idempotent, with $\|T\| = 1$. We will prove that $\ker T = (T(H))^\perp$. Let $x \in (\ker T)^\perp$. Since $T^2 = T$ it follows that $x - Tx \in \ker T$, so $0 = \langle x - Tx, x\rangle = \|x\|^2 - \langle Tx, x\rangle$, that is, $\langle Tx, x\rangle = \|x\|^2$. Then $\|x\|^2 \leq \|Tx\| \|x\| \leq \|x\| \|x\| = \|x\|^2$, i.e., $\|Tx\| = \|x\| = \sqrt{\langle Tx, x\rangle} \ \forall x \in H$. We deduce that $\|x - Tx\|^2 = \|x\|^2 - 2\Re \langle Tx, x\rangle + \|Tx\|^2 = 0$, i.e., $Tx = x$. So $(\ker T)^\perp \subseteq \ker(I - T) = T(H)$.

If $y \in T(H)$ then using that $H = (\ker T) \oplus (\ker T)^\perp$ it follows that there are $z \in \ker T$, $w \in (\ker T)^\perp$ with $y = z + w$. There is $h \in H$ such that $Th = w$, and then $T^2 h = Tw$, $Th = Tw$, that is, $w = Tw$. In the same way $y = Ty$. Then $z = Tz$ and as $z \in \ker T$ it follows that $z = 0$, i.e., $y \in (\ker T)^\perp$. Hence $\ker T = (T(H))^\perp$.

'(i) $\Rightarrow$ (iv)' If T is the orthogonal projection onto the closed linear subspace M in H then for any $x, y \in H$, $x = a + b$, $y = c + d$, $a, c \in M$, $b, d \in M^\perp$, we have $\langle Tx, y\rangle = \langle T(a+b), c+d\rangle = \langle a, c\rangle = \langle a+b, c\rangle = \langle a+b, T(c+d)\rangle = \langle x, Ty\rangle$, that is, $T = T^*$, i.e., (iv).

'(iv) $\Rightarrow$ (v)' It is obvious.

'(v) $\Rightarrow$ (ii)' Let us suppose that $T^2 = T$ and that $T^*T = TT^*$. Then $\|Tx\|^2 = \langle Tx, Tx\rangle = \langle x, T^*Tx\rangle = \langle x, TT^*x\rangle = \langle T^*x, T^*x\rangle = \|T^*x\|^2$. Hence $\|Tx\| = \|T^*x\|$ $\forall x \in H$. From here we easily deduce that $\ker T = \ker T^*$ and since we have $\ker T^* = (T(H))^\perp$ we obtain (ii).

'(i) $\Rightarrow$ (vi)' If T is the orthogonal projection onto the closed linear subspace M in H then for any $x \in H$, $x = a + b$, $a \in M$, $b \in M^\perp$ we have $\langle Tx, x\rangle = \langle T(a+b), a+b\rangle = \langle a, a+b\rangle = \|a\|^2 \geq 0$.

'(vi) $\Rightarrow$ (ii)' Let now $T \in L(H)$ such that $T^2 = T$ and $\langle Tx, x\rangle \geq 0 \ \forall x \in H$. Then $\langle T(x+y), x+y\rangle \geq 0 \ \forall x, y \in H$. Let $x \in T(H)$, $y \in \ker T$. Then using that $T(y) = 0$ we have $\langle Tx, x+y\rangle \geq 0$. Since $x \in T(H)$ we have $Tx = x$, hence $\langle x, y\rangle \geq -\|x\|^2$ $\forall x \in T(H), \forall y \in \ker T$. But for $y \in \ker T$ it follows that $ny \in \ker T \ \forall n \in \mathbb{N}$, so by above $\langle x, y\rangle \geq -\|x\|^2/n \ \forall n \in \mathbb{N}$, and passing to the limit for $n \to \infty$ it follows that $\langle x, y\rangle \geq 0$. Since also $-y \in \ker T$, we have $\langle x, -y\rangle \geq 0$ and then $\langle x, y\rangle = 0$. Hence $T(H) \perp \ker T$. We have $H = T(H) + \ker T$, therefore $H = T(H) \oplus \ker T$, hence $\ker T = (T(H))^\perp$, i.e., (ii).

35. See [6, exercise 18, chapter 9].

For any $x \in H$ we have $x - Tx \in (I-T)(H) \subseteq A^\perp$, $Tx \in T(H) \subseteq A$ by hypothesis, thus $(x - Tx) \perp Tx$, and so $\langle Tx, Tx\rangle = \langle Tx, x\rangle \ \forall x \in H$. If $Tx = 0$ then $\|Tx\| \leq \|x\|$. If $Tx \neq 0$, since $\langle Tx, Tx\rangle = \langle Tx, x\rangle$ it follows that $\|Tx\|^2 \leq \|Tx\| \|x\|$, hence again we obtain that $\|Tx\| \leq \|x\|$. So $T : H \to H$ is linear and continuous with $\|T\| \leq 1$.

11.2 Solutions

For any $x \in H$, $Tx - T^2x \in T(H) \cap (I-T)(H)$, so $Tx - T^2x \in A \cap A^\perp = \{0\}$. It follows that $T^2 = T$. If T is the null operator then $H \subseteq A^\perp$, so $A = \{0\}$, and then T is the orthogonal projection onto A. If there is $x \in H$ such that $Tx \neq 0$, then from $T(Tx) = Tx$ and $\|T\| \leq 1$ it follows that $\|T\| = 1$. Since $T^2 = T$ from exercise 34 it follows that T is the orthogonal projection onto $T(H)$, $T(H)$ being a closed linear subspace of H. But $T(H) \subseteq A$ and $(T(H))^\perp \subseteq A^\perp$, hence $T(H) \subseteq A$ and $A \subseteq A^{\perp\perp} \subseteq T(H)$. It follows that $T(H) = A$.

36. See [14, exercises 4, 6 and 7, chapter II, section 3].

i) Suppose that $P + Q$ is an orthogonal projection. Then $(P+Q)^2 = P+Q$, so $P^2 + PQ + QP + Q^2 = P + Q$. Using that $P^2 = P$ and $Q^2 = Q$, we obtain that $PQ + QP = 0$. For any $x \in Q(H)$ we have $x = y + z$ with $y \in P(H)$, $z \in (P(H))^\perp = \ker P$. Applying the operator P we obtain that $Px = y$. Applying the operator PQ and using that $PQ = -QP$ we obtain that $Px = -QP(y) - QP(z) = -Qy$. Hence $y + Qy = 0$ and applying again Q we deduce that $Qy = 0$ and then $y = 0$. It follows that $x = z \in (P(H))^\perp$, i.e., $Q(H) \subseteq (P(H))^\perp$, and then $P(H) \perp Q(H)$.

If $P(H) \perp Q(H)$, then $Q(H) \subseteq (P(H))^\perp = \ker P$, and analogously, $P(H) \subseteq \ker Q$. Then $PQ = QP = 0$ and so $(P+Q)^2 = P+Q$. Since $(P+Q)^* = P^* + Q^* = P + Q$, we can use exercise 34 to obtain that $P + Q$ is an orthogonal projection.

Let us suppose now that $P + Q$ is an orthogonal projection. Obviously, $(P+Q)(H) \subseteq P(H) + Q(H)$. Let $x \in P(H) + Q(H)$, $x = y + z$, $x \in P(H)$, $y \in Q(H)$. Then $(P+Q)(x) = (P+Q)(y) + (P+Q)(z) = y + z = x$, so $x \in (P+Q)(H)$. If $x \in \ker P \cap \ker Q$, then $x \in \ker(P+Q)$. Let now $x \in \ker(P+Q)$, i.e., $(P+Q)(x) = 0$. But $Px \perp Qx$, and then $Px = Qx = 0$, i.e., $x \in \ker P \cap \ker Q$.

ii) If PQ is an orthogonal projection, then $(PQ)^* = PQ$, so $Q^*P^* = PQ$, i.e., $QP = PQ$.

If $PQ = QP$, then $(PQ)^2 = PQPQ = P(PQ)Q = P^2Q^2 = PQ$, and $(PQ)^* = QP = PQ$. From exercise 34 it follows that PQ is an orthogonal projection.

If PQ is an orthogonal projection, then PQ is the orthogonal projection onto $(PQ)(H)$. Let $x \in (PQ)(H)$. Then $x \in P(H)$ and as $(PQ)(H) = (\ker(PQ))^\perp$, it follows that $x \in (\ker(PQ))^\perp$. But $\ker Q \subseteq \ker(PQ)$, hence $(\ker(PQ))^\perp \subseteq \ker(Q)^\perp$. Then $x \in \ker(Q)^\perp = Q(H)$, i.e., $x \in P(H) \cap Q(H)$, so $(PQ)(H) \subseteq P(H) \cap Q(H)$. Now if $x \in P(H) \cap Q(H)$, then $Px = x$, $Qx = x$, i.e., $PQx = x$, $x \in (PQ)(H)$, and so $(PQ)(H) = P(H) \cap Q(H)$.

Now we will prove that if A and B are two closed linear subspaces in a Hilbert space H, then $A \cap B = (A^\perp + B^\perp)^\perp$. Indeed, if $x \in A \cap B$ then $x \perp A^\perp$, $x \perp B^\perp$, so $x \perp (A^\perp + B^\perp)$, and therefore $x \in (A^\perp + B^\perp)^\perp$. If $x \in (A^\perp + B^\perp)^\perp$, since $0 \in A^\perp$ and $0 \in B^\perp$ it follows that $x \in (A^\perp)^\perp$ and $x \in (B^\perp)^\perp$, hence $x \in A \cap B$.

From $A \cap B = (A^\perp + B^\perp)^\perp$ we deduce that $(A \cap B)^\perp = \overline{(A^\perp + B^\perp)}$. Taking $A = P(H)$ and $B = Q(H)$ we obtain that

$$(P(H) \cap Q(H))^\perp = \overline{\left((P(H))^\perp + (Q(H))^\perp\right)},$$

so $((PQ)(H))^\perp = \overline{(\ker P + \ker Q)}$, i.e., $\ker(PQ) = \overline{(\ker P + \ker Q)}$.

iii) If $PQ = Q$ then $Q(H) \subseteq P(H)$. If $Q(H) \subseteq P(H)$, then $PQx = Qx \ \forall x \in H$, since $Qx \in P(H)$, and then (b) and (c) are equivalent.

Obviously (c) and (d) are equivalent, since if $PQ = Q$ then $(PQ)^* = Q^*$, so $QP = Q$, and conversely, if $QP = Q$ then $PQ = Q$.

If $P - Q$ is an orthogonal projection then $(P - Q)^2 = P - Q$, from whence $2Q = PQ+QP$. Multiplying to the left by P we obtain that $2PQ = PQ+PQP$, so $PQ = PQP$, and multiplying to the right by P we obtain that $QP = PQP$. So $PQ = QP$ and then $Q = PQ = QP$. If $Q = PQ$ then $QP = Q$, hence $2Q = PQ + QP$, from whence it follows that $(P - Q)^2 = P - Q$. But $(P - Q)^* = P - Q$, so from exercise 34 it follows that $P - Q$ is an orthogonal projection.

Let us suppose now that $P - Q$ is an orthogonal projection. Since $Q(H) \subseteq P(H)$ we can write $P(H) = Q(H) \oplus (P(H) \ominus Q(H))$ and then it follows that $(P - Q)(H) = P(H) \ominus Q(H)$, the orthogonal complement of $Q(H)$ as a linear subspace of $P(H)$. We have $H = \ker P \oplus P(H) = \ker P \oplus Q(H) \oplus (P(H) \ominus Q(H)) = \ker P \oplus Q(H) \oplus (P - Q)(H)$. Since $P - Q$ is an orthogonal projection then $\ker (P - Q) = ((P - Q)(H))^\perp = \ker P \oplus Q(H)$.

iv) If $P + Q - PQ$ is an orthogonal projection then $(P + Q - PQ)^* = P + Q - PQ$, so $PQ = QP$.

If $PQ = QP$ then $(P + Q - PQ)^* = P + Q - PQ$ and $(P + Q - PQ)^2 = P + Q - PQ$, hence by exercise 34 it follows that $P + Q - PQ$ is an orthogonal projection.

37. i) For $n = 2k + 1$, $k \in \mathbb{N}$, since $TS = ST$, $T^2 = T$, $S^2 = S$, we have

$$(T - S)^n = T - \binom{n}{1}TS + \binom{n}{2}TS - \binom{n}{3}TS + \cdots + \binom{n}{n-1}TS - S,$$

so $(T - S)^n = T - S$. Then $\|T - S\| = \|(T - S)^n\| \leq \|T - S\|^n \ \forall n = 2k + 1$, $k \in \mathbb{N}$. If $\|T - S\| < 1$ then passing to the limit for $k \to \infty$ we obtain that $\|T - S\| = 0$, i.e., $T = S$.

ii) See [40, exercise 8.40].

Let us denote by $P_i^\perp$ the orthogonal projection onto $L_i^\perp$. We have $I = P_i^\perp + P_i$, $i = 1, 2$, where $I : H \to H$ is the identity operator. Then $P_1 = P_1 I = P_1 P_2 + P_1 P_2^\perp$, $P_2 = IP_2 = P_1 P_2 + P_1^\perp P_2$, and so $P_1 - P_2 = P_1 P_2^\perp - P_1^\perp P_2$. For $x \in H$, $\|x\| \leq 1$, we have by the Pythagoras theorem

$$\begin{aligned} \|(P_1 - P_2)(x)\|^2 &= \|P_1(P_2^\perp(x)) - P_1^\perp(P_2(x))\|^2 = \|P_1(P_2^\perp(x))\|^2 + \|P_1^\perp(P_2(x))\|^2 \\ &\leq \|P_1\|^2 \|P_2^\perp(x)\|^2 + \|P_1^\perp\|^2 \|P_2(x)\|^2 \leq \|P_2^\perp(x)\|^2 + \|P_2(x)\|^2 \\ &= \|P_2^\perp(x) + P_2(x)\|^2 = \|x\|^2 \leq 1. \end{aligned}$$

Now passing to the supremum with respect to $x \in H$, $\|x\| \leq 1$, we obtain that $\|P_1 - P_2\| \leq 1$.

iii) We have $P^2 = P$, $Q^2 = Q$, and by exercise 36(ii) we have also $PQ = QP$. Now we can apply (i) and (ii).

Chapter 12

Linear and continuous operators on Hilbert spaces

Theorem. Let H_1, H_2 be two Hilbert spaces, and $U : H_1 \to H_2$ a linear and continuous operator. Then there is a unique linear and continuous operator $U^* : H_2 \to H_1$ such that $\langle Ux, y \rangle = \langle x, U^*y \rangle \; \forall x \in H_1, \forall y \in H_2$. U^* is called the *adjoint* of the operator U.

Definition. Let H be a Hilbert space, and $U : H \to H$ a linear and continuous operator. Then U is called *self-adjoint* if and only if $U^* = U$, i.e., if and only if $\langle Ux, y \rangle = \langle x, Uy \rangle \; \forall x, y \in H$.

For a Hilbert space H we denote $\mathcal{A}(H) = \{U \in L(H) \mid U \text{ is self-adjoint}\}$.

Theorem. If H is a Hilbert space and $U \in \mathcal{A}(H)$ then $\|U\| = \sup\limits_{\|x\| \leq 1} |\langle Ux, x \rangle|$.

Definition. Let H be a Hilbert space, and $U : H \to H$ a self-adjoint operator. U is called *positive*, and we write $U \geq 0$, if and only if $\langle Ux, x \rangle \geq 0 \; \forall x \in H$.

For a Hilbert space H we denote $\mathcal{A}_+(H) = \{U \in \mathcal{A}(H) \mid U \text{ is positive}\}$.

Definition. Let H be a Hilbert space, and $U, V : H \to H$ two self-adjoint operators. We write $U \leq V$ if and only if $V - U \geq 0$, i.e., if and only if $\langle Ux, x \rangle \leq \langle Vx, x \rangle \; \forall x \in H$.

Definition. Let H be a Hilbert space. A family $(U_i)_{i \in I} \subseteq \mathcal{A}(H)$ is called *upper bounded* (respectively, *lower bounded*) if and only if there is an operator $U \in \mathcal{A}(H)$ such that $U_i \leq U \; \forall i \in I$ (respectively, $U_i \geq U \; \forall i \in I$).

Definition. Let H be a Hilbert space. Given a family $(U_i)_{i \in I} \subseteq \mathcal{A}(H)$, an operator $U \in \mathcal{A}(H)$ is called the *supremum* (respectively, *infimum*) of the family $(U_i)_{i \in I}$ if and only if $U_i \leq U \; \forall i \in I$ (respectively, $U_i \geq U \; \forall i \in I$) and if $V \in \mathcal{A}(H)$ has the property that $U_i \leq V \; \forall i \in I$ (respectively, $U_i \geq V \; \forall i \in I$) then $U \leq V$ (respectively, $V \leq U$). In this case we write $U = \sup\limits_{i \in I} U_i$ (respectively, $U = \inf\limits_{i \in I} U_i$).

The Weierstrass convergence theorem. Let H be a Hilbert space. Then:

i) For any upper bounded increasing sequence $(U_n)_{n\in\mathbb{N}} \subseteq \mathcal{A}(H)$ there exists $U = \sup_{n\in\mathbb{N}} U_n$. Moreover, $U(x) = \lim_{n\to\infty} U_n(x) \ \forall x \in H$.

ii) For any lower bounded decreasing sequence $(U_n)_{n\in\mathbb{N}} \subseteq \mathcal{A}(H)$ there exists $U = \inf_{n\in\mathbb{N}} U_n$. Moreover, $U(x) = \lim_{n\to\infty} U_n(x) \ \forall x \in H$.

The square root of a positive operator. Let H be a Hilbert space, and $U \in \mathcal{A}_+(H)$. Then there is a unique positive operator denoted $\sqrt{U} \in \mathcal{A}_+(H)$ and called the *square root* of U, such that $(\sqrt{U})^2 = U$. $\sqrt{U}$ has also the property that it commutes with any operator that commutes with U.

Theorem. Let H be a Hilbert space, and $U, V \in \mathcal{A}_+(H)$ be such that $UV = VU$. Then $UV \in \mathcal{A}_+(H)$.

Definition. Let H_1, H_2 be two Hilbert spaces, and $U : H_1 \to H_2$ a linear and continuous operator. The *modulus* of the operator U, denoted $|U| : H_1 \to H_1$, is defined by $|U| = \sqrt{U^*U}$.

Definition. Let H be a Hilbert space, and $U \in L(H)$. Then the *positive part*, respectively, the *negative part* of U, is defined by $U_+ = (|U|+U)/2$, respectively, $U_- = (|U|-U)/2$.

Definition. Let H_1, H_2 be two Hilbert spaces. $A \in L(H_1, H_2)$ is called a *partial isometry* if and only if there is a closed linear subspace $G \subseteq H_1$ such that $\|A(x)\| = \|x\|$ $\forall x \in G$ and $A = 0$ on $G^\perp$. In this case G is called the *initial subspace* and $A(H_1)$ the *final subspace* of the partial isometry A.

The polar decomposition for an operator. Let H_1, H_2 be two Hilbert spaces, and $U : H_1 \to H_2$ a linear and continuous operator. Then:

i) $\ker |U| = \ker U$, $\overline{|U|(H_1)} = (\ker U)^\perp$ and $\||U|\| = \|U\|$;

ii) There exists a unique partial isometry $A \in L(H_1, H_2)$ such that $\ker A = \ker U$ and $U = A|U|$. In addition, the initial subspace of A is $(\ker U)^\perp$ and the final subspace of A is $\overline{U(H_1)}$.

The pair $(A, |U|)$ is called the *polar decomposition* of U.

Theorem. Let H_1, H_2 be two Hilbert spaces, $U : H_1 \to H_2$ a linear and continuous operator, and $(A, |U|)$ be the polar decomposition for U. Then $(A^*, |U^*|)$ is the polar decomposition for U^*.

Theorem. Let H be a Hilbert space, and $U \in L(H)$. The *spectral radius* of U is $\rho(U) = \sup\{|\lambda| \mid \lambda \in \mathbb{C}, \lambda I - U \text{ is not invertible}\} = \lim_{n\to\infty} \sqrt[n]{\|U^n\|}$.

12.1 Exercises

The norm for some operators between two Hilbert spaces

1. Let H_1, H_2 be two Hilbert spaces, $\{e_1, ..., e_n\} \subseteq H_1$ is an orthonormal system, $\{b_1, ..., b_n\} \subseteq H_2$ is an orthonormal system, $\lambda_1, ..., \lambda_n \in \mathbb{K}$, and $U : H_1 \to H_2$ is defined by $U(x) = \sum_{i=1}^n \lambda_i b_i \langle x, e_i \rangle$. Calculate $\|U\|$.

2. i) Let H be a Hilbert space, $a, b \in H \setminus \{0\}$ are two orthogonal elements and $U : H \to H$ is defined by $U(x) = a \langle x, b \rangle + b \langle x, a \rangle$. Calculate $\|U\|$.

12.1 Exercises

ii) Using (i) calculate $||U||$, where $U : L_2[0, \pi] \to L_2[0, \pi]$ is defined by

$$Uf(x) = \sin x \int_0^\pi f(t) \cos t\, dt + \cos x \int_0^\pi f(t) \sin t\, dt.$$

3. Let $A = \begin{pmatrix} a & b \\ c & d \end{pmatrix} \in \mathcal{M}_2(\mathbb{C})$. Prove that

$$\|A\|^2_{L(\mathbb{C}^2, \|\cdot\|_2)} = \frac{1}{2}N + \frac{1}{2}\sqrt{N^2 - 4|D|^2},$$

where $N = |a|^2 + |b|^2 + |c|^2 + |d|^2$ and $D = \det(A) = ad - bc$.

$L(H)$ is a Hilbert space if and only if $\dim_\mathbb{K} H = 1$

4. i) Let H be a Hilbert space, $e_1, e_2 \in H$ is an orthonormal system, $A = \begin{pmatrix} a & b \\ c & d \end{pmatrix}$ is a scalar square matrix and $U, V : H \to H$ are the operators defined by $U(x) = a\langle x, e_1 \rangle e_1 + b\langle x, e_2 \rangle e_2$ and $V(x) = c\langle x, e_1 \rangle e_1 + d\langle x, e_2 \rangle e_2$. Prove that

$$\|U + V\|^2 + \|U - V\|^2 = 2\left(\|U\|^2 + \|V\|^2\right)$$

if and only if

$$(\max(|a+c|, |b+d|))^2 + (\max(|a-c|, |b-d|))^2 = 2(\max(|a|, |b|)^2 + \max(|c|, |d|)^2).$$

ii) Using (i) prove that if H is a Hilbert space of dimension ≥ 2 then $L(H)$ is not a Hilbert space.

Infinite matrices and operators between separable infinite-dimensional Hilbert spaces

5. Let X and Y be two infinite-dimensional separable Hilbert spaces, $(x_n)_{n \in \mathbb{N}}$ an orthonormal basis in X and $(y_n)_{n \in \mathbb{N}}$ an orthonormal basis in Y.

i) Let $U : X \to Y$ be a linear and continuous operator. Let $U(x_n) = \sum_{m=1}^\infty \alpha_{mn} y_m$. Prove that $\sum_{m=1}^\infty |\alpha_{mn}|^2 \leq \|U\|^2\ \forall n \in \mathbb{N}$ and $\sum_{n=1}^\infty |\alpha_{mn}|^2 \leq \|U\|^2\ \forall m \in \mathbb{N}$.

ii) Give an example of a double-indexed sequence of scalars (α_{mn}) such that $\sum_{m=1}^\infty |\alpha_{mn}|^2 \leq 1\ \forall n \in \mathbb{N}$ and $\sum_{n=1}^\infty |\alpha_{mn}|^2 \leq 1\ \forall m \in \mathbb{N}$, but for which we cannot find $U \in L(X, Y)$ such that $\langle U(x_n), y_m \rangle = \alpha_{mn}\ \forall n, m \in \mathbb{N}$.

6. Let $(n_k)_{k \in \mathbb{N}}$ be a strictly increasing sequence in $\mathbb{N}$, and $(a_k)_{k \in \mathbb{N}}$ a sequence of scalars. Let us consider the $n_k \times n_k$ matrices $A_k = \begin{pmatrix} a_k & \cdots & a_k \\ \cdots & \cdots & \cdots \\ a_k & \cdots & a_k \end{pmatrix}$ and the double-indexed sequence (α_{mn}) defined by the matrix $\begin{pmatrix} A_1 & 0 & 0 & \cdots \\ 0 & A_2 & 0 & \cdots \\ 0 & 0 & A_3 & \cdots \\ \cdots & \cdots & \cdots & \cdots \end{pmatrix}$. Let X, Y be infinite-dimensional separable Hilbert spaces, $(x_n)_{n \in \mathbb{N}}$ an orthonormal basis in X and $(y_n)_{n \in \mathbb{N}}$ an orthonormal basis in Y. Prove that if there is $U \in L(X, Y)$ such that $\langle U(x_n), y_m \rangle = \alpha_{mn}$ $\forall n, m \in \mathbb{N}$, then the sequences $(n_k a_k^2)_{k \in \mathbb{N}}$ and $(n_k a_k)_{k \in \mathbb{N}}$ are bounded.

Linear symmetric operators are continuous operators

7. Let H be a Hilbert space and $T : H \to H$ a linear operator which is symmetric, i.e., $\langle Tx, y \rangle = \langle x, Ty \rangle \; \forall x, y \in H$. Prove that T is continuous.

Linear operators sending orthogonal vectors into orthogonal vectors

8. For $x = (x_1, ..., x_n)$, $y = (y_1, ..., y_n) \in \mathbb{C}^n$, consider $\langle x, y \rangle = \sum_{j=1}^{n} x_j \bar{y}_j$, the canonical inner product. Let $T : \mathbb{C}^n \to \mathbb{C}^n$ be a linear map such that $\langle Tx, Ty \rangle = 0$ if $\langle x, y \rangle = 0$. Prove that $T = kS$, for a scalar k and an unitary operator S (that is, $SS^* = S^*S = I$).

Decompositions for operators on a Hilbert space

9. i) Let H be a complex Hilbert space. Prove that any operator $T \in L(H)$ can be written in the form $T = U + iV$, where $U, V \in L(H)$ are self-adjoint operators.

ii) Let H be a Hilbert space. Prove that any operator $T \in L(H)$ can be written in the form $T = U - V$, where $U \in L(H)$ is self-adjoint and $V \in L(H)$ is antisymmetric, i.e., $V^* = -V$.

iii) Using (ii) prove that if H is a Hilbert space with $\dim_{\mathbb{K}} H \geq 2$ then the sets $\mathcal{A}(H) = \{U \in L(H) \mid U \text{ is self-adjoint}\}$ and $\{U \in L(H) \mid U \text{ is antisymmetric}\}$ are nowhere dense in $L(H)$.

Positive matrices

10. Let $A = (a_{ij}) \in \mathcal{M}_n(\mathbb{R})$. Prove that the following assertions are equivalent:

i) $A \in \mathcal{A}_+(\mathbb{R}^n)$, i.e., $\langle Ax, x \rangle \geq 0 \; \forall x \in \mathbb{R}^n$;

ii) All the diagonal minors of A are positive, i.e., for any $1 \leq k \leq n$ and any $1 \leq i_1 < \cdots < i_k \leq n$ we have

$$\Delta_{i_1,...,i_k} = \begin{vmatrix} a_{i_1 i_1} & a_{i_1 i_2} & \cdots & a_{i_1 i_k} \\ a_{i_2 i_1} & a_{i_2 i_2} & \cdots & a_{i_2 i_k} \\ \cdots & \cdots & \cdots & \cdots \\ a_{i_k i_1} & a_{i_k i_2} & \cdots & a_{i_k i_k} \end{vmatrix} \geq 0.$$

A rank one integral operator

11. i) Let $\varphi : [0, 1] \to \mathbb{R}$ be a continuous function and $A : L_2[0, 1] \to L_2[0, 1]$ defined by $(Af)(x) = \varphi(x) \int_0^1 \varphi(t) f(t) \, dt$.

a) Prove that A is self-adjoint and positive.

b) Prove that there is $\lambda \geq 0$ such that $A^2 = \lambda A$.

c) Calculate $\sqrt{A}$.

d) Calculate the spectral radius for the operator A.

e) Prove that A is a non-null orthogonal projector if and only if $\int_0^1 \varphi^2(t) dt = 1$.

ii) Let $A : L_2[0, 1] \to L_2[0, 1]$ be defined by

$$(Af)(x) = \frac{x}{1+x} \int_0^1 \frac{t}{1+t} f(t) dt.$$

12.1 Exercises

Deduce that A is a positive operator which is not an orthogonal projector, that $\sqrt{A} : L_2[0,1] \to L_2[0,1]$ is given by

$$\left(\sqrt{A}f\right)(x) = \frac{1}{\sqrt{3/2 - 2\ln 2}} \frac{x}{1+x} \int_0^1 \frac{t}{1+t} f(t)dt,$$

and that the spectral radius for A is $(3 - 4\ln 2)/2$.

The norm and the spectral radius for rank one operators

12. i) Let H be a Hilbert space, and $a, b \in H$. $U : H \to H$ is the rank one operator defined by $U(x) = \langle x, a \rangle b \ \forall x \in H$. Prove that $\|U\| = \|a\|\|b\|$, and that the spectral radius of U is $\rho(U) = |\langle a, b \rangle|$.

In particular, $\rho(U) = \|U\| \Leftrightarrow a, b$ are linearly dependent.

ii) Let X be a Banach space, $x^* \in X^*$, $y \in X$, and $U : X \to X$ is the rank one operator defined by $U(x) = x^*(x)y \ \forall x \in X$. Prove that $\|U\| = \|x^*\|\|y\|$, and that the spectral radius of U is $\rho(U) = |x^*(y)|$.

In particular, $\rho(U) = \|U\| \Leftrightarrow |x^*(y)| = \|x^*\|\|y\|$.

A characterization for one-dimensional real Hilbert spaces

13. i) If H is a Hilbert space and $A : H \to H$ is a self-adjoint operator such that $\langle Ax, x \rangle = 0 \ \forall x \in H$, prove that $A = 0$.

ii) Find a non-null matrix $A \in \mathcal{M}_2(\mathbb{R})$ such that $\langle Ax, x \rangle = 0 \ \forall x \in \mathbb{R}^2$.

iii) Using (ii) prove that if H is a real Hilbert space then the following assertions are equivalent:

a) For any $U \in L(H)$ with the property that $\langle Ux, x \rangle = 0 \ \forall x \in H$, it follows that $U = 0$;

b) $\dim_{\mathbb{R}} H = 1$;

c) The locally convex topology on $L(H)$ generated by the family of seminorms $(p_x)_{x \in H}$ is Hausdorff, where $p_x(U) = |\langle Ux, x \rangle|$.

Multiplication operators on l_2

14. Let $\lambda = (\lambda_n)_{n \in \mathbb{N}} \subseteq \mathbb{C}$ with $\sup_{n \in \mathbb{N}} |\lambda_n| < \infty$ and consider the multiplication operator $M_\lambda : l_2 \to l_2$ defined by $M_\lambda(x_1, x_2, \ldots) = (\lambda_1 x_1, \lambda_2 x_2, \ldots)$.

i) Calculate M_λ^* and prove that M_λ is self-adjoint $\Leftrightarrow \lambda_n \in \mathbb{R} \ \forall n \in \mathbb{N}$.

ii) Prove that M_λ is positive $\Leftrightarrow \lambda_n \geq 0 \ \forall n \in \mathbb{N}$, and in this case find $\sqrt{M_\lambda}$.

iii) Calculate $|M_\lambda|$ and the positive and negative parts M_λ^+, M_λ^- of M_λ.

Effective calculation of the square root for a positive operator

15. Let $A : \mathbb{R}^2 \to \mathbb{R}^2$ be defined by $A\begin{pmatrix} x_1 \\ x_2 \end{pmatrix} = \begin{pmatrix} 2x_1 + 3x_2 \\ 3x_1 + 5x_2 \end{pmatrix}$. Prove that A is self-adjoint and positive, and find $\sqrt{A}$.

16. Let $A : l_2 \to l_2$ be defined by $A(x_1, x_2, \ldots) = (0, 0, x_3, x_4, \ldots)$. Prove that A is linear, continuous, self-adjoint, positive, and find $\sqrt{A}$.

Monotony results for positive operators

17. Let H be a Hilbert space, and $A, B \in \mathcal{A}_+(H)$ with $AB = BA$. Prove that
$$A \leq B \Leftrightarrow \sqrt{A} \leq \sqrt{B}.$$

18. Let H be a Hilbert space, and $A, B \in \mathcal{A}_+(H)$ with $AB = BA$. Prove that
$$\frac{A+B}{2} \geq \sqrt{AB},$$
$$\sqrt{\frac{A+B}{2}} \geq \frac{\sqrt{A}+\sqrt{B}}{2}.$$

19. Let H be a Hilbert space and $A, B \in L(H)$ with $I \leq A \leq B$. Prove that A and B are invertible, with A^{-1}, B^{-1} continuous, and that $A^{-1} \geq B^{-1}$.

Convexity inequalities for operators

20. i) Let H_1, H_2 be two Hilbert spaces. Prove that:
a) $|A+B|^2 \leq 2\left(|A|^2 + |B|^2\right) \forall A, B \in L(H_1, H_2)$.
b) $|A_1 + \cdots + A_n|^2 \leq n(|A_1|^2 + \cdots + |A_n|^2) \, \forall A_1, ..., A_n \in L(H_1, H_2)$.
c) $|\alpha_1 A_1 + \cdots + \alpha_n A_n|^2 \leq (|\alpha_1|^2 + \cdots + |\alpha_n|^2)(|A_1|^2 + \cdots + |A_n|^2) \, \forall A_1, ..., A_n \in L(H_1, H_2), \forall \alpha_1, ..., \alpha_n \in \mathbb{K}$.

ii) Let H be a Hilbert space and $A_1, ..., A_n, B_1, ..., B_n \in L(H)$. Prove that
$$\left\|\sum_{i=1}^n A_i B_i\right\|^2 \leq \left\|\sum_{i=1}^n A_i A_i^*\right\| \left\|\sum_{i=1}^n B_i^* B_i\right\|.$$

iii) Let $A = \begin{pmatrix} 2 & 0 \\ 0 & 0 \end{pmatrix}$, $B = \begin{pmatrix} -1 & 1 \\ 1 & -1 \end{pmatrix} \in \mathcal{M}_2(\mathbb{R})$. Find $|A+B|, |A|, |B|$ and prove that the inequality $|A+B| \leq |A| + |B|$ is not true.

iv) Let H be a Hilbert space with $\dim_\mathbb{K} H \geq 2$. Using (iii) prove that we can find $U, V \in L(H)$ such that the inequality $|U+V| \leq |U| + |V|$ is not true.

21. Let H be a Hilbert space, $M : \mathcal{A}_+(H) \to \mathbb{R}_+$ be a positive homogeneous, subadditive and increasing mapping, and $n \in \mathbb{N}$. Prove that for any $A_1, ..., A_n \in \mathcal{A}_+(H)$ we have
$$M\left(\left(\sqrt{A_1} + \sqrt{A_2} + \cdots + \sqrt{A_n}\right)^2\right) \leq \left(\sqrt{M(A_1)} + \sqrt{M(A_2)} + \cdots + \sqrt{M(A_n)}\right)^2.$$

Linear and continuous operators associated to orthogonal systems

22. Let H be a Hilbert space, $(e_n)_{n \in \mathbb{N}} \subseteq H \setminus \{0\}$ an orthogonal system, $(a_n)_{n \in \mathbb{N}} \subseteq \mathbb{K}$ and $P_{(a_1, a_2, ...)} : H \to H$ be defined by $P_{(a_1, a_2, ...)}(x) = \sum_{n=1}^{\infty} a_n \langle x, e_n \rangle e_n$. Prove that $P_{(a_1, a_2, ...)}$ is well defined, linear, and continuous if and only if $\sup_{n \in \mathbb{N}} \left(|a_n| \|e_n\|^2\right) < \infty$, and in this case:

12.1 Exercises

i) $P^2_{(a_1,a_2,...)} = P_{(a_1^2\|e_1\|^2, a_2^2\|e_2\|^2,...)}$;

ii) $P^*_{(a_1,a_2,...)}(x) = \sum_{n=1}^{\infty} \overline{a_n} \langle x, e_n \rangle e_n$;

iii) If $a_n \geq 0 \ \forall n \in \mathbb{N}$, prove that

$$\sqrt{P_{(a_1,a_2,...)}}(x) = \sum_{n=1}^{\infty} \frac{\sqrt{a_n}}{\|e_n\|} \langle x, e_n \rangle e_n \quad \forall x \in H.$$

23. i) Let H_1 and H_2 be two Hilbert spaces, $(e_n)_{n\in\mathbb{N}} \subseteq H_1\setminus\{0\}$ and $(f_n)_{n\in\mathbb{N}} \subseteq H_2\setminus\{0\}$ two orthogonal systems, and $(\lambda_n)_{n\in\mathbb{N}} \subseteq \mathbb{K}$. Prove that the series $\sum_{n=1}^{\infty} \lambda_n \langle x, e_n \rangle f_n$ converges in H_2 for any $x \in H_1$ if and only if $\sup_{n\in\mathbb{N}} (|\lambda_n| \|e_n\| \|f_n\|) < \infty$. Deduce then that if $U : H_1 \to H_2$, $U(x) = \sum_{n=1}^{\infty} \lambda_n \langle x, e_n \rangle f_n$ is well defined, then U is linear and continuous, with $\|U\| = \sup_{n\in\mathbb{N}} (|\lambda_n| \|e_n\| \|f_n\|)$. If U is linear and continuous, prove that:

a) $U^*(x) = \sum_{n=1}^{\infty} \overline{\lambda_n} \langle x, f_n \rangle e_n \ \forall x \in H_2$;

b) $|U|(x) = \sum_{n=1}^{\infty} |\lambda_n| (\|f_n\| / \|e_n\|) \langle x, e_n \rangle e_n \ \forall x \in H_1$;

c) U is compact $\Leftrightarrow (\lambda_n \|e_n\| \|f_n\|)_{n\in\mathbb{N}} \in c_0$.

ii) Let $(\lambda_n)_{n\in\mathbb{N}} \subseteq \mathbb{K}$ and $U : L_2[0,1] \to l_2$ be the operator defined by

$$U(f) = \left(\lambda_n \int_{1/(n+1)}^{1/n} f(x)\, dx\right)_{n\in\mathbb{N}}.$$

Using (i) prove that U is well defined $\Leftrightarrow U$ linear and continuous $\Leftrightarrow \sup_{n\in\mathbb{N}}(|\lambda_n|/n) < \infty$, and in this case:

a) $U^*(x_1, x_2, ...) = \sum_{n=1}^{\infty} \overline{\lambda_n} x_n \chi_{(1/(n+1),1/n)}$.

b) $|U|(f) = \sum_{n=1}^{\infty} \left(\int_{1/(n+1)}^{1/n} f(x)\, dx\right) \sqrt{n^2+n} \, |\lambda_n| \chi_{(1/(n+1),1/n)}$.

c) U is compact $\Leftrightarrow \lambda_n/n \to 0$.

iii) Let (S_1, Σ_1, μ_1), (S_2, Σ_2, μ_2) be two measure spaces, $(A_n)_{n\in\mathbb{N}} \subseteq \Sigma_1$ and $(B_n)_{n\in\mathbb{N}} \subseteq \Sigma_2$ are two sequences of pairwise disjoint sets of strictly positive and finite measure, and $(\lambda_n)_{n\in\mathbb{N}} \subseteq \mathbb{K}$. Let $U : L_2(\mu_1) \to L_2(\mu_2)$ be the operator defined by

$$U(f) = \sum_{n=1}^{\infty} \lambda_n \left(\int_{A_n} f\, d\mu_1\right) \chi_{B_n}.$$

Using (i) prove that the operator U is well defined if and only if U is linear and continuous, or, if and only if $\sup_{n\in\mathbb{N}} \left(|\lambda_n| \sqrt{\mu_1(A_n)\mu_2(B_n)}\right) < \infty$, and that if U is continuous then $\|U\| = \sup_{n\in\mathbb{N}} \left(|\lambda_n| \sqrt{\mu_1(A_n)\mu_2(B_n)}\right)$. If U is continuous then

a) Calculate U^*, the adjoint of U;

b) Calculate $|U|$;

c) Prove that U is compact $\Leftrightarrow \lambda_n \sqrt{\mu_1(A_n)\mu_2(B_n)} \to 0$.

24. i) Let H be a Hilbert space, $(x_n)_{n\in\mathbb{N}} \subseteq H\setminus\{0\}$ is an orthogonal system, and $(\lambda_n)_{n\in\mathbb{N}} \subseteq \mathbb{K}$ is such that $\sup_{n\in\mathbb{N}} |\lambda_n|\,\|x_n\| < \infty$. Let us consider the operator $U : H \to l_2$ defined by $U(x) = (\lambda_n \langle x, x_n\rangle)_{n\in\mathbb{N}}$. Prove that U is compact $\Leftrightarrow \lambda_n x_n \to 0$ in norm.

ii) Then deduce that for any $\alpha \leq 1$ the operator $U : L_2[0,1] \to l_2$ defined by $U(f) = \left(n^\alpha \int_{1/(n+1)}^{1/n} f(x)dx\right)_{n\in\mathbb{N}}$ is linear and continuous, and that U is compact $\Leftrightarrow \alpha < 1$.

Linear and continuous operators associated to almost orthogonal systems

25. i) Let H_1 and H_2 be two Hilbert spaces, $(e_n)_{n\in\mathbb{N}} \subseteq H_1\setminus\{0\}$ and $(f_n)_{n\in\mathbb{N}} \subseteq H_2\setminus\{0\}$ are two orthogonal systems, and $(\lambda_n)_{n\in\mathbb{N}}$, $(\mu_n)_{n\in\mathbb{N}}$ are two scalar sequences. Prove that the series $\sum_{n=1}^\infty (\lambda_n \langle x, e_n\rangle f_n + \mu_n \langle x, e_{n+1}\rangle f_{n+1})$ converges in H_2 for any $x \in H_1$ if and only if $\sup_{n\in\mathbb{N}}(|\lambda_n|\,\|e_n\|\,\|f_n\|) < \infty$ and $\sup_{n\in\mathbb{N}}(|\mu_n|\,\|e_{n+1}\|\,\|f_{n+1}\|) < \infty$. Prove that in this case the operator $U : H_1 \to H_2$ defined by

$$U(x) = \sum_{n=1}^\infty (\lambda_n \langle x, e_n\rangle f_n + \mu_n \langle x, e_{n+1}\rangle f_{n+1})$$

is linear and continuous. If U is linear and continuous, prove that:

a) $U^*(x) = \sum_{n=1}^\infty (\overline{\lambda_n} \langle x, f_n\rangle e_n + \overline{\mu_n} \langle x, f_{n+1}\rangle e_{n+1})\ \forall x \in H_2$;

b) U is compact $\Leftrightarrow \lambda_n \|e_n\|\,\|f_n\| \to 0$ and $\mu_n \|e_{n+1}\|\,\|f_{n+1}\| \to 0$.

ii) Let $(\lambda_n)_{n\in\mathbb{N}} \subseteq \mathbb{K}$ and $U : L_2[1,\infty) \to l_2$ be the operator defined by

$$U(f) = \sum_{n=1}^\infty \left(\lambda_n \left(\int_n^{n+1} f(x)dx\right) f_n + \lambda_n \left(\int_{n+1}^{n+2} f(x)dx\right) f_{n+1}\right),$$

where $(f_n)_{n\in\mathbb{N}} \subseteq l_2$ is the standard basis. Using (i) prove that U is well defined $\Leftrightarrow U$ linear and continuous $\Leftrightarrow \sup_{n\in\mathbb{N}} |\lambda_n| < \infty$, and that in this case:

a) $U^*(x_1, x_2, \ldots) = \sum_{n=1}^\infty \overline{\lambda_n}(x_n\chi_{[n,n+1]} + x_{n+1}\chi_{[n+1,n+2]})$;

b) U is compact $\Leftrightarrow \lambda_n \to 0$.

Multiplication operators on $L_2(\mu)$

26. i) Let $\varphi \in L_\infty[0,1]$ and $M_\varphi : L_2[0,1] \to L_2[0,1]$ be the multiplication operator defined by φ, that is, $(M_\varphi f)(x) = \varphi(x)f(x)\ \forall f \in L_2[0,1]$. Prove that:

a) $M_\varphi^* = M_{\overline{\varphi}}$, i.e., $(M_\varphi^* f)(x) = \overline{\varphi(x)}f(x)\ \forall f \in L_2[0,1]$, where $\overline{\varphi}$ is the conjugate of φ. In particular, M_φ is self-adjoint if and only if φ takes real values a.e.;

b) $M_\varphi \geq 0$ if and only if $\varphi \geq 0$ a.e..

ii) Let (S, Σ, μ) be a σ-finite measure space, $\varphi \in L_\infty(\mu)$, and $M_\varphi : L_2(\mu) \to L_2(\mu)$ be the multiplication operator defined by φ, $(M_\varphi f)(x) = \varphi(x)f(x)\ \forall f \in L_2(\mu)$. Prove that the parts (a) and (b) from (i) are still true in this new context.

12.1 Exercises

iii) Let (S, Σ, μ) be a σ-finite measure space, $\varphi \in L_\infty(\mu)$, and $M_\varphi : L_2(\mu) \to L_2(\mu)$ be the multiplication operator defined by φ, $(M_\varphi f)(x) = \varphi(x)f(x)$ $\forall f \in L_2(\mu)$. If φ is positive a.e. prove that $\sqrt{M_\varphi} = M_{\sqrt{\varphi}}$, i.e., $(\sqrt{M_\varphi} f)(x) = \sqrt{\varphi(x)} f(x)$ $\forall f \in L_2(\mu)$.

Properties for positive operators

27. Let H be a Hilbert space and $A \in \mathcal{A}_+(H)$. Prove that the following assertions are equivalent:

i) $A(H) \subseteq H$ is dense;
ii) $\ker A = 0$;
iii) A is positive definite, i.e., $\langle Ax, x \rangle > 0$ $\forall x \neq 0$.

28. Let H be a Hilbert space.

i) Prove that a positive operator $T \in L(H)$ is bijective if and only if there is $\varepsilon > 0$ such that $T - \varepsilon I$ is positive definite.

ii) Prove that an operator $T \in L(H)$ is bijective if and only if there is $\varepsilon > 0$ such that $T^*T - \varepsilon I$ and $TT^* - \varepsilon I$ are positive definite.

An inner product associated to a positive operator

29. i) Let $(H, \langle \cdot, \cdot \rangle)$ be a Hilbert space, $A \in \mathcal{A}_+(H)$, and $\langle \cdot, \cdot \rangle_A : H \times H \to \mathbb{K}$ defined by $\langle x, y \rangle_A = \langle Ax, y \rangle$ $\forall x, y \in H$.

a) Prove that $\langle \cdot, \cdot \rangle_A$ is an inner product on $H \Leftrightarrow \langle Ax, x \rangle = 0$ implies $x = 0$;

b) If A satisfies the conditions from (a) then $(H, \langle \cdot, \cdot \rangle_A)$ is a Hilbert space if and only if there is $c > 0$ such that $\langle Ax, x \rangle \geq c \|x\|^2$ $\forall x \in H$.

ii) Let $a = (a_n)_{n \in \mathbb{N}} \subseteq [0, \infty)$ be a bounded sequence, and $\langle \cdot, \cdot \rangle_a : l_2 \times l_2 \to \mathbb{K}$ be defined by $\langle x, y \rangle_a = \sum_{n=1}^\infty a_n x_n \overline{y_n}$. Using (i) prove that:

a) $\langle \cdot, \cdot \rangle_a$ is an inner product on $l_2 \Leftrightarrow a_n > 0$ $\forall n \in \mathbb{N}$;

b) If $a_n > 0$ $\forall n \in \mathbb{N}$ then $(l_2, \langle \cdot, \cdot \rangle_a)$ is a Hilbert space if and only if $\inf_{n \in \mathbb{N}} a_n > 0$.

iii) Let (S, Σ, μ) be a σ-finite measure space, and $\varphi \in L_\infty(\mu)$ positive μ-a.e.. We consider the multiplication operator $M_\varphi : L_2(\mu) \to L_2(\mu)$ defined by $M_\varphi(f) = \varphi f$, and let $\langle \cdot, \cdot \rangle_\varphi : L_2(\mu) \times L_2(\mu) \to \mathbb{K}$ be defined by $\langle f, g \rangle_\varphi = \langle M_\varphi(f), g \rangle$. Using (i) prove that:

a) $\langle \cdot, \cdot \rangle_\varphi$ is an inner product on $L_2(\mu) \Leftrightarrow \varphi(x) > 0$, μ-a.e.;

b) If $\varphi(x) > 0$, μ-a.e., then $(L_2(\mu), \langle \cdot, \cdot \rangle_\varphi)$ is a Hilbert space if and only if there is $c > 0$ such that $\varphi(x) \geq c$, μ-a.e..

A Dini type of characterization for finite-dimensional Hilbert spaces

30. i) Let H be a Hilbert space, $(e_n)_{n \in \mathbb{N}} \subseteq H$ an orthonormal system, $(a_n)_{n \in \mathbb{N}} \subseteq [0, \infty)$ a bounded sequence, and for any $n \in \mathbb{N}$ define $U_n : H \to H$, $U_n(x) = \sum_{k=n}^\infty a_k \langle x, e_k \rangle e_k$. Prove that $0 \leq U_n$ $\forall n \in \mathbb{N}$, $U_n \geq U_{n+1}$ $\forall n \in \mathbb{N}$, $\inf_{n \in \mathbb{N}} U_n = 0$ and $\|U_n\| = \sup_{k \geq n} a_k$ $\forall n \in \mathbb{N}$. In particular $U_n \to 0$ in norm if and only if $a_n \to 0$.

ii) Let H be a Hilbert space. Using (i) prove that the following assertions are equivalent:

a) For any sequence $(U_n)_{n \in \mathbb{N}} \subseteq \mathcal{A}_+(H)$ with $U_n \searrow 0$ (i.e., $U_n \geq U_{n+1}$ $\forall n \in \mathbb{N}$ and $\inf_{n \in \mathbb{N}} U_n = 0$) it follows that $U_n \to 0$ in norm;

b) H is finite-dimensional.

Contractions on Hilbert spaces

31. Let H be a Hilbert space, and $T \in L(H)$ with $\|T\| \leq 1$.
i) Prove that $I - T^*T$ is positive.
ii) If $D_T := \sqrt{I - T^*T}$, prove that $\|x\|^2 = \|Tx\|^2 + \|D_T x\|^2$ $\forall x \in H$.
iii) Prove that $\{x \in H \mid \|x\| = \|Tx\|\} = \ker(D_T)$.

Compact positive operators

32. Let H be a Hilbert space, and $A \in \mathcal{A}_+(H)$. Prove that A is compact $\Leftrightarrow$ $\sqrt{A}$ is compact $\Leftrightarrow$ A_+ and A_- are compact $\Leftrightarrow$ $|A|$ is compact.

33. Let H be a Hilbert space. If $A, B : H \to H$ are self-adjoint operators with $0 \leq A \leq B$ and B is compact, prove that A is compact.

Pointwise convergence and uniform convergence in $L(H)$

34. Let H be a Hilbert space, and $(A_n)_{n \in \mathbb{N}} \subseteq L(H)$. If $\lim_{n \to \infty} |\langle A_n x, y \rangle| = 0$ $\forall x, y \in H$, does it follow that $\lim_{n \to \infty} \|A_n\| = 0$? Does it follow that the sequence $(\|A_n\|)_{n \in \mathbb{N}}$ is bounded?

35. Let $a = (a_n)_{n \in \mathbb{N}} \in l_\infty$ and define the sequence of operators $A_n : l_2 \to l_2$ by

$$A_n(x_1, x_2, \ldots) = (a_{n+1} x_{n+1}, a_{n+2} x_{n+2}, \ldots) \quad \forall (x_1, x_2, \ldots) \in l_2.$$

i) Prove that $A_n \in L(l_2)$ and $A_n(x) \to 0$ $\forall x \in l_2$.
ii) Calculate $A_n^* \in L(l_2)$.
iii) If $C = \{x \in l_2 \mid A_n^*(x) \to 0\}$ then prove that $C = l_2$ if and only if $a \in c_0$, and that $C = \{0\}$ if and only if $a \notin c_0$.

Two recurrence relations

36. i) Let $a, b \in \mathbb{R}$ be such that $a \geq 0$ and $|b| \leq a$. Define two sequences of real numbers $(x_n)_{n \in \mathbb{N}}$ and $(y_n)_{n \in \mathbb{N}}$ by the recurrence relations: $x_1 = a$, $y_1 = b$ and $x_{n+1} = a + x_n^2 + y_n^2$, $y_{n+1} = b + 2 x_n y_n$ $\forall n \in \mathbb{N}$. Prove that $(x_n)_{n \in \mathbb{N}}$ and $(y_n)_{n \in \mathbb{N}}$ are simultaneously convergent if and only if $0 \leq a \leq 1/4$ and $|b| \leq 1/4 - a$.

ii) Let a, b be two real numbers such that $a \in (0, 1)$ and $|b| < \min(a, 1 - a)$. Define two sequences $(a_n)_{n \geq 0}$ and $(b_n)_{n \geq 0}$ by the recurrence relations: $a_0 = a$, $b_0 = b$ and $a_{n+1} = a_n - a_n^2 - b_n^2$, $b_{n+1} = b_n - 2 a_n b_n$ $\forall n \geq 0$. Prove that

$$\lim_{n \to \infty} \frac{n}{\ln n} (n a_n - 1) = -1,$$

and that

$$\lim_{n \to \infty} \frac{n^2 b_n}{\ln n} = 0.$$

The supremum for a sequence of multiplication operators on l_2

37. i) Let $a = (a_n)_{n \in \mathbb{N}}, b = (b_n)_{n \in \mathbb{N}} \in l_\infty$, and $M_a, M_b : l_2 \to l_2$ be the multiplication operators given respectively by a and b. Prove that $M_a \leq M_b \Leftrightarrow a \leq b$ in l_∞, i.e., $a_n \leq b_n$ $\forall n \in \mathbb{N}$. (Here, l_∞ is considered as being defined over $\mathbb{R}$, and l_2 is defined over $\mathbb{K}$.)

ii) Let $(a_n)_{n \in \mathbb{N}} \subseteq l_\infty$ be an increasing sequence, i.e., $a_n \leq a_{n+1}$ $\forall n \in \mathbb{N}$ (in l_∞), and $M_{a_n} : l_2 \to l_2$ be the multiplication operator given by a_n. Prove that there exists $\sup_{n \in \mathbb{N}} M_{a_n}$ in $\mathcal{A}(l_2) \Leftrightarrow$ there exists $a \in l_\infty$ such that $a_n \leq a$ in l_∞ $\forall n \in \mathbb{N}$. In this case, $\sup_{n \in \mathbb{N}} M_{a_n} = M_a$, where $a = \sup_{n \in \mathbb{N}} a_n$ in l_∞.

iii) For every $n \in \mathbb{N}$ consider the operator $U_n : l_2 \to l_2$ defined by

$$U_n(x_1, x_2, \ldots) = \left(\frac{n}{n+1} x_1, \frac{2n}{n+2} x_2, \ldots, \frac{nk}{n+k} x_k, \ldots\right).$$

Using (ii) prove that $\sup_{n \in \mathbb{N}} U_n = I$ in the space $\mathcal{A}(l_2)$.

The supremum and the infimum for a sequence of multiplication operators on L_2

38. i) Let (S, Σ, μ) be a σ-finite measure space. Prove that if $\varphi, \psi \in L_\infty(\mu; \mathbb{R})$ and $M_\varphi, M_\psi : L_2(\mu) \to L_2(\mu)$ are the multiplication operators given by φ and ψ then $M_\varphi \leq M_\psi \Leftrightarrow \varphi \leq \psi$ in $L_\infty(\mu; \mathbb{R})$.

ii) Let (S, Σ, μ) be a σ-finite measure space and let $(\varphi_n)_{n \in \mathbb{N}} \subseteq L_\infty(\mu; \mathbb{R})$ be an increasing sequence, i.e., $\varphi_n \leq \varphi_{n+1}$ $\forall n \in \mathbb{N}$. Let $M_{\varphi_n} : L_2(\mu) \to L_2(\mu)$ be the multiplication operators given by φ_n. Prove that there exists $\sup_{n \in \mathbb{N}} M_{\varphi_n}$ in $\mathcal{A}(L_2(\mu)) \Leftrightarrow$ there exists $\varphi \in L_\infty(\mu; \mathbb{R})$ such that $\varphi_n \leq \varphi$ $\forall n \in \mathbb{N}$. In this case $\sup_{n \in \mathbb{N}} M_{\varphi_n} = M_\varphi$, where $\varphi = \sup_{n \in \mathbb{N}} \varphi_n$ in $L_\infty(\mu; \mathbb{R})$.

iii) Let $\varphi : [0, 1] \to [0, 1]$ be a continuous and increasing function, and for $n \in \mathbb{N}$ with $n \geq 2$ let $U_n : L_2[0, 1] \to L_2[0, 1]$ be defined by $(U_n f)(x) = \varphi(\sqrt[n]{x}) f(x)$. Prove that $\sup_{n \geq 2} U_n = \varphi(1) I$ in the space $\mathcal{A}(L_2[0, 1])$.

iv) Let $\varphi : [0, 1] \to [0, 1]$ be a continuous and increasing function, and for every $n \in \mathbb{N}$ let $U_n : L_2[0, 1] \to L_2[0, 1]$ be defined by $(U_n f)(x) = \varphi(x^n) f(x)$. Prove that $\inf_{n \in \mathbb{N}} U_n = \varphi(0) I$ in the space $\mathcal{A}(L_2[0, 1])$.

v) Let $(a_n)_{n \geq 2} \subseteq [0, \infty)$ be an increasing sequence, and for every $n \in \mathbb{N}$ with $n \geq 2$ let $U_n : L_2[0, 1] \to L_2[0, 1]$ be defined by $(U_n f)(x) = a_n \sqrt[n]{x} f(x)$. Prove that there exists $\sup_{n \geq 2} U_n$ in $\mathcal{A}(L_2[0, 1]) \Leftrightarrow \sup_{n \geq 2} a_n \in \mathbb{R}$, and in this case $\sup_{n \geq 2} U_n = \alpha I$, where $\alpha = \sup_{n \geq 2} a_n = \lim_{n \to \infty} a_n$.

Composition operators on the Hardy and the Bergman spaces

39. Let $\mathbb{D} \subseteq \mathbb{C}$ be the open unit disc in the complex plane, and $w \in \mathbb{D}$.

i) Let $T_w : H^2(\mathbb{D}) \to H^2(\mathbb{D})$ be the operator defined by $(T_w f)(z) = f(wz)$ $\forall z \in \mathbb{D}$. Prove that T_w is linear and continuous, $\|T_w\| = 1$, T_w is compact, and that T_w is self-adjoint $\Leftrightarrow w \in \mathbb{R}$.

ii) Let $T_w : \mathcal{A}^2(\mathbb{D}) \to \mathcal{A}^2(\mathbb{D})$ be the operator defined by $(T_w f)(z) = f(wz)\ \forall z \in \mathbb{D}$. Prove that T_w is linear and continuous, $\|T_w\| = 1$, T_w is compact, and that T_w is self-adjoint $\Leftrightarrow w \in \mathbb{R}$.

Compact operators on the Hardy space

40. Let $\mathbb{D} \subseteq \mathbb{C}$ be the open unit disc in the complex plane.

i) Let $(\xi_k)_{k\in\mathbb{N}} \subseteq \mathbb{D}$ and $\xi \in \mathbb{D}$ be such that $\xi_k \to \xi$. Consider the operator $U : H^2(\mathbb{D}) \to c_0$ defined by $U(f) = (f(\xi_k) - f(\xi))_{k\in\mathbb{N}}$. Prove that U is compact. In the case when $\xi = 0$ find the norm of U.

ii) Let $\xi \in \mathbb{D}$ and $U : H^2(\mathbb{D}) \to c_0$ be defined by $U(f) = (f(\xi^k) - f(0))_{k\in\mathbb{N}}$. Prove that U is compact. Calculate the norm of U.

iii) Let $U : H^2(\mathbb{D}) \to c_0$ be defined by $U(f) = (f(1/(k+1)) - f(0))_{k\in\mathbb{N}}$. Prove that U is compact. Calculate the norm of U.

Compact operators on the Bergman space

41. Let $\mathbb{D} \subseteq \mathbb{C}$ be the open unit disc in the complex plane.

i) Let $(\xi_k)_{k\in\mathbb{N}} \subseteq \mathbb{D}$ and $\xi \in \mathbb{D}$ such that $\xi_k \to \xi$. Consider the operator $U : \mathcal{A}^2(\mathbb{D}) \to c_0$ defined by $U(f) = (f(\xi_k) - f(\xi))_{k\in\mathbb{N}}$. Prove that U is compact. In the case when $\xi = 0$ find the norm of U.

ii) Let $\xi \in \mathbb{D}$ and $U : \mathcal{A}^2(\mathbb{D}) \to c_0$ be defined by $U(f) = (f(\xi^k) - f(0))_{k\in\mathbb{N}}$. Prove that U is compact. Calculate the norm of U.

iii) Let $U : \mathcal{A}^2(\mathbb{D}) \to c_0$ be defined by $U(f) = (f(1/(k+1)) - f(0))_{k\in\mathbb{N}}$. Prove that U is compact. Calculate the norm of U.

The polar decomposition for a rank one operator

42. Let H_1, H_2 be two Hilbert spaces, $a \in H_1\setminus\{0\}$, $b \in H_2\setminus\{0\}$, and $U : H_1 \to H_2$ defined by $Ux = \langle x, a\rangle b$.

i) Prove that U is linear and continuous.

ii) Calculate U^*.

iii) Calculate $|U|$.

iv) Find the polar decomposition of U.

v) If $H_1 = H_2 = H$, prove that U is self-adjoint if and only if there is $t \in \mathbb{R}\setminus\{0\}$ such that $b = ta$. Prove that U is normal if and only if there is $\lambda \in \mathbb{C}\setminus\{0\}$ such that $b = \lambda a$. Prove that U is an orthogonal projection if and only if $b = a/\|a\|^2$.

Effective calculation for the polar decomposition

43. Let $(A_n)_{n\in\mathbb{N}}$ be a sequence of pairwise disjoint Lebesgue measurable sets, with $\bigcup_{n\in\mathbb{N}} A_n = [0,\infty)$, $\mu(A_n) > 0\ \forall n \in \mathbb{N}$, and $\sup_{n\in\mathbb{N}} \mu(A_n) < \infty$. Consider the operator $U : l_2 \to L_2[0,\infty)$ defined by $U(x_1, x_2, ...) = \sum_{n=1}^{\infty} x_n \chi_{A_n}$.

i) Prove that U is linear and continuous.

ii) Calculate U^*, the adjoint of U.

iii) Calculate $|U|$.

12.1 Exercises

iv) Find the polar decomposition $(A, |U|)$ of U.

44. Let $U : L_2[0, \infty) \to l_2$ be defined by $Uf = \left(\int_n^{n+1} f(x)dx \right)_{n \geq 0}$.

i) Prove that U is linear and continuous.
ii) Calculate U^*, the adjoint of U.
iii) Calculate $|U|$.
iv) Find the polar decomposition of U.

The composition operator, the one variable case

45. Let $\varphi : [a, b] \to [c, d]$ be a C^1 diffeomorphism. Consider the composition operator $U : L_2[a, b] \to L_2[c, d]$ defined by $(Uf)(x) = f(\varphi^{-1}(x))\ \forall x \in [c, d]$. Prove that:

i) U is linear and continuous;
ii) $(U^*f)(x) = |\varphi'(x)|\, f(\varphi(x))$;
iii) $(|U| f)(x) = \sqrt{|\varphi'(x)|} f(x)$;
iv) The polar decomposition of U is $(A, |U|)$ where the operator A is given by the formula $(Af)(x) = \sqrt{|(\varphi^{-1})'(x)|} f(\varphi^{-1}(x))$.

46. Let $U : L_2[0, 1] \to L_2[0, 1]$ be defined by $(Uf)(x) = f(2x/(1+x))$.

i) Prove that U is linear and continuous.
ii) Calculate U^*, the adjoint of U.
iii) Calculate $|U|$.
iv) Find the polar decomposition of U.

47. Let $\varphi : [0, 1] \to [0, 1]$ be a C^1 diffeomorphism. Under what conditions on the function φ, is the composition operator $U : L_2[0, 1] \to L_2[0, 1]$ defined by $(Uf)(x) = f(\varphi^{-1}(x))$:

i) self-adjoint?
ii) normal?
iii) a projector?

48. i) Let $\varphi : [0, 1] \to [0, 1]$ be a C^1 diffeomorphism, and consider $h \in L_\infty[0, 1]$. Consider the operator $U : L_2[0, 1] \to L_2[0, 1]$ defined by $(Uf)(x) = h(x) f(\varphi^{-1}(x))$. Prove that $(U^*f) = \overline{h(\varphi(x))}\, |\varphi'(x)|\, f(\varphi(x))$ and that $(|U|f)(x) = |h(\varphi(x))| \sqrt{|\varphi'(x)|} f(x)$.

ii) Let $n \in \mathbb{N}$, and $U : L_2[0, 1] \to L_2[0, 1]$ defined by $(Uf)(x) = x^n f(1-x)$. Using (i) calculate U^* and $|U|$.

The composition operator, the several variables case

49. Let $n \in \mathbb{N}$, $A \subseteq \mathbb{R}^n$ is a bounded Lebesgue measurable set, U, V are open sets in $\mathbb{R}^n$ with $\overline{A} \subseteq U$, and $\varphi : U \to V$ is a C^1 diffeomorphism. Consider the composition operator $T : L_2(A) \to L_2(\varphi(A))$ defined by $(Tf)(x) = f(\varphi^{-1}(x))$. Prove that:

i) T is linear and continuous;
ii) $(T^*f)(x) = |\mathcal{J}_\varphi(x)|\, f(\varphi(x))$;
iii) $(|T| f)(x) = \sqrt{|\mathcal{J}_\varphi(x)|} f(x)$;
iv) The polar decomposition of T is $(A, |T|)$, where the operator A is given by the formula $(Af)(x) = \sqrt{|\mathcal{J}_{\varphi^{-1}}(x)|} f(\varphi^{-1}(x))$.
Here $\mathcal{J}_\varphi(x) = \det \varphi'(x)$ is the Jacobian of φ.

50. Let $n \in \mathbb{N}$, $A \subseteq \mathbb{R}^n$ is a bounded domain, U, V are open sets in $\mathbb{R}^n$ with $\overline{A} \subseteq U$, and $\varphi : U \to V$ is a C^1 diffeomorphism such that $\varphi(A) = A$. Consider the composition

operator $T : L_2(A) \to L_2(A)$ defined by $(Tf)(x) = f(\varphi^{-1}(x))$. Under what conditions on the function φ is the composition operator:
  i) self-adjoint?
  ii) normal?

51. Let $T : L_2\left([0,1]^2\right) \to L_2\left([0,1]^2\right)$ defined by
$$(Tf)(x,y) = f\left(\frac{2x}{1+x}, \frac{2y}{1+y}\right).$$

i) Prove that T is linear and continuous.

ii) Prove that
$$(T^*f)(x,y) = \frac{4}{(2-x)^2(2-y)^2} f\left(\frac{x}{2-x}, \frac{y}{2-y}\right).$$

iii) Prove that
$$(|T|f)(x,y) = \frac{2}{(2-x)(2-y)} f(x,y).$$

iv) Find the polar decomposition $(A, |T|)$ for T.

Extensions for operators

52. i) Let X, Y, Z be normed spaces, $S \in L(X, Z)$, $T \in L(X, Y)$ with T surjective, such that there is $M > 0$ with $\|Sx\| \leq M\|Tx\|$ $\forall x \in X$. Prove that there is a unique operator $R \in L(Y, Z)$ such that $R \circ T = S$ and $\|R\| \leq M$.

ii) If Y is a Hilbert space and Z is a Banach space, prove that we can obtain R with the above properties without the hypothesis that T is surjective. (However, in this case, we cannot obtain the uniqueness for R.)

A Hahn–Banach type of result in $L(H)$

53. Let H be a real Hilbert space.

i) Prove that there is $f : L(H) \to \mathbb{R}$, a linear and continuous functional of norm ≤ 1 such that
$$\inf_{\|x\|\leq 1} \langle Ux, x\rangle \leq f(U) \leq \sup_{\|x\|\leq 1} \langle Ux, x\rangle \quad \forall U \in \mathcal{A}(H).$$

ii) Prove that there is $g : L(H) \to \mathbb{R}$, a linear and continuous functional of norm equal to 1 such that
$$\inf_{\|x\|=1} \langle Ux, x\rangle \leq g(U) \leq \sup_{\|x\|=1} \langle Ux, x\rangle \quad \forall U \in \mathcal{A}(H).$$

12.2 Solutions

1. From the Pythagoras theorem and the Bessel inequality we have
$$\begin{aligned}\|Ux\|^2 &= \sum_{i=1}^n |\lambda_i|^2 \|b_i\|^2 |\langle x, e_i\rangle|^2 = \sum_{i=1}^n |\lambda_i|^2 |\langle x, e_i\rangle|^2 \\ &\leq M^2 \sum_{i=1}^n |\langle x, e_i\rangle|^2 \leq M^2 \|x\|^2,\end{aligned}$$

12.2 Solutions

where $M = \max_{i=\overline{1,n}}|\lambda_i|$. Hence $||Ux|| \leq M||x||\ \forall x \in H$, i.e., $||U|| \leq M$. Let $i = \overline{1,n}$. We have $||Ue_i|| \leq ||U||\,||e_i||$, and using that $Ue_i = \lambda_i b_i$ it follows that $|\lambda_i| \leq ||U||$, i.e., $\max_{i=\overline{1,n}}|\lambda_i| \leq ||U||$. Thus $||U|| = \max_{i=\overline{1,n}}|\lambda_i|$.

2. i) From the Pythagoras theorem we have $||Ux||^2 = ||a||^2\,|\langle x,b\rangle|^2 + ||b||^2\,|\langle x,a\rangle|^2$. Now from the Bessel inequality we deduce that

$$||Ux||^2 = ||a||^2||b||^2\left(|\langle x,b/\,||b||\rangle|^2 + |\langle x,a/\,||a||\rangle|^2\right) \leq ||a||^2||b||^2||x||^2.$$

Hence $||U|| \leq ||a||\,||b||$. But $U(a) = b||a||^2$, $||a||^2||b|| = ||U(a)|| \leq ||U||\,||a||$, i.e., $||U|| \geq ||a||\,||b||$.

ii) Since $\langle \sin x, \cos x\rangle = \int_0^\pi \sin x \cos x\,dx = 0$, $||U|| = ||\sin x||\,||\cos x||$. But $||\sin x||^2 = \int_0^\pi \sin^2 x\,dx = \pi/2$, and $||\cos x||^2 = \int_0^\pi \cos^2 x\,dx = \pi/2$, hence $||U|| = \pi/2$.

3. As is well known, we have $||A||^2_{L(\mathbb{C}^2,||\cdot||_2)} = \rho(A^*A)$, where

$$A^*A = \begin{pmatrix} |a|^2 + |c|^2 & \overline{a}b + \overline{c}d \\ a\overline{b} + c\overline{d} & |b|^2 + |d|^2 \end{pmatrix},$$

and ρ denotes the spectral radius. If λ_1 and λ_2 are the eigenvalues for the operator A^*A, then $\lambda_1, \lambda_2 \in \mathbb{R}_+$, since $A^*A \geq 0$. It follows that

$$\begin{aligned}\rho(A^*A) &= \max(\lambda_1,\lambda_2) = \frac{1}{2}(\lambda_1+\lambda_2) + \frac{1}{2}|\lambda_1-\lambda_2| = \frac{1}{2}(\lambda_1+\lambda_2) + \frac{1}{2}\sqrt{(\lambda_1-\lambda_2)^2}\\ &= \frac{1}{2}(\lambda_1+\lambda_2) + \frac{1}{2}\sqrt{(\lambda_1+\lambda_2)^2 - 4\lambda_1\lambda_2}\\ &= \frac{1}{2}\mathrm{tr}(A^*A) + \frac{1}{2}\sqrt{(\mathrm{tr}(A^*A))^2 - 4(\det(A^*A))^2}\\ &= \frac{1}{2}N + \frac{1}{2}\sqrt{N^2 - 4|D|^2},\end{aligned}$$

where $N = |a|^2 + |b|^2 + |c|^2 + |d|^2$ and $D = ad - bc$.

4. i) Using exercise 1 we have that $||U|| = \max(|a|,|b|)$, $||V|| = \max(|c|,|d|)$, $||U+V|| = \max(|a+c|,|b+d|)$ and $||U-V|| = \max(|a-c|,|b-d|)$. From here we obtain the statement.

ii) Since $\dim_{\mathbb{K}} H \geq 2$, we can find $x,y \in H$, two linearly independent elements. Using the Gram–Schmidt procedure on $\{x,y\}$ we can construct an orthonormal system $\{e_1,e_2\} \subseteq H$. Now construct U, V using the matrix $\begin{pmatrix} 2 & 1 \\ 0 & 2 \end{pmatrix}$, i.e., define $U, V : H \to H$ by $U(x) = 2\langle x,e_1\rangle e_1 + \langle x,e_2\rangle e_2$ and $V(x) = 2\langle x,e_2\rangle e_2$. If $L(H)$ were a Hilbert space, then by the parallelogram law we must have

$$||U+V||^2 + ||U-V||^2 = 2\left(||U||^2 + ||V||^2\right),$$

i.e., by (i)

$$[\max(2,3)]^2 + [\max(2,1)]^2 = 2([\max(2,1)]^2 + [\max(0,2)]^2),$$

and this is not true!

5. See [9, exercise 9, chapter V, section 2].

i) Since $U(x_n) \in Y$ and $(y_m)_{m \in \mathbb{N}}$ is an orthonormal basis in Y then from the Fourier formula we have $U(x_n) = \sum_{m=1}^{\infty} \langle U(x_n), y_m \rangle y_m$, i.e., $\alpha_{mn} = \langle U(x_n), y_m \rangle$. From the Parseval formula we have $\sum_{m=1}^{\infty} |\alpha_{mn}|^2 = \|Ux_n\|^2$. But $\|Ux_n\| \leq \|U\| \|x_n\| = \|U\|$, and therefore we obtain the first inequality.

For the second one we have

$$\sum_{n=1}^{\infty} |\alpha_{mn}|^2 = \sum_{n=1}^{\infty} |\langle U(x_n), y_m \rangle|^2 = \sum_{n=1}^{\infty} |\langle x_n, U^*(y_m) \rangle|^2 = \|U^*(y_m)\|^2,$$

where the last equality is true again by the Parseval formula. But $\|U^*(y_m)\| \leq \|U^*\| \|y_m\| = \|U^*\| = \|U\|$, i.e., we obtain the second inequality.

ii) Let us consider the double-indexed sequence defined by the infinite matrix

$$\begin{pmatrix} 1 & 0 & 0 & 0 & 0 & 0 & \cdots \\ 0 & \frac{1}{\sqrt{2}} & \frac{1}{\sqrt{2}} & 0 & 0 & 0 & \cdots \\ 0 & \frac{1}{\sqrt{2}} & \frac{1}{\sqrt{2}} & 0 & 0 & 0 & \cdots \\ 0 & 0 & 0 & \frac{1}{\sqrt{3}} & \frac{1}{\sqrt{3}} & \frac{1}{\sqrt{3}} & \cdots \\ 0 & 0 & 0 & \frac{1}{\sqrt{3}} & \frac{1}{\sqrt{3}} & \frac{1}{\sqrt{3}} & \cdots \\ 0 & 0 & 0 & \frac{1}{\sqrt{3}} & \frac{1}{\sqrt{3}} & \frac{1}{\sqrt{3}} & \cdots \\ \cdots & \cdots & \cdots & \cdots & \cdots & \cdots & \cdots \end{pmatrix}.$$

It is clear that this double-indexed sequence (α_{mn}) satisfies the conditions from the statement: we have $\sum_{m=1}^{\infty} |\alpha_{mn}|^2 = 1$ and $\sum_{n=1}^{\infty} |\alpha_{mn}|^2 = 1$. Let us suppose that there is $U \in L(X,Y)$ such that $\langle U(x_n), y_m \rangle = \alpha_{mn} \ \forall n, m \in \mathbb{N}$. If $y \in Y$ then $y = \sum_{m=1}^{\infty} \lambda_m y_m$, where $\lambda_m = \langle y, y_m \rangle$. Then

$$\langle U(x_n), y \rangle = \sum_{m=1}^{\infty} \langle Ux_n, \lambda_m y_m \rangle = \sum_{m=1}^{\infty} \overline{\lambda_m} \langle Ux_n, y_m \rangle = \sum_{m=1}^{\infty} \overline{\lambda_m} \alpha_{mn}$$

$$= \sum_{m=1}^{\infty} \langle y_m, y \rangle \alpha_{mn} = \left\langle \sum_{m=1}^{\infty} \alpha_{mn} y_m, y \right\rangle \ \forall y \in Y,$$

and then $U(x_n) = \sum_{m=1}^{\infty} \alpha_{mn} y_m \ \forall n \in \mathbb{N}$, i.e.,

$$U(x_1) = y_1, \ U(x_2) = U(x_3) = \frac{1}{\sqrt{2}} y_2 + \frac{1}{\sqrt{2}} y_3,$$

$$U(x_4) = U(x_5) = U(x_6) = \frac{1}{\sqrt{3}} y_4 + \frac{1}{\sqrt{3}} y_5 + \frac{1}{\sqrt{3}} y_6,$$

$$U(x_{k_n}) = \cdots = U(x_{k_{n+1}-1}) = \frac{1}{\sqrt{n}} y_{k_n} + \frac{1}{\sqrt{n}} y_{k_n+1} + \cdots + \frac{1}{\sqrt{n}} y_{k_{n+1}-1}, \ldots,$$

where $k_n = (n(n-1))/2 + 1\ \forall n \in \mathbb{N}$. Since U is linear and continuous we have

$$\|U(x_{k_n}) + U(x_{k_n+1}) + \cdots + U(x_{k_{n+1}-1})\|^2 \leq \|U\|^2 \|x_{k_n} + \cdots + x_{k_{n+1}-1}\|^2,$$

that is,

$$(k_{n+1} - k_n)^2 \|y_{k_n}/\sqrt{n} + \cdots + y_{k_{n+1}-1}/\sqrt{n}\|^2 \leq \|U\|^2 (k_{n+1} - k_n),$$

or $(k_{n+1} - k_n)^2 (k_{n+1} - k_n)/n \leq \|U\|^2 (k_{n+1} - k_n)$, i.e., $\|U\|^2 \geq (k_{n+1} - k_n)^2/n = n$, $\|U\|^2 \geq n\ \forall n \in \mathbb{N}$, contradiction.

Remark. It is clear that the part (ii) can be generalized in the way which is indicated in the next exercise.

6. Let us suppose that there is $U \in L(X, Y)$ as in the statement. Then by exercise 5(i) it follows that $\sum_{m=1}^{\infty} |\alpha_{mk}|^2 \leq \|U\|^2\ \forall k \in \mathbb{N}$, i.e., using the expression for the infinite matrix (α_{mk}), we obtain that $n_k |a_k|^2 \leq \|U\|^2\ \forall k \in \mathbb{N}$. For $k \geq 2$ let $t_k = n_1 + n_2 + \cdots + n_{k-1}$. We have $U(x_{t_k+1}) = a_k y_{t_k} + a_k y_{t_k+1} + \cdots + a_k y_{t_{k+1}}, \ldots, U(x_{t_{k+1}}) = a_k y_{t_k} + a_k y_{t_k+1} + \cdots + a_k y_{t_{k+1}}\ \forall k \geq 2$. Since $U \in L(X, Y)$ we have

$$\|U(x_{t_k+1}) + \cdots + U(x_{t_{k+1}})\|^2 \leq \|U\|^2 \|x_{t_k+1} + \cdots + x_{t_{k+1}}\|^2,$$

that is, $(t_{k+1} - t_k)^2 \|a_k y_{t_k+1} + \cdots + a_k y_{t_{k+1}}\|^2 \leq \|U\|^2 (t_{k+1} - t_k)$, i.e., using the orthonormality of the sequence $(y_n)_{n \in \mathbb{N}}$, $(t_{k+1} - t_k) |a_k| \leq \|U\|$, i.e., $n_k |a_k| \leq \|U\|\ \forall k \geq 2$.

7. See [35, exercise C, chapter 3].
Since H is a complete space and T is a linear operator, it is sufficient to prove that T is a closed graph operator. Let $(x_n)_{n \in \mathbb{N}}, x, y \in H$ such that $x_n \to x$ and $Tx_n \to y$. Since by hypothesis $\langle T(x_n - x), y \rangle = \langle x_n - x, Ty \rangle\ \forall y \in H$ we obtain that $\langle T(x_n - x), y \rangle \to 0$ $\forall y \in H$. Thus $T(x_n - x) \to 0$ *weak*, and since T is linear we obtain that $Tx_n \to Tx$ *weak*. But $Tx_n \to y$ in norm and therefore $Tx_n \to y$ *weak*. We obtain that $Tx = y$, i.e., T is a closed graph operator.

8. See [1, exercise 9, Fall 1978].
For $i, j \in \{1, 2, \ldots, n\}$ with $i \neq j$, since $\langle e_i, e_j \rangle = 0$ and $\langle e_i - e_j, e_i + e_j \rangle = \langle e_i, e_i \rangle + \langle e_i, e_j \rangle - \langle e_j, e_i \rangle - \langle e_j, e_j \rangle = 0$ then by hypothesis it follows that $\langle Te_i, Te_j \rangle = 0$ and $\langle T(e_i - e_j), T(e_i + e_j) \rangle = 0$, hence $\langle Te_i, Te_i \rangle = \langle Te_j, Te_j \rangle$. Thus there is $k \in \mathbb{C}$ such that $\langle Te_i, Te_i \rangle = k, i = \overline{1, n}$. Then for any $x = (x_1, \ldots, x_n), y = (y_1, \ldots, y_n)$ we obtain that

$$\langle Tx, Ty \rangle = \left\langle T\left(\sum_{i=1}^{n} x_i e_i\right), T\left(\sum_{j=1}^{n} y_j e_j\right) \right\rangle = \sum_{i=1}^{n} \sum_{j=1}^{n} x_i \bar{y}_j \langle Te_i, Te_j \rangle = k \langle x, y \rangle.$$

For any $x \in \mathbb{C}^n$, $\langle x, x \rangle \geq 0$, and therefore $k \geq 0$. If $k > 0$ we denote $S = T/\sqrt{k}$ and we obtain that $\langle Sx, Sy \rangle = (1/k) \langle Tx, Ty \rangle = \langle x, y \rangle\ \forall x, y \in \mathbb{C}^n$. Thus S is unitary and $T = \sqrt{k} S$, with $\sqrt{k} \in \mathbb{R}_+$. If $k = 0$ we obtain that $\langle Tx, Tx \rangle = 0\ \forall x \in \mathbb{C}^n$, i.e, $T = 0$, and then $T = 0 \cdot S$, where $S : \mathbb{C}^n \to \mathbb{C}^n$ is an arbitrary unitary operator.

9. i) Let $U = (T+T^*)/2$ and $V = (T-T^*)/(2i)$. Obviously $U, V \in L(H)$, $T = U + iV$, and

$$U^* = \left(\frac{T+T^*}{2}\right)^* = \frac{T^* + T^{**}}{2} = \frac{T^* + T}{2} = U,$$

$$V^* = \left(\frac{T-T^*}{2i}\right)^* = \frac{T^* - T}{-2i} = \frac{T - T^*}{2i} = V,$$

hence U and V are self-adjoint operators.

ii) Let $U = (T+T^*)/2$ and $V = (T^* - T)/2$. Obviously $U, V \in L(H)$, $T = U - V$ and $U^* = (T+T^*)/2 = U$, $V^* = (T^{**} - T^*)/2 = (T - T^*)/2 = -V$.

iii) Let $P : L(H) \to L(H)$ be the operator defined by $P(U) = (U^* - U)/2$. Clearly P is linear and continuous and then $\ker P = \{U \in L(H) \mid U \text{ is self-adjoint}\}$ is a closed linear subspace. Since $\dim_{\mathbb{K}} H \geq 2$ there exists an orthonormal system $\{e_1, e_2\} \subseteq H$, and then the operator $T : H \to H$ defined by $T(x) = \langle x, e_1 \rangle e_2$ is not self-adjoint, since $\langle Te_1, e_2 \rangle = 1 \neq 0 = \langle e_1, Te_2 \rangle$. Thus $\ker P$ is a proper subspace. From exercise 10(i) of chapter 9 it follows that $\ker P$ is a nowhere dense set. For the second set, one can take $S : L(H) \to L(H)$ defined by $S(U) = (U^* + U)/2$.

10. We will use the well known Sylvester theorem from the linear algebra, which says that if $A = (a_{ij}) \in \mathcal{M}_n(\mathbb{R})$, then A is positive definite (that is, $\langle Ax, x \rangle > 0, \forall x \in \mathbb{R}^n \setminus \{0\}$) if and only if

$$\begin{vmatrix} a_{11} & a_{12} & \cdots & a_{1k} \\ a_{21} & a_{22} & \cdots & a_{2k} \\ \cdots & \cdots & \cdots & \cdots \\ a_{k1} & a_{k2} & \cdots & a_{kk} \end{vmatrix} > 0,$$

for any $1 \leq k \leq n$.

For any $\varepsilon > 0$ we will write $A_\varepsilon = A + \varepsilon I_n$.

'(i) $\Rightarrow$ (ii)' Clearly $\langle A_\varepsilon(x), x \rangle = \langle A(x), x \rangle + \varepsilon \|x\|^2 \geq \varepsilon \|x\|^2 \geq 0 \; \forall x \in \mathbb{R}^n$, and if $\langle A_\varepsilon(x), x \rangle = 0$ then $\varepsilon \|x\|^2 = -\langle A(x), x \rangle \leq 0$, by (i), so since $\varepsilon > 0$ it follows that $\|x\|^2 \leq 0$, $\|x\| = 0$, i.e., $x = 0$. Hence A_ε is positive definite. Let D_k be a diagonal minor of order k for A. Then $D_k + \varepsilon I_k$ is a diagonal minor of order k for $A + \varepsilon I_n$. Since A_ε is positive definite then $D_k + \varepsilon I_k$ is positive definite, and from the Sylvester theorem we have that $\det(D_k + \varepsilon I_k) > 0 \; \forall \varepsilon > 0$, hence passing to the limit for $\varepsilon \to 0, \varepsilon > 0$, we obtain $\det(D_k) \geq 0$, i.e., (ii).

'(ii) $\Rightarrow$ (i)' We will prove by induction on $k = \overline{1, n}$ that under the hypothesis (ii), all the diagonal minors of order k for the matrix $A + \varepsilon I_n$ are positive definite, $\forall \varepsilon > 0$. For $k = 1$, we have $\triangle_{i_1}(\varepsilon) = a_{i_1 i_1} + \varepsilon \geq \varepsilon > 0$, since $a_{i_1 i_1} \geq 0$ by our hypothesis (ii). Let us suppose now that for any $\varepsilon > 0$ all the diagonal minors of order $\leq k - 1$ for the matrix $A + \varepsilon I_n$ are positive definite, and we will prove that all the diagonal minors of order k for the matrix $A + \varepsilon I_n$ are positive definite, $\forall \varepsilon > 0$. Let $\triangle_k(\varepsilon)$ be an arbitrary diagonal minor of order k for the matrix $A + \varepsilon I_n$. Say, for example, that

$$\triangle_k(\varepsilon) = \begin{vmatrix} a_{i_1 i_1} + \varepsilon & a_{i_1 i_2} & \cdots & a_{i_1 i_k} \\ a_{i_2 i_1} & a_{i_2 i_2} + \varepsilon & \cdots & a_{i_2 i_k} \\ \cdots & \cdots & \cdots & \cdots \\ a_{i_k i_1} & a_{i_k i_2} & \cdots & a_{i_k i_k} + \varepsilon \end{vmatrix}.$$

12.2 Solutions

Consider $\triangle_k$ as a function of $\varepsilon \in [0, \infty)$. Then

$$\triangle'_k(\varepsilon) = \begin{vmatrix} 1 & 0 & \cdots & 0 \\ a_{i_2 i_1} & a_{i_2 i_2}+\varepsilon & \cdots & a_{i_2 i_k} \\ \cdots & \cdots & \cdots & \cdots \\ a_{i_k i_1} & a_{i_k i_2} & \cdots & a_{i_k i_k}+\varepsilon \end{vmatrix} + \begin{vmatrix} a_{i_1 i_1}+\varepsilon & a_{i_1 i_2} & \cdots & a_{i_1 i_k} \\ 0 & 1 & \cdots & 0 \\ \cdots & \cdots & \cdots & \cdots \\ a_{i_k i_1} & a_{i_k i_2} & \cdots & a_{i_k i_k}+\varepsilon \end{vmatrix}$$

$$+ \cdots + \begin{vmatrix} a_{i_1 i_1}+\varepsilon & a_{i_1 i_2} & \cdots & a_{i_1 i_k} \\ a_{i_2 i_1} & a_{i_2 i_2}+\varepsilon & \cdots & a_{i_2 i_k} \\ \cdots & \cdots & \cdots & \cdots \\ 0 & 0 & \cdots & 1 \end{vmatrix}$$

$$= \begin{vmatrix} a_{i_2 i_2}+\varepsilon & \cdots & a_{i_2 i_k} \\ \cdots & \cdots & \cdots \\ a_{i_k i_2} & \cdots & a_{i_k i_k}+\varepsilon \end{vmatrix} + \cdots + \begin{vmatrix} a_{i_1 i_1}+\varepsilon & \cdots & a_{i_1 i_{k-1}} \\ \cdots & \cdots & \cdots \\ a_{i_{k-1} i_1} & \cdots & a_{i_{k-1} i_{k-1}}+\varepsilon \end{vmatrix}.$$

But $\begin{pmatrix} a_{i_2 i_2}+\varepsilon & \cdots & a_{i_2 i_k} \\ \cdots & \cdots & \cdots \\ a_{i_k i_2} & \cdots & a_{i_k i_k}+\varepsilon \end{pmatrix}, \ldots, \begin{pmatrix} a_{i_1 i_1}+\varepsilon & \cdots & a_{i_1 i_{k-1}} \\ \cdots & \cdots & \cdots \\ a_{i_{k-1} i_1} & \cdots & a_{i_{k-1} i_{k-1}}+\varepsilon \end{pmatrix}$ are all diagonal minors of order $k-1$ for the matrix $A + \varepsilon I_n$, and by our supposition they are all positive definite. Then $\triangle'_k(\varepsilon) > 0 \ \forall \varepsilon > 0$, i.e., $\triangle_k : [0, \infty) \to \mathbb{R}$ is a strictly increasing function, from whence $\triangle_k(\varepsilon) > \triangle_k(0) \ \forall \varepsilon > 0$. We also have

$$\triangle_k(0) = \begin{vmatrix} a_{i_1 i_1} & a_{i_1 i_2} & \cdots & a_{i_1 i_k} \\ a_{i_2 i_1} & a_{i_2 i_2} & \cdots & a_{i_2 i_k} \\ \cdots & \cdots & \cdots & \cdots \\ a_{i_k i_1} & a_{i_k i_2} & \cdots & a_{i_k i_k} \end{vmatrix} \geq 0,$$

by our hypothesis (ii). Hence $\triangle_k(\varepsilon) > 0 \ \forall \varepsilon > 0$. Then by the Sylvester theorem and using our hypothesis of induction we obtain that $\begin{pmatrix} a_{i_1 i_1}+\varepsilon & a_{i_1 i_2} & \cdots & a_{i_1 i_k} \\ a_{i_2 i_1} & a_{i_2 i_2}+\varepsilon & \cdots & a_{i_2 i_k} \\ \cdots & \cdots & \cdots & \cdots \\ a_{i_k i_1} & a_{i_k i_2} & \cdots & a_{i_k i_k}+\varepsilon \end{pmatrix}$ is positive definite $\forall \varepsilon > 0$.

Then, by induction, for $k = n$ we obtain that A_ε is positive definite, i.e.,

$$\langle A_\varepsilon(x), x \rangle = \langle A(x), x \rangle + \varepsilon \|x\|^2 > 0 \ \forall x \in \mathbb{R}^n \setminus \{0\},$$

hence passing to the limit for $\varepsilon \to 0, \varepsilon > 0$, we obtain that $\langle A(x), x \rangle \geq 0 \ \forall x \in \mathbb{R}^n \setminus \{0\}$, i.e., (i).

11. i) a) Obviously A is well defined, linear, and continuous. We prove that A is self-adjoint, i.e., $\langle Af, g \rangle = \langle f, Ag \rangle \ \forall f, g \in L_2[0, 1]$. Indeed, we have

$$\langle Af, g \rangle = \int_0^1 (Af)(x)\overline{g(x)}dx = \int_0^1 \varphi(x)\left(\int_0^1 \varphi(t)f(t)\,dt\right)\overline{g(x)}dx$$

$$= \left(\int_0^1 \varphi(x)\overline{g(x)}dx\right)\left(\int_0^1 \varphi(t)f(t)\,dt\right),$$

and, analogously,

$$\langle f, Ag \rangle = \int_0^1 f(x)\overline{(Ag)(x)}dx = \int_0^1 f(x)\varphi(x)\overline{\left(\int_0^1 \varphi(t)g(t)dt\right)}dx$$

$$= \left(\int_0^1 \varphi(t)\overline{g(t)}dt\right)\left(\int_0^1 \varphi(x)f(x)dx\right).$$

Also, A is a positive operator since $\langle Af, f \rangle = \left|\int_0^1 \varphi(x)f(x)dx\right|^2 \geq 0 \,\forall f \in L_2[0,1]$.

b) For $f \in L_2[0,1]$, let $g = Af \in L_2[0,1]$. Then $A^2 f = Ag$, where $(Ag)(x) = \varphi(x)\int_0^1 \varphi(t)g(t)dt$. Since $g(t) = \varphi(t)\int_0^1 \varphi(s)f(s)ds$, we obtain that

$$(Ag)(x) = \varphi(x)\int_0^1 \varphi(t)\varphi(t)\left(\int_0^1 \varphi(s)f(s)ds\right)dt$$

$$= \varphi(x)\left(\int_0^1 \varphi(s)f(s)ds\right)\left(\int_0^1 \varphi^2(t)dt\right) = (Af)(x)\left(\int_0^1 \varphi^2(t)dt\right),$$

i.e., $A^2 f = \lambda Af$, with $\lambda = \int_0^1 \varphi^2(t)dt \geq 0$.

c) If $\lambda > 0$, since $A \geq 0$ then $A/\sqrt{\lambda} \geq 0$, and using (ii) we deduce that $\left(A/\sqrt{\lambda}\right)^2 = A$. Now from the uniqueness of the square root it follows that $\sqrt{A} = A/\sqrt{\lambda}$, i.e.,

$$\left(\sqrt{A}f\right)(x) = \frac{\varphi(x)}{\sqrt{\int_0^1 \varphi^2(t)dt}}\int_0^1 \varphi(t)f(t)dt.$$

If $\lambda = 0$, i.e., $\int_0^1 \varphi^2(x)dx = 0$, then since φ is continuous it follows that $\varphi(x) = 0$ $\forall x \in [0,1]$. Then $\sqrt{A} = 0$, since $A = 0$.

d) Using the relation from (ii) we have $A^2 = \lambda A$, hence by induction we obtain that $A^n = \lambda^{n-1}A \,\forall n \in \mathbb{N}$, and from here

$$\|A^n\| = |\lambda|^{n-1}\|A\|, \quad \|A^n\|^{1/n} = |\lambda|^{(n-1)/n}\|A\|^{1/n} \to |\lambda|,$$

hence the spectral radius for the operator A is $\rho(A) = \lim_{n\to\infty}\|A^n\|^{1/n} = \int_0^1 \varphi^2(t)dt$.

e) Since A is self-adjoint, A is an orthogonal projector if and only if $A^2 = A$, or using the relation proved at (b), if and only if $\lambda A = A$, that is, $\int_0^1 \varphi^2(t)dt = 1$.

ii) We have $\varphi(x) = x/(x+1)$ and

$$\int_0^1 \varphi^2(x)dx = \int_0^1 \left(1 - \frac{1}{x+1}\right)^2 dx = \frac{3 - 4\ln 2}{2}.$$

12. i) We have $\|U(x)\| = |\langle x, a\rangle|\|b\| \leq \|a\|\|b\|\|x\|$ $\forall x \in H$, from whence $\|U\| \leq \|a\|\|b\|$. If $a = 0$ or $b = 0$ then $U = 0$. Let now $a \neq 0$ and $b \neq 0$. Then $\|U(a)\| = |\langle a, a\rangle|\|b\| \leq \|U\|\|a\|$, that is, $\|a\|^2\|b\| \leq \|U\|\|a\|$, and then $\|U\| \geq \|a\|\|b\|$.

For $x \in H$, let $y = U(x)$. Then $U^2(x) = U(y) = \langle y, a\rangle b$, where $\langle y, a\rangle = \langle Ux, a\rangle = \langle x, a\rangle \overline{\langle a, b\rangle}$, i.e., denoting with $\lambda = \overline{\langle a, b\rangle}$, we have $U^2 = \lambda U$, and from here, by induction, $U^n = \lambda^{n-1}U \,\forall n \in \mathbb{N}$. Now

$$\|U^n\|^{1/n} = |\lambda|^{(n-1)/n}\|U\|^{1/n} \to |\lambda|,$$

12.2 Solutions

and it follows that $\rho(U) = |\lambda| = |\langle a, b \rangle|$.

The last assertion from the statement is a consequence of the facts proved above and the fact that in the Cauchy–Buniakowski–Schwarz inequality we have equality if and only if the system $\{a, b\}$ is linearly dependent.

ii) We have $\|U(x)\| = |x^*(x)| \|y\| \leq \|x^*\| \|y\| \|x\| \; \forall x \in X$, so $\|U\| \leq \|x^*\| \|y\|$. If $y = 0$ then $U = 0$ and (ii) is true. Suppose now that $y \neq 0$. For $x \in X$ we have $\|U(x)\| \leq \|U\| \|x\|$, that is, $|x^*(x)| \|y\| \leq \|U\| \|x\|$, and then $|x^*(x)| \leq (\|U\| / \|y\|) \|x\|$, from whence $\|x^*\| \leq \|U\| / \|y\|$, i.e., $\|U\| \geq \|x^*\| \|y\|$.

For $x \in H$, let $z = U(x)$. Then $U^2(x) = U(z) = x^*(z)y$ and $z = U(x) = x^*(x)y$, $x^*(z) = x^*(x)x^*(y)$, and therefore $U^2(x) = x^*(y)x^*(x)y$. Denoting $\lambda = x^*(y)$ we have $U^2 = \lambda U$, and from here by induction we obtain that $U^n = \lambda^{n-1} U \; \forall n \in \mathbb{N}$. As above, we obtain $\rho(U) = |\lambda| = |x^*(y)|$.

13. i) Since $A : H \to H$ is self-adjoint, $\|A\| = \sup\limits_{\|x\| \leq 1} |\langle Ax, x \rangle|$ and hence by hypothesis $\|A\| = 0$, i.e., $A = 0$.

ii) Let $A = \begin{pmatrix} a & b \\ c & d \end{pmatrix} \in \mathcal{M}_2(\mathbb{R})$. From the definition for the inner product in $\mathbb{R}^2$ we have

$$\langle Ax, x \rangle = \left\langle \begin{pmatrix} ax_1 + bx_2 \\ cx_1 + dx_2 \end{pmatrix}, \begin{pmatrix} x_1 \\ x_2 \end{pmatrix} \right\rangle = ax_1^2 + (b+c)x_1 x_2 + dx_2^2 \; \forall x = \begin{pmatrix} x_1 \\ x_2 \end{pmatrix} \in \mathbb{R}^2.$$

For $a = d = 0$, $b = -c$, $c \in \mathbb{R} \setminus \{0\}$, we obtain that $A = \begin{pmatrix} 0 & -c \\ c & 0 \end{pmatrix} \in \mathcal{M}_2(\mathbb{R})$ is a non-null matrix with the property that $\langle Ax, x \rangle = 0 \; \forall x \in \mathbb{R}^2$. Obviously we do not contradict (i) since A is not symmetric, hence the associated operator from $\mathbb{R}^2$ to $\mathbb{R}^2$ is not self-adjoint.

iii) '(a) $\Rightarrow$ (b)' Suppose that $\dim_{\mathbb{R}} H \geq 2$. Then there is an orthonormal system $\{e_1, e_2\} \subseteq H$. Then, as in (ii), define $U : H \to H$ by $U(x) = \langle x, e_1 \rangle e_2 - \langle x, e_2 \rangle e_1$. Since H is a real Hilbert space we have

$$\langle U(x), x \rangle = \langle x, e_1 \rangle \langle x, e_2 \rangle - \langle x, e_2 \rangle \langle x, e_1 \rangle = 0 \; \forall x \in H.$$

Also $U \neq 0$, and this contradicts (a).

'(b) $\Rightarrow$ (a)' Since $\dim_{\mathbb{R}} H = 1$, let $e \in H$ with $\|e\| = 1$ such that $\{e\} \subseteq H$ is a basis. Let $U \in L(H)$ such that $\langle Ux, x \rangle = 0 \; \forall x \in H$. Then there is $\lambda \in \mathbb{R}$ such that $U(e) = \lambda e$. Since $\langle Ue, e \rangle = 0$ it follows that $\lambda = 0$. Now if $x \in H$ then there is $\alpha \in \mathbb{R}$ such that $x = \alpha e$, thus $U(x) = \alpha U(e) = 0$, i.e., $U = 0$.

'(c) $\Leftrightarrow$ (a)' The locally convex topology on $L(H)$ generated by the family of seminorms $(p_x)_{x \in H}$ is Hausdorff if and only if for every $U \in L(H)$ with $U \neq 0$ it exists $x \in H$ such that $p_x(U) \neq 0$.

14. i) The operator $M_\lambda : l_2 \to l_2$ is linear and from

$$\|M_\lambda x\|_2 = \left(\sum_{n=1}^{\infty} |\lambda_n|^2 |x_n|^2 \right)^{1/2} \leq \left(\sup_{n \in \mathbb{N}} |\lambda_n| \right) \|x\|_2 \; \forall x \in l_2,$$

it follows that M_λ is continuous. By definition we have $\langle M_\lambda x, y \rangle = \langle x, M_\lambda^* y \rangle$ $\forall x, y \in l_2$. For $y \in l_2$ let $z = M_\lambda^* y$. Then $\langle M_\lambda x, y \rangle = \sum_{n=1}^\infty \lambda_n x_n \overline{y_n}$ and $\langle x, M_\lambda^* y \rangle = \langle x, z \rangle = \sum_{n=1}^\infty x_n \overline{z_n}$. From $\langle M_\lambda x, y \rangle = \langle x, M_\lambda^* y \rangle$ it follows that $\sum_{n=1}^\infty \lambda_n x_n \overline{y_n} = \sum_{n=1}^\infty x_n \overline{z_n}$ $\forall x \in l_2$, from whence $\overline{z_n} = \lambda_n \overline{y_n}$ $\forall n \in \mathbb{N}$, i.e., $z_n = \overline{\lambda_n} y_n$ $\forall n \in \mathbb{N}$. Hence $M_\lambda^* : l_2 \to l_2$ is defined by $M_\lambda^*(y) = (\overline{\lambda_1} y_1, \overline{\lambda_2} y_2, ...)$ $\forall y \in l_2$, i.e., M_λ^* is the multiplication operator associated to the conjugate sequence $\overline{\lambda} = (\overline{\lambda_n})_{n \in \mathbb{N}}$. It is clear now that M_λ is self-adjoint if and only if $\lambda_n \in \mathbb{R}$ $\forall n \in \mathbb{N}$.

ii) M_λ is positive if and only if M_λ is self-adjoint and $\langle M_\lambda x, x \rangle \geq 0$ $\forall x \in l_2$, i.e., $\sum_{n=1}^\infty \lambda_n |x_n|^2 \geq 0$ $\forall x \in l_2$. Now if $\sum_{n=1}^\infty \lambda_n |x_n|^2 \geq 0$ $\forall x \in l_2$, then for $x = e_k$, $k \in \mathbb{N}$, we obtain that $\lambda_k \geq 0$ $\forall k \in \mathbb{N}$. Conversely, if $\lambda_k \geq 0$ $\forall k \in \mathbb{N}$ then M_λ is clearly positive.

Let $\lambda = (\lambda_n)_{n \in \mathbb{N}}$ and $\mu = (\mu_n)_{n \in \mathbb{N}}$ with $\sup_{n \in \mathbb{N}} |\lambda_n| < \infty$ and $\sup_{n \in \mathbb{N}} |\mu_n| < \infty$. Then a simple calculation shows that $M_\lambda \circ M_\mu = M_\nu$, where $\nu = (\lambda_n \mu_n)_{n \in \mathbb{N}}$. Suppose that $M_\lambda \geq 0$. Then there exists $\sqrt{M_\lambda}$. We try to find $\sqrt{M_\lambda}$ of the form M_μ, where $\mu = (\mu_n)_{n \in \mathbb{N}}$ with $\mu_n \geq 0$ $\forall n \in \mathbb{N}$ will be given by the condition $M_\mu^2 = M_\lambda$. Then, by the above, we have $\mu_n^2 = \lambda_n$ $\forall n \in \mathbb{N}$, thus $\mu_n = \sqrt{\lambda_n}$ $\forall n \in \mathbb{N}$, hence $\sqrt{M_\lambda} = M_{\sqrt{\lambda}}$, where $\sqrt{\lambda} = (\sqrt{\lambda_1}, \sqrt{\lambda_2}, ...)$, i.e.,

$$\sqrt{M_\lambda}(x) = \left(\sqrt{\lambda_1} x_1, \sqrt{\lambda_2} x_2, ...\right) \quad \forall x = (x_1, x_2, ...) \in l_2.$$

iii) By (i) we have $M_\lambda^* M_\lambda = M_{\overline{\lambda}} M_\lambda = M_\nu$, where $\nu_n = |\lambda_n|^2$ $\forall n \in \mathbb{N}$. Then by (ii) we have $|M_\lambda| = \sqrt{M_\lambda^* M_\lambda} = \sqrt{M_\nu} = M_{\sqrt{\nu}}$, i.e., $|M_\lambda|(x) = (|\lambda_1| x_1, |\lambda_2| x_2, ...)$. From here it follows that

$$M_\lambda^+(x) = \frac{|M_\lambda|(x) + M_\lambda(x)}{2} = \left(\frac{|\lambda_1| + \lambda_1}{2} x_1, \frac{|\lambda_2| + \lambda_2}{2} x_2, ...\right) \quad \forall x \in l_2,$$

and

$$M_\lambda^-(x) = \frac{|M_\lambda|(x) - M_\lambda(x)}{2} = \left(\frac{|\lambda_1| - \lambda_1}{2} x_1, \frac{|\lambda_2| - \lambda_2}{2} x_2, ...\right) \quad \forall x \in l_2.$$

Remark. For $\lambda \in \mathbb{R}$ if we denote $\lambda^+ = (|\lambda| + \lambda)/2$ and $\lambda^- = (|\lambda| - \lambda)/2$, and for a sequence of real numbers $\lambda = (\lambda_n)_{n \in \mathbb{N}}$ if we denote $|\lambda| = (|\lambda_n|)_{n \in \mathbb{N}}$, $\lambda_+ = (\lambda_n^+)_{n \in \mathbb{N}}$ and $\lambda_- = (\lambda_n^-)_{n \in \mathbb{N}}$, then it follows that if M_λ is the multiplication operator associated to $\lambda \in l_\infty$, then $|M_\lambda|$, M_λ^+ and M_λ^- are the multiplications operators associated to $|\lambda|$, λ_+, respectively λ_-.

15. See [40, exercise 18.41].

Identify the operator A with the matrix $A = \begin{pmatrix} 2 & 3 \\ 3 & 5 \end{pmatrix} \in \mathcal{M}_2(\mathbb{R})$. Since A is symmetric, the operator A is self-adjoint. The eigenvalues for A are $\lambda_1 = (7 + 3\sqrt{5})/2$ and $\lambda_2 = (7 - 3\sqrt{5})/2$. Since A is symmetric and with positive eigenvalues then A is positive, that is, the operator A is positive. The eigenvectors corresponding to the above eigenvalues

12.2 Solutions

are: for λ_1 we have $\left(1, \left(1+\sqrt{5}\right)/2\right)^t u$, $u \in \mathbb{R}$, and for λ_2, $\left(1, \left(1-\sqrt{5}\right)/2\right)^t u$, $u \in \mathbb{R}$. Considering in $\mathbb{R}^2$ the orthonormal basis

$$\left\{ \frac{1}{\sqrt{(10+2\sqrt{5})/4}} \left(1, \left(1+\sqrt{5}\right)/2\right)^t, \frac{1}{\sqrt{(10-2\sqrt{5})/4}} \left(1, \left(1-\sqrt{5}\right)/2\right)^t \right\}$$

the matrix of the operator A with respect to this basis is $\begin{pmatrix} \lambda_1 & 0 \\ 0 & \lambda_2 \end{pmatrix}$. Denoting by

$$B = \mathrm{col}\left(\frac{1}{\sqrt{(10+2\sqrt{5})/4}} \left(1, \left(1+\sqrt{5}\right)/2\right), \frac{1}{\sqrt{(10-2\sqrt{5})/4}} \left(1, \left(1+\sqrt{5}\right)/2\right) \right),$$

we have $B^{-1}AB = \begin{pmatrix} \lambda_1 & 0 \\ 0 & \lambda_2 \end{pmatrix}$. Then

$$A = B \begin{pmatrix} \lambda_1 & 0 \\ 0 & \lambda_2 \end{pmatrix} B^{-1} = \left(B \begin{pmatrix} \sqrt{\lambda_1} & 0 \\ 0 & \sqrt{\lambda_2} \end{pmatrix} B^{-1} \right)^2.$$

Let $C = B \begin{pmatrix} \sqrt{\lambda_1} & 0 \\ 0 & \sqrt{\lambda_2} \end{pmatrix} B^{-1}$. Then $C \in \mathcal{M}_2(\mathbb{R})$, C is positive, $C^2 = A$, and therefore $\sqrt{A} = C$.

16. See [40, exercise 18.39].

Obviously A is linear and continuous with $\|A\| = 1$, and $\langle Ax, y \rangle = \sum_{k=3}^{\infty} x_k \overline{y_k} = \langle x, Ay \rangle$ $\forall x, y \in l_2$, i.e., A is self-adjoint. Since $\langle Ax, x \rangle = \sum_{k=3}^{\infty} |x_k|^2 \geq 0$ $\forall x \in l_2$ it follows that A is positive. Then there exists the square root $\sqrt{A} : l_2 \to l_2$. Since

$$A^2(x) = A(A(x)) = A(0, 0, x_3, x_4, \ldots) = (0, 0, x_3, x_4, \ldots) = Ax \quad \forall x \in l_2,$$

we have $A^2 = A$, and hence by the uniqueness of the square root it follows that $\sqrt{A} = A$.

17. Let us suppose that $\sqrt{A} \leq \sqrt{B}$. Since $AB = BA$ then $A\sqrt{B} = \sqrt{B}A$, and from here $\sqrt{A}\sqrt{B} = \sqrt{B}\sqrt{A}$. Then $B - A = (\sqrt{B})^2 - (\sqrt{A})^2 = (\sqrt{B} - \sqrt{A})(\sqrt{B} + \sqrt{A})$. Since $\sqrt{B} - \sqrt{A} \geq 0$, $\sqrt{B} + \sqrt{A} \geq 0$, and the operators $\sqrt{B} - \sqrt{A}$, $\sqrt{B} + \sqrt{A}$ commute, it follows that their product is positive, i.e., $(\sqrt{B} - \sqrt{A})(\sqrt{B} + \sqrt{A}) \geq 0$, that is, $B - A \geq 0$, $B \geq A$.

Conversely, suppose that $A, B \in \mathcal{A}_+(H)$ with $AB = BA$ and $A \leq B$. We can suppose that $A \leq B \leq I$ (otherwise replace A with A/M and B with B/M, where $M \geq \|B\|$). We recall first that if $T_0 = 0$, $T_{n+1} = (A + T_n^2)/2$ $\forall n \geq 0$ then $T_n(x) \to T(x)$ $\forall x \in H$ where $T = \sup_{n \in \mathbb{N}} T_n$, and $T = I - \sqrt{I - A}$. In the same way, if $V_0 = 0$, $V_{n+1} = (B + V_n^2)/2$ $\forall n \geq 0$, then $V_n(x) \to V(x)$ $\forall x \in H$, where $V = \sup_{n \in \mathbb{N}} V_n$, and $V = I - \sqrt{I - B}$. Since $AB = BA$ it follows that for any $n \in \mathbb{N}$ we have $B^n - A^n = (B - A)(B^{n-1} + B^{n-2}A + \cdots$

$\cdots + BA^{n-2} + A^{n-1})$ and since both members from the right side are positive and commute it follows that $B^n - A^n \geq 0$, i.e., $B^n \geq A^n$. From here, if we observe that T_n, V_n are polynomials in A and respectively B, of degree 2^{n-1} with positive coefficients, we deduce that $V_n \geq T_n$ $\forall n \in \mathbb{N}$. Since for any $x \in H$ we have $T_n(x) \to T(x)$ and $V_n(x) \to V(x)$ then $\langle T_n(x), x \rangle \to \langle T(x), x \rangle$ and $\langle V_n(x), x \rangle \to \langle V(x), x \rangle$. But $\langle V_n(x), x \rangle \geq \langle T_n(x), x \rangle$ $\forall n \in \mathbb{N}$, hence passing to the limit for $n \to \infty$ we deduce that $\langle V(x), x \rangle \geq \langle T(x), x \rangle$, $V \geq T$, i.e., $I - \sqrt{I - B} \geq I - \sqrt{I - A}$, that is, $\sqrt{I - B} \leq \sqrt{I - A}$. Since $0 \leq A \leq B \leq I$ we can replace in the above reasoning $A \to I - B$ and $B \to I - A$ to deduce that $\sqrt{A} \leq \sqrt{B}$, i.e., the statement.

18. Since A and B commute it follows that $A\sqrt{B} = \sqrt{B}A$, and from here $\sqrt{A}\sqrt{B} = \sqrt{B}\sqrt{A}$, so $\left(\sqrt{A}\sqrt{B}\right)^2 = AB$. Also, $\left(\sqrt{AB}\right)^2 = AB$. By the uniqueness of the square root we deduce that $\sqrt{AB} = \sqrt{A}\sqrt{B}$. From this it follows that the operators from both sides of the first inequality from the statement commute. Using exercise 17, the inequality from the statement is equivalent with $((A+B)/2)^2 \geq \left(\sqrt{AB}\right)^2$, i.e., $(A^2 + B^2 + 2AB)/4 \geq AB$, $(A - B)^2 \geq 0$, which is obviously true.

In the same way the operators from both sides of the second inequality from the exercise commute. Again, using exercise 17 the inequality is equivalent to $\left(\sqrt{(A+B)/2}\right)^2 \geq \left(\left(\sqrt{A} + \sqrt{B}\right)/2\right)^2$, i.e., $(A+B)/2 \geq \left(A + B + 2\sqrt{A}\sqrt{B}\right)/4$, $(A+B)/2 \geq \sqrt{AB}$.

19. See [40, exercise 18.36].

Since $A \geq I$ it follows that $\langle Ax, x \rangle \geq ||x||^2$ $\forall x \in H$. Then A is injective and $A(H) \subseteq H$ is dense (see, for example, exercise 27 below). Now we will prove that $A(H)$ is closed, and then $A : H \to H$ will be bijective. Let $(y_n)_{n \in \mathbb{N}} \subseteq A(H)$ with $y_n \to y \in H$. Let $y_n = Ax_n$, $x_n \in H$. We have

$$||Ax_m - Ax_n|| \, ||x_m - x_n|| \geq \langle Ax_m - Ax_n, x_m - x_n \rangle \geq ||x_m - x_n||^2,$$

that is, $||Ax_m - Ax_n|| \geq ||x_m - x_n||$ $\forall m, n \in \mathbb{N}$. Since $(Ax_n)_{n \in \mathbb{N}}$ is Cauchy, being convergent, it follows that $(x_n)_{n \in \mathbb{N}}$ is Cauchy in H and hence there is $x \in H$ such that $x_n \to x$. Then $Ax_n \to Ax$ from whence $y = Ax$, i.e., $y \in A(H)$. We also have $\langle y, A^{-1}y \rangle \geq ||A^{-1}y||^2$ $\forall y \in H$, from whence $||y|| \, ||A^{-1}y|| \geq ||A^{-1}y||^2$ $\forall y \in H$, i.e., $||y|| \geq ||A^{-1}y||$ $\forall y \in H$, hence A^{-1} is linear and continuous with norm ≤ 1. Analogously B is bijective and $B^{-1} : H \to H$ is linear and continuous. Since A and B are self-adjoint operators, then A^{-1} and B^{-1} are also self-adjoint operators. We will prove now that $A^{-1} \geq B^{-1}$, i.e., $A^{-1} - B^{-1} \geq 0$. We have $A^{-1} - B^{-1} = A^{-1}(B - A)B^{-1}$ and then

$$\begin{aligned}
\langle (A^{-1} - B^{-1})x, x \rangle &= \langle A^{-1}(B-A)B^{-1}x, x \rangle = \langle (B-A)B^{-1}x, A^{-1}x \rangle \\
&= \langle (B-A)B^{-1}x, B^{-1}x \rangle + \langle (B-A)B^{-1}x, A^{-1}x - B^{-1}x \rangle \\
&\geq \langle (B-A)B^{-1}x, (A^{-1} - B^{-1})x \rangle \\
&= \langle (B-A)B^{-1}x, (A^{-1})(B-A)B^{-1}x \rangle \geq 0,
\end{aligned}$$

since $A : H \to H$ positive implies $A^{-1} : H \to H$ positive.

12.2 Solutions

20. i) See [31, Lemma 1.2.1].
Since $|A|^2 = A^*A$, then the relation (c) becomes

$$\left(\sum_{i=1}^n \alpha_i A_i\right)^* \left(\sum_{i=1}^n \alpha_i A_i\right) \le \left(\sum_{i=1}^n |\alpha_i|^2\right) \left(\sum_{i=1}^n A_i^* A_i\right),$$

that is,

$$\left\langle \left(\sum_{i=1}^n \alpha_i A_i\right)^* \left(\sum_{i=1}^n \alpha_i A_i\right) x, x \right\rangle \le \left(\sum_{i=1}^n |\alpha_i|^2\right) \left\langle \sum_{i=1}^n A_i^* A_i x, x \right\rangle \quad \forall x \in H_1,$$

or

$$\left\langle \sum_{i=1}^n \alpha_i A_i x, \sum_{i=1}^n \alpha_i A_i x \right\rangle \le \left(\sum_{i=1}^n |\alpha_i|^2\right) \left(\sum_{i=1}^n \langle A_i x, A_i x\rangle\right),$$

i.e.,

$$\left\| \sum_{i=1}^n \alpha_i A_i x \right\|^2 \le \left(\sum_{i=1}^n |\alpha_i|^2\right) \left(\sum_{i=1}^n \|A_i x\|^2\right).$$

But

$$\left\| \sum_{i=1}^n \alpha_i A_i x \right\| \le \sum_{i=1}^n |\alpha_i| \|A_i x\| \le \left(\sum_{i=1}^n |\alpha_i|^2\right)^{1/2} \left(\sum_{i=1}^n \|A_i x\|^2\right)^{1/2},$$

by the Cauchy–Buniakowski–Schwarz inequality, and from here we obtain (c).

ii) See [41, p. 37].
Let $x, y \in H$ with $\|x\| \le 1$ and $\|y\| \le 1$. Then we have:

$$\left|\left\langle \sum_{i=1}^n A_i B_i x, y \right\rangle\right| = \left|\sum_{i=1}^n \langle B_i x, A_i^* y\rangle\right| \le \sum_{i=1}^n |\langle B_i x, A_i^* y\rangle| \le \sum_{i=1}^n \|B_i x\| \|A_i^* y\|$$

$$\le \sqrt{\sum_{i=1}^n \|B_i x\|^2} \sqrt{\sum_{i=1}^n \|A_i^* y\|^2}$$

$$= \sqrt{\sum_{i=1}^n \langle B_i^* B_i x, x\rangle} \sqrt{\sum_{i=1}^n \langle A_i A_i^* y, y\rangle}$$

$$= \sqrt{\left\langle \left(\sum_{i=1}^n B_i^* B_i\right) x, x\right\rangle} \sqrt{\left\langle \left(\sum_{i=1}^n A_i A_i^*\right) y, y\right\rangle}.$$

But if $U \in \mathcal{A}_+(H)$ then $\|U\| = \sup\limits_{\|x\| \le 1} \langle Ux, x\rangle$. Now passing to the supremum over $\|x\| \le 1$ and $\|y\| \le 1$ and using this relation we obtain that

$$\left\|\sum_{i=1}^n A_i B_i\right\| \le \sqrt{\left\|\sum_{i=1}^n A_i A_i^*\right\|} \sqrt{\left\|\sum_{i=1}^n B_i^* B_i\right\|},$$

i.e., the statement.

iii) See [42, exercise 16, chapter VI].
We have
$$|A| = \sqrt{A^*A} = \begin{pmatrix} 2 & 0 \\ 0 & 0 \end{pmatrix}, B^*B = \begin{pmatrix} 2 & -2 \\ -2 & 2 \end{pmatrix}, |B| = \sqrt{B^*B} = \begin{pmatrix} 1 & -1 \\ -1 & 1 \end{pmatrix},$$

and
$$(A+B)^*(A+B) = \begin{pmatrix} 2 & 0 \\ 0 & 2 \end{pmatrix}, |A+B| = \sqrt{(A+B)^*(A+B)} = \begin{pmatrix} \sqrt{2} & 0 \\ 0 & \sqrt{2} \end{pmatrix}.$$

Then $|A+B|(e_2) = \sqrt{2}e_2$, $\langle |A+B|(e_2), e_2 \rangle = \sqrt{2}$, $|A|(e_2) = 0$, $|B|(e_2) = e_2 - e_1$, $\langle |B|(e_2), e_2 \rangle = 1$. If $|A+B| \leq |A| + |B|$, then we must have $\langle |A+B|(e_2), e_2 \rangle \leq \langle |A|(e_2), e_2 \rangle + \langle |B|(e_2), e_2 \rangle$, i.e., $\sqrt{2} \leq 1$, which is obviously false.

iv) Since $\dim_\mathbb{K} H \geq 2$, we can find an orthonormal system $\{e_1, e_2\} \subseteq H$. We consider $U, V \in L(H)$, $U(x) = 2\langle x, e_1 \rangle e_1$ and $V(x) = -\langle x, e_1 - e_2 \rangle (e_1 - e_2)$. Then we have $|U| = U$, $|V|(x) = \langle x, e_1 - e_2 \rangle (e_1 - e_2)$. Since $(U+V)(x) = \langle x, e_1 \rangle (e_1 + e_2) + \langle x, e_2 \rangle (e_1 - e_2)$, we obtain that $(U+V)^*(x) = \langle x, e_1 + e_2 \rangle e_1 + \langle x, e_1 - e_2 \rangle e_2$. Now $(U+V)^*(U+V)(x) = 2\langle x, e_1 \rangle e_1 + 2\langle x, e_2 \rangle e_2$ $\forall x \in H$, and from here by the orthonormality of the system $\{e_1, e_2\}$ we easily deduce that $|U+V|(x) = \sqrt{2}\langle x, e_1 \rangle e_1 + \sqrt{2}\langle x, e_2 \rangle e_2$ $\forall x \in H$. If $|U+V| \leq |U| + |V|$, then we must have $\langle |U+V| e_2, e_2 \rangle \leq \langle |U| e_2, e_2 \rangle + \langle |V| e_2, e_2 \rangle$, and by the orthonormality of the system $\{e_1, e_2\}$ we deduce that $\sqrt{2} \leq 1$, which is false.

Remark. The inequalities from (i) and (iv) and the inequalities from exercises 17–19 are well known for real numbers. Even if many of the analogies between the self-adjoint operators and the real numbers are true (the inequalities from 17, 18, 19 and 20 (i) for example), for the inequality from (iv) the analogy is no longer true.

21 Since M is positive homogeneous we may suppose that
$$M\left(\left(\sqrt{A_1} + \sqrt{A_2} + \cdots + \sqrt{A_n}\right)^2\right) = 1,$$

and then the inequality from the statement is
$$\left(\sqrt{M(A_1)} + \sqrt{M(A_2)} + \cdots + \sqrt{M(A_n)}\right)^2 \geq 1,$$

i.e., $(M(A_1), ..., M(A_n)) \in E$, where
$$E = \left\{(t_1, ..., t_n) \in [0, \infty)^n \mid \left(\sqrt{t_1} + \sqrt{t_2} + \cdots + \sqrt{t_n}\right)^2 \geq 1\right\}$$

is convex. If, for a contradiction, $(M(A_1), ..., M(A_n)) \notin E$, then using a separation theorem we can find $\alpha_1, ..., \alpha_n, t \in \mathbb{R}$ such that
$$\alpha_1 M(A_1) + \cdots + \alpha_n M(A_n) < t \leq \alpha_1 t_1 + \cdots + \alpha_n t_n \; \forall (t_1, ..., t_n) \in E. \tag{1}$$

12.2 Solutions

Exactly as in the solution for exercise 20 chapter 8 it follows that $\alpha_1 \geq 0, ..., \alpha_n \geq 0$, and therefore $t > 0$. Indeed, let $(t_1, ..., t_n) \in (0, \infty)^n$. Then $\lambda_0 = (\sqrt{t_1} + \sqrt{t_2} + \cdots + \sqrt{t_n})^{-2} > 0$ and then for $\lambda \geq \lambda_0$ we have

$$\left(\sqrt{\lambda t_1} + \sqrt{\lambda t_2} + \cdots + \sqrt{\lambda t_n}\right)^2 = \lambda \left(\sqrt{t_1} + \sqrt{t_2} + \cdots + \sqrt{t_n}\right)^2$$
$$\geq \lambda_0 \left(\sqrt{t_1} + \sqrt{t_2} + \cdots + \sqrt{t_n}\right)^2 = 1,$$

i.e., $(\lambda t_1, ..., \lambda t_n) \in E$, and from (1), $t \leq \alpha_1 \lambda t_1 + \cdots + \alpha_n \lambda t_n$, that is, $t/\lambda \leq \alpha_1 t_1 + \cdots + \alpha_n t_n$ $\forall \lambda \geq \lambda_0$. Passing to the limit for $\lambda \to \infty$ we obtain that $0 \leq \alpha_1 t_1 + \cdots + \alpha_n t_n$ $\forall (t_1, ..., t_n) \in (0, \infty)^n$ and for $t_2 \to 0, ..., t_n \to 0$, $\alpha_1 t_1 \geq 0$ $\forall t_1 > 0$, from whence $\alpha_1 \geq 0$. Analogously we obtain that $\alpha_2 \geq 0, ..., \alpha_n \geq 0$. Since $M(A_1) \geq 0, ..., M(A_n) \geq 0$, from (1) it follows that $t > 0$.

Let $x \in H$ and $\lambda = (\sqrt{\langle A_1 x, x \rangle} + \cdots + \sqrt{\langle A_n x, x \rangle})^2$. If $\lambda \neq 0$, then $\lambda > 0$, and

$$\left(\sqrt{\langle A_1 x, x \rangle / \lambda} + \cdots + \sqrt{\langle A_n x, x \rangle / \lambda}\right)^2 = 1,$$

i.e., $(\langle A_1 x, x \rangle / \lambda, ..., \langle A_n x, x \rangle / \lambda) \in E$, so by (1) it follows that

$$t \leq (\alpha_1 \langle A_1 x, x \rangle + \cdots + \alpha_n \langle A_n x, x \rangle) / \lambda,$$

i.e.,

$$t \left(\sqrt{\langle A_1 x, x \rangle} + \cdots + \sqrt{\langle A_n x, x \rangle}\right)^2 = \lambda t \leq \alpha_1 \langle A_1 x, x \rangle + \cdots + \alpha_n \langle A_n x, x \rangle.$$

If $\lambda = 0$, then the last inequality is clearly true, and then

$$t \left(\sqrt{\langle A_1 x, x \rangle} + \cdots + \sqrt{\langle A_n x, x \rangle}\right)^2 \leq \alpha_1 \langle A_1 x, x \rangle + \cdots + \alpha_n \langle A_n x, x \rangle \quad \forall x \in H.$$

We have

$$\left\langle \left(\sqrt{A_1} + \sqrt{A_2} + \cdots + \sqrt{A_n}\right)^2 x, x \right\rangle \leq \left(\sqrt{\langle A_1 x, x \rangle} + \cdots + \sqrt{\langle A_n x, x \rangle}\right)^2$$

if and only if

$$\left\| \left(\sqrt{A_1} + \sqrt{A_2} + \cdots + \sqrt{A_n}\right) x \right\|^2 \leq \left(\sqrt{\langle A_1 x, x \rangle} + \cdots + \sqrt{\langle A_n x, x \rangle}\right)^2,$$

that is, if and only if

$$\left\| \left(\sqrt{A_1} + \sqrt{A_2} + \cdots + \sqrt{A_n}\right) x \right\| \leq \sqrt{\langle A_1 x, x \rangle} + \cdots + \sqrt{\langle A_n x, x \rangle},$$

and using that

$$\langle Ax, x \rangle = \left\langle \sqrt{A}^2 x, x \right\rangle = \left\| \sqrt{A} x \right\|^2$$

for $A \in \mathcal{A}_+(H)$, this is equivalent to

$$\left\| \left(\sqrt{A_1} + \sqrt{A_2} + \cdots + \sqrt{A_n}\right) x \right\| \leq \left\| \sqrt{A_1} x \right\| + \cdots + \left\| \sqrt{A_n} x \right\|,$$

which is clearly true. So
$$t\left\langle \left(\sqrt{A_1} + \sqrt{A_2} + \cdots + \sqrt{A_n}\right)^2 x, x\right\rangle \leq \alpha_1 \langle A_1 x, x\rangle + \cdots + \alpha_n \langle A_n x, x\rangle \quad \forall x \in H$$
i.e.,
$$t\left(\sqrt{A_1} + \sqrt{A_2} + \cdots + \sqrt{A_n}\right)^2 \leq \alpha_1 A_1 + \cdots + \alpha_n A_n$$
in $\mathcal{A}_+(H)$. Then
$$tM\left(\left(\sqrt{A_1} + \sqrt{A_2} + \cdots + \sqrt{A_n}\right)^2\right) \leq \alpha_1 M(A_1) + \cdots + \alpha_n M(A_n).$$
But $M\left(\left(\sqrt{A_1} + \sqrt{A_2} + \cdots + \sqrt{A_n}\right)^2\right) = 1$, hence $t \leq \alpha_1 M(A_1) + \cdots + \alpha_n M(A_n)$, and this contradicts (1).

22. Suppose that the series $\sum_{k=1}^{\infty} a_k \langle x, e_k\rangle e_k$ converges $\forall x \in H$. For any $n \in \mathbb{N}$ let $U_n : H \to H$ be defined by $U_n(x) = \sum_{k=1}^{n} a_k \langle x, e_k\rangle e_k$. Then the sequence $(U_n)_{n \in \mathbb{N}}$ is pointwise convergent, hence by the Uniform Boundedness Principle we have $\sup_{n \in \mathbb{N}} \|U_n\| < \infty$. Using the results from exercise 1 we have $\|U_n\| = \max_{1 \leq k \leq n}\left(|a_k|\|e_k\|^2\right) \forall n \in \mathbb{N}$, and hence $\sup_{n \in \mathbb{N}}\left(|a_n|\|e_n\|^2\right) < \infty$. Conversely, if $M = \sup_{n \in \mathbb{N}}\left(|a_n|\|e_n\|^2\right) < \infty$ then for any $x \in H$ we have
$$\left\|\sum_{n=1}^{\infty} a_n \langle x, e_n\rangle e_n\right\|^2 = \sum_{n=1}^{\infty} |a_n|^2 \|e_n\|^2 |\langle x, e_n\rangle|^2 = \sum_{n=1}^{\infty} |a_n|^2 \|e_n\|^4 \left|\left\langle x, \frac{e_n}{\|e_n\|}\right\rangle\right|^2$$
$$\leq M^2 \sum_{n=1}^{\infty} \left|\left\langle x, \frac{e_n}{\|e_n\|}\right\rangle\right|^2 \leq M^2 \|x\|^2.$$

The last inequality is true by the Bessel inequality, the system $(e_n/\|e_n\|)_{n \in \mathbb{N}}$ being orthonormal. From here it follows that $P_{(a_1,a_2,\ldots)}$ is linear and continuous, with $\|P_{(a_1,a_2,\ldots)}\| \leq M$. The inequality $\|P_{(a_1,a_2,\ldots)}\| \geq M$ follows from
$$|a_n|\|e_n\|^3 = \|P_{(a_1,a_2,\ldots)}(e_n)\| \leq \|P_{(a_1,a_2,\ldots)}\|\|e_n\|$$
for any $n \in \mathbb{N}$.

i) For any sequences $(a_n)_{n \in \mathbb{N}}$ and $(b_n)_{n \in \mathbb{N}}$ with $\sup_{n \in \mathbb{N}}\left(|a_n|\|e_n\|^2\right) < \infty$ and $\sup_{n \in \mathbb{N}}\left(|b_n|\|e_n\|^2\right) < \infty$, we will prove that $P_{(a_1,a_2,\ldots)} = P_{(b_1,b_2,\ldots)}$ if and only if $a_n = b_n$ $\forall n \in \mathbb{N}$, and that $P_{(a_1,a_2,\ldots)} \circ P_{(b_1,b_2,\ldots)} = P_{(a_1 b_1 \|e_1\|^2, a_2 b_2 \|e_2\|^2, \ldots)}$.

The first assertion is clear, if we observe that from the orthogonality of the system $(e_k)_{k \in \mathbb{N}}$ we have $P_{(a_1,a_2,\ldots)}(e_n) = P_{(b_1,b_2,\ldots)}(e_n)$ if and only if $a_n = b_n$. For the second one, if $x \in H$ then let $y = P_{(b_1,b_2,\ldots)}(x)$. We have
$$(P_{(a_1,a_2,\ldots)} \circ P_{(b_1,b_2,\ldots)})(x) = P_{(a_1,a_2,\ldots)}(y) = \sum_{n=1}^{\infty} a_n \langle y, e_n\rangle e_n.$$

12.2 Solutions

By the separate continuity of the inner product and the orthogonality of the system $(e_n)_{n \in \mathbb{N}}$ we have

$$\langle y, e_n \rangle = \langle P_{(b_1,b_2,...)}(x), e_n \rangle = \sum_{k=1}^{\infty} b_k \langle x, e_k \rangle \langle e_k, e_n \rangle$$

$$= b_n \langle x, e_n \rangle \langle e_n, e_n \rangle = b_n \|e_n\|^2 \langle x, e_n \rangle,$$

and then

$$(P_{(a_1,a_2,...)} \circ P_{(b_1,b_2,...)})(x) = \sum_{n=1}^{\infty} a_n b_n \|e_n\|^2 \langle x, e_n \rangle e_n = P_{(a_1 b_1 \|e_1\|^2, a_2 b_2 \|e_2\|^2, ...)}(x).$$

From here we deduce that $P^2_{(a_1,a_2,...)} = P_{(a_1^2 \|e_1\|^2, a_2^2 \|e_2\|^2, ...)}$.

ii) We prove now that $P^*_{(a_1,a_2,...)} = P_{(\overline{a_1},\overline{a_2},...)}$. Indeed, for any $y \in H$ the series $\sum_{n=1}^{\infty} \overline{a_n} \langle y, e_n \rangle e_n$ converges (see (i)). Let $z = \sum_{n=1}^{\infty} \overline{a_n} \langle y, e_n \rangle e_n$, and then from the separate continuity of the inner product and the orthogonality of the system $(e_n)_{n \in \mathbb{N}}$ we have that

$$\langle P_{(a_1,a_2,...)}(x), y \rangle = \sum_{n=1}^{\infty} a_n \langle x, e_n \rangle \langle e_n, y \rangle = \langle x, z \rangle \quad \forall x \in H,$$

and therefore $P^*_{(a_1,a_2,...)}(y) = z$. From here it follows that $P_{(a_1,a_2,...)}$ is self-adjoint $\Leftrightarrow P^*_{(a_1,a_2,...)} = P_{(a_1,a_2,...)} \Leftrightarrow a_n = \overline{a_n} \; \forall n \in \mathbb{N} \Leftrightarrow a_n \in \mathbb{R} \; \forall n \in \mathbb{N}$.

iii) We have obviously $\langle P_{(a_1,a_2,...)} x, x \rangle = \sum_{n=1}^{\infty} a_n |\langle x, e_n \rangle|^2 \geq 0$, since $a_n \geq 0 \; \forall n \in \mathbb{N}$. Then we know that there exists $\sqrt{P_{(a_1,a_2,...)}}$. We search $\sqrt{P_{(a_1,a_2,...)}}$ of the form $P_{(b_1,b_2,...)}$, i.e., $\sqrt{P_{(a_1,a_2,...)}} = P_{(b_1,b_2,...)}$, where $b_n \geq 0 \; \forall n \in \mathbb{N}$ are to be calculated. Then $P_{(b_1,b_2,...)} \geq 0$, and from $P^2_{(b_1,b_2,...)} = P_{(a_1,a_2,...)}$ using (i) it follows that $P_{(a_1,a_2,...)} = P_{(b_1^2 \|e_1\|^2, b_2^2 \|e_2\|^2, ...)}$, i.e., $b_n^2 \|e_n\|^2 = a_n$, that is $b_n = \sqrt{a_n}/\|e_n\| \; \forall n \in \mathbb{N}$, i.e.,

$$\sqrt{P_{(a_1,a_2,...)}}(x) = \sum_{n=1}^{\infty} (\sqrt{a_n}/\|e_n\|) \langle x, e_n \rangle e_n \quad \forall x \in H.$$

23. i) From the Pythagoras theorem we have that the series $\sum_{n=1}^{\infty} \lambda_n \langle x, e_n \rangle f_n$ converges in H_2 for any $x \in H_1$ if and only if the series $\sum_{n=1}^{\infty} |\lambda_n|^2 \|f_n\|^2 |\langle x, e_n \rangle|^2$ converges $\forall x \in H_1$.

Suppose that the series $\sum_{n=1}^{\infty} \lambda_n \langle x, e_n \rangle f_n$ converges for any $x \in H_1$. Let $V : H \to l_2$ defined by $V(x) = (\lambda_n \|f_n\| \langle x, e_n \rangle)_{n \in \mathbb{N}}$. From the above V is well defined. Obviously, V is linear. It is easy to see that V has its graph closed, and since H and l_2 are Banach spaces, from the closed graph theorem it follows that V is continuous, and then $\|V(e_n)\| \leq \|V\| \|e_n\| \; \forall n \in \mathbb{N}$. Using that

$$V(e_n) = (0, ..., 0, \lambda_n \|f_n\| \langle e_n, e_n \rangle, 0, ...)$$

we obtain that $|\lambda_n| \|e_n\|^2 \|f_n\| \leq \|V\| \|e_n\|$, and since $e_n \neq 0$ it follows that $|\lambda_n| \|e_n\| \|f_n\| \leq \|V\|$ $\forall n \in \mathbb{N}$, i.e., $\sup_{n \in \mathbb{N}}(|\lambda_n| \|e_n\| \|f_n\|) \leq \|V\| < \infty$.

Suppose now that $M = \sup_{n \in \mathbb{N}}(|\lambda_n| \|e_n\| \|f_n\|) < \infty$. Then

$$\sum_{n=1}^\infty |\lambda_n|^2 \|f_n\|^2 |\langle x, e_n\rangle|^2 = \sum_{n=1}^\infty |\lambda_n|^2 \|e_n\|^2 \|f_n\|^2 \left|\left\langle x, \frac{e_n}{\|e_n\|}\right\rangle\right|^2$$
$$\leq M^2 \sum_{n=1}^\infty \left|\left\langle x, \frac{e_n}{\|e_n\|}\right\rangle\right|^2 \leq M^2 \|x\|^2.$$

The last inequality is true by the Bessel inequality, the system $(e_n/\|e_n\|)_{n \in \mathbb{N}}$ being orthonormal. Hence

$$\|Ux\| = \left\|\sum_{n=1}^\infty \lambda_n \langle x, e_n\rangle f_n\right\| = \left(\sum_{n=1}^\infty |\lambda_n|^2 \|f_n\|^2 |\langle x, e_n\rangle|^2\right)^{1/2} \leq M \|x\| \quad \forall x \in H,$$

and since U is linear it follows that $\|U\| \leq M = \sup_{n \in \mathbb{N}}(|\lambda_n| \|e_n\| \|f_n\|)$. We have $\|U(e_n)\| \leq \|U\| \|e_n\|$ $\forall n \in \mathbb{N}$, and since $U(e_n) = \lambda_n \langle e_n, e_n\rangle f_n = \lambda_n \|e_n\|^2 f_n$, we obtain that $|\lambda_n| \|e_n\| \|f_n\| \leq \|U\|$ $\forall n \in \mathbb{N}$, and then $\sup_{n \in \mathbb{N}}(|\lambda_n| \|e_n\| \|f_n\|) \leq \|U\|$.

a) Since $\sup_{n \in \mathbb{N}}(|\lambda_n| \|e_n\| \|f_n\|) < \infty$, using (i) the series $\sum_{n=1}^\infty \overline{\lambda_n} \langle x, f_n\rangle e_n$ converges $\forall x \in H_2$, and therefore U^* is well defined, linear, and continuous. We must prove that $\langle Ux, y\rangle = \langle x, U^*y\rangle$ $\forall x \in H_1$, $\forall y \in H_2$. For this we will use the separate continuity of the inner product. We have

$$\langle Ux, y\rangle = \left\langle \sum_{n=1}^\infty \lambda_n \langle x, e_n\rangle f_n, y\right\rangle = \sum_{n=1}^\infty \lambda_n \langle x, e_n\rangle \langle f_n, y\rangle,$$

and then

$$\langle x, U^*y\rangle = \left\langle x, \sum_{n=1}^\infty \overline{\lambda_n} \langle y, f_n\rangle e_n\right\rangle = \sum_{n=1}^\infty \langle x, \overline{\lambda_n}\langle y, f_n\rangle e_n\rangle$$
$$= \sum_{n=1}^\infty \lambda_n \overline{\langle y, f_n\rangle} \langle x, e_n\rangle = \sum_{n=1}^\infty \lambda_n \langle x, e_n\rangle \langle f_n, y\rangle = \langle Ux, y\rangle.$$

b) By definition we have $|U| = \sqrt{U^*U}$, i.e., we must calculate the square root of U^*U. For $x \in H$, let $y = U(x)$. Then we have $U^*Ux = U^*(y) = \sum_{n=1}^\infty \overline{\lambda_n} \langle y, f_n\rangle e_n$. Now again by the separate continuity of the inner product we obtain that

$$\langle y, f_n\rangle = \langle Ux, f_n\rangle = \left\langle \sum_{k=1}^\infty \lambda_k \langle x, e_k\rangle f_k, f_n\right\rangle$$
$$= \sum_{k=1}^\infty \lambda_k \langle x, e_k\rangle \langle f_k, f_n\rangle = \lambda_n \langle x, e_n\rangle \|f_n\|^2,$$

12.2 Solutions

and then
$$U^*Ux = \sum_{n=1}^{\infty} |\lambda_n|^2 \|f_n\|^2 \langle x, e_n \rangle e_n \quad \forall x \in H_1.$$

Let us observe that $U^*U = P_{(a_1, a_2, ...)}$, where $a_n = |\lambda_n|^2 \|f_n\|^2 \ \forall n \in \mathbb{N}$. From exercise 22,
$$\sqrt{U^*U} = \sqrt{P_{(a_1, a_2, ...)}} = P_{(\sqrt{a_1}/\|e_1\|, \sqrt{a_2}/\|e_2\|, ...)},$$

i.e.,
$$|U|(x) = \sqrt{U^*U}(x) = \sum_{n=1}^{\infty} |\lambda_n| (\|f_n\| / \|e_n\|) \langle x, e_n \rangle e_n \quad \forall x \in H_1.$$

c) Suppose that U is compact. Since $(e_n / \|e_n\|)_{n \in \mathbb{N}}$ is an orthonormal system then using exercise 11(i) of chapter 10 it follows that $e_n / \|e_n\| \to 0 \ weak$. Since U is compact from exercise 38 of chapter 14 it follows that $U(e_n / \|e_n\|) \to 0$ in norm, i.e., $\|U(e_n / \|e_n\|)\| \to 0$. Now from the orthogonality of the sequence $(e_n)_{n \in \mathbb{N}}$ we have
$$U(e_n / \|e_n\|) = \lambda_n \langle e_n / \|e_n\|, e_n \rangle f_n = \lambda_n \|e_n\| f_n,$$

and then $|\lambda_n| \|e_n\| \|f_n\| \to 0$.

Conversely, let us suppose that $|\lambda_n| \|e_n\| \|f_n\| \to 0$. In this case, for every $n \in \mathbb{N}$ let $U_n : H_1 \to H_2$ be defined by $U_n(x) = \sum_{k=1}^{n} \lambda_k \langle x, e_k \rangle f_k$. We prove that $U_n \to U$ in norm and then it follows that U is compact. We have
$$U(x) - U_n(x) = \sum_{k=n+1}^{\infty} \lambda_k \langle x, e_k \rangle f_k \quad \forall x \in H_1,$$

and from (a),
$$\|U - U_n\| = \sup_{k \geq n+1} (|\lambda_k| \|f_k\| \|e_k\|) \to 0.$$

ii) Take $e_n = \chi_{(1/(n+1), 1/n)} \in L_2[0, 1] \ \forall n \in \mathbb{N}$ and $f_n = (0, ..., 0, 1, 0, ...) \in l_2$ (1 on the nth position), and then apply (i). We have
$$U^*(x) = \sum_{n=1}^{\infty} \overline{\lambda_n} \langle x, f_n \rangle e_n = \sum_{n=1}^{\infty} \overline{\lambda_n} x_n \chi_{(1/(n+1), 1/n)} \quad \forall x = (x_n)_{n \in \mathbb{N}} \in l_2,$$

and
$$|U|(f) = \sum_{n=1}^{\infty} |\lambda_n| (\|f_n\| / \|e_n\|) \langle f, e_n \rangle e_n$$
$$= \sum_{n=1}^{\infty} \left(\int_{1/(n+1)}^{1/n} f(x) \, dx \right) |\lambda_n| \sqrt{n^2 + n} \chi_{(1/(n+1), 1/n)} \quad \forall f \in L_2[0, 1].$$

iii) Take $e_n = \chi_{A_n} \in L_2(\mu_1)$, $f_n = \chi_{B_n} \in L_2(\mu_2)$ and then apply (i). We obtain that
$$\|U\| = \sup_{n \in \mathbb{N}} \left(|\lambda_n| \sqrt{\mu_1(A_n) \mu_2(B_n)} \right),$$

that
$$U^*f = \sum_{n=1}^{\infty} \overline{\lambda_n} \left(\int_{B_n} f d\mu_2 \right) \chi_{A_n} \quad \forall f \in L_2(\mu_2)$$
and
$$|U|(f) = \sqrt{U^*U}(f) = \sum_{n=1}^{\infty} \left(|\lambda_n| \sqrt{\mu_2(B_n)} / \sqrt{\mu_1(A_n)} \right) \left(\int_{A_n} f d\mu_1 \right) \chi_{A_n} \quad \forall f \in L_2(\mu_1).$$

24. i) Take $e_n = x_n$ and $f_n = (0, ..., 0, 1, 0, ...) \in l_2$ $\forall n \in \mathbb{N}$ (($f_n)_{n \in \mathbb{N}}$ is the standard basis in l_2), and then apply the part (i) from exercise 23.

We give a second proof for the fact that $\lambda_n x_n \to 0$ in norm implies U compact. Let $A = U(B_H) \subseteq l_2$ (A is bounded). To prove that A is relatively compact we use the characterization for relatively compact sets in the space l_2: $A \subseteq l_2$ is relatively compact if and only if $\forall \varepsilon > 0$ $\exists n_\varepsilon \in \mathbb{N}$ such that $\sum_{k=n}^{\infty} |p_k(a)|^2 \leq \varepsilon$ $\forall n \geq n_\varepsilon$, $\forall a \in A$, where $p_k : l_2 \to \mathbb{K}$ are the canonical projections. For $a = U(x) \in A$ we have $p_k(a) = p_k(Ux) = \lambda_k \langle x, x_k \rangle$. Then
$$\sum_{k=n}^{\infty} |p_k(Ux)|^2 = \sum_{k=n}^{\infty} |\lambda_k|^2 |\langle x, x_k \rangle|^2 = \sum_{k=n}^{\infty} |\lambda_k|^2 \|x_k\|^2 |\langle x, x_k / \|x_k\| \rangle|^2$$
$$\leq \left(\sup_{k \geq n} |\lambda_k| \|x_k\| \right)^2 \sum_{k=n}^{\infty} |\langle x, x_k / \|x_k\| \rangle|^2.$$

From the Bessel inequality
$$\sum_{k=n}^{\infty} |\langle x, x_k / \|x_k\| \rangle|^2 \leq \|x\|^2.$$

So
$$\sum_{k=n}^{\infty} |p_k(Ux)|^2 \leq \left(\sup_{k \geq n} |\lambda_k| \|x_k\| \right)^2 \|x\|^2 \quad \forall x \in H.$$
Since $A = U(B_H)$ it follows that
$$\sum_{k=n}^{\infty} |p_k(a)|^2 \leq \left(\sup_{k \geq n} |\lambda_k| \|x_k\| \right)^2 \quad \forall a \in A.$$
Since $|\lambda_n| \|x_n\| \to 0$ it follows that $\sup_{k \geq n} |\lambda_k| \|x_k\| \to 0$. Hence A is relatively compact, i.e., U is compact.

ii) Let $x_n = \chi_{(1/(n+1), 1/n)} \in L_2[0, 1]$. From exercise 23 we know that U is continuous if and only if $\left(n^\alpha \|x_n\|_{L_2[0,1]} \right)_{n \in \mathbb{N}}$ is bounded, that is, if and only if $\alpha \leq 1$. Also, U is compact if and only if $n^\alpha \|x_n\|_{L_2[0,1]} \to 0$, i.e., if and only if $n^{\alpha-1} \to 0$, that is, $\alpha < 1$.

25. i) If the series from the statement converges, it follows that for any $x \in H_1$ we have $\|\lambda_n \langle x, e_n \rangle f_n + \mu_n \langle x, e_{n+1} \rangle f_{n+1}\| \to 0$, i.e., by the Pythagoras theorem,
$$|\lambda_n|^2 |\langle x, e_n \rangle|^2 \|f_n\|^2 + |\mu_n|^2 |\langle x, e_{n+1} \rangle|^2 \|f_{n+1}\|^2 \to 0,$$

12.2 Solutions

and from here $\lambda_n \langle x, e_n \rangle \|f_n\| \to 0 \; \forall x \in H_1$, i.e., $\lambda_n \|f_n\| e_n \to 0 \; weak$. Now from the Uniform Boundedness Principle it follows that $M_1 = \sup_{n \in \mathbb{N}} (|\lambda_n| \|e_n\| \|f_n\|) < \infty$. In the same way we obtain that $M_2 = \sup_{n \in \mathbb{N}} (|\mu_n| \|e_{n+1}\| \|f_{n+1}\|) < \infty$.

For the converse, let $M = \max(M_1, M_2) < \infty$ and observe that for $n, p \geq 1$ and for any $x \in H_1$ we have

$$\sum_{k=n}^{n+p} (\lambda_k \langle x, e_k \rangle f_k + \mu_k \langle x, e_{k+1} \rangle f_{k+1}) = \lambda_n \langle x, e_n \rangle f_n$$
$$+ \sum_{k=n}^{n+p-1} (\mu_k \langle x, e_{k+1} \rangle + \lambda_{k+1} \langle x, e_{k+1} \rangle) f_{k+1}$$
$$+ \mu_{n+p} \langle x, e_{n+p+1} \rangle f_{n+p+1}.$$

From the Pythagoras theorem, we have that

$$\left\| \sum_{k=n}^{n+p} (\lambda_k \langle x, e_k \rangle f_k + \mu_k \langle x, e_{k+1} \rangle f_{k+1}) \right\|^2 = |\lambda_n|^2 |\langle x, e_n \rangle|^2 \|f_n\|^2$$
$$+ \sum_{k=n}^{n+p-1} |\mu_k + \lambda_{k+1}|^2 |\langle x, e_{k+1} \rangle|^2 \|f_{k+1}\|^2$$
$$+ |\mu_{n+p}|^2 |\langle x, e_{n+p+1} \rangle|^2 \|f_{n+p+1}\|^2$$
$$= \|\lambda_n\|^2 \|e_n\|^2 \|f_n\|^2 |\langle x, e_n/\|e_n\| \rangle|^2$$
$$+ \sum_{k=n}^{n+p-1} |\mu_k + \lambda_{k+1}|^2 \|e_{k+1}\|^2 \|f_{k+1}\|^2$$
$$|\langle x, e_{k+1}/\|e_{k+1}\| \rangle|^2$$
$$+ |\mu_{n+p}|^2 \|e_{n+p+1}\|^2 \|f_{n+p+1}\|^2$$
$$|\langle x, e_{n+p+1}/\|e_{n+p+1}\| \rangle|^2$$
$$\leq 4M^2 \sum_{k=n}^{n+p+1} |\langle x, e_k/\|e_k\| \rangle|^2.$$

The inequality between the first term and the last one is true for $p = 0$ also, and now the Bessel inequality shows that the series from the statement converges $\forall x \in H_1$. It is easy to see now that if $M < \infty$ then U is linear and continuous. Using what we have just proved, the series $\sum_{n=1}^{\infty} (\overline{\lambda_n} \langle x, f_n \rangle e_n + \overline{\mu_n} \langle x, f_{n+1} \rangle e_{n+1})$ is convergent $\forall x \in H_2$, and then U^* defined as in the statement makes sense, and is linear and continuous. We must prove that $\langle Ux, y \rangle = \langle x, U^*y \rangle \; \forall x \in H_1, \forall y \in H_2$. For this we use the separate continuity of the inner product. We omit the details (see exercise 23).

b) Suppose that U is compact. Then, as in the solution for exercise 23(i)(c), it follows that $|\lambda_n| \|e_n\| \|f_n\| \to 0$ and $|\mu_n| \|e_{n+1}\| \|f_{n+1}\| \to 0$.

Conversely, let us suppose that $|\lambda_n| \|e_n\| \|f_n\| \to 0$ and $|\mu_n| \|e_{n+1}\| \|f_{n+1}\| \to 0$. Then

$|\mu_n + \lambda_{n+1}| \, \|e_{n+1}\| \, \|f_{n+1}\| \to 0$. In this case, let $U_n, V_n : H_1 \to H_2$ be defined by

$$U_n(x) = \sum_{k=1}^{n} (\lambda_k \langle x, e_k \rangle f_k + \mu_k \langle x, e_{k+1} \rangle f_{k+1}) = \lambda_1 \langle x, e_1 \rangle f_1$$
$$+ \sum_{k=1}^{n-1} (\mu_k \langle x, e_{k+1} \rangle f_{k+1} + \lambda_{k+1} \langle x, e_{k+1} \rangle f_{k+1})$$
$$+ \mu_n \langle x, e_{n+1} \rangle f_{n+1},$$

and

$$V_n(x) = \lambda_1 \langle x, e_1 \rangle f_1 + \sum_{k=1}^{n-1} (\mu_k \langle x, e_{k+1} \rangle f_{k+1} + \lambda_{k+1} \langle x, e_{k+1} \rangle f_{k+1}).$$

Since $\|U_n - V_n\| = |\mu_n| \, \|e_{n+1}\| \, \|f_{n+1}\| \to 0$ and $U_n(x) \to U(x) \,\forall x \in H_1$ it follows that $V_n(x) \to U(x) \,\forall x \in H_1$.

Now from the Pythagoras theorem and the Bessel inequality we have

$$\|V_{n+p}(x) - V_n(x)\|^2 = \sum_{k=n+1}^{n+p} |\mu_k + \lambda_{k+1}|^2 |\langle x, e_{k+1} \rangle|^2 \|f_{k+1}\|^2$$
$$= \sum_{k=n+1}^{n+p} |\mu_k + \lambda_{k+1}|^2 \|f_{k+1}\|^2 \|e_{k+1}\|^2 \left| \left\langle x, \frac{e_{k+1}}{\|e_{k+1}\|} \right\rangle \right|^2$$
$$\leq \sup_{k \geq n+1} (|\mu_k + \lambda_{k+1}| \, \|f_{k+1}\| \, \|e_{k+1}\|)^2 \sum_{k=n+1}^{n+p} \left| \left\langle x, \frac{e_{k+1}}{\|e_{k+1}\|} \right\rangle \right|^2$$
$$\leq \sup_{k \geq n+1} (|\mu_k + \lambda_{k+1}| \, \|f_{k+1}\| \, \|e_{k+1}\|)^2 \|x\|^2.$$

Passing to the limit for $p \to \infty$ we obtain that

$$\|U(x) - V_n(x)\|^2 \leq \sup_{k \geq n+1} (|\mu_k + \lambda_{k+1}| \, \|f_{k+1}\| \, \|e_{k+1}\|)^2 \|x\|^2,$$

i.e.,

$$\|U - V_n\| \leq \sup_{k \geq n+1} (|\mu_k + \lambda_{k+1}| \, \|f_{k+1}\| \, \|e_{k+1}\|) \to 0.$$

Since $\{V_n\}$ are finite rank operators and $V_n \to U$ in norm it follows that U is compact.

ii) Let $e_n = \chi_{[n,n+1)}$, $f_n = (0, ...0, 1, 0, ...) \in l_2$ (1 on the nth position) $\forall n \in \mathbb{N}$. Take $\mu_n = \lambda_n \,\forall n \in \mathbb{N}$. We have $\|e_n\| = 1$, $\|f_n\| = 1 \,\forall n \in \mathbb{N}$, and we apply (i).

26. i) a) Let $M > 0$ be such that $|\varphi(x)| \leq M$ a.e.. Then $|\varphi f|^2 \leq M^2 |f|^2$, and as φf is measurable and $|f|^2 \in L_1[0,1]$, it follows that $|\varphi f|^2 \in L_1[0,1]$, that is, $\varphi f \in L_2[0,1]$, and therefore U is well defined. Moreover,

$$\|M_\varphi f\|^2 = \int_0^1 |M_\varphi f(x)|^2 \, dx = \int_0^1 |\varphi(x)|^2 |f(x)|^2 \, dx \leq M^2 \|f\|^2,$$

i.e., $\|M_\varphi f\| \leq M \|f\| \,\forall f \in L_2[0,1]$, thus M_φ is linear and continuous, and $\|M_\varphi\| \leq M$.

12.2 Solutions

Let us calculate the norm for M_φ. First, if $s \in L_\infty[0,1]$ is a step function, i.e., $s = \alpha_1 \chi_{A_1} + \cdots + \alpha_n \chi_{A_n}$ where $\{A_1, ..., A_n\} \subseteq [0,1]$ are pairwise disjoint sets and $\mu(A_j) > 0$ for $1 \leq j \leq n$, then $\|M_s\| = \|s\|_\infty$. Indeed, we have $M_s(f) = \alpha_1 \chi_{A_1} f + \cdots + \alpha_n \chi_{A_n} f$ and

$$\|M_s(f)\|^2 = \sum_{k=1}^n |\alpha_k|^2 \int_{A_k} |f(x)|^2 \, dx \quad \forall f \in L_2[0,1].$$

For any $1 \leq j \leq n$ we have

$$\|M_s(\alpha_j \chi_{A_j})\|^2 = \sum_{k=1}^n |\alpha_k|^2 \int_{A_k} |\alpha_j|^2 \chi_{A_j} dx = |\alpha_j|^4 \mu(A_j)$$
$$\leq \|M_s\|^2 \|\alpha_j \chi_{A_j}\|^2 = \|M_s\|^2 |\alpha_j|^2 \mu(A_j),$$

i.e., $|\alpha_j|^2 \leq \|M_s\|^2$, hence $\|s\|_\infty = \max_{1 \leq j \leq n} |\alpha_j| \leq \|M_s\|$. We have proved above the reverse inequality, $\|M_s\| \leq \|s\|_\infty$, and then $\|M_s\| = \|s\|_\infty$ for any measurable step function s. But if $\varphi \in L_\infty[0,1]$ then there is a sequence of step functions $(s_n)_{n \in \mathbb{N}}$ such that $s_n \to \varphi$, μ-almost uniformly, i.e., $\|\varphi - s_n\|_\infty \to 0$ (see also the solution for exercise 11 of chapter 14). Then $\|M_\varphi - M_{s_n}\| = \|M_{\varphi - s_n}\| \leq \|\varphi - s_n\|_\infty \; \forall n \in \mathbb{N}$, hence $M_{s_n} \to M_\varphi$, and then $\|M_{s_n}\| \to \|M_\varphi\|$. But $\|M_{s_n}\| = \|s_n\|_\infty \; \forall n \in \mathbb{N}$, hence $\|s_n\|_\infty \to \|M_\varphi\|$. But $\|s_n\|_\infty \to \|\varphi\|_\infty$, and then $\|M_\varphi\| = \|\varphi\|_\infty$.

We have $M_\varphi^* = M_{\overline{\varphi}}$ if and only if $\langle M_\varphi f, g \rangle = \langle f, M_{\overline{\varphi}} g \rangle \; \forall f, g \in L_2[0,1]$, or $\int_0^1 (M_\varphi f)(x) \overline{g(x)} dx = \int_0^1 f(x) \overline{(M_{\overline{\varphi}} g)(x)} dx$, i.e., using the expression for M_φ and $M_{\overline{\varphi}}$, if and only if

$$\int_0^1 \varphi(x) f(x) \overline{g(x)} dx = \int_0^1 f(x) \varphi(x) \overline{g(x)} dx,$$

which is clearly true.

M_φ is self-adjoint if and only if $M_{\overline{\varphi}} = M_\varphi$. Let us suppose that M_φ is self-adjoint. As $[0,1]$ has finite measure the constant function 1 is in $L_2[0,1]$ and it follows that $M_{\overline{\varphi}}(1) = M_\varphi(1)$, i.e., $\overline{\varphi(x)} = \varphi(x)$ a.e., that is, φ takes real values a.e.. Conversely, if φ takes real values a.e. then $\overline{\varphi(x)} = \varphi(x)$ a.e., and then $\overline{\varphi(x)} f(x) = \varphi(x) f(x)$ a.e., i.e., $M_{\overline{\varphi}}(f) = M_\varphi(f) \; \forall f \in L_2[0,1]$, $M_{\overline{\varphi}} = M_\varphi$.

b) If $\varphi \geq 0$ a.e. then since $|f(x)|^2 \geq 0$ a.e. $\forall f \in L_2[0,1]$ it follows that $\varphi(x)|f(x)|^2 \geq 0$ a.e., and from the properties for the Lebesgue integral we have that $\int_0^1 \varphi(x) |f(x)|^2 \, dx \geq 0$, i.e., $\langle M_\varphi f, f \rangle \geq 0 \; \forall f \in L_2[0,1]$, that is, $M_\varphi \geq 0$.

Conversely, if $M_\varphi \geq 0$ then $\langle M_\varphi f, f \rangle \geq 0 \; \forall f \in L_2[0,1]$, i.e., $\int_0^1 \varphi(x) |f(x)|^2 \geq 0$ $\forall f \in L_2[0,1]$. Let $A = [\varphi < 0] \subseteq [0,1]$. Since $[0,1]$ has finite measure, $\chi_A \in L_2[0,1]$, and it follows that $\int_0^1 \varphi \chi_A dx \geq 0$, that is, $\int_0^1 (-\varphi) \chi_A dx \leq 0$. Since $-\varphi \chi_A \geq 0$ it follows that $\int_0^1 (-\varphi \chi_A) \, dx \geq 0$, and therefore $\int_0^1 (-\varphi \chi_A) \, dx = 0$. But $-\varphi \chi_A \geq 0$, so we obtain that $-\varphi \chi_A = 0$ a.e., i.e., $\varphi \chi_A = 0$ a.e., and therefore $\varphi \geq 0$ a.e..

ii) For this new context what we must justify is the fact that if $M_{\overline{\varphi}} = M_\varphi$ then φ takes real values a.e., and that if $M_\varphi \geq 0$ then $\varphi \geq 0$ a.e.. Since the measure space (S, Σ, μ) is σ-finite there is a sequence $(A_n)_{n \in \mathbb{N}} \subseteq \Sigma$ such that $S = \bigcup_{n \in \mathbb{N}} A_n$ and $\mu(A_n) < \infty \; \forall n \in \mathbb{N}$. Suppose that $M_{\overline{\varphi}} = M_\varphi$. We have $\chi_{A_n} \in L_2(\mu)$, from whence $M_{\overline{\varphi}}(\chi_{A_n}) = M_\varphi(\chi_{A_n})$, i.e., $\overline{\varphi} \chi_{A_n} = \varphi \chi_{A_n}$ a.e.. Let $E_n \in \Sigma$ with $\mu(E_n) = 0$ such that $\overline{\varphi(x)} \chi_{A_n}(x) = \varphi(x) \chi_{A_n}(x)$

$\forall x \in C_{E_n}$. Then for $E = \bigcup_{n \in \mathbb{N}} E_n$ we have $\mu(E) = 0$, and $\forall x \in S \backslash E$ there is $n \in \mathbb{N}$ such that $x \in A_n \backslash E$, and then $x \in \overline{A_n \backslash E_n}$. This implies that $\overline{\varphi(x) \chi_{A_n}}(x) = \varphi(x) \chi_{A_n}(x)$, and as $x \in A_n$, $\chi_{A_n}(x) = 1$, thus $\overline{\varphi(x)} = \varphi(x)$, i.e., $\varphi(x) \in \mathbb{R}$. So $\mu(E) = 0$ and $\forall x \in S \backslash E$, $\varphi(x) \in \mathbb{R}$, i.e. φ takes real values a.e.. Let us suppose now that $M_\varphi \geq 0$ and denote $A = [\varphi < 0]$. Then $A \in \Sigma$ and $A \cap A_n \subseteq A_n$, from whence $\mu(A \cap A_n) \leq \mu(A_n) < \infty$, i.e., $\chi_{A \cap A_n} \in L_2(\mu)$. Since $M_\varphi \geq 0$, then $\langle M_\varphi(\chi_{A \cap A_n}), \chi_{A \cap A_n} \rangle \geq 0$, i.e., $\int_S \varphi \chi_{A \cap A_n} d\mu \geq 0$, that is, $\int_S (-\varphi) \chi_{A \cap A_n} d\mu \leq 0$. Since $(-\varphi) \chi_{A \cap A_n} \geq 0$ it follows that $\int_S (-\varphi) \chi_{A \cap A_n} d\mu \geq 0$, and therefore $\int_S (-\varphi) \chi_{A \cap A_n} d\mu = 0$. Since $(-\varphi) \chi_{A \cap A_n} \geq 0$, it follows that $(-\varphi) \chi_{A \cap A_n} = 0$ a.e., i.e., $\varphi \chi_{A \cap A_n} = 0$ a.e., and as above we obtain that $\varphi \chi_A = 0$ a.e., and therefore $\varphi \geq 0$ a.e..

iii) Since $\varphi \geq 0$ a.e., then from (ii) $M_\varphi \geq 0$ and $M_{\sqrt{\varphi}} \geq 0$. We have $M_{\sqrt{\varphi}}^2 = M_\varphi$, hence using the uniqueness of the square root it follows that $\sqrt{M_\varphi} = M_{\sqrt{\varphi}}$, i.e.,

$$\left(\sqrt{M_\varphi} f\right)(x) = \sqrt{\varphi(x)} f(x) \quad \forall f \in L_2(\mu).$$

27. See [40, exercise 18.20].

'(i) $\Rightarrow$ (ii)' Suppose that $Ax = 0$. Then for any $y \in H$ we have $\langle Ax, y \rangle = 0$, and using that A is self-adjoint we obtain that $\langle x, Ay \rangle = 0$, i.e., $x \perp A(H)$, and then $x \perp \overline{A(H)}$. Using that $A(H) \subseteq H$ is dense we deduce that $x \perp H$, hence $x = 0$.

'(ii) $\Rightarrow$ (iii)' Since A is positive, let $\sqrt{A} = B : H \to H$ be the square root. If $x \in H$ has the property that $\langle Ax, x \rangle = 0$ then $\langle B^2 x, x \rangle = 0$, i.e., $\langle Bx, Bx \rangle = 0$, $\|Bx\|^2 = 0$, that is, $Bx = 0$. It follows that $Ax = B(Bx) = 0$, i.e., $x \in \ker A$, and by (ii) $x = 0$.

'(iii) $\Rightarrow$ (i)' If $A(H) \subseteq H$ is not dense then there is $x \in H \backslash \{0\}$ such that $x \perp A(H)$. In particular $x \perp Ax$, i.e., $\langle Ax, x \rangle = 0$. From (iii) it follows that $x = 0$, a contradiction.

28. i) Since T is positive let $S = \sqrt{T} \in L(H)$. Let $\varepsilon > 0$ be such that $T - \varepsilon I$ is positive definite. Then $\langle Tx, x \rangle \geq \varepsilon \|x\|^2$ $\forall x \in H$, hence $\langle S^2 x, x \rangle \geq \varepsilon \|x\|^2$ $\forall x \in H$, i.e., $\langle Sx, Sx \rangle \geq \varepsilon \|x\|^2$ $\forall x \in H$. We obtain that $\|Sx\| \geq \sqrt{\varepsilon} \|x\|$ $\forall x \in H$. From here it follows that S is injective and that $S(H) \subseteq H$ is closed. Indeed, if $Sx = 0$ then $x = 0$, and if $(Sx_n)_{n \in \mathbb{N}}$ converges then $(x_n)_{n \in \mathbb{N}}$ is a Cauchy sequence, which implies that $(x_n)_{n \in \mathbb{N}}$ is a Cauchy sequence, hence there is $x \in H$ such that $x_n \to x \in H$, which implies that $Sx_n \to Sx$. We prove now that S is surjective. If $x \in (S(H))^\perp$, then $\langle x, Sy \rangle = 0$ $\forall y \in H$, hence $\langle Sx, y \rangle = 0$ $\forall y \in H$, i.e., $Sx = 0$, and then $x = 0$. It follows that $(S(H))^\perp = \{0\}$, and since $S(H) \subseteq H$ is a closed linear subspace from the projection theorem it follows that $S(H) = H$. Since $S \in L(H)$ is bijective, from $S^2 = T$ it follows that T is bijective.

If T is bijective then S is bijective. Then from the inverse mapping theorem (see chapter 9) there is $m > 0$ such that $\|S^{-1} y\| \leq m \|y\|$ $\forall y \in H$, i.e., $\|x\|/m \leq \|Sx\|$ $\forall x \in H$. Denoting by $\varepsilon = 1/(2\sqrt{m})$ it follows that $\|Sx\|^2 > \varepsilon \|x\|^2$ $\forall x \in H \backslash \{0\}$. Then $\langle Sx, Sx \rangle > \varepsilon \|x\|^2$ $\forall x \in H \backslash \{0\}$, hence $\langle S^2 x, x \rangle > \varepsilon \|x\|^2$ $\forall x \in H \backslash \{0\}$, i.e., $\langle Tx, x \rangle > \varepsilon \|x\|^2$ $\forall x \in H \backslash \{0\}$, that is, $T - \varepsilon I$ is positive definite.

ii) If T is bijective it follows that T^* is also bijective. Indeed, $TT^{-1} = T^{-1} T = I$ implies that $(T^{-1})^* T^* = T^* (T^{-1})^* = I$, and therefore T^* is invertible. Hence $T^* T$ and TT^* are bijective. Since these operators are positive, from (i) it follows that we can find $\varepsilon_1, \varepsilon_2 > 0$ such that $T^* T - \varepsilon_1 I$ and $TT^* - \varepsilon_2 I$ are positive definite. Then for $\varepsilon = \min(\varepsilon_1, \varepsilon_2) > 0$ it follows that $T^* T - \varepsilon I$ and $TT^* - \varepsilon I$ are positive definite.

12.2 Solutions

Conversely, if there is $\varepsilon > 0$ such that $T^*T - \varepsilon I$ and $TT^* - \varepsilon I$ are positive definite, then by (i), T^*T and TT^* are bijective. T^*T being injective implies that T is injective, and TT^* being surjective implies that T is surjective. Hence T is bijective.

29. i) a) Since A is positive we have $\langle x, x \rangle_A = \langle Ax, x \rangle \geq 0 \ \forall x \in H$. From A being self-adjoint we have

$$\langle x, y \rangle_A = \langle Ax, y \rangle = \langle x, Ay \rangle = \overline{\langle Ay, x \rangle} = \overline{\langle y, x \rangle_A}.$$

The linearity of A implies that

$$\langle \alpha x + \beta y, z \rangle_A = \langle A(\alpha x + \beta y), z \rangle = \alpha \langle Ax, z \rangle + \beta \langle Ay, z \rangle_A = \alpha \langle x, z \rangle_A + \beta \langle y, z \rangle_A.$$

Therefore $\langle \cdot, \cdot \rangle_A$ is an inner product if and only if $\langle x, x \rangle_A = 0$ implies $x = 0$.

b) We will use the inverse mapping theorem (see chapter 9). Since A is linear and continuous,

$$\|x\|_A = \sqrt{\langle x, x \rangle_A} = \sqrt{\langle Ax, x \rangle} \leq \sqrt{\|A\|} \|x\| \ \forall x \in H,$$

i.e., the identity $I : (H, \|\cdot\|) \to (H, \|\cdot\|_A)$ is linear, continuous, and obviously bijective. Suppose that $(H, \langle \cdot, \cdot \rangle_A)$ is a Hilbert space. Since $(H, \langle \cdot, \cdot \rangle)$ is a Hilbert space, from the inverse mapping theorem it follows that $I^{-1} : (H, \|\cdot\|_A) \to (H, \|\cdot\|)$ is linear and continuous, i.e., there is $c > 0$ such that $\|x\| \leq c \|x\|_A \ \forall x \in H$, i.e., $\langle Ax, x \rangle \geq \|x\|^2 / c^2$ $\forall x \in H$. Conversely, suppose that there is $c > 0$ such that $\langle Ax, x \rangle \geq c \|x\|^2 \ \forall x \in H$, i.e. $\|x\| \leq \|x\|_A / \sqrt{c} \ \forall x \in H$. Then the two norms on H are equivalent. Since $\|\cdot\|$ is complete, from exercise 31(i) of chapter 1 it follows that $\|\cdot\|_A$ is also complete, hence $(H, \langle \cdot, \cdot \rangle_A)$ is a Hilbert space.

ii) a) Let $M_a : l_2 \to l_2$ be the multiplication operator given by a. Using exercise 14(ii) we have $M_a \in \mathcal{A}_+(l_2)$. Then

$$\langle x, y \rangle_{M_a} = \langle M_a x, y \rangle = \sum_{n=1}^{\infty} a_n x_n \overline{y_n} = \langle x, y \rangle_a.$$

From (i) we have that $\langle \cdot, \cdot \rangle_a$ is an inner product on l_2 if and only if from $\langle M_a x, x \rangle = 0$ it follows that $x = 0$.

Let us suppose that $\langle \cdot, \cdot \rangle_a$ is an inner product on l_2. Suppose that there is k such that $a_k = 0$. Then for $x = e_k$ we have $\langle M_a e_k, e_k \rangle = a_k = 0$, and since $\langle \cdot, \cdot \rangle_a$ is an inner product on l_2 it follows that $e_k = 0$, which is false. Hence $a_n \neq 0 \ \forall n \in \mathbb{N}$, and as $a_n \geq 0 \ \forall n \in \mathbb{N}$ it follows that $a_n > 0 \ \forall n \in \mathbb{N}$.

Conversely, suppose that $a_n > 0 \ \forall n \in \mathbb{N}$ and let $x \in l_2$ be such that $\langle M_a x, x \rangle = 0$, i.e., $\sum_{n=1}^{\infty} a_n |x_n|^2 = 0$. Then $a_n |x_n|^2 = 0 \ \forall n \in \mathbb{N}$, and as $a_n > 0 \ \forall n \in \mathbb{N}$ it follows that $x_n = 0$ $\forall n \in \mathbb{N}$, i.e., $x = 0$.

b) By (i) we know that $(l_2, \langle \cdot, \cdot \rangle_a)$ is a Hilbert space if and only if there is $c > 0$ such that $\langle M_a x, x \rangle \geq c \|x\|^2 \ \forall x \in l_2$. If the last inequality is true, then $\langle M_a e_n, e_n \rangle \geq c \|e_n\|^2$ $\forall n \in \mathbb{N}$, i.e., $a_n \geq c \ \forall n \in \mathbb{N}$, hence $\inf_{n \in \mathbb{N}} a_n \geq c > 0$. Conversely, if $c = \inf_{n \in \mathbb{N}} a_n > 0$ then for any $x \in l_2$ we have

$$\langle M_a x, x \rangle = \sum_{n=1}^{\infty} a_n |x_n|^2 \geq \left(\inf_{n \in \mathbb{N}} a_n \right) \sum_{n=1}^{\infty} |x_n|^2 = c \|x\|^2.$$

iii) We will use the following well known result from measure theory: for $g \in L_1(\mu)$ if $\int_S gf d\mu \leq 0 \ \forall f \in L_1(\mu)$ with $f \geq 0$, then $g \leq 0$ μ-a.e., and conversely, if $g \leq 0$ μ-a.e. then $\int_S gf d\mu \leq 0 \ \forall f \in L_1(\mu)$ with $f \geq 0$. Indeed, let $E \in \Sigma$ with $\mu(E) < \infty$, and take $h = g\chi_E$. Then $[h > 0] = \bigcup_{n \in \mathbb{N}} A_n$ where $A_n = E \cap [g \geq 1/n] \subseteq E$. We have $\chi_{A_n}/n \leq h\chi_{A_n}$. Since $\mu(A_n) \leq \mu(E) < \infty$ it follows that $\chi_{A_n} \in L_1(\mu)$, and then by hypothesis $\int_S h\chi_{A_n} d\mu \leq 0$, from whence $\int_S (\chi_{A_n}/n) d\mu \leq 0$, i.e., $\mu(A_n)/n \leq 0$, thus $\mu(A_n) = 0 \ \forall n \in \mathbb{N}$. Since $[h > 0] = \bigcup_{n \in \mathbb{N}} A_n$, it follows that $\mu([h > 0]) = 0$, i.e., $\mu([g\chi_E > 0]) = 0$. Now since (S, Σ, μ) is a σ-finite measure space there is a sequence $(E_k)_{k \in \mathbb{N}} \subseteq \Sigma$ such that $E_k \nearrow S$ and $\mu(E_k) < \infty \ \forall k \in \mathbb{N}$. From the above it follows that $\mu([g\chi_{E_k} > 0]) = 0 \ \forall k \in \mathbb{N}$. We have $[g > 0] = \bigcup_{k \in \mathbb{N}} [g\chi_{E_k} > 0]$, thus $\mu([g > 0]) = 0$, i.e., $g \leq 0$ μ-a.e.. The converse is obvious.

Suppose that $\langle \cdot, \cdot \rangle_\varphi$ is an inner product on $L_2(\mu)$. Let $A = \{s \in S \mid \varphi(s) = 0\}$. Since (S, Σ, μ) is a σ-finite measure space we can find $E_n \in \Sigma$ with $\mu(E_n) < \infty \ \forall n \in \mathbb{N}$ such that $S = \bigcup_{n \in \mathbb{N}} E_n$. For each $n \in \mathbb{N}$, $\chi_{A \cap E_n} \in L_2(\mu)$, and clearly

$$\langle \chi_{A \cap E_n}, \chi_{A \cap E_n} \rangle_\varphi = \int_{A \cap E_n} \varphi(s) d\mu(s) = 0.$$

Since $\langle \cdot, \cdot \rangle_\varphi$ is an inner product on $L_2(\mu)$, it follows that $\chi_{A \cap E_n} = 0$, i.e., $\mu(A \cap E_n) = 0$ $\forall n \in \mathbb{N}$, hence $\mu(A) = 0$, i.e., $\varphi(x) > 0$, μ-a.e..

Conversely, if $\varphi(x) > 0$, μ–a.e., then with the same notations as above if we take $h : S \to (0, \infty)$, $h(s) = \begin{cases} \varphi(s) & \text{if } s \notin A, \\ 1 & \text{if } s \in A, \end{cases}$ we have $h = \varphi$, μ-a.e., and hence $\langle \cdot, \cdot \rangle_\varphi = \langle \cdot, \cdot \rangle_h$. From (i) $\langle \cdot, \cdot \rangle_\varphi$ is an inner product on $L_2(\mu) \Leftrightarrow$ from $\langle M_\varphi f, f \rangle = 0$ it follows $f = 0$. If $f \in L_2(\mu)$ has the property that $\langle M_\varphi f, f \rangle = 0$, then $\int_S h|f|^2 d\mu = 0$, and then $h|f|^2 = 0$, μ-a.e., and using that $h(x) > 0 \ \forall x \in S$ it follows that $|f| = 0$, μ-a.e., i.e., $f = 0$ in $L_2(\mu)$.

From (i) $(L_2(\mu), \langle \cdot, \cdot \rangle_\varphi)$ is a Hilbert space if and only if there is $c > 0$ such that $\langle M_\varphi f, f \rangle \geq c \langle f, f \rangle \ \forall f \in L_2(\mu)$, that is, $\int_S (c - \varphi)|f|^2 d\mu \leq 0 \ \forall f \in L_2(\mu)$. Using the above proved remark this is equivalent to $c - \varphi \leq 0$, μ-a.e., i.e., $\varphi \geq c$, μ-a.e..

30. i) From exercise 22 it follows that U_n is well defined, linear, and continuous, with $\|U_n\| = \sup_{k \geq n} a_k \ \forall n \in \mathbb{N}$, and that U is self-adjoint. By the separate continuity of the inner product it follows that $\langle U_n(x), x \rangle = \sum_{k=n}^{\infty} a_k |\langle x, e_k \rangle|^2 \ \forall x \in H$. Since $a_n \geq 0 \ \forall n \in \mathbb{N}$ we have $\langle U_n(x), x \rangle \geq 0$ and $\langle U_n(x), x \rangle \geq \langle U_{n+1}(x), x \rangle \ \forall x \in H$. For any $x \in H$ by the Pythagoras theorem we obtain that

$$\|U_n(x)\|^2 = \sum_{k=n}^{\infty} a_k^2 |\langle x, e_k \rangle|^2 \leq \left(\sup_{k \in \mathbb{N}} a_k^2 \right) \sum_{k=n}^{\infty} |\langle x, e_k \rangle|^2 \to 0,$$

by the Bessel inequality, and so from the Weierstrass convergence theorem, $\inf_{n \in \mathbb{N}} U_n = 0$.

ii) '(a) $\Rightarrow$ (b)' Suppose that (b) is not true, i.e., H is infinite-dimensional. Then there is an infinite linearly independent system in H, and hence by the Gram–Schmidt procedure we can find an orthonormal system $(e_n)_{n\in\mathbb{N}} \subseteq H$. We define now $U_n : H \to H$, $U_n(x) = \sum_{k=n}^{\infty} \langle x, e_k \rangle e_k$. Then by (i) we have $0 \leq U_{n+1} \leq U_n$ $\forall n \in \mathbb{N}$, $\inf_{n\in\mathbb{N}} U_n = 0$, and $\|U_n\| = 1$ $\forall n \in \mathbb{N}$, hence U_n does not converge towards 0, which contradicts (a).

'(b) $\Rightarrow$ (a)' Suppose that H is finite-dimensional and let $(U_n)_{n\in\mathbb{N}} \subseteq \mathcal{A}_+(H)$, with $U_n \searrow 0$. Let $\varepsilon > 0$ be fixed. For any $x \in B_H$ by hypothesis there is $n \in \mathbb{N}$ (depending on ε and x) such that $\|U_n(x)\| < \varepsilon$. Then $\langle U_n(x), x \rangle < \varepsilon$, and therefore

$$B_H \subseteq \bigcup_{n\in\mathbb{N}} \{x \in H \mid \langle U_n(x), x \rangle < \varepsilon\}.$$

Moreover, the continuity for the operators U_n ensures that the sets from the right side are open. Since H is finite-dimensional B_H is compact, and so we can extract a finite cover, i.e., there is $n \in \mathbb{N}$ (depending on ε) such that

$$B_H \subseteq \bigcup_{k=1}^{n} \{x \in H \mid \langle U_k(x), x \rangle < \varepsilon\}.$$

Let $p \geq n$. For any $x \in B_H$ there is $1 \leq k \leq n$ such that $\langle U_k(x), x \rangle < \varepsilon$. Then $p \geq k$ and since $(U_n)_{n\in\mathbb{N}}$ is decreasing we have $\langle U_p(x), x \rangle \leq \langle U_k(x), x \rangle$, from whence $\langle U_p(x), x \rangle < \varepsilon$. Since $x \in B_H$ is arbitrary we obtain that $\|U_p\| = \sup_{\|x\|\leq 1} \langle U_p(x), x \rangle \leq \varepsilon$, i.e., $U_n \to 0$ in norm.

31. We have $(I - T^*T)^* = I - T^*(T^*)^* = I - T^*T$, hence $I - T^*T$ is self-adjoint, and for any $x \in H$ we have

$$\langle (I - T^*T)x, x \rangle = \langle x, x \rangle - \langle T^*Tx, x \rangle = \langle x, x \rangle - \langle Tx, Tx \rangle = \|x\|^2 - \|Tx\|^2 \geq 0,$$

since $\|T\| \leq 1$. Hence $I - T^*T$ is positive and we can define the operator $D_T = \sqrt{I - T^*T}$. Then

$$\|D_T x\|^2 = \langle D_T x, D_T x \rangle = \langle D_T^2 x, x \rangle = \langle (I - T^*T)x, x \rangle = \|x\|^2 - \|Tx\|^2,$$

i.e., $\|x\|^2 = \|Tx\|^2 + \|D_T x\|^2$ $\forall x \in H$. Then (iii) follows from (ii).

From here we deduce that if $\|T\| \leq 1$ then the space H can be written as $H = \ker(D_T) \oplus (\ker(D_T))^\perp$ such that on $\ker(D_T)$ the operator T is an isometry and for any $x \in (\ker(D_T))^\perp = \overline{D_T(H)}$ non-null we have $\|x\| > \|Tx\|$.

32. See [40, exercise 18.54].

From the ideal property for the class of compact operators, if $\sqrt{A}$ is compact, since $A = \sqrt{A}^2$ it follows that A is compact. Conversely, suppose that A is compact. Let $(x_n)_{n\in\mathbb{N}} \subseteq B_H$. Since A is compact there is a subsequence $(x_{k_n})_{n\in\mathbb{N}}$ for the sequence $(x_n)_{n\in\mathbb{N}}$ such that $(Ax_{k_n})_{n\in\mathbb{N}}$ converges. Using now that the square root of a positive operator is self-adjoint, we have

$$\left\|\sqrt{A}(x)\right\|^2 = \left\langle \sqrt{A}(x), \sqrt{A}(x) \right\rangle = \left\langle \sqrt{A}^2(x), x \right\rangle = \langle A(x), x \rangle \leq \|A(x)\| \|x\|$$

for all $x \in H$, and then

$$\left\|\sqrt{A}\left(x_{k_m}\right) - \sqrt{A}\left(x_{k_n}\right)\right\|^2 \leq \|A(x_{k_m} - x_{k_n})\| \|x_{k_m} - x_{k_n}\| \leq 2\|A(x_{k_m} - x_{k_n})\|.$$

Since the sequence $(Ax_{k_n})_{n\in\mathbb{N}}$ is Cauchy from the last inequality it follows that the sequence $(\sqrt{A}x_{k_n})_{n\in\mathbb{N}}$ is Cauchy, hence convergent, i.e., $\sqrt{A}$ is compact.

Suppose that A is compact. From the ideal property for the class of compact operators A^*A is compact, and then $\sqrt{A^*A}$ is compact, i.e., $|A|$ is compact. Conversely, suppose that $|A|$ is compact. Let $(U, |A|)$ be the polar decomposition of A. Then $A = U|A|$. Using again the ideal property it follows that $U|A|$ is compact, i.e., A is compact.

If A is compact then $|A|$ is compact, and using that $K(H)$ is a linear subspace it follows that $A_+ = (|A| + A)/2$ and $A_- = (|A| - A)/2$ are compact. Conversely, if A_+ and A_- are compact operators then $A = A_+ - A_-$ is compact.

33. See [40, exercise 18.55].

Since A is positive, let $\sqrt{A} = C$. Let $(x_n)_{n\in\mathbb{N}} \subseteq B_H$. Since B is compact there is a subsequence $(x_{k_n})_{n\in\mathbb{N}}$ for the sequence $(x_n)_{n\in\mathbb{N}}$ such that $(Bx_{k_n})_{n\in\mathbb{N}}$ converges. From the Cauchy–Buniakowski–Schwarz inequality we have $\langle Bx, x \rangle \leq \|Bx\| \|x\|$, and then

$$\langle Bx_{k_m} - Bx_{k_n}, x_{k_m} - x_{k_n}\rangle \leq \|Bx_{k_m} - Bx_{k_n}\|\|x_{k_m} - x_{k_n}\| \leq 2\|Bx_{k_m} - Bx_{k_n}\|.$$

As

$$\langle A(x_{k_m} - x_{k_n}), x_{k_m} - x_{k_n}\rangle \leq \langle B(x_{k_m} - x_{k_n}), x_{k_m} - x_{k_n}\rangle = \langle Bx_{k_m} - Bx_{k_n}, x_{k_m} - x_{k_n}\rangle,$$

we obtain that $\langle A(x_{k_m} - x_{k_n}), x_{k_m} - x_{k_n}\rangle \leq 2\|Bx_{k_m} - Bx_{k_n}\|$, so

$$\langle C(x_{k_m} - x_{k_n}), C(x_{k_m} - x_{k_n})\rangle \leq 2\|Bx_{k_m} - Bx_{k_n}\|,$$

that is, $\|Cx_{k_m} - Cx_{k_n}\|^2 \leq 2\|Bx_{k_m} - Bx_{k_n}\|$. From this it follows that the sequence $(Cx_{k_n})_{n\in\mathbb{N}}$ is Cauchy, hence convergent. We obtain that the operator $\sqrt{A}$ is compact. Since $A = (\sqrt{A})^2$ it follows that A is also compact.

34. For the first question the answer is: not necessarily. Let $H = l_2$ and for $n \in \mathbb{N}$ let $A_n \in L(l_2)$ be defined by $A_n((x_1, x_2, ..)) = (0, 0, ..., 0, x_1, x_2, ...)$, where 0 appears on the first n positions. Then $\|A_n\| = 1 \ \forall n \in \mathbb{N}$. Now let $x = (x_k)_{k\in\mathbb{N}}$ and $y = (y_k)_{k\in\mathbb{N}}$ in l_2. Then

$$|\langle A_n x, y\rangle| \leq \left(\sum_{k=n+1}^{\infty} |x_{k-n}|^2\right)^{1/2} \left(\sum_{k=n+1}^{\infty} |y_k|^2\right)^{1/2} = \|x\| \left(\sum_{k=n+1}^{\infty} |y_k|^2\right)^{1/2}.$$

But $y \in l_2$, i.e., $\sum_{k=1}^{\infty} |y_k|^2 < \infty$, and then $\lim_{n\to\infty} |\langle A_n x, y\rangle| = 0$.

We will prove now that under the same hypothesis the sequence $(\|A_n\|)_{n\in\mathbb{N}}$ is bounded. Let $x \in H$ and let us consider the sequence $(A_n x)_{n\in\mathbb{N}} \subseteq H^*$ (we have the identification of H with H^* given by the Riesz theorem). For any $y \in H$ the sequence $((A_n x)(y))_{n\in\mathbb{N}} = \left(\langle A_n x, y\rangle\right)_{n\in\mathbb{N}}$ is bounded. Using now the Uniform Boundedness Principle we obtain that

12.2 Solutions

the sequence $(A_n x)_{n\in\mathbb{N}} \subseteq H^*$ is bounded, hence the sequence $(A_n x)_{n\in\mathbb{N}} \subseteq H$ is bounded $\forall x \in H$. Applying again the Uniform Boundedness Principle we obtain that the sequence $(\|A_n\|)_{n\in\mathbb{N}}$ is bounded.

35. i) Let $n \in \mathbb{N}$. Obviously, A_n is linear and

$$\|A_n(x)\|_2 \leq \left(\sup_{k\geq n+1} |a_k|\right) \left(\sum_{k=n+1}^{\infty} |x_k|^2\right)^{1/2} \leq \left(\sup_{k\geq n+1} |a_k|\right) \|x\| \quad \forall x \in l_2,$$

hence A_n is continuous with $\|A_n\| \leq \sup_{k\geq n+1} |a_k|$. For $k \geq n+1$ we have

$$|a_k| = \|A_n(e_k)\| \leq \|A_n\| \, \|e_k\| = \|A_n\|,$$

hence $\sup_{k\geq n+1} |a_k| \leq \|A_n\|$, and therefore $\|A_n\| = \sup_{k\geq n+1} |a_k|$. For $x \in l_2$, we have

$$\|A_n(x)\|_2 \leq \|a\| \left(\sum_{k=n+1}^{\infty} |x_k|^2\right)^{1/2}. \text{ Since } \sum_{k=1}^{\infty} |x_k|^2 < \infty \text{ it follows that } \lim_{n\to\infty} A_n(x) = 0$$

$\forall x \in l_2$.

ii) We have $\langle A_n x, y\rangle = \langle x, A_n^* y\rangle \ \forall x, y \in l_2$. For $y \in l_2$, let $(\alpha_1, \alpha_2, ..., \alpha_n, ...) = A_n^* y$. Using the above relation we obtain that

$$\langle (a_{n+1}x_{n+1}, a_{n+2}x_{n+2}, ...), (y_1, y_2, ...)\rangle = \langle (x_1, x_2, ...), (\alpha_1, \alpha_2, ...)\rangle \quad \forall x \in l_2,$$

and then

$$a_{n+1}x_{n+1}\overline{y_1} + a_{n+2}x_{n+2}\overline{y_2} + \cdots = x_1\overline{\alpha_1} + x_2\overline{\alpha_2} + \cdots \quad \forall x \in l_2,$$

i.e., $\overline{\alpha_1} = \cdots = \overline{\alpha_n} = 0$, $\overline{\alpha_{n+1}} = a_{n+1}\overline{y_1}$, $\overline{\alpha_{n+2}} = a_{n+2}\overline{y_2}, \ldots$. Hence $A_n^*(y) = (0, ..., 0, \overline{a_{n+1}}y_1, \overline{a_{n+2}}y_2, ...)$, 0 on the first n positions.

iii) Let $x \in C$ with $x \neq 0$. Then $\|A_n^*(x)\| \to 0$. Since $x \neq 0$ there is $k \in \mathbb{N}$ such that $x_k \neq 0$. Then from $|a_{n+k}||x_k| \leq \|A_n^*(x)\| \ \forall n \in \mathbb{N}$ it follows that $a_n \to 0$, i.e., $a \in c_0$. This shows that if $C \neq \{0\}$ then $a \in c_0$. If $a \in c_0$ then by (i)

$$\|A_n\| = \sup_{k\geq n+1} |a_k| \to \limsup_{n\to\infty} |a_n| = 0,$$

and then $\|A_n^*(x)\| \leq \|A_n^*\| \, \|x\| = \|A_n\| \, \|x\| \to 0$ for any $x \in l_2$, i.e., $C = l_2$.

36. i) Let $A = \begin{pmatrix} a & b \\ b & a \end{pmatrix}$. Let us consider the sequence of matrices $(X_n)_{n\in\mathbb{N}}$ defined by $X_1 = A$, $X_{n+1} = A + X_n^2 \ \forall n \in \mathbb{N}$. It follows that $X_n = \begin{pmatrix} x_n & y_n \\ y_n & x_n \end{pmatrix} \ \forall n \in \mathbb{N}$, where $x_1 = a$, $y_1 = b$, $x_{n+1} = a + x_n^2 + y_n^2$, $y_{n+1} = b + 2x_n y_n \ \forall n \in \mathbb{N}$, i.e., the sequences $(x_n)_{n\in\mathbb{N}}$ and $(y_n)_{n\in\mathbb{N}}$ satisfy the same relations as in the statement. Let us observe that the sequences $(x_n)_{n\in\mathbb{N}}$ and $(y_n)_{n\in\mathbb{N}}$ are simultaneously convergent if and only if the sequence of matrices $(X_n)_{n\in\mathbb{N}}$ converges. That is, if and only if the sequence of linear and continuous operators associated to our matrices converges pointwise. Indeed, if $X_n \to X = \begin{pmatrix} x & y \\ y & x \end{pmatrix}$ pointwise it follows that $X_n \begin{pmatrix} 1 \\ 0 \end{pmatrix} \to X \begin{pmatrix} 1 \\ 0 \end{pmatrix}$, thus $x_n \to x$ and $y_n \to y$. Conversely, if $x_n \to x$ and $y_n \to y$ then

$$X_n \begin{pmatrix} u \\ v \end{pmatrix} = \begin{pmatrix} x_n u + y_n v \\ y_n u + x_n v \end{pmatrix} \to X \begin{pmatrix} u \\ v \end{pmatrix} \forall \begin{pmatrix} u \\ v \end{pmatrix} \in \mathbb{R}^2,$$

i.e., $X_n \to X$ pointwise. Hence the study for the convergence of the sequences $(x_n)_{n\in\mathbb{N}}$ and $(y_n)_{n\in\mathbb{N}}$ is equivalent with the study for the pointwise convergence for the sequence $(X_n)_{n\in\mathbb{N}} \subseteq L(\mathbb{R}^2)$. The idea is to use the Weierstrass convergence theorem for monotone and bounded sequences of operators. We have $A \geq 0$, by hypothesis. Since $X_n^2 \geq 0$ and $A \geq 0$ we obtain that $X_{n+1} = A + X_n^2 \geq 0$. Thus $X_n \geq 0 \ \forall n \in \mathbb{N}$. Also $X_n X_m = X_m X_n$ $\forall n, m \in \mathbb{N}$. We have $X_2 = A + A^2$ and then $X_2 - X_1 = A^2 \geq 0$, since A is self-adjoint. Suppose now that $X_{n+1} - X_n \geq 0$ and we will prove that $X_{n+2} - X_{n+1} \geq 0$. By induction we will obtain that $X_{n+1} - X_n \geq 0 \ \forall n \in \mathbb{N}$, i.e., $X_n \nearrow$. We have $X_{n+2} - X_{n+1} = A + X_{n+1}^2 - A - X_n^2 = X_{n+1}^2 - X_n^2 = (X_{n+1} - X_n)(X_{n+1} + X_n)$ (the last equality since X_n and X_{n+1} commute). But from the theory we know that if $A \geq 0$, $B \geq 0$ and $AB = BA$, then $AB \geq 0$. In our case $A = X_{n+1} - X_n \geq 0$ and $B = X_{n+1} + X_n \geq 0$, so $X_{n+2} - X_{n+1} \geq 0$.

Suppose that $(X_n)_{n\in\mathbb{N}}$ is pointwise convergent in $\mathcal{A}(\mathbb{R}^2)$. Then there is $M > 0$ such that $||X_n|| \leq M \ \forall n \in \mathbb{N}$, and then $\langle X_n x, x \rangle \leq M||x||^2 \ \forall n \in \mathbb{N}$, i.e., $X_n \leq MI \ \forall n \in \mathbb{N}$, i.e., the sequence $(X_n)_{n\in\mathbb{N}}$ is bounded in $\mathcal{A}(\mathbb{R}^2)$. Since $(X_n)_{n\in\mathbb{N}}$ is increasing from the Weierstrass convergence theorem for self-adjoint operators it follows that there is $U \in \mathcal{A}(\mathbb{R}^2)$ such that $X_n \to U$ pointwise, and, moreover, $\sup_{n\in\mathbb{N}} X_n = U$ in $\mathcal{A}(\mathbb{R}^2)$. $X_n \to U$ pointwise means that $X_n(x) \to U(x) \ \forall x \in \mathbb{R}^2$ and from here $X_n^2(x) \to U^2(x) \ \forall x \in \mathbb{R}^2$. Recall that $U^2 = U \circ U$, i.e., $U^2(x) = U(U(x)) \ \forall x \in \mathbb{R}^2$. Using this and passing to the limit in the recurrence relation we obtain that $U(x) = A(x) + U^2(x) \ \forall x \in \mathbb{R}^2$, i.e., $U = A + U^2$ in $\mathcal{A}(\mathbb{R}^2)$. We will prove that from this it follows that $A \leq I/4$. We have $U^2 - U + I/4 = I/4 - A$, or $(U - I/2)^2 = I/4 - A$. Since $(U - I/2)^2 \geq 0$ it follows that $A \leq I/4$.

Conversely, suppose that $A \leq I/4$. We will prove that in this case the sequence $(X_n)_{n\in\mathbb{N}}$ is upper bounded in $\mathcal{A}(\mathbb{R}^2)$. More precisely, we will prove that $X_n \leq I/2 - \sqrt{I/4 - A} \ \forall n \in \mathbb{N}$ (since $I/4 - A \geq 0$ then $\sqrt{I/4 - A}$ is well defined). To prove this property we will use the result from exercise 17: if $A, B \in \mathcal{A}_+(H)$ and $AB = BA$, then $A \leq B \Leftrightarrow \sqrt{A} \leq \sqrt{B}$. For $n = 1$, $X_1 = A \leq I/2 - \sqrt{I/4 - A}$ if and only if $\sqrt{I/4 - A} \leq I/2 - A$, that is, $I/4 - A \leq (I/2 - A)^2$, $I/4 - A \leq I/4 - A + A^2$, $A^2 \geq 0$, which is true. Let us suppose that $X_n \leq I/2 - \sqrt{I/4 - A}$ and we will prove that $X_{n+1} \leq I/2 - \sqrt{I/4 - A}$. We have

$$X_{n+1} = A + X_n^2 \leq A + \left(I/2 - \sqrt{I/4 - A}\right)^2 = A + I/4 - \sqrt{I/4 - A} + I/4 - A$$
$$= I/2 - \sqrt{I/4 - A}.$$

So, if $A \leq I/4$ it follows that $(X_n)_{n\in\mathbb{N}}$ is bounded in $\mathcal{A}(\mathbb{R}^2)$. Now from the Weierstrass convergence theorem it follows that the sequence $(X_n)_{n\in\mathbb{N}}$ is pointwise convergent.

12.2 Solutions

Therefore, the sequences $(x_n)_{n\in\mathbb{N}}$ and $(y_n)_{n\in\mathbb{N}}$ are simultaneously convergent if and only if $A \leq I/4$. But $I/4 - A \geq 0$ if and only if the matrix $\begin{pmatrix} 1/4 - a & -b \\ -b & 1/4 - a \end{pmatrix}$ is positive. Using the result from exercise 10 this is equivalent to $\begin{cases} 1/4 - a \geq 0, \\ (1/4 - a)^2 - b^2 \geq 0, \end{cases}$

that is, $\begin{cases} a \leq 1/4, \\ |b| \leq 1/4 - a, \end{cases}$ i.e., the statement.

Remarks. 1) Using the analogy between the self-adjoint (positive) operators and the real (positive) numbers, the exercise is a very natural one.

2) Also, let us observe that the verification of the commutativity is important in this new context.

3) From a recurrence relation in $\mathbb{R}$, we can construct a recurrence relation in a Hilbert space and then study the convergence of that sequence in the considered Hilbert space.

4) The recurrence relation has been suggested by [1, exercise 6, Spring 1984].

5) The general form for a real self-adjoint matrix is $\begin{pmatrix} a & b \\ b & c \end{pmatrix}$. Obviously, one can solve the same type of exercise for such a matrix. The statement is the following:

Let $a, b, c \in \mathbb{R}$, such that $a \geq 0$, $c \geq 0$ and $ac - b^2 \geq 0$. Define the sequences of real numbers $(x_n)_{n\geq 0}, (y_n)_{n\geq 0}$ and $(z_n)_{n\geq 0}$ by the recurrence relations: $x_0 = a$, $y_0 = b$, $z_0 = c$, and $x_{n+1} = a + x_n^2 + y_n^2$, $y_{n+1} = b + y_n(x_n + z_n)$, $z_{n+1} = c + y_n^2 + z_n^2$ $\forall n \geq 0$. Then these sequences are simultaneously convergent if and only if $a \leq 1/4$, $(a-1/4)(c-1/4) - b^2 \geq 0$ and $c \leq 1/4$.

ii) Recall first some notations which will be used below. If T, U are two symmetric matrices, we write $T \leq U$ if $\langle Tx, x \rangle \leq \langle Ux, x \rangle$ $\forall x \in \mathbb{R}^2$, where by $\langle \cdot, \cdot \rangle$ we denote the inner product on $\mathbb{R}^2$. A symmetric matrix T is positive definite if $\langle Tx, x \rangle \geq 0$ $\forall x \in \mathbb{R}^2$ and $\langle Tx, x \rangle = 0$ implies $x = 0$. By I we denote the unit matrix. We will also use that if two positive definite matrices commute then their product is a positive definite matrix. We define the sequence of matrices $(A_n)_{n\geq 0}$ by $A_0 = \begin{pmatrix} a & b \\ b & a \end{pmatrix}$ and $A_n = \begin{pmatrix} a_n & b_n \\ b_n & a_n \end{pmatrix}$ for $n \geq 1$. The relation from the exercise can be written in the form $A_{n+1} = A_n - A_n^2$. The hypothesis on a and b and the well known Sylvester theorem from linear algebra (see exercise 10) assures that A and $I - A$ are positive definite. From the recurrence relation it follows easily, by induction, that $0 \leq A_n \leq I$ and $A_n \geq A_{n+1}$ $\forall n \geq 0$. Now, from the Weierstrass convergence theorem for positive operators on a Hilbert space it follows that there exists $\lim_{n\to\infty} A_n \begin{pmatrix} x \\ y \end{pmatrix} \in \mathbb{R}^2$ $\forall x, y \in \mathbb{R}$. In particular there exists $\lim_{n\to\infty} A_n \begin{pmatrix} 1 \\ 0 \end{pmatrix} = \lim_{n\to\infty} \begin{pmatrix} a_n \\ b_n \end{pmatrix}$, i.e., there exists $\lim_{n\to\infty} a_n$ and $\lim_{n\to\infty} b_n$. From the first recurrence relation we deduce that $\lim_{n\to\infty} a_n = 0$ and $\lim_{n\to\infty} b_n = 0$, i.e. $A_n \to 0 = \begin{pmatrix} 0 & 0 \\ 0 & 0 \end{pmatrix}$. Let us observe that from $A_1 = A - A^2 = A(I - A)$ it follows that A_1 is positive definite. Also $I - A_1 = (I - A) + A^2 \geq I - A$, and since $I - A$ is positive definite it follows that $I - A_1$ is positive definite. In this way, by induction, we obtain that A_n and $I - A_n$ are positive definite for all $n \geq 0$ and so A_n and $I - A_n$ are invertible for all $n \geq 0$. Moreover, the matrices A_n

commute, hence their inverses commute. Let $T_n = A_n^{-1}$ for $n \geq 0$. We have

$$T_{n+1} - T_n = A_{n+1}^{-1} - A_n^{-1} = (A_n - A_{n+1})A_n^{-1}A_{n+1}^{-1} = A_n^2 A_n^{-1} A_{n+1}^{-1}$$
$$= A_n A_{n+1}^{-1} = (I - A_n)^{-1} \to I.$$

(We use that if $\|A\| < 1$ then $(I - A)^{-1} = I + A + A^2 + \cdots$, hence

$$\|(I - A)^{-1} - I\| \leq \|A\| + \|A\|^2 + \cdots = \frac{\|A\|}{1 - \|A\|}.$$

Here, $\|A_n\| \to 0$, $\|A_n\| < 1 \; \forall n \geq 0$, hence $\|(I - A_n)^{-1} - I\| \leq \|A_n\|/(1 - \|A_n\|)$ $\forall n \geq 0$, and then $(I - A_n)^{-1} \to I$.)

From the vectorial form of the Stolz–Cesaro lemma (see the end of our solution), it follows that

$$\lim_{n \to \infty} \frac{T_n}{n} = \lim_{n \to \infty} \frac{T_{n+1} - T_n}{n+1-n} = I,$$

i.e., $(nA_n)^{-1} \to I$, and then $nA_n \to I$, that is, $na_n \to 1$ and $nb_n \to 0$. Let now $Y_n = T_n - nI$. Then

$$Y_{n+1} - Y_n = T_{n+1} - T_n - I = (I - A_n)^{-1} - I = (I - A_n)^{-1}[I - (I - A_n)]$$
$$= (I - A_n)^{-1} A_n,$$

and then $n(Y_{n+1} - Y_n) = (I - A_n)^{-1}(nA_n) \to I$, that is, $(Y_{n+1} - Y_n)/(1/n) \to I$. Again, from the vectorial form of the Stolz–Cesaro lemma it follows that

$$\lim_{n \to \infty} \frac{Y_n}{1 + 1/2 + \cdots + 1/n} = I,$$

and therefore $\lim_{n \to \infty} (Y_n/\ln n) = I$, that is,

$$\lim_{n \to \infty} \frac{n}{\ln n}((nA_n)^{-1} - I) = I.$$

Then

$$\frac{n}{\ln n}(I - nA_n) = \frac{n}{\ln n}((nA_n)^{-1} - I)(nA_n) \to I \cdot I,$$

i.e., $(n/\ln n)(I - nA_n) \to I$, that is, $(n/\ln n)(1 - na_n) \to 1$ and $(n/\ln n)nb_n \to 0$, i.e. the statement.

Remark. The general form for a self-adjoint operator (matrix) on $\mathbb{R}^2$ is $\begin{pmatrix} a & b \\ b & c \end{pmatrix}$. One can solve the analogous exercise in this more general context.

A vectorial form for the Stolz–Cesaro lemma (the case $[\cdot/\infty]$)

Let X be a normed space, and $(x_n)_{n \in \mathbb{N}} \subseteq X$, $(a_n)_{n \in \mathbb{N}} \subseteq (0, \infty)$ such that $a_n \nearrow \infty$. If there exists

$$\lim_{n \to \infty} \frac{x_n - x_{n-1}}{a_n - a_{n-1}} = \alpha \in X$$

then there exists $\lim_{n \to \infty} (x_n/a_n)$ and $\lim_{n \to \infty} (x_n/a_n) = x$.

12.2 Solutions

Indeed, for $\varepsilon > 0$ we can find $n_\varepsilon \in \mathbb{N}$ such that

$$\left\| \frac{x_n - x_{n-1}}{a_n - a_{n-1}} - \alpha \right\| < \varepsilon \quad \forall n \geq n_\varepsilon.$$

Therefore

$$\|x_n - x_{n-1} - (a_n - a_{n-1})\alpha\| < \varepsilon(a_n - a_{n-1}) \quad \forall n \geq n_\varepsilon.$$

Let us denote $k = n_\varepsilon + 1$. For $n \geq k$, we have

$$\|x_{k+1} - x_k - (a_{k+1} - a_k)\alpha\| < \varepsilon(a_{k+1} - a_k),$$
$$\|x_{k+2} - x_{k+1} - (a_{k+2} - a_{k+1})\alpha\| < \varepsilon(a_{k+2} - a_{k+1}),$$
$$\ldots$$
$$\|x_n - x_{n-1} - (a_n - a_{n-1})\alpha\| < \varepsilon(a_n - a_{n-1}),$$

which by addition gives

$$\begin{aligned}\|x_n - x_k - (a_n - a_k)\alpha\| &\leq \|x_n - x_{n-1} - (a_n - a_{n-1})\alpha\| + \cdots \\ &\quad + \|x_{k+2} - x_{k+1} - (a_{k+2} - a_{k+1})\alpha\| \\ &\quad + \|x_{k+1} - x_k - (a_{k+1} - a_k)\alpha\| \\ &\leq \varepsilon(a_n - a_{n-1}) + \cdots + \varepsilon(a_{k+2} - a_{k+1}) + \varepsilon(a_{k+1} - a_k) \\ &= \varepsilon(a_n - a_k),\end{aligned}$$

i.e. $\|x_n - x_k - (a_n - a_k)\alpha\| \leq \varepsilon(a_n - a_k) \; \forall n \geq k$. Then

$$\left\| \frac{x_n}{a_n} - \alpha - \frac{x_k - a_k\alpha}{a_n} \right\| \leq \varepsilon - \varepsilon \frac{a_k}{a_n} \quad \forall n \geq k,$$

and then

$$\left\| \frac{x_n}{a_n} - \alpha \right\| \leq \frac{\|x_k - \alpha a_k\|}{a_n} + \varepsilon - \varepsilon \frac{a_k}{a_n} \quad \forall n \geq k.$$

Since $a_n \to \infty$ it follows that $\lim_{n \to \infty} \|x_k - \alpha a_k\|/a_n = 0$, thus there is $p_\varepsilon \in \mathbb{N}$ such that $\|x_k - \alpha a_k\|/a_n < \varepsilon \; \forall n \geq p_\varepsilon$. Then for $n \geq \max(k, p_\varepsilon,)$ we deduce that $\|x_n/a_n - \alpha\| < 2\varepsilon$, and therefore $\lim_{n \to \infty} x_n/a_n = \alpha$.

37. i) We have $M_a \leq M_b \Leftrightarrow \langle M_a x, x \rangle \leq \langle M_b x, x \rangle \; \forall x \in l_2$. If $M_a \leq M_b$ we have $\langle a_n e_n, e_n \rangle \leq \langle b_n e_n, e_n \rangle \; \forall n \in \mathbb{N}$, i.e., $a_n \leq b_n \; \forall n \in \mathbb{N}$. Conversely,

$$\langle M_a x, x \rangle = \sum_{n=1}^{\infty} a_n |x_n|^2 \leq \sum_{n=1}^{\infty} b_n |x_n|^2 = \langle M_b x, x \rangle \quad \forall x \in l_2.$$

ii) Since by (i) we have $M_{a_n} \leq M_{a_{n+1}} \; \forall n \in \mathbb{N}$, we can apply the Weierstrass convergence theorem for self-adjoint operators to deduce that $\sup_{n \in \mathbb{N}} M_{a_n}$ exists in $\mathcal{A}(l_2)$ if and only if the sequence $(M_{a_n})_{n \in \mathbb{N}}$ is bounded in $\mathcal{A}(l_2)$. Suppose that there is $U \in \mathcal{A}(l_2)$ such that $M_{a_n} \leq U \; \forall n \in \mathbb{N}$, i.e., $\langle M_{a_n} x, x \rangle \leq \langle Ux, x \rangle \; \forall x \in l_2, \forall n \in \mathbb{N}$. In particular, $\langle M_{a_n} e_k, e_k \rangle \leq \langle U e_k, e_k \rangle \; \forall k \in \mathbb{N}, \forall n \in \mathbb{N}$. If $a_n = (a_{n1}, a_{n2}, ..., a_{nk}, ...)$, we obtain that

$a_{nk} \leq a_k \ \forall k \in \mathbb{N}, \forall n \in \mathbb{N}$, where $a_k = \langle Ue_k, e_k \rangle$. But $|a_k| = |\langle Ue_k, e_k \rangle| \leq \|U\|$, i.e., $a = (a_1, a_2, ...) \in l_\infty$ and $a_n \leq a \ \forall n \in \mathbb{N}$. Conversely, if $a \in l_\infty$ and $a_n \leq a \ \forall n \in \mathbb{N}$, then by (i) we have $M_{a_n} \leq M_a \ \forall n \in \mathbb{N}$. By the Weierstrass convergence theorem for self-adjoint operators, there exists $\sup_{n \in \mathbb{N}} M_{a_n} = U$ in $\mathcal{A}(l_2)$ and, moreover, $M_{a_n}(x) \to U(x) \ \forall x \in l_2$. We must prove that $U = M_a$ where $a = \sup_{n \in \mathbb{N}} a_n$ in l_∞. Since $M_{a_n}(x) \to U(x) \ \forall x \in l_2$ then $\langle M_{a_n}(e_k), e_k \rangle \to \langle Ue_k, e_k \rangle$, i.e., $a_{nk} \to \langle Ue_k, e_k \rangle \ \forall k \in \mathbb{N}$. Let $a_k = \langle Ue_k, e_k \rangle$. Then we have $a_{nk} \to a_k$ and $a_{nk} \leq a_{n+1,k} \ \forall k \in \mathbb{N}$, and it follows that $a_k = \sup_{n \in \mathbb{N}} a_{nk} \ \forall k \in \mathbb{N}$. We have $\langle Ux, e_k \rangle = a_k x_k \ \forall x = (x_n)_{n \in \mathbb{N}} \in l_2, \ \forall k \in \mathbb{N}$. Then

$$Ux = \sum_{k=1}^\infty \langle Ux, e_k \rangle e_k = \sum_{k=1}^\infty a_k x_k e_k = (a_1 x_1, a_2 x_2, ...) = M_a(x) \ \ \forall x \in l_2.$$

iii) We take $a_n = (n/(1+n), 2n/(1+2n), ..., kn/(1+kn), ...)$. Then $a_{n+1} \geq a_n$ in $l_\infty \ \forall n \in \mathbb{N}$ and

$$\sup_{n \in \mathbb{N}} a_{nk} = \sup_{n \in \mathbb{N}} \frac{nk}{1+kn} = 1,$$

so by (ii) we have $\sup_{n \in \mathbb{N}} U_n = I$.

38. i) We have

$$M_\varphi \leq M_\psi \Leftrightarrow \int_S (\psi - \varphi)|f|^2 \, d\mu \geq 0 \ \ \forall f \in L_2(\mu).$$

Using the results from exercise 26 this is equivalent to $\varphi \leq \psi$, μ-a.e..

ii) Let us suppose that there exists $\sup_{n \in \mathbb{N}} M_{\varphi_n}$ in $\mathcal{A}(L_2(\mu))$. Then the sequence $(M_{\varphi_n})_{n \in \mathbb{N}}$ is bounded in $\mathcal{A}(L_2(\mu))$. Since by (i), $M_{\varphi_n} \leq M_{\varphi_{n+1}} \ \forall n \in \mathbb{N}$, we can apply the Weierstrass convergence theorem for self-adjoint operators to deduce that if $U = \sup_{n \in \mathbb{N}} M_{\varphi_n}$ then we have $\lim_{n \to \infty} M_{\varphi_n}(f) = U(f) \ \forall f \in L_2(\mu)$. Now we can use the Uniform Boundedness Principle ($L_2(\mu)$ is a Banach space) to deduce that the sequence $(M_{\varphi_n})_{n \in \mathbb{N}}$ is uniformly bounded, i.e., there is $c > 0$ such that $\|M_{\varphi_n}\| \leq c \ \forall n \in \mathbb{N}$, that is, (see exercise 26(i)) $\|\varphi_n\|_\infty \leq c \ \forall n \in \mathbb{N}$, i.e., there is $c > 0$ such that $\forall n \in \mathbb{N}, |\varphi_n| \leq c$, μ-a.e.. Since $\varphi_n \leq \varphi_{n+1}$ and $|\varphi_n| \leq c$, μ-a.e., it follows that there exists $\varphi = \sup_{n \in \mathbb{N}} \varphi_n$ and $|\varphi| \leq c$, μ-a.e., i.e., $\varphi \in L_\infty(\mu; \mathbb{R})$. Conversely, if there exists $\varphi = \sup_{n \in \mathbb{N}} \varphi_n \in L_\infty(\mu; \mathbb{R})$, then we have $M_{\varphi_n} \leq M_\varphi \ \forall n \in \mathbb{N}$. If $U \in \mathcal{A}(L_2(\mu))$ is such that $M_{\varphi_n} \leq U \ \forall n \in \mathbb{N}$, we obtain that $\int_S \varphi_n |f|^2 \, d\mu \leq \langle Uf, f \rangle$ $\forall n \in \mathbb{N}, \forall f \in L_2(\mu)$. Now we have $|\varphi_n| \leq \max(\|\varphi_1\|_\infty, \|\varphi\|_\infty) = M$, μ-a.e., and then $\varphi_n |f|^2 \to \varphi |f|^2$ and $|\varphi_n| |f|^2 \leq M |f|^2$. Hence by the Lebesgue dominated convergence theorem it follows that

$$\int_S \varphi |f|^2 \, d\mu = \lim_{n \to \infty} \int_S \varphi_n |f|^2 \, d\mu \leq \langle Uf, f \rangle \ \ \forall f \in L_2(\mu),$$

i.e., $M_\varphi \leq U$. Hence $\sup_{n \in \mathbb{N}} M_{\varphi_n} = M_\varphi$.

12.2 Solutions

iii) From exercise 26 it follows that U_n is self-adjoint and using that φ is increasing it follows that $U_n \leq U_{n+1}$. Also, $U_n \leq MI$ $\forall n \geq 2$, where $M = \sup_{x \in [0,1]} |\varphi(x)| < \infty$.
From the Weierstrass convergence theorem there exists $U = \sup_{n \geq 2} U_n$ and, moreover, $Uf = \lim_{n \to \infty} U_n f$ $\forall f \in L_2[0,1]$. We must calculate, for any $f \in L_2[0,1]$, $\lim_{n \to \infty} U_n f$ in the space $L_2[0,1]$, i.e., if $g_n(x) = \varphi(\sqrt[n]{x})f(x)$ we must find the limit for the sequence $(g_n)_{n \geq 2}$ in the space $L_2[0,1]$. But $\sqrt[n]{x} \to 1$ $\forall 0 < x \leq 1$, hence by the continuity of φ it follows that $\varphi(\sqrt[n]{x}) \to \varphi(1)$. Therefore $|\varphi(\sqrt[n]{x})f(x) - \varphi(1)f(x)|^2 \to 0$ $\forall 0 < x \leq 1$. Also, $|\varphi(\sqrt[n]{x}) - \varphi(1)|^2 |f(x)|^2 \leq 4M^2 |f(x)|^2$. Now we can apply the Lebesgue dominated convergence theorem to deduce that $\int_0^1 |g_n(x) - \varphi(1)f(x)|^2 \, dx \to 0$, i.e., $\|g_n - \varphi(1)f\| \to 0$ in $L_2[0,1]$, that is, $U_n f \to \varphi(1)f$ $\forall f \in L_2[0,1]$. Then $U = \varphi(1)I$, hence $\sup_{n \geq 2} U_n = \varphi(1)I$.

iv) We use the same reasoning as above. We omit the details.

v) We have $U_n \leq U_{n+1}$ $\forall n \geq 2$. We can apply the Weierstrass convergence theorem for self-adjoint operators to deduce that there exists $\sup_{n \geq 2} U_n$ in $\mathcal{A}(L_2[0,1])$ if and only if there exists $V \in \mathcal{A}(L_2[0,1])$ such that $U_n \leq V$ $\forall n \geq 2$. But $U_n \leq V$ $\forall n \geq 2$ if and only if $\langle U_n f, f \rangle \leq \langle Vf, f \rangle$ $\forall f \in L_2[0,1]$, $\forall n \geq 2$, that is,

$$a_n \int_0^1 \sqrt[n]{x} |f(x)|^2 \leq \int_0^1 (Vf)(x)\overline{f(x)}dx.$$

If $\sup_{n \geq 2} U_n$ exists then for $f(x) = 1$ we obtain that $a_n \int_0^1 \sqrt[n]{x}dx \leq M = \|V\|$ $\forall n \geq 2$, i.e., $a_n \leq M / \left(\int_0^1 \sqrt[n]{x}dx \right) \leq 2M$ $\forall n \geq 2$, i.e., $(a_n)_{n \geq 2}$ is bounded.

Conversely, suppose that $(a_n)_{n \geq 2}$ is also bounded and let $\alpha = \sup_{n \geq 2} a_n \in \mathbb{R}$. We have $U_n \leq \alpha I$ $\forall n \geq 2$, since

$$\langle U_n f, f \rangle = a_n \int_0^1 \sqrt[n]{x} |f(x)|^2 \leq \alpha \int_0^1 |f(x)|^2 \, dx = \alpha \langle f, f \rangle \quad \forall f \in L_2[0,1].$$

Let now $U \in \mathcal{A}(L_2[0,1])$, $U = \sup_{n \geq 2} U_n$. Then we know that $U_n f \to Uf$ $\forall f \in L_2[0,1]$. For $f \in L_2[0,1]$, let $g_n(x) = |a_n \sqrt[n]{x} - \alpha|^2 |f(x)|^2$. We have $g_n(x) \to 0$, $0 \leq g_n(x) \leq 4\alpha^2 |f(x)|^2$, $\forall x \in (0,1]$, and from the Lebesgue dominated convergence theorem it follows that $\int_0^1 g_n(x) \, dx \to 0$, i.e., $U_n f \to \alpha f$, and then $\sup_{n \geq 2} U_n = \alpha I$.

39. i) Let $h : l_2 \to H^2(\mathbb{D})$ be the isometric isomorphism from exercise 13(v) of chapter 10, i.e., $h(a_0, a_1, a_2, \ldots) = f \Leftrightarrow f(z) = a_0 + a_1 z + a_2 z^2 + \cdots \forall z \in \mathbb{D}$. We consider the diagram

$$\begin{array}{ccc} H^2(\mathbb{D}) & \xrightarrow{T_w} & H^2(\mathbb{D}) \\ h \uparrow & & \downarrow h^{-1} \\ l_2 & \xrightarrow{U_w} & l_2. \end{array}$$

Here, $U_w = h^{-1} \circ T_w \circ h$. We will prove that $U_w(a_0, a_1, a_2, ...) = (a_0, wa_1, w^2a_2, ...)$ $\forall (a_0, a_1, a_2, ...) \in l_2$. Indeed, we have

$$U_w(a_0, a_1, a_2, ...) = (a_0, wa_1, w^2a_2, ...) \Leftrightarrow T_w(h(a_0, a_1, a_2, ...)) = h(a_0, wa_1, w^2a_2, ...).$$

Now if $h(a_0, a_1, a_2, ...) = f$ and $h(a_0, wa_1, w^2a_2, ...) = g$, then $f(z) = \sum_{n=0}^{\infty} a_n z^n$ and $g(z) = \sum_{n=0}^{\infty} a_n w^n z^n$ $\forall z \in \mathbb{D}$, and the equality which we must prove is $T_w(f) = g$, i.e., $f(wz) = g(z)$ $\forall z \in \mathbb{D}$, which is clearly true. Hence U_w is a multiplication operator. From exercise 14, U_w is linear and continuous, with $\|U_w\| = 1$. Also U_w is self-adjoint if and only if $w \in \mathbb{R}$. Since $|w| < 1$ then $|w|^n \to 0$, and therefore using exercise 21 of chapter 5 it follows that U_w is compact.

Using the relation $U_w = h^{-1} \circ T_w \circ h$ and the fact that h is an isomorphic isometry of Hilbert spaces we obtain the assertions from the statement.

ii) Let $h : \mathcal{A}^2(\mathbb{D}) \to l_2$ be the isometric isomorphism from exercise 17(v) of chapter 10, i.e.,

$$h(f) = \left(\frac{\sqrt{\pi}}{\sqrt{n+1}} a_n\right)_{n \geq 0} \Leftrightarrow f(z) = \sum_{n=0}^{\infty} a_n z^n \ \forall z \in \mathbb{D}.$$

We observe that $h^{-1}(a_0, a_1, a_2, ...) = f \Leftrightarrow f(z) = \sum_{n=0}^{\infty} (\sqrt{n+1}/\sqrt{\pi}) a_n z^n$. We consider the diagram:

$$\begin{array}{ccc} \mathcal{A}^2(\mathbb{D}) & \xrightarrow{T_w} & \mathcal{A}^2(\mathbb{D}) \\ h^{-1} \uparrow & & \downarrow h \\ l_2 & \xrightarrow{U_w} & l_2. \end{array}$$

Here, $U_w = h \circ T_w \circ h^{-1}$. We will prove that $U_w(a_0, a_1, a_2, ...) = (a_0, wa_1, w^2a_2, ...)$ $\forall (a_0, a_1, a_2, ...) \in l_2$. Indeed, we have

$$U_w(a_0, a_1, a_2, ...) = (a_0, wa_1, w^2a_2, ...) \Leftrightarrow T_w(h^{-1}(a_0, a_1, a_2, ...)) = h^{-1}(a_0, wa_1, w^2a_2, ...).$$

But if $h^{-1}(a_0, a_1, a_2, ...) = f$ and $h^{-1}(a_0, wa_1, w^2a_2, ...) = g$ then we have that $f(z) = \sum_{n=0}^{\infty} (\sqrt{n+1}/\sqrt{\pi}) a_n z^n$ and $g(z) = \sum_{n=0}^{\infty} (\sqrt{n+1}/\sqrt{\pi}) a_n w^n z^n$ $\forall z \in \mathbb{D}$, so the equality which we must prove is $T_w(f) = g$, i.e., $f(wz) = g(z)$ $\forall z \in \mathbb{D}$, which is clearly true. From here we continue as in (i).

40. i) Let $\alpha, \beta \in \mathbb{D}$. We consider $x^* : H^2(\mathbb{D}) \to \mathbb{C}$ defined by $x^*(f) = f(\alpha) - f(\beta)$. From exercise 15 of chapter 10, x^* is a linear and continuous functional, and $\|x^*\|^2 = \sum_{n=0}^{\infty} |x^*(z^n)|^2$. But $x^*(z^n) = \alpha^n - \beta^n$ $\forall n \in \mathbb{N}$ and $x^*(1) = 0$, so $\|x^*\| = \left(\sum_{n=1}^{\infty} |\alpha^n - \beta^n|^2\right)^{1/2}$. Using the equality

$$\alpha^n - \beta^n = (\alpha - \beta)(\alpha^{n-1} + \alpha^{n-2}\beta + \cdots + \alpha\beta^{n-2} + \beta^{n-1}),$$

12.2 Solutions

we deduce that $|\alpha^n - \beta^n| \leq nM^{n-1}|\alpha - \beta|$, where $M = \max(|\alpha|, |\beta|) < 1$. Then

$$\sum_{n=1}^{\infty} |\alpha^n - \beta^n|^2 \leq |\alpha - \beta|^2 \sum_{n=1}^{\infty} n^2 M^{2(n-1)}.$$

But $\sum_{n=1}^{\infty} n^2 r^{n-1} = (1+r)/(1-r)^3$ for $0 < r < 1$, and then

$$\sum_{n=1}^{\infty} |\alpha^n - \beta^n|^2 \leq |\alpha - \beta|^2 \frac{1+M^2}{(1-M^2)^3}.$$

Hence $\|x^*\| \leq |\alpha - \beta|\sqrt{(1+M^2)/(1-M^2)^3}$. Let us observe that if $\beta = 0$, then

$$\|x^*\| = \left(\sum_{n=1}^{\infty} |\alpha|^{2n}\right)^{1/2} = \left(|\alpha|^2/(1-|\alpha|^2)\right)^{1/2}.$$

Let $x_k^* : H^2(\mathbb{D}) \to \mathbb{C}$ be defined by $x_k^*(f) = f(\xi_k) - f(\xi)$. Since $\xi_k \to \xi$ and $f \in H^2(\mathbb{D})$ is continuous it follows that $f(\xi_k) \to f(\xi)$, i.e., $x_k^*(f) \to 0 \ \forall f \in H^2(\mathbb{D})$. Thus, U is well defined. From the above proved relations it follows that x_k^* is a linear and continuous functional, with

$$\|x_k^*\| \leq |\xi_k - \xi|\left((1+M_k^2)/(1-M_k^2)^3\right)^{1/2} \ \forall k \in \mathbb{N},$$

where $M_k = \max(|\xi_k|, |\xi|) < 1$. Since $|\xi_k - \xi| \to 0$ and $M_k \to |\xi|$ it follows that $\|x_k^*\| \to 0$. From exercise 44(i) of chapter 5 it follows that U is a compact operator.

In the case in which $\xi = 0$ we have $\|x_k^*\| = \sqrt{|\xi_k|^2/(1-|\xi_k|^2)}$, and by exercise 42 of chapter 5 we have that

$$\|U\| = \sup_{k \in \mathbb{N}} \|x_k^*\| = \sup_{k \in \mathbb{N}} \sqrt{\frac{|\xi_k|^2}{1-|\xi_k|^2}} = \sqrt{\frac{M^2}{1-M^2}},$$

where $M = \sup_{k \in \mathbb{N}} |\xi_k|$, since the function $f : (0,1) \to (0,\infty)$ defined by $f(x) = \sqrt{x^2/(1-x^2)}$ is strictly increasing.

ii) The compactness of U follows from (i). Also, with the same notations as in (i), $\|U\| = \sup_{k \in \mathbb{N}} \|x_k^*\|$, and $\|x_k^*\| = \sqrt{|\xi|^{2k}/(1-|\xi|^{2k})}$. We obtain that

$$\|U\| = \sup_{k \in \mathbb{N}} \sqrt{\frac{|\xi|^{2k}}{1-|\xi|^{2k}}} = \sqrt{\frac{|\xi|^2}{1-|\xi|^2}}.$$

iii) The compactness of U follows from (i), and for the norm of U we have

$$\|U\| = \sup_{k \in \mathbb{N}} \sqrt{\sum_{n=1}^{\infty} \frac{1}{(k+1)^{2n}}} = \sup_{k \in \mathbb{N}} \frac{1}{\sqrt{k^2+2k}} = \frac{1}{\sqrt{3}}.$$

41. i) Let $\alpha, \beta \in \mathbb{D}$. We consider $x^* : \mathcal{A}^2(\mathbb{D}) \to \mathbb{C}$ defined by $x^*(f) = f(\alpha) - f(\beta)$. From exercise 18(iii) of chapter 10, x^* is a linear and continuous functional, and $\|x^*\| = \left(\sum_{n=0}^{\infty} |x^*(f_n)|^2\right)^{1/2}$, where $(f_n)_{n \in \mathbb{N}} \subseteq \mathcal{A}^2(\mathbb{D})$ is an orthonormal basis. Since in exercise 17(iv) of chapter 10 it is proved that

$$\left\{ f_n(z) = \left(\sqrt{n+1}/\sqrt{\pi}\right) z^n \mid n \geq 0 \right\} \subseteq \mathcal{A}^2(\mathbb{D})$$

is an orthonormal basis, we have $\|x^*\| = \left(\sum_{n=0}^{\infty} ((n+1)/\pi) |x^*(z^n)|^2\right)^{1/2}$. But $x^*(z^n) = \alpha^n - \beta^n \ \forall n \in \mathbb{N}$, and $x^*(1) = 0$, so

$$\|x^*\| = \left(\sum_{n=1}^{\infty} ((n+1)/\pi) |\alpha^n - \beta^n|^2\right)^{1/2}.$$

Using the inequality $|\alpha^n - \beta^n| \leq n M^{n-1} |\alpha - \beta|$, where $M = \max(|\alpha|, |\beta|) < 1$, we obtain that

$$\sum_{n=1}^{\infty} (n+1) |\alpha^n - \beta^n|^2 \leq |\alpha - \beta|^2 \sum_{n=1}^{\infty} (n+1) n^2 M^{2(n-1)}.$$

But we have that $\sum_{n=1}^{\infty} n^2 r^{n+1} = (r^2 + r^3)/(1-r)^3$, and then by derivation we obtain $\sum_{n=1}^{\infty} (n+1) n^2 r^{n-1} = (4r+2)/(1-r)^4$ for $0 < r < 1$. Then

$$\sum_{n=1}^{\infty} (n+1) |\alpha^n - \beta^n|^2 \leq |\alpha - \beta|^2 \frac{4M^2 + 2}{(1-M^2)^4},$$

and then

$$\|x^*\| \leq \frac{|\alpha - \beta|}{(1-M^2)^2 \sqrt{\pi}} \sqrt{4M^2 + 2}.$$

Let us observe that if $\beta = 0$ then

$$\|x^*\| = \sqrt{\sum_{n=1}^{\infty} \frac{n+1}{\pi} |\alpha|^{2n}} = \sqrt{\frac{2|\alpha|^2 - |\alpha|^4}{\pi (1 - |\alpha|^2)^2}},$$

since $\sum_{n=0}^{\infty} r^{n+1} = r/(1-r)$ and then by derivation $\sum_{n=0}^{\infty} (n+1) r^n = 1/(1-r)^2$, and therefore $\sum_{n=1}^{\infty} (n+1) r^n = (2r - r^2)/(1-r)^2$ for $0 < r < 1$.

Let $x_k^* : \mathcal{A}^2(\mathbb{D}) \to \mathbb{C}$ be defined by $x_k^*(f) = f(\xi_k) - f(\xi)$. Since $\xi_k \to \xi$ and $f \in \mathcal{A}^2(\mathbb{D})$ is continuous it follows that $f(\xi_k) \to f(\xi)$, i.e., $x_k^*(f) \to 0 \ \forall f \in \mathcal{A}^2(\mathbb{D})$. It follows that U is well defined. From the above proved relations it follows that x_k^* is a linear and continuous functional with

$$\|x_k^*\| \leq |\xi_k - \xi| \frac{\sqrt{4 M_k^2 + 2}}{(1 - M_k^2)^2 \sqrt{\pi}} \quad \forall k \geq 1,$$

12.2 Solutions

where $M_k = \max(|\xi_k|, |\xi|) < 1$. Since $|\xi_k - \xi| \to 0$ and $M_k \to |\xi|$ it follows that $\|x_k^*\| \to 0$. From exercise 44(i) of chapter 5 it follows that U is a compact operator.

In the case in which $\xi = 0$, we have

$$\|x_k^*\| = \sqrt{\frac{2|\xi_k|^2 - |\xi_k|^4}{\pi(1-|\xi_k|^2)^2}} = \frac{|\xi_k|}{1-|\xi_k|^2}\sqrt{\frac{2-|\xi_k|^2}{\pi}},$$

and by exercise 42 of chapter 5,

$$\|U\| = \sup_{k \in \mathbb{N}} \|x_k^*\| = \sup_{k \in \mathbb{N}} \sqrt{\frac{2|\xi_k|^2 - |\xi_k|^4}{\pi(1-|\xi_k|^2)^2}} = \frac{M}{1-M^2}\sqrt{\frac{2-M^2}{\pi}},$$

where $M = \sup_{k \in \mathbb{N}} |\xi_k|$, since the function $f(x) = (2x - x^2)/(1-x)^2$ is strictly increasing on $[0, 1)$.

ii) The compactness of U follows from (i). For the norm, by (1),

$$\|U\| = \frac{|\xi|}{1-|\xi|^2}\sqrt{\frac{2-|\xi|^2}{\pi}}.$$

iii) By (i) U is compact and

$$\|U\| = \frac{M}{1-M^2}\sqrt{\frac{2-M^2}{\pi}},$$

where $M = \sup_{k \in \mathbb{N}} \left|\frac{1}{k+1}\right| = 1/2$, hence $\|U\| = \sqrt{7}/(3\sqrt{\pi})$.

42. i) It is proved at exercise 12(i) that U is linear and continuous with $\|U\| = \|a\| \|b\|$.

ii) We have $\langle Ux, y \rangle = \langle x, U^*y \rangle \; \forall x \in H_1, \forall y \in H_2$. For $y \in H_2$ let $U^*y = w$. Then

$$\langle x, w \rangle = \langle Ux, y \rangle = \langle \langle x, a \rangle b, y \rangle = \langle x, a \rangle \langle b, y \rangle = \langle x, \langle y, b \rangle a \rangle \quad \forall x \in H_1,$$

and then $w = \langle y, b \rangle a$, i.e., $U^*y = \langle y, b \rangle a \; \forall y \in H_2$.

iii) We have

$$U^*Ux = \langle Ux, b \rangle a = \langle \langle x, a \rangle b, b \rangle a = \langle x, a \rangle \langle b, b \rangle a = \|b\|^2 \langle x, a \rangle a.$$

Let $P: H \to H$, $Px = \langle x, a \rangle a$. Then

$$P^2 x = P(Px) = \langle x, a \rangle Pa = \langle x, a \rangle \langle a, a \rangle a = \|a\|^2 Px,$$

and $\langle Px, x \rangle = \langle x, a \rangle \langle a, x \rangle = |\langle x, a \rangle|^2 \geq 0$, i.e., $P \geq 0$. Since $(P/\|a\|)^2 = P$ from the uniqueness of the square root we obtain that $\sqrt{P} = P/\|a\|$. Since $U^*U = \|b\|^2 P$ it follows that

$$|U| = \sqrt{U^*U} = \|b\| \sqrt{P} = (\|b\|/\|a\|) P,$$

i.e., $|U|(x) = (\|b\|/\|a\|) \langle x, a \rangle a \; \forall x \in H_1$.

iv) Let $(A, |U|)$ be the polar decomposition for U. We have $\ker A = \ker |U|$. But $x \in \ker |U| \Leftrightarrow \langle x, a \rangle = 0 \Leftrightarrow x \perp a \Leftrightarrow x \in \{a\}^\perp$, and therefore $\ker A = \{a\}^\perp$. Since $|U|(H_1) = \text{Sp}\{a\}$ and $A : |U|(H_1) \to H_2$ is defined by $Ax = Uz \Leftrightarrow x = |U|z$, then

$$A(a) = A\left(|U|\left(\frac{a}{\|a\| \|b\|}\right)\right) = U\left(\frac{a}{\|a\| \|b\|}\right),$$

and therefore $A(a) = (\|a\|/\|b\|) b$. Hence $A : H_1 \to H_2$ is defined by

$$A(x) = \begin{cases} 0, & \text{if } x \in \{a\}^\perp, \\ \lambda (\|a\|/\|b\|) b, & \text{if } x = \lambda a, \lambda \in \mathbb{K}. \end{cases}$$

v) We have $U : H \to H$, $Ux = \langle x, a \rangle b$. U is self-adjoint $\Leftrightarrow U^* = U \Leftrightarrow Ux = U^*x$ $\forall x \in H \Leftrightarrow \langle x, a \rangle b = \langle x, b \rangle a$ $\forall x \in H$. Suppose that U is self-adjoint. Then $\|a\|^2 b = \langle a, b \rangle a$, i.e., $b = ta$, with $t \in \mathbb{C} \setminus \{0\}$. We obtain that $\langle x, a \rangle ta = \overline{t} \langle x, a \rangle a$ $\forall x \in H$, and from here $t \in \mathbb{R} \setminus \{0\}$, i.e., $b = ta$ with $t \in \mathbb{R} \setminus \{0\}$. Conversely, if there is $t \in \mathbb{R} \setminus \{0\}$ such that $b = ta$ then $\langle x, a \rangle b = \langle x, b \rangle a$ $\forall x \in H$ and therefore U is self-adjoint.

U is normal $\Leftrightarrow U^*U = UU^*$. We have that $U^*U(x) = \langle x, a \rangle \|b\|^2 a$ and $UU^*(x) = \langle x, b \rangle \|a\|^2 b$ $\forall x \in H$. Hence U is normal $\Leftrightarrow \|b\|^2 \langle x, a \rangle a = \|a\|^2 \langle x, b \rangle b$ $\forall x \in H$. If U is normal, it follows that $\|b\|^2 \langle b, a \rangle a = \|a\|^2 \|b\|^2 b$, and therefore $b = \langle b, a \rangle a / \|a\|^2$, i.e., there is $\lambda \in \mathbb{C} \setminus \{0\}$ such that $b = \lambda a$. Conversely, if there is $\lambda \in \mathbb{C} \setminus \{0\}$ such that $b = \lambda a$ then $\forall x \in H$ we have

$$\|a\|^2 \langle x, b \rangle b = \|a\|^2 \overline{\lambda} \langle x, a \rangle \lambda a = |\lambda|^2 \|a\|^2 \langle x, a \rangle a = \|\lambda a\|^2 \langle x, a \rangle a = \|b\|^2 \langle x, a \rangle a,$$

and therefore U is a normal operator.

U is an orthogonal projector $\Leftrightarrow U$ is self-adjoint and $U^2 = U$. U is self-adjoint if and only if there is $t \in \mathbb{R} \setminus \{0\}$ such that $b = ta$. Also, $U^2 = U$ if and only if

$$\begin{aligned} U^2 x &= U(Ux) = \langle Ux, a \rangle b = \langle \langle x, a \rangle b, a \rangle b = \langle x, a \rangle \langle b, a \rangle b \\ &= Ux = \langle x, a \rangle b \ \forall x \in H, \end{aligned}$$

i.e. if and only if $\langle a, b \rangle = 1$. Hence U is an orthogonal projector $\Leftrightarrow b = ta$ and $\langle a, b \rangle = 1 \Leftrightarrow b = a/\|a\|^2$, and therefore $Ux = \langle x, a \rangle a/\|a\|^2$ $\forall x \in H$.

43. i) Let $U(x) = f$, where $f = \sum_{n=1}^\infty x_n \chi_{A_n}$. Since the sets A_n are pairwise disjoint, $|f|^2 = \sum_{n=1}^\infty |x_n|^2 \chi_{A_n}$, and then $\int_0^\infty |f|^2 d\mu = \sum_{n=1}^\infty |x_n|^2 \mu(A_n)$. Let $M = \sup_{n \in \mathbb{N}} \mu(A_n) < \infty$. Then

$$\|Ux\| = \|f\| = \left(\int_0^\infty |f|^2 d\mu\right)^{1/2} \leq \sqrt{M} \|x\| \ \forall x \in l_2.$$

From here it follows that U is well defined. Obviously U is linear, and therefore U is continuous, with $\|U\| \leq \sqrt{M}$.

ii) U^* is defined by $\langle Ux, f \rangle = \langle x, U^*f \rangle$ $\forall x \in l_2, \forall f \in L_2[0, \infty)$. For $f \in L_2[0, \infty)$, let $U^* f = y = (y_n)_{n \in \mathbb{N}}$. Then

$$\langle Ux, f \rangle = \int_0^\infty (Ux) \overline{f} d\mu = \int_0^\infty \sum_{n=1}^\infty x_n \chi_{A_n} \overline{f} d\mu = \sum_{n=1}^\infty x_n \overline{\int_{A_n} f d\mu},$$

12.2 Solutions

and $\langle x, y \rangle = \sum_{n=1}^{\infty} x_n \overline{y_n}$. Thus

$$\sum_{n=1}^{\infty} \overline{x_n} \int_{A_n} f d\mu = \sum_{n=1}^{\infty} \overline{x_n} y_n \quad \forall x = (x_n)_{n \in \mathbb{N}} \in l_2,$$

and then $y_n = \int_{A_n} f d\mu \ \forall n \in \mathbb{N}$, i.e., $U^* f = \left(\int_{A_n} f d\mu \right)_{n \in \mathbb{N}}$.

iii) By definition $|U| = \sqrt{U^*U}$, i.e., we must calculate the square root of U^*U. But

$$U^*Ux = U^* \left(\sum_{n=1}^{\infty} x_n \chi_{A_n} \right) = \sum_{n=1}^{\infty} x_n U^* (\chi_{A_n}),$$

i.e., we must evaluate $U^* (\chi_{A_n})$. We have

$$U^* (\chi_{A_n}) = \left(\int_{A_k} \chi_{A_n} d\mu \right)_{k \in \mathbb{N}} = (\mu(A_k \cap A_n))_{k \in \mathbb{N}} = \mu(A_n) e_n,$$

since the sets $(A_k)_{k \in \mathbb{N}}$ are pairwise disjoint. Then

$$U^*Ux = \sum_{n=1}^{\infty} x_n \mu(A_n) e_n = (x_1 \mu(A_1), x_2 \mu(A_2), ...),$$

i.e., U^*U is a multiplication operator. By exercise 14(iii) we obtain

$$|U|(x) = \left(x_1 \sqrt{\mu(A_1)}, x_2 \sqrt{\mu(A_2)}, ... \right) \quad \forall x = (x_n)_{n \in \mathbb{N}} \in l_2.$$

vi) We have $\|U\| = \||U|\| = \sup_{n \in \mathbb{N}} \sqrt{\mu(A_n)} = \sqrt{M}$. Also $\ker U = \ker |U| = \{0\}$, since $\mu(A_n) > 0 \ \forall n \in \mathbb{N}$. From the theory, we know that $A : |U|(l_2) \to U(l_2)$ is defined by $Ax = Uz \Leftrightarrow x = |U|z$. But $x = |U|z \Leftrightarrow x_n = z_n \sqrt{\mu(A_n)} \ \forall n \in \mathbb{N}$, i.e., $z_n = x_n / \sqrt{\mu(A_n)} \ \forall n \in \mathbb{N}$, from whence $A : l_2 \to L_2[0, \infty)$ is defined by

$$A(x_1, x_2, ...) = U \left(\frac{x_1}{\sqrt{\mu(A_1)}}, \frac{x_2}{\sqrt{\mu(A_2)}}, ... \right) = \sum_{n=1}^{\infty} \frac{x_n}{\sqrt{\mu(A_n)}} \chi_{A_n}.$$

44. i) By the Cauchy–Buniakowski–Schwarz inequality we have

$$\left| \int_n^{n+1} f(x) dx \right|^2 \leq \left(\int_n^{n+1} |f(x)|^2 dx \right) \left(\int_n^{n+1} 1 dx \right) = \int_n^{n+1} |f(x)|^2 \quad \forall n \geq 0.$$

Then

$$\sum_{n=0}^{\infty} \left| \int_n^{n+1} f(x) dx \right|^2 \leq \sum_{n=0}^{\infty} \int_n^{n+1} |f(x)|^2 = \int_0^{\infty} |f(x)|^2 dx < \infty.$$

Hence U is well defined, and $\|Uf\| \leq \|f\| \ \forall f \in L_2[0, \infty)$. Since obviously U is linear it follows that U is continuous.

ii) $U^* : l_2 \to L_2[0, \infty)$ is defined by the relation $\langle Uf, x \rangle = \langle f, U^*x \rangle$ $\forall f \in L_2[0, \infty)$, $\forall x \in l_2$. For $x = (x_n)_{n \geq 0} \in l_2$, let $U^*x = g$. Then

$$\langle Uf, x \rangle = \sum_{n=0}^{\infty} \overline{x_n} \int_n^{n+1} f(t)dt = \sum_{n=0}^{\infty} \overline{x_n} \int_0^{\infty} \chi_{[n,n+1)} f dt = \int_0^{\infty} \left(\sum_{n=0}^{\infty} \overline{x_n} \chi_{[n,n+1)} \right) f dt,$$

and $\langle f, U^*x \rangle = \langle f, g \rangle = \int_0^{\infty} f(t)\overline{g(t)}dt$. Hence

$$\int_0^{\infty} \left(\sum_{n=0}^{\infty} \overline{x_n} \chi_{[n,n+1)} \right) f dt = \int_0^{\infty} f \overline{g} dt \quad \forall f \in L_2[0, \infty),$$

and then it follows that $g = \sum_{n=0}^{\infty} x_n \chi_{[n,n+1)}$, i.e., $U^*(x) = \sum_{n=0}^{\infty} x_n \chi_{[n,n+1)}$ $\forall x = (x_n)_{n \geq 0} \in l_2$.

iii) Let $f \in L_2[0, \infty)$. Then

$$(U^*U)f = U^* \left(\left(\int_n^{n+1} f dt \right)_{n \geq 0} \right) = \sum_{n=0}^{\infty} \left(\int_n^{n+1} f dt \right) \chi_{[n,n+1)} = Vf.$$

We will prove that $V^2 = V$. Indeed, let $Vf = g$, i.e., $g = \sum_{n=0}^{\infty} \left(\int_n^{n+1} f dt \right) \chi_{[n,n+1)}$. Then $V^2 f = Vg = \sum_{n=0}^{\infty} \left(\int_n^{n+1} g dt \right) \chi_{[n,n+1)}$. Since the intervals $([n, n+1))_{n \geq 0}$ are pairwise disjoint it follows that $g\chi_{[n,n+1)} = \left(\int_n^{n+1} f dt \right) \chi_{[n,n+1)}$, and therefore

$$\int_n^{n+1} g dt = \int_0^{\infty} g\chi_{[n,n+1)} dt = \int_n^{n+1} f dt.$$

Then

$$V^2 f = Vg = \sum_{n=0}^{\infty} \left(\int_n^{n+1} f dt \right) \chi_{[n,n+1)} = Vf.$$

We have V positive and $V^2 = V$, so $\sqrt{V} = V$, i.e., $|U| = \sqrt{U^*U} = \sqrt{V} = V$, that is,

$$|U|(f) = \sum_{n=0}^{\infty} \left(\int_n^{n+1} f(t)\,dt \right) \chi_{[n,n+1)} \quad \forall f \in L_2[0, \infty).$$

iv) It is proved in (ii) that $U^*(x) = \sum_{n=0}^{\infty} x_n \chi_{[n,n+1)}$. Then by exercise 43 the polar decomposition of U^* is $(B, |U^*|)$, where $B : l_2 \to L_2[0, \infty)$ is defined by

$$B(x_1, x_2, \ldots) = \sum_{n=0}^{\infty} \frac{x_n}{\sqrt{\mu([n, n+1))}} \chi_{[n,n+1)} = \sum_{n=0}^{\infty} x_n \chi_{[n,n+1)}.$$

Now from the theory we know that if $(B, |W|)$ is the polar decomposition of W then $(B^*, |W^*|)$ is the polar decomposition for W^*. Therefore if $(B, |U^*|)$ is the polar decomposition for U^* then $(B^*, |U^{**}|) = (B^*, |U|)$ is the polar decomposition for U. Hence

12.2 Solutions

to find the polar decomposition for the operator U from our exercise, we must calculate B^*. We apply the same exercise and we obtain that $B^*f = \left(\int_{A_n} f(t)\,dt\right)_{n\geq 0}$, where $A_n = [n, n+1)$, that is, $B^*f = \left(\int_n^{n+1} f(t)\,dt\right)_{n\geq 0}$. Hence the polar decomposition for U is given by $U = A|U|$, where

$$|U|(f) = \sum_{n=0}^{\infty}\left(\int_n^{n+1} f(t)\,dt\right)\chi_{[n,n+1)}, \quad A(f) = \left(\int_n^{n+1} f(t)\,dt\right)_{n\geq 0} = U(f).$$

45. We have that if $f \in L_1[c,d]$ and $\varphi : [a,b] \to [c,d]$ is a C^1 diffeomorphism, then

$$\int_c^d f(x)\,dx = \int_a^b |\varphi'(x)| f(\varphi(x))\,dx$$

(the change of variables formula for the Lebesgue integral).

i) We have $\int_c^d |f(\varphi^{-1}(x))|^2\,dx = \int_a^b |\varphi'(t)|\,|f(t)|^2\,dt$. Hence

$$\|Uf\|^2 = \int_a^b |\varphi'(t)|\,|f(t)|^2\,dt \leq \left(\sup_{t\in[a,b]}|\varphi'(t)|\right)\|f\|^2,$$

and then $\|Uf\| \leq \left(\sup_{t\in[a,b]}|\varphi'(t)|\right)^{1/2}\|f\|$, i.e. U is well defined. Since U is obviously linear, U is continuous and $\|U\| \leq \left(\sup_{t\in[a,b]}|\varphi'(t)|\right)^{1/2}$.

ii) We have $\langle Uf, g\rangle = \langle f, U^*g\rangle\ \forall f \in L_2[a,b],\ \forall g \in L_2[c,d]$. For $g \in L_2[c,d]$, let $h = U^*g$. Then

$$\langle Uf, g\rangle = \int_c^d f(\varphi^{-1}(x))\overline{g(x)}\,dx = \int_a^b |\varphi'(t)|\,f(t)\overline{g(\varphi(t))}\,dt.$$

Also $\langle f, U^*g\rangle = \langle f, h\rangle = \int_a^b f(t)\overline{h(t)}\,dt$. Since $\langle Uf, g\rangle = \langle f, U^*g\rangle$, it follows that

$$\int_a^b |\varphi'(t)|\,f(t)\overline{g(\varphi(t))}\,dt = \int_a^b f(t)\overline{h(t)}\,dt\ \ \forall f \in L_2[a,b],$$

so $h(x) = |\varphi'(x)|\,g(\varphi(x))$ a.e., i.e.,

$$(U^*g)(x) = |\varphi'(x)|\,g(\varphi(x))\ \ \forall g \in L_2[c,d].$$

iii) For $f \in L_2[a,b]$, let $Uf = g$. Then $(U^*g)(x) = |\varphi'(x)|\,g(\varphi(x))$. Now

$$g(\varphi(x)) = (Uf)(\varphi(x)) = f(\varphi^{-1}(\varphi(x))) = f(x),$$

and then $(U^*g)(x) = |\varphi'(x)| f(x)$, i.e., $(U^*Uf)(x) = |\varphi'(x)| f(x)$, that is, $U^*U = M_{|\varphi'|}$, where $M_{|\varphi'|}$ is the multiplication operator given by $|\varphi'|$ (see exercise 26). Then $\sqrt{M_{|\varphi'|}} = M_{\sqrt{|\varphi'|}}$, and therefore

$$|U| = \sqrt{U^*U} = \sqrt{M_{|\varphi'|}} = M_{\sqrt{|\varphi'|}}.$$

iv) We have $f \in \ker |U| \Leftrightarrow |U|f = 0 \Leftrightarrow \sqrt{\varphi'(x)} f(x) = 0$ a.e. $\Leftrightarrow f(x) = 0$ a.e., i.e., $f = 0$ in $L_2[a,b]$. Therefore $\ker |U| = \{0\}$. If $(A, |U|)$ is the polar decomposition for U we know that $(\ker A)^\perp = (\ker |U|)^\perp = \{0\}^\perp = L_2[a,b]$ is the initial subspace for A and that the final subspace for A is $\overline{U(L_2[a,b])} = L_2[c,d]$. Moreover, $A : |U|(L_2[a,b]) \to L_2[c,d]$ is defined by $Af = Ug \Leftrightarrow f = |U|g$. But $f = |U|g \Leftrightarrow f(x) = \sqrt{|\varphi'(x)|} g(x)$ a.e. $\Leftrightarrow g(x) = f(x)/\sqrt{|\varphi'(x)|}$ a.e.. Hence

$$(Af)(x) = (Ug)(x) = g(\varphi^{-1}(x)) = \frac{f(\varphi^{-1}(x))}{\sqrt{|\varphi'(\varphi^{-1}(x))|}}.$$

Since $\varphi(\varphi^{-1}(x)) = x \; \forall x \in [c,d]$ we have $\varphi'(\varphi^{-1}(x))(\varphi^{-1})'(x) = 1$, and then

$$(Af)(x) = \sqrt{|(\varphi^{-1})'(x)|} f(\varphi^{-1}(x)).$$

46. Let $\varphi : [0,1] \to [0,1]$ be defined by $\varphi(x) = x/(2-x)$. Then $\varphi^{-1}(x) = 2x/(1+x)$, i.e., $(Uf)(x) = f(\varphi^{-1}(x))$. From exercise 45 it follows that

$$(U^*f)(x) = |\varphi'(x)| f(\varphi(x)) = \frac{2}{(2-x)^2} f\left(\frac{x}{2-x}\right),$$

$$(|U|f)(x) = \sqrt{|\varphi'(x)|} f(x) = \frac{\sqrt{2}}{2-x} f(x),$$

and the polar decomposition for U is $(A, |U|)$, where

$$(Af)(x) = \sqrt{|(\varphi^{-1})'(x)|} f(\varphi^{-1}(x)) = \frac{\sqrt{2}}{1+x} f\left(\frac{2x}{1+x}\right).$$

47. We will use the results from exercise 45.

i) U is self-adjoint $\Leftrightarrow U = U^* \Leftrightarrow (Uf)(x) = (U^*f)(x) \Leftrightarrow f(\varphi^{-1}(x)) = |\varphi'(x)| f(\varphi(x)) \; \forall f \in L_2[0,1]$. Suppose that U is self-adjoint. For $f(x) = 1 \in L_2[0,1]$ it follows that $|\varphi'(x)| = 1$, equality in $L_2[0,1]$, i.e., $|\varphi'(x)| = 1$ a.e.. Since both members are continuous functions then $|\varphi'(x)| = 1 \; \forall x \in [0,1]$. Since φ' has the Darboux property it follows that $\varphi'(x) = 1 \; \forall x \in [0,1]$, or $\varphi'(x) = -1 \; \forall x \in [0,1]$. In the first case it follows that there is a constant $k \in \mathbb{R}$ such that $\varphi(x) = x + k \; \forall x \in [0,1]$. Since $\varphi : [0,1] \to [0,1]$ is bicontinuous it follows that $k = 0$, i.e., $\varphi(x) = x \; \forall x \in [0,1]$, so $U = I$. In the second case it follows that there is a constant $k \in \mathbb{R}$ such that $\varphi(x) = -x + k \; \forall x \in [0,1]$. Since $\varphi : [0,1] \to [0,1]$ is bicontinuous it follows that $k = 1$, i.e., $\varphi(x) = 1 - x \; \forall x \in [0,1]$, and then $Uf(x) = f(1-x)$. Conversely, it is clear that the above two operators are self-adjoint.

12.2 Solutions

ii) We have $(U^*Uf)(x) = |\varphi'(x)| f(x)$, and

$$(UU^*f)(x) = (Ug)(x) = g(\varphi^{-1}(x)) = |\varphi'(\varphi^{-1}(x))| f(x),$$

where $g(x) = (U^*f)(x) = |\varphi'(x)| f(\varphi(x))$. Hence U is normal $\Leftrightarrow |\varphi'(x)| f(x) = |\varphi'(\varphi^{-1}(x))| f(x)$ $\forall f \in L_2[0,1]$ $\Leftrightarrow |\varphi'(x)| = |\varphi'(\varphi^{-1}(x))|$ a.e. $\Leftrightarrow |\varphi'(x)| = |\varphi'(\varphi^{-1}(x))|$ $\forall x \in [0,1]$, by continuity. For $x \to \varphi(x)$, it follows that U is normal if and only if $|\varphi'(\varphi(x))| = |\varphi'(x)|$ $\forall x \in [0,1]$.

iii) $U \in L(H)$ is an orthogonal projector if and only if U is self-adjoint and $U^2 = U$ (see exercise 34 of chapter 11). If $U^2 = U$ we obtain that $f(\varphi^{-1}(\varphi^{-1}(x))) = f(\varphi^{-1}(x))$ $\forall f \in L_2[0,1]$ and taking $f(x) = x$ it follows that $\varphi^{-1}(\varphi^{-1}(x)) = \varphi^{-1}(x)$ a.e., and therefore $\varphi^{-1}(\varphi^{-1}(x)) = \varphi^{-1}(x)$ $\forall x \in [0,1]$, that is, $\varphi(\varphi(x)) = \varphi(x)$ $\forall x \in [0,1]$, and since φ is injective, $\varphi(x) = x$ $\forall x \in [0,1]$, in which case $(Uf)(x) = f(x)$, i.e., $U = I$, the identity operator, which is obviously an orthogonal projector. Hence U is an orthogonal projector if and only if $\varphi(x) = x$ $\forall x \in [0,1]$.

48. i) We have $\langle Uf, g \rangle = \langle f, U^*g \rangle$ $\forall f, g \in L_2[0,1]$. Let $U^*g = v$. Then

$$\int_0^1 h(x) f(\varphi^{-1}(x)) \overline{g(x)} dx = \int_0^1 f(x) \overline{v(x)} dx.$$

Now by a change of variables we have

$$\int_0^1 h(x) f(\varphi^{-1}(x)) \overline{g(x)} dx = \int_0^1 h(\varphi(t)) f(t) \overline{g(\varphi(t))} |\varphi'(t)| dt \quad \forall f \in L_2[0,1].$$

Then $\overline{v(x)} = h(\varphi(x)) |\varphi'(x)| \overline{g(\varphi(x))}$ a.e., i.e., $v(x) = \overline{h(\varphi(x))} |\varphi'(x)| g(\varphi(x))$, i.e.,

$$(U^*g)(x) = \overline{h(\varphi(x))} |\varphi'(x)| g(\varphi(x)).$$

We also have

$$\begin{aligned}(U^*Uf)(x) &= (U^*g)(x) = \overline{h(\varphi(x))} |\varphi'(x)| g(\varphi(x)) \\ &= \overline{h(\varphi(x))} |\varphi'(x)| h(\varphi(x)) f(x) = |h(\varphi(x))|^2 |\varphi'(x)| f(x),\end{aligned}$$

i.e., $U^*U = M_u$, where $u(x) = |h(\varphi(x))|^2 |\varphi'(x)|$. Using exercise 26 we have that $|U| = \sqrt{U^*U} = \sqrt{M_u} = M_{\sqrt{u}}$, i.e.,

$$(|U|f)(x) = (M_{\sqrt{u}} f)(x) = |h(\varphi(x))| \sqrt{|\varphi'(x)|} f(x).$$

ii) Take $h(x) = x^n$, $\varphi(x) = 1 - x$ and then apply (i). We obtain that $(U^*g)(x) = (1-x)^n g(1-x)$ and $(|U|f)(x) = (1-x)^n f(x)$.

49. We have that if $f \in L_1(\varphi(A))$ then

$$\int_{\varphi(A)} f(x) dx = \int_A f(\varphi(x)) |\mathcal{J}_\varphi(x)| dx,$$

where $\mathcal{J}_\varphi(x) = \det \varphi'(x)$ is the Jacobian of φ (the change of variables formula for the Lebesgue integral in $\mathbb{R}^n$).

i) For $f \in L_2(A)$, we have $\int_{\varphi(A)} |f(\varphi^{-1}(x))|^2 \, dx = \int_A |\mathcal{J}_\varphi(x)| \, |f(x)|^2 \, dx$. Hence

$$\|Tf\|^2 = \int_A |f(\varphi^{-1}(x))|^2 \, dx \leq \left(\sup_{x \in A} |\mathcal{J}_\varphi(x)| \right) \|f\|^2,$$

and then $\|Tf\| \leq \sqrt{\sup_{x \in A} |\mathcal{J}_\varphi(x)|} \, \|f\| < \infty$, i.e., T is well defined. Since T is obviously linear then T is continuous and $\|T\| \leq \sqrt{\sup_{x \in A} |\mathcal{J}_\varphi(x)|}$. ($\mathcal{J}_\varphi$ is continuous on $\overline{A}$, and therefore bounded on $\overline{A}$, since $\overline{A}$ is compact.)

ii) We have by the definition that $\langle Tf, g \rangle = \langle f, T^*g \rangle$ $\forall f \in L_2(A), \forall g \in L_2(\varphi(A))$. For $g \in L_2(\varphi(A))$ let $h = T^*g$. Then

$$\langle Tf, g \rangle = \int_{\varphi(A)} f(\varphi^{-1}(x)) \overline{g(x)} dx = \int_A |\mathcal{J}_\varphi(t)| \, f(t) \overline{g(\varphi(t))} dt.$$

Also $\langle f, T^*g \rangle = \langle f, h \rangle = \int_A f(t) \overline{h(t)} dt$. Using that $\langle Tf, g \rangle = \langle f, T^*g \rangle$ we obtain

$$\int_A |\mathcal{J}_\varphi(t)| \, f(t) \overline{g(\varphi(t))} dt = \int_A f(t) \overline{h(t)} dt \ \ \forall f \in L_2(A),$$

and then $h(x) = |\mathcal{J}_\varphi(x)| \, g(\varphi(x))$ a.e., i.e., $(T^*g)(x) = |\mathcal{J}_\varphi(x)| \, g(\varphi(x))$.

iii) For $f \in L_2(A)$, let $Tf = g$. We have $(T^*g)(x) = |\mathcal{J}_\varphi(x)| \, g(\varphi(x))$. But

$$g(\varphi(x)) = (Tf)(\varphi(x)) = f(\varphi^{-1}(\varphi(x))) = f(x),$$

and therefore $(T^*Tf)(x) = |\mathcal{J}_\varphi(x)| \, f(x)$, i.e. $T^*T = M_h$, where $h(x) = |\mathcal{J}_\varphi(x)| \geq 0$. From exercise 26 we have $\sqrt{M_h} = M_{\sqrt{h}}$, and then $|T| = M_{\sqrt{h}}$, i.e., $(|T|f)(x) = \sqrt{|\mathcal{J}_\varphi(x)|} f(x)$.

iv) We have $f \in \ker |T| \Leftrightarrow |T|f = 0 \Leftrightarrow \sqrt{|\mathcal{J}_\varphi(x)|} f(x) = 0$ a.e. $\Leftrightarrow f(x) = 0$ a.e. $\Leftrightarrow f = 0$ in $L_2(A)$, i.e. $\ker |T| = \{0\}$. If $(A, |T|)$ is the polar decomposition for T, we know that $(\ker A)^\perp = (\ker |T|)^\perp = \{0\}^\perp = L_2(A)$ is the initial subspace for A and $\overline{T(L_2(A))} = L_2(\varphi(A))$ is the final subspace for A. In addition, $A : |T|(L_2(A)) \to L_2(\varphi(A))$ is defined by $Af = Tg \Leftrightarrow f = |T|g$. Using what we have proved, $f = |T|g \Leftrightarrow f(x) = \sqrt{|\mathcal{J}_\varphi(x)|} g(x)$ a.e. $\Leftrightarrow g(x) = f(x)/\sqrt{|\mathcal{J}_\varphi(x)|}$ a.e., i.e.,

$$(Af)(x) = (Tg)(x) = g(\varphi^{-1}(x)) = \frac{f(\varphi^{-1}(x))}{\sqrt{|\mathcal{J}_\varphi(\varphi^{-1}(x))|}}.$$

Since $\varphi(\varphi^{-1}(x)) = x$ by differentiation $\varphi'(\varphi^{-1}(x))(\varphi^{-1})'(x) = I_{\mathbb{R}^n}$ and passing to the determinant we obtain that $\mathcal{J}_\varphi(\varphi^{-1}(x)) = 1/\mathcal{J}_{\varphi^{-1}}(x)$, and therefore

$$(Af)(x) = \sqrt{|\mathcal{J}_{\varphi^{-1}}(x)|} f(\varphi^{-1}(x)).$$

12.2 Solutions

50. We will use exercise 49.

i) T is self-adjoint if and only if $T = T^*$, that is, if and only if

$$f(\varphi^{-1}(x)) = |\mathcal{J}_\varphi(x)| \, f(\varphi(x)) \quad \forall f \in L_2(A).$$

Suppose that T is self-adjoint. Since A is bounded, hence of finite Lebesgue measure, $f(x) = 1 \in L_2(A)$, and we obtain that $|\mathcal{J}_\varphi(x)| = 1$ a.e.. By continuity we have $|\mathcal{J}_\varphi(x)| = 1 \; \forall x \in A$, and therefore $f(\varphi^{-1}(x)) = f(\varphi(x)) \; \forall f \in L_2(A)$. For $i = \overline{1,n}$ the canonical projections $p_i : A \to \mathbb{R}$ defined by $p_i(x_1, x_2, ..., x_n) = x_i$ belong to $L_2(A)$, and then $p_i(\varphi^{-1}(x)) = p_i(\varphi(x))$ a.e.. Again from the continuity it follows that $p_i(\varphi^{-1}(x)) = p_i(\varphi(x)) \; \forall x \in A$, and therefore $\varphi^{-1}(x) = \varphi(x) \; \forall x \in A$. Since $\varphi(A) = A$ for $x \longmapsto \varphi(x)$ we obtain that $\varphi(\varphi(x)) = x \; \forall x \in A$. Hence if T is self-adjoint then $|\mathcal{J}_\varphi(x)| = 1 \; \forall x \in A$ and $\varphi(\varphi(x)) = x \; \forall x \in A$. Conversely, from these two conditions it follows easily that T is self-adjoint. Now if we use the Darboux property, the connectedness of A and the continuity for φ' and for the determinant function, it follows that T is self-adjoint if and only if $\mathcal{J}_\varphi(x) = 1 \; \forall x \in A$ and $\varphi(\varphi(x)) = x \; \forall x \in A$, or $\mathcal{J}_\varphi(x) = -1 \; \forall x \in A$ and $\varphi(\varphi(x)) = x \; \forall x \in A$.

ii) We have $(T^*Tf)(x) = |\mathcal{J}_\varphi(x)| \, f(x)$. For $f \in L_2(A)$ let $g = T^*f$. Then $g(x) = |\mathcal{J}_\varphi(x)| \, f(\varphi(x))$ and then

$$Tg(x) = g(\varphi^{-1}(x)) = |\mathcal{J}_\varphi(\varphi^{-1}(x))| \, f(\varphi(\varphi^{-1}(x))) = |\mathcal{J}_\varphi(\varphi^{-1}(x))| \, f(x),$$

i.e., $(TT^*f)(x) = |\mathcal{J}_\varphi(\varphi^{-1}(x))| \, f(x)$.

T is normal if and only if $TT^* = T^*T$, that is, if and only if

$$|\mathcal{J}_\varphi(x)| \, f(x) = |\mathcal{J}_\varphi(\varphi^{-1}(x))| \, f(x) \quad \forall f \in L_2(A),$$

which is equivalent with $|\mathcal{J}_\varphi(x)| = |\mathcal{J}_\varphi(\varphi^{-1}(x))|$ a.e.. From the continuity this is equivalent with $|\mathcal{J}_\varphi(x)| = |\mathcal{J}_\varphi(\varphi^{-1}(x))| \; \forall x \in A$, i.e.,

$$\left|\det \varphi'(x)\right| = \left|\det \varphi'(\varphi^{-1}(x))\right| \quad \forall x \in A \subseteq \mathbb{R}^n.$$

Since $\varphi(A) = A$ for $x \longmapsto \varphi(x)$ we obtain that

$$\left|\det \varphi'(\varphi(x))\right| = \left|\det \varphi'(x)\right| \quad \forall x \in A,$$

and therefore $\det \varphi'(\varphi(x)) = \det \varphi'(x) \; \forall x \in A$, or $\det \varphi'(\varphi(x)) = -\det \varphi'(x) \; \forall x \in A$.

51. Let $\varphi : (-\infty, 2)^2 \to (-1, \infty)^2$ be a C^1 diffeomorphism such that $\varphi^{-1} : (-1, \infty)^2 \to (-\infty, 2)^2$ is defined by

$$\varphi^{-1}(x, y) = (2x/(1+x), 2y/(1+y)).$$

The Jacobian for φ^{-1} is

$$\mathcal{J}_{\varphi^{-1}}(x, y) = \begin{vmatrix} 2/(1+x)^2 & 0 \\ 0 & 2/(1+y)^2 \end{vmatrix} = \frac{4}{(1+x)^2 (1+y)^2},$$

and $\varphi : (-\infty, 2)^2 \to (-1, \infty)^2$ is given by $\varphi(x,y) = (x/(2-x), y/(2-y))$. The Jacobian is

$$\mathcal{J}_\varphi(x,y) = \begin{vmatrix} 2/(2-x)^2 & 0 \\ 0 & 2/(2-y)^2 \end{vmatrix} = \frac{4}{(2-x)^2(2-y)^2}.$$

Using exercise 49 we deduce that

$$(U^*f)(x,y) = \frac{4}{(2-x)^2(2-y)^2} f\left(\frac{x}{2-x}, \frac{y}{2-y}\right),$$

$$(|U|f)(x,y) = \frac{2}{(2-x)(2-y)} f(x,y),$$

and

$$(Af)(x,y) = \frac{2}{(1+x)(1+y)} f\left(\frac{2x}{1+x}, \frac{2y}{1+y}\right).$$

52. i) We define $R : Y \to Z$ by $R(y) = S(x)$ if and only if $y = T(x)$. If $y = T(x_1)$ and $y = T(x_2)$ then $T(x_1 - x_2) = 0$, and then, by hypothesis, $||S(x_1 - x_2)|| = 0$, i.e., $S(x_1) = S(x_2)$, so $R : Y \to Z$ is well defined (T is surjective, therefore R is defined on Y). Let now $\alpha, \beta \in \mathbb{K}$, $y, z \in Y$. Let $y = T(x)$, $z = T(w)$, $x, w \in X$. Then by the definition of R and the linearity of S we have $\alpha R(y) + \beta R(z) = \alpha S(x) + \beta S(w) = S(\alpha x + \beta w)$. Since $T(\alpha x + \beta w) = \alpha y + \beta z$, it follows again by the definition of R that $R(\alpha y + \beta z) = S(\alpha x + \beta w)$, hence $R(\alpha y + \beta z) = \alpha R(y) + \beta R(z)$, i.e., R is linear. For any $y \in Y$ we have $||Ry|| = ||Sx|| \leq M ||Tx|| = M ||y||$, ($y = Tx$), and so $R \in L(Y, Z)$ with $||R|| \leq M$. By construction, $R \circ T = S$. The uniqueness of R follows from the fact that T is surjective.

ii) Let $S : X \to Z$ and $T : X \to H$ be linear and continuous, where Z is a Banach space and H a Hilbert space. Using the same technique as in (i) we can define $\widetilde{R} : T(X) \subseteq H \to Z$, a linear and continuous operator, such that $\widetilde{R} \circ T = S$, $||\widetilde{R}|| \leq M$. Since Z is a Banach space, $\widetilde{R}$ has a unique linear extension to $\overline{T(X)}$, i.e., there is $\overline{\widetilde{R}} : \overline{T(X)} \subseteq H \to Z$ linear and continuous, unique with $||\overline{\widetilde{R}}|| = ||\widetilde{R}||$ and $\overline{\widetilde{R}}\big|_{T(X)} = \widetilde{R}$. Since $\overline{T(X)} \subseteq H$ is a closed linear subspace in a Hilbert space, let $P : H \to \overline{T(X)}$ be the orthogonal projection. We define then $R : H \to Z$ by $R = \overline{\widetilde{R}} \circ P$. Then $R \in L(H, Z)$ and $||R|| \leq ||\overline{\widetilde{R}}|| \, ||P|| \leq ||\overline{\widetilde{R}}|| = ||\widetilde{R}|| \leq M$. For any $x \in X$ we have

$$S(x) = \widetilde{R}(T(x)) = \overline{\widetilde{R}}(T(x)) = \overline{\widetilde{R}}(P((T(x))),$$

since $T(x) \in T(X) \subseteq \overline{T(X)}$, hence $S = R \circ T$.

53. i) Let $q : L(H) \to \mathbb{R}$ be defined by $q(U) = \inf_{||x|| \leq 1} \langle Ux, x \rangle$. We will prove that q is supra-linear. Let $U, V \in L(H)$. Then $\langle (U+V)x, x \rangle = \langle Ux, x \rangle + \langle Vx, x \rangle \geq q(U) + q(V)$, $\forall ||x|| \leq 1$, and therefore passing to the infimum over $||x|| \leq 1$ we deduce that $q(U+V) \geq q(U) + q(V)$. For $\lambda \geq 0$ we have $q(\lambda U) = \inf_{||x|| \leq 1} \langle \lambda Ux, x \rangle = \lambda \inf_{||x|| \leq 1} \langle Ux, x \rangle = \lambda q(U)$.

12.2 Solutions

Also if $U \in \mathcal{A}(H)$ for $x \in B_H$ we have $\langle Ux, x \rangle \leq |\langle Ux, x \rangle| \leq \sup_{\|y\|\leq 1} |\langle Uy, y \rangle| = \|U\|$, and then $q(U) \leq \|U\|$ $\forall U \in \mathcal{A}(H)$. Let $p : L(H) \to \mathbb{R}$ be defined by $p(U) = \|U\|$. Since p is a norm, p is sublinear. So we have the following situation: $p, q : X \to \mathbb{R}$, with p sublinear and q supra-linear, $G \subseteq X$ is a convex cone, and $q(x) \leq p(x)$ $\forall x \in G$. From exercise 3(ii) of chapter 7 it follows that there is $f : X \to \mathbb{R}$, a linear functional, such that $q(x) \leq f(x)$ $\forall x \in G$ and $f(x) \leq p(x)$ $\forall x \in X$. In our case we deduce that there is $f : L(H) \to \mathbb{R}$ linear such that $\inf_{\|x\|\leq 1} \langle Ux, x \rangle \leq f(U)$ $\forall U \in \mathcal{A}(H)$ and $f(U) \leq \|U\|$ $\forall U \in L(H)$. From here we deduce that f is continuous and $\|f\| \leq 1$. Also from $\inf_{\|x\|\leq 1} \langle -Ux, x \rangle \leq f(-U)$ $\forall U \in \mathcal{A}(H)$ it follows that $-\sup_{\|x\|\leq 1} \langle Ux, x \rangle \leq -f(U)$ $\forall U \in \mathcal{A}(H)$, and therefore $f(U) \leq \sup_{\|x\|\leq 1} \langle Ux, x \rangle$ $\forall U \in \mathcal{A}(H)$.

ii) As above, taking $q : L(H) \to \mathbb{R}$ defined by $q(U) = \inf_{\|x\|=1} \langle Ux, x \rangle$ we deduce that there is a linear and continuous functional $g : L(H) \to \mathbb{R}$ with norm ≤ 1 such that

$$\inf_{\|x\|=1} \langle Ux, x \rangle \leq g(U) \leq \sup_{\|x\|=1} \langle Ux, x \rangle \quad \forall U \in \mathcal{A}(H).$$

Since $\inf_{\|x\|=1} \langle x, x \rangle \leq g(I) \leq \sup_{\|x\|=1} \langle x, x \rangle$ it follows that $g(I) = 1$ and therefore $\|g\| = 1$.

Part III

General topological spaces

Chapter 13

Linear topological and locally convex spaces

Definition. A *linear topological* space is a linear space X with a topology τ on it such that each of the following maps is continuous:
  i) $+ : X \times X \to X, (x, y) \to x + y$;
  ii) $\cdot : \mathbb{K} \times X \to X, (\lambda, y) \to \lambda x$.

In this case the topology τ is called a *linear topology* on X.

Definition. Let X be a linear topological space. A subset $A \subseteq X$ is called:

  i) *balanced* if and only if $\forall \lambda \in \mathbb{K}$ with $|\lambda| \leq 1$ and $\forall x \in A$ it follows that $\lambda x \in A$;
  ii) *absorbing* if and only if $\forall x \in X$ $\exists \delta > 0$ such that $\forall \lambda \in \mathbb{K}$ with $\lambda \neq 0$ and $|\lambda| < \delta$ it follows that $\lambda x \in A$;
  iii) *bounded* if and only if $\forall V$ a neighborhood of 0 it exists $\lambda \in \mathbb{R}$ such that $A \subseteq \lambda V$;
  iv) *convex* if and only if $\forall x, y \in A, \forall \lambda \in [0, 1]$ it follows that $\lambda x + (1 - \lambda) y \in A$;
  v) *convex cone* if and only if $\forall x, y \in A$ it follows that $x + y \in A$ and $\forall x \in A, \forall \lambda \geq 0$ it follows that $\lambda x \in A$.

Theorem. Let X be a linear topological space and $A \subseteq X$ a balanced subset. Then A is absorbing if and only if $\forall x \in X$ $\exists \delta > 0$ such that $\delta x \in A$.

Theorem. A linear topological space X is Hausdorff if and only if $\forall x \neq 0$ $\exists W$ a neighborhood of 0 such that $x \notin W$.

Theorem. Let X be a linear topological space. Then there exists a basis $\mathcal{B}$ for the system of neighborhoods at 0 such that:

  i) $\forall U, V \in \mathcal{B}$ $\exists W \in \mathcal{B}$ such that $W \subseteq U \cap V$;
  ii) U is balanced and absorbing $\forall U \in \mathcal{B}$;
  iii) $\forall U \in \mathcal{B}, \forall \lambda \in \mathbb{K}$ with $\lambda \neq 0$ it follows that $\lambda U \in \mathcal{B}$;
  iv) $\forall U \in \mathcal{B}$ $\exists V \in \mathcal{B}$ such that $V + V \subseteq U$.

Conversely, let X be a linear space and $\mathcal{B}$ a non-empty family of subsets of X which satisfies (i), (ii), (iii) and (iv), and let

$$\tau = \{W \subseteq X \mid \forall x \in W \; \exists U \in \mathcal{B} \text{ such that } x + U \subseteq W\}.$$

Then τ is a linear topology for X having $\mathcal{B}$ as a basis of neighborhoods for 0.

Definition. Let X be a linear topological space. The dual of X is $X^* = \{x^* : X \to \mathbb{K} \mid x^* \text{ is linear and continuous}\}$.

Definition. Let X be a linear topological space and $A \subseteq X$ an absorbing set. The functional $p_A : X \to \mathbb{R}$ defined by $p_A(x) = \inf\{\mu > 0 \mid x \in \mu A\}$ is called the *Minkowski functional* associated with A.

Theorem. Let X be a linear topological space, $A \subseteq X$ an absorbing set, and $p_A : X \to \mathbb{R}$ the Minkowski functional associated with A.

i) We have $p_A(\lambda x) = \lambda p_A(x) \; \forall x \in X, \forall \lambda > 0$. If $0 \in A$ then $p_A(0) = 0$ and in this case p_A is positive homogeneous.

ii) If A is convex then p_A is subadditive, i.e., $p_A(x + y) \leq p_A(x) + p_A(y) \; \forall x, y \in X$.

iii) If A is balanced then $p_A(\lambda x) = |\lambda| p_A(x) \; \forall x \in X, \forall \lambda \in \mathbb{K}$.

If A is also convex and balanced then p_A is a seminorm, called the *Minkowski seminorm* associated with A.

Definition. A linear topological space X is called *locally convex* if and only if there is a basis of neighborhoods $\mathcal{B}$ for 0 consisting only of convex sets.

Theorem. Let X be a linear space and $(p_i)_{i \in I}$ a family of seminorms on X. For every $\varepsilon > 0$, $n \in \mathbb{N}$, and $i_1, ..., i_n \in I$, we define $W(0; p_{i_1}, ..., p_{i_n}; \varepsilon) = \{x \in X \mid p_{i_1}(x) < \varepsilon, ..., p_{i_n}(x) < \varepsilon\}$. Then the family $\mathcal{B} = \{W(0; p_{i_1}, ..., p_{i_n}; \varepsilon) \mid i_1, ..., i_n \in I, \varepsilon > 0, n \in \mathbb{N}\}$ is a basis of neighborhoods of 0 for a locally convex topology on X. This topology is called the locally convex topology generated by the family of seminorms $(p_i)_{i \in I}$.

Conversely, if X is a locally convex space then there exists a family of seminorms $(p_i)_{i \in I}$ on X such that the linear topology on X coincides with the locally convex topology generated by the family of seminorms $(p_i)_{i \in I}$.

Theorem. Let X be a locally convex space and $(p_i)_{i \in I}$ a family of seminorms on X which generates the topology of X. Then:

i) X is a Hausdorff space if and only if the family $(p_i)_{i \in I}$ separates the points of X, i.e., $\forall x \neq 0 \; \exists i \in I$ such that $p_i(x) \neq 0$.

ii) A net $(x_\delta)_{\delta \in \Delta} \subseteq X$ converges to an element $x \in X$ if and only if $p_i(x_\delta - x) \to 0$ $\forall i \in I$.

Definition. A locally convex space is called an *F-space* or a *Fréchet space* if and only if its topology is generated by a complete metric.

13.1 Exercises

Examples of linear spaces with topologies that are not linear

1. Let $X \neq \{0\}$ be a linear space on which we consider the discrete topology, i.e., the topology given by all the subsets of X. Prove that the addition $+ : X \times X \to X$ is continuous, but the multiplication $\cdot : \mathbb{K} \times X \to X$ is not continuous.

2. Let $X = C((0,1); \mathbb{R})$. For $f \in X$ and $r > 0$ we denote by $V(f, r) = \{\gamma \in X \mid |f(x) - \gamma(x)| < r \; \forall x \in (0,1)\}$. We consider on X the topology generated by the family of sets $\{V(f, r) \mid f \in X, r > 0\}$. Prove that $+ : X \times X \to X$ is continuous, but $\cdot : \mathbb{R} \times X \to X$ is not continuous.

3. Let X be a linear space of dimension ≥ 2, and $Y \subseteq X$ a proper linear subspace. On X we consider the topology τ given by: $\tau = \{\emptyset, Y \setminus \{0\}, X\}$. Prove that $+ : X \times X \to X$ is not continuous, but $\cdot : \mathbb{K} \times X \to X$ is continuous.

4. Let $X \neq \{0\}$ be a linear space, $X^{\#} = \{x^{\#} : X \to \mathbb{K} \mid x^{\#} \text{ is linear}\}$ and let τ be the topology on X generated by the family $\mathcal{A} \subseteq \mathcal{P}(X)$, $\mathcal{A} = \{\ker x^{\#} \mid x^{\#} \in X^{\#}\}$. Prove that:

i) $+ : X \times X \to X$ is continuous at $(x_0, y_0) \in X \times X$ if and only if:
   (a) the system $\{x_0, y_0\} \subseteq X$ is linearly dependent and $x_0 + y_0 \neq 0$;

or

   (b) $x_0 = y_0 = 0$;

ii) $\cdot : \mathbb{K} \times X \to X$ is continuous at $(\lambda_0, y_0) \in \mathbb{K} \times X$ if and only if $(\lambda_0, y_0) \in ((\mathbb{K} \setminus \{0\}) \times X) \cup (0, 0)$.

Computations in linear topological spaces

5. Let X be a linear space and $A \subseteq X$ a convex subset which contains 0. Prove that A is balanced if and only if $\forall x \subset A$, $\forall \lambda \in \mathbb{K}$ with $|\lambda| = 1$ it follows that $\lambda x \in A$.

6. Let A be a convex subset in a real linear space X. Prove that A contains a convex, balanced, and absorbing set if and only if for each line d passing through $0 \in X$ the set $d \cap A$ contains an open segment, which contains $0 \in X$ (i.e., for each $x \in X$, $x \neq 0$, we can find $\alpha < 0 < \beta$ such that $\lambda x \in A \; \forall \lambda \in (\alpha, \beta)$).

7. i) Let X be a linear space and $A \subseteq X$ a non-empty and balanced set. Prove that the convex hull of A is balanced.

ii) Let X be a linear space and $A \subseteq X$ a non-empty set. Prove that the convex hull of the balanced hull of A is the convex balanced hull of A.

iii) Let $A = \{(0,0), (0,1), (1,0)\} \subseteq \mathbb{R}^2$. Prove that the balanced hull of the convex hull of A is not convex.

8. Let X be a linear space.

i) If $B_i \subseteq X$ is balanced for every $i \in I$, prove that $B = \bigcup_{i \in I} B_i$ is balanced.

ii) If $A \subseteq X$ is a set which contains the origin, let $B = \bigcup \{C \mid C \text{ balanced}, C \subseteq A\}$. Then B is the biggest balanced set included in A and prove that $B = \bigcap_{|\lambda| \geq 1} (\lambda A)$.

iii) Give an example of a non-empty set $A \subseteq X$, $0 \notin A$, such that $\bigcap_{|\lambda| \geq 1} (\lambda A) = \emptyset$.

9. i) Let X be a linear topological space, and $B \subseteq X$ a balanced set. Prove that $\overline{B}$ is also balanced. If $0 \in \overset{\circ}{B}$, prove that $\overset{\circ}{B}$ is balanced.

ii) Let $B \subseteq (\mathbb{C}^2, \|\cdot\|_2)$, $B = \{(z_1, z_2) \in \mathbb{C}^2 \mid |z_1| \leq |z_2|\}$. Prove that B is balanced, but $\overset{\circ}{B}$ is not balanced.

10. Let X be a linear topological space. Prove that:

i) The convex hull of a non-empty open subset of X is open;
ii) If $A, B \subseteq X$ are bounded, prove that $A + B \subseteq X$ is bounded;
iii) If $A, B \subseteq X$ are compact sets, prove that $A + B \subseteq X$ is compact;

13.1 Exercises

iv) If $A \subseteq X$ is compact and $B \subseteq X$ is closed, prove that $A + B \subseteq X$ is closed;
v) Not always the sum of two closed sets is closed.

Midpoint closed sets

11. Let X be a linear space and $A \subseteq X$ a non-empty set.
i) Prove that if $A + A \subseteq 2A$ then $\alpha A + \beta A \subseteq A$ for every $\alpha > 0$ and $\beta > 0$, α and β in $\mathbb{Q}$, $\alpha + \beta = 1$, each α and β being a finite sum of negative powers of 2.
ii) Prove that if A is closed and $A + A \subseteq 2A$ then A is convex.

The interior and the closure for a convex set

12. Let X be a linear topological space and $A \subseteq X$ a convex set.
i) Prove that if $x \in \overset{\circ}{A}$ and $y \in \overline{A}$ then $\{\lambda x + (1-\lambda) y \mid \lambda \in (0,1]\} \subseteq \overset{\circ}{A}$.
ii) Prove that $\overset{\circ}{A}, \overline{A} \subseteq X$ are convex sets.
iii) If $\overset{\circ}{A} \neq \varnothing$ prove that $\overset{\circ}{A} = \overset{\circ}{\overline{A}}$, $\overline{A} = \overline{\overset{\circ}{A}}$.

The Kakutani lemma and the Stone theorem

13. Let X be a linear space and $A, B \subseteq X$ two non-empty convex and disjoint sets.
i) Prove that for every $x \in X \setminus (A \cup B)$ we have $\operatorname{co}(A \cup \{x\}) \cap B = \varnothing$ or $\operatorname{co}(B \cup \{x\}) \cap A = \varnothing$.
ii) Using (i) deduce that there exist $C, D \subseteq X$ convex and disjoint sets such that $A \subseteq C$, $B \subseteq D$ and $C \cup D = X$.

Helly's theorem for convex sets

14. i) Let $n \in \mathbb{N}$ and $r \in \mathbb{N}$ with $r > n + 1$. Let $A_1, ..., A_r \subseteq \mathbb{R}^n$ be convex sets such that $\bigcap_{j=\overline{1,r},\, j \neq k} A_j \neq \varnothing$, for any $k = 1, 2, ..., r$. Prove that $\bigcap_{j=1}^{r} A_j \neq \varnothing$.
ii) Let $n \in \mathbb{N}$ and $\mathcal{C}$ be a family of compact and convex subsets of $\mathbb{R}^n$ such that for every choice of $n+1$ elements from $\mathcal{C}$ their intersection is non-empty. Prove that $\bigcap_{C \in \mathcal{C}} C \neq \varnothing$.

A linear topology need not be sequential

15. For $0 \in \mathbb{R}$ we consider a base of neighborhoods as being given by the sets of the form $(-\varepsilon, \varepsilon)$, with $\varepsilon > 0$, from which we eliminate a countable set of non-zero elements. We consider on $\mathbb{R}$ the linear topology given by this base of neighborhoods for $0 \in \mathbb{R}$. Find a set $A \subseteq \mathbb{R}$ such that $0 \in \overline{A}$, and yet we have no null convergent sequence in A.

Characterizations for linear and continuous functionals

16. Let X be a real linear topological space.
i) If $f : X \to \mathbb{R}$ is linear and continuous at $0 \in X$, prove that f is continuous on X.
ii) If $f : X \to \mathbb{R}$ is linear, prove that the following assertions are equivalent:
a) f is continuous;
b) $\ker f \subseteq X$ is closed;
c) f is bounded on a neighborhood V of $0 \in X$.

Connected open non-empty sets in a linear topological space are arcwise connected

17. Let X be a linear topological space and $G \subseteq X$ a non-empty connected and open subset. Prove that G is arcwise connected.

Additive operators continuous at 0 are linear operators

18. Let X and Y be two real linear topological spaces, Y being Hausdorff, and $T : X \to Y$ an additive operator (i.e., $T(x+y) = T(x) + T(y)$, $\forall x, y \in X$), continuous at $0 \in X$. Prove that T is linear.

The extension by continuity for a linear and continuous operator

19. Let X be a linear topological space, $M \subseteq X$ a dense linear subspace, Y an F-space and $T : M \to Y$ a linear and continuous operator (on M we have the topology given by X). Prove that T has a unique linear and continuous extension $\widetilde{T} : X \to Y$.

A non-metrizable linear topological space

20. On the linear space $C[0,1]$, we consider two topologies. The first, σ, given by the metric
$$d(f,g) = \int_0^1 \frac{|f(x) - g(x)|}{1 + |f(x) - g(x)|} dx,$$
and the second, τ_p, the pointwise convergence topology given by the family of seminorms $(p_x)_{x \in [0,1]}$, where $p_x(f) = |f(x)|$.

i) Prove that the identity $I : (C[0,1], \tau_p) \to (C[0,1], \sigma)$ is sequentially continuous but not continuous.

ii) Deduce that the pointwise convergence topology τ_p is not metrizable.

The form for some particular linear and continuous functionals

21. i) Let X be a normed space and $\Gamma \subseteq X^*$ a non-empty set. We consider on X the topology τ as being the locally convex topology given by the family of seminorms $\{p_{x^*} \mid x^* \in \Gamma\}$, where $p_{x^*} : X \to \mathbb{R}_+$, $p_{x^*}(x) = |x^*(x)|$. Prove that if $f : (X, \tau) \to \mathbb{K}$ is linear and continuous, we can then find $n \in \mathbb{N}$, $\alpha_1, ..., \alpha_n \in \mathbb{K}$ and $x_1^*, ..., x_n^* \in \Gamma$ such that $f = \sum_{i=1}^n \alpha_i x_i^*$.

ii) Deduce the form for the linear and continuous functionals on $(C[0,1], \tau_p)$, where τ_p is the pointwise convergence topology.

A linear topological space for which the dual is the null space

22. Let $X = C([0,1]; \mathbb{R})$. For every $\varepsilon > 0$ and $\delta \in (0,1)$ let $V_{\varepsilon,\delta}$ be the set of functions f in X for which there is an open set $A \subseteq [0,1]$, $A = \bigcup_{i=1}^\infty (\alpha_i, \beta_i)$, $((\alpha_i, \beta_i))_{i \in \mathbb{N}}$ pairwise disjoint intervals with $\sum_{i=1}^\infty (\beta_i - \alpha_i) \leq \delta$ such that $|f(t)| \leq \varepsilon \ \forall t \notin A$. (We can have $\alpha_i = \beta_i$, in this case $(\alpha_i, \beta_i) = \varnothing$.)

i) Prove that there is a linear topology τ on X such that $(V_{\varepsilon,\delta})_{\varepsilon > 0, 0 < \delta < 1}$ form a basis of neighborhoods for $0 \in X$ in the τ topology.

ii) Prove that (X, τ) is not a locally convex space.

iii) Prove that if f is a linear and continuous functional from (X, τ) to $\mathbb{R}$ then $f = 0$.

Effective calculation for the Minkowski seminorm

23. Let $(a_n)_{n \in \mathbb{N}}$ be a sequence of scalars, $1 \leq p < \infty$, and

$$A = \left\{ (x_n)_{n \in \mathbb{N}} \in l_p \;\middle|\; \sum_{n=1}^{\infty} |a_n| \, |x_n|^p \leq 1 \right\} \subseteq l_p.$$

i) Prove that A is balanced and convex.
ii) Prove that A is an absorbing set if and only if $(a_n)_{n \in \mathbb{N}} \in l_\infty$.
iii) If $(a_n)_{n \in \mathbb{N}} \in l_\infty$, calculate the Minkowski seminorm given by A.

24. i) Let X be a linear space, $p_n : X \to \mathbb{R}_+$, $n \in \mathbb{N}$, be a sequence of seminorms, and $A = \{x \in X \mid p_n(x) \leq 1 \; \forall n \in \mathbb{N}\}$.
a) Prove that A is convex and balanced.
b) Prove that A is absorbing if and only if $\sup_{n \in \mathbb{N}} p_n(x) < \infty \; \forall x \in X$, i.e., if and only if the family $(p_n)_{n \in \mathbb{N}}$ is pointwise bounded. In this case, find the Minkowski seminorm associated with A.
c) If, in addition, X is a Banach space and every seminorm p_n is continuous, then A is absorbing if and only if $\exists M > 0$ such that $\sup_{n \in \mathbb{N}} p_n(x) \leq M \|x\| \; \forall x \in X$.

ii) Let $\mathcal{R}[0,1] = \{f : [0,1] \to \mathbb{R} \mid f \text{ is Riemann integrable}\}$, which is a linear space with respect to the usual operations of addition and scalar multiplication. A linear functional $x^* : \mathcal{R}[0,1] \to \mathbb{R}$ is called positive if $x^*(f) \geq 0 \; \forall f \in \mathcal{R}[0,1]$ with $f \geq 0$. Let $x_n^* : \mathcal{R}[0,1] \to \mathbb{R}$, $n \in \mathbb{N}$, be a sequence of positive linear functionals and let $(a_n)_{n \in \mathbb{N}} \subseteq \mathbb{R}$.
a) We define

$$A = \{f \in \mathcal{R}[0,1] \mid |a_n x_n^*(f)| \leq 1 \; \forall n \in \mathbb{N}\}.$$

Using (i) prove that A is absorbing if and only if $(a_n x_n^*(1))_{n \in \mathbb{N}}$ belongs to l_∞, and in this case find the Minkowski seminorm associated with A. Then deduce that if $\varphi_n : [0,1] \to [0,\infty)$, $n \in \mathbb{N}$, is a sequence of Riemann integrable functions and

$$A = \left\{ f \in \mathcal{R}[0,1] \;\middle|\; \left| \int_0^1 \varphi_n(x) f(x) dx \right| \leq 1 \; \forall n \in \mathbb{N} \right\}$$

then A is absorbing if and only if the sequence $\left(\int_0^1 \varphi_n(x) dx \right)_{n \in \mathbb{N}}$ is upper bounded, and in this case find the Minkowski seminorm associated with A.

b) We define

$$A = \left\{ f \in \mathcal{R}[0,1] \;\middle|\; \sum_{n=1}^{\infty} |a_n x_n^*(f)| \leq 1 \right\}.$$

Using (i) prove that A is absorbing if and only if the sequence $(a_n x_n^*(1))_{n \in \mathbb{N}}$ belongs to l_1, and in this case find the Minkowski seminorm associated with A. In particular, if $(a_n)_{n \geq 0} \subseteq \mathbb{R}$ and

$$A = \left\{ f \in \mathcal{R}[0,1] \;\middle|\; \sum_{n=0}^{\infty} \left| a_n \int_0^1 x^n f(x) dx \right| \leq 1 \right\},$$

prove that A is absorbing if and only if the series $\sum_{n=0}^{\infty} \dfrac{a_n}{n+1}$ is absolutely convergent, and in this case find the Minkowski seminorm associated with A.

iii) Let X be a Banach space, $(x_n^*)_{n\in\mathbb{N}} \subseteq X^*$, and

$$A = \left\{ x \in X \mid \sum_{n=1}^\infty |x_n^*(x)| \leq 1 \right\}.$$

Using (i) prove that A is absorbing if and only if

$$\sup\left\{ \left\| \sum_{n\in F} x_n^* \right\| \mid F \text{ finite}, F \subseteq \mathbb{N} \right\} < \infty,$$

and in this case find the Minkowski seminorm associated with A.

iv) Let H be a Hilbert space, $(e_n)_{n\in\mathbb{N}} \subseteq H$ is an orthonormal system, $(a_n)_{n\in\mathbb{N}} \subseteq \mathbb{K}$ and

$$A = \left\{ x \in H \mid \sum_{n=1}^\infty |a_n \langle x, e_n \rangle| \leq 1 \right\}.$$

Using (iii) prove that A is absorbing if and only if the sequence $(a_n)_{n\in\mathbb{N}}$ belongs to l_2, and in this case find the Minkowski seminorm associated with A.

v) Let H be a Hilbert space, $(e_n)_{n\in\mathbb{N}} \subseteq H$ an orthonormal system, $(a_n)_{n\in\mathbb{N}} \subseteq \mathbb{K}$, and

$$A = \left\{ U \in L(H) \mid \sum_{n=1}^\infty |a_n \langle U(e_n), e_n \rangle| \leq 1 \right\}.$$

Using (iii) prove that A is absorbing if and only if $(a_n)_{n\in\mathbb{N}} \in l_1$, and in this case find the Minkowski seminorm associated with A.

vi) For a fixed $1 < p < \infty$ and $(a_n)_{n\in\mathbb{N}} \subseteq \mathbb{K}$, we consider the set

$$A = \left\{ f \in C[0,1] \mid \sum_{n=1}^\infty \left| a_n \int_{1/(n+1)}^{1/n} f(x)dx \right|^p \leq 1 \right\}.$$

Using (i) prove that A is absorbing if and only if $(a_n/(n(n+1)))_{n\in\mathbb{N}} \in l_p$, and in this case find the Minkowski seminorm associated with A.

vii) Let $(a_n)_{n\in\mathbb{N}} \subseteq \mathbb{K}$ and

$$A = \{ f \in C[0,1] \mid |a_n f(1/n)| \leq 1 \; \forall n \in \mathbb{N} \}.$$

Using (i) prove that A is absorbing if and only if $(a_n)_{n\in\mathbb{N}} \in l_\infty$, and in this case find the Minkowski seminorm associated with A.

Separation in real linear topological spaces

25. Let X be a real linear topological space, and $A \subseteq X$ a convex subset with $0 \in \overset{\circ}{A}$.

i) If $p_A : X \to \mathbb{R}_+$ is the Minkowski functional for A prove that $p_A : X \to \mathbb{R}$ is continuous at 0. Moreover, if A is symmetric, i.e., $A = -A$, then $p_A : X \to \mathbb{R}_+$ is continuous.

ii) Prove that for any $x_0 \in X \setminus A$ there is a linear and continuous functional $f : X \to \mathbb{R}$ such that $f(x) \leq 1 \; \forall x \in A$ and $f(x_0) = 1$.

13.1 Exercises

26. i) Let X be a real normed space, $n \in \mathbb{N}$, $\{x_1^*, ..., x_n^*\} \subseteq X^*$ a linearly independent system, $A = \{x \in X \mid |x_1^*(x)| \leq 1, ..., |x_n^*(x)| \leq 1\}$ and $x_0 \notin A$. Prove that $0 \in \overset{\circ}{A}$ and that any $f \in X^*$ with the properties $|f(x)| \leq 1 \; \forall x \in A$ and $f(x_0) = 1$ is of the form $f = \sum_{i=1}^{n} \lambda_i x_i^*$, where $\lambda_1 \in \mathbb{R}, ..., \lambda_n \in \mathbb{R}$ satisfy the conditions $\sum_{i=1}^{n} \lambda_i x_i^*(x_0) = 1$ and $\sum_{i=1}^{n} |\lambda_i| \leq 1$.

ii) Let $A = \{U \in L(l_2) \mid |\langle Ue_1, e_2 \rangle| \leq 1, \; |\langle Ue_2, e_2 \rangle| \leq 1\}$ and $U_0 \in L(l_2)$ defined by $U_0(x) = \langle x, 2e_1 + 3e_2 \rangle e_2$. Using (i) prove that any $f \in (L(l_2))^*$ with the properties $|f(U)| \leq 1 \; \forall U \in A$ and $f(U_0) = 1$ is of the form

$$f(U) = \frac{3\lambda \langle Ue_1, e_2 \rangle + (1 - 2\lambda)\langle Ue_2, e_2 \rangle}{3},$$

where $\lambda \in [-2/5, 4/5]$.

Algebraic closure for convex sets

27. Let X be a real linear topological space and let $A \subseteq X$.
i) Prove that $\bigcap_{r>1} (rA) \subseteq \overline{A}$.
ii) If $0 \in \overset{\circ}{A}$ and $A \subseteq X$ is convex, prove that $\bigcap_{r>1} (rA) = \overline{A}$.

The space $L_p[0, 1]$ for $0 < p < 1$

28. Let μ be the Lebesgue measure on $[0, 1] \subseteq \mathbb{R}$, and $0 < p < 1$. For a measurable function $f : [0, 1] \to \mathbb{R}$, let $\|f\|_p = \int_0^1 |f(t)|^p \, dt$. Then

$$\mathcal{L}_p[0, 1] = \left\{ f : [0, 1] \to \mathbb{R} \mid f \text{ measurable and } \|f\|_p < \infty \right\}$$

is a linear space with respect to the usual operations of addition and scalar multiplication. Factorizing this space by the equivalence relation given by the equality μ-a.e., we obtain the space $L_p[0, 1]$.

i) Prove that $\|f\|_p = 0 \Leftrightarrow f = 0$ in $L_p[0, 1]$; $\|\alpha f\|_p = |\alpha|^p \|f\|_p \; \forall \alpha \in \mathbb{R}, \forall f \in L_p[0, 1]$; $\|f + g\|_p \leq \|f\|_p + \|g\|_p \; \forall f, g \in L_p[0, 1]$.

ii) Define $d : L_p[0, 1] \times L_p[0, 1] \to [0, \infty)$, $d(f, g) = \|f - g\|_p \; \forall f, g \in L_p[0, 1]$. Prove that d is a translation invariant metric on $L_p[0, 1]$.

iii) If we consider on $L_p[0, 1]$ the topology τ given by the metric d, prove that the only open convex sets in $(L_p[0, 1], \tau)$ are $\emptyset$ and $L_p[0, 1]$, and then deduce that for each non-empty open set in $(L_p[0, 1], \tau)$ its convex hull is the space $L_p[0, 1]$.

iv) Prove that if X is a locally convex Hausdorff space and $T : (L_p[0, 1], \tau) \to X$ is a linear and continuous operator, then $T = 0$.

v) Prove that the topology τ on $L_p[0, 1]$ is not a locally convex one and that the dual of $(L_p[0, 1], \tau)$ is the null space.

The space $C(\Omega)$ for a non-empty open set $\Omega \subseteq \mathbb{C}$

29. Let $\Omega \subseteq \mathbb{C}$ be a non-empty open set and $C(\Omega)$ the linear space of all continuous functions from Ω to $\mathbb{C}$.

i) For each $n \in \mathbb{N}$, let $V_n = \{z \in \mathbb{C} \mid |z| > n\} \cup \bigcup_{z \notin \Omega} D(z, 1/n) \subseteq \mathbb{C}$, where $D(z, r)$ is the open disc of center z and radius r in the complex plane. Denoting $K_n = \complement_{V_n} \ \forall n \in \mathbb{N}$ prove that $\bigcup_{n=1}^{\infty} K_n = \Omega$, $K_n \subseteq \overset{\circ}{K}_{n+1} \ \forall n \in \mathbb{N}$, that K_n is a compact set for each $n \in \mathbb{N}$ and that if $K \subseteq \Omega$ is a compact set then there is an $n \in \mathbb{N}$ such that $K \subseteq K_n$.

ii) For any $n \in \mathbb{N}$ let $p_n : C(\Omega) \to [0, \infty)$, $p_n(f) = \sup_{x \in K_n} |f(x)|$. Prove that p_n is a seminorm for any $n \in \mathbb{N}$, $p_n \leq p_{n+1}$ for any $n \in \mathbb{N}$ and that the family $(p_n)_{n \in \mathbb{N}}$ separates the elements of $C(\Omega)$.

iii) If τ is the locally convex topology on $C(\Omega)$ defined by the family $(p_n)_{n \in \mathbb{N}}$, prove that the sets $\{W_n \mid n \in \mathbb{N}\}$, where $W_n = \{f \in C(\Omega) \mid p_n(f) \leq 1/n\}$ form a basis of convex neighborhoods for the origin in the τ topology.

iv) Prove that the topology τ is compatible with the distance $d : C(\Omega) \times C(\Omega) \to [0, \infty)$,
$$d(f, g) = \sum_{n=1}^{\infty} \frac{p_n(f-g)}{2^n(1 + p_n(f-g))},$$
i.e., the topology associated to the distance d coincide with τ, and that d is a translation invariant complete metric.

v) Prove that $f_n \to f$ in $(C(\Omega), \tau)$ if and only if for each compact $K \subseteq \Omega$ we have $f_n \to f$ uniformly on K.

vi) Prove that $(C(\Omega), d)$ is not normable.

The space $\mathcal{H}(\Omega)$ for a non-empty open set $\Omega \subseteq \mathbb{C}$

30. Let $\Omega \subseteq \mathbb{C}$ be a non-empty open set and consider $\mathcal{H}(\Omega)$, the linear space of all holomorphic functions on Ω, considered as a linear subspace of $(C(\Omega), \tau)$, where τ is the topology considered at exercise 29.

i) Prove that $\mathcal{H}(\Omega) \subseteq (C(\Omega), \tau)$ is a closed linear subspace and then deduce that $(\mathcal{H}(\Omega), \tau)$ is a Fréchet space.

ii) Prove that any closed bounded subset in $(\mathcal{H}(\Omega), \tau)$ is compact in $(\mathcal{H}(\Omega), \tau)$.

iii) Prove that $(\mathcal{H}(\Omega), \tau)$ is not normable.

The space of all entire functions

31. Let $\mathcal{I}(\mathbb{C}) = \{f : \mathbb{C} \to \mathbb{C} \mid f \text{ entire}\}$, which is a linear space over $\mathbb{C}$ with respect to the usual operations of addition and scalar multiplication. For any $n \in \mathbb{N}$ consider
$$\|\cdot\|_n : \mathcal{I}(\mathbb{C}) \to \mathbb{R}, \ \|f\|_n = \sup_{|z| \leq n} |f(z)| \ \forall f \in \mathcal{I}(\mathbb{C}).$$

i) Prove that $(\mathcal{I}(\mathbb{C}), \|\cdot\|_n)_{n \in \mathbb{N}}$ is a Hausdorff locally convex space.

ii) For $f \in \mathcal{I}(\mathbb{C})$ and $\xi \in \mathbb{C}$ we define $f_\xi : \mathbb{C} \to \mathbb{C}$, $f_\xi(z) = f(\xi z)$. Prove that $f_\xi \in \mathcal{I}(\mathbb{C})$ and that $f_\xi^{(n)}(z) = f^{(n)}(\xi z)\xi^n \ \forall z \in \mathbb{C}, \forall n \geq 0$.

13.2 Solutions   **377**

iii) Let $x^* : \mathcal{I}(\mathbb{C}) \to \mathbb{C}$ be a linear and continuous functional, $f \in \mathcal{I}(\mathbb{C})$, and consider $F : \mathbb{C} \to \mathbb{C}$, $F(\xi) = x^*(f_\xi)$. Prove that F is analytic on $\mathbb{C}$ and that $F^{(n)}(0) = f^{(n)}(0) x^*(z^n) \, \forall n \geq 0$.

iv) Deduce that if $f \in \mathcal{I}(\mathbb{C})$ has the property that $f^{(n)}(0) \neq 0 \, \forall n \geq 0$ then the set $\{f_\xi \mid \xi \in \mathbb{C}\} \subseteq \mathcal{I}(\mathbb{C})$ is fundamental.

A locally convex space analogous to l_1

32. Consider two sets A and B, and let $(c_{\alpha\beta})_{(\alpha,\beta) \in A \times B} \subseteq [0, \infty)$. Denote by

$$X = \left\{ x = (x_\alpha)_{\alpha \in A} \subseteq \mathbb{K} \,\bigg|\, p_\beta(x) = \sum_{\alpha \in A} c_{\alpha\beta} |x_\alpha| < \infty \, \forall \beta \in B \right\},$$

which is a linear space with the usual operations of addition and scalar multiplication.

i) Prove that $(p_\beta)_{\beta \subset B}$ is a family of seminorms on X and that X with the locally convex topology generated by the family $(p_\beta)_{\beta \in B}$ is Hausdorff if and only if $\forall \alpha \in A \, \exists \beta \in B$ such that $c_{\alpha\beta} > 0$.

ii) Prove that if in addition the family $(c_{\alpha\beta})_{(\alpha,\beta) \in A \times B}$ is uniformly filtered with respect to A, i.e., $\forall \beta_1, \beta_2 \in B \, \exists \beta \in B$ such that $c_{\alpha\beta_1}, c_{\alpha\beta_2} \leq c_{\alpha\beta} \, \forall \alpha \in A$, then each linear and continuous functional $x^* : X \to \mathbb{K}$ can be represented by

$$x^*\big((x_\alpha)_{\alpha \in A}\big) = \sum_{\alpha \in A} u_\alpha x_\alpha \quad \forall (x_\alpha)_{\alpha \in A} \in X,$$

where the family $(u_\alpha)_{\alpha \in A} \subseteq \mathbb{K}$ verifies the property: $\exists \beta \in B, \exists a > 0$ such that $|u_\alpha| \leq a c_{\alpha\beta} \, \forall \alpha \in A$.

Thus if $c_{\alpha\beta} > 0 \, \forall (\alpha, \beta) \in A \times B$ then X^* can be identified with the space

$$\left\{ (u_\alpha)_{\alpha \in A} \subseteq \mathbb{K} \,\bigg|\, \exists \beta \in B \text{ such that } \sup_{\alpha \in A} \frac{|u_\alpha|}{c_{\alpha\beta}} < \infty \right\}.$$

13.2 Solutions

1. The discrete topology is $\tau = \mathcal{P}(X)$. Consider $y \in X$, $z \in X$ and denote $x = y + z$. For V, an open neighborhood of x, we consider $V_1 = \{y\}$, an open neighborhood of y, and $V_2 = \{z\}$, an open neighborhood of z, and we have $V_1 + V_2 = \{x\} \subseteq V$.

We now prove that the multiplication with scalars is not continuous. Consider $x_0 \in X$, $x_0 \neq 0$. Then, since $1/n \to 0$ in $\mathbb{K}$, if the multiplication were continuous at $(0, x_0)$ we would have that $(1/n) \cdot x_0 \to 0$ in X. Since $\{0\} \in \tau = \mathcal{P}(X)$ there is an $n \in \mathbb{N}$ such that $(1/n) \cdot x_0 \in \{0\}$, i.e, $(1/n) \cdot x_0 = 0$, that is, $x_0 = 0$, contradiction.

2. See [44, exercise 23, chapter 1].

We recall first some well known results of general topology.

a) Let X be a set. Let $\mathcal{B} = (B_i)_{i \in I} \subseteq \mathcal{P}(X)$ be a family of sets with the properties: $\bigcup_{i \in I} B_i = X$, and $B_1, B_2 \in \mathcal{B}$ implies $B_1 \cap B_2 \in \mathcal{B}$. Then $\tau = \{\bigcup_{i \in J} B_i \mid J \subseteq I\}$ is a topology on X named the topology generated by $\mathcal{B}$. In this case we say that $\mathcal{B}$ is a basis for

τ (if $J = \emptyset$, then by convention $\bigcup_{i \in \emptyset} B_i = \emptyset$), and $\forall x \in X$, $\{B \in \mathcal{B} \mid x \in B\}$ is a basis of neighborhoods for x in the τ topology.

b) Let X be a set, and $\mathcal{G} = (G_i)_{i \in I} \subseteq \mathcal{P}(X)$ a family of subsets. We denote by $\mathcal{B} = \{\bigcap_{i \in J} G_i \mid J \text{ finite} \subseteq I\}$ (by convention, if $J = \emptyset$ then $\bigcap_{i \in \emptyset} G_i = X$). Then $\mathcal{B}$ is a basis for a topology τ on X, i.e., it satisfies the conditions from (a). Therefore $\forall x \in X$

$$\left\{ \bigcap_{k=1}^{n} G_{i_k} \;\middle|\; G_{i_1} \in \mathcal{G}, ..., G_{i_n} \in \mathcal{G}, x \in \bigcap_{k=1}^{n} G_{i_k}, n \in \mathbb{N} \right\}$$

is a basis of neighborhoods for x in the τ topology. In this case the family $\mathcal{G}$ is called a sub-basis for τ and τ is called the topology generated by the family $\mathcal{G} = (G_i)_{i \in I} \subseteq \mathcal{P}(X)$.

Using (b) we know that the set $\mathcal{G} \subseteq \mathcal{P}(X)$, $\mathcal{G} = \{V(f, r) \mid f \in X, r > 0\}$ generates a topology τ on X and that for any $f \in X$,

$$\left\{ \bigcap_{k=1}^{n} V(f_k, r_k) \;\middle|\; f_1, ..., f_n \in X, r_1 > 0, ..., r_n > 0, n \in \mathbb{N}, f \in \bigcap_{k=1}^{n} V(f_k, r_k) \right\}$$

is a basis of neighborhoods for f in the τ topology.

We prove now that the addition $+ : X \times X \to X$ is continuous. Let $u \in X$, $v \in X$ and denote by $h = u + v \in X$. Let G be a neighborhood of h from the basis of neighborhoods for h in the τ topology. We can find $n \in \mathbb{N}$, $f_1, ..., f_n \in X$, $r_1, ..., r_n > 0$ such that $h \in G = \bigcap_{k=1}^{n} V(f_k, r_k)$, i.e., $h \in V(f_i, r_i)$ $\forall i = \overline{1, n}$. For every $1 \leq i \leq n$ we denote by $u_i = u + (f_i - h)/2$, $v_i = v + (f_i - h)/2$. Let $z \in V(u_i, r_i/2)$, $w \in V(v_i, r_i/2)$. We obtain that

$$\begin{aligned} |z(x) + w(x) - f_i(x)| &= \left| z(x) - \left(u + \frac{f_i - h}{2}\right)(x) + w(x) - \left(v + \frac{f_i - h}{2}\right)(x) \right| \\ &\leq |z(x) - u_i(x)| + |w(x) - v_i(x)| < \frac{r_i}{2} + \frac{r_i}{2} = r_i \quad \forall x \in (0, 1), \end{aligned}$$

therefore $z + w \in V(f_i, r_i)$, and $V(u_i, r_i/2) + V(v_i, r_i/2) \subseteq V(f_i, r_i)$. If we denote by $D_1 = \bigcap_{i=1}^{n} V(u_i, r_i/2)$ and $D_2 = \bigcap_{i=1}^{n} V(v_i, r_i/2)$ then $u \in D_1$, $v \in D_2$, and then D_1 is an open neighborhood of u and D_2 is an open neighborhood of v. Also $D_1 + D_2 \subseteq \bigcap_{i=1}^{n} V(f_i, r_i)$. We have proved that $+ : X \times X \to X$ is continuous.

We now prove that the multiplication from $\mathbb{R} \times X$ to X is not continuous. Consider $f : (0, 1) \to \mathbb{R}$, $f(x) = 1/(x(x-1))$ $\forall x \in (0, 1)$. We prove that the multiplication is not continuous at $(0, f)$. We suppose, for a contradiction, that it is continuous. Then there exist $\varepsilon > 0$, $n \in \mathbb{N}$, $f_1, ..., f_n \in X$ and $r_1 > 0, ..., r_n > 0$ such that $f \in \bigcap_{i=1}^{n} V(f_i, r_i)$, and that $\forall \lambda \in (-\varepsilon, \varepsilon)$ it follows that $\lambda \cdot \bigcap_{i=1}^{n} V(f_i, r_i) \subseteq V(0 \cdot f, 1)$. Then $\forall \lambda \in (-\varepsilon, \varepsilon)$ we have $\lambda f \in V(0, 1)$, i.e., $|\lambda f(x)| < 1$ $\forall x \in (0, 1)$, $\forall \lambda \in (-\varepsilon, \varepsilon)$. In particular, we have $\varepsilon/(2x(1-x)) < 1$ $\forall x \in (0, 1)$, that is, $\varepsilon < 2x(1-x)$ $\forall x \in (0, 1)$ and, for $x \to 0$, $x > 0$, we obtain that $\varepsilon \leq 0$, contradiction.

13.2 Solutions

3. Consider $x_0 \in X \setminus Y$ and $y_0 \in Y \setminus \{0\}$. Then $(x_0 + y_0) + (-x_0) = y_0$. If the addition were continuous, using that $y_0 \in Y \setminus \{0\}$ we obtain that there exist V_1 a neighborhood of $(x_0 + y_0)$ and V_2 a neighborhood of $-x_0$ such that $x + y \in Y \setminus \{0\}$ $\forall x \in V_1, \forall y \in V_2$. Since V_1 is a neighborhood of $x_0 + y_0$ there is a set $A \in \tau$ such that $(x_0 + y_0) \in A \subseteq V_1$. If $A = Y \setminus \{0\}$ then $x_0 + y_0 \in Y$ and, since $y_0 \in Y$ and Y is a linear subspace, $(x_0 + y_0) - y_0 = x_0 \in Y$, contradiction! We obtain that $A = X$, therefore $V_1 = X$ and, since $V_1 + V_2 \subseteq Y \setminus \{0\}$, we obtain that $Y \setminus \{0\} = X$, a contradiction!

Let $\lambda_0 \in \mathbb{K}$, $x_0 \in X$ and consider $(\lambda_0, x_0) \in \mathbb{K} \times X$. Let V be a neighborhood of $\lambda_0 x_0 \in X$. Then there is a set $A \in \tau$ such that $\lambda_0 x_0 \in A \subseteq V$. If $\lambda_0 = 0$ then $\lambda_0 x_0 = 0$ and $A = X$. We obtain that $V = X$ and, for every neighborhood V_1 of λ_0 in $\mathbb{K}$ and for every neighborhood V_2 of x_0 in X we have $V_1 V_2 \subseteq V$. If $\lambda_0 \neq 0$ then $\lambda_0 x_0$ could be non-zero, and then A could be X or $Y \setminus \{0\}$. If $A = X$ we obtain the same conclusion as before. If $A = Y \setminus \{0\}$ then $(\lambda_0 x_0) \in Y \setminus \{0\} \subseteq V$. Therefore as $\lambda_0 \neq 0$ we obtain that $x_0 \in Y \setminus \{0\}$ and, for $V_1 = \mathbb{K} \setminus \{0\}$, $V_2 = Y \setminus \{0\}$, we have $V_1 V_2 \subseteq V$, where V_1 is an open neighborhood of $\lambda_0 \neq 0$ and V_2 is an open neighborhood of x_0.

Remark. We observe that (X, τ) is not a Hausdorff space.

4. We shall use (a) and (b) from exercise 2.

Suppose that the addition is continuous at (x_0, y_0). For any $x^\# \in X^\#$, with $x^\# (x_0 + y_0) = 0$, the set $V = \ker x^\#$ is a τ-neighborhood for $z_0 = x_0 + y_0$. Therefore there exist $V_1 = \bigcap_{i=1}^{n} \ker y_i^\#$, $y_i^\# (x_0) = 0$, $i = \overline{1, n}$, $V_2 = \bigcap_{i=1}^{p} \ker z_i^\#$, $z_i^\# (y_0) = 0$, $i = \overline{1, p}$ such that $V_1 + V_2 \subseteq V$. Since $0 \in V_1, 0 \in V_2$ and $x_0 \in V_1, y_0 \in V_2$, we obtain that $x_0, y_0 \in V$, thus $x^\# (x_0) = x^\# (y_0) = 0$. Therefore if $x^\# \in X^\#$ and $x^\# (z_0) = 0$, we obtain that $x^\# (x_0) = x^\# (y_0) = 0$. For any $x \in X$ we denote by $\widehat{x} : X^\# \to \mathbb{K}$, $\widehat{x}(x^\#) = x^\# (x)$ $\forall x^\# \in X^\#$. Then $\widehat{x} \in X^{\#\#}$ $\forall x \in X$, and the above relation shows that $\widehat{z_0}(x^\#) = 0$ implies $\widehat{x_0}(x^\#) = \widehat{y_0}(x^\#) = 0$, that is, $\ker \widehat{z_0} \subseteq \ker \widehat{x_0}$, $\ker \widehat{z_0} \subseteq \ker \widehat{y_0}$. Using exercise 16(i) of chapter 14 we know that there are $\alpha, \beta \in \mathbb{K}$ such that $\widehat{x_0} = \alpha \widehat{z_0}$, $\widehat{y_0} = \beta \widehat{z_0}$, that is, $x_0 = \alpha z_0$, $y_0 = \beta z_0$, and therefore $\beta x_0 - \alpha y_0 = 0$. Suppose that $x_0 + y_0 = 0$, that is, $z_0 = 0$. Then $\forall x^\# \in X^\#$, $z_0 \in \ker x^\#$. There exist V_1 a neighborhood for x_0, V_2 a neighborhood for y_0, finite intersections of kernels of linear functionals, with $V_1 + V_2 \subseteq \ker x^\#$. Since $0 \in V_1, x_0 \in V_1, 0 \in V_2, y_0 \in V_2$, we obtain that $x^\# (x_0) = x^\# (y_0) = 0$ $\forall x^\# \in X^\#$, i.e., $x_0 = y_0 = 0$.

If $x_0 = y_0 = 0$ it is immediate that the addition is continuous at $(0, 0)$. Suppose now that $x_0, y_0 \in X$ are such that there exist $\alpha, \beta \in \mathbb{K}$, not both zero, with $\alpha x_0 + \beta y_0 = 0$. Suppose, for example, that $\beta \neq 0$, and then $y_0 = \gamma x_0$. We obtain that $(1 + \gamma) x_0 = x_0 + y_0 =: z_0$. Suppose that $\gamma \neq -1$, and then $\gamma + 1 \neq 0$. If $x_0 = 0$, then $y_0 = 0$, and we have already seen this case. If $x_0 \neq 0$ then $y_0 \neq 0$. Let V be a neighborhood of z_0, $V = \bigcap_{i=1}^{n} \ker x_i^\#$, $x_i^\# (z_0) = 0$, $i = \overline{1, n}$. Since $x_0 = z_0 / (1 + \gamma)$ we obtain that $x_i^\# (x_0) = 0$, $i = \overline{1, n}$, and since $y_0 = \gamma x_0$ we obtain that $x_i^\# (y_0) = 0$, $i = \overline{1, n}$. Then V is a neighborhood for x_0 and y_0, and also $V + V \subseteq V$. Thus, the addition is continuous at (x_0, y_0).

(ii) Suppose that $\cdot : \mathbb{K} \times X \to X$ is continuous at $(0, x_0)$, and that $x_0 \neq 0$. Then $z_0 := 0 \cdot x_0 = 0$. For any $x^\# \in X^\#$, $V := \ker x^\#$ is a neighborhood for z_0, and therefore there are $\varepsilon > 0$ and V_1 a neighborhood for x_0 such that $(-\varepsilon, \varepsilon) \cdot V_1 \subseteq V$, and then $\lambda x_0 \in V$ $\forall \lambda \in (-\varepsilon, \varepsilon)$, thus $x^\# (x_0) = 0$ $\forall x^\# \in X^\#$. Since $X^\#$ separates the elements of X we obtain that $x_0 = 0$, contradiction!

We now prove that the multiplication is continuous at (λ_0, x_0), $\lambda_0 \neq 0$, and at $(0,0) \in \mathbb{K} \times X$. Let $\lambda_0 \in \mathbb{K}$, $\lambda_0 \neq 0$ and $x_0 \in X$. Denote $z_0 = \lambda_0 x_0$. Let $V = \bigcap_{i=1}^{n} \ker x_i^{\#}$ be a neighborhood for z_0, i.e., $x_i^{\#}(z_0) = 0$, $i = \overline{1,n}$. Then since $x_0 = z_0/\lambda_0$ we obtain that $x_i^{\#}(x_0) = 0$, $i = \overline{1,n}$, and therefore V is a neighborhood for x_0. We have $\mathbb{K} \cdot V \subseteq V$ and therefore the multiplication with scalars is continuous at (λ_0, x_0). For the continuity at $(0,0)$, if V is a neighborhood for $0 = 0 \cdot 0$ in the τ topology, with the form given above, again $\mathbb{K} \cdot V \subseteq V$, and therefore the multiplication with scalars is continuous at $(0,0)$.

5. If $A \subseteq X$ is a balanced set then it is trivial that $\forall x \in A$, $\forall \lambda \in \mathbb{K}$ with $|\lambda| = 1$ it follows that $\lambda x \in A$. Suppose now that $0 \in A$, $A \subseteq X$ is a convex set and that $\forall x \in A$, $\forall \lambda \in \mathbb{K}$ with $|\lambda| = 1$ it follows that $\lambda x \in A$. Let $\lambda \in \mathbb{K}$, $|\lambda| \leq 1$ and $x \in A$. Then $|\lambda| x = (1 - |\lambda|) 0 + |\lambda| x$ and since A is convex, $|\lambda| x \in A$. If $\lambda = 0$ then $\lambda x = 0 \in A$, by our hypothesis. If $\lambda \neq 0$ then since $|\lambda/|\lambda|| = 1$, using the hypothesis we obtain that $(\lambda/|\lambda|)(|\lambda| x) \in A$, thus $\lambda x \in A$.

6. If $B \subseteq A$ is a balanced and absorbing set then for any $x \in X$ there is $\delta_x > 0$ such that $\lambda x \in B$, $\forall \lambda \in \mathbb{R}$ with $0 \leq |\lambda| < \delta_x$. Then for any $x \in X$, $x \neq 0$, there is a $\delta_x > 0$ such that $\lambda x \in A$ $\forall \lambda \in (-\delta_x, \delta_x)$.

Suppose now that for any $x \in X$, $x \neq 0$, there are $\alpha < 0 < \beta$ such that $\lambda x \in A$ $\forall \lambda \in (\alpha, \beta)$. Let $B = A \cap (-A)$. For any $x \in X$, $x \neq 0$, denote by $\delta_x = \min(-\alpha, \beta) > 0$, where α and β (depending on x) are given by our hypothesis. Then for any $\lambda \in \mathbb{R}$ with $|\lambda| \leq \delta_x$ we have that $\lambda \in (\alpha, \beta)$, and therefore $\lambda x \in A$. Also $-\lambda \in (\alpha, \beta)$, and therefore $-\lambda x \in A$, that is, $\lambda x \in -A$. Thus $\lambda x \in B$ for every $\lambda \in \mathbb{R}$, $|\lambda| \leq \delta_x$. We obtain that $B \subseteq X$ is absorbing. Clearly $B \subseteq X$ is a convex set. Using exercise 5, in order to prove that $B \subseteq X$ is a balanced set, it is sufficient to prove that $\forall x \in B$, $\forall \lambda \in \mathbb{R}$ with $|\lambda| = 1$ ($\lambda = \pm 1$) it follows that $\lambda x \in B$. But $B = -B$, and therefore the last assertion is true.

7. i) Let $A \subseteq X$ be a non-empty balanced set and $x \in \text{co}(A)$. Then there are $n \in \mathbb{N}$, $\lambda_1 \geq 0, ..., \lambda_n \geq 0$, with $\sum_{i=1}^{n} \lambda_i = 1$ and $a_1, ..., a_n \in A$ such that $x = \sum_{i=1}^{n} \lambda_i a_i$. Let $\lambda \in \mathbb{K}$, $|\lambda| \leq 1$. Then $\lambda x = \sum_{i=1}^{n} \lambda_i (\lambda a_i) = \sum_{i=1}^{n} \lambda_i b_i$, with $b_i = \lambda a_i$, $i = \overline{1,n}$. Since A is balanced it follows that $b_i \in A$, $i = \overline{1,n}$, and therefore $\lambda x \in \text{co}(A)$.

ii) Using (i) the convex hull of the balanced hull of A is a balanced, and, obviously, a convex set. Then the convex and balanced hull of A is contained in $\text{co}(\text{ec}(A))$. Consider now $x \in \text{co}(\text{ec}(A))$. We obtain that there are $n \in \mathbb{N}$, $\alpha_1 \geq 0, ..., \alpha_n \geq 0$, with $\sum_{i=1}^{n} \alpha_i = 1$, $x_1, ..., x_n \in \text{ec}(A)$, such that $x = \sum_{i=1}^{n} \alpha_i x_i$. But $\text{ec}(A) = \bigcup_{|\lambda| \leq 1} \lambda A$, and therefore, for each $1 \leq i \leq n$, we can find $a_i \in A$, $\lambda_i \in \mathbb{K}$, $|\lambda_i| \leq 1$, such that $x_i = \lambda_i a_i$. We obtain that $x = \sum_{i=1}^{n} \lambda_i \alpha_i a_i$. Then if $C \subseteq X$ is a balanced and convex set which contains A we obtain that $\lambda_i a_i \in C$ for each $1 \leq i \leq n$, and therefore $\sum_{i=1}^{n} \lambda_i \alpha_i a_i \in C$, that is, $x \in C$. We obtain

13.2 Solutions

that x belongs to the convex and balanced hull of A. (For a set $W \subseteq X$, co (W) is the convex hull for W and ec (W) is the balanced hull for W.)

iii) co $(A) = \{(x,y) \in \mathbb{R}^2 \mid x \geq 0, \ y \geq 0, \ x+y \leq 1\}$. Since $(0,1) \in$ co (A) we obtain that $(0,-1) \in$ ec $($co $(A))$. Also $(1,0) \in A \subseteq$ ec $($co $(A))$. If ec $($co $(A))$ were convex, then $(1/2, -1/2) \in$ ec $($co $(A))$. It results that there are $\lambda \in \mathbb{K}$, $|\lambda| \leq 1$ and $(x,y) \in$ co (A), such that: $(1/2, -1/2) = \lambda(x,y)$. Then $x = -y$ and, since $x, y \geq 0$, we obtain that $x = y = 0$, a contradiction.

8. i) Let $x \in B$ and $\lambda \in \mathbb{K}$ with $|\lambda| \leq 1$. Then there is $i \in I$ such that $x \in B_i$ and using that B_i is balanced we have $\lambda x \in B_i$, and therefore $\lambda x \in B$.

ii) We observe first that we have at least one set in our union, since $\{0\} \subseteq A$ is a balanced set. Let $D = \bigcap_{|\lambda| \geq 1} (\lambda A)$. Then $x \in D \Leftrightarrow x \in \lambda A \ \forall |\lambda| \geq 1 \Leftrightarrow \mu x \in A \ \forall |\mu| \leq 1$. We obtain that $D \subseteq X$ is balanced. Since $D \subseteq A$ we obtain that $D \subseteq B$. But if $x \in B$ then there is a balanced set $C \subseteq A$ such that $x \in C$. Then $\mu x \in C \subseteq A \ \forall |\mu| \leq 1$, and therefore $x \in D$.

iii) Consider $A = \{x\}$, with $x \in X$, $x \neq 0$.

9. i) See [35, proposition 5.2, chapter 2].

Let $B \subseteq X$ balanced and $\lambda \in \mathbb{C}$ with $|\lambda| \leq 1$. For $0 < |\lambda| \leq 1$, $\lambda \overline{B} = \overline{\lambda B}$, and therefore $\lambda \overline{B} \subseteq \overline{B}$. If $\lambda = 0$ then $\lambda \overline{B} = \{0\} \in B$, since B is balanced. We have proved that $\overline{B}$ is balanced. Suppose now that $0 \in \overset{\circ}{B}$. For any $\lambda \in \mathbb{C}$ with $0 < |\lambda| \leq 1$ we have $\lambda \cdot \overset{\circ}{B} = \widehat{\lambda B} \subseteq \overset{\circ}{B}$. If $\lambda = 0$ then $\lambda \cdot \overset{\circ}{B} = \{0\} \in \overset{\circ}{B}$. Therefore $\overset{\circ}{B}$ is balanced.

ii) See [44, exercise 4, chapter 1].

Evidently if $\lambda \in \mathbb{C}$ with $|\lambda| \leq 1$ and $|z_1| \leq |z_2|$, then $|\lambda z_1| \leq |\lambda z_2|$, and therefore $B \subseteq \mathbb{C}^2$ is balanced. We prove now that $\overset{\circ}{B} = \{(z_1, z_2) \in \mathbb{C}^2 \mid |z_1| < |z_2|\}$. Indeed let A the set from the right side. Since A is open (to see this we can use the continuity of the function $f(z_1, z_2) = |z_1| - |z_2|$) and $A \subseteq B$ it follows that $A \subseteq \overset{\circ}{B}$. Now if $(z_1, z_2) \in \overset{\circ}{B}$ there are $a, b \in (0, 1)$ such that $D(z_1, a) \times D(z_2, b) \subseteq B$ and therefore $|z| \leq |w| \ \forall z \in D(z_1, a)$, $\forall w \in D(z_2, b)$. If $z_2 = 0$, since $|z_1| \leq |z_2|$ we obtain that $z_1 = 0$, and therefore $|z| \leq 0$ $\forall z \in D(0, a)$, contradiction. Therefore $z_2 \neq 0$. If $z_1 = 0$, then $|z_1| < |z_2|$. If $z_1 \neq 0$ then $z_1 + z_1 a/(2|z_1|) \in D(z_1, a)$, and then $|z_1|(1 + a/(2|z_1|)) \leq |z_2|$. We obtain that $|z_1| < |z_2|$. We have proved that $\overset{\circ}{B} = \{(z_1, z_2) \in \mathbb{C}^2 \mid |z_1| < |z_2|\}$. If we suppose, for a contradiction, that $\overset{\circ}{B}$ is a balanced set, we obtain that $0 \cdot \overset{\circ}{B} \subseteq \overset{\circ}{B}$ and, since $\overset{\circ}{B} \neq \emptyset$, we obtain that $(0, 0) \in \overset{\circ}{B}$, a contradiction.

10. See [9, exercise 17, chapter II, section 1], or [44, exercise 3, chapter 1].

i) Let $A \subseteq X$ be an open and non-empty set. We know that

$$\text{co}(A) = \left\{ \sum_{i=1}^{n} t_i x_i \ \middle|\ n \in \mathbb{N}, \ t_i \in [0,1], \ x_i \in A \ \forall 1 \leq i \leq n, \ \sum_{i=1}^{n} t_i = 1 \right\}.$$

Therefore we have the equality:

$$\text{co}(A) = \bigcup \left\{ (t_1 A + \cdots + t_n A) \ \middle|\ n \in \mathbb{N}, \ (t_1, ..., t_n) \in [0,1]^n, \ \sum_{i=1}^{n} t_i = 1 \right\}.$$

Since A is open then for any scalar $\alpha \neq 0$ we have that $\alpha A \subseteq X$ is an open set. Then for each $n \in \mathbb{N}$ the set $(t_1 A + \cdots + t_n A) \subseteq X$ is open, because at least one of the t_i is not zero, and in a linear topological space the summation of a non-empty set and a non-empty open set is always an open set (we use the relation $U + V = \bigcup_{u \in U} (u + V)$, the fact that the translation $x \longmapsto u + x$ is an homeomorphism and the fact that a union of open sets is an open set). Therefore $co(A) \subseteq X$ is an open set, being a union of open sets.

ii) Let V be a neighborhood of 0. There exist two balanced neighborhoods of 0, V_A and V_B, such that $V_A + V_B \subseteq V$. Since A is bounded, for V_A there is $t_A \in \mathbb{R}$ such that $A \subseteq t_A V_A$. For B and V_B there is $t_B \in \mathbb{R}$ such that $B \subseteq t_B V_B$. Let then $t = \max(|t_A|, |t_B|)$. Since V_A and V_B are balanced and $t \geq |t_A|, |t_B|$ we have the inclusions $t_A V_A \subseteq t V_A$ and $t_B V_B \subseteq t V_B$. Then $A + B \subseteq t_A V_A + t_B V_B \subseteq t V_A + t V_B = t(V_A + V_B) \subseteq tV$, and therefore $A + B$ is bounded in X.

iii) Since A and B are compact sets in X, we obtain that $A \times B \subseteq X \times X$ is a compact set if we consider the product topology on $X \times X$. Since $+ : X \times X \to X$ is continuous, and the range of a compact set by a continuous function is compact, we obtain that $A + B \subseteq X$ is a compact set.

iv) Let $A \subseteq X$ be a compact set and $B \subseteq X$ be a closed set. Let $x \in X$, $x \in \overline{A + B}$. There is a net $(x_\delta)_{\delta \in \Delta} \subseteq (A + B)$ such that $x_\delta \to x$. For any $\delta \in \Delta$ there are $a_\delta \in A$ and $b_\delta \in B$ such that $x_\delta = a_\delta + b_\delta$. We obtain the nets $(a_\delta)_{\delta \in \Delta} \subseteq A$ and $(b_\delta)_{\delta \in \Delta} \subseteq B$. Since A is compact the net $(a_\delta)_{\delta \in \Delta}$ has a subnet which converges, i.e., there are $a \subset A$ and $\left(a_{\varphi(\tau)}\right)_{\tau \in \Delta'} \subseteq (a_\delta)_{\delta \in \Delta}$ such that $a_{\varphi(\tau)} \to a$. Since $b_{\varphi(\tau)} = x_{\varphi(\tau)} - a_{\varphi(\tau)}$ for any $\tau \in \Delta'$ we obtain that $b_{\varphi(\tau)} \to x - a$. But $\left(b_{\varphi(\tau)}\right)_\tau \subseteq B$ and B is a closed set. We obtain that $x - a =: b \in B$, and therefore $x = a + b$, with $a \in A$ and $b \in B$, i.e., $x \in A + B$.

v) Consider $\mathbb{R}^2$ with the Euclidean norm. Let $A = \{(x, 1/x) \mid x \in \mathbb{R} \backslash \{0\}\}$, $B = \{(y, 0) \mid y \in \mathbb{R}\}$. Then $A, B \subseteq \mathbb{R}^2$ are closed sets and

$$A + B = \{(x + y, 1/x) \mid x \in \mathbb{R} \backslash \{0\}, y \in \mathbb{R}\} = \mathbb{R} \times (\mathbb{R} \backslash \{0\})$$

is not closed in $\mathbb{R} \times \mathbb{R}$.

11. See [35, chapter I, A].

i) For any $x, y \in A$ we have $x + y = 2a$, with $a \in A$, and then $x/2 + y/2 \in A$. For any α and β as in the statement of our exercise we can write $\alpha = \sum_{i=1}^{n} \alpha_i / 2^i$, $n \in \mathbb{N}$, $\alpha_i \in \{0, 1\}$, $i = \overline{1, n}$, $\alpha_n = 1$, and $\beta = \sum_{i=1}^{n} \beta_i / 2^i$, with $\beta_i = 1 - \alpha_i$, $i = \overline{1, n-1}$, $\beta_n = 1$. We prove the assertion by induction on $n \in \mathbb{N}$. For $n = 1$, $\alpha = \beta = 1/2$, and the assertion is true. Suppose the assertion true for $n - 1$ and we prove it for n. Let $x, y \in A$ and let α and β as above. We suppose, for example, that $\alpha_1 = 0$. Then $\beta_1 = 1$. Let $\gamma = \sum_{i=2}^{n} \alpha_i / 2^{i-1}$, $\delta = \sum_{i=2}^{n} \beta_i / 2^{i-1}$. Then $\gamma + \delta = 2(\alpha + \beta - 1/2) = 1$. Using the hypothesis of induction we obtain that $\gamma x + \delta y \in A$. Since $y \in A$ we deduce that $y/2 + (\gamma x + \delta y)/2 \in A$, that is, $\alpha x + \beta y \in A$.

ii) Let $x, y \in A$ and $\alpha \geq 0, \beta \geq 0, \alpha + \beta = 1$. We write α and β as a sum of negative powers of 2 and we obtain that we can find sequences $(\alpha_n)_{n \in \mathbb{N}}$ and $(\beta_n)_{n \in \mathbb{N}}$ in $\mathbb{Q}_+ \backslash \{0\}$,

13.2 Solutions

$\alpha_n + \beta_n = 1 \ \forall n \in \mathbb{N}$, each α_n and each β_n being a finite sum of negative powers of 2 such that $\alpha_n \to \alpha$, $\beta_n \to \beta$. Using (i) we have $\alpha_n x + \beta_n y \in A$ for each $n \in \mathbb{N}$. Since the topology is linear then $\alpha_n x + \beta_n y \to \alpha x + \beta y$. Thus $\alpha x + \beta y \in \overline{A} = A$.

12. See [11, proposition 16, chapter II, section 2]

i) If $\lambda = 1$ then the assertion is obvious. Suppose now that $\lambda \neq 1$. Let $t = \lambda x + (1-\lambda) y$. Then using, if necessary, a translation, we can suppose that $t = 0$. Then $y = \lambda x / (\lambda - 1)$. Since $\overset{\circ}{A}$ is an open set and $\lambda/(\lambda-1) \in \mathbb{R} \setminus \{0\}$ using the fact that the mapping $u \longmapsto \lambda u / (\lambda - 1)$ is a homeomorphism we obtain that $(\lambda / (\lambda - 1)) \overset{\circ}{A}$ is a neighborhood of y. Since $y \in \overline{A}$ there is a $z \in A \cap (\lambda / (\lambda - 1)) \overset{\circ}{A}$. Let $w \in \overset{\circ}{A}$ be such that $z = \lambda w / (\lambda - 1)$. Then $w \in \overset{\circ}{A}$ and $(\lambda / (\lambda - 1)) w \in A$. Since $\lambda \neq 0$, the mapping $u \longmapsto \lambda (u - w)$ is a homeomorphism which takes w into 0. We obtain that $U = \{\lambda (u - w) \mid u \in \overset{\circ}{A}\}$ is a neighborhood for the origin. Let $v \in U$. Then $v = \lambda (u - w)$, $u \in \overset{\circ}{A}$, and therefore $v = \lambda u + (1 - \lambda)(\lambda/(\lambda - 1)) w = \lambda u + (1 - \lambda) z$, with $u \in \overset{\circ}{A} \subseteq A$, $z \in A$, and then $v \in A$. We have obtained that $U \subseteq A$ and therefore $0 \in \overset{\circ}{A}$.

ii) The fact that $\overset{\circ}{A}$ is a convex set follows from (i). To show that $\overline{A}$ is also a convex set, we define $f : X \times X \times \mathbb{R} \to X$, $f(x, y, \lambda) = \lambda x + (1 - \lambda) y$. Then f is continuous and $f(A \times A \times [0,1]) \subseteq A$, since A is a convex set. We obtain that $f\left(\overline{A} \times \overline{A} \times [0,1]\right) \subseteq \overline{A}$, that is, $\lambda \overline{A} + (1 - \lambda) \overline{A} \subseteq \overline{A} \ \forall \lambda \in [0,1]$. Therefore $\overline{A}$ is a convex set.

iii) Obviously $\overset{\circ}{A} \subseteq A$ implies $\overset{\circ}{A} \subseteq \overline{A}$, and $A \subseteq \overline{A}$ implies $\overset{\circ}{A} \subseteq \overset{\circ}{\overline{A}}$. Let $y \in \overset{\circ}{\overline{A}}$ and $x \in \overset{\circ}{A}$. For any $\lambda \in (0,1)$ let $z_\lambda = \lambda x + (1 - \lambda) y$. Then $z_\lambda \in \overset{\circ}{A}$ (we use (i)) and $y = \lim_{\lambda \to 0} z_\lambda$. Therefore $y \in \overline{\overset{\circ}{A}}$. We have proved that $\overset{\circ}{\overline{A}} \subseteq \overline{\overset{\circ}{A}}$ and then $\overline{\overset{\circ}{A}} = \overline{A}$. We prove now that $\overset{\circ}{\overline{A}} \subseteq \overset{\circ}{A}$. Using, if necessary, a translation, it is sufficient to prove that $0 \in \overset{\circ}{\overline{A}}$ implies $0 \in \overset{\circ}{A}$. Let $W \subseteq \overline{A}$ be a symmetric neighborhood of the origin ($W = -W$). We can choose $W \subseteq \overline{A}$ because 0 is an interior point for $\overline{A}$. Since $0 \in \overline{A}$ we obtain that $0 \in \overset{\circ}{A}$ and then $W \cap \overset{\circ}{A} \neq \emptyset$. Let $z \in W \cap \overset{\circ}{A}$. Then $z \in \overset{\circ}{A}$, $-z \in (-W) = W \subseteq \overline{A}$, and using (i), $0 = z/2 + (-z)/2 \in \overset{\circ}{A}$.

13. i) (Kakutani's lemma). See [35, exercise B, chapter I].

Suppose that $\text{co}(A \cup \{x\}) \cap B \neq \emptyset$ and $\text{co}(B \cup \{x\}) \cap A \neq \emptyset$. Therefore there are $\alpha, \beta \in [0,1]$ and $a, a' \in A$, $b, b' \in B$, such that $\alpha a + (1 - \alpha) x = b'$, $\beta b + (1 - \beta) x = a'$. Since $x \notin A$ and $x \notin B$ we obtain that $\alpha, \beta \neq 0$. Then

$$a + \frac{1-\alpha}{\alpha} x = \frac{b'}{\alpha}, \quad b + \frac{1-\beta}{\beta} x = \frac{a'}{\beta}.$$

If $\alpha = \beta$ then $a + a'/\alpha = b + b'/\alpha$. Let $t > 0$ be such that $t(1 + 1/\alpha) = 1$. Since $ta + ta'/\alpha = tb + tb'/\alpha$ using the convexity of A and B we obtain that $ta + ta'/\alpha \in A \cap B$, contradiction. For $\alpha \neq \beta$, if we eliminate x we obtain that

$$\frac{1-\beta}{\beta} a + \frac{1-\alpha}{\alpha \beta} a' = \frac{1-\alpha}{\alpha} b + \frac{1-\beta}{\alpha \beta} b'.$$

Again, we consider $t > 0$ such that

$$t\left(\frac{1-\beta}{\beta} + \frac{1-\alpha}{\alpha\beta}\right) = t\left(\frac{1-\alpha}{\alpha} + \frac{1-\beta}{\alpha\beta}\right) = 1,$$

and we obtain that $A \cap B \neq \emptyset$, a contradiction. (Since $A \cap B = \emptyset$ we have that $\alpha, \beta \neq 1$ and therefore $(1-\beta)/\beta + (1-\alpha)/(\alpha\beta) \neq 0$.)

ii) (Stone's theorem). See [9, exercise 3, chapter II, section 1].

Let $\mathcal{S} = \{(C,D) \subseteq X \times X \mid C, D \text{ convex}, A \subseteq C, B \subseteq D, C \cap D = \emptyset\}$. We observe that $\mathcal{S} \neq \emptyset$, since $(A,B) \in \mathcal{S}$. We consider on $\mathcal{S}$ the partial order relation '$\prec$' given by: $(C_1, D_1) \prec (C_2, D_2) \Leftrightarrow C_1 \subseteq C_2, D_1 \subseteq D_2$. We prove that every totally ordered family of sets in $(\mathcal{S}, \prec)$ has an upper bound. Indeed, let $\mathcal{T} = \{(C_i, D_i)\}_{i \in I} \subseteq \mathcal{S}$ be a totally ordered subset. Let then $C = \bigcup_{i \in I} C_i$, $D = \bigcup_{i \in I} D_i$. Consider $x, y \in C$. There are $i, j \in I$ such that $x \in C_i$, $y \in C_j$. But, for example, $C_i \subseteq C_j$, and therefore $x, y \in C_j$. Then for every $\alpha \in [0,1]$, $\alpha x + (1-\alpha) y \in C_j$, and therefore $\alpha x + (1-\alpha) y \in C$. We have proved that $C \subseteq X$ is a convex set and, analogously, $D \subseteq X$ is a convex set. Obviously $A \subseteq C$, $B \subseteq D$. Suppose that there is an $x \in C \cap D$, and therefore there are $i, j \in I$ such that $x \in C_i$, $x \in D_j$. But, for example, $(C_i, D_i) \prec (C_j, D_j)$, and then $x \in C_j \cap D_j$, contradiction. Thus (C,D) belongs to $\mathcal{S}$. Using Zorn's lemma we obtain that $(\mathcal{S}, \prec)$ admits a maximal element, denoted by (C,D). Then C and D are convex disjoint sets and $A \subseteq C$, $B \subseteq D$. Suppose that $C \cup D \neq X$, and let $x \in X \setminus (C \cup D)$. Using (i) we can suppose, for example, that $\operatorname{co}(C \cup \{x\}) \cap D = \emptyset$. If we denote by $C' = \operatorname{co}(C \cup \{x\})$, we obtain that $(C', D) \in \mathcal{S}$, and $(C, D) \subseteq (C', D)$. Since (C, D) is maximal, we obtain that $C = C'$, that is, $x \in C$, contradiction.

14. See [9, exercise 16, chapter II, section 1].

i) Let $x_k \in \bigcap_{j=\overline{1,r}, j \neq k} A_j$, $k = 1, \ldots, r$. There are then $\alpha_1, \ldots, \alpha_r$ in $\mathbb{R}$, not all zero, such that $\sum_{i=1}^{r} \alpha_i = 0$, $\sum_{i=1}^{r} \alpha_i x_i = 0$ (any homogeneous system with m equations and p unknowns has a non-zero solution if $m < p$). It cannot happen that all α_i be ≤ 0, or that all α_i be ≥ 0, since the relation $\sum_{i=1}^{r} \alpha_i = 0$ would imply that $\alpha_i = 0\ \forall 1 \leq i \leq r$. Suppose, for example, that $\alpha_1, \ldots, \alpha_m < 0$, and $\alpha_{m+1}, \ldots, \alpha_r \geq 0$, where $1 \leq m < r$, and not all $\alpha_{m+1}, \ldots, \alpha_r$ are zero. Then $\sum_{i=1}^{m} (-\alpha_i) x_i = \sum_{i=m+1}^{r} \alpha_i x_i$. We denote this common value with $x \in \mathbb{R}^n$. Since $\sum_{i=1}^{r} \alpha_i = 0$, we obtain that $\sum_{i=1}^{m} (-\alpha_i) = \sum_{i=m+1}^{r} \alpha_i$, and we denote this common value with $t \in \mathbb{R}$. Let us consider the element $x/t \in \mathbb{R}^n$. For each $j \in \{m+1, \ldots, r\}$, $x_i \in A_j\ \forall i \in \{1, \ldots, m\}$, and then $x/t \in A_j$, for each $j \in \{m+1, \ldots, r\}$. For each $j \in \{1, \ldots, m\}$, $x_i \in A_j\ \forall i \in \{m+1, \ldots, r\}$, and then $x/t \in A_j$, for each $j \in \{1, \ldots, m\}$. Thus $x/t \in \bigcap_{j=1}^{r} A_j$.

ii) We shall prove that for any finite set of elements of $\mathcal{C}$ their intersection is non-empty. Induction on k, the number of sets that we want to intersect. If $k \in \{1, \ldots, n+1\}$, any intersection of k elements from $\mathcal{C}$ is non-empty by our hypothesis. Suppose now that any intersection of k elements from $\mathcal{C}$ is non-empty and that $k \geq n+1$. We shall prove that

13.2 Solutions

any intersection of $k+1$ elements from $\mathcal{C}$ is non-empty. Let $A_1, ..., A_{k+1} \in \mathcal{C}$. Then $\bigcap_{j=\overline{1,k+1}, j \neq i} A_j \neq \emptyset$, for each $1 \leq i \leq k+1$. Using (i) we obtain that $\bigcap_{j=\overline{1,k+1}} A_j \neq \emptyset$ and this is what we wanted to prove.

Suppose now that $\mathcal{C} = \{A_i\}_{i \in I}$ and that $\bigcap_{i \in I} A_i = \emptyset$. Then $\bigcup_{i \in I} \mathbf{C}_{A_i} = \mathbb{R}^n$ and then for an $i_0 \in I$ we have that $A_{i_0} \subseteq \bigcup_{i \in I} \mathbf{C}_{A_i}$. Since A_{i_0} is a compact set and the sets $\mathbf{C}_{A_i}$ are open we obtain that there are $i_1, ..., i_p \in I$ such that $A_{i_0} \subseteq \bigcup_{j=1}^{p} \mathbf{C}_{A_{i_j}}$ and therefore $\bigcap_{j=0}^{p} A_{i_j} = \emptyset$, a contradiction.

15. Let $A = (-1,1) \setminus \{0\}$. For any V a neighborhood of 0 from the fundamental system of neighborhoods given by our hypothesis, we have $V \cap A \neq \emptyset$ and therefore $0 \in \overline{A}$. Suppose that there is a sequence $(x_n)_{n \in \mathbb{N}} \subseteq A$ such that $x_n \to 0$ in our topology. Since $(x_n)_{n \in \mathbb{N}} \subseteq A$ we obtain that $x_n \neq 0 \; \forall n \in \mathbb{N}$. We consider $V = (-1,1) \setminus \{x_n \mid n \in \mathbb{N}\}$. Then V is a neighborhood of 0 in this topology and since $x_n \to 0$ we can find elements from the sequence $(x_n)_{n \in \mathbb{N}}$ which belong to V and we obtain a contradiction.

16. See [35, proposition 5.4, chapter 2].

i) Consider $x \in X$ and $\varepsilon > 0$. Since f is continuous at 0, there is a neighborhood W of 0 in X such that $|f(y)| < \varepsilon \; \forall y \in W$. Then for any $z \in x+W$, $z = x+y$, $y \in W$, we have $|f(z) - f(x)| = |f(y)| < \varepsilon$. Since the topology on X is linear, we obtain that $x + W$ is a neighborhood for x and therefore f is continuous at x.

ii) If f is continuous, since $\ker f = f^{-1}(\{0\})$ and $\{0\} \subseteq \mathbb{R}$ is a closed set we obtain that $\ker f \subseteq X$ is a closed set, and therefore (a) implies (b). Suppose now that (b) is true. If $\ker f = X$ then $f = 0$, and f is continuous. If $\ker f \neq X$ we can find $x_0 \in X$, $x_0 \notin \ker f$. Since $\ker f$ is a closed set there is a neighborhood V for $0 \in X$, balanced, such that $(x_0 + V) \cap \ker f = \emptyset$. Since V is balanced and f is linear we obtain that $f(V) \subseteq \mathbb{R}$ is balanced. Also $0 = f(0) \in f(V)$. Therefore $f(V)$ is a non-empty balanced set in $\mathbb{R}$. If $f(V)$ is unbounded since for each $\lambda \in f(V)$ all real numbers of modulus $\leq |\lambda|$ are in $f(V)$, we obtain that $f(V) = \mathbb{R}$ and therefore there is $x \in V$ such that $f(x) = -f(x_0)$. If we denote by $y = x + x_0$ then $y \in (x_0 + V) \cap \ker f$, a contradiction. Therefore (b) implies (c). If there is a neighborhood V of $0 \in X$ and an $M \geq 0$ such that $|f(x)| \leq M$ $\forall x \in V$, then, for any $\varepsilon > 0$ if we denote by $W = (\varepsilon/(M+1))V$, for every $y \in W$ we have $|f(x)| \leq \varepsilon$, and therefore f is continuous at $0 \in X$. Using (i) we obtain that f is continuous on X, and therefore (c) implies (a).

17. Let $x_0 \in G$. We denote by $D = \{x \in G \mid \exists f : [0,1] \to G \text{ continuous}, f(0) = x_0, f(1) = x\}$. We shall prove that $D \subseteq G$ is a closed and open set with respect to the topology obtained on G by restricting the topology of X. Since $x_0 \in D$, and therefore D is non-empty, using the connectedness of G we will obtain that $D = G$, and then G will be arcwise connected. Let $x \in D$. Since $D \subseteq G$ we obtain that $x \in G$ and, since G is open, we obtain that there is V a balanced neighborhood of the origin such that $x + V \subseteq G$. For any $y \in x + V$, let $g : [0,1] \to X$, $g(t) = (1-t)x + ty$. Then g is continuous, $g(0) = x$, $g(1) = y$. For every $t \in [0,1]$, $g(t) = x + t(y-x)$. But $y - x = v \in V$. Since V is balanced we obtain that $\lambda v \in V \; \forall \lambda \in \mathbb{K}$ with $|\lambda| \leq 1$. Then $t(y-x) \in V$, $g(t) \in x + V \; \forall t \in [0,1]$, and therefore $g(t) \in G$, $\forall t \in [0,1]$. We obtain that any

$y \in x + V$ can be connected with x by a continuous path, contained in G. Since there is a continuous path in G from x_0 to x, we obtain that there is a continuous path in G from y to x_0, and therefore $x + V \subseteq D$, that is, $D \subseteq G$ is an open set. Consider now x belonging to the closure of D in G. Then $x \in G$ and for any neighborhood V of $0 \in X$ we have $D \cap ((x + V) \cap G) \neq \varnothing$. Let V be a balanced neighborhood of $0 \in X$ such that $x + V \subseteq G$. Then $D \cap (x + V) \neq \varnothing$ and let $z \in D \cap (x + V)$. Using the results proved above, x and z can be connected by a continuous path contained in $x + V$, and therefore contained in G. Since x_0 and z can be connected by a continuous path contained in G, we obtain that x and x_0 can be connected by a continuous path contained in G, and therefore $x \in D$.

18. Since T is additive, $T(0) = 2T(0)$, $T(0) = 0$. Let $x \in X$ and $(x_n)_{n \in \mathbb{N}} \subseteq X$ be such that $x_n \to x$. Since the topology on X is linear we obtain that $x_n - x \to 0$. Using the continuity of T at $0 \in X$ we obtain that $T(x_n - x) \to T(0) = 0$. Then $T(x_n - x) + T(x) \to T(x)$. But $T(x_n - x) + T(x) = T(x_n)$ because T is additive, and therefore $T(x_n) \to T(x)$ in Y.

Let $x \in X$ and $\lambda \in \mathbb{R}$. Then $0 = T(0) = T(x - x) = T(x) + T(-x)$, and therefore $T(-x) = -T(x)$. Using the fact that T is additive we obtain that $T(mx) = mT(x)$, for any $m \in \mathbb{Z}$. Let now $n \in \mathbb{N}$ and $m \in \mathbb{Z}$. Then: $mT(x) = T(mx) = T(n((m/n)x)) = nT((m/n)x)$, that is, $T((m/n)x) = (m/n)T(x)$. Therefore $T(qx) = qT(x)\ \forall q \in \mathbb{Q}$. Let $\lambda \in \mathbb{R}$. Then there is a sequence $(q_n)_{n \in \mathbb{N}} \subseteq \mathbb{Q}$ such that $q_n \to \lambda$. Since the topology on X is linear we obtain that $q_n x \to \lambda x$, and therefore $T(q_n x) \to T(\lambda x)$. Then $q_n T(x) \to T(\lambda x)$. Also since the topology on X is linear we obtain that $q_n T(x) \to \lambda T(x)$. But Y is Hausdorff, hence $T(\lambda x) = \lambda T(x)$. Therefore $T(\lambda x) = \lambda T(x)\ \forall x \in X$ and $\forall \lambda \in \mathbb{R}$, i.e., $T : X \to Y$ is a linear operator.

19. See [44, exercise 19, chapter 1].

Let d be the translation invariant complete metric which generates the linear topology on Y. Let V_0 be a neighborhood of $0 \in X$. The operator $T : M \to Y$ is continuous at $0 \in M$. Therefore there is a neighborhood W_1 of $0 \in X$ such that $d(0, Tx) \leq 1/2$ $\forall x \in W_1 \cap M$. For W_1 neighborhood of $0 \in X$ there is a neighborhood G_1 of $0 \in X$ such that $G_1 + G_1 \subseteq W_1 \cap V_0$. For G_1 there is a balanced neighborhood V_1 of 0 such that $V_1 \subseteq G_1$. Thus there is a balanced neighborhood V_1 of $0 \in X$ such that $V_1 + V_1 \subseteq V_0$ and $d(0, Tx) \leq 1/2\ \forall x \in V_1 \cap M$. In this way, by induction, we construct a sequence of balanced neighborhoods of $0 \in X$, $(V_n)_{n \in \mathbb{N}}$, such that $V_n + V_n \subseteq V_{n-1}$ for every $n \in \mathbb{N}$ and $d(0, Tx) \leq 1/2^n\ \forall x \in V_n \cap M, \forall n \in \mathbb{N}$.

Consider now $x \in X$ arbitrary. Since $M \subseteq X$ is dense, for every $n \in \mathbb{N}$ there is an $x_n \in (V_n + x) \cap M$. Then for every $n \geq m \geq 1$ we have $d(Tx_m, Tx_n) = d(0, Tx_n - Tx_m) = d(0, T(x_n - x_m))$. But $x_n - x_m \in V_n - V_m \subseteq V_m + V_m \subseteq V_{m-1}$ and, since x_n and x_m are from M and M is a linear subspace in X we have that $x_n - x_m \in M$. Therefore $x_n - x_m \in V_{m-1} \cap M$ and therefore $d(0, T(x_n - x_m)) \leq 1/2^{m-1}$. From here it follows that the sequence $(Tx_n)_{n \in \mathbb{N}} \subseteq (Y, d)$ is Cauchy. Using the completeness it results that there is a unique element in Y, denoted $\widetilde{T}x$, such that $Tx_n \to \widetilde{T}x$ in (Y, d). We obtain the operator $\widetilde{T} : X \to Y$. We must prove that $\widetilde{T}$ is well defined, that $\widetilde{T}$ is an extension of T, and that $\widetilde{T}$ is linear and continuous.

First, we observe that $\widetilde{T}$ is well defined, that is, if we consider two sequences $(x_1^n)_{n \in \mathbb{N}}$

and $(x_2^n)_{n\in\mathbb{N}}$, with $x_1^n \in M \cap (V_n + x)$, $x_2^n \in M \cap (V_n + x)$ $\forall n \in \mathbb{N}$, then $\lim_{n\to\infty} Tx_1^n = \lim_{n\to\infty} Tx_2^n$. Indeed, we consider the sequence $(x_3^n)_{n\in\mathbb{N}} = (x_1^1, x_2^2, x_1^3, x_2^4, ...)$. Then $x_3^n \in M \cap (V_n + x)$ $\forall n \in \mathbb{N}$ and using what we have proved above we obtain that the sequence $(Tx_3^n)_{n\in\mathbb{N}}$ converges, and then, evidently, $\lim_{n\to\infty} Tx_1^n = \lim_{n\to\infty} Tx_2^n = \lim_{n\to\infty} Tx_3^n$. Thus $\widetilde{T} : X \to Y$ is well defined. For $x \in M$ if we consider $x_n = x$ $\forall n \in \mathbb{N}$ we obtain that $\widetilde{T}$ is an extension of T. We now prove that $\widetilde{T}$ is additive. Let $x, y \in X$ and we consider, respectively, the sequences $(x_n)_{n\in\mathbb{N}}$ and $(y_n)_{n\in\mathbb{N}}$ with which we define, respectively, $\widetilde{T}x$ and $\widetilde{T}y$. Let $z_n = x_n + y_n$ for every $n \in \mathbb{N}$ and $z = x + y$. Then $z_n \in M$ for every $n \in \mathbb{N}$ and $z_n \in z + V_n + V_n \subseteq z + V_{n-1}$, for every $n \in \mathbb{N}$. Thus $z_n \in M \cap (z + V_{n-1})$ for all $n \in \mathbb{N}$ and therefore $\widetilde{T}z = \lim_{n\to\infty} Tz_n = \lim_{n\to\infty} (Tx_n + Ty_n) = \lim_{n\to\infty} Tx_n + \lim_{n\to\infty} Ty_n = \widetilde{T}x + \widetilde{T}y$. Thus $\widetilde{T} : X \to Y$ is additive. Consider now $x \in X$ and $\alpha \in \mathbb{K}$ with $|\alpha| \le 1$. Let $(x_n)_{n\in\mathbb{N}}$ be a sequence with which we define $\widetilde{T}x$. Then $x_n \in M \cap (V_n + x)$ $\forall n \in \mathbb{N}$. Since $M \subseteq X$ is a linear subspace and $V_n \subseteq X$ is balanced, we obtain that $\alpha x_n \in M \cap (V_n + \alpha x)$ $\forall n \in \mathbb{N}$. Therefore we can consider the sequence $(\alpha x_n)_{n\in\mathbb{N}}$ for the definition of $\widetilde{T}(\alpha x)$, and we have $\widetilde{T}(\alpha x) = \lim_{n\to\infty} T(\alpha x_n) = \lim_{n\to\infty} \alpha T(x_n) = \alpha \widetilde{T}(x)$. Thus, for any $x \in X$ and for any $\alpha \in \mathbb{K}$ with $|\alpha| \le 1$ we have $\widetilde{T}(\alpha x) = \alpha \widetilde{T}(x)$. Consider now $x \in X$ and $\alpha \in \mathbb{K}$ arbitrary. Then there is $n \in \mathbb{N}$ such that $|\alpha/n| \le 1$. Then by the above $\widetilde{T}(\alpha x) = \widetilde{T}(n((\alpha/n)x)) = n\widetilde{T}((\alpha/n)x) = n(\alpha/n)\widetilde{T}(x) = \alpha \widetilde{T}(x)$. Thus $\widetilde{T} : X \to Y$ is also homogeneous, and therefore $\widetilde{T} : X \to Y$ is linear. It remains to be proved that $\widetilde{T}$ is continuous. Since $\widetilde{T} : X \to Y$ is linear and the topologies on X and Y are linear it is sufficient to prove that $\widetilde{T}$ is continuous at $0 \in X$. Let $\varepsilon > 0$. Since $T : M \to Y$ is continuous, there is an open neighborhood V of $0 \in X$ such that $d(0, Tx) < \varepsilon/2$ $\forall x \in M \cap V$. Let $x \in V$ arbitrary. Then there is a neighborhood W for $0 \in X$ such that $x + W \subseteq V$. Let $W_0 = W$ and then we can find a sequence $(W_n)_{n\in\mathbb{N}}$ of neighborhoods of $0 \in X$ such that $W_n + W_n \subseteq W_{n-1}$ for each $n \in \mathbb{N}$. Consider now $z_n \in (W_n + x) \cap (V_n + x) \cap M$ for each $n \in \mathbb{N}$, where the sets V_n are the neighborhoods used for the definition of $\widetilde{T}$. We can always choose z_n because $(W_n + x) \cap (V_n + x)$ is a neighborhood for $x \in X$ and $M \subseteq X$ is dense. Using the definition for $\widetilde{T}$ we obtain that $\widetilde{T}x = \lim_{n\to\infty} Tz_n$. But $z_n \in (W_n + x) \subseteq (W_{n-1} + x) \subseteq \cdots \subseteq (W + x) \subseteq V$ $\forall n \in \mathbb{N}$ and, since $z_n \in M$ $\forall n \in \mathbb{N}$, we obtain that $d(0, Tz_n) < \varepsilon/2$ $\forall n \in \mathbb{N}$. Then $d\left(0, \widetilde{T}x\right) \le \varepsilon/2$ and therefore for each $x \in V$ we have that $\widetilde{T}x \in B_d(0, \varepsilon)$. Thus $\widetilde{T}$ is continuous at $0 \in X$.

20. See [44, exercise 13, chapter 1].

i) Let $f \in C[0,1]$ and $(f_n)_{n\in\mathbb{N}} \subseteq C[0,1]$ such that $f_n \xrightarrow{\tau_p} f$. We obtain that $f_n(x) \to f(x)$ $\forall x \in [0,1]$. Let $(g_n)_{n\in\mathbb{N}} \subseteq C[0,1]$,

$$g_n(x) = \frac{|f_n(x) - f(x)|}{1 + |f_n(x) - f(x)|} \quad \forall x \in [0,1], \forall n \in \mathbb{N}.$$

Then $g_n(x) \to 0$ $\forall x \in [0,1]$ and $|g_n(x)| \le 1$ $\forall n \in \mathbb{N}$, $\forall x \in [0,1]$. Using the Lebesgue dominated convergence theorem we obtain that $\int_0^1 g_n(x)\,dx \to 0$, and therefore $d(f_n, f) \to 0$. We have proved that the identity $I : (C[0,1], \tau_p) \to (C[0,1], \sigma)$ is sequentially continuous.

Suppose now that $I : (C[0,1], \tau_p) \to (C[0,1], \sigma)$ is continuous at $0 \in C[0,1]$. Then for any $\varepsilon \in (0,1)$ there are $\delta > 0$, $n \in \mathbb{N}$ and $x_1, ..., x_n \in [0,1]$ such that $\{f \in C[0,1] \mid |f(x_i)| \leq \delta, i = 1, ..., n\} \subseteq B_d(0, \varepsilon)$. For each $k \in \mathbb{N}$, let $f_k(x) = k(x - x_1) \cdots (x - x_n)$. Then $f_k \in C[0,1]$ and for each $1 \leq i \leq n$, $|f_k(x_i)| = 0 \leq \delta$, and therefore $f_k \in B_d(0, \varepsilon)$, i.e., $\int_0^1 (|f_k(x)|/(1 + |f_k(x)|))\, dx \leq \varepsilon \ \forall k \in \mathbb{N}$ (1). If we denote now by $h_k(x) = |f_k(x)|/(1 + |f_k(x)|) \ \forall x \in [0,1], \ \forall k \in \mathbb{N}$, then we have $h_k(x) \to 1 \ \forall x \in [0,1] \setminus \{x_1, ..., x_n\}$, i.e., $h_k \to 1$ almost everywhere. Also $|h_k(x)| \leq 1$ $\forall k \in \mathbb{N}, \ \forall x \in [0,1]$. Using the Lebesgue dominated convergence theorem we obtain that $\int_0^1 h_k(x)\, dx \to \int_0^1 1\, dx = 1$. Then passing to the limit in (1) we obtain that $\varepsilon \geq 1$, which contradicts our choice for ε.

ii) The fact that the topology τ_p is not metrizable is now obvious, if we use (i), because for a function between two metric spaces the continuity is equivalent to the sequential continuity.

21. i) Let $f : (X, \tau) \to \mathbb{K}$ be linear and continuous. Using the continuity of f at $0 \in X$, for $\varepsilon > 0$ we obtain that there are $n \in \mathbb{N}$, $x_1^*, ..., x_n^* \in \Gamma$ and $\delta > 0$ such that if $x \in X$ is such that $|x_i^*(x)| \leq \delta, i = 1, ..., n$, then $|f(x)| \leq \varepsilon$. Let $x \in \bigcap_{i=1}^n \ker x_i^*$. Then $|x_i^*(kx)| = 0 \leq \delta, i = 1, ..., n, \ \forall k \in \mathbb{N}$, and therefore $|f(kx)| \leq \varepsilon \ \forall k \in \mathbb{N}$, that is, $|f(x)| \leq \varepsilon/k \ \forall k \in \mathbb{N}$ and passing to the limit for $k \to \infty$ we obtain that $f(x) = 0$. We obtain that $\bigcap_{i=1}^n \ker x_i^* \subseteq \ker f$ and, using exercise 16(i) of chapter 14 we obtain that there are $\alpha_1, ..., \alpha_n \in \mathbb{K}$ such that $f = \sum_{i=1}^n \alpha_i x_i^*$.

In the case in which $\Gamma \subseteq X^*$ is a linear subspace we observe that if $f : (X, \tau) \to \mathbb{K}$ is linear and continuous then $f \in \Gamma$. Conversely, if $x^* \in \Gamma$ then $x^* : (X, \tau) \to \mathbb{K}$ is linear and continuous. Thus we have the identification $\Gamma = (X, \tau)^*$.

ii) See [44, exercise 13 (c), chapter 1].

The topology τ_p on $C[0,1]$ is given by $\Gamma \subseteq (C[0,1])^*$, $\Gamma = \{\delta_x \mid x \in [0,1]\}$, where, for $x \in [0,1]$, $\delta_x(f) = |f(x)| \ \forall f \in C[0,1]$. By (i) we obtain that any linear and continuous functional on $(C[0,1], \tau_p)$ is of the form $f \mapsto \sum_{i=1}^n \alpha_i f(x_i)$, with $n \in \mathbb{N}$, $\alpha_1, ..., \alpha_n \in \mathbb{K}$, $x_1, ..., x_n \in [0,1]$.

22. See [9, exercise 2 (a), (c), chapter I, section 2].

i) We have $V_{\varepsilon/2, \delta/2} + V_{\varepsilon/2, \delta/2} \subseteq V_{\varepsilon, \delta}$. Indeed, let $f, g \in V_{\varepsilon/2, \delta/2}$. Then we can find an open set $A \subseteq [0,1]$, $A = \bigcup_{i=1}^\infty (\alpha_i, \beta_i)$, $((\alpha_i, \beta_i))_{i \in \mathbb{N}}$ pairwise disjoint intervals with $\sum_{i=1}^\infty (\beta_i - \alpha_i) \leq \delta/2$ such that $|f(t)| \leq \varepsilon/2 \ \forall t \notin A$, and an open set $B \subseteq [0,1]$, $B = \bigcup_{i=1}^\infty (\xi_i, \eta_i)$, $((\xi_i, \eta_i))_{i \in \mathbb{N}}$ pairwise disjoint intervals with $\sum_{i=1}^\infty (\eta_i - \xi_i) \leq \delta/2$ such that $|g(t)| \leq \varepsilon/2 \ \forall t \notin B$. Let $C = A \cup B$. Then C is open, and therefore C is the union of its component intervals: $C = \bigcup_{i=1}^\infty (u_i, v_i)$, $((u_i, v_i))_{i \in \mathbb{N}}$ pairwise disjoint intervals. If μ is the Lebesgue measure on $[0,1]$ then $\mu(C) \leq \mu(A) + \mu(B)$, and therefore $\sum_{i=1}^\infty (v_i - u_i) \leq \delta$.

13.2 Solutions

Also $\forall t \notin C$ we have $|f(t) + g(t)| \leq |f(t)| + |g(t)| \leq \varepsilon$. Therefore $f + g \in V_{\varepsilon,\delta}$.

If $\varepsilon_1 < \varepsilon_2$ and $\delta_1 < \delta_2$ then $V_{\varepsilon_1,\delta_1} \subseteq V_{\varepsilon_2,\delta_2}$. Indeed, if $f \in V_{\varepsilon_1,\delta_1}$ then we can find an open set $A \subseteq [0,1]$, $A = \bigcup_{i=1}^{\infty}(\alpha_i, \beta_i)$, $((\alpha_i, \beta_i))_{i \in \mathbb{N}}$ pairwise disjoint intervals with $\sum_{i=1}^{\infty}(\beta_i - \alpha_i) \leq \delta_1$, such that $|f(t)| \leq \varepsilon_1 \; \forall t \notin A$. Then $\sum_{i=1}^{\infty}(\beta_i - \alpha_i) \leq \delta_2$ and $|f(t)| \leq \varepsilon_2$ $\forall t \notin A$, and we obtain that $f \in V_{\varepsilon_2,\delta_2}$. Then $V_{\min\{\varepsilon',\varepsilon''\},\min\{\delta',\delta''\}} \subseteq V_{\varepsilon',\delta'} \cap V_{\varepsilon'',\delta''}$.

Clearly, for $\varepsilon > 0$ and $0 < \delta < 1$, $V_{\varepsilon,\delta}$ is absorbing and balanced. Also, if $\lambda \neq 0$ then $\lambda V_{\varepsilon,\delta} = V_{|\lambda|\varepsilon,\delta}$. Therefore there is a linear topology τ on X such that $(V_{\varepsilon,\delta})_{\varepsilon,\delta}$ form a basis of neighborhoods for $0 \in X$ in the τ topology.

ii) It is sufficient to prove that if $V_{\varepsilon',\delta'} \subseteq U \subseteq V_{\varepsilon,\delta}$ where $\varepsilon, \varepsilon' > 0$, $\delta, \delta' \in (0, 1/3)$, then U is not a convex set. Indeed, there is $n \in \mathbb{N}$ such that $1 > n\delta' > 2\delta$. Also, let $I_i \subseteq [0,1]$, $i = \overline{1,n}$, n pairwise disjoint intervals, each of length δ'. For each $1 \leq i \leq n$, let J_i be the interval of length $\delta'/2$ centered in the middle of I_i and we consider $f_i \in X$ with the properties $f_i = 0$ outside I_i, $f_i > n\varepsilon$ on J_i. Then $f_i \in V_{\varepsilon',\delta'} \; \forall 1 \leq i \leq n$ and therefore $f_i \in U \; \forall 1 \leq i \leq n$. If U were convex then $(1/n)\sum_{i=1}^{n} f_i \in U$, and therefore $(1/n)\sum_{i=1}^{n} f_i \in V_{\varepsilon,\delta}$, a contradiction, since $\left((1/n)\sum_{i=1}^{n} f_i\right)(t) > \varepsilon \; \forall t \in \bigcup_{i=1}^{n} J_i$, and the sum of the lengths for the intervals J_i is $n(\delta'/2) > \delta$.

iii) Let $\varphi : (X, \tau) \to \mathbb{R}$ be linear and continuous. From the continuity, there are $\varepsilon > 0$ and $\delta \in (0,1)$ such that $|\varphi(f)| \leq 1 \; \forall f \in V_{\varepsilon,\delta}$. We consider a covering of $[0,1]$ with open intervals U_i, $i = \overline{1,n}$, such that every interval U_i has its length lesser then δ. Using the partition of unity theorem there are $f_1, ..., f_n \in X$ such that $\sum_{i=1}^{n} f_i = 1$, and $f_i = 0$ on $[0,1] \setminus U_i \; \forall 1 \leq i \leq n$. Then for any $f \in X$ we have $f = \sum_{i=1}^{n} ff_i$ and since φ is linear, $\varphi(f) = \sum_{i=1}^{n} \varphi(ff_i)$. But $(ff_i)(t) = 0$ for every $t \in [0,1]$ with the exception of an interval of length lesser than δ. We obtain that $k(ff_i) \in V_{\varepsilon,\delta}$ for every $k \in \mathbb{N}$, and then $|\varphi(k(ff_i))| \leq 1$ for every $k \in \mathbb{N}$. Then $|\varphi(ff_i)| \leq 1/k$ for every $k \in \mathbb{N}$, and passing to the limit for $k \to \infty$ we deduce that $\varphi(ff_i) = 0$ for each $1 \leq i \leq n$. Then $\varphi(f) = 0$ $\forall f \in X$, and $\varphi = 0$.

23. i) Let $|\lambda| \leq 1$ and $x \in A$. Then $\sum_{n=1}^{\infty} |a_n| |x_n|^p \leq 1$ and therefore

$$\sum_{n=1}^{\infty} |a_n| |\lambda x_n|^p = |\lambda|^p \sum_{n=1}^{\infty} |a_n| |x_n|^p \leq |\lambda|^p \leq 1,$$

i.e., $\lambda x \in A$. Since $p \geq 1$ the function $t \longmapsto |t|^p$ is convex and then

$$|\lambda x_n + (1-\lambda) y_n|^p \leq \lambda |x_n|^p + (1-\lambda) |y_n|^p \; \forall n \in \mathbb{N}, \; \forall 0 \leq \lambda \leq 1.$$

If $x = (x_n)_{n \in \mathbb{N}} \in A$, $y = (y_n)_{n \in \mathbb{N}} \in A$, and $0 \leq \lambda \leq 1$, then

$$\sum_{n=1}^{\infty} |a_n| |\lambda x_n + (1-\lambda) y_n|^p \leq \lambda \sum_{n=1}^{\infty} |a_n| |x_n|^p + (1-\lambda) \sum_{n=1}^{\infty} |a_n| |y_n|^p \leq \lambda + (1-\lambda) = 1,$$

i.e., $\lambda x + (1 - \lambda)y \in A$.

ii) Suppose that A is absorbing. Then $\forall x \in l_p \; \exists \lambda > 0$ such that $\lambda x \in A$, that is, $\forall \sum_{n=1}^{\infty} |x_n|^p < \infty \; \exists \lambda > 0$ such that $\sum_{n=1}^{\infty} |a_n| \lambda^p |x_n|^p \leq 1$, that is, $\sum_{n=1}^{\infty} |a_n| |x_n|^p \leq 1/\lambda^p$. But $\forall x = (x_n)_{n \in \mathbb{N}} \in l_1$, we have $\left(|x_n|^{1/p} \right)_{n \in \mathbb{N}} \in l_p$, and then we can find λ such that $\sum_{n=1}^{\infty} |a_n| |x_n| \leq 1/\lambda^p$, i.e., for every $x = (x_n)_{n \in \mathbb{N}} \in l_1$ the series $\sum_{n=1}^{\infty} a_n x_n$ converges absolutely. Using exercise 24(iii) of chapter 9 we obtain that $(a_n)_{n \in \mathbb{N}} \in l_\infty$.

Conversely, if $(a_n)_{n \in \mathbb{N}} \in l_\infty$, then $\forall x \in l_p \; \exists \lambda = (M \|x\|^p + 1)^{-1/p}$ (here $M = \sup_{n \in \mathbb{N}} |a_n| < \infty$) such that $\lambda x \in A$, since

$$\sum_{n=1}^{\infty} |a_n| \lambda^p |x_n|^p \leq \lambda^p M \|x\|^p \leq \frac{M \|x\|^p}{M \|x\|^p + 1} \leq 1.$$

Since A is balanced we obtain that A is absorbing.

iii) Since (ii) is true we can calculate the Minkowski seminorm. We have

$$p_A(x) = \inf\{\mu > 0 \mid x \in \mu A\} = \inf \left\{ \mu > 0 \; \Big| \; \sum_{n=1}^{\infty} |a_n| \frac{|x_n|^p}{\mu^p} \leq 1 \right\}$$

$$= \inf \left\{ \mu > 0 \; \Big| \; \left(\sum_{n=1}^{\infty} |a_n| |x_n|^p \right)^{1/p} \leq \mu \right\} = \left(\sum_{n=1}^{\infty} |a_n| |x_n|^p \right)^{1/p}.$$

24. i) a) If $p : X \to \mathbb{R}_+$ is a seminorm then a simple calculation shows that the set $\{x \in X \mid p(x) \leq 1\}$ is convex and balanced, hence A is convex and balanced, as an intersection of convex and balanced sets.

b) Suppose that A is absorbing. Then $\forall x \in X \; \exists \varepsilon > 0$ such that $\varepsilon x \in A$, i.e., $p_n(\varepsilon x) \leq 1$ $\forall n \in \mathbb{N}$, i.e., since p_n are seminorms, $p_n(x) \leq 1/\varepsilon$ $\forall n \in \mathbb{N}$. Conversely, if $\sup_{k \in \mathbb{N}} p_k(x) < \infty$ $\forall x \in X$ then $\forall x \in X \; \exists \varepsilon = \left(1 + \sup_{k \in \mathbb{N}} p_k(x)\right)^{-1} > 0$ such that

$$p_n(\varepsilon x) = \varepsilon p_n(x) = \frac{p_n(x)}{1 + \sup_{k \in \mathbb{N}} p_k(x)} \leq 1 \; \forall n \in \mathbb{N},$$

and then A is absorbing. If A is absorbing we have

$$p_A(x) = \inf \{\mu > 0 \mid x \in \mu A\} = \inf \{\mu > 0 \mid p_n(x) \leq \mu \; \forall n \in \mathbb{N}\} = \sup_{n \in \mathbb{N}} p_n(x).$$

c) Suppose that A is absorbing. Using (b) we can define $p : X \to \mathbb{R}$, $p(x) = \sup_{n \in \mathbb{N}} p_n(x)$, and p is a seminorm. Since p takes finite values we have $X = \bigcup_{k \in \mathbb{N}} E_k$, where

$$E_k = \{x \in X \mid p(x) \leq k\} = \bigcap_{n \in \mathbb{N}} \{x \in X \mid p_n(x) \leq k\}.$$

13.2 Solutions

Since the seminorms p_n are continuous then the sets E_k are closed. By the Baire category theorem (X is a Banach space) it follows that there is $k \in \mathbb{N}$ such that $\overset{\circ}{E_k} \neq \emptyset$. Then there are $\varepsilon > 0$ and $a \in X$ such that $\overline{B}(a, \varepsilon) \subseteq E_k$. Let now $x \in X$, $x \neq 0$. Then $a + \varepsilon x / \|x\| \in \overline{B}(a, \varepsilon)$, hence $a + \varepsilon x / \|x\| \in E_k$, i.e., $p(a + \varepsilon x / \|x\|) \leq k$, and then $(\varepsilon / \|x\|) p(x) \leq k + p(a)$, i.e., $p(x) \leq ((k + p(a))/\varepsilon) \|x\|$.

The converse is clear by (b).

ii) a) Let $p_n : \mathcal{R}[0,1] \to \mathbb{R}$ be defined by $p_n(f) = |a_n x_n^*(f)|$. Clearly all p_n are seminorms and $A = \{f \in \mathcal{R}[0,1] \mid p_n(f) \leq 1 \ \forall n \in \mathbb{N}\}$. If A is absorbing then by (i) the family $(p_n)_{n \in \mathbb{N}}$ is pointwise bounded, in particular $\sup_{n \in \mathbb{N}} p_n(1) < \infty$, i.e., $\sup_{n \in \mathbb{N}} |a_n x_n^*(1)| < \infty$. For the converse we remark first that if $x^* : \mathcal{R}[0,1] \to \mathbb{R}$ is linear and positive, and $a \in \mathbb{R}$, then we have

$$|a x^*(f)| \leq |a| \, x^*(1) \sup_{x \in [0,1]} |f(x)| \quad \forall f \in \mathcal{R}[0,1].$$

Then for any $f \in \mathcal{R}[0,1]$ and any $n \in \mathbb{N}$ we have by the above remark that

$$p_n(f) = |a_n x_n^*(f)| \leq |a_n| \, x_n^*(1) \sup_{x \in [0,1]} |f(x)|,$$

hence

$$\sup_{n \in \mathbb{N}} p_n(f) \leq \left(\sup_{n \in \mathbb{N}} |a_n| \, x_n^*(1) \right) \left(\sup_{x \in [0,1]} |f(x)| \right) < \infty$$

and this implies that A is absorbing.

If A is absorbing then $p_A(f) = \sup_{n \in \mathbb{N}} p_n(f) = \sup_{n \in \mathbb{N}} |a_n x_n^*(f)|, \forall f \in \mathcal{R}[0,1]$

For the functions φ_n we take $x_n^* : \mathcal{R}[0,1] \to \mathbb{R}$ defined by $x_n(f) = \int_0^1 \varphi_n(x) f(x) dx$ and we apply the above calculations.

b) Let $p_n : \mathcal{R}[0,1] \to \mathbb{R}$ be defined by $p_n(f) = \sum_{k=1}^n |a_k x_k^*(f)|$. Clearly all the p_n are seminorms and $A = \{f \in \mathcal{R}[0,1] \mid p_n(f) \leq 1 \ \forall n \in \mathbb{N}\}$. If A is absorbing then the family $(p_n)_{n \in \mathbb{N}}$ is pointwise bounded, in particular $\sup_{n \in \mathbb{N}} p_n(1) < \infty$, i.e., $\sum_{k=1}^\infty |a_k x_k^*(1)| < \infty$. For the converse we use the same remark: if $x^* : \mathcal{R}[0,1] \to \mathbb{R}$ is linear and positive and $a \in \mathbb{R}$, then we have

$$|a x^*(f)| \leq |a| \, x^*(1) \sup_{x \in [0,1]} |f(x)| \quad \forall f \in \mathcal{R}[0,1].$$

Then for any $f \in \mathcal{R}[0,1]$ and any $n \in \mathbb{N}$ we have by the above remark that

$$p_n(f) = \sum_{k=1}^n |a_k x_k^*(f)| \leq \sup_{x \in [0,1]} |f(x)| \cdot \sum_{k=1}^n |a_k x_k^*(1)|$$

hence

$$\sup_{n \in \mathbb{N}} p_n(f) \leq \sup_{x \in [0,1]} |f(x)| \cdot \sum_{k=1}^\infty |a_k x_k^*(1)| < \infty,$$

and hence A is absorbing. We also have that if A is absorbing then $p_A(f) = \sum_{k=1}^\infty |a_k x_k^*(f)|$ $\forall f \in \mathcal{R}[0,1]$.

For the particular case we take $x_n^* : \mathcal{R}[0,1] \to \mathbb{R}$ defined by $x_n(f) = \int_0^1 x^n f(x)dx$ and we apply the above calculations.

iii) For every $n \in \mathbb{N}$ consider $p_n : X \to \mathbb{R}$, $p_n(x) = \sum_{k=1}^{n} |x_k^*(x)|$. Then

$$A = \{x \in X \mid p_n(x) \leq 1\ \forall n \in \mathbb{N}\}.$$

Since X is a Banach space from the part (c) of (i) it follows that A is absorbing if and only if $\exists M > 0$ such that $\sup_{n \in \mathbb{N}} p_n(x) \leq M \|x\|\ \forall x \in X$, i.e., $\exists M > 0$ such that $\sum_{n=1}^{\infty} |x_n^*(x)| \leq M$ $\forall x \in B_X$. Now since for a finite system of scalars $\lambda_1, ..., \lambda_n$ we have that

$$\sup\left\{\left|\sum_{i \in F} \lambda_i\right| \mid F \subseteq \{1, ..., n\}\right\} \leq \sum_{k=1}^{n} |\lambda_k| \leq 6 \sup\left\{\left|\sum_{i \in F} \lambda_i\right| \mid F \subseteq \{1, ..., n\}\right\}$$

(see [45, Lemma 6.3]) this is equivalent to

$$\sup\left\{\left\|\sum_{n \in F} x_n^*\right\| \mid F \text{ finite, } F \subseteq \mathbb{N}\right\} < \infty.$$

Also if A is absorbing then $p_A(x) = \sup_{n \in \mathbb{N}} p_n(x) = \sum_{n=1}^{\infty} |x_n^*(x)|$.

iv) Taking $x_n^* : H \to \mathbb{K}$, $x_n^*(x) = a_n \langle x, e_n \rangle$, by (iii) we obtain that A is absorbing if and only if $\sup\left\{\left\|\sum_{n \in F} x_n^*\right\| \mid F \text{ finite, } F \subseteq \mathbb{N}\right\} < \infty$ (1). Also for any finite set $F \subseteq \mathbb{N}$ we have

$$\left(\sum_{n \in F} x_n^*\right)(x) = \left\langle x, \sum_{n \in F} \overline{a_n} e_n \right\rangle \quad \forall x \in H,$$

so $\left\|\sum_{n \in F} x_n^*\right\| = \left\|\sum_{n \in F} a_n e_n\right\| = \left(\sum_{n \in F} |a_n|^2\right)^{1/2}$ (for the last equality we have used the Pythagoras theorem). From here it follows easily that (1) is equivalent with the fact that $\sum_{k=1}^{\infty} |a_k|^2 < \infty$.

If A is absorbing then $p_A(x) = \sum_{n=1}^{\infty} |a_n \langle x, e_n \rangle|\ \forall x \in H$.

v) Taking $x_n^* : L(H) \to \mathbb{K}$, $x_n^*(U) = a_n \langle U(e_n), e_n \rangle$, by (iii) we obtain that A is absorbing if and only if $\sup\left\{\left\|\sum_{n \in F} x_n^*\right\| \mid F \text{ finite, } F \subseteq \mathbb{N}\right\} < \infty$ (2). Also for any finite set $F \subseteq \mathbb{N}$ we have $\left\|\sum_{n \in F} x_n^*\right\| = \sum_{n \in F} |a_n|$. Indeed for any $U \in L(H)$ we have

$$\left|\left(\sum_{n \in F} x_n^*\right)(U)\right| \leq \sum_{n \in F} |a_n| |\langle U(e_n), e_n \rangle| \leq \sum_{n \in F} |a_n| \|U\|,$$

so

$$\left\|\sum_{n \in F} x_n^*\right\| \leq \sum_{n \in F} |a_n|.$$

For the converse, let $U \in L(l_2)$ defined by $U(x) = \sum_{n \in F} \operatorname{sgn}(a_n) \langle x, e_n \rangle e_n$. Then $\left|\left(\sum_{n \in F} x_n^*\right)(U)\right| \leq \|U\| \left\|\sum_{n \in F} x_n^*\right\|$. We have $\left(\sum_{n \in F} x_n^*\right)(U) = \sum_{n \in F} |a_n|$, and by the Pythagoras theorem and the Bessel inequality

$$\|U(x)\|^2 = \sum_{n \in F} |\operatorname{sgn}(a_n) \langle x, e_n \rangle|^2 \leq \sum_{n \in F} |\langle x, e_n \rangle|^2 \leq \|x\|^2 \quad \forall x \in H,$$

13.2 Solutions

thus $\|U\| \leq 1$. We obtain that $\left\|\sum_{n\in F} x_n^*\right\| \geq \sum_{n\in F} |a_n|$. From here it follows easily that (2) is equivalent with the fact that $\sum_{k=1}^{\infty} |a_k| < \infty$.

If A is absorbing then $p_A(U) = \sum_{n=1}^{\infty} |a_n \langle U(e_n), e_n \rangle| \; \forall U \in L(H)$

vi) For every $n \in \mathbb{N}$ let us consider the set $E_n = [1/(n+1), 1/n)$, and $p_n : C[0,1] \to \mathbb{R}$, $p_n(f) = \left(\sum_{k=1}^{n} \left| a_k \int_{E_k} f(x)dx \right|^p \right)^{1/p}$. Then all p_n are seminorms and

$$A = \{x \in X \mid p_n(x) \leq 1 \; \forall n \in \mathbb{N}\}.$$

By (i), A is absorbing if and only if $\exists M > 0$ such that $\sup_{n \in \mathbb{N}} p_n(f) \leq M \|f\| \; \forall f \in C[0,1]$ (3). Using the Hölder inequality we have

$$p_n(f) = \sup \left\{ \left| \sum_{k=1}^{n} a_k b_k \int_{E_k} f(x)dx \right| \; \middle| \; \left(\sum_{k=1}^{n} |b_k|^q \right)^{1/q} \leq 1 \right\}$$

(q is the conjugate of p). It follows that the condition (3) is equivalent with the fact that

$$\left| \sum_{k=1}^{n} a_k b_k \int_{E_k} f(x)dx \right| \leq M \; \forall \|f\| \leq 1, \; \forall \sum_{k=1}^{n} |b_k|^q \leq 1, \; \forall n \in \mathbb{N}.$$

But

$$\sup_{f \in C[0,1], \|f\| \leq 1} \left| \sum_{k=1}^{n} a_k b_k \int_{E_k} f(x)dx \right| = \sup_{f \in C[0,1], \|f\| \leq 1} \left| \int_0^1 \sum_{k=1}^{n} a_k b_k \chi_{E_k} f(x)dx \right|$$

$$= \int_0^1 \left| \sum_{k=1}^{n} a_k b_k \chi_{E_k} \right| dx = \sum_{k=1}^{n} |a_k b_k| \mu(E_k)$$

$$= \sum_{k=1}^{n} \frac{1}{k(k+1)} |a_k b_k|,$$

since the sets E_k are pairwise disjoint. Thus the last inequality is equivalent with

$$\sum_{k=1}^{n} \frac{1}{k(k+1)} |a_k b_k| \leq M \; \forall \sum_{k=1}^{n} |b_k|^q \leq 1, \; \forall n \in \mathbb{N}.$$

Using again the Hölder inequality this is equivalent with the fact that

$$\sum_{k=1}^{n} \frac{1}{k^p (k+1)^p} |a_k|^p \leq M^p, \; \forall n \in \mathbb{N},$$

i.e., $((1/(n(n+1))) a_n)_{n \in \mathbb{N}} \in l_p$.

If A is absorbing then

$$p_A(f) = \left(\sum_{k=1}^{\infty} \left| a_k \int_{1/(k+1)}^{1/k} f(x)dx \right|^p \right)^{1/p} \quad \forall f \in C[0,1].$$

vii) Let $x_n^*(f) = a_n f(1/n)$. Then $\|x_n^*\| = |a_n| \, \forall n \in \mathbb{N}$. Observing that

$$A = \{f \in C[0,1] \mid |x_n^*(f)| \leq 1 \, \forall n \in \mathbb{N}\}$$

by (i) it follows that A is absorbing if and only if $\sup_{n \in \mathbb{N}} \|x_n^*\| < \infty$, i.e., $\sup_{n \in \mathbb{N}} |a_n| < \infty$. Also if A is absorbing then

$$p_A(f) = \sup_{n \in \mathbb{N}} |x_n^*(f)| = \sup_{n \in \mathbb{N}} |a_n f(1/n)|.$$

25. i) Observe first that since $0 \in \overset{\circ}{A}$, A is a neighborhood of 0, hence absorbing, and therefore p_A has a meaning. Let $\varepsilon > 0$. Then $V := (\varepsilon/2) A \cap (-(\varepsilon/2) A)$ is a neighborhood of 0, and for each $x \in V$ we have that $p_A(x) \leq \varepsilon/2 < \varepsilon$ and $p_A(-x) \leq \varepsilon/2 < \varepsilon$. We obtain that $x \longmapsto p_A(x)$ is continuous at 0, and also $x \longmapsto \max(p_A(x), p_A(-x))$ is continuous at 0. If A is symmetric then $p_A(-x) = p_A(x) \, \forall x \in X$ and then since $p_A(x+y) \leq p_A(x) + p_A(y) \, \forall x, y \in X$ we obtain that

$$|p_A(x) - p_A(y)| \leq p_A(x-y) \quad \forall x, y \in X.$$

Since p_A is continuous at $0 \in X$ from here it follows that p_A is continuous.

ii) See [42, chapter V, the proof of Theorem V.4].

Since $x_0 \in X \setminus A$, $x_0 \neq 0$. Let $M = \mathrm{Sp}\{x_0\} = \{\lambda x_0 \mid \lambda \in \mathbb{R}\}$ and $g : M \to \mathbb{R}$, $g(\lambda x_0) = \lambda \, \forall \lambda \in \mathbb{R}$. For any $\lambda \leq 0$ we have $g(\lambda x_0) = \lambda \leq 0 \leq p_A(\lambda x_0)$. If $\lambda > 0$ then since $p_A(x_0) \geq 1$ ($x_0 \notin A$) we have $g(\lambda x_0) = \lambda \leq \lambda p_A(x_0) = p_A(\lambda x_0)$. Then $g \leq p_A$ on M and by the Hahn–Banach theorem there is $f : X \to \mathbb{R}$ linear such that $f|_M = g$ and $f \leq p_A$ on X. Then $f(x) \leq p_A(x) \leq 1 \, \forall x \in A$ and $f(x_0) = g(x_0) = 1$. We must prove now that $f : X \to \mathbb{R}$ is continuous. We have $f(x) \leq p_A(x)$, $-f(x) = f(-x) \leq p_A(-x)$ $\forall x \in X$, and therefore $|f(x)| \leq \max(p_A(x), p_A(-x)) \, \forall x \in X$. Then f is continuous at 0, because it is proved at (i) that $x \longmapsto \max(p_A(x), p_A(-x))$ is continuous at 0. Using the linearity of f it follows that f is continuous on X.

Remark. If X is a linear topological space which contains a convex neighborhood of the origin, strictly contained in X, then the dual of X is not the null space. Conversely, if for a linear topological space X its dual is not the null space, then X has a convex neighborhood of the origin strictly contained in X. For example, $V = f^{-1}((-1,1))$, where $f \in X^*$ is not zero.

If X is a locally convex space, we obtain that its dual is not the null space.

26. i) We have $\{x \in X \mid |x_1^*(x)| < 1, ..., |x_n^*(x)| < 1\} \subseteq A$ and since the set from the left side is open and contains the origin it follows that $0 \in \overset{\circ}{A}$. Let $f \in X^*$ with the properties $|f(x)| \leq 1 \, \forall x \in A$ and $f(x_0) = 1$ (by exercise 25 there is at least one such f). We will prove that $\bigcap_{i=1}^n \ker x_i^* \subseteq \ker f$. Indeed, let $x \in \bigcap_{i=1}^n \ker x_i^*$. Then for any $n \in \mathbb{N}$ the element nx is in A, hence $|f(nx)| \leq 1$, therefore $|f(x)| \leq 1/n$ and passing to the limit for $n \to \infty$ it follows that $|f(x)| \leq 0$, i.e., $f(x) = 0$, $x \in \ker f$. Using exercise 16(i) of chapter 14, we can find $\lambda_1 \in \mathbb{R}, ..., \lambda_n \in \mathbb{R}$ such that $f = \sum_{i=1}^n \lambda_i x_i^*$. Since $\{x_1^*, ..., x_n^*\} \subseteq X^*$ is a linearly

13.2 Solutions

independent system by a well known result from the linear algebra we obtain that we can find $\{x_1, ..., x_n\} \subseteq X$ such that $x_i^*(x_j) = \delta_{ij}\ \forall 1 \leq i, j \leq n$. Form here it follows that there is $x \in X$ ($x = \sum_{i=1}^{n} \operatorname{sgn}(\lambda_i) x_i$) such that $x_i^*(x) = \operatorname{sgn}(\lambda_i)$, for any $1 \leq i \leq n$. This means that $x \in A$, hence $f(x) \leq 1$, i.e., $\sum_{i=1}^{n} \lambda_i x_i^*(x) \leq 1$, so $\sum_{i=1}^{n} |\lambda_i| \leq 1$.

ii) We have $U_0(e_1) = \langle e_1, 2e_1 + 3e_2 \rangle e_2 = 2e_2$ and $U_0(e_2) = \langle e_2, 2e_1 + 3e_2 \rangle e_2 = 3e_2$, hence $U_0 \notin A$. If $f \in (L(l_2))^*$ has the properties $|f(U)| \leq 1\ \forall U \in A$ and $f(U_0) = 1$, then by (i) we can find $\lambda, \mu \in \mathbb{R}$ such that $f(U) = \lambda \langle Ue_1, e_2 \rangle + \mu \langle Ue_2, e_2 \rangle\ \forall U \in L(l_2)$, $|\lambda| + |\mu| \leq 1$ and $\lambda \langle U_0 e_1, e_2 \rangle + \mu \langle U_0 e_2, e_2 \rangle = 1$. From this we obtain that $2\lambda + 3\mu = 1$ and $|\lambda| + |\mu| \leq 1$, i.e., $\mu = (1 - 2\lambda)/3$ and $|\lambda| + |1 - 2\lambda|/3 \leq 1$. If $\lambda \leq 0$ we have $-\lambda + (1 - 2\lambda)/3 \leq 1$, and then $\lambda \in [-2/5, 0]$. If $0 < \lambda \leq 1/2$ we have $\lambda + (1 - 2\lambda)/3 \leq 1$, $\lambda \leq 2$, so $0 < \lambda \leq 1/2$. For $\lambda > 1/2$, we have $\lambda + (2\lambda - 1)/3 \leq 1$, $\lambda \leq 4/5$, so $1/2 < \lambda \leq 4/5$. Hence $\lambda \in [-2/5, 4/5]$.

27. See [35, exercise G, chapter 2].

i) Let $x \in \bigcap_{r>1}(rA)$. Then for any $n \in \mathbb{N}$ there is an $a_n \in A$ such that $x = (1 + 1/n)a_n$. Then $a_n = x/(1 + 1/n)$ for every $n \in \mathbb{N}$ and therefore $a_n \to x$. We obtain that $x \in \overline{A}$.

ii) From (i), $\bigcap_{r>1}(rA) \subseteq \overline{A}$. Suppose that $x_0 \in \overline{A}$ and $x_0 \notin \bigcap_{r>1}(rA)$. Hence there is $r > 1$ such that $x_0 \notin (rA)$. Since rA is a convex set and 0 belongs to the interior of rA then by exercise 25 we obtain that there is a linear and continuous functional $f : X \to \mathbb{R}$ such that $f(x) \leq 1\ \forall x \in rA$ and $f(x_0) = 1$. Then $f(ra) \leq 1\ \forall a \in A$, and therefore $f(a) \leq 1/r\ \forall a \in A$. Since $x_0 \in \overline{A}$ then by the continuity of f we obtain that $f(x_0) \leq 1/r$. We obtain that $1 \leq 1/r$, $r \leq 1$, contradiction.

28. See [32, chapter 2], for more general results.

i) We have:

$$\|f\|_p = 0 \Leftrightarrow \int_0^1 |f(t)|^p\, dt = 0 \Leftrightarrow f(t) = 0\ \mu\text{-a.e.} \Leftrightarrow f = 0 \text{ in } L_p[0, 1].$$

For $\lambda \in \mathbb{R}$ and $f \in L_p[0, 1]$,

$$\|\lambda f\|_p = \int_0^1 |\lambda f(t)|^p\, dt = |\lambda|^p \int_0^1 |f(t)|^p\, dt = |\lambda|^p \|f\|_p,$$

and for f and g in $L_p[0, 1]$,

$$\|f + g\|_p = \int_0^1 |f(t) + g(t)|^p\, dt \leq \int_0^1 (|f(t)|^p + |g(t)|^p)dt = \|f\|_p + \|g\|_p.$$

We observe that $\|\cdot\|_p$ is not a norm on $L_p[0, 1]$.

ii) We have

$$d(f, g) = 0 \Leftrightarrow \|f - g\|_p = 0 \Leftrightarrow f = g \text{ in } L_p[0, 1].$$

Also

$$d(f, g) = \|f - g\|_p = |-1|^p \|g - f\|_p = d(g, f)\ \forall f, g \in L_p[0, 1],$$

and $\forall f, g, h \in L_p[0,1]$ we have
$$d(f,g) = \|f-g\|_p \le \|f-h\|_p + \|h-g\|_p = d(f,h) + d(h,g),$$
and therefore d is a metric on $L_p[0,1]$. We observe that d is translation invariant. Analogously with the case $p \ge 1$ one can prove that d is a complete metric and therefore $(L_p[0,1], \tau)$ is an F-space.

iii) Let $V \subseteq L_p[0,1]$ be a non-empty open and convex set. Using, if necessary, a translation, we can suppose that $0 \in V$. Since V is open there is $\varepsilon > 0$ such that $B_d(0, \varepsilon) \subseteq V$. Let $f \in L_p[0,1]$. Since $p < 1$, there is an $n \in \mathbb{N}$ such that $\|f\|_p n^{p-1} < \varepsilon$. If we consider $h : [0,1] \to \mathbb{R}$, $h(x) = \int_0^x |f(t)|^p dt$, then h is continuous, $h(0) = 0$, $h(1) = \|f\|_p$ and therefore by the continuity of h we can find $0 = x_0 < x_1 < \cdots < x_n = 1$ such that $\int_{x_{i-1}}^{x_i} |f(t)|^p dt = \|f\|_p / n$ for each $1 \le i \le n$. For any $1 \le i \le n$ we define $g_i : [0,1] \to \mathbb{R}$,
$$g_i(t) = \begin{cases} nf(t), & t \in (x_{i-1}, x_i], \\ 0, & t \notin (x_{i-1}, x_i]. \end{cases}$$
Then for any $1 \le i \le n$ we have $\|g_i\|_p = n^p \int_{x_{i-1}}^{x_i} |f(t)|^p = \|f\|_p n^{p-1} < \varepsilon$ and therefore $g_i \in B_d(0, \varepsilon) \subseteq V$. But $f(t) = (1/n) \sum_{i=1}^n g_i(t)$ $\forall t \in (0,1]$, and therefore $f = (1/n) \sum_{i=1}^n g_i$ in $L_p[0,1]$. Since V is convex we obtain that $f \in V$ and therefore $L_p[0,1] - V$.

If $W \subseteq L_p[0,1]$ is an open non-empty set using exercise 10(i) we obtain that co$(W) \subseteq L_p[0,1]$ is an open and convex set and then co$(W) = L_p[0,1]$.

iv) Consider now any locally convex Hausdorff space (X, σ) and $T : (L_p[0,1], \tau) \to (X, \sigma)$ a linear and continuous operator. Using the local convexity of (X, σ) we obtain that there is a basis of neighborhoods $\mathcal{B}$ for $0 \in X$ containing only convex sets. Consider $V \in \mathcal{B}$. Since V is a neighborhood for 0, there is an open neighborhood W for $0 \in X$ such that $W \subseteq V$. Let $D = $ co(W). Since V is convex and $W \subseteq V$ we obtain that co$(W) \subseteq V$. Since $W \subseteq V$ is an open set, using exercise 10(i) we obtain that co(W) is an open set. Since $T : (L_p[0,1], \tau) \to (X, \sigma)$ is linear and continuous it follows that $T^{-1}($co$(W)) \subseteq L_p[0,1]$ is a non-empty open and convex set. Using (iii) we obtain that $T^{-1}($co$(W)) = L_p[0,1]$, i.e., $T(L_p[0,1]) \subseteq $ co(W) and, since co$(W) \subseteq V$ we obtain that $T(L_p[0,1]) \subseteq V$. Then
$$T(L_p[0,1]) \subseteq \bigcap_{V \in \mathcal{B}} V = \{0\},$$
since (X, σ) is Hausdorff. Thus $T = 0$.

v) If we consider now $X = \mathbb{R}$ with the usual topology then by (iv) we obtain that the only linear and continuous functional on $(L_p[0,1], \tau)$ is the zero functional.

Since the identity $I : (L_p[0,1], \tau) \to (L_p[0,1], \tau)$ is a linear and continuous operator, if, for a contradiction, $(L_p[0,1], \tau)$ is a locally convex space, then by (iv) for $(X, \sigma) = (L_p[0,1], \tau)$ we obtain that I must be the null operator, which is false! Hence $(L_p[0,1], \tau)$ is not a locally convex space.

29. See [44, example 1.44, chapter 1].

i) Let $n \in \mathbb{N}$. Then V_n is an open set, being a union of open sets, and therefore K_n is closed. Since $K_n \subseteq \overline{D}(0, n)$ we obtain that K_n is a compact set. If $z \in \Omega$ and $z \notin \bigcup_{n=1}^{\infty} K_n$,

13.2 Solutions

then $z \in \bigcap_{n=1}^{\infty} C_{K_n} = \bigcap_{n=1}^{\infty} V_n$. Let $m \in \mathbb{N}$ be such that $|z| \leq m$. Then $z \in \bigcup_{w \notin \Omega} D(w, 1/n)$ for any $n \geq m$ and therefore there is a sequence $(w_n)_{n \geq m}$ in C_Ω such that $w_n \to z$. Since C_Ω is closed we obtain that $z \notin \Omega$, a contradiction! Thus $\bigcup_{n=1}^{\infty} K_n = \Omega$. If $z \in K_n$ and $r = 1/(n(n+1))$, then for any $w \in D(z, r)$ we have $|w| \leq |z| + |w - z| \leq n + 1$. If $w \in \bigcup_{t \notin \Omega} D(t, 1/(n+1))$, then there is $t \in C_\Omega$ such that $|w - t| < 1/(n+1)$. Then $|z - t| \leq |z - w| + |w - t| < 1/n$, and then $z \in \bigcup_{t \notin \Omega} D(t, 1/n)$, $z \in V_n$, contradiction.

Thus $D(z, r) \subseteq K_{n+1}$ and therefore $K_n \subseteq \overset{\circ}{K}_{n+1} \forall n \in \mathbb{N}$. We obtain that

$$\Omega = \bigcup_{n=1}^{\infty} K_n \subseteq \bigcup_{n=1}^{\infty} \overset{\circ}{K}_{n+1} \subseteq \bigcup_{n=1}^{\infty} \overset{\circ}{K}_n,$$

and therefore $\Omega = \bigcup_{n=1}^{\infty} \overset{\circ}{K}_n$. If $K \subseteq \Omega$ is a compact set since $K \subseteq \bigcup_{n=1}^{\infty} \overset{\circ}{K}_n$ we obtain that there are $i_1, \ldots, i_m \in \mathbb{N}$ such that $K \subseteq \bigcup_{j=1}^{m} \overset{\circ}{K}_{i_j}$. Taking $n = \max_{1 \leq j \leq m} i_j$, then $K \subseteq K_n$.

ii) Since a continuous function on a compact set is bounded, we obtain that all the p_n are well defined. For any $n \in \mathbb{N}$ we have $p_n(\lambda f) = |\lambda| p_n(f) \ \forall \lambda \in \mathbb{C}, \forall f \in C(\Omega)$, and $p_n(f + g) \leq p_n(f) + p_n(g) \ \forall f, g \in C(\Omega)$. Thus p_n is a seminorm for every $n \in \mathbb{N}$. Since $K_n \subseteq K_{n+1}$ for each $n \in \mathbb{N}$, we obtain that $p_n \leq p_{n+1}$. If $f \in C(\Omega)$ is such that $p_n(f) = 0 \ \forall n \in \mathbb{N}$ then f is zero on every K_n and, since $\Omega = \bigcup_{n=1}^{\infty} K_n$, we obtain that $f = 0$. Thus, the family of seminorms $(p_n)_{n \in \mathbb{N}}$ is a total one (it separates the elements of $C(\Omega)$).

iii) Since $p_n \leq p_{n+1} \ \forall n \in \mathbb{N}$ we obtain that a basis of convex neighborhoods for the origin of $C(\Omega)$ in the locally convex topology τ generated by this family of seminorms is given by the sets W_n.

iv) The fact that d is a distance follows from the fact that the family $(p_n)_{n \in \mathbb{N}}$ is a total one. We prove now that the topology generated by d is τ. It is sufficient to prove that $\forall r > 0$ $\exists n \in \mathbb{N}$ such that $W_n \subseteq B_d(0, r)$, and $\forall n \in \mathbb{N}$ $\exists r > 0$ such that $B_d(0, r) \subseteq W_n$. Let $r > 0$. Since $\sum_{n=1}^{\infty} 1/2^n < \infty$ there is $m \in \mathbb{N}$ such that $\sum_{n=m+1}^{\infty} 1/2^n < r/2$. Let $k \in \mathbb{N}$ be such that $1/k < r/2$ and let $n = \max(k, m)$. Consider $f \in W_n$. Then $p_n(f) < 1/n$ and then $p_n(f) < r/2$. We obtain that $p_i(f) < r/2 \ \forall 1 \leq i \leq n$ and then

$$d(f, 0) = \sum_{i=1}^{\infty} \frac{p_i(f)}{2^i(1 + p_i(f))} = \sum_{i=1}^{n} \frac{p_i(f)}{2^i(1 + p_i(f))} + \sum_{i=n+1}^{\infty} \frac{p_i(f)}{2^i(1 + p_i(f))}$$

$$\leq \frac{r}{2} \sum_{i=1}^{n} \frac{1}{2^i} + \sum_{i=n+1}^{\infty} \frac{1}{2^i} < r.$$

Thus $W_n \subseteq B_d(0, r)$. Consider now $n \in \mathbb{N}$, and let $r = 1/(2^n(n+1))$. If $f \in B_d(0, r)$, then

$$\sum_{i=1}^{\infty} \frac{p_i(f)}{2^i(1 + p_i(f))} < \frac{1}{2^n(n+1)}, \quad \frac{p_n(f)}{2^n(1 + p_n(f))} < \frac{1}{2^n(n+1)},$$

and therefore $p_n(f)/(1+p_n(f)) < 1/(n+1)$. Since the function $t \longmapsto t/(t+1)$ is increasing on $(0, \infty)$, and $1/(n+1) = (1/n)/(1+1/n)$ we obtain that $p_n(f) < 1/n$, that is, $f \in W_n$. We have proved that $B_d(0, r) \subseteq W_n$. Therefore the topology τ on $C(\Omega)$ coincides with the topology generated by d. Evidently d is translation invariant. We prove now that d is complete. Consider $(f_k)_{k \in \mathbb{N}} \subseteq (C(\Omega), d)$, a Cauchy sequence. Then for each $n \in \mathbb{N}$ the sequence $(f_k|_{K_n})_{k \in \mathbb{N}} \subseteq (C(K_n), \|\cdot\|_\infty)$ is Cauchy. We obtain that for each $n \in \mathbb{N}$ there is $g_n \in C(K_n)$ such that $(f_k)_{k \in \mathbb{N}}$ converges uniformly to g_n on K_n. Since $K_n \subseteq K_{n+1}$ for every $n \in \mathbb{N}$, we obtain that $g_{n+1}|_{K_n} = g_n$ for every $n \in \mathbb{N}$, and we can define $g : \Omega = \bigcup_{n=1}^{\infty} K_n \to \mathbb{C}$, $g(x) = g_n(x)$, if $x \in K_n$. Then $g \in C(\Omega)$ and we must prove that $d(f_k, g) \to 0$. Let $\varepsilon > 0$. There is $n \in \mathbb{N}$ such that $\sum_{i=n+1}^{\infty} 1/2^i < \varepsilon/2$. Since $(f_k)_{k \in \mathbb{N}}$ converges uniformly to g on K_n it results that there is $m \in \mathbb{N}$ such that $p_n(f_k - g) < \varepsilon/2 \ \forall k \geq m$. Then $p_i(f_k - g) < \varepsilon/2 \ \forall k \geq m, \forall i = 1, ..., n$, and therefore for any $k \geq m$ we have

$$d(f_k, p) = \sum_{i=1}^{n} \frac{p_i(f_k - g)}{2^i(1 + p_i(f_k - g))} + \sum_{i=n+1}^{\infty} \frac{p_i(f_k - g)}{2^i(1 + p_i(f_k - g))} < \varepsilon.$$

Thus $d(f_k, g) \to 0$ and we obtain that $(C(\Omega), \tau)$ is an F-space.

v) Suppose that $f_n \to f$ in $(C(\Omega), d)$. Then since for any $i \in \mathbb{N}$ we have

$$\frac{p_i(f_n - f)}{2^i(1 + p_i(f_n - f))} \leq d(f_n, f) \to 0,$$

it follows that for any $i \in \mathbb{N}$ we have $p_i(f_n - f) \to 0$, i.e., $f_n \to f$ uniformly on every K_i. If $K \subseteq \Omega$ is a compact set, then by (i) there is $p \in \mathbb{N}$ such that $K \subseteq K_p$ and by above it follows that $f_n \to f$ uniformly on K_p, thus on K also.

Suppose now that for any compact set $K \subseteq \Omega$ we have $f_n \to f$ uniformly on K. Consider $0 < \varepsilon < 1$. There is $n \in \mathbb{N}$ such that $\sum_{i=n+1}^{\infty} 1/2^i < \varepsilon/2$. For $K = \bigcup_{i=1}^{n} K_i$, which is compact, by hypothesis $f_n \to f$ uniformly on K, and therefore we can find $m_\varepsilon \in \mathbb{N}$ such that $|f_k(x) - f(x)| \leq \varepsilon/2 \ \forall k \geq m_\varepsilon, \forall x \in K$. Then $p_i(f_k - f) \leq \varepsilon/2$ $\forall k \geq m_\varepsilon, \forall i = \overline{1, n}$. Now, for any $k \geq m_\varepsilon$ we have

$$d(f_k, f) = \sum_{i=1}^{n} \frac{p_i(f_k - f)}{2^i(1 + p_i(f_k - f))} + \sum_{i=n+1}^{\infty} \frac{p_i(f_k - f)}{2^i(1 + p_i(f_k - f))} \leq \frac{\varepsilon}{2} \sum_{i=1}^{n} \frac{1}{2^i} + \frac{\varepsilon}{2} < \varepsilon,$$

i.e., $f_n \to f$ in $(C(\Omega), d)$.

vi) We prove now that $(C(\Omega), \tau)$ is not normable. Suppose, for a contradiction, that there is a norm $\|\cdot\|$ on $C(\Omega)$ which generates the topology τ on $C(\Omega)$. Let $\varepsilon > 0$. Since the sets $(W_n)_{n \in \mathbb{N}}$ form a basis of neighborhoods for the origin in the τ topology and $B(0, \varepsilon)$ is a neighborhood of the origin in the topology τ it follows that there is $n \in \mathbb{N}$ such that $W_n \subseteq B(0, \varepsilon) = \{f \in C(\Omega) \mid \|f\| < \varepsilon\}$ (1). We cannot have $K_n = \Omega$ because Ω is open and K_n is compact. Let $z_0 \in \Omega \setminus K_n$. Then there is $f \in C(\Omega)$ such that $f|_{K_n} = 0$ and $f(z_0) = 1$ (for example, one may take $f(z) = \text{dist}(z, K_n)/\text{dist}(z_0, K_n) \ \forall z \in \Omega)$. For any $m \in \mathbb{N}$ we have $p_n(mf) = 0 < 1/n$ and therefore $mf \in W_n \ \forall m \in \mathbb{N}$. Then by (1)

13.2 Solutions

$mf \in B(0, \varepsilon) \; \forall m \in \mathbb{N}$, i.e., $\|f\| < \varepsilon/m \; \forall m \in \mathbb{N}$ and passing to the limit for $m \to \infty$ and using the fact that $\|\cdot\|$ is a norm it follows that $f = 0$. But $f(z_0) = 1$, and we obtain a contradiction.

30. See [44, example 1.45, chapter 1].

i) Evidently $\mathcal{H}(\Omega) \subseteq C(\Omega)$ is a linear subspace. Since by exercise 29(iv) τ is metrizable on $C(\Omega)$ in order to prove that $\mathcal{H}(\Omega) \subseteq (C(\Omega), \tau)$ is closed it is sufficient to prove that if $f \in C(\Omega)$ and $(f_n)_{n \in \mathbb{N}} \subseteq \mathcal{H}(\Omega)$ with $f_n \to f \in C(\Omega)$ in the τ topology then $f \in \mathcal{H}(\Omega)$. Using exercise 29(v) the fact that $(f_n)_{n \in \mathbb{N}}$ converges towards f in the τ topology is equivalent with the fact that $f_n \to f$ uniformly on every compact set $K \subseteq \Omega$. Using the properties for holomorphic functions we obtain that $f \in \mathcal{H}(\Omega)$. Thus $\mathcal{H}(\Omega) \subseteq (C(\Omega), \tau)$ is a closed linear subspace. We obtain that the metric d which generates the τ topology on $C(\Omega)$ gives on $\mathcal{H}(\Omega)$ a structure of an F-space.

ii) Let $A \subseteq (\mathcal{H}(\Omega), \tau)$ be a closed bounded set. In order to prove that A is a compact set, since τ is metrizable it is sufficient to prove that A is sequentially compact. Let $n \in \mathbb{N}$. Since A is bounded and W_n is a neighborhood for the origin (see exercise 29(iii)) it follows that there is a $\lambda > 0$ such that $A \subseteq \lambda W_n$, that is, $p_n(f) \leq \lambda/n \; \forall f \in A$. We obtain that A is uniformly bounded on every K_n. Let $K \subseteq \Omega$ be an arbitrary compact set. Using exercise 29(i), we obtain that there is $n \in \mathbb{N}$ such that $K \subseteq K_n$ and therefore A is uniformly bounded on K. From Montel's theorem it follows that the family $A \subseteq \mathcal{H}(\Omega)$ is a normal one, and thus for any sequence $(f_k)_{k \in \mathbb{N}} \subseteq A$ there is a subsequence $(f_{n_k})_{k \in \mathbb{N}}$ which converges uniformly on every compact subset of Ω. From (i) it follows that there is $f \in \mathcal{H}(\Omega)$ such that $f_{n_k} \to f$ uniformly on every compact subset of Ω. Using exercise 29(v) we obtain that $f_{n_k} \to f$ in the τ topology. Since $A \subseteq (\mathcal{H}(\Omega), \tau)$ is closed it follows that $f_{n_k} \to f \in A$ in the τ topology and therefore A is a sequentially compact set.

iii) If we suppose that $(\mathcal{H}(\Omega), \tau)$ is normable since in a normed space the closed unit ball is closed and bounded then by (ii) it follows that the closed unit ball in $(\mathcal{H}(\Omega), \tau)$ is compact which by the Riesz theorem would imply that $\dim_{\mathbb{C}} \mathcal{H}(\Omega) < \infty$. But every polynomial is a holomorphic function, and the space of all polynomials is not finite-dimensional.

31. See [21, exercise 2.4, chapter 2].

i) For every $n \in \mathbb{N}$ the map $\|\cdot\|_n : \mathcal{I}(\mathbb{C}) \to [0, \infty)$ is a norm. Indeed, if $\|f\|_n = 0$, $f \in \mathcal{I}(\mathbb{C})$, then $f = 0$ on $D(0, n)$ and using the identity theorem for analytic functions we obtain that $f = 0$. Evidently $\|\lambda f\|_n = |\lambda| \|f\|_n$ and $\|f + g\|_n \leq \|f\|_n + \|g\|_n \; \forall \lambda \in \mathbb{C}$, $\forall f, g \in \mathcal{I}(\mathbb{C})$. We consider now $\{\|\cdot\|_n \mid n \in \mathbb{N}\}$ as a family of seminorms on $\mathcal{I}(\mathbb{C})$ and this family generates a Hausdorff locally convex topology on $\mathcal{I}(\mathbb{C})$.

ii) Evidently for any $\xi \in \mathbb{C}$ the function $z \longmapsto \xi z$ is analytic from $\mathbb{C}$ to $\mathbb{C}$ and then $f_\xi \in \mathcal{I}(\mathbb{C})$ as a composition of entire functions. Also $\forall n \in \mathbb{N}$ we have $f_\xi^{(n)}(z) = f^{(n)}(\xi z) \xi^n$ $\forall z \in \mathbb{C}$.

iii) Let $x^* : \mathcal{I}(\mathbb{C}) \to \mathbb{C}$ be a linear and continuous functional, where on $\mathcal{I}(\mathbb{C})$ we consider the topology given at (i), and let $f \in \mathcal{I}(\mathbb{C})$. Since x^* is linear and continuous, we can find $M > 0$ and $m \in \mathbb{N}$ such that $|x^*(g)| \leq M \|g\|_m \; \forall g \in \mathcal{I}(\mathbb{C})$, that is $|x^*(g)| \leq M \sup_{|z| \leq m} |g(z)| \; \forall g \in \mathcal{I}(\mathbb{C})$ (1). For any $n \in \mathbb{N}$ let $g_n \in \mathcal{I}(\mathbb{C})$, $g_n(z) = z^n \; \forall z \in \mathbb{C}$.

For $f \in \mathcal{I}(\mathbb{C})$ considered in the statement, let $f(z) = \sum_{n=0}^{\infty} a_n z^n$ on $\mathbb{C}$. We obtain that

$f_\xi(z) = \sum_{n=0}^{\infty} a_n \xi^n z^n$ and therefore $f_\xi(z) = \sum_{n=0}^{\infty} a_n \xi^n g_n(z) \; \forall \xi, z \in \mathbb{C}$. We fix $\xi \in \mathbb{C}$. For any $N \in \mathbb{N}$ let $s_N = \sum_{n=0}^{N} a_n \xi^n g_n$, $s_N \in \mathcal{I}(\mathbb{C})$. Then

$$\sup_{|z| \leq m} |(f_\xi - s_N)(z)| = \sup_{|z| \leq m} \left| \sum_{n=N+1}^{\infty} a_n (\xi z)^n \right|.$$

If we denote $w = z\xi$ we obtain that

$$\sup_{|z| \leq m} |(f_\xi - s_N)(z)| = \sup_{|w| \leq m|\xi|} \left| \sum_{n=N+1}^{\infty} a_n w^n \right|.$$

Since the series $\sum_{n=0}^{\infty} a_n w^n$ has ∞ as its radius of convergence, we obtain that it converges on a closed disc of radius $\geq m|\xi|$, and therefore it converges uniformly on this disc. We obtain that for any $\varepsilon > 0$ there is an $N_\varepsilon \in \mathbb{N}$ such that

$$\left| \sum_{n=p}^{\infty} a_n w^n \right| < \varepsilon \; \forall w \in \overline{D}(0, m|\xi|), \; \forall p \geq N_\varepsilon.$$

From here it follows that

$$\sup_{|z| \leq m} |(f_\xi - s_N)(z)| \leq \varepsilon \; \forall N \geq N_\varepsilon$$

and, using (1), we obtain that $x^*(f_\xi) = \sum_{n=0}^{\infty} x^*(g_n) a_n \xi^n$. Thus, for any $\xi \in \mathbb{C}$ we have

$$F(\xi) = x^*(f_\xi) = \sum_{n=0}^{\infty} x^*(g_n) a_n \xi^n.$$

Using again (1) we obtain that $|x^*(g_n)| \leq Mm^n$ for every $n \in \mathbb{N}$. Since the series $\sum_{n=0}^{\infty} a_n \xi^n$ defines an analytic function, it converges absolutely on $\mathbb{C}$. We obtain that the series which defines $F(\xi)$ is uniformly convergent on every bounded set of $\mathbb{C}$ and therefore $F \in \mathcal{I}(\mathbb{C})$. We also have
$$F^{(n)}(0) = n! a_n x^*(g_n) = f^{(n)}(0) x^*(g_n) \; \forall n \geq 0.$$

iv) Let $x^* \in (\mathcal{I}(\mathbb{C}))^*$ be such that $x^*(f_\xi) = 0 \; \forall \xi \in \mathbb{C}$. Using (iii) we obtain that $x^*(g_n) = 0$ for any $n \in \mathbb{N}$. But if $g \in \mathcal{I}(\mathbb{C})$, $g(z) = \sum_{n=0}^{\infty} c_n z^n \; \forall z \in \mathbb{C}$, as in (iii) one can easily see that $x^*(g) = \sum_{n=0}^{\infty} c_n x^*(g_n)$, therefore $x^*(g) = 0$, and then $x^* = 0$. This means (see chapter 8) that the set $\{f_\xi \mid \xi \in \mathbb{C}\} \subseteq \mathcal{I}(\mathbb{C})$ is fundamental in the locally convex space $\mathcal{I}(\mathbb{C})$.

13.2 Solutions

32. See [9, exercise 1, chapter IV, section 1].

Recall that if $(x_\alpha)_{\alpha \in A} \subseteq \mathbb{K}$, then this family is summable if and only if the net $(s_F)_{F \in \mathcal{F}(A)}$ is convergent, where $\mathcal{F}(A) = \{F \subseteq A \mid F \text{ finite}\}$ is directed by inclusion and $s_F = \sum_{\alpha \in F} x_\alpha \ \forall F \text{ finite} \subseteq A$. In this case we write $\sum_{\alpha \in A} x_\alpha = \lim_{F \in \mathcal{F}(A)} s_F$, or simply $\lim_F s_F$.

i) The fact that $p_\beta : X \to [0, \infty)$ are seminorms is obvious. Suppose that X is Hausdorff. Let $\alpha \in A$. Let $e_\alpha = (\delta_{i\alpha})_{i \in A}$, where $\delta_{i\alpha} = \begin{cases} 1 & \text{if } i = \alpha, \\ 0 & \text{if } i \neq \alpha. \end{cases}$ We have $e_\alpha \in X$, because $p_\beta(e_\alpha) = c_{\alpha\beta} < \infty \ \forall \beta \in B$. Observe that $e_\alpha \neq 0$. Since X is Hausdorff there is $\beta \in B$ such that $p_\beta(e_\alpha) \neq 0$, i.e., there is $\beta \in B$ such that $c_{\alpha\beta} \neq 0$ and, since $c_{\alpha\beta} \geq 0$ we obtain that $c_{\alpha\beta} > 0$, i.e., $\forall \alpha \in A \ \exists \beta \in B$ such that $c_{\alpha\beta} > 0$.

Conversely, suppose that $\forall \alpha \in A \ \exists \beta \in B$ such that $c_{\alpha\beta} > 0$. We want to show that X is Hausdorff, i.e., $\forall x \in X, x \neq 0, \exists \beta \in B$ such that $p_\beta(x) \neq 0$. Consider $x \in X$, $x \neq 0$, $x = (x_\alpha)_{\alpha \in A}$. Since $x \neq 0$ there is an $\alpha \in A$ such that $x_\alpha \neq 0$. For this $\alpha \in A$, by hypothesis there is $\beta \in B$ such that $c_{\alpha\beta} > 0$. Then $p_\beta(x) = \sum_{\gamma \in A} c_{\gamma\beta} |x_\gamma| \geq c_{\alpha\beta} |x_\alpha| > 0$.

ii) Consider $x^* : \left(X, (p_\beta)_{\beta \in B}\right) \to \mathbb{K}$, a linear and continuous functional. From our hypothesis the family of seminorms $(p_\beta)_{\beta \in B}$ is filtered: $\forall \beta_1, \beta_2 \in B \ \exists \beta \in B$ such that $p_{\beta_1}, p_{\beta_2} \leq p_\beta$. Then since x^* is continuous we can find $a > 0, \beta \in B$ such that $|x^*(x)| \leq a p_\beta(x) \ \forall x \in X$ (1). For $\alpha \in A, e_\alpha \in X$. We denote $u_\alpha = x^*(e_\alpha) \ \forall \alpha \in A$. Then $|u_\alpha| = |x^*(e_\alpha)| \leq a p_\beta(e_\alpha) = a c_{\alpha\beta}$, i.e., $|u_\alpha| \leq a c_{\alpha\beta} \ \forall \alpha \in A$. Consider now $x \in X$. We know that $p_\beta(x) = \sum_{\alpha \in A} c_{\alpha\beta} |x_\alpha| < \infty$, and therefore $\lim_{F \in \mathcal{F}(A)} p_\beta(x - s_F) = 0$, where $s_F = \sum_{\alpha \in F} x_\alpha e_\alpha \ \forall F \in \mathcal{F}(A)$. Using (1) we obtain that

$$|x^*(x) - x^*(s_F)| \leq a p_\beta(x - s_F) \ \forall F \in \mathcal{F}(A),$$

hence $x^*(x) = \lim_{F \in \mathcal{F}(A)} x^*(s_F)$. But x^* is linear and F finite, and therefore

$$x^*(s_F) = \sum_{\alpha \in F} x_\alpha x^*(e_\alpha) = \sum_{\alpha \in F} u_\alpha x_\alpha.$$

Then

$$x^*(x) = \lim_{F \in \mathcal{F}(A)} \sum_{\alpha \in F} u_\alpha x_\alpha = \sum_{\alpha \in A} u_\alpha x_\alpha.$$

Conversely, from $|u_\alpha| \leq a c_{\alpha\beta} \ \forall \alpha \in A$ we obtain for $(x_\alpha)_{\alpha \in A} \in X$ that $|u_\alpha x_\alpha| \leq a c_{\alpha\beta} |x_\alpha| \ \forall \alpha \in A$. Since $\sum_{\alpha \in A} c_{\alpha\beta} |x_\alpha|$ is summable we obtain that $\sum_{\alpha \in A} |u_\alpha x_\alpha|$ is also summable, i.e., $\sum_{\alpha \in A} u_\alpha x_\alpha$ is absolutely summable, and therefore summable. Let

$$x^*(x) = \sum_{\alpha \in A} u_\alpha x_\alpha \ \forall x = (x_\alpha)_{\alpha \in A} \in X.$$

Then x^* is clearly well defined and linear. By our hypothesis, $\exists a > 0, \beta \in B$ such that

$$|x^*(x)| \leq \sum_{\alpha \in A} |u_\alpha x_\alpha| \leq \sum_{\alpha \in A} a c_{\alpha\beta} |x_\alpha| = a p_\beta(x) \ \forall x \in X,$$

and then $x^* \in X^*$.

If $c_{\alpha\beta} > 0\ \forall\,(\alpha,\beta) \in A \times B$ then we can identify the space X^* with the space

$$\left\{ (u_\alpha)_{\alpha \in A} \subseteq \mathbb{K} \;\middle|\; \exists \beta \in A \text{ such that } \sup_{\alpha \in A} \frac{|u_\alpha|}{c_{\alpha\beta}} < \infty \right\}.$$

Chapter 14

The weak topologies

Definition. Let X be a normed space or, more generally, a locally convex Hausdorff space, and X^* its dual. The *weak topology* on X is the locally convex topology generated by the family of seminorms $(p_{x^*})_{x^* \in X^*}$, where $p_{x^*} : X \to \mathbb{R}$ is defined by $p_{x^*}(x) = |x^*(x)|$ $\forall x \in X$.

Proposition. Let X be a normed space or, more generally, a locally convex Hausdorff space. For $n \in \mathbb{N}$, $x_1^*, ..., x_n^* \in X^*$, $\varepsilon > 0$, define
$$W(0; x_1^*, ..., x_n^*; \varepsilon) = \{x \in X \mid |x_1^*(x)| < \varepsilon, \; ..., \; |x_n^*(x)| < \varepsilon\}.$$
Then the family of sets $\{W(0; x_1^*, ..., x_n^*; \varepsilon) \mid x_1^*, ..., x_n^* \in X^*, \; \varepsilon > 0, \; n \in \mathbb{N}\}$ is a basis of neighborhoods for 0 in the *weak* topology. Also, for any $x \in X$, the sets $W(x; x_1^*, ..., x_n^*; \varepsilon) = x + W(0; x_1^*, ..., x_n^*; \varepsilon)$ form a basis of neighborhoods for x in the *weak* topology.

Definition. Let X be a normed space or, more generally, a locally convex Hausdorff space, and X^* its dual. The *weak* topology* on X^* is the locally convex topology generated by the family of seminorms $(p_x)_{x \in X}$, where $p_x : X^* \to \mathbb{R}$ is defined by $p_x(x^*) = |x^*(x)|$ $\forall x^* \in X^*$.

Proposition. Let X be a normed space or, more generally, a locally convex Hausdorff space, and X^* its dual. For $n \in \mathbb{N}$, $x_1, ..., x_n \in X$, $\varepsilon > 0$, define
$$W(0; x_1, ..., x_n; \varepsilon) = \{x^* \in X^* \mid |x^*(x_1)| < \varepsilon, \; ..., \; |x^*(x_n)| < \varepsilon\}.$$
Then the family of sets $\{W(0; x_1, ..., x_n; \varepsilon) \mid x_1, ..., x_n \in X, \; \varepsilon > 0, \; n \in \mathbb{N}\}$ is a basis of neighborhoods for 0 in the *weak** topology. Also, for any $x^* \in X^*$, the sets $W(x^*; x_1, ..., x_n; \varepsilon) = x^* + W(0; x_1, ..., x_n; \varepsilon)$ form a basis of neighborhoods for x^* in the *weak** topology.

Proposition. Let X be a normed space or, more generally, a locally convex Hausdorff space, and X^* its dual. Then the *weak* and *weak** topologies are Hausdorff.

Proposition. Given a normed or, more generally, a locally convex Hausdorff space X, then:

i) a sequence $x_n \to x$ *weak* $\Leftrightarrow x^*(x_n) \to x^*(x) \ \forall x^* \in X^*$;
ii) a sequence $x_n^* \to x^*$ *weak*$^* \Leftrightarrow x_n^*(x) \to x^*(x) \ \forall x \in X$;
iii) a net $x_\delta \to x$ *weak* $\Leftrightarrow x^*(x_\delta) \to x^*(x) \ \forall x^* \in X^*$;
iv) a net $x_\delta^* \to x^*$ *weak*$^* \Leftrightarrow x_\delta^*(x) \to x^*(x) \ \forall x \in X$.

The Alaoglu–Bourbaki Theorem. i) Let X be a normed space. Then $B_{X^*} \subseteq X^*$ is *weak** compact.

ii) Let X be a Banach space. Then X is a reflexive space if and only if the closed unit ball B_X is *weak* compact.

The Goldstine Theorem. Let X be a normed space and $K_X : X \to X^{**}$ the canonical embedding into the bidual. Then $K_X(B_X)$ is *weak** dense in $B_{X^{**}}$, and so $K_X(X)$ is *weak** dense in X^{**}.

The Mazur Theorem. Let X be a normed space or, more generally, a locally convex Hausdorff space, X^* its dual, and $A \subseteq X$ a convex set. Then $\overline{A} = \overline{A}^{\text{weak}}$. In particular, if A is a closed convex set then A is *weak* closed.

The Eberlein–Smulian Theorem. Let X be a Banach space and consider a subset $A \subseteq X$. Then the following assertions are equivalent:

i) A is *weak* compact;
ii) A is *weak* sequentially compact, i.e., for each sequence $(x_n)_{n\in\mathbb{N}} \subseteq A$ there exist a subsequence $(x_{k_n})_{n\in\mathbb{N}}$ and an element $x \in A$ such that $x_{k_n} \to x$ *weak*.

The Helly Theorem. Let X be a normed space, $n \in \mathbb{N}$, $c_1, ..., c_n \in \mathbb{K}$, $x_1^*, ..., x_n^* \in X^*$ and $M > 0$. Then the following assertions are equivalent:

i) $\forall \varepsilon > 0 \ \exists x_\varepsilon \in X$ such that $\|x_\varepsilon\| < M + \varepsilon$ and $x_1^*(x_\varepsilon) = c_1, ..., x_n^*(x_\varepsilon) = c_n$;
ii) $\forall \lambda_1, ..., \lambda_n \in \mathbb{K}$ we have

$$\left| \sum_{i=1}^n \lambda_i c_i \right| \leq M \left\| \sum_{i=1}^n \lambda_i x_i^* \right\|.$$

Definition. Let X, Y be Banach spaces. An operator $U \in L(X, Y)$ is called *weak* compact if and only if $U(B_X) \subseteq Y$ is relatively *weak* compact (that is, $\overline{U(B_X)}^{\text{weak}} = \overline{U(B_X)}^{\|\cdot\|}$ is *weak* compact).

The Gantmacher Theorem. Let X, Y be Banach spaces. An operator $U \in L(X, Y)$ is *weak* compact if and only if its dual $U^* \in L(Y^*, X^*)$ is *weak* compact.

Theorem. Let X, Y, Z, T be Banach spaces. If $A : X \to Y$, $B : Y \to Z$, $C : Z \to T$ are linear and continuous operators and B is *weak* compact, then the composition CBA is a *weak* compact operator (the ideal property for *weak* compact operators).

The Krein Theorem. Let X be a Banach space and $A \subseteq X$ a subset. Then the following assertions are equivalent:

i) A is relatively *weak* compact;
ii) co(A) is relatively *weak* compact;
iii) ec(A) is relatively *weak* compact;
iv) eco(A) is relatively *weak* compact.

14.1 Exercises

Weak convergent sequences in c_0

1. i) Let $x \in c_0$ and $(x_n)_{n \in \mathbb{N}} \subseteq c_0$. Prove that $x_n \to x$ *weak* if and only if the sequence $(x_n)_{n \in \mathbb{N}}$ is norm bounded in c_0 and $p_k(x_n) \to p_k(x)$ $\forall k \in \mathbb{N}$, where $p_k : c_0 \to \mathbb{K}$ are the canonical projections.

ii) Let $\varphi : [1, 2] \to \mathbb{K}$ be a continuous function and consider the sequences
$$x_n = \left(0, ..., 0, \varphi\left(\frac{n}{n}\right), \varphi\left(\frac{n+1}{n}\right), ..., \varphi\left(\frac{n+n}{n}\right), 0, ...\right) \in c_0 \ \forall n \in \mathbb{N}$$
($\varphi(n/n)$ on the nth position). Prove that $x_n \to 0$ *weak* and that $(x_n)_{n \in \mathbb{N}}$ is norm convergent if and only if $\varphi = 0$.

*Weak** convergent sequences in l_1

2. i) Let $x_0 \in l_1$, $(x_n)_{n \in \mathbb{N}} \subseteq l_1 = c_0^*$. Prove that $x_n \to x_0$ *weak** if and only if the sequence $(x_n)_{n \in \mathbb{N}}$ is norm bounded in l_1 and $p_k(x_n) \to p_k(x_0)$ $\forall k \in \mathbb{N}$, where $p_k : l_1 \to \mathbb{K}$ are the canonical projections.

ii) Let $(a_n)_{n \in \mathbb{N}} \subseteq \mathbb{K}$. Define $x_n^* \in c_0^*$, $x_n^*(x_1, x_2, ...) = a_n x_n$ $\forall (x_1, x_2, ..) \in c_0$. Prove that the sequence $(x_n^*)_{n \in \mathbb{N}}$ is *weak** convergent if and only if $(a_n)_{n \in \mathbb{N}} \in l_\infty$. In this case prove that $x_n^* \to 0$ *weak**.

iii) Let $(a_n)_{n \in \mathbb{N}} \subseteq \mathbb{K}$. Define $x_n^* \in c_0^*$, $x_n^*(x_1, x_2, ...) = a_1 x_1 + a_2 x_2 + \cdots + a_n x_n$. Prove that the sequence $(x_n^*)_{n \in \mathbb{N}}$ is *weak** convergent if and only if $(a_n)_{n \in \mathbb{N}} \in l_1$. In this case prove that $x_n^* \to x^*$ *weak**, where $x^*(x_1, x_2, ...) = \sum_{n=1}^{\infty} a_n x_n$ $\forall (x_1, x_2, ...) \in c_0$.

iv) Let $(a_n)_{n \in \mathbb{N}} \subseteq \mathbb{K}$. Define $x_n^* \in c_0^*$, $x_n^*(x_1, x_2, ...) = a_n x_1 + a_{n-1} x_2 + \cdots + a_1 x_n$ $\forall (x_1, x_2, ..) \in c_0$. Prove that $(x_n^*)_{n \in \mathbb{N}}$ is *weak** convergent if and only if $(a_n)_{n \in \mathbb{N}} \in l_1$. In this case prove that $x_n^* \to 0$ *weak**.

v) Let $\varphi : [0, 1] \to \mathbb{K}$ be a continuous function, and consider
$$x_n = \left(\frac{1}{n}\varphi\left(\frac{1}{n}\right), \frac{1}{n}\varphi\left(\frac{2}{n}\right), ..., \frac{1}{n}\varphi\left(\frac{n}{n}\right), 0, ...\right) \in l_1 \ \forall n \in \mathbb{N}.$$
Prove that $x_n \to 0$ *weak** and that $(x_n)_{n \in \mathbb{N}}$ is norm convergent if and only if $\varphi = 0$.

vi) Let $\varphi : [0, 1] \to \mathbb{K}$ be a continuous function and $U : c_0 \to c_0$,
$$U((a_k)_{k \in \mathbb{N}}) = \left(\frac{1}{n}\sum_{k=1}^{n} a_k \varphi\left(\frac{k}{n}\right)\right)_{n \in \mathbb{N}}.$$
Prove that U is linear and continuous and that U is compact if and only if $\varphi = 0$.

The *weak* sequential convergence coincides with the norm convergence in l_1

3. i) Prove that $(B_{l_\infty}, weak^*)$ is metrizable.

ii) Let $x \in l_1$ and $(x_n)_{n \in \mathbb{N}} \subseteq l_1$ such that $x_n \to x$ weak. For $\varepsilon > 0$, we define
$$B_m := \bigcap_{n \geq m} \{x^* \in B_{l_\infty} \mid |x^*(x_n) - x^*(x)| \leq \varepsilon/3\} \quad \forall m \geq 1.$$
Prove that B_m is weak* closed in l_∞ for all $m \in \mathbb{N}$, and that $B_{l_\infty} = \bigcup_{m \geq 1} B_m$.

iii) If $x_n \to x$ weak prove that $x_n \to x$ in the norm topology of l_1.

iv) **Schur's theorem.** In l_1 weak and norm convergence of sequences coincide.

v) Does the part (iv) remain true for nets? I.e., for $(x_\delta)_{\delta \in \Delta} \subseteq l_1$ with $x_\delta \to 0$ weak, does it follow that $x_\delta \to 0$ in norm?

Weak convergent sequences in l_p $(1 < p < \infty)$

4. Let $1 < p < \infty$.

i) Let $x \in l_p$ and $(x_n)_{n \in \mathbb{N}} \subseteq l_p$. Prove that $x_n \to x$ weak if and only if the sequence $(x_n)_{n \in \mathbb{N}}$ is norm bounded in l_p and $p_k(x_n) \to p_k(x)$ $\forall k \in \mathbb{N}$, where $p_k : l_p \to \mathbb{K}$ are the canonical projections.

ii) Let $\varphi : [0, 1] \to \mathbb{K}$ be a continuous function and consider the sequences
$$x_n = n^{-1/p} \left(\varphi\left(\frac{1}{n}\right), \varphi\left(\frac{2}{n}\right), ..., \varphi\left(\frac{n}{n}\right), 0, ... \right) \in l_p \quad \forall n \in \mathbb{N}.$$
Using (i) prove that $x_n \to 0$ weak and that $(x_n)_{n \in \mathbb{N}}$ is norm convergent if and only if $\varphi = 0$.

iii) Let $\varphi : [0, 1] \to \mathbb{K}$ be a continuous function, $1 < q < \infty$ be the conjugate of p, and $U : l_q \to c_0$,
$$U\left((a_k)_{k \in \mathbb{N}}\right) = \left(n^{-1/p} \sum_{k=1}^{n} a_k \varphi\left(\frac{k}{n}\right) \right)_{n \in \mathbb{N}}.$$
Prove that U is linear and continuous and that U is compact if and only if $\varphi = 0$.

A converse of the Schur theorem

5. Let $1 \leq p < \infty$. Prove that the following assertions are equivalent:

i) For any sequence $(x_n)_{n \in \mathbb{N}} \subseteq l_p$ with the property that $x_n \to 0$ weak it follows that $x_n \to 0$ in the norm of l_p.

ii) $p = 1$.

Examples of weak* convergent sequences in l_∞ and l_∞^*

6. i) For $(a_n)_{n \in \mathbb{N}} \in l_\infty$ let $x_n = (0, 0, ..., 0, a_{n+1}, a_{n+2}, ..) \in l_\infty$ $\forall n \in \mathbb{N}$ (n times 0). Prove that $x_n \to 0$ weak* in l_∞.

ii) Let $(a_n)_{n \in \mathbb{N}}$ be a sequence of scalars. Define $x_n^* \in l_\infty^*$, $x_n^*(x_1, x_2, ...) = a_n x_n$. Prove that $(x_n^*)_{n \in \mathbb{N}}$ is weak* convergent in l_∞^* if and only if $(a_n)_{n \in \mathbb{N}} \in c_0$. In this case prove that $x_n^* \to 0$ in norm.

iii) Let $(a_n)_{n \in \mathbb{N}}$ be a sequence of scalars. Define $x_n^* \in l_\infty^*$, $x_n^*(x_1, x_2, ...) = a_1 x_1 + \cdots + a_n x_n$. Prove that the sequence $(x_n^*)_{n \in \mathbb{N}}$ is weak* convergent if and only if $(a_n)_{n \in \mathbb{N}} \in l_1$. In this case prove that $x_n^* \to x^*$ weak*, where $x^*(x_1, x_2, ...) = \sum_{n=1}^{\infty} a_n x_n$ $\forall (x_1, x_2, ...) \in l_\infty$.

Weak convergent sequences in $C(T)$

7. i) Let T be a compact Hausdorff space, $f \in C(T)$ and $(f_n)_{n \in \mathbb{N}} \subseteq C(T)$. Then $f_n \to f$ *weak* if and only if $\sup_{n \in \mathbb{N}} \|f_n\|_\infty < \infty$ and $f_n(t) \to f(t)$ $\forall t \in T$.

ii) Let $\varphi : [0, 1] \to \mathbb{K}$ be a continuous function and $f_n : [0, 1] \to \mathbb{K}$, $n \in \mathbb{N}$, be defined by $f_n(x) = \varphi(x^n)$. Using (i) prove that $(f_n)_{n \in \mathbb{N}} \subseteq C[0, 1]$ is *weak* convergent if and only if $\varphi(0) = \varphi(1)$.

iii) Let $(a_n)_{n \in \mathbb{N}}$ be a sequence scalars and $f_n : [0, 1] \to \mathbb{K}$, $n \in \mathbb{N}$, be defined by $f_n(x) = a_n x^n (1 - x^n)$. Prove that $(f_n)_{n \in \mathbb{N}} \subseteq C[0, 1]$ is *weak* convergent if and only if $(a_n)_{n \in \mathbb{N}} \in l_\infty$, and that $(f_n)_{n \in \mathbb{N}} \subseteq C[0, 1]$ is norm convergent if and only if $(a_n)_{n \in \mathbb{N}} \in c_0$.

8. Let T be a compact Hausdorff space and let $(f_n)_{n \in \mathbb{N}} \subseteq C(T)$, $f \in C(T)$, such that $\sup_{n \in \mathbb{N}} \|f_n\| < \infty$ and $f_n(t) \to f(t)$ $\forall t \in T$. Prove that there is a sequence $(g_n)_{n \in \mathbb{N}} \subseteq C(T)$, each g_n being a convex combination of elements from $(f_k)_{k \in \mathbb{N}}$ such that $g_n \to f$ uniformly.

Pointwise convergence in $C(T)$

9. Let (T, ρ) be a compact metric space and τ_p the topology of the pointwise convergence on $C(T)$.

i) Prove that the topology τ_p is metrizable on each τ_p compact subset of $C(T)$.

ii) Prove that each τ_p compact subset of $C(T)$ which is uniformly bounded is *weak* compact in $C(T)$.

Fundamental sets and the *weak* convergence

10. i) Let X be a normed space, $x \in X$ and $(x_n)_{n \in \mathbb{N}} \subseteq X$. Prove that $x_n \to x$ *weak* if and only if the sequence $(x_n)_{n \in \mathbb{N}}$ is bounded and there is a fundamental subset $Y \subseteq X^*$ (that is, $\overline{\text{Sp}(Y)} = X^*$) such that $x^*(x_n) \to x^*(x)$ $\forall x^* \in Y$.

ii) Let H be a Hilbert space, $E \subseteq H$ an orthonormal basis, $x \in H$ and $(x_n)_{n \in \mathbb{N}} \subseteq H$. Using (i) prove that $x_n \to x$ *weak* if and only if the sequence $(x_n)_{n \in \mathbb{N}}$ is bounded and $\langle x_n, a \rangle \to \langle x, a \rangle$ $\forall a \in E$.

Weak convergent sequences in $L_1[0, 1]$

11. Consider the space $L_1[0, 1]$ of scalar integrable functions defined on $[0, 1]$, the measure used being the Lebesgue measure. Let $f \in L_1[0, 1]$ and $(f_n)_{n \in \mathbb{N}} \subseteq L_1[0, 1]$. Prove that $f_n \to f$ *weak* if and only if there is an $M > 0$ such that $\int_0^1 |f_n(t)| dt \le M$ $\forall n \in \mathbb{N}$, and $\int_E f_n(t) dt \to \int_E f(t) dt$ $\forall E \subseteq [0, 1]$ Lebesgue measurable.

Weak convergent sequences in $L_p[0, 1]$ ($1 < p < \infty$)

12. For $1 < p < \infty$ we consider the space $L_p[0, 1]$ of scalar p-integrable functions defined on $[0, 1]$, the measure used being the Lebesgue measure.

i) Let $f \in L_p[0, 1]$ and $(f_n)_{n \in \mathbb{N}} \subseteq L_p[0, 1]$. Prove that $f_n \to f$ *weak* if and only if $\int_0^x f_n(t) dt \to \int_0^x f(t) dt$ $\forall x \in [0, 1]$ and there exists $M > 0$ such that $\int_0^1 |f_n(t)|^p dt \le M$ $\forall n \in \mathbb{N}$.

ii) Let $(a_n)_{n\in\mathbb{N}} \subseteq \mathbb{K}$ and $(b_n)_{n\in\mathbb{N}} \subseteq [0,1]$ such that $\left(|a_n| b_n^{1/p}\right)_{n\in\mathbb{N}}$ is bounded and $a_n b_n \to 0$.

a) Using (i) prove that the sequence of functions $(f_n)_{n\in\mathbb{N}}$, where $f_n = a_n \chi_{(0,b_n)}$, is *weak* null in $L_p[0,1]$.

b) Let $1 < q < \infty$ be the conjugate of p. Prove that the operator $U : L_q[0,1] \to c_0$, $U(f) = \left(a_n \int_0^{b_n} f(x)\,dx\right)_{n\in\mathbb{N}}$ is linear and continuous and that U is compact if and only if $\left(|a_n| b_n^{1/p}\right)_{n\in\mathbb{N}} \in c_0$.

13. Let $1 < p < \infty$ and $f_n : [0,1] \to \mathbb{K}$, $n \in \mathbb{N}$, be a sequence of functions in $L_p[0,1]$ with $\|f_n\| = 1$ $\forall n \in \mathbb{N}$ and such that: $\forall t \in [0,1]$ there is at most one $n \in \mathbb{N}$ such that $f_n(t) \neq 0$. Prove that $f_n \to 0$ *weak*. What about the case $p = 1$?

Examples of *weak* and *weak** continuous functions

14. Let T be a compact Hausdorff space. For each $t \in T$ we define the Dirac functional $\delta_t : C(T) \to \mathbb{K}$, $\delta_t(f) = f(t)$ $\forall f \in C(T)$. Prove that the function $\varphi : T \to C(T)^*$, $\varphi(t) = \delta_t$, is continuous if we consider the *weak** topology on $C(T)^*$.

15. For a given normed space X prove that the canonical mapping $K_X : X \to K_X(X) \subseteq X^{**}$ is a *weak*-to-*weak** homeomorphism.

An algebraic property for infinite-dimensional linear spaces

16. i) Let X be a linear space, $n \in \mathbb{N}$, and $f, f_1, f_2, ..., f_n : X \to \mathbb{K}$ be linear functionals such that $\bigcap_{i=1}^{n} \ker f_i \subseteq \ker f$. Prove that f is a linear combination of $f_1, f_2, ..., f_n$.

ii) Let X be a linear space with $\dim_{\mathbb{K}} X = \infty$, $n \in \mathbb{N}$, and $f_1, ..., f_n : X \to \mathbb{K}$ be linear functionals. Prove that there is an $x_0 \in X$, $x_0 \neq 0$, such that $f_i(x_0) = 0$ $\forall i = \overline{1,n}$.

Continuous and *weak* continuous functionals

17. i) Let X be a normed space and $f : X \to \mathbb{K}$ a linear functional. Prove that $f : (X, \|\cdot\|) \to \mathbb{K}$ is continuous if and only if $f : (X, weak) \to \mathbb{K}$ is continuous.

ii) If Ω is a topological space and $f : \Omega \to X$ a function, prove that f is *weak* continuous (continuous from Ω with values in $(X, weak)$) if and only if $x^* \circ f : \Omega \to \mathbb{K}$ is continuous for every $x^* \in X^*$.

The norm continuity and the *weak* continuity coincide for linear operators

18. Let X and Y be two normed spaces and $T : X \to Y$ a linear operator. Then T is norm-to-norm continuous if and only if it is *weak*-to-*weak* continuous.

A linear *weak**-to-*weak** continuous operator is a dual operator

19. Let X and Y be two normed spaces and $S : Y^* \to X^*$ a linear operator. Prove that S is *weak**-to-*weak** continuous if and only if there is $T \in L(X,Y)$ such that $T^* = S$.

14.1 Exercises

The *weak* topology restricted to a linear subspace

20. Let X be a normed space and $Y \subseteq X$ a linear subspace. Prove that the *weak* topology on Y (given by its dual space Y^*) is the restriction on $Y \subseteq X$ of the *weak* topology from X.

A separation result for the *weak* and *weak** topologies

21. Let X be a real normed space.

i) Prove that if $A \subseteq X$ is a convex *weak* compact set and $C \subseteq X^*$ is a *weak** closed convex cone with the property that $\forall x^* \in C \, \exists x \in A$ such that $x^*(x) \leq 0$, then there exists $x_0 \in A$ such that $x^*(x_0) \leq 0 \, \forall x^* \in C$.

ii) Prove that if $A \subseteq X^*$ is a convex *weak** compact set and $C \subseteq X$ is a *weak* closed convex cone with the property that $\forall x \in C \, \exists x^* \in A$ such that $x^*(x) \leq 0$, then there exists $x_0^* \in A$ such that $x_0^*(x) \leq 0 \, \forall x \in C$.

The *weak* topology and finite-dimensional normed spaces

22. If $(X, \|\cdot\|)$ is a finite-dimensional normed space prove that the *weak* topology on X coincide with the norm topology. Thus in this case $x_n \to x$ *weak* if and only if $x_n \to x$ in norm.

23. Prove that if the *weak* topology on a normed space X is metrizable then X is finite-dimensional.

The *weak* and the norm topology on infinite-dimensional normed spaces

24. If $(X, \|\cdot\|)$ is an infinite-dimensional normed space prove that the *weak* topology is smaller than the norm topology, that is there are sets which are open (respectively closed) in the norm topology, and not open (respectively closed) in the *weak* topology.

For example, if $(X, \|\cdot\|)$ is an infinite-dimensional normed space, then $\overline{S_X}^{weak} = B_X$.

25. i) If $(X, \|\cdot\|)$ is an infinite-dimensional normed space prove that the interior for $B(0, 1)$ with respect to the *weak* topology on X is the empty set.

Consequently, the open unit ball $B(0, 1)$ is an open set in the norm topology, but it is not open in the *weak* topology.

ii) Using (i) prove that if X is an infinite-dimensional normed space then $(X, weak)$ is of the first Baire category.

The *weak* topology need not be sequential

26. i) Let $A \subseteq l_2$, $A = \{e_m + m e_n \mid m, n \in \mathbb{N}, \, 1 \leq m < n < \infty\}$. Then $0 \in \overline{A}^{weak}$ but we do not have in A a sequence which converges towards 0 in the *weak* topology.

ii) Let $S = \{n e_n \mid n \in \mathbb{N}\} \subseteq c_0$. Show that $0 \in \overline{S}^{weak}$ but we do not have in A a sequence which converges towards 0 in the *weak* topology.

The *weak* topology on infinite-dimensional normed spaces is never complete

27. Let X be an infinite-dimensional normed space.

i) Prove that there is a linear functional φ on X^* which is not continuous.

ii) Using Helly's lemma and the Hahn–Banach theorem, build a *weak* Cauchy net in X indexed by the finite-dimensional subspaces of X^* with φ the only possible *weak* limit point.

iii) Deduce that the *weak* topology on infinite-dimensional spaces is never complete, i.e., we can find *weak* Cauchy nets which are not *weak* convergent.

Separately continuous bilinear operators on infinite-dimensional normed spaces which are not continuous

28. i) Let H be an infinite-dimensional Hilbert space. Prove that the inner product $\langle \cdot, \cdot \rangle : (H, weak) \times (H, weak) \to \mathbb{K}$ is separately continuous but is not continuous as a function of two variables.

ii) Let X be an infinite-dimensional normed space. Prove that the duality bracket $\langle \cdot, \cdot \rangle : (X^*, weak^*) \times (X, weak) \to \mathbb{K}$, $\langle x^*, x \rangle = x^*(x)$, is separately continuous but is not continuous as a function of two variables.

iii) Let X be an infinite-dimensional normed space, and on $L(X)$ we consider the pointwise convergence topology, i.e., the locally convex topology given by the family of seminorms $(p_x)_{x \in X}$, where $p_x(U) = \|U(x)\|$. Prove that the map $E : L(X) \times X \to X$ defined by $E(U, x) = U(x)$ is separately continuous but is not continuous as a function of two variables (on X we consider the norm topology and on $L(X)$ the pointwise convergence topology).

Metrizability for the *weak* and the *weak** topologies and separable Banach spaces

29. Let $(X, \|\cdot\|)$ be a separable normed space, and consider a dense subset $(x_n)_{n \in \mathbb{N}} \subseteq X \setminus \{0\}$. We define $d : B_{X^*} \times B_{X^*} \longrightarrow \mathbb{R}_+$,

$$d(x^*, y^*) = \sum_{n=1}^{\infty} \frac{|(x^* - y^*)(x_n)|}{2^n \|x_n\|} \quad \forall x^*, y^* \in B_{X^*}.$$

i) Prove that d is a metric on B_{X^*}.

ii) Prove that the identity $I : (B_{X^*}, weak^*) \longrightarrow (B_{X^*}, d)$ is a homeomorphism.

iii) Deduce that the *weak** topology on B_{X^*} is metrizable.

30. Let X be a Banach space. Prove that $(B_X, weak)$ is metrizable if and only if X^* is separable.

31. Let X be a Banach space. Prove that if $(B_X, weak)$ is metrizable then X^* has a countable total set. (A subset $A \subseteq X^*$ is called total if for any $x \in X$ with the property that $x^*(x) = 0 \ \forall x^* \in A$ it follows that $x = 0$.)

32. Let X be a Banach space.

i) If X^* has a countable total set prove that on every *weak* compact subset of X the *weak* topology is metrizable.

ii) Prove that if X is reflexive and X^* has a countable total set then X^* is separable.

33. Let X be a Banach space such that $(B_{X^*}, weak^*)$ is a compact metric space. Prove that X is separable.

Dual $weak^*$-to-norm continuous operators are finite rank operators

34. Let X and Y be normed spaces, and $T: X \to Y$ a bounded linear operator. Prove that if $T^*: Y^* \to X^*$ is $weak^*$-to-norm continuous then T is a finite rank operator.

A $weak^*$-to-$weak^*$ bicontinuous linear operator

35. Let X be a Banach space and $Y \subseteq X$ a dense linear subspace.
i) Prove that the operator $T: X^* \to Y^*$, $T(x^*) = x^*|_Y$ is an isometric isomorphism.
ii) Prove that T is a $weak^*$-to-$weak^*$ homeomorphism if and only if $X = Y$.

A characterization for injective operators

36. Let X and Y be two normed spaces and $T: X \to Y$ a linear bounded operator. Then T is injective if and only if $T^*(Y^*) \subseteq X^*$ is $weak^*$ dense.

Conditions for a Banach space to be a dual space

37. Let X be a Banach space and let $E \subseteq X^*$. Suppose that E separates the elements of X and that B_X is compact in the topology of pointwise convergence on E. Prove that X is the dual of a Banach space (X is isometrically isomorphic with $(\overline{\mathrm{Sp}(E)})^*$).

$Weak$ sequential convergence and compact operators on Hilbert spaces

38. Let H be a complex Hilbert space and $T \in L(H)$.
i) Prove that T is compact if and only if $(x_n)_{n \in \mathbb{N}} \subseteq H$, $x_n \to 0$ weak, imply $Tx_n \to 0$ in the norm topology of H.
ii) Prove that T is compact if and only if $(x_n)_{n \in \mathbb{N}} \subseteq H$, $x_n \to 0$ weak, imply $\langle Tx_n, x_n \rangle \to 0$.

A compact operator on a reflexive space attains its norm

39. Let X, Y be two Banach spaces, $Y \neq \{0\}$. Prove that X is reflexive if and only if for each compact operator $A: X \to Y$ there is an $x \in X$, $\|x\| \leq 1$, such that $\|A\| = \|Ax\|$.

A characterization for compact operators between two Banach spaces

40. Let X, Y be two Banach spaces, and consider a bounded linear operator $T: X \to Y$. Prove that T is compact if and only if T^* is $weak^*$-to-norm continuous on $weak^*$ compact subsets of Y^*.

$Weak$ compact sets and norm closed and bounded sets

41. Let X be a normed space and $K \subseteq X$ a $weak$ compact subset. Then $K \subseteq X$ is norm closed and norm bounded.

42. i) Prove that $B_{c_0} \subseteq c_0$ is norm closed and norm bounded and that it is not $weak$ compact.
ii) Prove that $B_{l_1} \subseteq l_1$ is norm closed, norm bounded, but it is not $weak$ compact.

The summation operator is not $weak$ compact

43. Prove that the summation operator $\Sigma: l_1 \to l_\infty$,

$$\Sigma((t_n)_{n \in \mathbb{N}}) = \left(\sum_{i=1}^{n} t_i\right)_{n \in \mathbb{N}}$$

is not a $weak$ compact operator.

The failure of the Eberlein–Smulian theorem in the *weak** topology

44. Let Γ be a non-empty set and denote by $l_1(\Gamma)$ the set of all functions $x : \Gamma \to \mathbb{K}$ for which $\sum_{\gamma \in \Gamma} |x(\gamma)| < \infty$. Analogously to the case l_1 we obtain that $(l_1(\Gamma), \|.\|_1)$ is a Banach space, where $\|x\|_1 = \sum_{\gamma \in \Gamma} |x(\gamma)|$ $\forall x \in l_1(\Gamma)$, and $(l_1(\Gamma))^* = l_\infty(\Gamma)$, where $l_\infty(\Gamma)$ is the set of all functions $\varphi : \Gamma \to \mathbb{K}$, with $\|\varphi\|_\infty := \sup_{\gamma \in \Gamma} |\varphi(\gamma)| < \infty$. The action of $\varphi \in l_\infty(\Gamma)$ on $x \in l_1(\Gamma)$ is given by $\varphi(x) = \sum_{\gamma \in \Gamma} \varphi(\gamma) x(\gamma)$.

Let now $\Gamma = \mathcal{P}(\mathbb{N})$, thus Γ is the set of all the subsets of $\mathbb{N}$. For each $n \in \mathbb{N}$, let $\delta_n : \Gamma \to \mathbb{K}$, $\delta_n(A) = \begin{cases} 1, & \text{if } n \in A, \\ 0, & \text{if } n \notin A. \end{cases}$ Prove that $\delta_n \in B_{l_\infty(\Gamma)}$ $\forall n \in \mathbb{N}$ and that we cannot find a *weak** convergent subsequence for $(\delta_n)_{n \in \mathbb{N}}$.

45. Define $x_n^* \in l_\infty^*$, $x_n^*(x_1, x_2, \ldots) = x_n$ $\forall (x_1, x_2, \ldots) \in l_\infty$. Prove that $\|x_n^*\| = 1$ $\forall n \in \mathbb{N}$, but $(x_n^*)_{n \in \mathbb{N}}$ does not have a *weak** convergent subsequence.

Weak entire functions are entire functions

46. Let X be a complex Banach space and $f : \mathbb{C} \to X$ a function such that $x^* \circ f : \mathbb{C} \to \mathbb{C}$ is entire, for each $x^* \in X^*$.

i) Prove that f is continuous.

ii) Prove that there is a sequence $(x_n)_{n \geq 0} \subseteq X$ such that $f(z) = \sum_{n=0}^\infty x_n z^n$ for all $z \in \mathbb{C}$, and $\sum_{n=0}^\infty \|x_n\| R^n < \infty$ for every $R > 0$.

Thus, if $f : \mathbb{C} \to X$ is '*weak* entire' then f is entire.

14.2 Solutions

1. We observe that $x_n \to x$ *weak* if and only if $x_n - x \to 0$ *weak*, and then we will prove the equivalence from the exercise only for $x = 0$.

i) One implication is immediate. If $x_n \to 0$ *weak* using that a *weak* convergent sequence in a normed space is norm bounded we obtain that the set $\{\|x_n\|_\infty \mid n \in \mathbb{N}\} \subseteq \mathbb{R}$ is bounded. Since for each $k \in \mathbb{N}$ the canonical projection $p_k : c_0 \to \mathbb{K}$, $p_k((\alpha_n)_{n \in \mathbb{N}}) = \alpha_k$ is linear and continuous it follows that $p_k(x_n) \to p_k(0) = 0$ $\forall k \in \mathbb{N}$.

Suppose now that the set $\{\|x_n\|_\infty \mid n \in \mathbb{N}\} \subseteq \mathbb{R}$ is bounded and that $p_k(x_n) \to 0$ $\forall k \in \mathbb{N}$. Let $\varphi \in c_0^* = l_1$. Then there is $(\lambda_n)_{n \in \mathbb{N}} \in l_1$ such that $\varphi((\alpha_n)_{n \in \mathbb{N}}) = \sum_{n=1}^\infty \lambda_n \alpha_n$, $\forall \alpha = (\alpha_n)_{n \in \mathbb{N}} \in c_0$. Let $\varepsilon > 0$. Since $(\lambda_n)_{n \in \mathbb{N}} \in l_1$, $\sum_{n=1}^\infty |\lambda_n| < \infty$, and therefore there is an $N_\varepsilon \in \mathbb{N}$ such that $\sum_{n=N_\varepsilon}^\infty |\lambda_n| \leq \varepsilon/(2M)$, where $M := 1 + \sup_{n \in \mathbb{N}} \|x_n\|_\infty$. Since $p_1(x_n) \to 0$, $\ldots, p_{N_\varepsilon}(x_n) \to 0$ for $n \to \infty$, there is an $n_\varepsilon \in \mathbb{N}$ such that $\forall n \geq n_\varepsilon$ we have

$$|p_1(x_n)| \leq \frac{\varepsilon}{2\left(1 + \sum_{i=1}^{N_\varepsilon} |\lambda_i|\right)}, \ldots, |p_{N_\varepsilon}(x_n)| \leq \frac{\varepsilon}{2\left(1 + \sum_{i=1}^{N_\varepsilon} |\lambda_i|\right)}.$$

14.2 Solutions

Consequently,

$$|\varphi(x_n)| \leq \sum_{i=1}^{N_\varepsilon} |\lambda_i| |p_i(x_n)| + \sum_{i=N_\varepsilon+1}^{\infty} |\lambda_i| |p_i(x_n)|$$

$$\leq \sum_{i=1}^{N_\varepsilon} |\lambda_i| \frac{\varepsilon}{2\left(1+\sum_{i=1}^{N_\varepsilon}|\lambda_i|\right)} + \sum_{i=N_\varepsilon+1}^{\infty} |\lambda_i| \|x_n\|$$

$$\leq \frac{\varepsilon}{2} + \frac{\varepsilon}{2M} M = \varepsilon \ \forall n \geq n_\varepsilon.$$

Thus, $\varphi(x_n) \to 0 \ \forall \varphi \in c_0^*$, i.e., $x_n \to 0$ weak.

ii) We have $\|x_n\| = \sup_{n \leq k \leq 2n} |\varphi(k/n)| \leq \sup_{x \in [1,2]} |\varphi(x)| < \infty \ \forall n \in \mathbb{N}$. Also, for a fixed $k \in \mathbb{N}$ we have $p_k(x_n) = 0 \ \forall n \geq k+1$, and therefore $p_k(x_n) \to 0$. From (i) it follows that $x_n \to 0$ weak. Suppose now that $x_n \to 0$ in norm. Then $\forall \varepsilon > 0 \ \exists n_\varepsilon \in \mathbb{N}$ such that $\forall n \geq n_\varepsilon$ it follows that $\sup_{n \leq k \leq 2n} |\varphi(k/n)| < \varepsilon$. Consider now a fixed $1 \leq x \leq 2$. Then for $n \geq n_\varepsilon$ let $k = [nx]$, the integer part of nx. Then $k \leq nx < k+1$, so $n \leq k \leq 2n$ and therefore $|\varphi(k/n)| < \varepsilon$, i.e., $|\varphi([nx]/n)| < \varepsilon$. Resuming, we have proved that $|\varphi([nx]/n)| < \varepsilon$ $\forall n \geq n_\varepsilon$ and passing to the limit for $n \to \infty$, using that $[nx]/n \to x$, and using the continuity of φ, it follows that $|\varphi(x)| \leq \varepsilon$. Since $\varepsilon > 0$ is arbitrary we deduce that $\varphi = 0$.

2. i) If $x_n \to x_0$ weak* then for each $x \in c_0$, $x_n(x) \to x_0(x)$. Taking $x = e_k \in c_0$ we obtain that $\lim_{n \to \infty} x_n^k = x_0^k$. That the sequence $(x_n)_{n \in \mathbb{N}}$ is norm bounded is obtained by applying the Uniform Boundedness Principle.

Suppose now that there is an $M > 0$ such that $\|x_n\| \leq M \ \forall n \geq 0$, and that $\lim_{n \to \infty} x_n^k = x_0^k \ \forall k \geq 1$. Let $x \in c_0$, $x = (\alpha_1, \alpha_2, \alpha_3, ...)$. Then $x_0(x) = \sum_{i=1}^{\infty} \alpha_i x_0^i$, $x_n(x) = \sum_{i=1}^{\infty} \alpha_i x_n^i$. For any $\varepsilon > 0$ there is an $N_\varepsilon \in \mathbb{N}$ such that $|\alpha_i| \leq \varepsilon/(4M) \ \forall i \geq N_\varepsilon$. For this N_ε, there is an $n_\varepsilon \in \mathbb{N}$ such that

$$\left|x_0^i - x_n^i\right| \leq \frac{\varepsilon}{2N_\varepsilon (\|x\|+1)} \quad i = \overline{1, N_\varepsilon}, \ \forall n \geq n_\varepsilon.$$

Then for $n \geq n_\varepsilon$ we have

$$|x_n(x) - x_0(x)| \leq \sum_{i=1}^{N_\varepsilon} |\alpha_i| \left|x_n^i - x_0^i\right| + \sum_{i=N_\varepsilon+1}^{\infty} |\alpha_i| \left|x_n^i - x_0^i\right|$$

$$\leq \|x\| \sum_{i=1}^{N_\varepsilon} \left|x_n^i - x_0^i\right| + \frac{\varepsilon}{4M} \sum_{i=N_\varepsilon+1}^{\infty} (|x_n^i| + |x_0^i|)$$

$$\leq \|x\| \sum_{i=1}^{N_\varepsilon} \frac{\varepsilon}{2N_\varepsilon (\|x\|+1)} + \frac{\varepsilon}{4M} \left(\sum_{i=1}^{\infty} |x_n^i| + \sum_{i=1}^{\infty} |x_0^i|\right)$$

$$\leq \frac{\varepsilon}{2} + \frac{\varepsilon}{4M} 2M = \varepsilon.$$

Thus $x_n(x) \to x_0(x) \ \forall x \in c_0$, i.e., $x_n \to x_0$ weak*.

ii) By the Uniform Boundedness Principle a $weak^*$ convergent sequence is norm bounded. Thus, if $(x_n^*)_{n\in\mathbb{N}}$ is $weak^*$ convergent then $\sup_{n\in\mathbb{N}} \|x_n^*\| < \infty$. But $\|x_n^*\| = |a_n|$ $\forall n \in \mathbb{N}$, from whence $\sup_{n\in\mathbb{N}} \|a_n\| < \infty$, i.e., $(a_n)_{n\in\mathbb{N}} \in l_\infty$. Conversely, for $(x_n)_{n\in\mathbb{N}} \in c_0$, if $(a_n)_{n\in\mathbb{N}} \in l_\infty$ then $a_n x_n \to 0$, i.e., $x_n^*(x) \to 0$ $\forall x \in c_0$.

iii) It is easy to see (see the part (iv) below) that $\|x_n^*\| = |a_1| + \cdots + |a_n|$ $\forall n \in \mathbb{N}$. Now we continue as above.

iv) If $(x_n^*)_{n\in\mathbb{N}}$ is $weak^*$ convergent then $\sup_{n\in\mathbb{N}} \|x_n^*\| < \infty$. But

$$|x_n^*(x_1, x_2, ...)| \leq (|a_1| + \cdots + |a_n|)\sup_{k\in\mathbb{N}} |x_k|,$$

i.e., $\|x_n^*\| \leq |a_1| + \cdots + |a_n|$. Since

$$\begin{aligned} x_n^*(\operatorname{sgn}(a_n), ..., \operatorname{sgn}(a_1), 0, 0, ...) &= |a_1| + \cdots + |a_n| \\ &\leq \|x_n^*\| \|(\operatorname{sgn}(a_n), ..., \operatorname{sgn}(a_1), 0, 0, ...)\| \\ &\leq \|x_n^*\|, \end{aligned}$$

it follows that $\|x_n^*\| = |a_1| + \cdots + |a_n|$. Therefore

$$\sup_{n\subset\mathbb{N}}(|a_1| + \cdots + |a_n|) = \sup_{n\in\mathbb{N}} \|x_n^*\| < \infty,$$

i.e., the series $\sum_{n=1}^{\infty} |a_n|$ converges, i.e., $(a_n)_{n\in\mathbb{N}} \in l_1$. Conversely, if $(a_n)_{n\in\mathbb{N}} \in l_1$ then a classical result in analysis (proved at exercise 14 of chapter 6) assures that $\forall (x_n)_{n\in\mathbb{N}} \in c_0$ it follows that

$$a_1 x_n + a_2 x_{n-1} + \cdots + a_{n-1} x_2 + a_n x_1 \to 0,$$

i.e., $x_n^*(x_1, x_2, ...) \to 0$, $x_n^* \to 0$ $weak^*$.

v) We have $\|x_n\| = (1/n)\sum_{k=1}^{n} |\varphi(k/n)|$, and therefore $(\|x_n\|)_{n\in\mathbb{N}}$ is bounded, since $(1/n)\sum_{k=1}^{n} |\varphi(k/n)| \to \int_0^1 |\varphi(x)|\,dx$. Also, for any fixed $k \in \mathbb{N}$ we have $|p_k(x_n)| = (1/n)|\varphi(k/n)| \leq (1/n)\sup_{x\in[0,1]} |\varphi(x)|$ $\forall n \geq k$, hence $p_k(x_n) \to 0$. Now we apply (i) and we obtain that $x_n \to 0$ $weak^*$. If $(x_n)_{n\in\mathbb{N}}$ is norm convergent then $(x_n)_{n\in\mathbb{N}}$ must converge towards 0. Then $\|x_n\| \to 0$, and then $\int_0^1 |\varphi(x)|\,dx = 0$, $\varphi = 0$.

vi) The idea is to use exercises 42 and 44 from chapter 5. Let $x_n^* : c_0 \to \mathbb{K}$,

$$x_n^*((a_k)_{k\in\mathbb{N}}) = \frac{1}{n}\sum_{k=1}^{n} a_k \varphi\left(\frac{k}{n}\right).$$

Then $x_n^* \in c_0^* = l_1$ is defined by the sequence

$$x_n = \left(\frac{1}{n}\varphi\left(\frac{1}{n}\right), \frac{1}{n}\varphi\left(\frac{2}{n}\right), ..., \frac{1}{n}\varphi\left(\frac{n}{n}\right), 0, ...\right) \in l_1$$

and then by (v) $x_n^* \to 0$ $weak^*$. Using exercise 42 of chapter 5 it follows that U is linear and continuous. Since $\|x_n\| = \|x_n^*\|$ $\forall n \in \mathbb{N}$, if U is compact then by exercise 44 of chapter 5 we have $\|x_n^*\| \to 0$, i.e., $\|x_n\| \to 0$, and using (v) it follows that $\varphi = 0$.

14.2 Solutions

3. See [19, chapter I, Schur's Theorem].

i) Since l_1 is separable using exercise 29 we obtain that there is a distance d on B_{l_∞} which generates the $weak^*$ topology on B_{l_∞}.

ii) For every $x, y \in l_1$ and $\varepsilon > 0$, the set $\{x^* \in B_{l_\infty} \mid |x^*(x) - x^*(y)| \leq \varepsilon/3\}$ is $weak^*$ closed, because the function $x - y : (l_\infty, weak^*) \to (\mathbb{K}, |\cdot|)$ is continuous, and $B_{l_\infty} \subseteq l_\infty$ is $weak^*$ closed. We obtain that B_m is $weak^*$ closed in l_∞, for each $m \geq 1$.

If $x_n \to x$ $weak$ then for each $x^* \in l_1^* = l_\infty$ we have $x^*(x_n) \to x^*(x)$, and consequently for each $x^* \in B_{l_\infty}$, $x^*(x_n) - x^*(x) \to 0$. Let $\varepsilon > 0$ and let $x^* \in B_{l_\infty}$. There is an $m \in \mathbb{N}$ such that for each $n \geq m$, $|x^*(x_n) - x^*(x)| \leq \varepsilon/3$, and consequently $x^* \in B_m$. We obtain that $B_{l_\infty} = \bigcup_{m \geq 1} B_m$.

iii) If we consider (B_{l_∞}, d) we have a compact metric space because the metric d generates the $weak^*$ topology on B_{l_∞} and $(B_{l_\infty}, weak^*)$ is a compact topological space, by the Alaoglu–Bourbaki theorem. Using Baire's category theorem and (ii) we obtain that there is $m_0 \in \mathbb{N}$ such that $\overset{\circ}{\overline{B_{m_0}}} \neq \emptyset$, where the closure and the interior are taken with respect to the $weak^*$ topology on B_{l_∞}. Since B_{m_0} is $weak^*$ closed (in $(l_\infty, weak^*)$, and then in $(B_{l_\infty}, weak^*)$ also), we obtain that $\overset{\circ}{B_{m_0}} \neq \emptyset$, and consequently we can find $x_0^* \in B_{l_\infty}$, $y_1, ..., y_k \in l_1$, $\tau > 0$ such that

$$W(x_0^*; y_1, ..., y_k; \tau) \cap B_{l_\infty} = \{x^* \in l_1^* \mid |(x^* - x_0^*)(y_i)| < \tau, \, i = \overline{1, k}\} \cap B_{l_\infty} \subseteq B_{m_0}.$$

We will show that there are $p \in \mathbb{N}$ and $\delta > 0$ such that

$$B_{l_\infty} \cap W(x_0^*; e_1, ..., e_p; \delta) \subseteq W(x_0^*; y_1, ..., y_k; \tau) \cap B_{l_\infty}.$$

Let $p \in \mathbb{N}$ such that $\sum_{i=p+1}^{\infty} |y_j^i| < \tau/4$, $j = \overline{1, k}$. For $x^* \in B_{l_\infty}$, if

$$|(x^* - x_0^*)(e_i)| \leq \delta = \frac{\tau}{2(\max\{\|y_j\|_1 \mid j = \overline{1, k}\} + 1)}, \, i = \overline{1, p},$$

then

$$|(x^* - x_0^*)(y_j)| = \left|\sum_{i=1}^{\infty} ((x^*)_i - (x_0^*)_i) y_j^i\right| \leq \sum_{i=1}^{p} |(x^*)_i - (x_0^*)_i| |y_j^i|$$

$$+ \sum_{i=p+1}^{\infty} |y_j^i| (\|x_0^*\| + \|x^*\|)$$

$$\leq \delta \sum_{i=1}^{\infty} |y_j^i| + \frac{\tau}{4} 2 \leq \delta \max_{s=\overline{1,k}} \|y_s\| + \frac{\tau}{2} \leq \tau, \, j = \overline{1, k},$$

thus $x^* \in W(x_0^*; y_1, ..., y_k; \tau)$.

Since $x_n \to x$ $weak$ we obtain that $p_k(x_n) \to p_k(x)$ $\forall k \in \mathbb{N}$, i.e., $x_n^k \to x^k$ $\forall k \in \mathbb{N}$, with obvious notations. Thus there is an $m > m_0$ such that $\sum_{k=1}^{p} |x_n^k - x^k| \leq \varepsilon/3$ $\forall n \geq m$.

Then, for $n \geq m$ (m depends on ε), we have

$$\begin{aligned}
\|x_n - x\|_1 &= \sum_{k=1}^{\infty} |x_n^k - x^k| = \sum_{k=1}^{p} |x_n^k - x^k| + \sum_{k=p+1}^{\infty} |x_n^k - x^k| \\
&= \sum_{k=1}^{p} |x_n^k - x^k| - \sum_{k=1}^{p} (x_0^*)_k (x_n^k - x^k) \\
&\quad + \sum_{k=p+1}^{\infty} |x_n^k - x^k| + \sum_{k=1}^{p} (x_0^*)_k (x_n^k - x^k) \\
&\leq 2 \sum_{k=1}^{p} |x_n^k - x^k| + |x^*(x_n - x)|,
\end{aligned}$$

where $x^* \in l_1^*$ is given by $\left((x_0^*)_1, ..., (x_0^*)_p, \operatorname{sgn}(x_n^{p+1} - x_0^{p+1}), ...\right) \in B_{l_\infty}$. But $(x^* - x_0^*)(e_j) = 0, j = \overline{1,p}$, and therefore

$$x^* \in W(x_0^*; e_1, ..., e_p; \delta) \cap B_{l_\infty} \subseteq B_{m_0} \subseteq B_n \ \forall n \geq m_0.$$

Thus $x^*(x_n - x) < \varepsilon/3 \ \forall n \geq m_0$, and consequently $\forall n \geq \max(m_0, m)$, $\|x_n - x\| \leq 2(\varepsilon/3) + \varepsilon/3 = \varepsilon$. We obtain that $x_n \to x$ in the norm of l_1.

iv) It follows from (iii).

v) The answer is no! From exercise 24 it follows that if X is an infinite-dimensional normed space then $B_X = \overline{S_X}^{\text{weak}}$. Since $0 \in B_X$ from the well known characterization of the closure in topological spaces it follows that there is a net $(x_\delta)_{\delta \in \Delta} \subseteq X$ such that $\|x_\delta\| = 1 \ \forall \delta \in \Delta$ and $x_\delta \to 0$ *weak*.

4. i) We can suppose that $x = 0$. We use the fact that $l_p^* = l_q$, $1/p + 1/q = 1$, with the usual identification. Since every *weak* convergent sequence is bounded, $(x_n)_{n \in \mathbb{N}} \subseteq l_p$ is bounded, and since $p_k : l_p \to \mathbb{K}$ are linear and continuous it follows that $p_k(x_n) \to 0$ $\forall k \in \mathbb{N}$.

Suppose now that $\{\|x_n\|_p \mid n \in \mathbb{N}\}$ is bounded and that $p_k(x_n) \to 0 \ \forall k \in \mathbb{N}$. Let $\varphi \in l_p^*$. There is a sequence $(\lambda_n)_{n \in \mathbb{N}} \in l_q$ such that $\varphi\left((\alpha_n)_{n \in \mathbb{N}}\right) = \sum_{n=1}^{\infty} \alpha_n \lambda_n \ \forall (\alpha_n)_{n \in \mathbb{N}} \in l_p$.

Let $\varepsilon > 0$. Since $(\lambda_n)_{n \in \mathbb{N}} \in l_q$, there is $N_\varepsilon \in \mathbb{N}$ such that $\left(\sum_{n=N_\varepsilon+1}^{\infty} |\lambda_n|^q\right)^{1/q} \leq \varepsilon/(2M)$ where $M := 1 + \sup_{n \in \mathbb{N}} \|x_n\|_p$. Since $p_1(x_n) \to 0, ..., p_{N_\varepsilon}(x_n) \to 0$, there is $n_\varepsilon \in \mathbb{N}$ such that $|p_1(x_n)| \leq \varepsilon/(2L), ..., |p_{N_\varepsilon}(x_n)| \leq \varepsilon/(2L) \ \forall n \geq n_\varepsilon$, where $L := \sum_{n=1}^{N_\varepsilon} |\lambda_n| + 1$.

14.2 Solutions

Consequently for every $\varepsilon > 0$ there is $n_\varepsilon \in \mathbb{N}$ such that

$$|\varphi(x_n)| \leq \sum_{k=1}^{\infty} |\lambda_k p_k(x_n)| = \sum_{k=1}^{N_\varepsilon} |\lambda_k| \, |p_k(x_n)| + \sum_{k=N_\varepsilon+1}^{\infty} |\lambda_k| \, |p_k(x_n)|$$

$$\leq \frac{\left(\sum_{k=1}^{N_\varepsilon} |\lambda_k|\right) \varepsilon}{2L} + \left(\sum_{k=N_\varepsilon+1}^{\infty} |\lambda_k|^q\right)^{1/q} \left(\sum_{k=N_\varepsilon+1}^{\infty} |p_k(x_n)|^p\right)^{1/p}$$

$$\leq \frac{\varepsilon}{2} + \frac{\varepsilon}{2M} \|x_n\|_p \leq \frac{\varepsilon}{2} + \frac{\varepsilon}{2M} M = \varepsilon \quad \forall n \geq n_\varepsilon.$$

We have proved that $\varphi(x_n) \to 0 \; \forall \varphi \in l_p^*$, i.e., $x_n \to 0$ weak.

ii) We have $\|x_n\| = \left((1/n) \sum_{k=1}^{n} |\varphi(k/n)|^p\right)^{1/p}$, and therefore the sequence $(\|x_n\|)_{n \in \mathbb{N}}$ is bounded, since $(1/n) \sum_{k=1}^{n} |\varphi(k/n)|^p \to \int_0^1 |\varphi(x)|^p \, dx$. Also for any fixed $k \in \mathbb{N}$ we have $|p_k(x_n)| = (1/n)^{1/p} |\varphi(k/n)| \leq (1/n)^{1/p} \sup_{x \in [0,1]} |\varphi(x)| \; \forall n \geq k$, hence $p_k(x_n) \to 0$. Now we apply (i). (See also the solution for the part (v) from exercise 2.)

iii) The idea is the same as the one for exercise 2(vi), i.e., to use exercises 42 and 44 from chapter 5. Let $x_n^* : l_q \to \mathbb{K}$,

$$x_n^*\left((a_k)_{k \in \mathbb{N}}\right) = n^{-1/p} \sum_{k=1}^{n} a_k \varphi\left(\frac{k}{n}\right).$$

Then $x_n^* \in l_q^* = l_p$ is defined by the sequence

$$x_n = n^{-1/p} \left(\varphi\left(\frac{1}{n}\right), \varphi\left(\frac{2}{n}\right), \ldots, \varphi\left(\frac{n}{n}\right), 0, \ldots\right) \in l_p$$

and since by (ii) $x_n \to 0$ *weak*, this means that $x_n^* \to 0$ *weak**. By exercise 42 of chapter 5 it follows that U is linear and continuous. Since $\|x_n\| = \|x_n^*\| \; \forall n \in \mathbb{N}$, if U is compact then by exercise 44 of chapter 5 we have $\|x_n^*\| \to 0$, i.e., $\|x_n\| \to 0$, and using (ii) it follows that $\varphi = 0$.

5. '(i) $\Rightarrow$ (ii)' Suppose, for a contradiction, that $p \neq 1$, i.e., $1 < p < \infty$. Consider the standard basis $(e_n)_{n \in \mathbb{N}} \subseteq l_p$. Since $p > 1$ then $e_n \to 0$ weak. Indeed, if $x^* \in l_p^*$, then x^* is given by $(\xi_n)_{n \in \mathbb{N}} \in l_q$, $1/p + 1/q = 1$, and then $x^*(e_n) = \xi_n \to 0$. Our hypothesis implies now that $e_n \to 0$ in the norm of l_p, and this is false since $\|e_n\|_p = 1 \; \forall n \in \mathbb{N}$.

'(ii) $\Rightarrow$ (i)' This is the Schur theorem proved at exercise 3.

6. i) We have $l_1^* = l_\infty$ and let $\varphi \in l_1$, $\varphi = (\lambda_k)_{k \in \mathbb{N}}$, $\sum_{k=1}^{\infty} |\lambda_k| < \infty$. Since $|a_k \lambda_k| \leq \left(\sup_{n \in \mathbb{N}} |a_n|\right) |\lambda_k| \; \forall k \in \mathbb{N}$ and the series $\sum_{k=1}^{\infty} \lambda_k$ converges, by the comparison test it follows that the series $\sum_{k=1}^{\infty} a_k \lambda_k$ converges, and therefore $\sum_{k=n+1}^{\infty} a_k \lambda_k \to 0$, i.e.,

$$x_n(\varphi) = \sum_{k=n+1}^{\infty} a_k \lambda_k \to 0 \text{ for } n \to \infty.$$

ii) Let $x^* \in l_\infty^*$ such that $x_n^* \to x^*$ $weak^*$, i.e., $x_n^*(x_1, x_2, ...) \to x^*(x_1, x_2, ...)$ $\forall (x_1, x_2, ..) \in l_\infty$. In particular, $x_n^*(0, 1, 0, 1, ...) \to x^*(0, 1, 0, 1, ...)$, from whence $x_{2n}^*(0, 1, 0, 1, ...) \to x^*(0, 1, 0, 1...)$ and $x_{2n+1}^*(0, 1, 0, 1, ...) \to x^*(0, 1, 0, 1...)$. But $x_{2n}^*(0, 1, 0, 1, ...) = a_{2n}$ and $x_{2n+1}^*(0, 1, 0, 1, ...) = 0$, hence $a_{2n} \to 0$. Analogously $a_{2n+1} \to 0$. Then $a_n \to 0$, i.e., $(a_n)_{n \in \mathbb{N}} \in c_0$. The converse is clear. We observe that if $(a_n)_{n \in \mathbb{N}} \in c_0$ then $\|x_n^*\| = |a_n| \to 0$.

iii) Let $x^* \in l_\infty^*$ be such that $x_n^* \to x^*$ $weak^*$, i.e., $x_n^*(x_1, x_2, ..) \to x^*(x_1, x_2, ...)$ $\forall (x_1, x_2, ...) \in l_\infty$. In particular, for $(\text{sgn}(a_1), \text{sgn}(a_2), ...) \in l_\infty$,

$$x_n^*(\text{sgn}(a_1), \text{sgn}(a_2), ...) \to x^*(\text{sgn}(a_1), \text{sgn}(a_2), ...) =: s \in \mathbb{K}.$$

But

$$x_n^*(\text{sgn}(a_1), \text{sgn}(a_2), ...) = a_1 \text{sgn}(a_1) + \cdots + a_n \text{sgn}(a_n)$$
$$= |a_1| + \cdots + |a_n|.$$

Hence $|a_1| + |a_2| + \cdots + |a_n| \to s$, i.e., the series $\sum_{n=1}^\infty a_n$ converges absolutely, that is, $(a_n)_{n \in \mathbb{N}} \in l_1$. Conversely, suppose that $\sum_{n=1}^\infty a_n$ converges absolutely. Let $x^* : l_\infty \to \mathbb{K}$, $x^*(x_1, x_2, ..) = \sum_{n=1}^\infty a_n x_n$. Then

$$\sum_{n=1}^\infty |a_n x_n| \leq \left(\sup_{n \in \mathbb{N}} |x_n|\right) \sum_{n=1}^\infty |a_n|,$$

and therefore x^* is well defined, linear, and continuous. Then

$$x_n^* \to x^* \ weak^* \Leftrightarrow x_n^*(x_1, x_2, ...) \to x^*(x_1, x_2, ...) \ \forall (x_1, x_2, ...) \in l_\infty$$
$$\Leftrightarrow \sum_{k=1}^n a_k x_k \to \sum_{k=1}^\infty a_k x_k \ \forall (x_1, x_2, ...) \in l_\infty,$$

and the last relation is true since the series $\sum_{k=1}^\infty a_k x_k$ converges.

7. i) If $f_n \to f$ $weak$ then $(f_n)_{n \in \mathbb{N}}$ is bounded in $C(T)$ and therefore $\sup_{n \in \mathbb{N}} \|f_n\|_\infty < \infty$. For each $t \in T$ we consider the Dirac functional $\delta_t(f) = f(t)$ $\forall f \in C(T)$. Then $\delta_t \in (C(T))^*$ $\forall t \in T$, and thus $\delta_t(f_n) \to \delta_t(f)$ $\forall t \in T$. We obtain that $f_n(t) \to f(t)$ $\forall t \in T$.

Suppose now that $f_n(t) \to f(t)$ $\forall t \in T$ and that $\sup_{n \in \mathbb{N}} \|f_n\|_\infty = M < \infty$. Let $\varphi \in (C(T))^*$. Using the Riesz representation theorem we obtain a finite Borel regular measure μ on T such that $\varphi(f) = \int_T f d\mu$ $\forall f \in C(T)$. Since $f_n(t) \to f(t)$ $\forall t \in T$, $|f_n(t)| \leq M$ $\forall n \in \mathbb{N}$, $\forall t \in T$, and $|\mu|(T) < \infty$, by the Lebesgue dominated convergence theorem we obtain that $\int_T f_n d\mu \to \int_T f d\mu$, i.e., $\varphi(f_n) \to \varphi(f)$, and thus $f_n \to f$ $weak$.

ii) Since φ is continuous it follows that $\|f_n\| \leq \sup_{x \in [0,1]} |\varphi(x)| < \infty$ $\forall n \in \mathbb{N}$. Also, we have $\lim_{n \to \infty} f_n(x) = \lim_{n \to \infty} \varphi(x^n) = \begin{cases} \varphi(0) & \text{if } 0 \leq x < 1, \\ \varphi(1) & \text{if } x = 1. \end{cases}$ If the sequence $(f_n)_{n \in \mathbb{N}} \subseteq$

14.2 Solutions

$C[0,1]$ is *weak* convergent then there is $f \in C[0,1]$ such that $f_n \to f$ weak. By (i) and the above calculations we must have $f(x) = \begin{cases} \varphi(0) & \text{if } 0 \leq x < 1, \\ \varphi(1) & \text{if } x = 1. \end{cases}$ Since f is continuous we obtain that $\varphi(0) = \varphi(1)$. The converse is clear.

iii) Suppose that $(f_n)_{n\in\mathbb{N}}$ is *weak* convergent. Then by (i) there is $M > 0$ such that $|f_n(x)| \leq M \; \forall x \in [0,1], \forall n \in \mathbb{N}$. In particular, $|f_n(1 - 1/n)| \leq M \; \forall n \in \mathbb{N}$, i.e.,

$$|a_n| \leq \frac{M}{(1-1/n)^n \left(1-(1-1/n)^n\right)} \quad \forall n \in \mathbb{N}.$$

Since $(1-1/n)^n \to 1/e$ it follows that the sequence $(a_n)_{n\in\mathbb{N}}$ is bounded, i.e., $(a_n)_{n\in\mathbb{N}} \in l_\infty$. Conversely, if $(a_n)_{n\in\mathbb{N}} \in l_\infty$ then since $x^n \to 0 \; \forall x \in (0,1)$ it follows that $f_n(x) \to 0$ $\forall x \in (0,1)$. Also $f_n(0) = f_n(1) = 0 \; \forall n \in \mathbb{N}$, and $|f_n(x)| \leq 2 \sup_{k\in\mathbb{N}} |a_k| \; \forall x \in [0,1]$, $\forall n \in \mathbb{N}$. Now we use (i) to deduce that $f_n \to 0$ *weak*.

Suppose that there is $f \in C[0,1]$ such that $f_n \to f$ in norm. Then $(f_n)_{n\in\mathbb{N}}$ is *weak* convergent and by the above $f_n \to 0$ *weak*. Then $f = 0$, and then $f_n \to 0$ in norm, i.e., $\forall \varepsilon > 0 \; \exists n_\varepsilon \in \mathbb{N}$ such that $\forall n \geq n_\varepsilon$ it follows that $|f_n(x)| < \varepsilon \; \forall x \in [0,1]$. In particular, $\forall \varepsilon > 0 \; \exists n_\varepsilon \in \mathbb{N}$ such that $\forall n \geq n_\varepsilon$ it follows that $|f_n(1-1/n)| < \varepsilon$, that is,

$$|a_n| \leq \frac{\varepsilon}{(1-1/n)^n \left(1-(1-1/n)^n\right)}.$$

From here we deduce that $\limsup_{n\to\infty} |a_n| \leq (e^2\varepsilon)/(e-1) \; \forall \varepsilon > 0$, and therefore $\limsup_{n\to\infty} |a_n| = 0$, i.e., $(a_n)_{n\in\mathbb{N}} \in c_0$. The converse is clear since $\|f_n\| \leq 2|a_n| \; \forall n \in \mathbb{N}$.

8. See [20, exercise 21, chapter V, section 7].

From exercise 7 we obtain that $f_n \to f \in C(T)$ *weak*. We use now the Mazur theorem to solve the exercise.

9. i) Let $A \subseteq (C(T), \tau_p)$ be a compact subset. For each $k \in \mathbb{N}$ using the compactness of T we obtain that there are $x_{1,k}, ..., x_{n_k,k} \in T$ such that $T \subseteq \bigcup_{i=1}^{n_k} B_\rho(x_{i,k}; 1/k)$. We enumerate the set $(x_{i,k})_{k\in\mathbb{N},\; i=\overline{1,n_k}}$ as a sequence $(x_n)_{n\in\mathbb{N}}$, and we obtain $(x_n)_{n\in\mathbb{N}} \subseteq T$ such that: $\forall k \in \mathbb{N}, \forall x \in T$, there is $n \in \mathbb{N}$ such that $\rho(x_n, x) < 1/k$. Let $d : A \times A \to \mathbb{R}_+$,

$$d(f,g) = \sum_{n=1}^{\infty} \frac{p_n(f-g)}{2^n (1 + p_n(f-g))} \quad \forall f, g \in A,$$

where $p_n(f) = |f(x_n)| \; \forall f \in A$. That the sequence $(x_n)_{n\in\mathbb{N}}$ is dense in T implies that d is a distance on A.

We will show now that the topology generated by d on A and the restriction of τ_p on A coincide. Without loss of generality we can suppose that $0 \in A$. Let $\varepsilon \in (0,1)$. Let $m \in \mathbb{N}$ be such that $\sum_{n=m+1}^{\infty} 1/2^n \leq \varepsilon/2$. Let

$$f \in W(0; p_1, ..., p_m; \varepsilon/(2-\varepsilon)) = \{g \in C(T) \mid p_i(g) < \varepsilon/(2-\varepsilon),\; i = \overline{1,m}\}.$$

Then
$$d(f,0) \leq \left(\sum_{n=1}^{m} \frac{1}{2^n}\right)\frac{\varepsilon}{2} + \sum_{n=m+1}^{\infty} \frac{1}{2^n} < \varepsilon,$$

and therefore $W(0; p_1, ..., p_m; \varepsilon/(2-\varepsilon)) \subseteq B_d(0, \varepsilon)$.

Let us consider now $\varepsilon > 0$ and $y_1, ..., y_m \in T$. We must show that there is an $r > 0$ such that
$$B_d(0, r) \cap A \subseteq W(0; p_{y_1}, ..., p_{y_m}; \varepsilon),$$

where $p_{y_i}(f) = |f(y_i)| \, \forall f \in C(T)$. For each $f \in C(T)$, we can find $\left\{i_1^{(f)}, ..., i_m^{(f)}\right\} \subseteq \mathbb{N}$ such that $|f(x_{i_j^{(f)}}) - f(y_j)| < \varepsilon/4$, $j = \overline{1, m}$. Since

$$A \subseteq \bigcup_{f \in A} W\left(f; p_{i_1^{(f)}}, ..., p_{i_m^{(f)}}, p_{y_1}, ..., p_{y_m}; \varepsilon/4\right),$$

using the τ_p compactness of A and the fact that the sets from the union are τ_p open we obtain that there are $k \in \mathbb{N}$ and $f_1, ..., f_k \in A$ such that

$$A \subseteq \bigcup_{j=1}^{k} W\left(f_j; p_{i_1^{(f_j)}}, ..., p_{i_m^{(f_j)}}, p_{y_1}, ..., p_{y_m}; \varepsilon/4\right).$$

Consider now the set $W\left(0; p_{i_1^{(f_1)}}, ..., p_{i_m^{(f_1)}}, ..., p_{i_1^{(f_k)}}, ..., p_{i_m^{(f_k)}}; \varepsilon/4\right)$. Using the definition for d, we obtain that there is $r > 0$ such that

$$B_d(0, r) \subseteq W\left(0; p_{i_1^{(f_1)}}, ..., p_{i_m^{(f_1)}}, ..., p_{i_1^{(f_k)}}, ..., p_{i_m^{(f_k)}}; \varepsilon/4\right).$$

Then for each $f \in B_d(0, r) \cap A$ we have

$$f \in W\left(0; p_{i_1^{(f_1)}}, ..., p_{i_m^{(f_1)}}, ..., p_{i_1^{(f_k)}}, ..., p_{i_m^{(f_k)}}; \varepsilon/4\right) \cap A.$$

Since $f \in A$ there is $1 \leq j \leq k$ such that $f \in W\left(f_j; p_{i_1^{(f_j)}}, ..., p_{i_m^{(f_j)}}, p_{y_1}, ..., p_{y_m}; \varepsilon/4\right)$. Then for each $1 \leq s \leq m$ we have

$$\begin{aligned}p_{y_s}(f) &= |f(y_s)| \leq |f(y_s) - f_j(y_s)| + \left|f_j(y_s) - f_j\left(x_{i_s}^{(f_j)}\right)\right| \\ &\quad + \left|f_j\left(x_{i_s}^{(f_j)}\right) - f\left(x_{i_s}^{(f_j)}\right)\right| + \left|f\left(x_{i_s}^{(f_j)}\right)\right| \\ &< 4\frac{\varepsilon}{4} = \varepsilon,\end{aligned}$$

and we obtain that $B_d(0, r) \cap A \subseteq W(0; p_{y_1}, ..., p_{y_m}; \varepsilon)$.

We obtain that the neighborhood basis for $0 \in A$ with respect to the two topologies we have on A are equivalent. Analogously we obtain the same assertion for every element $f \in A$. Thus the two topologies we have on A coincide.

ii) Let $A \subseteq C(T)$, a τ_p compact and uniformly bounded set. Using (i) we obtain the fact that the τ_p topology on A is metrizable. If we consider a sequence $(g_n)_{n \in \mathbb{N}} \subseteq A$ then by the compactness and the metrizability of (A, τ_p) we obtain that there are a subsequence $(n_k)_{k \in \mathbb{N}}$ of $\mathbb{N}$ and an element $g \in A$ such that $g_{n_k} \to g$ in the τ_p topology, thus $g_{n_k} \to g$ pointwise. Since $A \subseteq C(T)$ is uniformly bounded we obtain the fact that $(g_{n_k})_{k \in \mathbb{N}}$ is uniformly bounded. Using exercise 7(i) we obtain that $g_{n_k} \to g$ *weak* in $C(T)$. Thus $A \subseteq C(T)$ is sequentially *weak* compact and by the Eberlein–Smulian theorem we obtain that $A \subseteq C(T)$ is *weak* compact.

10. i) If $x_n \to x$ *weak* then the sequence $(x_n)_{n \in \mathbb{N}}$ is bounded and we can consider $Y = X^*$. Conversely, suppose that there are $M > 0$ and $Y \subseteq X^*$ with $\overline{\mathrm{Sp}(Y)} = X^*$ such that $\|x_n\| \leq M \ \forall n \in \mathbb{N}$ and $x^*(x_n) \to x^*(x) \ \forall x^* \in Y$. If we denote $Z = \mathrm{Sp}(Y)$, then $Z \subseteq X^*$, $\overline{Z} = X^*$ and $x^*(x_n) \to x^*(x) \ \forall x^* \in Z$. Consider $x^* \in X^*$. Since $\overline{Z} = X^*$ we can find a sequence $(z_n^*)_{n \in \mathbb{N}} \subseteq Z$ such that $z_n^* \to x^*$ in X^*. For every $\varepsilon > 0$ there is an $n_\varepsilon \in \mathbb{N}$ such that $\|z_n^* - x^*\| \leq \varepsilon \ \forall n \geq n_\varepsilon$. But $z_{n_\varepsilon}^* \in Z$, thus $z_{n_\varepsilon}^*(x_n) \to z_{n_\varepsilon}^*(x)$, and therefore there is an $N_\varepsilon \in \mathbb{N}$ such that $\left|z_{n_\varepsilon}^*(x_n) - z_{n_\varepsilon}^*(x)\right| \leq \varepsilon \ \forall n \geq N_\varepsilon$. Then, for $n \geq N_\varepsilon$, we have

$$\begin{aligned} |x^*(x_n) - x^*(x)| &\leq \left|x^*(x_n) - z_{n_\varepsilon}^*(x_n)\right| + \left|z_{n_\varepsilon}^*(x_n) - z_{n_\varepsilon}^*(x)\right| + \left|z_{n_\varepsilon}^*(x) - x^*(x)\right| \\ &\leq \left\|x^* - z_{n_\varepsilon}^*\right\| \|x_n\| + \varepsilon + \left\|x^* - z_{n_\varepsilon}^*\right\| \|x\| \leq \varepsilon(2M+1), \end{aligned}$$

and therefore $x^*(x_n) \to x^*(x)$. We obtain that $x_n \to x$ *weak*.

ii) We apply (i), Riesz's representation theorem, and that if $E \subseteq H$ is an orthonormal basis then E is fundamental.

11. See [20, exercise 25, chapter IV, section 13].

If $f_n \to f$ *weak* in $L_1[0,1]$ we obtain that the sequence $(f_n)_{n \in \mathbb{N}} \subseteq L_1[0,1]$ is bounded, i.e., there is an $M > 0$ such that $\int_0^1 |f_n(t)| dt \leq M \ \forall n \in \mathbb{N}$. Using the fact that $(L_1[0,1])^* = L_\infty[0,1]$, since $\chi_E \in L_\infty[0,1] \ \forall E \subseteq [0,1]$ Lebesgue measurable we obtain that $\chi_E(f_n) \to \chi_E(f)$, thus $\int_E f_n(t)dt \to \int_E f(t)dt \ \forall E \subseteq [0,1]$ Lebesgue measurable.

Suppose now that the sequence $(f_n)_{n \in \mathbb{N}} \subseteq L_1[0,1]$ is bounded and that $\int_E f_n(t)dt \to \int_E f(t)dt \ \forall E \subseteq [0,1]$ Lebesgue measurable. We obtain that for each step function $g : [0,1] \to \mathbb{K}$, $\lim_{n \to \infty} \int_0^1 g(t)f_n(t)dt = \int_0^1 g(t)f(t)dt$. We will prove that the set of all step functions is dense in $L_\infty[0,1]$, and then we apply exercise 10 to obtain that $f_n \to f$ *weak*.

Let $h \in L_\infty[0,1]$. Suppose that $h(t) \in \mathbb{R}$ a.e. $t \in [0,1]$. Without loss of generality we can suppose that there is $k > 0$ such that $-k < h(t) < k \ \forall t \in [0,1]$. Let $\varepsilon > 0$ and choose $n \in \mathbb{N}$ such that $2k/n \leq \varepsilon$. For $i = \overline{1,n}$, let

$$E_i = \{t \in [0,1] \mid -k + (2k/n)(i-1) \leq h(t) < -k + (2k/n)i\}.$$

Since h is measurable we obtain that $E_i \subseteq [0,1]$ is Lebesgue measurable. Let $t_i \in E_i$, $i = \overline{1,n}$, and define $g : [0,1] \to \mathbb{R}$, $g = \sum_{i=1}^n h(t_i)\chi_{E_i}$. For each $t \in [0,1]$ there is an i such that $t \in E_i$. Then $|g(t) - h(t)| = |h(t_i) - h(t)|$. Suppose, for example, that $h(t_i) \geq h(t)$, and then

$$|g(t) - h(t)| = h(t_i) - h(t) \leq -k + (2k/n)i + k - (2k/n)(i-1) = (2k/n) \leq \varepsilon.$$

We obtain that $\|g - h\|_\infty \leq \varepsilon$, where $g : [0,1] \to \mathbb{R}$ is a step function. Now, if $h : [0,1] \to \mathbb{K}$ is bounded then $h = u + iv$ with $u, v : [0,1] \to \mathbb{R}$ bounded. Then we can find two step functions $u_1, v_1 : [0,1] \to \mathbb{R}$ such that $\|u - u_1\|_\infty \leq \varepsilon/2$ and $\|v - v_1\|_\infty \leq \varepsilon/2$, and then, for $h_1 = u_1 + iv_1$, we have $\|h - h_1\|_\infty \leq \varepsilon$, and h_1 is a step function.

12. See [20, exercise 23, chapter IV, section 13].

i) If $f_n \to f$ weak in $L_p[0,1]$ we obtain that the sequence $(f_n)_{n \in \mathbb{N}}$ is bounded, i.e., there is $M > 0$ such that $\int_0^1 |f_n(t)|^p \, dt \leq M$ $\forall n \in \mathbb{N}$. Since $(L_p[0,1])^* = L_q[0,1]$, $1 < q < \infty$, $1/p + 1/q = 1$, and $\chi_{[0,x]} \in L_q[0,1]$ $\forall x \in [0,1]$, we obtain that $\chi_{[0,x]}(f_n) \to \chi_{[0,x]}(f)$ $\forall x \in [0,1]$, that is, $\int_0^x f_n(t)\,dt \to \int_0^x f(t)\,dt$ $\forall x \in [0,1]$.

Suppose now that $\int_0^x f_n(t)\,dt \to \int_0^x f(t)\,dt$ $\forall x \in [0,1]$ and that $(f_n)_{n \in \mathbb{N}} \subseteq L_p[0,1]$ is norm bounded. For every $0 \leq x < y \leq 1$ we have

$$\lim_{n \to \infty} \int_x^y f_n(t)\,dt = \lim_{n \to \infty} \left(\int_0^y f_n(t)\,dt - \int_0^x f_n(t)\,dt \right)$$
$$= \lim_{n \to \infty} \int_0^y f_n(t)\,dt - \lim_{n \to \infty} \int_0^x f_n(t)\,dt$$
$$= \int_0^y f(t)\,dt - \int_0^x f(t)\,dt = \int_x^y f(t)\,dt.$$

We obtain that for every $g : [0,1] \to \mathbb{K}$, $g = \sum_{i=1}^N \alpha_i \chi_{[a_i, b_i]}$, we have

$$\int_0^1 f_n(t) g(t) \to \int_0^1 f(t) g(t)\,dt.$$

Let us consider $X \subseteq L_q[0,1]$ as being the set of all function g with the above form: we obtain that $g(f_n) \to g(f)$ $\forall g \in X$. Using exercise 10, if we prove that $X \subseteq L_q[0,1]$ is dense we will have $f_n \to f$ weak.

Since the set of all step functions is dense in $L_q[0,1]$ it is enough to prove that X is dense in the set of all step functions in $L_q[0,1]$. Let $h \in L_q[0,1]$, $h = \sum_{i=1}^N \alpha_i \chi_{E_i}$, $N \in \mathbb{N}$, $\alpha_i \in \mathbb{K}$, E_i μ-measurable, $i = \overline{1, N}$, and consider $\varepsilon > 0$. For each $i = \overline{1, N}$, since E_i is Lebesgue measurable from the regularity of the Lebesgue measure we obtain that there is a set $A_i \subseteq [0,1]$, $A_i = \bigcup_{j=1}^{N_i} I_i^j$, where I_i^j, $j = \overline{1, N_i}$, are pairwise disjoint intervals such that

$$\mu(E_i \Delta A_i) < \frac{\varepsilon^q}{\left(\sum_{i=1}^N |\alpha_i| + 1 \right)^q}.$$

We obtain that

$$\left(\int_0^1 |\chi_{E_i} - \chi_{A_i}|^q \, dt \right)^{1/q} < \frac{\varepsilon}{\left(\sum_{i=1}^N |\alpha_i| + 1 \right)}.$$

Let $g = \sum_{i=1}^{N} \alpha_i \sum_{j=1}^{N_i} \chi_{I_i^j}$. We obtain that

$$\|g-h\|_q = \left(\int_0^1 |g(t)-h(t)|^q \, dt\right)^{1/q} = \left(\int_0^1 \left|\sum_{i=1}^{N} \alpha_i \left(\sum_{j=1}^{N_i} \chi_{I_i^j}(t) - \chi_{E_i}(t)\right)\right|^q dt\right)^{1/q}$$

$$= \left(\int_0^1 \left|\sum_{i=1}^{N} \alpha_i \left(\chi_{A_i}(t) - \chi_{E_i}(t)\right)\right|^q dt\right)^{1/q} = \left\|\sum_{i=1}^{N} \alpha_i(\chi_{A_i} - \chi_{E_i})\right\|_q$$

$$\leq \sum_{i=1}^{N} |\alpha_i| \, \|\chi_{E_i} - \chi_{A_i}\|_q = \sum_{i=1}^{N} |\alpha_i| \left(\int_0^1 |\chi_{E_i}(t) - \chi_{A_i}(t)|^q \, dt\right)^{1/q}$$

$$\leq \sum_{i=1}^{N} |\alpha_i| \frac{\varepsilon}{\left(\sum_{i=1}^{N} |\alpha_i| + 1\right)} < \varepsilon.$$

ii) a) Let $a \in \mathbb{K}, b \in [0,1]$, and let $f = a\chi_{(0,b)}$. Then $\|f\|_p = \left(\int_0^1 |f(x)|^p \, dx\right)^{1/p} = |a| \, b^{1/p}$ and for $x \in [0,1]$,

$$\int_0^x f(t)dt = \int_0^x a\chi_{(0,b)} dt = a\mu\left((0,x) \cap (0,b)\right) = \begin{cases} ax & \text{if } 0 < x < b, \\ ab & \text{if } x \geq b. \end{cases}$$

We have $\|f_n\| = |a_n| b_n^{1/p} \; \forall n \in \mathbb{N}$. Also $\int_0^x f_n(t)dt = \begin{cases} a_n x & \text{if } 0 < x < b_n, \\ a_n b_n & \text{if } x \geq b_n, \end{cases}$
$\forall x \in [0,1], \, \forall n \in \mathbb{N}$, from whence

$$\left|\int_0^x f_n(t)dt\right| = \begin{cases} |a_n| x & \text{if } 0 < x < b_n \\ |a_n| b_n & \text{if } x \geq b_n \end{cases} \leq |a_n| b_n \to 0.$$

Using the hypothesis and the part (i), the statement follows.

b) We use the same ideas as in the solutions for exercises 2(vi) and 4(iii). We omit the details.

13. For each $n \in \mathbb{N}$ we denote $A_n = \{t \in [0,1] \mid f_n(t) \neq 0\}$. Then A_n is μ-measurable for every $n \in \mathbb{N}$, and by hypothesis $A_n \cap A_m = \emptyset$ for $m \neq n$. If $1 < p < \infty$, using exercise 12(i) we know that it is sufficient to show that $\int_0^x f_n(t)dt \to 0 \; \forall x \in [0,1]$. We have

$$\left|\int_0^x f_n(t)dt\right| \leq \int_0^x |f_n(t)| \, dt = \int_0^1 |f_n(t)| \, |\chi_{[0,x]}(t)| \, dt = \int_{A_n} |f_n(t)| \, \chi_{[0,x]}(t) dt$$

$$\leq \left(\int_{A_n} |f_n(t)|^p \, dt\right)^{1/p} \left(\int_{A_n} |\chi_{[0,x]}(t)|^q \, dt\right)^{1/q}$$

$$= \left(\int_{A_n} \chi_{[0,x]}(t) dt\right)^{1/q} = (\mu(A_n \cap [0,x]))^{1/q} \leq (\mu(A_n))^{1/q},$$

where q is the conjugate of p. We shall prove that $\lim_{n\to\infty} \mu(A_n) = 0$. Suppose that the relation is not true, and we obtain $\varepsilon > 0$ and a subsequence $(n_k)_{k\in\mathbb{N}}$ of $\mathbb{N}$ such that $\mu(A_{n_k}) \geq$

$\varepsilon\ \forall k\in\mathbb{N}$. Denote by $A = \bigcup_{k\in\mathbb{N}} A_{n_k}$. Since the sets A_{n_k} are pairwise disjoint and μ is countably additive we obtain that $\mu(A) = \sum_{k=1}^{\infty} \mu(A_{n_k}) = \infty$, a contradiction!

Consider now the case $p = 1$. For $n \in \mathbb{N}$ consider the function $f_n : [0, 1] \to \mathbb{K}$,
$$f_n(t) = \begin{cases} n(n+1), & t \in (1/(n+1), 1/n], \\ 0, & t \notin (1/(n+1), 1/n]. \end{cases}$$ Then $f_n \in L_1[0, 1]$, $\|f_n\|_1 = 1\ \forall n \in \mathbb{N}$ and $A_n = (1/(n+1), 1/n]\ \forall n \in \mathbb{N}$, thus $A_n \cap A_m = \varnothing$ if $m \neq n$. Suppose that $f_n \to 0$ $weak$. Then for the linear and continuous functional $x^* : L_1[0, 1] \to \mathbb{K}$, $x^*(f) = \int_0^1 f(x)\,dx$ we must have $x^*(f_n) \to 0$. But $x^*(f_n) = \int_0^1 f_n(t)dt = 1\ \forall n \in \mathbb{N}$, and we have a contradiction.

14. See [44, exercise 29, chapter3].
First solution. Evidently $\delta_t \in C(T)^*\ \forall t \in T$. Let $t_0 \in T$ and consider the set
$$W(\delta_{t_0}; f_1, f_2, ..., f_n; \varepsilon) = \{x^* \in C(T)^* \mid |(x^* - \delta_{t_0})(f_i)| < \varepsilon,\ i = \overline{1, n}\} \subseteq C(T)^*.$$

If $x^* \in W(\delta_{t_0}; f_1, ..., f_n; \varepsilon)$ then $|x^*(f_i) - f_i(t_0)| < \varepsilon$, $i = \overline{1, n}$. Define $U = \bigcap_{i=1}^n U_i$, where $U_i = \{t \in T \mid |f_i(t) - f_i(t_0)| < \varepsilon\}$. Since f_i is continuous, $U_i \subseteq T$ is open $\forall 1 \leq i \leq n$, and then it follows that U is open. Since $t_0 \in U_i\ \forall i = \overline{1, n}$ we obtain that $t_0 \in \bigcap_{i=1}^n U_i = U$, i.e., U is an open neighborhood of t_0 in the topology of T. If we can prove that $\forall t \in U$ we have $\varphi(t) = \delta_t \in W(\delta_{t_0}; f_1, ...f_n; \varepsilon)$, the exercise will be solved. But if $t \in U$ then $|\delta_t(f_i) - f_i(t_0)| = |f_i(t) - f_i(t_0)| < \varepsilon$, $i = \overline{1, n}$ and then $\delta_t \in W(\delta_{t_0}; f_1, ...f_n; \varepsilon)$.

Second solution. In order to show that $\varphi : T \to C(T)^*$ is continuous from the topology of T to the $weak^*$ topology on $C(T)^*$, it is sufficient to prove that for each net $(t_\tau)_{\tau \in \Delta}$, $t_\tau \to t$ in T, we have $\varphi(t_\tau) \to \varphi(t)$ $weak^*$ in $C(T)^*$, that is, $\varphi(t_\tau)(f) \to \varphi(t)(f)\ \forall f \in C(T)$. This means that $f(t_\tau) \to f(t)$, and this is true because $t_\tau \to t$ in T and $f : T \to \mathbb{K}$ is continuous.

15. Since K_X is an isometry, $K_X : X \to X^{**}$ is injective, and therefore $K_X : X \to K_X(X)$ is bijective. Let $x^{**} \in X^{**}$, $x^{**} \in K_X(X)$. Then there is $x \in X$ such that $K_X(x) = x^{**}$, that is, $x^{**} = \widehat{x}$. Consider a $weak^*$ neighborhood of $\widehat{x}$ from the usual basis, $U = W(\widehat{x}; x_1^*, ..., x_n^*; \varepsilon)$. We take then $V = W(x; x_1^*, ..., x_n^*; \varepsilon)$, neighborhood of $x \in X$ in the $weak$ topology of X. For every $y \in V$, $|x_i^*(x-y)| < \varepsilon$, $i = \overline{1, n}$, thus $|(\widehat{x} - \widehat{y})(x_i^*)| < \varepsilon$, $i = \overline{1, n}$, that is, $K_X(y) \in U$, and therefore K_X is $weak$-to-$weak^*$ continuous. In order to prove that $K_X^{-1} : K_X(X) \to X$ is $weak^*$-to-$weak$ continuous we consider a net $(\widehat{x}_\delta)_{\delta \in \Delta}$ in $K_X(X)$, with $\widehat{x}_\delta \to \widehat{x}$ $weak^*$, $x_\delta \in X\ \forall \delta \in \Delta$, $x \in X$. Then $\widehat{x}_\delta(x^*) \to \widehat{x}(x^*)\ \forall x^* \in X^*$, that is, $x^*(x_\delta) \to x^*(x)\ \forall x^* \in X^*$, i.e., $x_\delta \to x$ $weak$. We obtain that $K_X^{-1}(\widehat{x}_\delta) \to K_X^{-1}(\widehat{x})$ $weak$.

16. i) See [18, Lemma, p. 10].
Induction on $n \in \mathbb{N}$. Let $n = 1$ and let $f, g : X \to \mathbb{K}$ be linear functionals such that $\ker g \subseteq \ker f$. If $g = 0$ then $\ker g = X$, $\ker f = X$, and therefore $f = 0$, i.e., $f = \alpha g\ \forall \alpha \in \mathbb{K}$. Suppose now that there is $x_0 \in X$ such that $g(x_0) \neq 0$. For every $x \in X$, $x - x_0 g(x)/g(x_0) \in \ker g$ and then by hypothesis $x - x_0 g(x)/g(x_0) \in \ker f$. We obtain that

14.2 Solutions

$f(x - x_0 g(x)/g(x_0)) = 0$, and therefore $f(x) = \alpha g(x) \ \forall x \in X$, where $\alpha := f(x_0)/g(x_0)$ does not depend on x.

Suppose the assertion true for $n \in \mathbb{N}$ and suppose that $\bigcap_{i=1}^{n+1} \ker f_i \subseteq \ker f$. Then

$$\bigcap_{i=1}^{n} \ker \left(f_i \big|_{\ker f_{n+1}} \right) \subseteq \ker \left(f \big|_{\ker f_{n+1}} \right).$$

Applying the hypothesis of induction for $f \big|_{\ker f_{n+1}}$, $f_1 \big|_{\ker f_{n+1}}, \ldots, f_n \big|_{\ker f_{n+1}}$ we obtain that, on $\ker f_{n+1}$, f is a linear combination of $f_1, f_2, \ldots, f_n$, i.e., $f = \sum_{i=1}^{n} \alpha_i f_i$ on $\ker f_{n+1}$, i.e.,

$$\ker f_{n+1} \subseteq \ker \left(f - \sum_{i=1}^{n} \alpha_i f_i \right).$$

We apply what we have already showed and we obtain that $f = \sum_{i=1}^{n+1} \alpha_i f_i$.

ii) Suppose that $\bigcap_{i=1}^{n} \ker f_i = \{0\}$. Then for every linear functional $f : X \to \mathbb{K}$ we have

$$\{0\} = \bigcap_{i=1}^{n} \ker f_i \subseteq \ker f.$$

Using (i) we obtain that $f \in \operatorname{Sp}\{f_1, \ldots, f_n\}$, i.e., the dimension for the algebraic dual of X is finite, therefore $\dim_{\mathbb{K}} X < \infty$, a contradiction!

17. i) Obviously, if f is continuous it is *weak* continuous.

Suppose now that f is *weak* continuous. Then $V = \{x \in X \mid |f(x)| < 1\}$ is a neighborhood of $0 \in X$ in the *weak* topology. Therefore there are $x_1^*, \ldots, x_n^* \in X^*$, $\varepsilon > 0$ such that $W(0; x_1^*, \ldots, x_n^*; \varepsilon) \subseteq V$, i.e., $|x_i^*(x)| < \varepsilon \ \forall i = \overline{1,n}$ implies $|f(x)| < 1$. If $x \in \bigcap_{i=1}^{n} \ker x_i^*$, then $tx \in W(0; x_1^*, \ldots, x_n^*; \varepsilon) \ \forall t \in \mathbb{K}$ and therefore $|f(tx)| < 1, \ \forall t \in \mathbb{K}$, that is, $|f(x)| < 1/|t| \ \forall t \in \mathbb{K}\setminus\{0\}$, and for $|t| \to \infty$, $|f(x)| \leq 0$, i.e., $f(x) = 0$. We obtain that $\bigcap_{i=1}^{n} \ker x_i \subseteq \ker f$ and using exercise 16 we deduce that f is a linear combination of $x_1^*, \ldots, x_n^*$, that is, $f \in X^*$.

ii) If $f : \Omega \to X$ is *weak* continuous and $x^* \in X^*$, since x^* is *weak* continuous we obtain that the composition $x^* \circ f : \Omega \to \mathbb{K}$ is continuous.

Let now $f : \Omega \to X$ such that $x^* \circ f : \Omega \to \mathbb{K}$ is continuous $\forall x^* \in X^*$. We want to prove that $f : \Omega \to X$ is *weak* continuous and for that consider $t_0 \in \Omega$ and

$$W(f(t_0); x_1^*, \ldots, x_n^*; \varepsilon) = \{x \in X \mid |x_i^*(x - f(t_0))| < \varepsilon, \ i = \overline{1,n}\}.$$

We shall prove that

$$f^{-1}(W(f(t_0); x_1^*, \ldots, x_n^*; \varepsilon)) = \bigcap_{i=1}^{n} (x_i^* \circ f)^{-1}(D(x_i^*(f(t_0)), \varepsilon)),$$

and then $f^{-1}(W(f(t_0); x_1^*, ..., x_n^*; \varepsilon))$ is open in the topology of Ω, being an intersection of open sets, as the function $x^* \circ f : \Omega \to \mathbb{K}$ is continuous for every $x^* \in X^*$.

Let $t \in f^{-1}(W(f(t_0); x_1^*, ..., x_n^*; \varepsilon))$, i.e., $f(t) \in W(f(t_0); x_1^*, ...x_n^*; \varepsilon)$. Then $|x_i^*(f(t) - f(t_0))| < \varepsilon \; \forall 1 \leq i \leq n$, that is, $|x_i^*(f(t)) - x_i^*(f(t_0))| < \varepsilon \; \forall 1 \leq i \leq n$, i.e., equivalently, $t \in \bigcap_{i=1}^{n}(x_i^* \circ f)^{-1}(D(x_i^*(f(t_0)), \varepsilon))$ ($D(z, \varepsilon)$ is the open disc of center z and radius ε in the complex plane).

Remark. One can solve the same type of exercise working with the *weak** topology on X^* instead of the *weak* topology on X.

18. Suppose that $T : X \to Y$ is continuous if we consider on X and Y the norm topologies. Using exercise 17 we know that T is *weak*-to-*weak* continuous if and only if for every $y^* \in Y^*$, $y^* \circ T : X \to \mathbb{K}$ is a *weak* continuous functional. By the same exercise, this is equivalent to $y^* \circ T : X \to \mathbb{K}$ being a continuous functional (with the norm topology on X), which is true by our hypothesis.

Suppose now that $T : X \to Y$ is *weak*-to-*weak* continuous. Consider $K_Y : Y \to Y^{**}$ and the family of linear and continuous functionals $T_x : Y^* \to \mathbb{K}$, $x \in B_X$, where $T_x = K_Y(T(x)) = \widehat{T(x)}$. Observe that $\|T_x\| = \|T(x)\| \; \forall x \in B_X$. We will prove that the family $(T_x)_{x \in B_X}$ is pointwise bounded. Let $y^* \in Y^*$. Using our hypothesis and exercise 17 we obtain that $y^* \circ T : X \to \mathbb{K}$ is a continuous functional, and we have

$$|(y^* \circ T)(x)| \leq \|y^* \circ T\| \|x\| \leq \|y^* \circ T\| \; \forall x \in B_X.$$

Now by the definition of T_x we have $T_x(y^*) = \widehat{T(x)}(y^*) = (y^* \circ T)(x)$ from whence $\sup_{x \in B_X} |T_x(y^*)| \leq \|y^* \circ T\| < \infty$. From the Uniform Boundedness Principle (Y^* is a Banach space) it follows that $\sup_{x \in B_X} \|T_x\| < \infty$, that is, $\sup_{\|x\| \leq 1} \|T(x)\| < \infty$, i.e., $T(B_X) \subseteq Y$ is bounded and therefore $T : X \to Y$ is norm-to-norm continuous.

19. See [44, exercise 6, chapter 4].

We prove that if $T : X \to Y$ is a linear bounded operator, then $T^* : Y^* \to X^*$ is *weak**-to-*weak** continuous. Using exercise 17 it is sufficient to show that $\widehat{x} \circ T^* : Y^* \to \mathbb{K}$ is *weak** continuous. By the same exercise it is sufficient to prove that $\widehat{x} \circ T^*$ is continuous if we consider the norm topology on Y^*, and this is obvious.

Now let $S : Y^* \to X^*$ be linear and *weak**-to-*weak** continuous. Then for each $x \in X$ the function $\widehat{x} \circ S : Y^* \to \mathbb{K}$ is *weak** continuous. Using the form for the *weak** continuous linear functionals, we obtain that there is a unique $y_x \in Y$ such that $\widehat{x} \circ S = \widehat{y_x}$. Define $T : X \to Y$, $T(x) = y_x$. S being linear implies that T is linear. Indeed, let $x_1, x_2 \in X$, $\alpha, \beta \in \mathbb{K}$, $x = \alpha x_1 + \beta x_2$. There are $y_1, y_2 \in Y$ such that $\widehat{x_1} \circ S = \widehat{y_1}$, $\widehat{x_2} \circ S = \widehat{y_2}$, and an element $y \in Y$ such that $\widehat{x} \circ S = \widehat{y}$. Then $(\widehat{\alpha x_1 + \beta x_2}) \circ S = \alpha(\widehat{x_1} \circ S) + \beta(\widehat{x_2} \circ S) = \alpha \widehat{y_1} + \beta \widehat{y_2}$, therefore $\widehat{x} \circ S = \widehat{\alpha y_1 + \beta y_2}$, and then $y = \alpha y_1 + \beta y_2$, that is, $T(\alpha x_1 + \beta x_2) = \alpha T(x_1) + \beta T(x_2)$, T is linear. We have $(S(y^*))(x) = y^*(T(x)) \; \forall x \in X, \forall y^* \in Y$. Since $B_{Y^*} \subseteq Y^*$ is a *weak** compact set we obtain that $S(B_{Y^*}) \subseteq X^*$ is a *weak** compact set and therefore bounded in norm: there is $M > 0$ such that $\|S(y^*)\| \leq M \; \forall y^* \in B_{Y^*}$. Then for any $x \in X$ we have

$$\|Tx\| = \sup_{y^* \in B_{Y^*}} |y^*(Tx)| = \sup_{y^* \in B_{Y^*}} |(S(y^*))(x)| \leq M \|x\|,$$

14.2 Solutions

and therefore $T \in L(X, Y)$. The operator $T^* \in L(Y^*, X^*)$ is defined by

$$T^* y^* = x^* \Leftrightarrow x^*(x) = y^*(Tx) \quad \forall x \in X.$$

Using that $(S(y^*))(x) = y^*(Tx) \; \forall x \in X, \forall y^* \in Y^*$ we obtain that $S = T^*$.

20. For $x_1^*, ..., x_n^* \in X^*$ and $\varepsilon > 0$, let $y_k^* = x_k^* \circ i$, $k = 1, ..., n$, where $i : Y \to X$ is the inclusion. We obtain that $y_k^* \in Y^*$. We observe that

$$\begin{aligned} W(0; y_1^*, ..., y_n^*; \varepsilon) &= \{y \in Y \mid |y_k^*(y)| < \varepsilon, \; k = \overline{1, n}\} \\ &= \{y \in Y \mid |x_k^*(y)| < \varepsilon, \; k = \overline{1, n}\} \\ &= Y \cap \{x \in X \mid |x_k^*(x)| < \varepsilon, \; k = \overline{1, n}\} \\ &= Y \cap W(0; x_1^*, ..., x_n^*; \varepsilon). \end{aligned}$$

Let now $y_1^*, ..., y_n^* \in Y^*$ and $\varepsilon > 0$, and let $(x_k^*)_{1 \le k \le n} \subseteq X^*$ be, respectively, extensions for $(y_k^*)_{1 \le k \le n}$ to the whole space X given by the Hahn–Banach theorem. As above we obtain that $\widetilde{W}(0; y_1^*, ..., y_n^*; \varepsilon) = Y \cap W(0; x_1^*, ..., x_n^*; \varepsilon)$. Therefore the two *weak* topologies coincide on Y.

21. See [9, exercise 8 (a), chapter IV, section 1].

We will prove the following more general result. Let X, Y be a dual pair of real linear spaces and $\langle \cdot, \cdot \rangle$ the duality bracket. Suppose that $A \subseteq X$ is a convex *weak* compact set and that $C \subseteq Y$ is a *weak* closed convex cone with the property that $\forall y \in C \; \exists x \in A$ such that $\langle x, y \rangle \le 0$. Then there exists $x_0 \in A$ such that $\langle x_0, y \rangle \le 0 \; \forall y \in C$. Applying this general result to the pair (X, X^*) we obtain (i), and for the pair (X^*, X) we obtain (ii).

Recall that if X, Y is a pair of real linear spaces with the duality bracket $\langle \cdot, \cdot \rangle$ and $P \subseteq X$ is a convex cone, then for the polar of P we have that $P^0 = \{y \in Y \mid \langle x, y \rangle \le 0 \; \forall x \in P\}$. Indeed, let $y \in P^0$. Then by the definition for the polar of a set, $\langle x, y \rangle \le 1 \; \forall x \in P$. Let now $x \in P$ and $\lambda > 0$. Since P is a cone, $\lambda x \in P$, from whence $\langle \lambda x, y \rangle \le 1$, or $\langle x, y \rangle \le 1/\lambda \; \forall \lambda > 0$, hence for $\lambda \to \infty$, $\langle x, y \rangle \le 0$. Conversely is trivial since if $\langle x, y \rangle \le 0 \; \forall x \in P$ then $\langle x, y \rangle \le 1 \; \forall x \in P$, i.e., $y \in P^0$.

Let us suppose, for a contradiction, that the conclusion from the statement is not true. Then $\forall x \in A \; \exists y \in C$ such that $\langle x, y \rangle > 0$. As $C^0 = \{x \in X \mid \langle x, y \rangle \le 0 \; \forall y \in C\}$ then $A \cap C^0 = \emptyset$. But the polar for a set is *weak* closed, convex, and contains the origin 0. Using a separation theorem, we can find $f : X \to \mathbb{R}$ *weak* continuous and $t \in \mathbb{R}$ such that $f(a) > t > f(b) \; \forall a \in A, \forall b \in C^0$. Since f is *weak* continuous there exists $y \in Y$ such that $f(x) = \langle x, y \rangle \; \forall x \in X$, so $\langle a, y \rangle > t > \langle b, y \rangle \; \forall a \in A, \forall b \in C^0$. Since $0 \in C^0$, $t > \langle 0, y \rangle = 0$, i.e., $t > 0$. Then $\langle b, y/t \rangle < 1 \; \forall b \in C^0$, i.e., $y/t \in C^{00}$. Now by the bipolar theorem, $C^{00} = \overline{\mathrm{co}}^{\mathrm{weak}}(C \cup \{0\})$. Since $0 \in C$ and C is convex and *weak* closed we have $\overline{\mathrm{co}}^{\mathrm{weak}}(C) = C$, i.e., $C^{00} = C$, from whence $y/t \in C$. But by hypothesis there exists $x \in A$ such that $\langle x, y/t \rangle \le 0$, that is, $\langle x, y \rangle \le 0$. We also have $\langle a, y \rangle > t \; \forall a \in A$ and since $x \in A$, $\langle x, y \rangle > t > 0$, i.e., $\langle x, y \rangle > 0$, a contradiction!

22. Since every *weak* neighborhood for an element $x_0 \in X$, of the form

$$W(x_0; x_1^*, ..., x_n^*; \varepsilon) = \{x \in X \mid |x_i^*(x - x_0)| < \varepsilon, \; i = \overline{1, n}\},$$

is open in the norm topology, and the sets of the form $W(x_0; x_1^*, ..., x_n^*; \varepsilon)$ with $n \in \mathbb{N}$, $x_1^*, ..., x_n^* \in X^*$, $\varepsilon > 0$, form a basis of neighborhoods for $x_0 \in X$ in the *weak* topology, to show that the two topologies coincide it is sufficient to show that for every open neighborhood U of x_0 in the norm topology there is a neighborhood V of x_0 in the *weak* topology of X such that $V \subseteq U$. Since U is an open neighborhood of x_0 in the norm topology there is an $r > 0$ such that $B(x_0, r) \subseteq U$. Let $\{x_1, ..., x_n\} \subseteq X$ be a basis of X. We can suppose that $\|x_i\| = 1$ $\forall 1 \leq i \leq n$. For every $1 \leq i \leq n$, we define $e_i^* : X \to \mathbb{K}$, $e_i^*(x) = \alpha_i$ $\forall x = \sum_{j=1}^{n} \alpha_j x_j$. Since $\{x_1, ..., x_n\} \subseteq X$ is a basis, e_i^* is well defined and linear for every $1 \leq i \leq n$. Since X is finite-dimensional, $e_i^* \in X^*$, $i = \overline{1, n}$. Let us consider $V = W(x_0; e_1^*, ..., e_n^*; r/n)$. For every $x \in V$, $|e_i^*(x - x_0)| < r/n$, and therefore

$$\|x - x_0\| = \left\| \sum_{i=1}^{n} e_i^*(x - x_0) x_i \right\| \leq \sum_{i=1}^{n} |e_i^*(x - x_0)| \|x_i\| < \sum_{i=1}^{n} \frac{r}{n} = r,$$

i.e., $V \subseteq B(x_0, r) \subseteq U$.

23. See [18, chapter II].

Suppose that the *weak* topology on X is metrizable: there exists a metric d on X which generates the *weak* topology on X. Then $\{B(0, \varepsilon) \mid \varepsilon \in \mathbb{Q}_+ \setminus \{0\}\}$, the balls of center 0 and radius $\varepsilon > 0$ with respect to d form a basis of neighborhoods for $0 \in X$ with respect to the *weak* topology. For each $\varepsilon \in \mathbb{Q}_+ \setminus \{0\}$ we can find $x_{1,\varepsilon}^*, ..., x_{n(\varepsilon),\varepsilon}^* \in X^* \setminus \{0\}$ and $\delta_\varepsilon > 0$ such that $W(0; x_{1,\varepsilon}^*, ..., x_{n(\varepsilon),\varepsilon}^*; \delta_\varepsilon) \subseteq B(0, \varepsilon)$. The set $\{x_{1,\varepsilon}^*, ..., x_{n(\varepsilon),\varepsilon}^*\}_{\varepsilon \in \mathbb{Q}_+ \setminus \{0\}}$ is countable and we will consider it as a sequence $(x_n^*)_{n \in \mathbb{N}} \subseteq X^*$. For every $x^* \in X^*$ and $\delta > 0$, the set $W(0; x^*; \delta) = \{x \in X \mid |x^*(x)| < \delta\}$ is a *weak* neighborhood of 0 and therefore there is an $\varepsilon \in \mathbb{Q}_+ \setminus \{0\}$ such that $B(0; \varepsilon) \subseteq W(0; x^*; \delta)$. We obtain that we can find $\{x_{k_1}^*, ..., x_{k_n}^*\} \subseteq (x_j^*)_{j \in \mathbb{N}}$ such that $W(0; x_{k_1}^*, ..., x_{k_n}^*; \delta_\varepsilon) \subseteq B(0, \varepsilon) \subseteq W(0; x^*; \delta)$. We will show that $\bigcap_{i=1}^{n} \ker x_{k_i}^* \subseteq \ker x^*$. Let $x \in \bigcap_{i=1}^{n} \ker x_{k_i}^*$. Then $x_{k_i}^*(\alpha x) = 0$, $i = 1, ..., n$, $\forall \alpha \in \mathbb{N}$, and therefore $\alpha x \in W(0; x_{k_1}^*, ..., x_{k_n}^*; \delta_\varepsilon)$ $\forall \alpha \in \mathbb{N}$. We obtain that $\alpha x \in W(0; x^*; \delta)$ $\forall \alpha \in \mathbb{N}$ and therefore $|x^*(x)| < \delta/\alpha$ $\forall \alpha \in \mathbb{N}$, so passing to the limit for $\alpha \to \infty$ we deduce that $x^*(x) = 0$, i.e., $x \in \ker x^*$. Using exercise 16(i) we obtain that $x^* \in \text{Sp}\{x_{k_1}^*, ..., x_{k_n}^*\}$. If we denote by $X_n = \text{Sp}\{x_1^*, ..., x_n^*\}$ $\forall n \in \mathbb{N}$, we have proved that $X^* = \bigcup_{n \in \mathbb{N}} X_n$. For every $n \in \mathbb{N}$ the subspace X_n is closed. Since X^* is a Banach space, applying Baire's theorem we obtain that we can find $n_0 \in \mathbb{N}$, $r > 0$, and $x_0^* \in X_{n_0}$ such that $B(x_0^*, r) \subseteq X_{n_0}$. Using exercise 15 of chapter 1 we obtain that $X^* = X_{n_0}$, i.e., $\dim_\mathbb{K} X^* < \infty$, and therefore $\dim_\mathbb{K} X < \infty$.

Remark. One can solve the same type of exercise working with the *weak** topology on X^* instead of the *weak* topology on X.

24. See [14, exercise 10, chapter V, section 1].

Since B_X is convex and closed in the norm topology and $S_X \subseteq B_X$ we obtain that $\overline{S_X}^{\text{weak}} \subseteq \overline{B_X}^{\text{weak}} = B_X$. We will show now that $B_X \subseteq \overline{S_X}^{\text{weak}}$. It is sufficient to show that for every $x_0 \in X$, with $\|x_0\| < 1$, we have $x_0 \in \overline{S_X}^{\text{weak}}$.

Let $\varepsilon > 0$, $n \in \mathbb{N}$, and $x_1^*, ..., x_n^* \in X^*$. It is sufficient to show that $V \cap S_X \neq \emptyset$, where $V = W(x_0; x_1^*, ..., x_n^*; \varepsilon)$. Using exercise 16(ii) we obtain that there is $z \in X \setminus \{0\}$ such

14.2 Solutions

that $x_i^*(z) = 0 \ \forall 1 \leq i \leq n$. We define $\varphi : [0, \infty) \to \mathbb{R}$, $\varphi(t) = \|x_0 + tz\| \ \forall t \geq 0$. Then φ is continuous, $\varphi(0) = \|x_0\| < 1$, and $\lim_{t \to \infty} \varphi(t) = \infty$ since $z \neq 0$ ($\|x_0 + tz\| \geq t\|z\| - \|x_0\| \to \infty$ for $t \to \infty$). From the Darboux property we obtain that there is $t_0 \geq 0$ such that $\|x_0 + t_0 z\| = 1$. Let $w = x_0 + t_0 z$. Then $\|w\| = 1$, and $x_i^*(w) - x_i^*(x_0) = t_0 x_i(z) = 0$ $\forall i = \overline{1, n}$ and therefore $w \in V$. Then $w \in S_X \cap V$, and therefore $S_X \cap V \neq \varnothing$.

25. See [42, exercise 40, chapter IV] or [46, exercise 22, chapter II].

i) Suppose that we can find x_0, an interior point of $B(0, 1)$ with respect to the *weak* topology. Then we can find $\varepsilon > 0$, $n \in \mathbb{N}$, and $x_1^*, ..., x_n^* \in X^*$ such that $x_0 \in W(x_0; x_1^*, ..., x_n^*; \varepsilon) \subseteq B(0, 1)$. Using exercise 16(ii) we obtain that there is an $y_0 \in X \setminus \{0\}$ such that $x_i^*(y_0) = 0 \ \forall i = \overline{1, n}$. Then, for every $t \in \mathbb{R}$, $x_0 + t y_0 \in W(x_0; x_1^*, ..., x_n^*; \varepsilon) \subseteq B(0, 1)$, i.e., $\|x_0 + t y_0\| < 1 \ \forall t \in \mathbb{R}$. But $y_0 \neq 0$, and therefore $\|x_0 + t y_0\| \to \infty$ for $t \to \infty$, a contradiction.

In the same way one can show that the interior for B_X with respect to the *weak* topology is $\varnothing$.

ii) For any $x \in X$ there is $n \in \mathbb{N}$ such that $\|x\| \leq n$, i.e., $x \in n B_X$. Thus $X = \bigcup_{n \in \mathbb{N}} (n B_X)$. We have $\overline{B_X}^{\text{weak}} = \overline{B_X}^{\|\cdot\|} = B_X$. Since by (i) the interior for B_X with respect to the *weak* topology is $\varnothing$ it follows that B_X is a nowhere dense set with respect to the *weak* topology and hence $(X, weak)$ is of the first Baire category.

26. i) See [20, exercise 38, chapter V, section 7], the von Neumann example.

Since the sets of the form $W(0; x_1^*, ..., x_n^*; \varepsilon) = \{x \in l_2 \mid |x_i^*(x)| < \varepsilon, i = \overline{1, n}\}$ with $n \in \mathbb{N}$, $\varepsilon > 0$, $x_1^*, ..., x_n^* \in l_2^*$ form a basis of neighborhoods for the origin, in order to prove that $0 \in \overline{A}^{\text{weak}}$ it is sufficient to prove that $A \cap W(0; x_1^*, ..., x_n^*; \varepsilon) \neq \varnothing$ for any $n \in \mathbb{N}$, $x_1^*, ..., x_n^* \in l_2^*$, $\varepsilon > 0$. Let $n \in \mathbb{N}$, $\varepsilon > 0$ and $x_1^*, ..., x_n^* \in l_2^*$. Using Riesz's representation theorem we can find $(\lambda_k^i)_{k \in \mathbb{N}} \in l_2$, $i = \overline{1, n}$, such that $x_i^*(x) = \langle (x_k)_{k \in \mathbb{N}}, (\lambda_k^i)_{k \in \mathbb{N}} \rangle_{l_2}$ $\forall x = (x_k)_{k \in \mathbb{N}} \in l_2$, for $1 \leq i \leq n$. Since $\lim_{k \to \infty} \lambda_k^i = 0$, $i = \overline{1, n}$, there is $s_\varepsilon \in \mathbb{N}$ such that $|\lambda_{s_\varepsilon}^i| < \varepsilon/2$ for $1 \leq i \leq n$. Since $\lim_{k \to \infty} s_\varepsilon \lambda_k^i = 0$, $i = \overline{1, n}$, there is $t_\varepsilon \in \mathbb{N}$, $t_\varepsilon > s_\varepsilon$, such that $|s_\varepsilon \lambda_{t_\varepsilon}^i| < \varepsilon/2$ for $1 \leq i \leq n$. Then

$$|x_i^*(e_{s_\varepsilon} + s_\varepsilon e_{t_\varepsilon})| = |\lambda_{s_\varepsilon}^i + s_\varepsilon \lambda_{t_\varepsilon}^i| \leq |\lambda_{s_\varepsilon}^i| + |s_\varepsilon \lambda_{t_\varepsilon}^i| < \varepsilon, \ i = \overline{1, n},$$

that is, $e_{s_\varepsilon} + s_\varepsilon e_{t_\varepsilon} \in A \cap W(0; x_1^*, ..., x_n^*; \varepsilon)$.

Suppose now that there is a sequence $(x_k)_{k \in \mathbb{N}} \subseteq A$ such that $x_k \to 0$ *weak*. Let $x_k = e_{m_k} + m_k e_{n_k}$, $1 \leq m_k < n_k < \infty \ \forall k \in \mathbb{N}$. Since $x_k \to 0$ *weak*, the sequence $(x_k)_{k \in \mathbb{N}}$ is bounded, that is, the sequence $(m_k)_{k \in \mathbb{N}}$ is bounded. Using Cesaro's lemma we can find a convergent subsequence $(m_{k_s})_{s \in \mathbb{N}}$. Therefore there is an $s_0 \in \mathbb{N}$ and an $n \in \mathbb{N}$ such that $m_{k_s} = n \ \forall s \geq s_0$ (we have $m_{k_s} \in \mathbb{Z} \ \forall s \in \mathbb{N}$). Then

$$\langle e_n, x_{k_s} \rangle = \langle e_n, e_{m_{k_s}} + m_{k_s} e_{n_{k_s}} \rangle = \langle e_n, e_n \rangle + n \langle e_n, e_{n_{k_s}} \rangle = 1 \ \forall s \geq s_0,$$

a contradiction, since $\langle e_n, x_{k_s} \rangle \to 0$ for $s \to \infty$.

ii) See [50, exercise 8.4.120].

Let $W(0; x_1^*, ..., x_n^*; \varepsilon) = \{x \in c_0 \mid |x_k^*(x)| < \varepsilon, \ k = 1, ..., n\}$ be a *weak* neighborhood for $0 \in c_0$. We have $x_k^* \in c_0^*$, $k = 1, ..., n$, and therefore we can find

$\left(\lambda_j^{(1)}\right)_{j\in\mathbb{N}},...,\left(\lambda_j^{(n)}\right)_{j\in\mathbb{N}} \in l_1$ such that $x_k^*\left((x_j)_{j\in\mathbb{N}}\right) = \sum_{j=1}^{\infty} \lambda_j^{(k)} x_j \ \forall (x_j)_{j\in\mathbb{N}} \in c_0$, $k = 1,...,n$. Let $a_j = |\lambda_j^{(1)}| + \cdots + |\lambda_j^{(n)}| \ \forall j \in \mathbb{N}$. Then $a_j \geq 0 \ \forall j \in \mathbb{N}$ and $\sum_{j=1}^{\infty} a_j < \infty$. Then we can find $j \in \mathbb{N}$ such that $ja_j < \varepsilon$. Indeed, suppose that we cannot find such j. Then $a_j \geq \varepsilon/j \ \forall j \in \mathbb{N}$, and then $\sum_{j=1}^{\infty} a_j \geq \sum_{j=1}^{\infty} \varepsilon/j = \infty$, a contradiction. If $ja_j < \varepsilon$ then $|x_k^*(je_j)| = |j\lambda_j^{(k)}| \leq ja_j < \varepsilon$, $k = 1,...,n$, and therefore $je_j \in W(0; x_1^*,...,x_n^*; \varepsilon)$. This implies that $S \cap W(0; x_1^*,...,x_n^*; \varepsilon) \neq \emptyset$, and then $0 \in \overline{S}^{\text{weak}}$.

Suppose now that we can find a subsequence $(n_k)_{k\in\mathbb{N}}$ of $\mathbb{N}$ such that $n_k e_{n_k} \to 0$ weak in c_0. Choosing, if necessary, a subsequence for $(n_k)_{k\in\mathbb{N}}$, without loss of generality we can suppose that $\sum_{k=1}^{\infty} 1/n_k < \infty$ (for example, we can choose a subsequence $\left(n_{k_p}\right)_{p\in\mathbb{N}}$ of $(n_k)_{k\in\mathbb{N}}$ such that $n_{k_p} \geq 2^p \ \forall p \in \mathbb{N}$ and then $n_{k_p} e_{n_{k_p}} \to 0$ weak in c_0 and $\sum_{p=1}^{\infty} 1/n_{k_p} \leq \sum_{p=1}^{\infty} 1/2^p < \infty$). Then consider the sequence $(\lambda_n)_{n\in\mathbb{N}} \in l_1$, $\lambda_n = 0$ if $n \notin \{n_k \mid k \in \mathbb{N}\}$, $\lambda_n = 1/n_k$ if $n = n_k$, $k \in \mathbb{N}$. $(\lambda_n)_{n\in\mathbb{N}}$ gives an element $x^* \in c_0^*$, and therefore $x^*(n_k e_{n_k}) \to 0$. But $x^*(n_k e_{n_k}) = 1 \ \forall k \in \mathbb{N}$, and we obtain a contradiction.

27. See [18, exercise 3 (ii), chapter II].

i) If X^* were finite-dimensional X^{**} would be finite-dimensional, and we obtain that X is finite-dimensional, since $X \hookrightarrow X^{**}$. Contradiction! Therefore X^* is infinite-dimensional and, using exercise 2 of chapter 3, we obtain that we can find $\varphi : X^* \to \mathbb{K}$ linear which is not continuous.

ii) On the set of all finite-dimensional subspaces of X^*, denoted by $\text{Fin}(X^*)$, we define the following relation: $Y \leq Z \Leftrightarrow Y \subseteq Z$. We obtain that $(\text{Fin}(X^*), \leq)$ is a directed set, since for $Y, Z \in \text{Fin}(X^*)$, if $W = \text{Sp}(Y \cup Z)$, then $W \in \text{Fin}(X^*)$ and $Y \leq W, Z \leq W$. Let now $Y \in \text{Fin}(X^*)$. Since $\varphi : X^* \to \mathbb{K}$ is linear, $\varphi \mid_Y : Y \to \mathbb{K}$ is linear. Since Y is finite-dimensional, $\varphi \mid_Y : Y \to \mathbb{K}$ is linear and continuous. Using the Hahn–Banach theorem, we obtain $\widetilde{\varphi \mid_Y} : X^* \to \mathbb{K}$ linear and continuous such that

$$\widetilde{\varphi \mid_Y}(y) = \varphi(y) \ \forall y \in Y, \ \left\|\widetilde{\varphi \mid_Y}\right\|_{X^{**}} = \|\varphi \mid_Y\|.$$

Using Helly's lemma we obtain that $\forall \varepsilon > 0 \ \forall Y \in \text{Fin}(X^*)$ there is an $x_Y \in X$ such that

$$\|x_Y\| \leq \left\|\widetilde{\varphi \mid_Y}\right\| + \varepsilon, \ y^*(x_Y) = (\widetilde{\varphi \mid_Y})(y^*) \ \forall y^* \in Y.$$

In this way we obtain a net $(x_Y)_{Y\in\text{Fin}(X^*)} \subseteq X$ such that

$$y^*(x_Y) = \varphi(y^*) \ \forall y^* \in Y, \ \forall Y \in \text{Fin}(X^*).$$

We show that $(x_Y)_{Y\in\text{Fin}(X^*)}$ is a *weak* Cauchy net. Consider $W(0; x_1^*,...,x_n^*; \varepsilon) = \{x \in X \mid |x_i^*(x)| < \varepsilon, i = \overline{1,n}\}$. Denote by $Z = \text{Sp}\{x_1^*,...,x_n^*\}$, $Z \in \text{Fin}(X^*)$. Then, for $Y_1, Y_2 \in \text{Fin}(X^*)$, $Y_1, Y_2 \geq Z$, we have $x_{Y_1} - x_{Y_2} \in W(0; x_1^*,...,x_n^*; \varepsilon)$. Indeed, if $Y_1 \geq Z$, $Y_2 \geq Z$ then $Z \subseteq Y_1$, $Z \subseteq Y_2$, and therefore $x^*(x_{Y_1}) = \varphi(x^*) = x^*(x_{Y_2}) \ \forall x^* \in Z$, that is, $x^*(x_{Y_1} - x_{Y_2}) = 0 \ \forall x^* \in Z$, and then $x_{Y_1} - x_{Y_2} \in W(0; x_1^*,...,x_n^*; \varepsilon)$.

14.2 Solutions

Suppose that there is an $x \in X$ such that $(x_Y)_{Y \in \text{Fin}(X^*)} \to x$ $weak$. Then $(x^*(x_Y))_{Y \in \text{Fin}(X^*)} \to x^*(x)$ $\forall x^* \in X^*$. Let now $x^* \in X^*$ be fixed. We consider the set $\Delta = \{Y \in \text{Fin}(X^*) \mid x^* \in Y\}$. Then $(\Delta, \leq)$ is a directed set and considering the inclusion $i : \Delta \hookrightarrow \text{Fin}(X^*)$ we have: $\forall Y_0 \in \text{Fin}(X^*)$ $\exists Z_0 = \text{Sp}(Y_0 \cup \{x^*\}) \in \Delta$ such that if $Y \geq Z_0$ then $i(Y) \geq Y_0$. Therefore $(x_{i(Z)})_{Z \in \Delta}$ is a subnet of $(x_Y)_{Y \in \text{Fin}(X^*)}$ and then $(x_{i(Z)})_{Z \in \Delta} \to x$ $weak$. We obtain that $(x^*(x_{i(Z)}))_{Z \in \Delta} \to x^*(x)$ and since $x^* \in Z$ $\forall Z \in \Delta$ we obtain that $x^*(x_{i(Z)}) = \varphi(x^*)$ $\forall Z \in \Delta$, and then $\varphi(x^*) = x^*(x)$. Therefore if $(x_Y)_{Y \in \text{Fin}(X^*)}$ is $weak$ convergent towards x, we have $x^*(x) = \varphi(x^*)$ $\forall x^* \in X^*$.

iii) If, for a contradiction, the $weak$ topology is complete, then since the net $(x_Y)_{Y \in \text{Fin}(X^*)}$ is $weak$ Cauchy there is $x \in X$ such that $(x_Y)_{Y \in \text{Fin}(X^*)} \to x$ $weak$. By (ii) we have $\varphi(x^*) = x^*(x)$ $\forall x^* \in X^*$. Since $x \in X$, the function $x^* \to x^*(x)$ from X^* to $\mathbb{K}$ is continuous, i.e., $\varphi : X^* \to \mathbb{K}$ is continuous, which contradicts (i).

28. i) See [44, exercise 30, chapter 12].

For every $x \in H$ we must show that $\langle \cdot, x \rangle : (H, weak) \to \mathbb{K}$ is continuous. Using Riesz's representation theorem, $\langle \cdot, x \rangle \in H^*$, and therefore $\langle \cdot, x \rangle : H \to \mathbb{K}$ is $weak$ continuous. For the function $\langle x, \cdot \rangle : H \to \mathbb{K}$, we have $\langle x, \cdot \rangle = \overline{\langle \cdot, x \rangle}$, and therefore $\langle x, \cdot \rangle : (H, weak) \to \mathbb{K}$ is continuous. We have proved that $\langle \cdot, \cdot \rangle : (H, weak) \times (H, weak) \to \mathbb{K}$ is separately continuous.

Suppose now that $\langle \cdot, \cdot \rangle : (H, weak) \times (H, weak) \to \mathbb{K}$ is continuous as a function of two variables. Then the function is continuous at $(0,0) \in H \times H$, and therefore there are $\delta > 0, \tau > 0, x_1^*, ..., x_n^* \in H^*, y_1^*, ..., y_m^* \in H^*$, such that

$$W(0; x_1^*, ..., x_n^*; \delta) \times W(0; y_1^*, ..., y_m^*; \tau) \subseteq \{(x,y) \in H \times H \mid |\langle x, y \rangle| < 1\}, \quad (1)$$

with the usual notations. Using Riesz's theorem we can find $x_1, ..., x_n, y_1, ..., y_m \in H$ such that $x_i^*(x) = \langle x, x_i \rangle$, $i = 1, ..., n$, $y_j^*(x) = \langle x, y_j \rangle$, $j = 1, ..., m$, $\forall x \in H$. Then the relation (1) becomes

$$\{(x,y) \mid |\langle x, x_i \rangle| < \delta, \, i = \overline{1, n}, \, |\langle y, y_j \rangle| < \tau, \, j = \overline{1, m}\} \subseteq \{(x,y) \mid |\langle x, y \rangle| < 1\}. \quad (2)$$

Let $X = \text{Sp}\{x_1, ..., x_n\}$ and $Y = \text{Sp}\{y_1, ..., y_m\}$. We will prove that

$$X^\perp \times \{y \in H \mid |\langle y, y_j \rangle| < \tau, \, j = 1, ..., m\} \subseteq \{(x,y) \in H \times H \mid \langle x, y \rangle = 0\} \quad (3)$$

Let $x \in X^\perp$ and $y \in H$ such that $|\langle y, y_j \rangle| < \tau, j = 1, ..., m$. Then $\lambda x \in X^\perp$ $\forall \lambda \in \mathbb{K}$, and therefore $|\langle \lambda x, x_i \rangle| = 0 < \delta$ $i = 1, ..., n$. Using (2) we obtain that $|\langle \lambda x, y \rangle| < 1$ $\forall \lambda \in \mathbb{K}$, i.e., $|\langle x, y \rangle| \leq 1/|\lambda|$ $\forall \lambda \in \mathbb{K} \setminus \{0\}$, and for $\lambda \to \infty$, $\langle x, y \rangle = 0$, i.e., (3). We prove now that

$$X^\perp \times Y^\perp \subseteq \{(x,y) \in H \times H \mid \langle x, y \rangle = 0\}. \quad (4)$$

Let $x \in X^\perp$ and $y \in Y^\perp$. Then $|\langle y, y_j \rangle| = 0 < \tau, j = 1, ..., m$, and using (3) we obtain that $\langle x, y \rangle = 0$, i.e., (4).

We prove now that $(X+Y)^\perp = \{0\}$, and then $H = X+Y+(X+Y)^\perp = X+Y$, which contradicts that the dimension of H is infinite. Let $x \in (X+Y)^\perp$, i.e., $x \perp (X+Y)$. Then $x \perp X$ and $x \perp Y$, i.e., $(x, x) \in X^\perp \times Y^\perp$ and, using (4), $\langle x, x \rangle = 0$, i.e., $x = 0$.

ii) For every $x^* \in X^*$, we must show that $\langle x^*, \cdot \rangle : (X, weak) \to \mathbb{K}$ is continuous, which is obvious. For any $x \in X$ the function $\langle \cdot, x \rangle : X^* \to \mathbb{K}$, $x^* \to x^*(x)$ is obviously $weak^*$ continuous, and therefore $\langle \cdot, \cdot \rangle : (X^*, weak^*) \times (X, weak) \to \mathbb{K}$ is separately continuous.

Suppose now that $\langle \cdot, \cdot \rangle : (X^*, weak^*) \times (X, weak) \to \mathbb{K}$ is continuous as a function of two variables. Then the function is continuous at $(0,0) \in X^* \times X$ and therefore there are $\delta > 0, \tau > 0, x_1, ..., x_n \in X, x_1^*, ..., x_m^* \in X^*$, such that

$$W(0; x_1, ..., x_n; \delta) \times W(0; x_1^*, ..., x_m^*; \tau) \subseteq \{(x^*, x) \in X^* \times X \mid |\langle x^*, x \rangle| < 1\}$$

Let $p : X^* \to \mathbb{R}, p(x^*) = \max_{i=\overline{1,n}} |x^*(x_i)|$, and $q : X \to \mathbb{R}, q(x) = \max_{j=\overline{1,m}} |x_j^*(x)|$. We will prove that

$$|x^*(x)| \leq \frac{1}{\delta \tau} p(x^*) q(x) \quad \forall x^* \in X^*, \forall x \in X. \tag{5}$$

Let $x \in X, x^* \in X^*$ and $\varepsilon > 0$. Then

$$\frac{\delta x^*}{p(x^*) + \varepsilon} \in W(0; x_1, ..., x_n; \delta), \quad \frac{\tau x}{q(x) + \varepsilon} \in W(0; x_1^*, ..., x_m^*; \tau)$$

and using (5) we obtain that

$$\left| \left\langle \frac{\delta x^*}{p(x^*) + \varepsilon}, \frac{\tau x}{q(x) + \varepsilon} \right\rangle \right| < 1,$$

that is, $|x^*(x)| \leq (1/\delta\tau)(p(x^*) + \varepsilon)(q(x) + \varepsilon)$. For $\varepsilon \to 0, \varepsilon > 0$, we obtain that $|x^*(x)| \leq (1/\delta\tau) p(x^*) q(x)$. Therefore (5) is true. Let $x^* \in X^*$ such that $x^*(x_i) = 0$ $\forall 1 \leq i \leq n$, i.e., $p(x^*) = 0$. Then by (5) it follows that $x^*(x) = 0$ $\forall x \in X$. Thus we obtain that the set $\{x_1, ..., x_n\}$ is fundamental, i.e., $\overline{\mathrm{Sp}\{x_1, ..., x_n\}} = X$, and then X is finite-dimensional, a contradiction.

iii) The fact that E is separately continuous is obvious. Suppose now that $E : L(X) \times X \to X$ is continuous as a function of two variables. Then, as in (ii), we can find $M > 0$ and a finite system $\{x_1, ..., x_n\} \subseteq X$ such that

$$\|U(x)\| \leq M \sup_{1 \leq i \leq n} \|U(x_i)\| \|x\| \quad \forall U \in L(X), \forall x \in X. \tag{6}$$

Let $x^* \in X^*$ such that $x^*(x_i) = 0$ $\forall 1 \leq i \leq n$. Take any $a \in X \setminus \{0\}$. Then for the operator $U(x) = x^*(x) a$ we have $U(x_i) = 0$ $\forall 1 \leq i \leq n$, and hence by (6) we have $U(x) = 0$ $\forall x \in X$, i.e., $x^*(x) = 0$ $\forall x \in X$. Then $x^* = 0$, and therefore the set $\{x_1, ..., x_n\}$ is fundamental in X, i.e., $\overline{\mathrm{Sp}\{x_1, ..., x_n\}} = X$, and then X is finite-dimensional, a contradiction.

29. i) We first observe that if $x^*, y^* \in B_{X^*}$, then

$$\sum_{n=1}^{\infty} \frac{|(x^* - y^*)x_n|}{2^n \|x_n\|} \leq \sum_{n=1}^{\infty} \frac{\|x^* - y^*\| \|x_n\|}{2^n \|x_n\|} = \|x^* - y^*\| \sum_{n=1}^{\infty} \frac{1}{2^n} = \|x^* - y^*\|,$$

and therefore d is well defined. Let now $x^*, y^* \in B_{X^*}$ such that $d(x^*, y^*) = 0$. Then $(x^* - y^*)(x_n) = 0$ $\forall n \in \mathbb{N}$, and using the continuity of x^* and y^* and the density of $(x_n)_{n \in \mathbb{N}}$ in X we obtain that $(x^* - y^*)(x) = 0$ $\forall x \in X$, that is, $x^* = y^*$. The other axioms for a metric are easy to verify.

ii) Consider the identity $I : (B_{X^*}, weak^*) \to (B_{X^*}, d)$, and $x_0^* \in B_{X^*}$. We shall prove that I is continuous at x_0^*. Let $\varepsilon > 0$. There is an $N \in \mathbb{N}$ such that $1/2^{N-1} < \varepsilon/2$ and there is an $r > 0$ such that

$$r \sum_{n=1}^{N} \frac{1}{2^n \|x_n\|} < \frac{\varepsilon}{2}.$$

14.2 Solutions

We consider $U = W(x_0^*; x_1, ..., x_N; r) \cap B_{X^*}$. For any $x^* \in U$, we have

$$d(x^*, x_0^*) = \sum_{n=1}^{N} \frac{|(x^* - x_0^*)(x_n)|}{2^n \|x_n\|} + \sum_{n=N+1}^{\infty} \frac{|(x^* - x_0^*)(x_n)|}{2^n \|x_n\|}$$

$$\leq r \sum_{n=1}^{N} \frac{1}{2^n \|x_n\|} + \sum_{n=N+1}^{\infty} \frac{\|x^* - x_0^*\| \|x_n\|}{2^n \|x_n\|}$$

$$< \frac{\varepsilon}{2} + \sum_{n=N+1}^{\infty} \frac{1}{2^{n-1}} = \frac{\varepsilon}{2} + \frac{1}{2^{N-1}} < \varepsilon,$$

and therefore $I(U) \subseteq B_d(x_0^*, \varepsilon)$, that is, I is continuous at x_0^*. Since I is bijective, $(B_{X^*}, weak^*)$ is a compact topological space, and (B_{X^*}, d) is a Hausdorff space, we obtain that I is a homeomorphism.

iii) Indeed, using (ii) we obtain that the $weak^*$ topology on B_{X^*} is generated by d.

Remark. If (X, τ) is a compact topological space and (Y, σ) is a Hausdorff topological space then any continuous bijective function $f : (X, \tau) \to (Y, \sigma)$ is a homeomorphism. Indeed, let $f^{-1} = g : (Y, \sigma) \to (X, \tau)$. For any closed set $F \subseteq X$ since X is compact we obtain that F is compact. Since $f : (X, \tau) \to (Y, \sigma)$ is continuous, $f(F) \subseteq Y$ is a compact set. Since (Y, σ) is Hausdorff we obtain that $f(F) \subseteq Y$ is closed, and thus $\forall F \subseteq X$ closed, $g^{-1}(F) = f(F) \subseteq Y$ is closed, i.e., g is continuous.

30. Suppose that $(B_X, weak)$ is a metric space. Similarly to the solution for exercise 23, we can find $(x_n^*)_{n \in \mathbb{N}} \subseteq X^*$ and $(\varepsilon_n)_{n \in \mathbb{N}} \subseteq (0, \infty)$ such that

$$\{0\} = \bigcap_{n=1}^{\infty} (W(0; x_n^*; \varepsilon_n) \cap B_X).$$

We shall prove that $\text{Sp}\{x_n^* \mid n \in \mathbb{N}\}$ is dense in X^*, and we shall obtain that X^* is separable. Suppose that $\overline{\text{Sp}\{x_n^* \mid n \in \mathbb{N}\}}^{\|\cdot\|} \neq X^*$, and consider $x_0^* \in X^* \setminus \overline{\text{Sp}\{x_n^* \mid n \in \mathbb{N}\}}^{\|\cdot\|}$. We obtain that $\delta = d(x_0^*, \text{Sp}\{x_n^* \mid n \in \mathbb{N}\}) > 0$ and by a corollary of the Hahn–Banach theorem there is an $x^{**} \in X^{**}$, $\|x^{**}\| = 1/\delta$, such that $x^{**} = 0$ on $\text{Sp}\{x_n^* \mid n \in \mathbb{N}\}$ and $x^{**}(x_0^*) = 1$. Let $U = W(0; x_0^*; \delta/2) \cap B_X$. U is a neighborhood of 0 in the $weak$ topology on B_X. Since $(B_X, weak)$ is a metric space there is an open ball with center at 0 (with respect to the distance on B_X which generates the $weak$ topology) included in U. Similarly to the solution for exercise 23 it results that there is a finite set $A \subseteq \mathbb{N}$ such that

$$\bigcap_{n \in A} W(0; x_n^*; \varepsilon_n) \cap B_X \subseteq U.$$

Since $\delta x^{**} \in B_{X^{**}}$ and from the Goldstine theorem $K_X(B_X) \subseteq B_{X^{**}}$ is a $weak^*$ dense set, there is an $x_0 \in B_X$ such that

$$\widehat{x}_0 \in \bigcap_{n \in A} W(\delta x^{**}; x_n^*; \varepsilon_n) \cap W(\delta x^{**}; x_0^*; \delta/2).$$

Then $|\widehat{x}_0(x_n^*) - \delta x^{**}(x_n^*)| < \varepsilon_n$, that is, $|x_n^*(x_0)| < \varepsilon_n \ \forall n \in A$, and therefore $x_0 \in \bigcap_{n \in A} W(0; x_n^*; \varepsilon_n)$. Since $x_0 \in B_X$ we obtain that $x_0 \in U$. We have $|\widehat{x}_0(x_0^*) - \delta x^{**}(x_0^*)| <$

$\delta/2$, and therefore $|x_0^*(x_0) - \delta| < \delta/2$. We obtain that $|x_0^*(x_0)| \geq \delta - |x_0^*(x_0) - \delta| > \delta/2$, that is, $x_0 \notin U$, a contradiction!

Suppose now that X^* is separable. Using exercise 29 we obtain that $(B_{X^{**}}, weak^*)$ is a metrizable space. Let d be the metric which generates the $weak^*$ topology on $B_{X^{**}}$. For $x, y \in B_X$, we consider $\rho(x,y) := d(K_X(x), K_X(y))$. Since d is a metric on $B_{X^{**}}$ we obtain that ρ is a metric on B_X. One can easily see that $K_X : (B_X, \rho) \to (K_X(B_X), d)$ is a homeomorphism. Using exercise 15 we obtain that $K_X : (B_X, weak) \to (K_X(B_X), weak^*)$ is a homeomorphism. Then, for the identity $I : B_X \to B_X$, we have the decomposition

$$I : (B_X, \rho) \xrightarrow{K_X} (K_X(B_X), d) = (K_X(B_X), weak^*) \xrightarrow{(K_X)^{-1}} (B_X, weak),$$

and therefore I is a homeomorphism, i.e., the topology generated by ρ on B_X is the $weak$ topology.

31. See [20, exercise 35, chapter V, section 7].

Using exercise 30 we obtain that X^* is separable, i.e., there is $(x_n^*)_{n \in \mathbb{N}} \subseteq X^*$, a countable dense set. We shall prove that $A = \{x_n^* \mid n \in \mathbb{N}\} \subseteq X^*$ is a total set. Let $x \in X$ such that $x_n^*(x) = 0\ \forall n \in \mathbb{N}$. Then for $\widehat{x} \in X^{**}$ we have $\widehat{x}(x_n^*) = 0\ \forall n \in \mathbb{N}$. Using the continuity of $\widehat{x}$ on X^* we obtain that $\widehat{x} = 0$ on $\overline{A}$. But $\overline{A} = X^*$, and therefore $\widehat{x} = 0$, i.e., $x = 0$.

32. i) Let $A \subseteq X$ be a $weak$ compact set and $(x_n^*)_{n \in \mathbb{N}} \subseteq X^*$ a total set. Define $d : A \times A \to \mathbb{R}$,

$$d(x,y) = \sum_{n=1}^{\infty} \frac{|x_n^*(x-y)|}{2^n (\|x_n^*\| + 1)}.$$

For any $x, y \in A$,

$$\sum_{n=1}^{\infty} \frac{|x_n^*(x-y)|}{2^n (\|x_n^*\| + 1)} \leq \sum_{n=1}^{\infty} \frac{\|x_n^*\|(\|x\| + \|y\|)}{2^n (\|x_n^*\| + 1)} \leq 2M \sum_{n=1}^{\infty} \frac{1}{2^n} = 2M < \infty,$$

where $M = \sup_{x \in A} \|x\|$, $M < \infty$ since any $weak$ compact set is norm bounded. Therefore $d(x,y) \in \mathbb{R}$. If $d(x,y) = 0$ then $x_n^*(x-y) = 0\ \forall n \in \mathbb{N}$ and using that $(x_n^*)_{n \in \mathbb{N}} \subseteq X^*$ is a total set we obtain that $x - y = 0$, $x = y$. The other axioms for a metric are easy to verify. Similarly to the solution for exercise 29 we obtain that the $weak$ topology on A is generated by d.

ii) See [20, exercise 36, chapter V, section 7].

Since X is a reflexive space, $B_X \subseteq X$ is a $weak$ compact set. Since X^* has a countable total set, by (i) we obtain that the $weak$ topology on B_X is metrizable and then, using exercise 30, we obtain that X^* is separable.

33. Let d be the metric which generates the $weak^*$ topology on B_{X^*}. In any metric space we have

$$\{0\} = \bigcap_{n=1}^{\infty} B(0, 1/n),$$

14.2 Solutions

where $B(0, 1/n)$ is the open ball of center 0 and radius $1/n$ with respect to d. For any $n \in \mathbb{N}$ there are $x_{1,n}, \ldots, x_{k_n,n} \in X$, $\varepsilon_n > 0$, such that

$$0 \in W(0; x_{1,n}, \ldots, x_{k_n,n}; \varepsilon_n) \cap B_{X^*} \subseteq B(0, 1/n).$$

We deduce that there is a sequence $(x_n)_{n \in \mathbb{N}} \subseteq X$ and a sequence $(\varepsilon_n)_{n \in \mathbb{N}} \subseteq (0, \infty)$ such that

$$\{0\} = \bigcap_{n=1}^{\infty} (W(0; x_n; \varepsilon_n) \cap B_{X^*}).$$

Let now $x^* \in X^*$ such that $x^*(x_n) = 0 \; \forall n \in \mathbb{N}$. Then

$$\frac{x^*}{1 + \|x^*\|} \in \bigcap_{n=1}^{\infty} (W(0; x_n; \varepsilon_n) \cap B_{X^*}),$$

and therefore $x^* = 0$. This means that the set $\{x_n \mid n \in \mathbb{N}\} \subseteq X$ is fundamental, i.e., $\mathrm{Sp}\{x_n \mid n \in \mathbb{N}\} \subseteq X$ is a dense linear subspace, and this implies that X is separable.

34. Let $T^* : Y^* \to X^*$ be the dual of T, which, by our hypothesis, is continuous if we consider the $weak^*$ topology on Y^* and the norm topology on X^*. Let us consider $B(0,1) = \{x^* \in X^* \mid \|x^*\| < 1\}$. Then $(T^*)^{-1}(B(0,1)) \subseteq Y^*$ is an open set with respect to the $weak^*$ topology which contains the zero functional on Y. We obtain that there are $y_1, \ldots, y_n \in Y$ and $\varepsilon > 0$ such that $T^*(W(0; y_1, \ldots, y_n; \varepsilon)) \subseteq B(0,1) \subseteq X^*$. (1) Therefore $\forall y^* \in Y^*$ such that $|y^*(y_1)| < \varepsilon, \ldots, |y^*(y_n)| < \varepsilon$, we have $T^*y^* \in B(0,1) \subseteq X^*$, i.e., $\|T^*y^*\| < 1$. Consider $K_Y : Y \to Y^{**}$ and let

$$y^* \in \bigcap_{i=1}^{n} \ker(K_Y(y_i)).$$

Then $\forall m \in \mathbb{N}$ we have $(my^*)(y_i) = (K_Y(y_i))(my^*) = m K_Y(y_i)(y^*) = 0$, $i = \overline{1,n}$, which implies $my^* \in W(0; y_1, \ldots, y_n; \varepsilon)$, and then, by (1), $\|T^*(my^*)\| < 1$, and therefore $\|T^*y^*\| < 1/m \; \forall m \in \mathbb{N}$. Passing to the limit for $m \to \infty$ we obtain that $T^*y^* = 0$. Then $\forall x \in X$ we have $(T^*y^*)(x) = 0$, that is, $\widehat{x}(T^*y^*) = 0 \; \forall x \in X$. In fact, we have proved that

$$\bigcap_{i=1}^{n} \ker(\widehat{y_i}) \subseteq \ker(\widehat{x} \circ T^*) \quad \forall x \in X.$$

Using exercise 16(i) we obtain that for any $x \in X$ there are $\alpha_1, \ldots, \alpha_n \in \mathbb{K}$, which depend on x, such that $\widehat{x} \circ T^* = \sum_{i=1}^{n} \alpha_i \widehat{y_i}$. Therefore $\widehat{x}(T^*y^*) = \sum_{i=1}^{n} \alpha_i \widehat{y_i}(y^*) \; \forall y^* \in Y^*$, that is, $(T^*y^*)(x) = \sum_{i=1}^{n} \alpha_i y^*(y_i) \; \forall y^* \in Y^*$, i.e., $y^*(Tx) = y^* \left(\sum_{i=1}^{n} \alpha_i y_i \right) \; \forall y^* \in Y^*, Tx = \sum_{i=1}^{n} \alpha_i y_i$, i.e., $T(X) \subseteq \mathrm{Sp}\{y_1, \ldots, y_n\} \subseteq Y$, and therefore T is a finite rank operator.

35. See [20, exercise 40, chapter V, section 4].
i) For every $x^* \in X^*$,

$$\|Tx^*\|_{Y^*} = \sup_{\|y\| \leq 1, \, y \in Y} |x^*(y)| \leq \sup_{\|x\| \leq 1, \, x \in X} |x^*(x)| = \|x^*\|_{X^*}.$$

Therefore T is well defined, linear, and, from the above inequality, $\|Tx^*\|_{Y^*} \leq \|x^*\|_{X^*}$ $\forall x^* \in X^*$. In fact, as $Y \subseteq X$ is dense,

$$\sup_{\|x\|\leq 1,\, x\in X} |x^*(x)| = \sup_{\|y\|\leq 1,\, y\in Y} |x^*(y)|,$$

and therefore $\|Tx^*\|_{Y^*} = \|x^*\|_{X^*}$ $\forall x^* \in X^*$. We obtain that T injective. T is also surjective, because every $y^* \in Y^* = L(Y,\mathbb{K})$ can be uniquely extended to $x^* \in L(\overline{Y},\mathbb{K}) = L(X,\mathbb{K}) = X^*$, $\mathbb{K}$ being a complete field.

ii) One implication is obvious. Suppose now that $T : X^* \to Y^*$ is a $weak^*$-to-$weak^*$ homeomorphism. Let $x \in X$. We will show that $x \in Y$. Consider $U = W(0;x;1) = \{x^* \in X^* \mid |x^*(x)| < 1\} \subseteq X^*$. U is an open neighborhood of $0 \in X^*$ in the $weak^*$ topology of X^*. Since T is a $weak^*$-to-$weak^*$ homeomorphism we obtain that $T(U) \subseteq Y^*$ is an open neighborhood of $T(0) = 0 \in Y^*$ in the $weak^*$ topology of Y^*. Therefore there are $y_1, \ldots, y_n \in Y$ and $\varepsilon > 0$ such that

$$V = W(0;y_1,\ldots,y_n;\varepsilon) = \{y^* \in Y^* \mid |y^*(y_i)| < \varepsilon,\ i = \overline{1,n}\} \subseteq T(U).$$

We obtain that $\bigcap_{i=1}^n \ker \widehat{y_i} \subseteq \ker \widehat{x}$. Indeed, let $x^* \in \bigcap_{i=1}^n \ker \widehat{y_i}$. Then $\widehat{y_i}(x^*) = 0$, $i = \overline{1,n}$, and therefore $x^*(y_i) = 0$, $i = \overline{1,n}$. For every $m \in \mathbb{N}$, $(mx^*)(y_i) = 0$, $i = \overline{1,n}$, and therefore $T(mx^*) \in V$. We obtain that $T(mx^*) \in T(U)$ $\forall m \in \mathbb{N}$ and therefore $mx^* \in U$ $\forall m \in \mathbb{N}$, i.e., $|x^*(x)| < 1/m$ $\forall m \in \mathbb{N}$. We obtain that $x^*(x) = 0$ and therefore $x^* \in \ker \widehat{x}$. Applying exercise 16(i) we obtain that there are $\alpha_1, \ldots, \alpha_n \in \mathbb{K}$ such that $\widehat{x} = \sum_{i=1}^n \alpha_i \widehat{y_i}$, i.e.,

$$K_X\left(x - \sum_{i=1}^n \alpha_i y_i\right) = 0,$$

that is, $x = \sum_{i=1}^n \alpha_i y_i$, and then $x \in Y$.

36. Suppose that T is injective. If $T^*(Y^*) \subseteq X^*$ is not $weak^*$ dense by a corollary of the Hahn–Banach theorem we can find $\varphi : (X^*, weak^*) \to \mathbb{K}$ linear and continuous such that $\varphi \neq 0$ and $\varphi|_{T^*(Y^*)} = 0$. Using the form for the linear and $weak^*$ continuous functionals we obtain that there is an $x \in X$, $x \neq 0$, such that $\varphi(x^*) = x^*(x)$ $\forall x^* \in X^*$. The fact that $\varphi(T^*y^*) = 0$ $\forall y^* \in Y^*$ implies that $(T^*y^*)(x) = 0$ $\forall y^* \in Y^*$, i.e., $y^*(Tx) = 0$ $\forall y^* \in Y^*$, that is, $Tx = 0$. Since T is injective and linear we obtain that $x = 0$, a contradiction!

Suppose now that $T^*(Y^*) \subseteq X^*$ is $weak^*$ dense. If, for a contradiction, there is an $x \in X$, $x \neq 0$, such that $Tx = 0$ then $y^*(Tx) = 0$ $\forall y^* \in Y^*$, and therefore $(T^*y^*)(x) = 0$ $\forall y^* \in Y^*$. We have obtained that the linear and continuous functional $\widehat{x} : (X^*, weak^*) \to \mathbb{K}$ is equal to zero on the set $T^*(Y^*) \subseteq X^*$, which is $weak^*$ dense. Then $\widehat{x} : X^* \to \mathbb{K}$ is the null functional, and therefore $\widehat{x} = 0$, $x = 0$, a contradiction!

37. See [18, exercise 5, chapter II].

Any element $x \in X$ generates an element $\widehat{x} \in X^{**}$. If we restrict $\widehat{x}$ to $\overline{\mathrm{Sp}(E)}$ we obtain the canonical operator $T : X \to \left(\overline{\mathrm{Sp}(E)}\right)^*$, $Tx = \widehat{x}\,|_{\overline{\mathrm{Sp}(E)}}$. Then T is linear and for any

14.2 Solutions

$x \in X$ we have the fact that

$$\|Tx\| = \left\|\widehat{x}\,|_{\overline{\mathrm{Sp}(E)}}\right\| \le \|\widehat{x}\| = \|x\|,$$

i.e., T is continuous, with $\|T\| \le 1$. Suppose that $Tx = 0$. Then $\widehat{x}\,|_{\overline{\mathrm{Sp}(E)}} = 0$, and then $x^*(x) = 0 \; \forall x^* \in E$, and by hypothesis this implies $x = 0$. Therefore $T : X \to \left(\overline{\mathrm{Sp}(E)}\right)^*$ is a linear bounded injective operator. We shall prove that T is also surjective and for that we will use the compactness hypothesis. We denote by τ_p the topology of pointwise convergence on E, i.e., the topology generated by the family of seminorms $(p_{x^*})_{x^* \in E}$, $p_{x^*} : X \to \mathbb{R}_+$, $p_{x^*}(x) = |x^*(x)| \; \forall x \in X$. We shall prove that $T : (B_X, \tau_p) \to \left(\left(\overline{\mathrm{Sp}(E)}\right)^*, weak^*\right)$ is continuous. Let $x_0 \in B_X$ and $(x_\delta)_{\delta \in \Delta}$ be a net in B_X such that $x_\delta \to x_0$ in the τ_p topology. Equivalently, $x^*(x_\delta) \to x^*(x_0) \; \forall x^* \in E$, and we obtain that $x^*(x_\delta) \to x^*(x_0) \; \forall x^* \in \mathrm{Sp}(E)$. Let now $x^* \in \overline{\mathrm{Sp}(E)}$. Then for $\varepsilon > 0$ there is an $y^* \in \mathrm{Sp}(E)$ such that $\|y^* - x^*\|_{X^*} < \varepsilon/3$, and therefore $|y^*(x_\delta) - x^*(x_\delta)| < \varepsilon/3$ $\forall \delta \in \Delta \; (\|x_\delta\| \le 1)$. But $y^*(x_\delta) \to y^*(x_0)$ and therefore there is $\delta_\varepsilon \in \Delta$ such that $|y^*(x_\delta) - y^*(x_0)| < \varepsilon/3 \; \forall \delta \ge \delta_\varepsilon$. Then for any $\delta \ge \delta_\varepsilon$ we have

$$|x^*(x_\delta) - x^*(x_0)| \le |x^*(x_\delta) - y^*(x_\delta)| + |y^*(x_\delta) - y^*(x_0)| + |y^*(x_0) - x^*(x_0)| < \varepsilon.$$

We have obtained that $x^*(x_\delta) \to x^*(x_0) \; \forall x^* \in \overline{\mathrm{Sp}(E)}$, i.e., $Tx_\delta \to Tx_0$ weak* in $\left(\overline{\mathrm{Sp}(E)}\right)^*$, and this is what we wanted to show. (B_X, τ_p) is compact by our hypothesis, and therefore $T(B_X) \subseteq \left(\left(\overline{\mathrm{Sp}(E)}\right)^*, weak^*\right)$ is a weak* compact set. Using the Goldstine theorem, $\{\widehat{x} \mid x \in B_X\} \subseteq B_{X^{**}}$ is weak* dense. Then we obtain that $\{\widehat{x}|_{\overline{\mathrm{Sp}(E)}} \mid x \in B_X\} \subseteq B_{\left(\overline{\mathrm{Sp}(E)}\right)^*}$ is weak* dense, that is, $T(B_X)$ is weak* dense in $B_{\left(\overline{\mathrm{Sp}(E)}\right)^*}$. Since $T(B_X)$ is weak* compact and the weak* topology is Hausdorff, we obtain that $T(B_X)$ is weak* closed in $B_{\left(\overline{\mathrm{Sp}(E)}\right)^*}$ and, using the density we obtain that $T(B_X) = B_{\left(\overline{\mathrm{Sp}(E)}\right)^*}$. Since T is linear we obtain that T is surjective. Therefore $T : X \to \left(\overline{\mathrm{Sp}(E)}\right)^*$ is a bijective linear and bounded operator. It remains to be proved that $\|Tx\| = \|x\| \; \forall x \in X$. We have already proved that $\|Tx\| \le \|x\| \; \forall x \in X$, and suppose that we can find $x_0 \in X$ such that $\|Tx_0\| < \|x_0\|$. Let $r \in (\|Tx_0\|, \|x_0\|)$ and consider $y_0 = x_0/r$. Then $\|Ty_0\| = \|Tx_0\|/r < 1$, therefore $Ty_0 \in B_{\left(\overline{\mathrm{Sp}(E)}\right)^*}$, and then since $T(B_X) = B_{\left(\overline{\mathrm{Sp}(E)}\right)^*}$ and T is bijective we obtain that $T^{-1}(Ty_0) \in B_X$, that is, $\|y_0\| \le 1$, $\|x_0\| \le r$, and this contradicts our choice for r.

38. i) Suppose that $x_n \to 0$ weak in H implies $Tx_n \to 0$ in the norm topology of H. We shall prove that T is a compact operator. Consider $(x_n)_{n \in \mathbb{N}} \subseteq B_H$. A Hilbert space is always reflexive, and using the Eberlein–Smulian theorem there is a subsequence $(n_k)_{k \in \mathbb{N}}$ of $\mathbb{N}$ and $x \in H$ such that $x_{n_k} \to x$ weak. Then $x_{n_k} - x \to 0$ weak and, using the hypothesis, $\|T(x_{n_k} - x)\| \to 0$, i.e., $Tx_{n_k} \to Tx$, and therefore T is compact.

Conversely, suppose that T is compact. Let $(x_n)_{n \in \mathbb{N}} \subseteq H$, $x_n \to 0$ weak in H. Then the set $\{x_n \mid n \in \mathbb{N}\} \subseteq H$ is bounded. Suppose that Tx_n does not converge towards 0 in the norm of H. Then there is an $\varepsilon > 0$ and a subsequence $(n_k)_{k \in \mathbb{N}}$ of $\mathbb{N}$ such that $\|Tx_{n_k}\| \ge \varepsilon \; \forall k \in \mathbb{N}$. But $x_{n_k} \to 0$ weak and since T is norm-to-norm continuous, using exercise 18, $T : (H, weak) \to (H, weak)$ is continuous. We obtain that $Tx_{n_k} \to 0$ weak. But T is compact and therefore there is a subsequence $(n_{k_p})_{p \in \mathbb{N}}$ of $(n_k)_{k \in \mathbb{N}}$ such that $Tx_{n_{k_p}} \to y \in H$ in the norm of H. Since the weak topology is Hausdorff we obtain that $y = 0$, i.e., $\|Tx_{n_{k_p}}\| \to 0$, and therefore there is a $p \in \mathbb{N}$ such that $\|Tx_{n_{k_p}}\| < \varepsilon$, contradiction.

ii) Suppose that T is a compact operator. Let $(x_n)_{n \in \mathbb{N}} \subseteq H$, $x_n \to 0$ weak. Then $(x_n)_{n \in \mathbb{N}}$ is bounded. Also, by (i), $Tx_n \to 0$. Then $\langle Tx_n, x_n \rangle \to 0$.

Conversely, suppose that $x_n \to 0$ weak in H implies $\langle Tx_n, x_n \rangle \to 0$. Using the polarization identity,

$$4 \langle Tx, y \rangle = \langle T(x+y), x+y \rangle - \langle T(x-y), x-y \rangle + i \langle T(x+iy), x+iy \rangle \\ - i \langle T(x-iy), x-iy \rangle$$

for $x, y \in H$, we obtain that if $x_n \to 0$, $y_n \to 0$ weak in H, then $\langle Tx_n, y_n \rangle \to 0$. But $T : (H, weak) \to (H, weak)$ is continuous (see exercise 18) and therefore for any sequence $(x_n)_{n \in \mathbb{N}} \subseteq H$, $x_n \to 0$ weak, we have $Tx_n \to 0$ weak, and then $\langle Tx_n, Tx_n \rangle \to 0$, i.e., $\|Tx_n\| \to 0$. The fact that $x_n \to 0$ weak in H implies the fact that $\|Tx_n\| \to 0$. Using (i) we obtain that T is a compact operator.

39. Suppose that X is a reflexive space and let $A : X \to Y$ be a compact operator. There is a sequence $(x_n)_{n \in \mathbb{N}} \subseteq X$, $\|x_n\| \leq 1$ $\forall n \in \mathbb{N}$, such that $\|Ax_n\| \to \|A\|$. Since A is compact we obtain that the set $\{Ax_n \mid n \in \mathbb{N}\} \subseteq Y$ is relatively compact and therefore there is a subsequence $(n_k)_{k \in \mathbb{N}}$ of $\mathbb{N}$ such that $Ax_{n_k} \to y \in Y$ in norm. Consider now the set $\{x_{n_k} \mid k \in \mathbb{N}\}$. This set is included in the closed unit ball of X, which is weak compact, by our hypothesis. Applying the Eberlein–Smulian theorem we obtain that there is a subsequence $(x_{n_{k_s}})_{s \in \mathbb{N}} \subseteq (x_{n_k})_{k \in \mathbb{N}}$ and an $x \in B_X$ such that $x_{n_{k_s}} \to x$ weak. Since $A : X \to Y$ is continuous, using exercise 18 we obtain that A is weak-to-weak continuous, and therefore $Ax_{n_{k_s}} \to Ax$ weak. But $Ax_{n_{k_s}} \to y$ in norm, and then $Ax_{n_{k_s}} \to y$ weak. Since the weak topology is Hausdorff we obtain that $y = Ax$, i.e., $Ax_{n_{k_s}} \to Ax$ in norm, and therefore $\|Ax_{n_{k_s}}\| \to \|Ax\|$, i.e., $\|A\| = \|y\| = \|Ax\|$, with $x \in X$, $\|x\| \leq 1$.

Conversely, suppose that for any compact operator $A : X \to Y$ there is an $x \in X$, $\|x\| \leq 1$ such that $\|A\| = \|Ax\|$. Since $Y \neq \{0\}$, there is $y \in Y$ with $\|y\| = 1$. Consider $x^* \in X^*$. Let $A : X \to Y$, $A(x) = x^*(x)y$, i.e., $A = x^* \otimes y$ is a rank 1 operator. Then A is compact and therefore by hypothesis there is an $x \in X$, $\|x\| \leq 1$, such that $\|A\| = \|Ax\|$. Since $\|y\| = 1$ we obtain that $\|x^*\| = |x^*(x)|$. We apply the James theorem (the difficult part of the James theorem, see exercise 14 of chapter 3) and we obtain that X is reflexive.

40. See [18, exercise 4 (iv), chapter II].

Suppose that $T^* : Y^* \to X^*$ is weak*-to-norm continuous on the weak* compact sets from Y^*. Using the Alaoglu–Bourbaki theorem, $B_{Y^*} \subseteq Y^*$ is weak* compact, and then since T^* is weak*-to-norm continuous it follows that $T^*(B_{Y^*}) \subseteq X^*$ is compact in the norm topology. This means that T^* is compact and we apply the Schauder theorem to obtain that $T : X \to Y$ is compact.

Conversely, let $T : X \to Y$ be a linear, continuous, and compact operator. Let $E = \overline{T(X)} \subseteq Y$. Since $T(B_X)$ is separable (its closure is a compact set) and $T(X) = \bigcup_{n \in \mathbb{N}} nT(B_X)$ we obtain that E is a separable Banach space (with the norm from

14.2 Solutions

Y). The co-restriction of T, $\widetilde{T} : X \to E$, is a compact operator ($\widetilde{T}(x) = T(x) \; \forall x \in X$). Consider $\widetilde{T}^* : E^* \to X^*$, which is also compact by the Schauder theorem. We shall prove that $\widetilde{T}^* : B_{E^*} \to X^*$ is $weak^*$-to-norm continuous. Since E is separable using exercise 29 we obtain that the $weak^*$ topology on B_{E^*} is metrizable. Since B_{E^*} and X^* are both metric spaces and $\widetilde{T}^*$ is linear, it is sufficient to prove that $\widetilde{T}^*$ is sequentially continuous at 0, i.e., if $(y_n^*)_{n\in\mathbb{N}} \subseteq B_{E^*}$, $y_n^* \to 0$ $weak^*$, then $\widetilde{T}^* y_n^* \to 0$ in the norm of X^*. Let then $(y_n^*)_{n\in\mathbb{N}} \subseteq B_{E^*}$ such that $y_n^* \to 0$ $weak^*$. Suppose, for a contradiction, that $\widetilde{T}^* y_n^*$ does not converge towards 0 in the norm of X^*. Passing if necessary to a subsequence we may suppose that there is an $\varepsilon > 0$ such that $\|\widetilde{T}^* y_n^*\| \geq \varepsilon \; \forall n \in \mathbb{N}$. Since $\widetilde{T}^* : E^* \to X^*$ is a bounded linear operator, it is $weak^*$-to-$weak^*$ continuous, and therefore $\widetilde{T}^* y_n^* \to 0$ $weak^*$. Since the closure of $\widetilde{T}^*(B_{E^*}) \subseteq X^*$ is a compact set there is a subsequence $(n_k)_{k\in\mathbb{N}}$ of $\mathbb{N}$ and an $x^* \in X^*$ such that $\|\widetilde{T}^* y_{n_k}^* - x^*\| \to 0$. Then $\widetilde{T}^* y_{n_k}^* \to x^*$ $weak^*$ and since the $weak^*$ topology is Hausdorff it follows that $x^* = 0$. Therefore $\|\widetilde{T}^* y_{n_k}^*\| \to 0$, which contradicts our choice for ε.

Consider now $T^* : Y^* \to X^*$. We want to show that T^* is $weak^*$-to-norm continuous on the $weak^*$ compact subsets of Y^*, and, as the $weak^*$ compacts sets are bounded it is sufficient to show that $T^* : B_{Y^*} \to X^*$ is $weak^*$-to-norm continuous. Let $(y_\delta^*)_{\delta\in\Delta} \subseteq B_{Y^*}$ be a net such that $y_\delta^* \to y_0^* \in B_{Y^*}$ $weak^*$. Define $\widetilde{y_\delta^*}, \widetilde{y_0^*} : E \to \mathbb{K}$, $\widetilde{y_\delta^*}(y) = y_\delta^*(y)$, $\widetilde{y_0^*}(y) = y_0^*(y) \; \forall y \in E$. Then $\widetilde{y_0^*} \in B_{E^*}$, $\widetilde{y_\delta^*} \in B_{E^*} \; \forall \delta \in \Delta$. Clearly, $\widetilde{y_\delta^*} \to \widetilde{y_0^*}$ $weak^*$ in E^* and using that $\widetilde{T}^*$ is $weak^*$-to-norm continuous we obtain that $\widetilde{T}^* \widetilde{y_\delta^*} \to \widetilde{T}^* \widetilde{y_0^*}$ in the norm of X^*. But $\widetilde{T}^* \widetilde{y^*} = T^* y^* \; \forall y^* \in B_{Y^*}$ (if $\widetilde{y}^*$ is defined as above). Indeed, let $x \in X$ and $y^* \in B_{Y^*}$. Since $\widetilde{T}$ is the co-restriction of T we have $\widetilde{T} x = Tx$. Using the definition for $\widetilde{y}^*$ we obtain that $\widetilde{y}^*(\widetilde{T} x) = y^*(Tx)$, and therefore $\left(\widetilde{T}^* \widetilde{y}^*\right)(x) = (T^* y^*)(x)$. We obtain that $T^* y_\delta^* \to T^* y_0^*$ in the norm of X^*.

41. See [18, p. 17].

For any $x^* \in X^*$, $x^* : (X, weak) \to (\mathbb{K}, |\cdot|)$ is continuous. Since $K \subseteq X$ is a $weak$ compact set we obtain that $x^*(K) \subseteq (\mathbb{K}, |\cdot|)$ is a compact set (continuous functions between topological spaces map compact sets into compact sets), and therefore bounded. For $x \in X$ we have $\widehat{x} \in X^{**}$, $\|x\|_X = \|\widehat{x}\|_{X^{**}}$, where $\widehat{x}$ is defined by the relation $\widehat{x}(x^*) = x^*(x)$ $\forall x^* \in X^*$. We have obtained that the set $(\widehat{x})_{x\in K} \subseteq L(X^*, \mathbb{K})$ is pointwise bounded. We apply the Uniform Boundedness Principle and we obtain that the set $(\widehat{x})_{x\in K} \subseteq L(X^*, \mathbb{K})$ is uniformly bounded, i.e., there is an $M > 0$ such that $\|\widehat{x}\|_{X^{**}} \leq M \; \forall x \in K$. Since $\|x\|_X = \|\widehat{x}\|_{X^{**}} \; \forall x \in X$ we obtain that $\|x\| \leq M \; \forall x \in K$, i.e., K is bounded in norm. Since $K \subseteq X$ is $weak$ compact and the $weak$ topology is Hausdorff, we obtain that $K \subseteq X$ is $weak$ closed.

We prove now that any $weak$ closed set $A \subseteq X$ is closed with respect to the norm topology on X. Indeed, let $x_0 \in \overline{A}^{\|\cdot\|}$. For every $\varepsilon > 0$ and for every $x_1^*, ..., x_n^* \in X^*$, the set $W(x_0; x_1^*, ..., x_n^*; \varepsilon)$ is open in $(X, \|\cdot\|)$ and $x_0 \in W(x_0; x_1^*, ..., x_n^*, \varepsilon)$. We obtain that $W(x_0; x_1^*, ..., x_n^*; \varepsilon) \cap A \neq \varnothing$, and then $x_0 \in \overline{A}^{weak} = A$, since $A \subseteq X$ is $weak$ closed.

42. See [18, p. 17].

i) Obviously B_{c_0} is norm closed and norm bounded. Suppose that B_{c_0} is $weak$ compact. We consider the sequence $(\sigma_n)_{n\in\mathbb{N}} \subseteq B_{c_0}$, $\sigma_n = e_1 + \cdots + e_n \; \forall n \in \mathbb{N}$, where $(e_n)_{n\in\mathbb{N}}$ is the standard basis in c_0. Then there is a subnet $(\sigma_{\varphi(\delta)})_{\delta \in \Delta}$ of $(\sigma_n)_{n \in \mathbb{N}}$ and $\sigma \in B_{c_0}$ such that $\sigma_{\varphi(\delta)} \to \sigma$ weak. Then for any $n \in \mathbb{N}$ we have $p_n(\sigma_{\varphi(\delta)}) \longrightarrow p_n(\sigma)$, where $p_n \in c_0^*$, $p_n(x_1, x_2, ..., x_n, ...) = x_n$. By the definition for a subnet, $\varphi : \Delta \to \mathbb{N}$ has the following property: $\forall n \in \mathbb{N} \, \exists \delta_0 \in \Delta$ such that $\varphi(\delta) \geq n \, \forall \delta \geq \delta_0$. Then $\forall n \in \mathbb{N} \, \exists \delta_0 \in \Delta$ such that $p_n(\sigma_{\varphi(\delta)}) = 1 \, \forall \delta \geq \delta_0$. Taking the limit with respect to $\delta \in \Delta$, $\delta \geq \delta_0$ we obtain that $p_n(\sigma) = 1 \, \forall n \in \mathbb{N}$, therefore $\sigma = (1, 1, 1, ...)$, $\sigma \notin c_0$, contradiction!

ii) Suppose that $B_{l_1} \subseteq l_1$ is weak compact. Consider the standard basis $(e_n)_{n \in \mathbb{N}}$, $e_n \in B_{l_1} \, \forall n \in \mathbb{N}$. We prove then that $e_n \to 0$ weak. Suppose, for a contradiction, that the assertion is not true, and we obtain $k \in \mathbb{N}$, $\varepsilon > 0$, $x_1^*, ..., x_k^* \in l_1^*$ such that $\forall n \in \mathbb{N}$ there is an $m \geq n$ such that $e_m \notin W(0; x_1^*, ..., x_k^*; \varepsilon)$. Therefore we can construct a subsequence $(e_{k_n})_{n \in \mathbb{N}}$ of $(e_n)_{n \in \mathbb{N}}$ such that $\forall n \in \mathbb{N}$, $e_{k_n} \notin W(0; x_1^*, ..., x_k^*; \varepsilon)$. Since $(e_{k_n})_{n \in \mathbb{N}} \subseteq B_{l_1}$ and B_{l_1} is supposed weak compact, there is a a subnet $(e_{\varphi(\delta)})_{\delta \in \Delta}$ of $(e_{k_n})_{n \in \mathbb{N}}$ and $e \in B_{l_1}$ such that $e_{\varphi(\delta)} \to e$ weak. Then $\lambda(e_{\varphi(\delta)}) \to \lambda(e)$ for any $\lambda \in c_0 \subseteq l_\infty = l_1^*$. But $\lambda(e_{\varphi(\delta)}) = \lambda_{\varphi(\delta)}$ $\forall \delta \in \Delta$, where $\lambda = (\lambda_1, \lambda_2, ..., \lambda_n, ...)$, and since $\lambda \in c_0$ we have $\lambda_{\varphi(\delta)} \to 0$. Then $\lambda(e) = 0 \, \forall \lambda \in c_0$. Taking $\lambda = e_n \in c_0$ we obtain that $e = 0$. Therefore $\forall \delta \in \Delta$ we have $e_{\varphi(\delta)} \notin W(0; x_1^*, ..., x_k^*; \varepsilon)$, and $e_{\varphi(\delta)} \to 0$ weak. This is contradictory, and therefore $e_n \to 0$ weak. Using the Mazur theorem we obtain that there is a sequence $(\sigma_n)_{n \in \mathbb{N}}$ of convex combinations of elements from $(e_n)_{n \in \mathbb{N}}$ such that $\sigma_n \to 0$ in the norm topology of l_1. But $\|\sigma_n\|_1 = 1 \, \forall n \in \mathbb{N}$, and we obtain a contradiction.

43. See [18, exercise 6 (i), chapter VII] or [39, chapter 7, 7.6.3].

It is proved in the solution of exercise 22(i) of chapter 5 that $\Sigma : l_1 \to l_\infty$ is linear and continuous, with $\|\Sigma\| = 1$. For any $n \in \mathbb{N}$, $\Sigma(e_n) = (0, 0, ..., 0, 1, 1, 1, ...)$ ($n - 1$ occurrences of 0) and in order to prove that Σ is not weak compact we will prove that the set $\{\Sigma(e_n) \mid n \in \mathbb{N}\} \subseteq l_\infty$ is not relatively compact in the weak topology of l_∞. Suppose, for a contradiction, that it is relatively compact. Then by the Eberlein–Smulian theorem we can find $x \in l_\infty$ and a subsequence $(n_k)_{k \in \mathbb{N}}$ of $\mathbb{N}$ such that $\Sigma(e_{n_k}) \to x$ weak. But $\Sigma(e_{n_k}) \to 0$ weak* in l_∞ and therefore $x = 0$. We obtain that $\Sigma(e_{n_k}) \to 0$ weak in l_∞. Using the Mazur theorem we obtain that there is a sequence $(y_n)_{n \in \mathbb{N}} \subseteq l_\infty$, each y_n being a convex combination of elements from $\{\Sigma(e_{n_k}) \mid k \in \mathbb{N}\}$, such that $y_n \to 0$ in the norm of l_∞. But $\|y_n\|_\infty = 1 \, \forall n \in \mathbb{N}$, and we obtain a contradiction.

44. See [18, exercise 1, chapter III].

Obviously $\|\delta_n\|_\infty = 1 \, \forall n \in \mathbb{N}$, and then $\delta_n \in B_{l_\infty(\Gamma)} \, \forall n \in \mathbb{N}$. Suppose that we can find a subsequence $(k_n)_{n \in \mathbb{N}} \subseteq \mathbb{N}$ and $\delta \in l_\infty(\Gamma)$ such that $\delta_{k_n} \to \delta$ weak*. Then $\delta_{k_n}(x) \to \delta(x) \, \forall x \in l_1(\Gamma)$. Put $A \subseteq \mathbb{N}$, $A = \{k_0, k_2, k_4, ...\}$, and define $x : \Gamma \to \mathbb{K}$,
$$x(B) = \begin{cases} 1 & \text{if } B = A, \\ 0 & \text{if } B \neq A. \end{cases}$$
Then $x \in l_1(\Gamma)$ and therefore the sequence $(\delta_{k_n}(x))_{n \in \mathbb{N}} \subseteq \mathbb{K}$ converges. But

$$\delta_{k_n}(x) = \sum_{B \subseteq \mathbb{N}} \delta_{k_n}(B) x(B) = \delta_{k_n}(A) = \begin{cases} 1 & \text{if } k_n \in A, \\ 0 & \text{if } k_n \notin A, \end{cases}$$

and therefore $\delta_{k_n}(x) = \begin{cases} 1 & \text{if } n \text{ is even}, \\ 0 & \text{if } n \text{ is odd}. \end{cases}$ Evidently $(\delta_{k_n}(x))_{n \in \mathbb{N}} \subseteq \mathbb{K}$ is not a convergent sequence.

14.2 Solutions

45. See [9, exercise 13, chapter IV, section 2].

Let $n \in \mathbb{N}$ be fixed. We have

$$|x_n^*(x_1, x_2, ...)| = |x_n| \leq \sup_{k \in \mathbb{N}} |x_k| \quad \forall (x_1, x_2, ...) \in l_\infty,$$

hence $\|x_n^*\| \leq 1$. Also,

$$1 = |x_n^*(e_n)| \leq \|x_n^*\| \|(e_n)\| = \|x_n^*\|.$$

Suppose that there exists a subsequence $(n_k)_{k \in \mathbb{N}}$ of $\mathbb{N}$ such that $(x_{n_k}^*)_{k \in \mathbb{N}}$ converges *weak** towards $x^* \in l_\infty^*$. Then $\lim_{k \to \infty} x_{n_k}^*(x) = x^*(x) \; \forall x \in l_\infty$. Consider $x = (x_n)_{n \in \mathbb{N}}$, where $x_n = \begin{cases} 1 & \text{if } n \in \{n_1, n_3, n_5, ...\}, \\ 2 & \text{if } n \notin \{n_1, n_3, n_5, ...\}. \end{cases}$ Then $x \in l_\infty$, and therefore the sequence $(x_{n_k}^*(x))_{k \in \mathbb{N}}$ converges. But $x_n^*(x) = \begin{cases} 1 & \text{if } n \in \{n_1, n_3, n_5, ...\}, \\ 2 & \text{if } n \notin \{n_1, n_3, n_5, ...\}, \end{cases}$ and we obtain a contradiction.

46. i) See [44, theorem 3.31 (i), chapter 3].

We prove, for example, that f is continuous at $0 \in \mathbb{C}$. Consider D, the closed unit disc in $\mathbb{C}$, and Γ, the unit circle. Let $x^* \in X^*$. For $0 < |z| < 1$,

$$\frac{1}{2\pi i} \int_\Gamma \frac{(x^* \circ f)(\xi)}{(\xi - z)\xi} d\xi = \frac{1}{z} \left(\frac{1}{2\pi i} \int_\Gamma \frac{(x^* \circ f)(\xi)}{\xi - z} d\xi - \frac{1}{2\pi i} \int_\Gamma \frac{(x^* \circ f)(\xi)}{\xi} d\xi \right)$$

$$= \frac{1}{z} ((x^* \circ f)(z) - (x^* \circ f)(0)).$$

Let $M_{x^*} = \sup_{z \in D} |(x^* \circ f)(z)| < \infty$, since $x^* \circ f$ is continuous and D is a compact set. Then for $0 < |z| < 1/2$ we have $|\xi - z| \geq 1/2 \; \forall \xi \in \Gamma$, and then

$$\left| \frac{1}{2\pi i} \int_\Gamma \frac{(x^* \circ f)(\xi)}{(\xi - z)\xi} d\xi \right| \leq 2 M_{x^*}.$$

We obtain that the set $\{x^* ((f(z) - f(0))/z) \mid 0 < |z| < 1/2\}$ is bounded $\forall x^* \in X^*$. Using the Uniform Boundedness Principle it follows that the set $\{((f(z) - f(0))/z) \mid 0 < |z| < 1/2\}$ is norm bounded in X and then there is an $M > 0$ such that $\|((f(z) - f(0))/z)\| \leq M \; \forall 0 < |z| < 1/2$, i.e., $\|f(z) - f(0)\| \leq M \|z\|$ $\forall 0 < |z| < 1/2$, hence is f is continuous at 0.

ii) For every $x^* \in X^*$, $(x^* \circ f)(z) = \sum_{n=0}^\infty a_n^{x^*} z^n \; \forall z \in \mathbb{C}$. Since the series $\sum_{n=0}^\infty a_n^{x^*} z^n$ has its radius of convergence equal to infinity, and therefore bigger than 1, we obtain that $\sum_{n=0}^\infty |a_n^{x^*}| < \infty \; \forall x^* \in X^*$. Thus we can define the operator $T : X^* \to l_1$, $T(x^*) = (a_n^{x^*})_{n \geq 0}$. Obviously T is linear. Let $M = \sup_{|z| \leq 2} \|f(z)\|$. Then $M < \infty$ (f is continuous), and $\sup_{|z| \leq 2} |(x^* \circ f)(z)| \leq M \|x^*\| \; \forall x^* \in X^*$. Using Cauchy's estimates, $\forall x^* \in X^*$ and $\forall n \in \mathbb{N}$ we have

$$|a_n^{x^*}| = \left| \frac{(x^* \circ f)^{(n)}(0)}{n!} \right| \leq \frac{1}{2^n} \sup_{|z| \leq 2} |(x^* \circ f)(z)| \leq \frac{M \|x^*\|}{2^n}.$$

Then
$$\|Tx^*\| = \sum_{n=0}^{\infty} |a_n^{x^*}| \leq \sum_{n=0}^{\infty} \frac{M\|x^*\|}{2^n} = 2M\|x^*\|,$$
and therefore $T \in L(X^*, l_1)$.

More than that, we shall prove that $T : X^* \to l_1 = c_0^*$ is a $weak^*$-to-$weak^*$ continuous operator. Indeed, let $(x_\delta^*)_{\delta \in \Delta} \subseteq X^*$ be a net such that $x_\delta^* \to 0 \in X^*$ $weak^*$. Then $\forall z \in D_2$, we have $(x_\delta^* \circ f)(z) \to 0 \in \mathbb{C}$. We shall prove that $(x_\delta^* \circ f)'(z) \to 0 \in \mathbb{C}$ $\forall z \in D_{3/2}$ and, by induction, that $(x_\delta^* \circ f)^{(n)}(z) \to 0 \in \mathbb{C}$ $\forall z \in D_{(n+2)/(n+1)}$ $\forall n \in \mathbb{N}$ (we have written D_r for the closed disc in $\mathbb{C}$ of center 0 and radius r, and Γ_r for the circles). Since $x_\delta^* \to 0$ $weak^*$ we obtain that there is an $N > 0$ such that $\|x_\delta^*\| \leq N$ $\forall \delta \in \Delta$. Since $\Gamma_2 \subseteq \mathbb{C}$ is a compact set and f is continuous we obtain that $f(\Gamma_2) \subseteq X$ is a compact set. Let $\varepsilon > 0$. Then there are $\zeta_1, ..., \zeta_k \in \Gamma_2$ such that
$$f(\Gamma_2) \subseteq \bigcup_{m=1}^{k} B(f(\zeta_m), \varepsilon).$$
Since $(x_\delta^* \circ f)(\zeta_m) \to 0$, $m = 1, ..., k$, there is a $\delta_\varepsilon \in \Delta$ such that
$$|(x_\delta^* \circ f)(\zeta_m)| < \varepsilon, \ m = 1, ..., k, \ \forall \delta \geq \delta_\varepsilon.$$
Consider $\delta \geq \delta_\varepsilon$. For each $\xi \in \Gamma_2$ there is an m such that $f(\xi) \in B(f(\zeta_m), \varepsilon)$. Then
$$|(x_\delta^* \circ f)(\xi)| \leq |(x_\delta^* \circ f)(\xi) - (x_\delta^* \circ f)(\zeta_m)| + |(x_\delta^* \circ f)(\zeta_m)|$$
$$\leq \|x_\delta^*\| |f(\xi) - f(\zeta_m)| + \varepsilon \leq (N+1)\varepsilon.$$
Then for a fixed $z \in D_{3/2}$ and for any $\delta \geq \delta_\varepsilon$ we have:
$$|(x_\delta^* \circ f)'(z)| = \left| \frac{1}{2\pi i} \int_{\Gamma_2} \frac{(x_\delta^* \circ f)(\xi)}{(\xi - z)^2} d\xi \right| \leq 8(N+1)\varepsilon,$$
and therefore $(x_\delta^* \circ f)'(z) \to 0$ $\forall z \in D_{3/2}$. It is easy to see now that, by induction, we obtain for any $n \geq 0$ that $(x_\delta^* \circ f)^{(n)}(z) \to 0$ $\forall z \in D_{(n+2)/(n+1)}$. Therefore for any fixed $n \geq 0$, $\lim_{\delta \in \Delta} a_n^{x_\delta^*} = 0$. Also, since the net $(x_\delta^*)_{\delta \in \Delta} \subseteq X^*$ is norm bounded we obtain that the net
$$(Tx_\delta^*)_{\delta \in \Delta} = \left\{ \left(a_n^{x_\delta^*} \right)_{n \geq 0} \right\}_{\delta \in \Delta} \subseteq l_1$$
is norm bounded. Similarly to the solution for exercise 2 we obtain that $Tx_\delta^* \to 0$ $weak^*$ in l_1 and therefore T is a $weak^*$-to-$weak^*$ continuous operator.

Using exercise 19, we obtain that there is a bounded linear operator $S : c_0 \to X$ such that $S^* = T$. Let $x_n = Se_n$ $\forall n \in \mathbb{N}$. We obtain that
$$a_n^{x^*} = (Tx^*)_n = (Tx^*)(e_n) = x^*(Se_n) = x^*(x_n) \ \forall n \in \mathbb{N},$$
and therefore $\forall x^* \in X^*$ and $\forall z \in \mathbb{C}$, $(x^* \circ f)(z) = \sum_{n=0}^{\infty} x^*(x_n) z^n$. Let $R > 0$ and consider $0 < r < R$. Let $M_R = \sup_{|z| \leq R} \|f(z)\|$. Then $\forall x^* \in B_{X^*}$, $\sup_{|z| \leq R} |(x^* \circ f)(z)| \leq M_R$. Using again Cauchy's estimates, for any $x^* \in B_{X^*}$ and for any $n \in \mathbb{N}$ we have
$$|x^*(x_n)| = \left| \left((x^* \circ f)^{(n)}(0) \right) / (n!) \right| \leq M_R / R^n.$$

14.2 Solutions

Since M_R depends only on f and R we obtain that $\|x_n\| \leq M_R/R^n$ $\forall n \in \mathbb{N}$. Then

$$\|x_n\| r^n \leq \frac{M_R}{(R/r)^n} \quad \forall n \in \mathbb{N}.$$

Since $R/r > 1$ the series $\sum_{n=0}^{\infty} \|x_n\| r^n$ converges. Since $R > 0$ was arbitrary we obtain that $\sum_{n=0}^{\infty} \|x_n\| R^n < \infty$ $\forall R > 0$. From here it follows that the series $\sum_{n=0}^{\infty} x_n z^n$ is absolutely convergent in X for all $z \in \mathbb{C}$, hence convergent in X (X is a Banach space). We have

$$x^* \left(\sum_{n=0}^{\infty} x_n z^n \right) = \sum_{n=0}^{\infty} x^*(x_n) z^n = x^*(f(z)) \quad \forall z \in \mathbb{C}, \ \forall x^* \in X^*,$$

thus $f(z) = \sum_{n=0}^{\infty} x_n z^n$ $\forall z \in \mathbb{C}$. We also have $\sum_{n=0}^{\infty} \|x_n\| R^n < \infty$ $\forall R > 0$.

Bibliography

[1] *Berkeley Preliminary Exams (http://math.berkeley.edu/ ~ desouza/pb.html)*.

[2] D. Amir (1986). *Characterizations of inner product spaces.* Birkhauser Verlag, Basel–Boston–Stuttgart.

[3] B. Aupetit (1991). *A primer on spectral theory.* Springer Verlag.

[4] A. V. Balakrishnan (1976). *Applied Functional Analysis.* Springer Verlag.

[5] S. Banach (1932). *Théorie des operations linéaires.* Monografie matematyczne, Warszawa.

[6] B. Bollobas (1992). *Linear analysis, an introductory course.* Cambridge Press.

[7] F. F. Bonsall (1957). The decomposition of continuous linear functionals into non-negative components. *Proc. Univ. Durham Phil. Soc.,*, Ser. A 13, No. 2, 6–12.

[8] F. F. Bonsall and J. Duncan (1973). *Complete normed algebras.* Springer Verlag.

[9] N. Bourbaki (1953), (1958). *Espaces vectoriels topologiques.* Actualités Sci. et Ind. 1189 and 1229, Hermann, Paris.

[10] N. Bourbaki (1953), (1958). *Intégration.* Actualités Sci. et Ind. 1189 and 1229, Hermann, Paris.

[11] N. Bourbaki (1965). *Espaces vectoriels topologiques.* Actualités Sci. et Ind. 1189 and 1229, Hermann, Paris.

[12] N. Bourbaki (1965). *Intégration.* Actualités Sci. et Ind. 1189 and 1229, Hermann, Paris.

[13] N. Bourbaki (1966). *Espaces vectoriels topologiques.* 2-ème edition, Hermann, Paris.

[14] J. Conway (1985). *A Course in Functional Analysis.* Springer Verlag.

[15] M. Cwikel and N. Kalton (1995). Interpolation of compact operators by the methods of Calderón and Gustavsson–Peetre. *Proc. Edin. Math. Soc.*, II. Ser. 38, No. 2, 261–276.

[16] A. Defant and K. Floret (1993). *Tensor Norms and Operator Ideals.* North-Holland.

[17] R. Deville, G. Godefroy, and V. Zizler (1993). *Smoothness and renormings in Banach spaces*. Pitman Monographs, Longman Scientific and Technical.

[18] J. Diestel (1984). *Sequences and series in Banach spaces*. Springer Verlag.

[19] J. Diestel, H. Jarchow, and A. Tonge (1995). *Absolutely Summing Operators*. Cambridge University Press.

[20] N. Dunford and J. Schwartz (1958). *Linear operators*, volume 1. New York, London, Interscience Publishers.

[21] R. E. Edwards (1965). *Functional Analysis; Theory and Applications*. Holt, Rinehart and Winston, Inc., New York.

[22] R. E. Edwards (1979). *Fourier series–A modern introduction*, volume 1. Springer Verlag.

[23] K. Floret (1980). *Weakly compact sets*. Springer Verlag.

[24] H. G. Garnir, M. De Wilde, and J. Schmets (1968). *Analyse fonctionnelle*, volume III. Birkhauser Verlag.

[25] I. M. Glazman and Y. I. Lyubich (1974). *Finite-dimensional linear analysis: A systematic presentation in problem form*. Cambridge, Mass. - London: The M.I.T. Press. XIV. Translated and edited by G. P. Barker and G. Kuerti (English).

[26] I. Gohberg and S. Goldberg (2001). *Basic operator theory*. Birkhauser, Boston–Basel–Berlin.

[27] P. Halmos (1950). *Measure theory*. Van Nostrand, Princeton, New Jersey.

[28] G. Hardy, J. Littlewood, and G. Polya (1964). *Inequalities*. Cambridge University Press.

[29] E. Hewitt and K. Stromberg (1965). *Real and Abstract Analysis*. Springer Verlag.

[30] E. Hille and R. S. Phillips (1957). *Functional Analysis and Semigroups*. A.M.S. Coll. Publ., XXXI.

[31] R. Jajte (1991). *Strong Limit Theorems in Noncommutative L_2-spaces*, volume 1477. Lecture Notes in Math., Springer Verlag, Berlin.

[32] N. Kalton, N. Peck, and J. Roberts (1984). *An F-space sampler*. Cambridge University Press.

[33] L. V. Kantorovich and G. P. Akilov (1982). *Functional Analysis*. Pergamon Press XIV.

[34] J. Kelley (1955). *General topology*. Van Nostrand, Princeton, New Jersey.

[35] J. Kelley and I. Namioka (1976). *Linear topological spaces*. Springer Verlag.

[36] H. Elton Lacey (1974). *The isometric theory of classical Banach spaces.* Springer Verlag.

[37] J. Lindenstrauss and L. Tzafriri (1977). *Classical Banach Spaces*, volume I. Springer Verlag.

[38] A. Pietsch (1980). *Operator Ideals.* North-Holland.

[39] A. Pietsch and J. Wenzel (1998). *Orthogonal Systems and Banach space Geometry.* Cambridge University Press.

[40] B. M. Pisarevski, T. S. Sobolev, and V. A. Trenoghin (1984). *Problems and exercises in functional analysis.* In Russian.

[41] G. Pisier (1996). *The operator Hilbert space OH, complex interpolation and tensor norms*, volume 122. Memoirs of A.M.S.

[42] M. Reed and B. Simon (1972). *Methods of modern mathematical physics*, volume 1. Academic Press, New York–London.

[43] J. R. Ringrose (1972). *Lectures on the trace in finite von Neumann algebras*, volume 247. Lecture Notes in Math., Springer Verlag, Berlin.

[44] W. Rudin (1973). *Functional analysis.* McGraw-Hill.

[45] W. Rudin (1974). *Real and complex analysis.* McGraw-Hill.

[46] H. Schaefer (1966). *Topological vector spaces.* The MacMillan Company, New York, Collier-MacMillan Limited, London.

[47] I. Singer (1981). *Basis in Banach spaces*, volume 2. Springer Verlag.

[48] E. Titchmarsh (1932). *The theory of functions.* Oxford, Clarendon Press.

[49] N. Tomczack-Jaegermann (1989). *Banach–Mazur distances and finite-dimensional operator ideals.* Longman Scientific and Technical.

[50] A. Wilansky (1978). *Modern methods in topological vector spaces.* McGraw-Hill, New York.

[51] K. Yoshida (1965). *Functional Analysis.* Springer Verlag.

[52] K. Zhu (1990). *Operator theory in function spaces.* Marcel Dekker, Inc., New York and Basel.

List of Symbols

$\mathbb{N}$ set of natural numbers.
$\mathbb{Z}$ set of integers.
$\mathbb{R}$ set of real numbers.
$\mathbb{C}$ set of complex numbers.
$\mathbb{K} = \mathbb{R}$ or $\mathbb{C}$.
$\mathbb{R}_+ = [0, \infty)$.
$\mathbb{K}^n$ for $n \in \mathbb{N}$.
$\Re(z), \Im(z), |z|, \bar{z}$, real and imaginary part, modulus and conjugate of $z \in \mathbb{C}$.
$\operatorname{sgn}(z) = \begin{cases} |z|/z, & \text{if } z \neq 0 \\ 0, & \text{if } z = 0 \end{cases}$, for $z \in \mathbb{K}$.
$[a,b] = \{x \in \mathbb{R} \mid a \leq x \leq b\}$, for $a, b \in \mathbb{R}$, $a \leq b$.
$[a,b) = \{x \in \mathbb{R} \mid a \leq x < b\}$, for $a, b \in \mathbb{R}$, $a < b$.
$(x_n)_{n \in \mathbb{N}} \subseteq X$, $x_n \in X \ \forall n \in \mathbb{N}$.
$\mathbf{C}_A = \{x \in X \mid x \notin A\}$ complement of a subset $A \subseteq X$.
$\varnothing$ empty set.
$[f \geq \alpha] = \{x \in S \mid f(x) \geq \alpha\}$ for $f : S \to \mathbb{R}$.
χ_A, characteristic function of a subset $A \subseteq S$ defined by $\chi_A(s) = 1$ if $s \in S$ and $\chi_A(s) = 0$ if $s \notin S$.
$\mathbb{D} = \{z \in \mathbb{C} \mid |z| < 1\}$ the open unit disc in the complex plane.
B_X closed unit ball $\{x \in X \mid \|x\| \leq 1\}$ in a normed space, 3.
S_X closed unit sphere $\{x \in X \mid \|x\| = 1\}$ in a normed space, 3.
$B(x, \varepsilon) = \{y \in X \mid \|y - x\| < \varepsilon\}$, the open ball of center x and radius ε in a normed space X, 3.
$\overline{B}(x, \varepsilon) = \{y \in X \mid \|y - x\| \leq \varepsilon\}$, the closed ball of center x and radius ε in a normed space X, 3.
X/Y, quotient space, 4.
$\langle \cdot, \cdot \rangle$, inner product, 243.
$l_p, 1 \leq p \leq \infty, \|\cdot\|_p$, ix.
$l_p^n, 1 \leq p \leq \infty, n \in \mathbb{N}$, ix.
c_0, c the space of null convergent scalar sequences and the space of convergent scalar sequences, ix.

$X = l_p$, with $1 \leq p \leq \infty$, or c_0, $p_n : X \to \mathbb{K}$ the canonical projections, ix.
$\mathcal{L}_p(\mu), L_p(\mu), 1 \leq p \leq \infty, \|\cdot\|_p$, x.
q conjugate of $1 \leq p < \infty$, satisfying $1/p + 1/q = 1$.
μ-a.e. = a.e., almost everywhere.
$H^p(\mathbb{D}), N_p, 1 \leq p < \infty$, Hardy space, 248.
$A^p(\Omega), A_p, 1 \leq p < \infty$, Bergman space, 248.
$C[a,b], C(T)$, space of continuous functions $f : T \to \mathbb{K}$, where T is a compact Hausdorff space, x.
$L(X, Y) = \{U : X \to Y \mid U \text{ linear and continuous}\}, L(X) = L(X, X), 36$.
$K(X, Y) = \{U \in L(X, Y) \mid U \text{ is compact}\}, K(X) = K(X, X), 108$.
$\mathcal{A}(H) = \{U \in L(H) \mid U \text{ is self-adjoint}\}$, where H is a Hilbert space, 305.
$\mathcal{A}_+(H) = \{U \in \mathcal{A}(H) \mid U \geq 0\}$, where H is a Hilbert space, 305.
$\ker U = \{x \in X \mid U(x) = 0\}$ the kernel of $U \in L(X, Y)$.
$U(X) = \{y \in Y \mid \exists x \in X \text{ such that } y = U(x)\}$ range of $U \in L(X, Y)$.
I or $I_X : X \to X, I(x) = x$, identity operator.
M_φ, M_λ multiplication operator, 309.
p_A, Minkowski functional, 369.
$X^* = L(X, \mathbb{K})$, dual of X, $x^* \in X^*$, 68.
$X^{**} = L(X^*, \mathbb{K})$, bidual of X, $x^{**} \in X^{**}$, 68.
$K_X : X \to X^{**}, K_X(x) = \hat{x}$, the canonical embedding into bidual, 68.
$\hat{x} : X^* \to \mathbb{K}, \hat{x}(x^*) = x^*(x)$, 68.
p_{x^*}, p_x, 403.
$W(0; x_1^*, ..., x_n^*; \varepsilon) = \{x \in X \mid |x_1^*(x)| < \varepsilon, ..., |x_n^*(x)| < \varepsilon\}, W(x; x_1^*, ..., x_n^*; \varepsilon)$, 403.
$W(0; x_1, ..., x_n; \varepsilon) = \{x^* \in X^* \mid |x^*(x_1)| < \varepsilon, ..., |x^*(x_n)| < \varepsilon\}, W(x^*; x_1, ..., x_n; \varepsilon)$, 403.
$\|U\| = \sup \{\|U(x)\| \mid \|x\| \leq 1\}$ norm of U, 36.
U^*, dual of U for normed spaces, adjoint for Hilbert spaces, 36.
$U \geq 0, U \leq V$, 305.
$|U|, U_+, U_-$, modulus, positive and negative part of an operator, 306.

$\sqrt{U}$, square root of a positive operator, 306.
$\sup_{i \in I} U_i$, $\inf_{i \in I} U_i$, 305.
$f|_G$ restriction of a function.
LIM Banach Limit, 180.
$x \perp y$, $\langle x, y \rangle = 0$ in an inner product space H.
$A^\perp = \{x \in H \mid \langle x, a \rangle = 0, \forall a \in A\}$ orthogonal complement in an inner product space H, 244.
$\Pr_A$ orthogonal projection in a Hilbert space, 271.
S_c, space of convergent scalar series, 7.
c_{00}, space of scalar sequences with finite support, 9.
$\mathcal{M}_n(\mathbb{K})$, space of $n \times n$ scalar matrices.
$\text{tr}(A)$ trace for $A \in \mathcal{M}_n(\mathbb{K})$.
$\mathcal{R}[0, 1]$, space of Riemann integrable functions $f : [0, 1] \to \mathbb{R}$, 153.
$\text{Lip}_\alpha[0, 1]$, $0 < \alpha \leq 1$, space of α-Lipschitz functions $f : [0, 1] \to \mathbb{K}$, 214.
$C^k[a, b]$, $k \in \mathbb{N} \bigcup \{\infty\}$, space of functions $f : [a, b] \to \mathbb{K}$ which are of class C^k, 214.
$\mathcal{I}(\mathbb{C})$, space of entire functions, 376.
$\mathcal{H}(\Omega)$, space of holomorphic functions $f : \Omega \to \mathbb{C}$, where $\Omega \subseteq \mathbb{C}$ is a non-empty open set, 376.

$C(\Omega)$, space of continuous functions $f : \Omega \to \mathbb{C}$, where $\Omega \subseteq \mathbb{C}$ is a non-empty open set, 376.
τ_p, pointwise convergence topology, 109.
$l_p(X)$, $1 \leq p \leq \infty$, 165.
$c_0(X)$, 155.
$w_p(X)$, $1 \leq p < \infty$, 115.
$\overset{\circ}{A}$, $\overline{A}$, A', $\text{Fr}(A) = \overline{A} - \overset{\circ}{A}$, interior, closure, derivative, border of a subset in a topological space.
$\dim_\mathbb{K} X$ dimension of a linear space X.
Let A be a subset in a linear space
$\text{Sp}(A)$ linear hull of A.
$\text{co}(A)$ convex hull of A.
$\text{ec}(A)$ balanced hull of A.
$\text{eco}(A)$ absolutely convex hull of A.
Let A be a subset in a linear topological space
$\overline{\text{co}}(A)$ closed convex hull of A.
$\overline{\text{ec}}(A)$ closed balanced hull of A.
$\overline{\text{eco}}(A)$ closed absolutely convex hull of A.
$d(x, A) = \text{dist}(x, A)$ distance from a point to a non-empty subset in a metric space, 86.
$d(A, B) = \text{dist}(A, B)$ distance between two non-empty subsets in a metric space, 86.
$\widehat{(x, y)}, \widehat{(x, G)}, \widehat{(L, M)}$ angles in a Hilbert space, 277.

Index

A
Additive,
  functional, 68.
  operator, 372.
Angle in a Hilbert space, 277.

B
Ball,
  closed, 3.
  open, 3.
Banach limit, 180.
Basis,
  orthonormal, 244.
  Schauder, 4.
Bergman kernel, 248.
Bessel inequality, 244.
Bidual, 68.
Bounded pointwise, 147.

C
Canonical embedding into the bidual, 68.
Cauchy–Buniakowski–Schwarz inequality, 243.
Closure, 4.
Convergence,
  pointwise, 372.
  $weak$, 403.
  $weak^*$, 403.
convex cone, 368.

D
Dual, 68.
Distance = metric,
  associated with a norm, 4.
  from an element to a subset, 86.
  between two sets, 86.

E
ε-net, 107.
Equicontinuous family of functions, 107.
Equivalent norms, 4.
Extension,
  Hahn–Banach, 175.

Philips, 42, 197.

F
Formula,
  Fourier, 244.
  Parseval, 244.
Functional,
  positive homogeneous, 3.
  subadditive, 3.
  sublinear, 3.
  supra-linear, 3.

G
Gram determinant, 274.

H
Hilbertian product, 243.

I
Ideal property,
  for compact operators, 108.
  for $weak$ compact operators, 404.
Infimum for a family of operators, 305.
Inner product,
  space, 243.
  associated with a positive operator, 313.
Interior, 4.
Isometrically isomorphic, 36.
Isometry, 36.
Isomorphism, 36.

K
Kernel
  Bergman, 249.

L
Lemma,
  Auerbach, 69.
  Cantor, 155.
  Kakutani, 383.
  Riesz, 86.

M
Metric associated with a norm, 4.
Minkowski,

functional, 369.
seminorm, 369.

N
Norm,
    associated with an inner product, 243.
    quotient, 4.

O
Operator,
    adjoint, 36.
    compact, 107.
    composition, 37.
    dual, 36.
    finite rank, 108.
    Hardy, 38, 111.
    Hilbert, 39.
    inversion, 246.
    linear and continuous, 36.
    modulus, 306.
    multiplication, 111.
    negative part, 306.
    norm, 36.
    positive, 305.
    positive part, 306.
    Riemann–Liouville modified, 39.
    Schur, 39.
    self-adjoint, 305.
    summation, 111.
    Volterra, 111.
    weak compact, 404.
Orthogonal complement, 244.

P
Parallelogram law, 243.
Partial isometry, 306.
Polar decomposition, 306.
Projection = projector,
    canonical, ix.
    orthogonal, 271.

R
Relative compactness in l_p, c_0, 107.
Relatively compact, 107.

S
Seminorm, 3.
Sequence,
    Cauchy, 3.
    convergent, 3.
Series,
    convergent, 4.
    absolutely convergent, 4.
Set,
    absorbing, 368.
    balanced, 368.
    bounded, 368.
    convex, 368.
    dense, 4.
    first Baire category, 213.
    fundamental, 195.
    nowhere dense, 213.
    orthogonal, 244.
    orthonormal, 244.
    second Baire category, 213.
    total, 407.
Space,
    Banach, 3.
    Bergman, 248.
    Fréchet, 369.
    Hardy, 248.
    Hilbert, 243.
    linear topological, 368.
    locally convex, 369.
    normed, 3.
    quotient linear, 4.
    separable, 4.
Spectral radius, 306.
Square root of a positive operator, 306.
Subspace,
    initial, 306.
    final, 306.
Supremum for a family of operators, 305.

T
Theorem,
    Alaoglu–Bourbaki, 404.
    Arzela–Ascoli, 107.
    Baire category, 213.
    Banach–Steinhaus, 147.
    closed graph, 213.
    Eberlein–Smulian, 404.
    Gantmacher, 404.
    Goldstine, 404.
    Grothendieck, 108.
    Hahn–Banach, 175.
    Helly, 404.
    inverse mapping, 213.
    James, 70.
    von Neumann, 243.
    Krein, 404.
    Mazur, 108, 404.
    open mapping, 213.
    orthogonal projection, 271.
    Pythagoras, 244.
    Riesz, 89.
    Riesz representation, 244.
    Schauder, 108.
    Schur, 406.
    Stone, 384.
    Weierstrass convergence, 305.
Totally bounded, 107.
Topology,
    generated by a family of seminorms, 369.

linear, 368.
locally convex, 369.
pointwise convergence, 372.
weak, 404.
weak*, 404.

U

Uniformly bounded, 147.

Uniform boundedness principle, 147.
Unit,
  ball, 3.
  sphere, 3.

W

Weak Cauchy series, 113.

Kluwer Texts in the Mathematical Sciences

1. A.A. Harms and D.R. Wyman: *Mathematics and Physics of Neutron Radiography.* 1986
   ISBN 90-277-2191-2
2. H.A. Mavromatis: *Exercises in Quantum Mechanics.* A Collection of Illustrative Problems and Their Solutions. 1987   ISBN 90-277-2288-9
3. V.I. Kukulin, V.M. Krasnopol'sky and J. Horácek: *Theory of Resonances.* Principles and Applications. 1989   ISBN 90-277-2364-8
4. M. Anderson and Todd Feil: *Lattice-Ordered Groups.* An Introduction. 1988
   ISBN 90-277-2643-4
5. J. Avery: *Hyperspherical Harmonics.* Applications in Quantum Theory. 1989
   ISBN 0-7923-0165-X
6. H.A. Mavromatis: *Exercises in Quantum Mechanics.* A Collection of Illustrative Problems and Their Solutions. Second Revised Edition. 1992   ISBN 0-7923-1557-X
7. G. Micula and P. Pavel: *Differential and Integral Equations through Practical Problems and Exercises.* 1992   ISBN 0-7923-1890-0
8. W.S. Anglin: *The Queen of Mathematics.* An Introduction to Number Theory. 1995
   ISBN 0-7923-3287-3
9. Y.G. Borisovich, N.M. Bliznyakov, T.N. Fomenko and Y.A. Izrailevich: *Introduction to Differential and Algebraic Topology.* 1995   ISBN 0-7923-3499-X
10. J. Schmeelk, D. Takacqi and A. Takacqi: *Elementary Analysis through Examples and Exercises.* 1995   ISBN 0-7923-3597-X
11. J.S. Golan: *Foundations of Linear Algebra.* 1995   ISBN 0-7923-3614-3
12. S.S. Kutateladze: *Fundamentals of Functional Analysis.* 1996   ISBN 0-7923-3898-7
13. R. Lavendhomme: *Basic Concepts of Synthetic Differential Geometry.* 1996
    ISBN 0-7923-3941-X
14. G.P. Gavrilov and A.A. Sapozhenko: *Problems and Exercises in Discrete Mathematics.* 1996
    ISBN 0-7923-4036-1
15. R. Singh and N. Singh Mangat: *Elements of Survey Sampling.* 1996   ISBN 0-7923-4045-0
16. C.D. Ahlbrandt and A.C. Peterson: *Discrete Hamiltonian Systems.* Difference Equations, Continued Fractions, and Riccati Equations. 1996   ISBN 0-7923-4277-1
17. J. Engelbrecht: *Nonlinear Wave Dynamics.* Complexity and Simplicity. 1997
    ISBN 0-7923-4508-8
18. E. Pap, A. Takači and D. Takači: *Partial Differential Equations through Examples and Exercises.* 1997   ISBN 0-7923-4724-2
19. O. Melnikov, V. Sarvanov, R. Tyshkevich, V. Yemelichev and I. Zverovich: *Exercises in Graph Theory.* 1998   ISBN 0-7923-4906-7
20. G. Călugăreanu and P. Hamburg: *Exercises in Basic Ring Theory.* 1998
    ISBN 0-7923-4918-0
21. E. Pap: *Complex Analysis through Examples and Exercises.* 1999   ISBN 0-7923-5787-6
22. G. Călugăreanu: *Lattice Concepts of Module Theory.* 2000   ISBN 0-7923-6488-0
23. P.M. Gadea and J. Muñoz Masqué: *Analysis and Algebra on Differentiable Manifolds: A Workbook for Students and Teachers.* 2002   ISBN 1-4020-0027-8; Pb 1-4020-0163-0
24. U. Faigle, W. Kern and G. Still: *Algorithmic Principles of Mathematical Programming.* 2002
    ISBN 1-4020-0852-X
25. G. Călugăreanu, S. Breaz, C. Modoi, C. Pelea and D. Vălcan: *Exercises in Abelian Group Theory.* 2003   ISBN 1-4020-1183-1

Kluwer Texts in the Mathematical Sciences

26. C. Costara and D. Popa: *Exercises in Functional Analysis.* 2003　　　ISBN 1-4020-1560-7

KLUWER ACADEMIC PUBLISHERS – DORDRECHT / BOSTON / LONDON